Resorcinol

Its Uses and Derivatives

TOPICS IN APPLIED CHEMISTRY

Current volumes in the series:

BIOCATALYSTS FOR INDUSTRY
Edited by Jonathan S. Dordick

CHEMICAL TRIGGERING
Reactions of Potential Utility in Industrial Processes
Gebran J. Sabongi

THE CHEMISTRY AND APPLICATION OF DYES
Edited by David R. Waring and Geoffrey Hallas

HIGH-TECHNOLOGY APPLICATIONS OF ORGANIC COLORANTS
Peter Gregory

INFRARED ABSORBING DYES
Edited by Masaru Matsuoka

LEAD-BASED PAINT HANDBOOK
Jan W. Gooch

ORANOFLUORINE CHEMISTRY
Principles and Commercial Applications
Edited by R. E. Banks, B. E. Smart, and J. C. Tatlow

POLY (ETHYLENE GLYCOL) CHEMISTRY
Biotechnical and Biomedical Applications
Edited by J. Milton Harris

RADIATION CURING
Science and Technology
Edited by S. Peter Pappas

RESORCINOL
Its Uses and Derivatives
Hans Dressler

TARGET SITES FOR HERBICIDE ACTION
Edited by Ralph C. Kirkwood

Resorcinol

Its Uses and Derivatives

Hans Dressler
Formerly of INDSPEC Chemical Corporation
Pittsburgh, Pennsylvania

Plenum Press • New York and London

Library of Congress Cataloging-in-Publication Data

Dressler, Hans.
Resorcinol : its uses and derivatives / Hans Dressler.
p. cm. -- (Topics in applied chemistry)
Includes bibliographical references and index.
ISBN 0-306-44850-5
1. Resorcinol. I. Title. II. Series.
QD341.P5D67 1994
547'.633--dc20 94-30813
CIP

ISBN 0-306-44850-5

A Division of Plenum Publishing Corporation
233 Spring Street, New York, N.Y. 10013

Printed in the United States of America

Preface

The chemistry, manufacture, and uses of the specialty chemical resorcinol (1,3-dihydroxybenzene) have a rich history, variety, and current interest. This book is intended to be the first detailed review of this product. The numerous references and the Index given in the book are intended to help users of resorcinol and resorcinol derivatives in their work, perhaps even to help in the development of new ideas.

The chemical industry in the United States, as well as in many other parts of the world, is being challenged on many fronts, old and new, due to global competitive, environmental, governmental, political, and technological trends that seem more variable than ever before. The case of resorcinol mirrors this array of challenges and demonstrates the broad range of successes and some of the problems of the industry.

It is shown how a sprout of a chemical grew over time into a sizable, still growing item of commerce. In the description of this development it becomes apparent how many different talents and trades it takes to develop, manufacture, and market a product.

Also given are indications of how academic research and applied research/development quite often, intentionally or not, influence each other and can spur the creative process to the benefit of both.

Areas are outlined where the current, often intense, work employing resorcinol is aimed. Thus, the old chemical is shown to be involved in an intriguing tomorrow.

Hans Dressler

Monroeville, Pennsylvania

Acknowledgments

The help and input of my former colleagues at INDSPEC Chemical Corp., and the former Koppers Co., on select aspects of this book is acknowledged gratefully. The permission to use an office/facilities at INDSPEC's research and development department in Harmarville, Pennsylvania, during the preparation of the manuscript was particularly helpful. Special thanks and mention are due: At Harmarville, to Fred M. Covelli—Vice President for R&D, Ray Bender, Harv Daily, Raj Durairaj, Valerie Kennedy, and Alex Peterson; and at INDSPEC's Pittsburgh, Pennsylvania, office to Frank Spinola—President, H. James Michaels, and James Kozler.

A great contribution was made by Dr. T. David Westmoreland of Wesleyan University, Middletown, Connecticut, who did the structure drawings. His help is acknowledged with relief and thorough appreciation.

Last but not least, a load of thanks is due to my wife, Bernice, for understanding the travails of this author in the time-consuming process of preparing a manuscript.

Contents

1. Introduction 1

2. The Properties and Chemistry of Resorcinol

2.1. Structure, Names, and Numbering of Ring Positions 5
2.2. Physical Properties, Including Spectra 5
2.3. Toxicology/Occupational Safety and Health Information 6
2.4. Environmental Concerns/Waste Disposal and Handling of Spills into Water 11
2.5. Chemistry 12
2.5.1. Introductory Remarks 12
2.5.2. Resorcinolsulfonic Acids 13
2.5.3. 2,4-Dihydroxybenzoic Acid (beta-Resorcylic Acid) 14
2.5.4. 3,5-Dihydroxybenzoic Acid (alpha-Resorcylic Acid) 15
2.5.5. 2,6-Dihydroxybenzoic Acid (gamma-Resorcylic Acid) 16
2.5.6. Resorcinol-4,6-dicarboxylic Acid (4,6-Dihydroxyisophthalic Acid) 17
2.5.7. Hydrogenated Resorcinols 17
2.5.8. Alkylresorcinols 19
2.5.9. Acylresorcinols 21
2.5.10. Haloresorcinols 21
2.5.11. Other Examples of Regioselective Reactions 22
2.6. Methods of Analysis 23
2.7. References 23

3. Processes for Making Resorcinol

3.1. Resorcinol by the Disulfonation/Caustic Fusion of Benzene 27

3.2. Resorcinol from the Hydroperoxidation/Cleavage of *m*-Diisopropylbenzene 31
3.3. Resorcinol from the Acid Hydrolysis of *m*-Phenylenediamine 40
3.4. Resorcinol from the Acetone/Acrylic Compound Condensation/Cyclization/Dehydrogenation 42
3.4.1. Stamicarbon's Version 43
3.4.2. Hoechst's Version 44
3.4.3. Koppers's Version 45
3.5. Resorcinol from Low-Temperature Coal Pyrolyzates, Shale Oil, and Plant Sources 47
3.6. Resorcinol from Haloaromatic Precursors 49
3.7. Routes Starting with Oxygenated Cyclohexanes/Cyclohexenes 49
3.8. Resorcinol Derivatives from Noncyclic Precursors Other than Acetone/Acrylic Compounds 51
3.9. Miscellaneous Routes 54
3.10. References 55

4. The Use of Resorcinol in Rubber Compositions

4.1. Introduction and Overview 59
4.2. Examples of Actual Application Results 62
4.3. Selections from the Patent Literature, 1983–1993 65
4.4. Nontire Applications of Resorcinol in Rubber Compounds 73
4.5. Resotropin 75
4.6. Thio- and Dithio-bis-Resorcinol 76
4.7. Concluding Remarks 77
4.8. References 78

5. Resorcinol/Formaldehyde Resins—Adhesives for Wood and Other Nonrubber Applications

5.1. Introduction 85
5.2. Adhesives for Wood 85
5.2.1. Overview 85
5.2.2. Selections from the Literature 91
5.3. Fire-Retardant, Fiber-Reinforced Composites 95
5.3.1. Introduction and Selections from the Literature 95
5.3.2. Specific Examples of Applications 97
5.4. Ion-Exchange Resins 102
5.5. Soil Stabilizers and Drilling Aids 106
5.6. Binders for Foundry Cores, Molds, Refractories, and Friction Materials 107

5.7. Phenolic Resin Foams, Aerogels, and Related Polymer Foams 109
5.8. Additives for Concrete 111
5.9. Tanning Agents for Animal Hides 112
5.10. Microcapsules 112
5.11. Coatings for Polymerization Reactors 114
5.12. Lithography, Printed Circuit Boards, Magnetic Tapes, and Encapsulated Semiconductors 115
5.13. Miscellany 116
5.14. References 119

6. *m*-Aminophenol

6.1. Introduction 125
6.2. *m*-Aminophenol from Resorcinol 126
6.3. *N*-Alkyl- and *N,N*-Dialkyl-*m*-Aminophenols from Resorcinol or *m*-Aminophenol 130
6.4. *m*-Aminophenol(s) from 1,3-Cyclohexanedione(s) 132
6.5. Stabilization of *m*-Aminophenol 133
6.6. References 133

7. Agricultural Chemicals, Including Veterinary Products

7.1. Introduction 135
7.2. Commercial Products 136
7.3. A Sampling of Developments from 1980–1992 141
7.3.1. Cyclohexanediones 142
7.3.2. Phenyl Ethers 157
7.3.3. Resorcylic Acid Derivatives 163
7.3.4. Heterocyclic Derivatives of Resorcinol 165
7.3.5. Miscellaneous Resorcinol Derivatives 167
7.3.6. Derivatives of *m*-Aminophenol 168
7.4. References 171

8. Pharmaceuticals, Over-the-Counter Medications, and Diagnostic Aids

8.1. Commercial Products 179
8.1.1. Bronchospasmolytics, Antiasthmatics, and Respiratory Stimulants . . 179
8.1.2. Coronary Vasodilators and Betablockers 181
8.1.3. An Ulcer Medication 181
8.1.4. An Antiseptic and Anthelminthic Agent 182

8.1.5. Choleretic Agents . . . 182
8.1.6. An Antiadrenergic Agent . . . 183
8.1.7. A Tuberculostatic Drug . . . 183
8.1.8. Antidepressants and Antipsychotics . . . 183
8.1.9. Reversible Cholinesterase Inhibitors . . . 183
8.1.10. Antibiotics, Antibacterials, and Antifungal Agents . . . 184
8.1.11. Antiemetics . . . 185
8.1.12. Miscellaneous Pharmaceuticals . . . 185
8.1.13. Acne and Antiseborrheic Preparations, Dandruff Control . . . 185
8.1.14. Ultraviolet Light Screens for Skin Protection . . . 187
8.1.15. Diagnostic Aids and Reagents . . . 187
8.1.16. Outlook . . . 188
8.2. Selection of Developments Since 1980 . . . 189
8.3. References . . . 220

9. The Uses of Resorcinol/Derivatives in Polymers

9.1. Introduction . . . 229
9.2. Monomers Based on Resorcinol or *m*-Aminophenol . . . 230
9.2.1. Resorcinol Diacetate . . . 230
9.2.2. Resorcinol Di-(2-hydroxyethyl) Ether [HER] . . . 230
9.2.3. (*m*-Phenylenedioxy)diacetic Acid . . . 231
9.2.4. 3,4′-Diaminodiphenyl Ether . . . 232
9.2.5. 4,6-Diaminoresorcinol . . . 233
9.2.6. 1,3-bis-(4-aminophenoxy)benzene, 1,3-bis-(3-aminophenoxy)-5-chlorobenzene, and 1,3-bis-(3-aminophenoxy)benzene . . . 234
9.2.7. Diamines of Aromatic Ether-Ketones . . . 234
9.2.8. Diamines from 2,6-Dichloropyridine . . . 235
9.2.9. Diamines of Aromatic Ether-Sulfones . . . 236
9.2.10. 3,8-Dihydroxy-5*a*,10*b*-diphenylcumarano-2,3,2′,3′-cumaran . . . 237
9.3. Polyurethanes . . . 237
9.3.1. HER as a Chain Extender . . . 237
9.3.2. Resorcinol Di-(hydroxypropyl) Ether . . . 238
9.3.3. Resorcinol/Formaldehyde Resins . . . 238
9.4. Epoxy Resins . . . 238
9.4.1. Resorcinol Diglycidyl Ether . . . 239
9.4.2. Glycidyl Ethers of Resorcinol Derivatives and Resins . . . 241
9.4.3. The Use of Resorcinol, Resorcinol Derivatives, or Resins as Curing Agents for Epoxy Resins . . . 245
9.5. Polyesters . . . 248
9.5.1. Partially Aromatic Polyesters . . . 248
9.5.2. Polyarylates . . . 249

9.5.3. Thermotropic Liquid-Crystal Polymers (LCPs) 251
9.6. Polyamides 252
9.7. Polyimides, Including Maleimide-Type Polymers 253
9.8. Polyamides/Imides and Other Multifunctional Polymers 258
9.9. Polyaryl Ethers 260
9.9.1. Polyphenyl Ethers 260
9.9.2. Poly(cyanoaryl) Ethers 262
9.9.3. Polyaryl Ether/Sulfones and Polyaryl Ether/Ketones 265
9.10. Dicyanates 266
9.11. Polymers Containing Other Heterocyclic Units 266
9.12. Miscellaneous Polymers 268
9.13. References 269

10. Dyes, Fluorescent Chemicals, Optical Bleaches, Laser Dyes, and Imaging/Recording Technologies

10.1. Introduction 279
10.2. Dyes and Related Chemicals in Commercial Use 280
10.2.1. Azo Dyes 280
10.2.2. Coumarins, Xanthenes, and Oxazines 281
10.2.3. Plant Extracts as Dyes 286
10.2.4. Chemicals for Diazotype Reprography 286
10.2.5. Chemicals for Carbonless Paper Copying 287
10.2.6. Chemicals for Thermographic Direct and Transfer Recording . . 289
10.2.7. Applications in Other Imaging Techologies 290
10.3. A Sampling of Developments from 1980 On 291
10.3.1. Dyestuff Intermediates and Processes 292
10.3.2. Azo Dyes 293
10.3.3. Fluorescent 7-Hydroxy- and 7-Aminocoumarins, Xanthenes, and Fluorans 295
10.3.4. Chemicals for Diazotypy 297
10.3.5. Leuco Dyes for Carbonless and Thermal Copying Technologies . . 299
10.3.6. Developers and Sensitizers for Carbonless and Thermal Printing Paper 310
10.3.7. Silver Halide Photographic Materials 313
10.3.8. Lithography and Photoresists 316
10.3.9. Liquid Crystals 320
10.3.10. Miscellaneous Applications in High Technology 321
10. 4. References 324

11. Additives of Many Types

11.1. Introduction . . . 331
11.2. Commercial Products . . . 331
11.2.1. Ultraviolet Light Absorbers . . . 331
11.2.2. Flame Retardants . . . 335
11.3. Selection of Developments in Additives . . . 336
11.3.1. Adhesion Promoters . . . 336
11.3.2. Antioxidants/Heat and Processing Stabilizers for Organic Materials . . . 336
11.3.3. Catalysts . . . 341
11.3.4. Corrosion Inhibitors . . . 343
11.3.5. Electrolytes . . . 343
11.3.6. Flame Retardants . . . 344
11.3.7. Additives for Functional Fluids . . . 348
11.3.8. Ultraviolet Light Stabilizers . . . 349
11.3.9. Food Additives . . . 364
11.3.10. Miscellaneous Additives . . . 365
11.4. References . . . 367

12. Other Uses for Resorcinol

12.1. Explosives . . . 375
12.2. Beauty Aids and Flavorants . . . 378
12.2.1. Hair Dyes . . . 378
12.2.2. Perfumes and Aromas . . . 380
12.2.3. Other Applications in Cosmetics . . . 382
12.3. Miscellaneous Other Uses . . . 383
12.4. References . . . 389

13. Selected Business Aspects

13.1. History of Actual/Potential Producers of Resorcinol . . . 393
13.2. Markets/Marketing . . . 398
13.3. Management Tasks and Challenges . . . 400
13.3.1. General . . . 400
13.3.2. Environmental and Health Regulations . . . 401
13.3.3. Research & Development and Patents . . . 402
13.4. References . . . 404

14. Occurrence in Nature—A Domain of Academic Researchers

14.1. Introduction 405
14.2. Resorcinol and Relatively Simple Derivatives 406
14.3. Cashew Nut Shell Liquid 410
14.4. Coumarins 411
14.5. Precocene-1, An Insect Antijuvenile Hormone 414
14.6. Lichen Constituents 416
14.7. Cannabis/Marijuana 420
14.8. Psoralens and Related Compounds 423
14.9. Macrolides 424
14.10. Others 426
14.11. References 431

15. Other Examples of Mostly Academic Work with Resorcinol

15.1. Introduction 437
15.2. Host–Guest Chemistry and Molecular Recognition 437
15.2.1. Calixarenes, Baskets, Carcerands, Cavitands, Hemicarcerands, Velcraplexes, Enzyme Mimics, and Related Topics in Supramolecular Chemistry 437
15.2.2. Other Macrocyclic Hosts 445
15.2.3. Noncyclic Hosts 448
15.3. Physical and Analytical Chemistry 450
15.3.1. Studies in Physical Chemistry 450
15.3.2. Topics in Analytical Chemistry 455
15.4. 1,3-Cyclohexanedione in Organic Synthesis 460
15.5. References 478

16. Postscript 485

Index 489

1

Introduction

Resorcinol has been known for about 125 years, yet, to the nonspecialist and the general public, the mention of resorcinol does not ring a bell. In part, the reason for this gap in recognition is that there have been only few, relatively short reviews of resorcinol in technical encyclopedias and most chemistry textbooks barely mention it, if at all. Yet, resorcinol is a significant contributor to many technologies, enhancing the properties of a wide range of materials or being the starting material for high-value-added products. Resorcinol plays a significant part in academic research, but rarely is the focus of attention.

This book is intended to fill this information gap by seeing the world of chemistry, in broad perspective, from the resorcinol standpoint in a purely descriptive manner. To this end, a more detailed, yet selective, picture of this chemical is given. To keep the book to a reasonable size, equations for syntheses and fine details are presented only occasionally to serve the specialist or to interest the novice, but over 1800 references are given for unearthing further details. The literature before 1980 is used but sparingly. The majority of the references are from the 1980–1992 period, with a sprinkling of 1993 publications.

The overall approach has been to attempt to integrate the whole chemical enterprise, from the technical (chemical and interdisciplinary) and business aspects to the admirable academic work involving resorcinol and its derivatives, to encourage mutual understanding. Once again it is demonstrated that chemistry plays an important part in many older areas of technology and in newer, high-technology fields. Both the joys and the pains of being involved with chemicals nowadays will become apparent.

After this introduction, the book proceeds with a chapter on the properties and chemistry of resorcinol. Much information on the toxicity and safety in the handling of resorcinol is presented. The reader will come to understand that massive studies have been conducted on these aspects, in good part voluntarily by industry and in part mandated by government regulations, at great expense to industry and the public, indirectly. The presentation of the chemistry of resorcinol is concentrated on ways by which (regio)selective reactions can be obtained in the face of multiple possible points of substitution.

Next, a chapter on processes, both commercial and standby, for making resorcinol goes into some detail to show the intensive competition that exists worldwide, even on this one specialty chemical, to develop the lowest cost process—low raw-materials cost, high yield/little waste, safety, low energy use, and low investment cost in a new plant. The new process concepts are presented as a review of the literature.

Chapters 4 to 12 describe the many uses of resorcinol in industry, covering commercial applications first, followed by a selection of patents used to indicate newer ideas and trends. It will become apparent that old uses constantly require modified products to fit evolutionary needs, while new uses may be based on old chemistry but require new application technology. Indeed, it will become clear that each application category requires specialized testing techniques, equipment, and expertise. There are hints of rapid growth for new products and maturity, if not decline for some older products. There are some good examples of the value-added concept, with higher prices if not profits needed and obtainable for the performance obtained from derivatives of resorcinol.

Included in the applied chemistry treatment is a chapter on the preparation of *m*-aminophenol from resorcinol, because this route has been commercialized in Japan. *m*-Aminophenol is then treated as a resorcinol derivative in this book; many of its applications, mostly for high-value-added products, are included in the descriptions of uses for resorcinol.

The applied chemistry sections are followed by Chapter 13 on selected business aspects, showing that many have tried but only few have succeeded in making a profit from the manufacture of resorcinol. This chapter demonstrates that making a profit is not obscene but necessary for survival and as a reward for successfully managing the old and new complexities—from global competition to environmental realities to operational efficiency.

Chapter 14 sketches the occurrence of resorcinol in nature, necessarily, within the limited space of a book of this scope, as a collection of snapshots but adequate to show the fascinating diversity of the compounds evolved by nature. They range from simple compounds to complex molecules and their isolation, structure elucidation, and proof of structure by synthesis are shown to be largely in the academic domain. There are examples of fragrances, dyes, folk medicines, and oily plant extracts in this review which are commercially used as such or serve as models for the development of new drugs and other useful materials.

The penultimate Chapter 15 presents a few, selected examples of other, mostly academic work involving resorcinol chemistry. Mentioned are the massive work on macrocycles and host–guest chemistry, studies in physical and analytical chemistry, and the use of 1,3-cyclohexanedione (dihydroresorcinol) in the synthesis of many interesting compounds.

The placement of Chapters 14 and 15 near the end of the book is not intended as an indication of the lesser importance of these topics, but rather as an intent to finish with a note of enthusiasm, to show the burst of ideas and the connection to the whole fabric.

The interconnection and cross-fertilization between applied chemistry and academia are apparent throughout the whole text. There are learned papers from industry and application ideas coming from academic researchers and teachers. Referrals are given in each chapter to topics in other chapters, to connect the parts to the whole. An ample index is provided to further help the reader to find the connections.

The postscript contains musings on resorcinol as a mirror of the current state of chemistry and its future, concluding on an optimistic note.

2

The Properties and Chemistry of Resorcinol

2.1. STRUCTURE, NAMES, AND NUMBERING OF RING POSITIONS

Other names which have been used are 1,3-benzenediol, resorcin, *m*-dihydroxybenzene, 1,3-dihydroxybenzene, and 1,3-dioxybenzene. The conventional numbering of the carbon atoms of resorcinol is shown inside structure **2-1**. The alpha (α), beta (β), gamma (γ) designation was used in the older literature and is still used occasionally today.

2.2. PHYSICAL PROPERTIES, INCLUDING SPECTRA

Resorcinol is a colorless solid with a mild odor. In nature, resorcinol appears in the form of both simple and complex derivatives (see Chapter 14). The solid resorcinol occurs in two crystalline forms, the alpha form which is obtained by crystallization of resorcinol from ethanol, and the beta form which is obtained on crystallizing resorcinol from benzene or on cooling of molten resorcinol. (For a recent, intensive study of these phenomena, see Chapter 15, Ref. 81.)

Table 2.1 lists the important physical properties of resorcinol. Figures 2.1, 2.2, and 2.3 present the IR, ^{1}H-NMR, and ^{13}C-NMR spectra, respectively, of resorcinol.

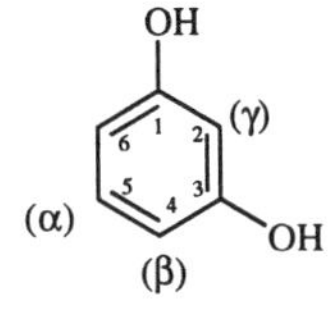

2 - 1

Resorcinol

CAS Registry No. [108-46-3]

UN Number 2876

Table 2.1. Physical Properties of Resorcinol

Molecular weight	110.11
Boiling point	276.7 °C/760 mm Hg (280.8 °C/1013 mbar)
	210.0 °C/100 mm Hg (209.8 °C/133 mbar)
	152.3 °C/10 mm Hg (152.1 °C/13.3 mbar)
	108.4 °C/1 mm Hg (110.8 °C/1.3 mbar)
Melting point	110.8 °C
Density, solid at 20 °C	1.292 g/cm^3
molten at 150 °C	1.1506 g/cm^3
Viscosity, 140 °C	38 millipoises
150 °C	31 millipoises
pH, 10% solution in water	4.5
Dissociation constants in water at 30 °C	$K_1 = 7.11 \times 10^{-10}$; $K_2 = 4.78 \times 10^{-12}$
Specific gravity (15 °C/4 °C)	1.272
Solubility, wt%, in water	58.3/20 °C; 83.3/60 °C; 96.8/100 ° C
in ethanol	61/20 °C; 73/60 °C
in acetone	67/20 °C; 83/60 °C
in benzene	2.2/20 °C; 14/60 °C
in carbon tetrachloride	very low/20 °C; 0.3/60 °C
Flash point, tag, open cup	168 °C
closed cup	127 °C
Heat capacity, at 25 °C	0.28 cal/g °C
at 110 °C	0.43 cal/g °C
at 130 °C	0.57 cal/g °C
at 150 °C	0.58 cal/g °C
Heat of crystallization	5.09 kcal/mole
Heat of combustion	683.90 kcal/mole

2.3. TOXICOLOGY/OCCUPATIONAL SAFETY AND HEALTH INFORMATION

The toxic effects of resorcinol are related to dose, as seen from the data below. Resorcinol is relatively easily absorbed by the skin and mucosa, and, after gluconoride formation and sulfation, is predominantly eliminated from the body via the kidneys. The acute toxicity of resorcinol is significantly less than that of phenol.[1]

General toxicity data are given in Table 2.2. Acute animal toxicity data are given in Table 2.3. These data are from brochures published by INDSPEC Chemical Corp. and are included here with their permission.

In 1992, a report from the National Toxicology Program appeared, giving the results of a two-year gavage (feeding) study of resorcinol: there was no evidence of carcinogenicity in male rats (F 334/N strain) fed 112 or 225 mg/kg/day or female rats (344/N strain) fed 50, 100, or 150 mg/kg/day; no signs of carcinogenicity were found in male or female mice (B6C3F$_1$ strain) fed resorcinol at 112 or 225 mg/kg/day; there were some signs of chemical-related neurotoxicity.[6]

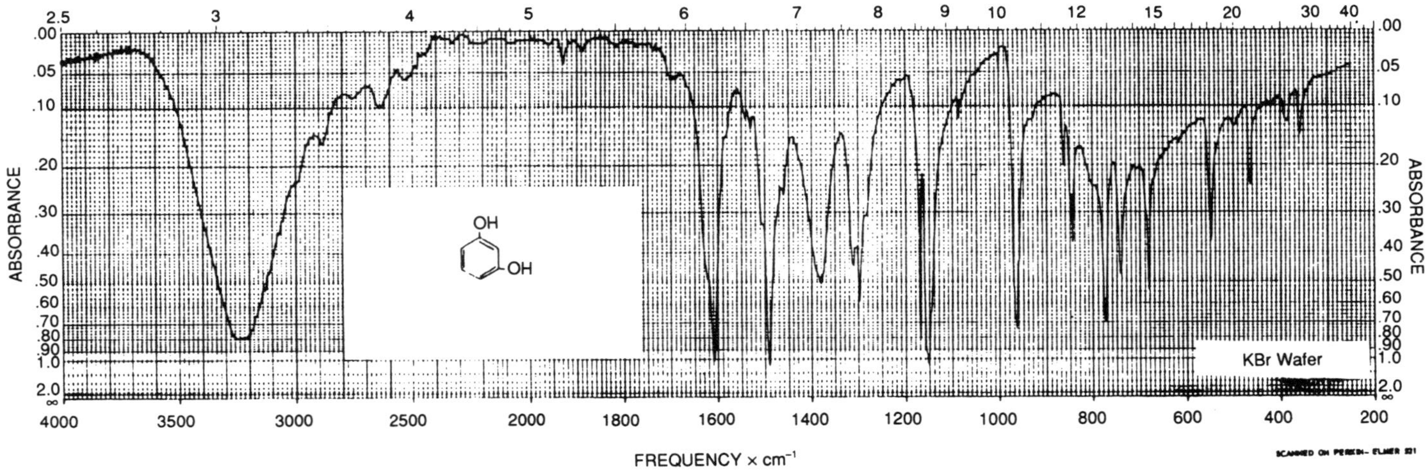

Figure 2.1. IR spectrum of resorcinol. Permission for the publication herein of Sadtler Standard Spectra® has been granted and all rights reserved by Sadtler Research Laboratories, Division of Bio-Rad Laboratories, Inc.

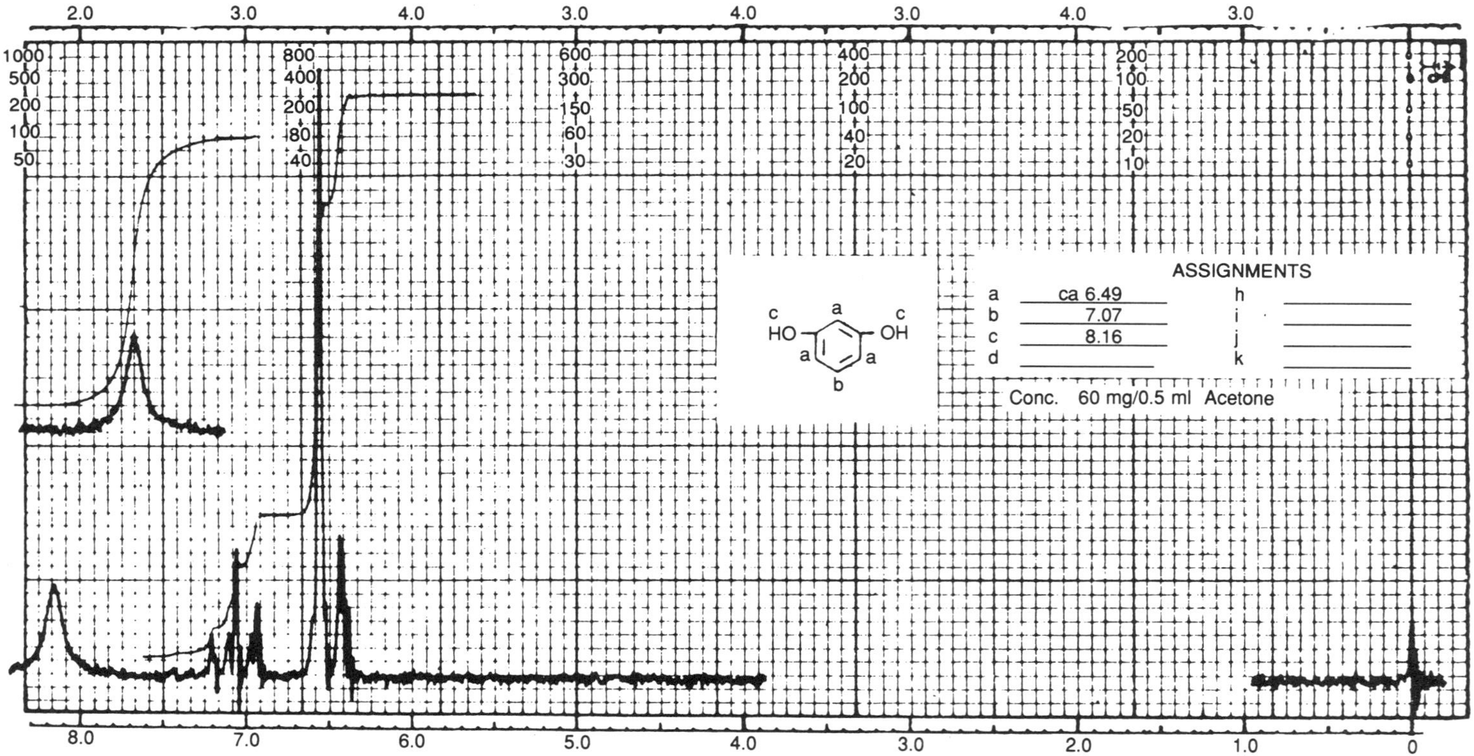

Figure 2.2. ^{1}H-NMR spectrum of resorcinol. Permission for the publication herein of Sadtler Standard Spectra has been granted and all rights reserved by Sadtler Research Laboratories, Division of Bio-Rad Laboratories, Inc.

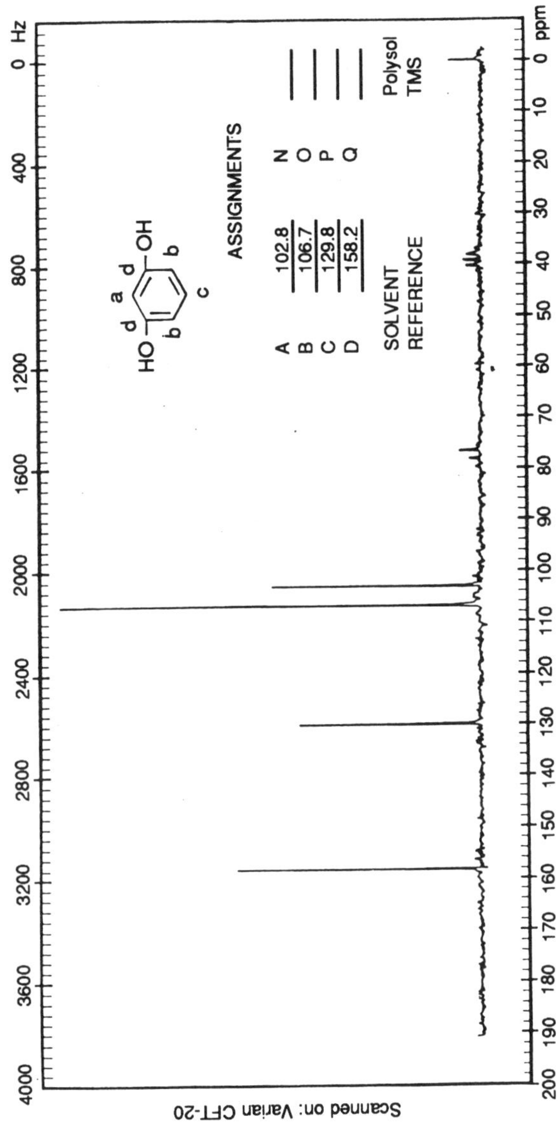

Figure 2.3. ^{13}C-NMR spectrum of resorcinol. Permission for the publication herein of Sadtler Standard Spectra has been granted and all rights reserved by Sadtler Research Laboratories, Division of Bio-Rad Laboratories, Inc.

Table 2.2 General Toxicity Data

Threshold Limit Value—Time Weighted Average (TLV-TWA)[2] for substances in workroom air, adopted by ACGIH	10 ppm (45 mg/m^3)
Qualifying Toxic Dose, Oral—human, LD_{LO}[b], [3a]	29 mg/kg
Oral—rat, LD_{50}[c]	301 mg/kg
Subcutaneous—mouse, LD_{LO}[b]	340 mg/kg
Ames Salmonella/Microsome Test for Mutagenicity[4a]	Negative
Subacute Animal Toxicity Data[5a]—Throat Spray Tests with guinea pigs and rats; 3 daily spraying of the throat for 2 weeks with a 1% water solution of resorcinol	No gross evidence of respiratory damage
Inhalation Tests—guinea pigs, rats, and rabbits: exposed to a resorcinol concentration of 34 mg/m^3, 6 hr daily, for two weeks	No evidence of toxic effects to lungs and trachea were noted when the animals were sacrificed

[a]This number refers to a reference at the end of the chapter.
[b]Lowest published lethal dose.
[c]Lethal dose, 50% kill.

It is appropriate to point out here that the greatest potential of exposure to resorcinol would be in the production facility of the only U.S. producer, INDSPEC Chemical Corp. With the observance of proper precautions, this plant has had a good record as to the safety and health of its employees for many years.

A massive and valuable evaluation, published in 1986 by the Cosmetics Ingredients Review (CIR) Expert Panel, was the "Final Report on the Safety Assessment of 2-Methylresorcinol and Resorcinol."[7] This report considered literature published through 1984 as well as unpublished safety test data by the cosmetics industry on these

Table 2.3. Acute Animal Toxicity Data[a]

Test	Dose	Remarks
Single Dose Oral—LD_{50}[b]	980 mg/kg	Toxic as defined by the Federal Hazardous Substances Act (FHSA)
Single Dose Skin Penetration—LD_{50}, 24 hr contact, intact and abraded skin, rabbit	3360 mg/kg	Not classified by FHSA at this low range of toxicity
Primary skin irritation—24 hr contact, intact and abraded skin, rabbit	500 mg	Not a primary skin irritant by FHSA definition
Eye injury—24 hr exposure, rabbit	100 mg	Corrosive to the eye by FHSA definition
Inhalation—resorcinol/water aerosol,		
rat, 1 hr	7800 mg/m^3[c]	No toxic signs or symptoms
rat, 8 hr	2800 mg/mg^3	No toxic signs or symptoms
rat, 8 hr daily for 90 days	957 mg/m^3	Hyperplasia of the thyroid gland but no significant effect on thyroid weight

[a]Results of tests sponsored by INDSPEC Chemical Corp, reproduced here with permission.
[b]Lethal dose, 50% kill.
[c]Maximum concentration that could be obtained with the Dautrebonde aerosol generator.

ingredients. The major use of resorcinol and 2-methylresorcinol there is in cosmetic hair dye formulations at concentrations between 1% and 5%. A gamut of tests with mice, rabbits, and guinea pigs, and skin irritation tests with human subjects concluded that, based on acute oral toxicity studies, resorcinol is slightly to moderately toxic and 2-methylresorcinol is moderately toxic; resorcinol was a mild skin irritant and a rare sensitizer in clinical testing, but not when tested on nonclinical groups; both 2-methylresorcinol and resorcinol were considered safe as cosmetic ingredients in the present practices of use and concentration.

In addition, in 1991 the FDA (Food and Drug Administration) gave the GRAS (generally recognized as safe) designation to over-the-counter acne treatments containing about 2% resorcinol (see Chapter 8, Section 8.1.13 and Ref. 3 therein) in combination with sulfur and other ingredients.

2.4. ENVIRONMENTAL CONCERNS/WASTE DISPOSAL AND HANDLING OF SPILLS INTO WATER

The information given below is from brochures of INDSPEC Chemical Corp. and published here with their permission.

Resorcinol is covered by the Hazardous Waste Management System, 40 CFR 261.33 regulation. The Hazardous Waste Number for resorcinol is U201.

These regulations provide specific requirements which must be complied with whenever resorcinol is discarded, intended to be discarded, or is spilled into or on any land or water. The disposal of such waste and any residue or contaminated soil, water, or other debris resulting from the cleanup of a spill of resorcinol product must be in accordance with current regulations of Title C of the Solid Waste Disposal Act, as amended by the Resource Conservation and Recovery Act (RCRA) of 1976, as well as pertinent state and local regulations.

Resorcinol is also covered by the regulations of The Clean Water Act, CFR Title 40, Parts 116 and 117. These regulations stipulate that the U.S. Coast Guard National Response Center must be notified immediately when a Reportable Quantity of resorcinol is spilled into or upon the navigable water of the United States or adjoining shorelines and contiguous zones. The Reportable Quantity of resorcinol as defined by the Clean Water Act is 2270 kg (5000 lb.). It should be noted that under these regulations the spill must be ≥2270 kg (5000 lb.).

Disposal methods for resorcinol include complete incineration, land (soil) farming, and decomposition in activated sludge-type wastewater treatment plants. All disposal practices should be carefully evaluated for compliance with applicable local, state, and federal regulations.

Resorcinol is biodegradable, as confirmed by tests conducted by Koppers Co. Environmental Resources Department following the Test Methods for Oxygen Uptake proposed by ASTM Task Group D19.24.06.02 in both resorcinol- and phenol-acclimated activated sludges. The rate of biodegradation of resorcinol in an activated sludge treatment facility is dependent upon many factors, e.g., concentration and temperature.

Table 2.4. Resorcinol—Water Quality Criteria

Threshold Concentration for odor[8a]	40 mg/l
Fresh Water Critical Concentration, toxic threshold for bleak and carp	35 mg/l
LC_{50}[b], fathead minnows[9a]	71 mg/l

[a]This number refers to a reference at the end of the chapter.
[b]Lethal concentration, 50% kill.

Information related to the requirements of water quality in regards to odor and effects on fish is presented in Table 2.4.

2.5. CHEMISTRY

2.5.1. Introductory Remarks

Resorcinol behaves much like phenol, since the aromatic ring is activated by the hydroxyl groups due to resonance and inductive effects. Thus, resorcinol can be readily substituted at the 2-, 4-, and/or 6-positions by halogenation, alkylation, acylation, sulfonation, nitration, nitrosation, or coupling with aryldiazonium compounds; it can be carboxylated by the Kolbe reaction or its variations (e.g., the Kolbe-Schmitt reaction); it can be (poly)condensed with aldehydes and ketones; it can be converted to resorcinate salts by reaction with inorganic or organic bases; e.g., a 1:1 salt, commonly called resotropin, is formed by the reaction of resorcinol with hexamethylenetetramine (the preparation and a use of resotropin is described in Chapter 4); it can be aminated by ammonia or amines with the replacement of hydroxyl groups, an example of the Bucherer reaction (this is described in detail in Chapter 6); and resorcinol can be hydrogenated to give nonaromatic compounds. The chemistry of resorcinol is well dispersed in the literature. A thorough summary of this chemistry through 1959 is contained in Beilstein's Handbuch der Organischen Chemie.[10] Additional literature on resorcinol chemistry to the early 1970s is given in *Houben-Weyl's Methoden der Organischen Chemie.*[11] The general chemistry of phenols, with references to resorcinol as well, is also summarized in *Rodd's Chemistry of Carbon Compounds.*[12]

As a dihydric phenol, resorcinol has much higher reactivity than phenol; almost any reaction of resorcinol gives several products. In ring-substitution reactions, regiospecificity is often not easily achieved but may be attained in many cases by special procedures. In reactions at the hydroxyl groups, such as etherification and esterification, mono-substitution may be difficult to achieve in high yield; pure products may also be difficult to obtain, even when di-substituted derivatives are made. From the industrial viewpoint, high in-hand yields of the desired products, of high purity or readily purifiable, and the minimization of byproducts have become more and more important for reasons such as lower production cost, demands/needs of the customer (e.g., 99+% pure monomers), and the minimization of waste disposal problems, such as the potentially high cost of waste disposition. The use of less toxic and/or hazardous reagents for highly toxic/hazardous reagents has also gained urgency.

The following subsections present examples of special techniques that have been developed to make the reactions of resorcinol more selective, if not different from the chemistry of phenols.

2.5.2. Resorcinolsulfonic Acids

The seemingly simple sulfonation reaction is difficult to study because of multiple products, difficulties of analysis (especially before the advent of NMR spectroscopy and HPLC), and the difficulties in preparing high-purity analytical standards. Therefore, few studies are available.

Podstata and Allan in the 1960s studied the sulfonation of resorcinol with 70% H_2SO_4 which gave predominantly the resorcinol-4-sulfonic acid (2,4-dihydroxybenzenesulfonic acid), **2-2**, while 90% H_2SO_4 gave resorcinol-4,6-disulfonic acid, **2-3** (the *Chemical Abstract* entries of the paper said resorcinol-2,6-disulfonic acid was obtained as the main product); and with 20% oleum the resorcinol-2,4,6-trisulfonic acid, **2-4**, was produced. The formation of byproduct sulfones was recognized. The analysis of the products was by paper chromatography and color reactions (i.e., coupling with aryldiazonium halides). The equilibration of the sulfonation products at 130 °C was also studied.[13]

Recent work by H. Cerfontain, the contemporary grandmaster of mechanistic and kinetics studies of aromatic compound sulfonation, and his associates showed that, in the sulfonation of resorcinol and its mono- and di-methyl ether with a large excess of conc. H_2SO_4, protonation occurred initially at the C-4 and C-6 positions; subsequent sulfonation of resorcinol and its dimethyl ether gave the 4-sulfonic acid, and sulfonation of resorcinol mono-methyl ether (3-methoxyphenol) gave the 4- and 6-sulfonic acids. The sulfonation of resorcinol-4-sulfonic acid gave a mixture of the 2,4- and 4,6-disulfonic acids, whereas the sulfonation of resorcinol-4-sulfonic acid mono- and di-methyl ether gave the corresponding 4,6-disulfonic acids; sulfonation of 3-methoxyphenol-6-sulfonic acid gave a mixture of the corresponding 2,6- and 4,6-disulfonic acids. Resorcinol-2,4-disulfonic acid and 3-methoxyphenol-2,6-disulfonic acid slowly isomerized to the corresponding 4,6-disulfonic acids. With sulfuric acid of >82.8% concentration, resorcinol-4,6-disulfonic acid was slowly converted to the 2,4,6-trisulfonic acid; the disulfonic acids of the resorcinol mono- and di-methyl ether

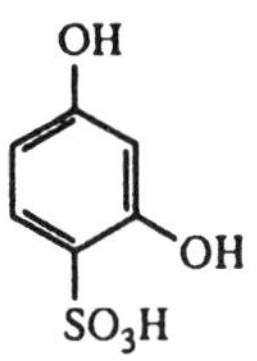

2-2

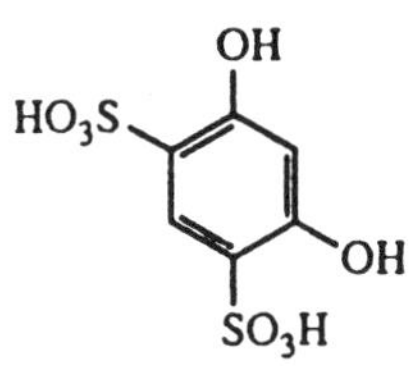

2-3

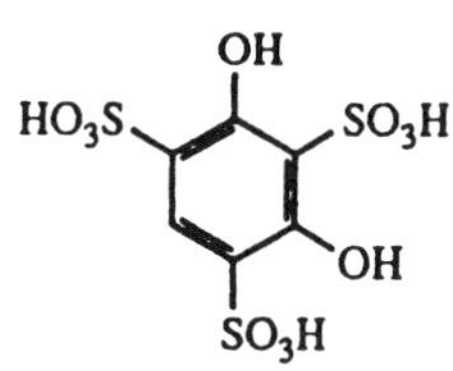

2-4

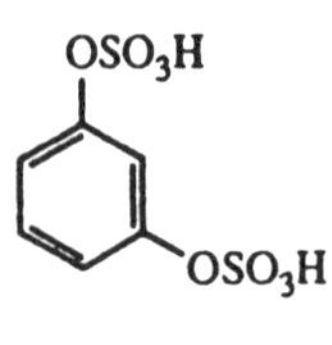

2 - 5

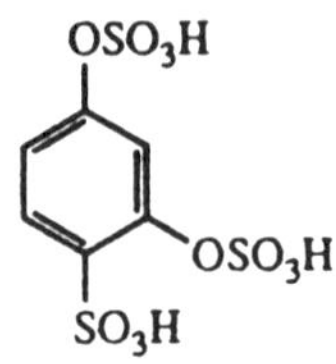

2 - 6

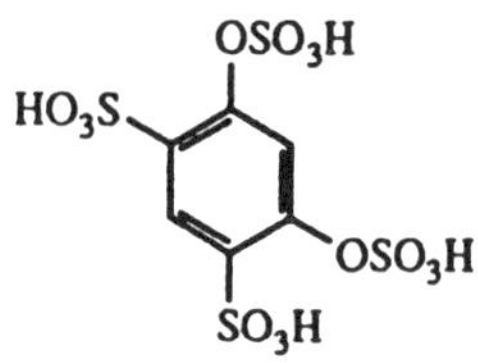

2 - 7

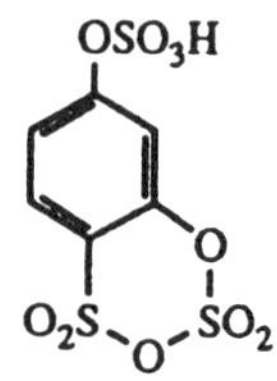

2 - 8

were not further sulfonated. ^{1}H-NMR analysis was the major analytical tool. The kinetics of resorcinol sulfonation was studied with 73.5%, 82.8%, 88.5%, 95.5%, and 98.5% sulfuric acid.[14]

Cerfontain and his associates also studied the sulfonation of resorcinol (and the other dihydroxybenzenes) and its methyl ethers with SO_3 in nitromethane at 0 °C. Resorcinol with 1 equiv. SO_3 gave the resorcinol-4-sulfonic acid; but resorcinol with 4 equiv. SO_3 first gave the disulfate of resorcinol, **2-5**, which transformed subsequently into the resorcinol-4-sulfonic acid disulfate, **2-6**, and finally into a mixture of resorcinol-4,6-disulfonic acid disulfate, **2-7**, and the cyclic anhydride, **2-8**; small amounts of trans-sulfonation products appeared. The sulfonation of the mono- and di-methyl ethers of resorcinol with SO_3 also showed such complexities.[15]

Nonetheless, single compounds of good purity have been prepared. Thus, when an equimolar amount of chlorosulfonic acid, $ClSO_3$, in nitrobenzene was rapidly added to a solution of resorcinol in nitrobenzene at 15–75 °C, pure resorcinol-4-sulfonic acid (2,4-dihydroxybenzenesulfonic acid) was said to be precipitated in almost quantitative yield.[16]

When resorcinol was dissolved in sulfuric acid and sulfonated with 30% oleum (30 wt% SO_3 in sulfuric acid) for 2 hr/100 °C, then cooled, neutralized with sodium hydroxide, and refrigerated for 2 days, crystalline disodium resorcinol-4,6-disulfonate was obtained (purity not given).[17]

2.5.3. 2,4-Dihydroxybenzoic Acid (beta-Resorcylic Acid), 2-9

This compound, **2-9**, is obtained by the reaction of resorcinol in aqueous solution with sodium or potassium bicarbonate. For example, a good laboratory process was described in *Organic Synthesis*[18]: Resorcinol was reacted with $KHCO_3$ in 1/5.5 mole

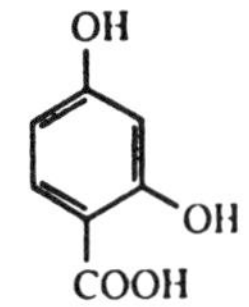

2-9

ratio in water for 4 hr/100 °C, then refluxed for 30 min while passing a rapid stream of CO_2 through the solution; after workup by acidification, etc., and purification by recrystallization/charcoal treatment, a 60% yield of pure 2,4-dihydroxybenzoic acid was obtained.

Since beta-resorcylic acid is important industrially, e.g., to make dyes and explosive materials, attempts to improve this process have been made. Thus, resorcinol was carboxylated in water with an excess of alkali carbonate at 80–100 °C while passing CO_2 through the solution from the start, followed by cooling to precipitate alkali carbonate, filtration (the solid alkali carbonate was recycled), and acidification of the filtrate to obtain a 50–60% conversion of resorcinol and ≥95% selectivity to 2,4-dihydroxybenzoic acid; the net amount of alkali carbonate required was said to be one fourth of that required by previous methods.[19]

In another variation, resorcinol/$KHCO_3$/$NaHCO_3$ in 1.0/1.67/4.0 mole ratio was reacted in an intensive mixer apparatus for 3 hr/120 °C under a stream of CO_2, then cooled and acidified with conc. HCl to pH 3, cooled to 10–15 °C, and filtered; the cake was washed and dried to give an 81% yield of 2,4-dihydroxybenzoic acid of 99.5% purity by HPLC.[20]

Finally, resorcinol/sodium hydroxide in 1.0/1.7 mole ratio was reacted in a stirred autoclave in 3-methoxy-1-butanol for 3 hr/140–150 °C under 3 atm of CO_2, then cooled, vented, and filtered; the cake was dissolved in water and acidified with conc. HCl to give a 95% yield of 2,4-dihydroxybenzoic acid.[21]

Overall, it should be noted that the process with the highest yield of beta-resorcylic acid is not necessarily the most cost effective; the production cost may be higher than in a process with a lower per-pass yield, as long as the unreacted resorcinol can be recovered, e.g., by extraction, and recycled.

2.5.4. 3,5-Dihydroxybenzoic Acid (alpha-Resorcylic Acid), 2-10

This product cannot be prepared by the carboxylation of resorcinol. Rather, it is best prepared by the sulfonation/caustic fusion of benzoic acid. An example of such a process is described in *Organic Synthesis*, using 30% oleum for the disulfonation of benzoic acid, $BaCO_3$ to neutralize the disulfonic acid, and a 1/1 mixture of KOH/NaOH for the fusion step; after drowning of the fusion mass in water and acidification, a 58–65% yield of the 3,5-dihydroxybenzoic acid was obtained.[22]

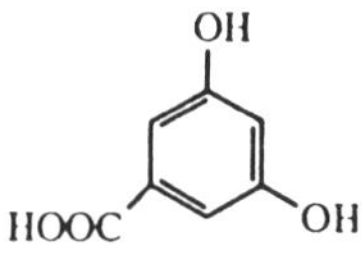

2-10

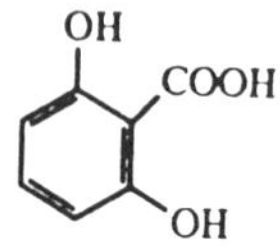

2-11

2.5.5. 2,6-Dihydroxybenzoic Acid (gamma-Resorcylic Acid), 2-11

This acid was prepared, e.g., by reacting resorcinol/$KHCO_3$ in 1/4 mole ratio in water in a stirred autoclave for 3.5 hr/130 °C under 27 atm CO_2, then venting, cooling, and acidification to obtain a 97% yield of a 52/48 mixture of 2,4-/2,6-dihydroxybenzoic acids; the mixture was then separated by ion-exchange chromatography.[23] This method is not suitable for industrial use.

Russian workers developed a method of selective precipitation/separation of the 2,4- and 2,6-dihydroxybenzoic acid from alkaline solutions by acidification, the 2,4-isomer being obtained at pH 2.6 and the 2,6-isomer at pH 1.0–1.5.[24]

A further advance came from Japan: The carboxylation of resorcinol was performed in a stirred autoclave at a 1.0/1.38 resorcinol/$KHCO_3$ mole ratio in ethanol at 140 °C under CO_2 pressure (10 kg/cm^2 total); the alcohol was then stripped off, the residue was taken up in water, and the solution was acidified to pH 4 with 75% H_2SO_4 to obtain a 36% yield of 2,6-dihydroxybenzoic acid (the byproduct must have been the 2,4-isomer).[25]

2,6-Dimethoxybenzoic acid, **2-12**, is one of the starting materials for the preparation of methicillin, a semisynthetic penicillin antibiotic (see Chapter 8). This compound can be made directly from resorcinol dimethyl ether by regioselective metalation at the 2-position in diethyl ether, treatment with CO_2, and acidification. Thus, Wittig *et al.* had done this in modest (14%) yield, starting with the metalation of 1,3-dimethoxybenzene with phenyllithium.[26]

The work was improved upon by Gilman and his group, who obtained 2,6-dimethoxybenzoic acid in 55% yield, starting by metalating 1,3-dimethoxybenzene with *n*-butyllithium and treatment with solid CO_2.[27] A big improvement, by Italian scientists, was the treatment of 2,6-dimethoxybenzene with BuLi and CO_2 in the presence of $Me_2NCH_2CH_2NMe_2$; a 99% yield of **2-12** and a 1% yield of the 4-isomer were obtained.[28] The metalation of 1,3-dimethoxybenzene can also be achieved with other organometallics. Thus, phenylsodium gave 2,6-dimethoxyphenyl sodium, which was

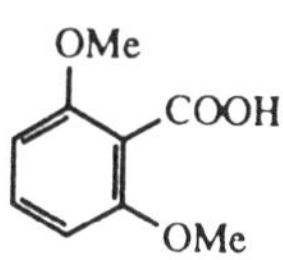

2-12

treated with solid CO_2; after acidification, an 80% yield of **2-12** was obtained.[29] Most recently, 1,3-dimethoxybenzene in $Me_2NCH_2CH_2NMe_2$ was metalated in the 2-position with potassium dialkylamide and treated with CO_2 to give a 69% yield of **2-12**.[30] Alternatively, 1,3-dimethoxybenzene in $Me_2NCH_2CH_2NMe_2$ and in the presence of alpha-methylstyrene was metalated by a dispersion of potassium metal in triethylamine (containing a drop of chlorobenzene as a dispersing agent) at 70 °C, followed by treatment with CO_2, stripping of volatiles, and dissolution in water/acidification to give an 85% yield of 2,6-dimethoxybenzoic acid.[31]

2.5.6. Resorcinol-4,6-dicarboxylic Acid (4,6-Dihydroxyisophthalic Acid), 2-13

This dicarboxylic acid can be obtained by the carbonation of resorcinol with K_2CO_3/CO_2 or $KHCO_3$ at appropriate ratios at high temperature. For instance, in a rotating ball mill/autoclave 1.0 m resorcinol was reacted with 4.4 m $KHCO_3$ (prepared in the reactor from calcined K_2CO_3 and the theoretical amount of water under CO_2 pressure) for 4 hr/180 °C, then cooled; the charge was dissolved in water and acidified to give **2-13** as the only product.[32]

Although the resorcinol-4,6-dicarboxylic acid is somewhat difficult to work with because of its high mp, 312 °C, and its low solubility in water and common organic solvents, its diesters have lower melting points and good solubility in organic solvents. For example, the dimethyl ester of **2-13** was prepared by treatment of the diacid with diazomethane and had mp 148 °C.[33]

Resorcinol-4,6-dicarboxylic acid is an interesting monomer and chelating agent.

2.5.7. Hydrogenated Resorcinols

To produce a specific hydroresorcinol in good yield and purity from resorcinol by catalytic hydrogenation, care must be taken to avoid over-hydrogenation or hydrogenolysis with a loss of keto or carbinol groups.

2.5.7.1. Dihydroresorcinol (1,3-Cyclohexanedione), 2-14

This compound exists in equilibrium with, or can react as, the enol-one, **2-15**. It is an important intermediate for agricultural chemicals and drugs (see Chapters 7 and 8) and for organic syntheses (academic work on this is discussed in Chapter 15). Since both names, dihydroresorcinol and 1,3-cyclohexanedione, have been oft-used in the literature, both names are being used interchangeably in this book to allow the reader to recognize them by either name.

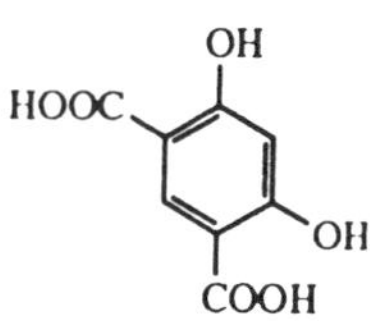

2-13

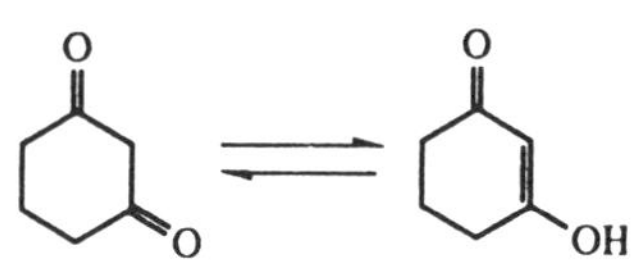

2-14 2-15

Dihydroresorcinol itself is still best prepared as described in an early volume of *Organic Synthesis*: Monosodium resorcinate in water was hydrogenated (1000–1500 psig was prescribed in this paper, but lower pressures can be used) for several hr/15 °C in the presence of a supported or Raney nickel catalyst until 1.0 m H_2 was absorbed per 1.0 m resorcinol. Then the reactor was vented and the solution filtered to remove the catalyst, the filtrate was acidified with HCl to precipitate crude dihydroresorcinol which was filtered off, dissolved in hot benzene, and filtered again to remove any sodium chloride; the filtrate was cooled/filtered to obtain dihydroresorcinol in 85–95% yield.[34] 1,3-Cyclohexanedione can also be made by the hydrogenation of resorcinol in aqueous, alkaline solution at room temperature and 50 psi H_2 pressure using 5% Rh-on-Al_2O_3 catalyst; after 16–18 hr, the yield of the desired product was 85%.[35]

For industrial purposes, the choice of catalyst will again depend on the total cost, determined by the initial cost of the catalyst (rhodium being much more expensive than nickel), the number of recycles before catalyst deactivation becomes pronounced, and the cost of disposal and/or recovery/reactivation/refabrication of the catalyst. The simple reaction may require a fairly extended process development study, including evaluation of batch and continuous reactors.

At any rate, attempts to improve the preparation of 1,3-cyclohexanedione continue. A recent Japanese patent describes the hydrogenation of resorcinol in aqueous solution in the presence of dibutylamine and 5% Pd/C catalyst at 15 kg/cm^2 hydrogen at 100 °C/1.5 hr to obtain the dione in 79% yield at essentially complete conversion of resorcinol.[36]

Most substituted dihydroresorcinols are best made from noncyclic precursors, i.e., by the Mannich condensation of ketones or esters with activated olefins, followed by Claisen cyclization of the adduct; examples of such syntheses are shown in Chapters 3 and 12. Many 2-substituted dihydroresorcinols can be directly made from its alkali salts by alkylation or acylation (both O- and/or C-substitution is possible).

2.5.7.2. Hexahydroresorcinol (1,3-Cyclohexanediol), **2-16**

1,3-Cyclohexanediol has no known, sizable commercial use but has potential as a monomer. For example, unsaturated (co)polyesters incorporating this compound showed good thermal resistance up to 220 °C and had good chemical resistance.[37]

The hydrogenation of resorcinol in ethanol, using precious metal catalysts, was studied at Engelhard Corp. With 5% Pd-on-C or 5% Pt-on-C catalyst, extensive hydrogenolysis was observed—substantial amounts of cyclohexenol, cyclohexanol,

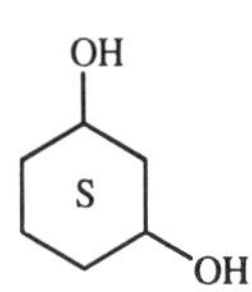

2 - 16

(mixture of *cis*- and *trans*-isomers)

and/or cyclohexanone were formed under various conditions; with a 5% Rh-on-Al_2O_3 catalyst, a 93% yield of 1,3-cyclohexanediol was obtained (7% cyclohexanol as a byproduct). Up to 90% yields of 1,3-cyclohexanediol were also obtained under similar conditions, but using a nickel catalyst.[38]

2.5.8. Alkylresorcinols

The Friedel-Crafts alkylation of resorcinol with alkyl halides or olefins usually yields mixtures of mono-, di-, and tri-substitution products which are difficult to separate; the alkyl groups are secondary-alkyl even when starting with alpha-olefins. Therefore, to prepare 4-*n*-hexylresorcinol, a well known bacteriostat (see Chapter 8), resort has to be taken to making 4-*n*-hexanoylresorcinol first and then reducing the ketone.

An exception is the acid-catalyzed *t*-alkylation of resorcinol with *t*-butanol or isobutylene, where the bulky *t*-alkyl group is prevented from entering the 2-position and, at a 2/1 mole ratio of isobutylene/resorcinol, a high yield of 4,6-di-*t*-butylresorcinol, **2-17**, can be obtained. Similarly, the alkylation of resorcinol with alpha-methylstyrene in 1/2 mole ratio gave a high in-hand yield of 4,6-di(α,α-dimethylbenzyl)resorcinol, **2-18**.[39]

2-Methylresorcinol, **2-19**, is useful in hair dyes and 2-*n*-propylresorcinol is a pharmaceutical intermediate. The synthesis of these and other 2-alkylresorcinols has been of much interest in the last 15 years. An unusual process was developed in Japan for 2-methylresorcinol (2,6-dihydroxytoluene); the patent claimed that the alkylation of resorcinol with methanol in the presence of ammonium chloride in a stirred autoclave at 300–400 °C under an inert atmosphere gave **2.19** in high yield.[40] Much work was done in Japan on alkylating 4,6-di-*t*-butylresorcinol in the 2-position, followed by debutylation to obtain 2-alkylresorcinols. For instance, to 0.1 m 4,6-di-*t*-butylresorcinol in toluene in the presence of 0.2 m NaOH (as a 12% aqueous solution) 0.11 m MeI at 70 °C was slowly added and this charge was reacted for 2 hr/70 °C to give a 71% yield of 2-methyl-4,6-di-*t*-butylresorcinol which was debutylated in nitrobenzene with conc. H_2SO_4 as the catalyst for 1 hr/200 °C to give 2-methylresorcinol in 80% yield.[41] In another, related process, 4,6-di-*t*-butylresorcinol, **2-17**, was reacted with formaldehyde and dimethylamine to give 2-dimethylaminomethyl-4,6-di-*t*-butylresorcinol, **2-20**, which was submitted to catalytic hydrogenolysis to produce 2-methyl-4,6-di-*t*-butylresorcinol, **2-21**, finally debutylated using Amberlyst 15 (sul-

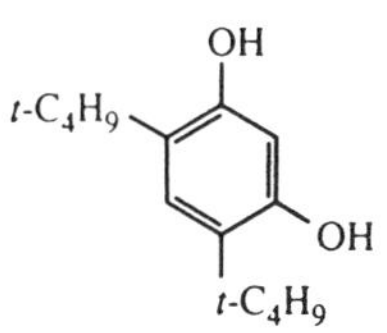

2-17

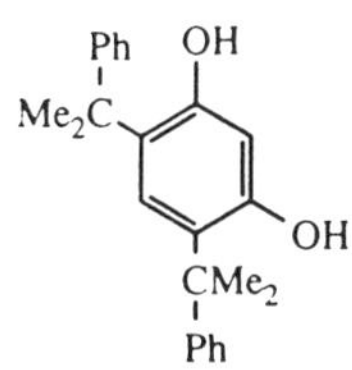

2-18

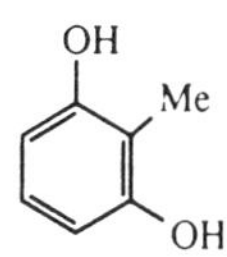

2-19

fonated ion-exchange resin) as the catalyst for 3 hr/120 °C to give an 88% yield (based on **2-17**) of 2-methylresorcinol, as shown in Eq. (2.1).[42]

(2.1)

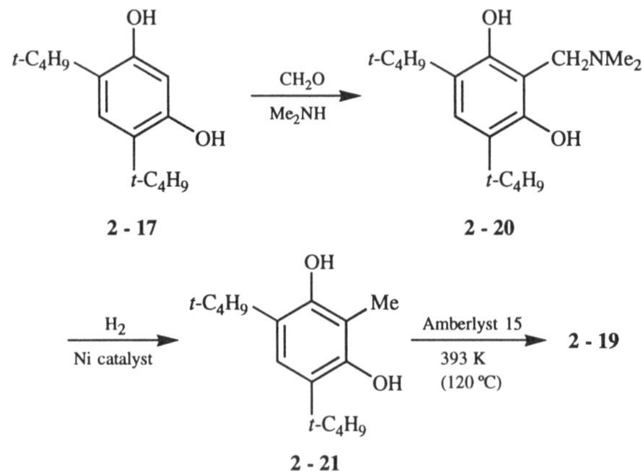

In another variation, 4,6-di-*t*-butylresorcinol, **2-17**, was reacted with an aldehyde and caustic in the presence of H_2/5% Pd-on-C catalyst to produce 2-alkyl- or 2-aralkyl-4,6-di-*t*-butylresorcinol, which was debutylated as before to 2-(ar)alkylresorcinol.[43] Separately claimed was the use of a heteropolyacid, such as tungstophosphoric acid, as the catalyst for debutylating **2-21** at 210 °C to obtain 2-methylresorcinol in 97% yield.[44]

A good route to other alkylresorcinols starts with noncyclic precursors such as ketones or aldehydes and activated olefins, as mentioned briefly above. A sophisticated variation of this route is shown in Eq. (2.2), consisting of the reaction of methyl phenylsulfinylacetate, **2-22**, with the alpha,beta-unsaturated ketone, **2-23**, in the presence of magnesium methoxide to give the substituted 1,3-cyclohexanedione, **2-24** (isolation not necessary) which, on heating, eliminated benzenesulfinic acid to produce 5-*n*-pentylresorcinol (olivetol), **2-25**, in 48% yield.[45]

(2.2)

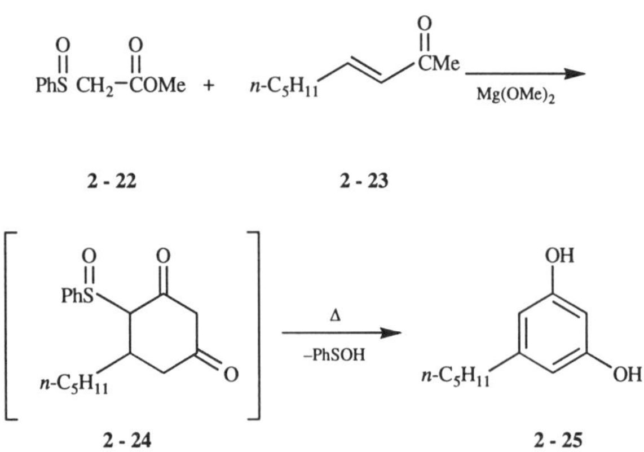

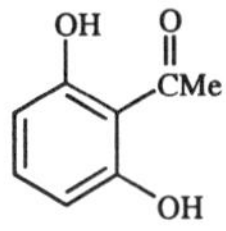

2 - 26

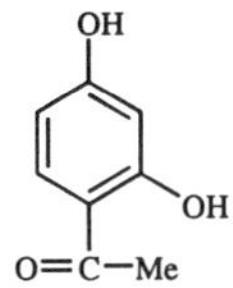

2 - 27

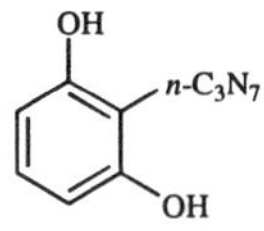

2 - 28

2.5.9. Acylresorcinols

The difficulties in the Friedel-Crafts acylation of resorcinol are mediocre yield, impure products, and, nowadays, problems of catalyst disposal.

Two examples of efforts to attain particular products/purity are now stated.

2-Acetylresorcinol, **2-26**, was prepared from dihydroresorcinol by acetylation with acetic anhydride in the presence of a base and dehydrogenation of the 2-acetyl-1,3-cyclohexanedione in *p*-isopropyltoluene (cymene) in the presence of a Pd-on-carbon catalyst for 9 hr/175 °C; the yield of compound **2-26** was 66% (based on 2-acetyl-1,3-cyclohexanedione).[46]

4-Acetylresorcinol is a starting material for a series of promising drugs. From Lilly's laboratories came two recent processes which do not require the disposal of a standard Friedel-Crafts catalyst. In one, resorcinol was reacted with acetic acid containing 5% HBr at 80 °C for 4–6 hr to obtain a 53% yield of **2-27**.[47] In the other, 2-alkyl-1,3-dimethoxybenzenes were reacted with acetic acid containing 48% HBr as the catalyst to give the 2-alkyl-4-acetylresorcinol; e.g., 2-*n*-propyl-1,3-dimethoxybenzene gave 2-*n*-propyl-4-acetylresorcinol [3′-(*n*-propyl)-2′,4′-dihydroxyacetophenone], **2-28**, in 73% yield.[48] This compound is also an important intermediate for drugs.

2.5.10. Haloresorcinols

4-Chloro-, 4,6-dichloro-, and 2,4,6-trichlororesorcinols have been prepared in good yields and purity by the reaction of resorcinol with SO_2Cl_2 (sulfuryl chloride) in the required amounts in suspension or solution in an appropriate solvent, such as chlorobenzene.

2-Chlororesorcinol, an intermediate for the preparation of medicinals, was prepared indirectly. Resorcinol was disulfonated first with 96% sulfuric acid to produce largely resorcinol-4,6-disulfonic acid, which was drowned in ice/37% HCl and chlorinated by the slow addition of $KClO_4$ at –15 °C, then warmed to 25 °C, and treated with 50% aq. KOH; after 12 hr, the suspension was filtered to obtain a cake of the dipotassium 2-chlororesorcinol-4,6-disulfonate. This K-sulfonate was then desulfonated by dissolving it in dilute sulfuric acid and refluxing for 24 hr, to obtain the 2-chlororesorcinol.[49]

An example of regiospecific reactions for iodinating resorcinol (this reaction usually gives crude products which are difficult to purify) came in a recent paper from Denmark.[50] 2-Iodoresorcinol was prepared in 77% yield by adding I_2/$NaHCO_3$ to a solution of resorcinol in ice water, with suitable workup/recrystallization of the crude from water. The 4-iodoresorcinol was obtained in 85% yield by adding ICl to resorcinol in diethyl ether at 0–25 °C; the crude product was dissolved in $CHCl_3$/AcOH and purified by chromatography on silica. 4,6-Diiodoresorcinol was made in 90% yield by adding 20% more than the required ICl to resorcinol in ether at 0–25 °C. 2,4,6-Triiodoresorcinol was obtained in 57% yield by adding $NaHCO_3$ and iodine suspended in water, with 20 hr reaction at 25 °C. Similar methods were applied to mono- and di-iodinate beta-resorcylic acid.

4-Fluororesorcinol was prepared in 60% yield by fluorinating 1,3-dimethoxybenzene in $CFCl_3$ at –78 °C with CF_3OF, then cleaving the methoxy groups by refluxing the 4-fluoro-1,3-dimethoxybenzene in 48% HBr/AcOH.[51]

2.5.11. Other Examples of Regioselective Reactions

In the coupling of aryldiazonium salts with resorcinol, the first substitution takes place in the 4-position; a second equivalent of the diazonium salt reacts at the 2-position on the resorcinol ring at acid pH, and at the 6-position at alkaline pH.[52]

2,4-Dihydroxybenzaldehyde (beta-resorcaldehyde), **2-29**, was obtained in 96% yield when resorcinol was reacted with $CHCl_3$ and NaOH (the Reimer–Tiemann reaction) in the presence of beta-cyclodextrin. Under the same conditions, but in the absence of beta-cyclodextrin, a 24% yield of 2,4-dihydroxybenzaldehyde plus a 40% yield of resorcyl-2,6-dialdehyde, **2-30**, was obtained.[53]

Resorcinol monobenzoate, **2-31**, was prepared in 85–99% yield by the slow, simultaneous addition of 1 equiv. benzoyl chloride and 1 equiv. aqueous sodium hydroxide solution to a solution of 15% resorcinol in water at 20 °C, controlling the pH in the 7.5–8.5 range. Under these conditions, the product precipitated from the solution and thus was removed from further benzoylation. Deviations from the exact conditions described in this process can promote undesirable side reactions that form resorcinol dibenzoate and/or convert benzoyl chloride to benzoic acid.[54]

A final example is the development of reagents for the preparation of resorcinol alkyl ethers (and ethers of other phenolics) which are less toxic than the dialkyl sulfates and which minimize ring alkylation; these reagents are the dialkyl carbonates. Thus, 22 parts of resorcinol, 18 parts of dimethyl carbonate, and 0.5 parts of 4-dimethylaminopyridine (catalyst) were held for 10 hr at 180 °C. The conversion of resorcinol was 46%, and an 80% yield of resorcinol mono-methyl ether plus a 20% yield of resorcinol dimethyl ether were obtained and can be separated by distillation. Trialkyl- and triarylphosphines were also suitable catalysts for this reaction.[55]

In conclusion, the above examples of the chemistry of resorcinol have sketched some of the problems encountered and how they were solved. Subsequent treatment in Chapters 3–12 on production and applications of resorcinol and in Chapters 14 and

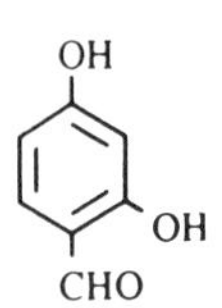

2-29

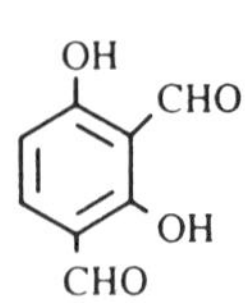

2-30

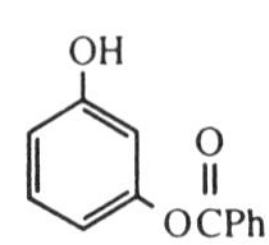

2-31

15 on natural products and academic work with resorcinol fill out the picture of the richness of this chemistry.

2.6. METHODS OF ANALYSIS

For the quantitative analysis of resorcinol, HPLC and GC are suitable. For the determination of the purity of resorcinol, the freezing point method or DSC (differential scanning calorimetry) can be used.

For a quick estimate of the amount of resorcinol in a given sample, the old bromination method is quite suitable. Excess bromine is reacted with KI solution to liberate iodine, which is titrated with standardized thiosulfate solution. A suitable version of this method is described in "The United States Pharmacopeia".[56]

2.7. REFERENCES

1. A. Hopp and W. Bunge, in *Ullmann's Encyklopaedie der Technischen Chemie*, 4th ed., Vol. 20, p. 190, Verlag Chemie, Weinheim, 1981, and references cited therein.
2. *Threshold Limit Values for Chemical Substances in the Workroom Air*, Adopted by ACGIH and recognized by OSHA.
3. *Registry of Toxic Effects of Chemical Substances.*
4. B. N. Ames *et al.*, *Proc. Natl. Acad. Sci. U.S.A., 72*, 5135 (1975).
5. C. W. Flickinger, *Am. Ind. Hygiene Assoc. J., 37*, 596 (1976).
6. *Natl. Toxicol. Program Tech. Rep. Ser., 1992*, 403; *C.A., 118*, 96019 (1993).
7. Cosmetics Ingredients Review (CIR) Expert Panel, *J. Am. Coll. Toxicology, 5(3)*, 167–223 (1986).
8. J. E. McKee and H. W. Wolf, *State of California Water Quality Control Board, Publication No.3A* (1963).
9. Based on tests conducted by Koppers Co. Environmental Resources and Occupational Health Department following EPA 600/4-78-012 "Methods for Measuring the Acute Toxicity of Effluents to Aquatic Organisms."
10. *Beilstein's Handbuch der Organischen Chemie*, Vol. 6 and supplementary volumes, Springer-Verlag, Berlin; resorcinol itself is cataloged under Syst. Nr. 554.
11. *Houben-Weyl's Methoden Der Organischen Chemie*, 4th edition (E. Mueller, ed.), Vol. VI/1c, Parts 1 and 2, G. Thieme Verlag, Stuttgart, 1976.
12. *Rodd's Chemistry of Carbon Compounds*, 2nd edition (S. Coffey, ed.), Vol. IIIA, p. 395, Elsevier, Amsterdam, 1971.
13. J. Podstata and Z. J. Allan, *Collect. Czech. Chem. Commun., 32(8)*, 3004 (1967).
14. H. Cerfontain and A. Koeberg-Telder, *Recl. Trav. Chim. Pays-Bas, 107(9)*, 543 (1988).
15. H. Cerfontain, N. J. Coenjaarts, and A. Koeberg-Telder, *Recl. Trav. Chim. Pays-Bas, 108(1)*, 7 (1989).
16. U.S. Pat. 2,487,586 (Nov. 8, 1949; W. D. Peterson to General Aniline and Film Corp.).
17. U.S. Pat. 3,772,379 (Nov. 13, 1973; P. E. Woodgate to Eastman Kodak Co.).
18. N. Nierenstein and D. A. Clibbens, *Organic Synthesis,* Coll. Vol. 2, p. 557, Wiley, New York, 1943.
19. Jpn. Kokai Tokkyo Koho JP 59,184,149 [84,184,149] (Oct. 19, 1984; to Sumitomo Chemical Co. K.K.); *C.A., 102*, 113072 (1985).
20. Ger. Offen. DE 3,832,076 (March 22, 1990; P. Neumann and U. Eichenauer to BASF A.G.); *C.A., 113*, 114824 (1990).
21. Eur. Pat. Appl. EP 436,379 (July 10, 1991; T. K. Choi, H. J. Park, and H. W. Kim to Kolon Industries, Korea); *C.A., 115*, 158725 (1991).
22. A. W. Weston and C. M. Suter, *Organic Synthesis*, Coll. Vol. 3, p. 288, Wiley, New York, 1955.

23. D. K. Hale, A. R. Hawdon, J. I. Jones, and D. I. Packham, *J. Chem. Soc., 1952*, 3503.
24. I. T. Strukov and V. V. Vantsova, *Khim.-Farm. Zh., 3(6)*, 58 (1969); *C.A., 72*, 91014 (1969).
25. Jpn. Kokai Tokkyo Koho JP62,000,440 [87,000,440] (Jan. 6, 1987) and JP62,010,043 [87,010,043] (Jan. 19, 1987), both to Mitsui Petrochemical Industries K.K.; *C.A., 107*, 39406 and 39412 (1987).
26. G. Wittig and U. Pockels, *Chem. Ber., 72*, 89 (1939).
27. H. Gilman, H. B. Willis, T. H. Cook, F. J. Webb, and R. N. Meals, *J. Am. Chem. Soc., 62*, 667 (1940).
28. S. Cabiddu, A. Maccioni, P. P. Piras, and A. Plumitallo, *Gazz. Chim. Ital., 111(3–4)*, 123 (1981); *C.A., 95*, 132423 (1981).
29. Japan. Pat. 68,022,969 (Oct. 3, 1968; K. Nakagawa and I. Matsumoto to Banyu Pharmaceutical Co.); *C.A., 72*, 77600 (1969).
30. U.S. Pat. 4,845,276 (July 4, 1989; L.-C. Yu to Mine Safety Appliances Co.).
31. U.S. Pat. 4,845,277 (July 4, 1989; L.-C. Yu to Mine Safety Appliances Co.).
32. U.S.Pat. 2,703,812 (March 8, 1955; R. F. Bann and R. C. Conn to American Cyanamid Co.).
33. T. S. Gore, P. K. Inamdar, and P. M. Nagarkar, *Indian J. Chem., 12*, 946 (1974); *C.A., 82*, 139584 (1975).
34. R. B. Thompson, *Organic Synthesis*, Coll. Vol., 3, p. 278, Wiley, New York, 1955; some more details on the preparation are given in a procedure for making 2-methyl-1,3-cyclohexanedione by A. B. Mekler, S. Ramachandran, S. Swaminathan, and M. S. Newman, *Organic Synthesis*, Coll. Vol. 5, p. 743, Wiley, New York, 1973.
35. J. C. Sircar and A. I. Meyers, *J. Org. Chem., 30*, 3206 (1965).
36. Jpn. Kokai Tokkyo Koho JP 04,013,644 [92, 013,644] (Jan. 17, 1992; M. Kondo and H. Mikito to Mitsui Petrochemical Industries, Ltd.); *C.A., 116*, 193783 (1992).
37. Unpublished information of Koppers Co., Inc.
38. P. N. Rylander and N. Himelstein, in *Engelhard Industries, Inc., Techn. Bull., 5(2)*, 43 (1964); U.S. Pat. 3,193,584 (July 6, 1965; P. N. Rylander and J. F. Kreidl to Engelhard Industries, Inc.).
39. U.S. Pat. 4,599,464 (July 8, 1986; H. Dressler to Koppers Co., Inc.).
40. Japan. Pat. 54,088,227 (Oct. 10, 1979; to Seitetsu Chemical Industries K.K.); *C.A., 91*, 157443 (1979).
41. Ger. Offen. DE3,042,121 (May 14, 1981; Y. Takisawa and S. Hasegawa to Sumitomo Chemical Co., Ltd.); *C.A., 95*, 150166 (1981).
42. Jpn. Kokai Tokkyo Koho JP 57,142,935 [82,142,935] (Sept. 3, 1982; to Showa Denko K.K.); *C.A., 98*, 53390 (1983).
43. Jpn. Kokai Tokkyo Koho JP61,007,226 [86,007,226] (Jan. 13, 1986; to Showa Denko K.K.); similar is the process given in Jpn. Kokai Tokkyo Koho JP 63,222,138 [88,222,138] (Sept. 16, 1988; A. Takeshita *et al.* to Sumitomo Chemical Co. K.K.); *C.A., 110*, 75043 (1989).
44. Jpn. Kokai Tokkyo Koho JP63,130,546 [88,130,546] (June 2, 1988; A. Takeshita *et al.* to Sumitomo Chemical Co., Ltd.); *C.A., 109*, 190009 (1988).
45. A. A. Jaxa-Chamiec, P. G. Sammes, and P. D. Kennewell, *J. Chem. Soc., Perkin Trans. 1*, 1980, 170.
46. Eur. Pat. Appl. EP 219,282 (Oct. 11, 1985; K. Tanaka to Sumitomo Chemical Co., Ltd.); *C.A., 107*, 175666 (1987).
47. U.S. Pat. 4,777,298 (Oct. 11, 1988; W. S. Marshall, S. K. Sigmund, and C. A. Whitesitt to Eli Lilly and Co.).
48. U.S. Pat. 4,777,299 (Oct. 11, 1988; W. S. Marshall, S. K. Sigmund, and C. A. Whitesitt to Eli Lilly and Co.).
49. Eur. Pat. Appl. EP 146,923 (July 3, 1985; F. della Valle and A. Romeo to FIDIA SpA.); *C.A., 103*, 214974 (1985).
50. I. Thomsen and K. B. G. Torsell, *Acta Chem. Scand., 45*, 539 (1991).
51. P. C. Belanger, C. K. Lau, H. W. R. Williams, C. Dufresne, and J. Scheigetz, *Can. J. Chem., 66(6)*, 1479 (1988); the paper is from the laboratories of Merck Frosst Canada Inc.
52. K. Hunger, P. Mischke, and W. Rieper, "Azofarbstoffe", in *Ullmann's Encyklopaedie der Technischen Chemie*, 4th ed., Vol. 8, p. 250, Verlag Chemie, Weinheim, 1974.
53. M. Komiyama and H. Hirai, *Makromol. Chem., Rapid Commun., 2(11)*, 707 (1981).

54. U.S. Pat. 2,571,703 (Oct. 16, 1951; W. M. Gearhart, R. O. Hill, Jr., and M. H. Broyles to Eastman Kodak Co.).
55. U.S. Pat. 4,192,948 (March 11, 1980; F. Merger, F. Towae, and L. Schroff to BASF A.-G.).
56. *The United States Pharmacopeia*, Twenty-first Revision, USP XXI, p. 941, The United States Pharmacopeial Convention, Inc., Rockville, MD, 1984.

3

Processes for Making Resorcinol

At present, only the benzene sulfonation/fusion and the hydroperoxidation/cleavage of diisopropylbenzene routes are used commercially on a large scale. The remainder of the routes described below are used on a small scale only, if at all.

3.1. RESORCINOL BY THE DISULFONATION/CAUSTIC FUSION OF BENZENE

This old process, much improved over time, is being used in the U.S. by INDSPEC Chemical Corp., which has about 50% of the world capacity for resorcinol. There are smaller producers in China and India using this process as well.

The sulfonation/caustic fusion of benzene has been used since the 1870s to make resorcinol.[1] Usually, the process is described in textbooks similar to Eqs. (3.1)–(3.3). This is quite a simplification, since many byproducts are formed.

In the 1940s, in Germany benzene, **3-1**, was mono-sulfonated by addition to 100% sulfuric acid (called monohydrate) in cast iron, jacketed reactors, first at 25–60 °C,

(3.1)

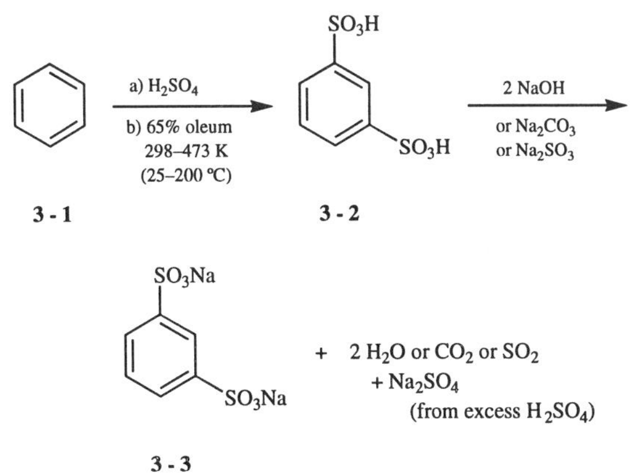

(3.2) 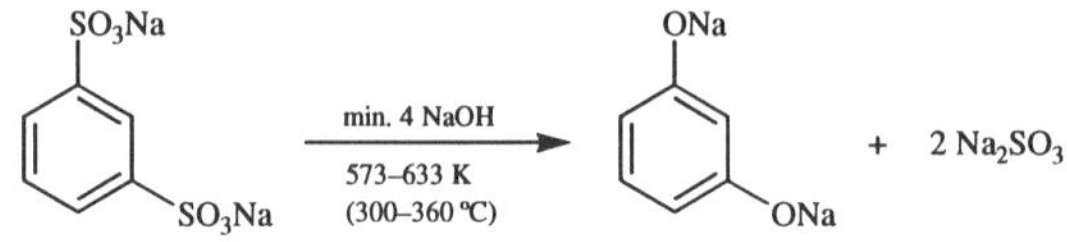

3 - 3 3 - 4

(3.3) 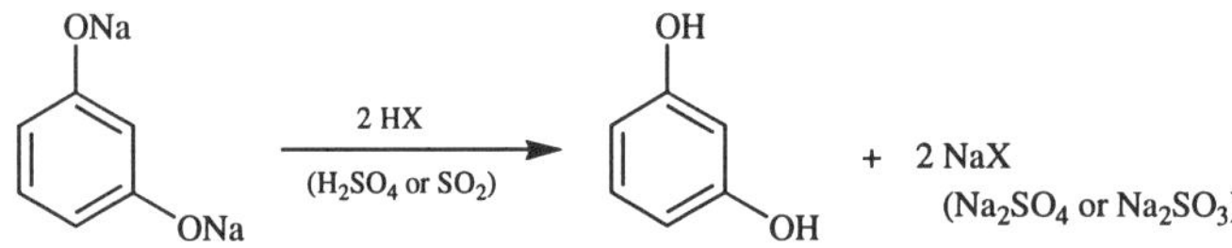

3 - 4 Resorcinol

then heated to 100 °C, and held at 100 °C for 4 hr. The benzenesulfonic acid produced was then cooled and slowly added to 65% oleum (65% SO_3 in sulfuric acid), with the temperature slowly rising from 30 °C to 80 °C, then held at 80–85 °C for 2 hr to produce *m*-benzenedisulfonic acid, **3-2**. This diacid was then added to water containing sodium sulfate, the solution was heated to 100 °C and neutralized with a suspension containing 25% lime, to convert the disulfonic acid to the calcium salt and excess sulfuric acid to calcium sulfate. The precipitated calcium sulfate was filtered off, washed with hot water and discarded. The filtrate/wash was heated to 95 °C and treated with soda ash (sodium carbonate). This precipitated calcium carbonate which was filtered off, and the filtrate, were evaporated to leave the disodium *m*-benzenedisulfonate, **3-3**, in 93% yield. This disodium salt was then added to a melt of an excess over the stoichiometrically required sodium hydroxide at *ca* 330 °C in a stirred reactor to give, after completion of the fusion, a powdery product which was dropped into water and neutralized with hydrochloric acid. A suspension was produced which contained undissolved sodium sulfite, and this was filtered off. The filtrate was treated with charcoal, filtered again, and the filtrate was extracted with diethyl ether. The extract was concentrated and the ether was recovered. The residue was crude resorcinol, which was dehydrated and vacuum-distilled to give up to an 82% yield of resorcinol.[2] As needed, the distilled resorcinol was further purified by crystallization from benzene or by vacuum-sublimation.

Obviously, the process as described above would not be viable today in most countries for environmental reasons and byproduct disposal problems, as well as the high cost of production.

The complications of this process are numerous. For example, in the benzene sulfonation step there are byproducts, including traces of *o*-benzenedisulfonic acid, several percent of *p*-benzenedisulfonic acid (which yields only sodium *p*-phenolsulfonic acid on alkali fusion), monosulfonated diphenylsulfones, and disulfonated diphenylsulfones, which need to be minimized. Indeed, the formation of sulfones can reach 15–20% unless a sulfone inhibitor, such as sodium sulfate, is present from the start of the sulfonation to minimize this byproduct. Much effort has been invested at optimizing sulfone inhibitors, controlling the sulfonation to reduce other byproducts, and reducing the amount of oleum used to minimize the byproduced inorganic salts; much of this effort has been summarized in two valuable books which examine the literature up to *ca* 1965.[3,4] Other possible complications in the sulfonation of benzene, particularly with SO_3, include the formation of pyrosulfonic acids and of sulfonic anhydrides. Because of the aggressive, hygroscopic media and the complex species involved, the kinetics of sulfonation of benzene is difficult to study. Actual kinetic studies may be limited to monosulfonation and usually do not extend beyond the first 20% of the reaction, while several of the undesirable byproducts form mainly in the later stages, at higher temperatures. An overview of the literature through 1965 on the kinetics of benzene sulfonation demonstrates well the complexities in this area.[5]

The fusion of sodium *m*-benzenedisulfonate with caustic at *ca* 330–350 °C is also not a clean reaction. Difficulties which have been observed include foaming (hydrogen generation is possible), "burning" (oxidation of sodium resorcinate), and the formation of sulfite addition products to the resorcinate ring in later stages. The consistency and mixing of the fusion mass is largely dependent on the excess of caustic which is used, but also on the amount of inorganics which are present. The crude resorcinol obtained after extraction/extract stripping of the drowned, neutralized fusion mass contains some phenol and higher molecular weight products; the main component of the latter is 2,4,4′-trihydroxybiphenyl.

Some improvements, from the 1960s onward, in the old process are sketched below.

The continuous disulfonation of benzene by injecting benzene and SO_3 in 1.0/2.2 mole ratio at 120–140 °C into a heel of finished disulfonate containing sodium sulfate or sodium benzenedisulfonate as the sulfone inhibitor was said to give high quality (90%) *m*-benzenedisulfonic acid, using either a cascade reactor[6] or a loop reactor.[7]

An interesting disulfonation process for benzene involved the use of 96% sulfuric acid for the sulfonation, with continuous removal of water and excess sulfuric acid at 200–235 °C/1 mm Hg in a thin film evaporator.[8] A similar process involved stripping sulfuric acid from the disulfonation product of benzene with oleum in a wiped film evaporator.[9]

Another variation for removing the *ca* 25 wt% sulfuric acid present in an in-process disulfonic acid consisted of injecting benzene at 130–140 °C, without the necessity of removing water by the slow, azeotropic distillation with benzene; essentially all of the excess sulfuric acid was converted to benzene(mono)sulfonic acid.[10]

Useful inhibitors of sulfone formation in the disulfonation of benzene, other than sodium sulfate, were shown to be sodium *m*-benzenedisulfonate, acetic acid or anhydride, aliphatic and aromatic ketones, clays and phosphoric acid, but none is as good as sodium sulfate.

Among the improvements described for the caustic fusion step was a continuous process where sodium benzenedisulfonate and sodium hydroxide were fed first to a thin film evaporator and then to an intensive mixer.[11]

Another variation of the caustic fusion process claimed that the use of a reaction medium of aliphatic, alicyclic, or aromatic, high-boiling hydrocarbons or aromatic

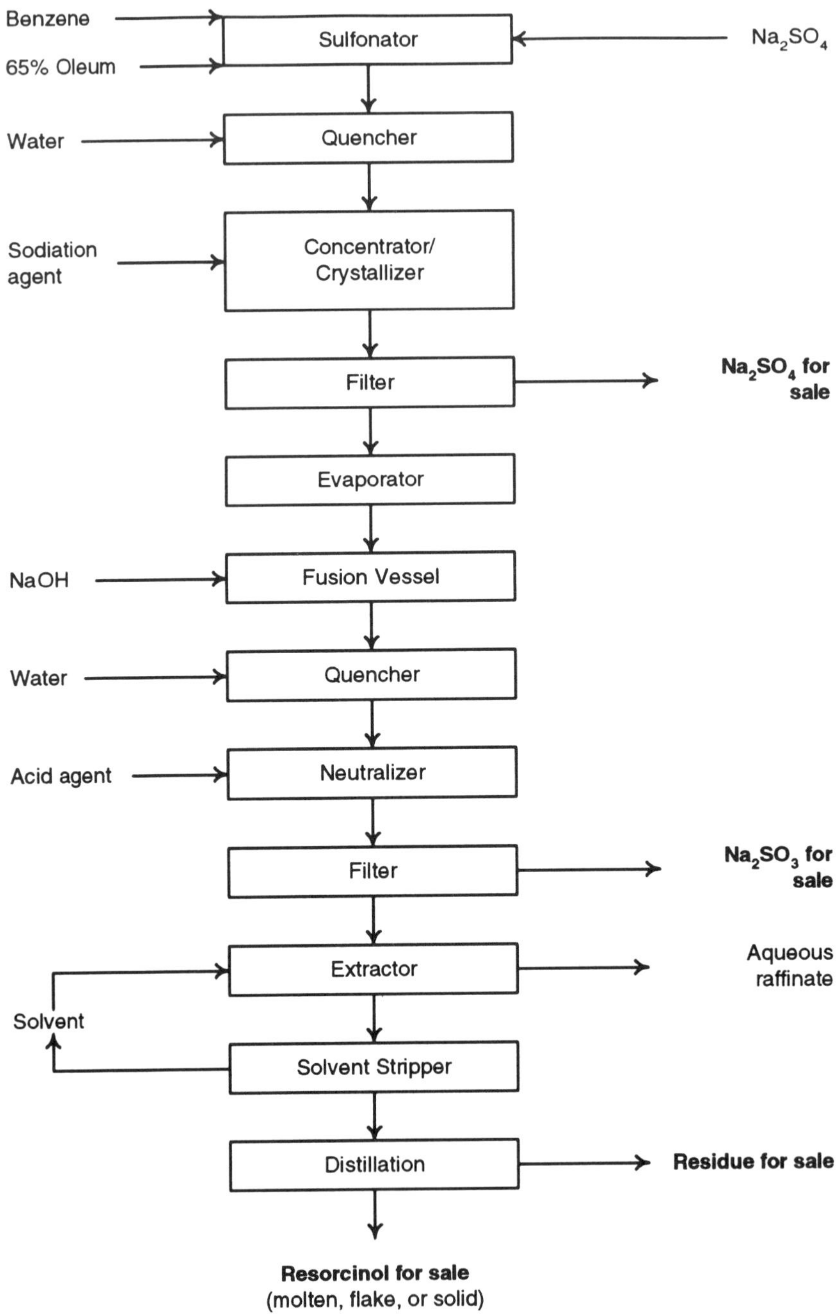

Figure 3.1. Resorcinol by the disulfonation/caustic fusion of benzene.

ethers was beneficial. Thus, disodium *m*-benzenedisulfonate, 50% aqueous sodium hydroxide, and a white (petroleum) oil were mixed and heated to 300 °C under a stream of nitrogen for dehydration. The dehydrated mix was stirred at 320 °C for 1 hr and worked-up to give a 94.4% yield of resorcinol.[12]

There have been reports that the simultaneous caustic fusion of sodium *m*-benzenedisulfonate and another arenesulfonate gave good yields of resorcinol and hydroxyarene. Thus, benzene was disulfonated with oleum. Benzene vapor was blown into the crude disulfonate containing sulfuric acid to produce benzenesulfonic acid; the mixed sulfonates were converted to the sodium salts with alkali, then fused with caustic to produce resorcinol and phenol.[13] Similarly, the caustic fusion of a mixture of disodium *m*-benzenedisulfonate and sodium 2-naphthalenesulfonate gave a 75% yield of resorcinol plus an 80% yield of 2-naphthol.[14] Also, caustic fusion of mixtures of disodium *m*-benzenedisulfonate and sodium toluenesulfonate gave high yields of resorcinol and *p*-cresol;[15] and the caustic fusion of mixtures of disodium *m*-benzenedisulfonate and sodium xylenesulfonate gave resorcinol and xylenol.[16] The mixed fusion masses were said to be quite fluid-favorable for good mixing. However, it is unlikely that any of these processes are practiced commercially because of the additional processing complications which are introduced, and because of the difficulty of balancing production and marketing of the coproducts.

Finally, today's commercial practice of the benzene sulfonation/caustic fusion process for resorcinol is described in general terms in the block flow diagram presented in Fig. 3.1.

Needless to say, the process details are proprietary information of INDSPEC Chemical Co. and others. The sulfonation is done in a continuous, temperature/dose-programmed fashion. Both sodium sulfate and sodium sulfite are produced for sale as high-purity byproducts. The current process is the result of continuous process development and improvements in materials of construction, process equipment, process modeling, and so on. All waste streams are handled in the plant. Statistical quality control is employed. The successful and safe operation of a process of this complexity requires the complete attention of engineering and production personnel. Certainly, this is a major engineering achievement.

3.2. RESORCINOL FROM THE HYDROPEROXIDATION/ CLEAVAGE OF *m*-DIISOPROPYLBENZENE

This process was developed diligently in the 1950s by Distillers Chemicals & Plastics in England and Hercules in the U.S. The viability of the process according to their pioneering patents remained doubtful, because of only fair yields of resorcinol and questions of how to handle and recycle byproducts.

The general chemistry of this process is shown in Eqs. (3.4), (3.5), and (3.6); a new, improved version of the last step is shown in Eq. (3.6a).

(3.4)

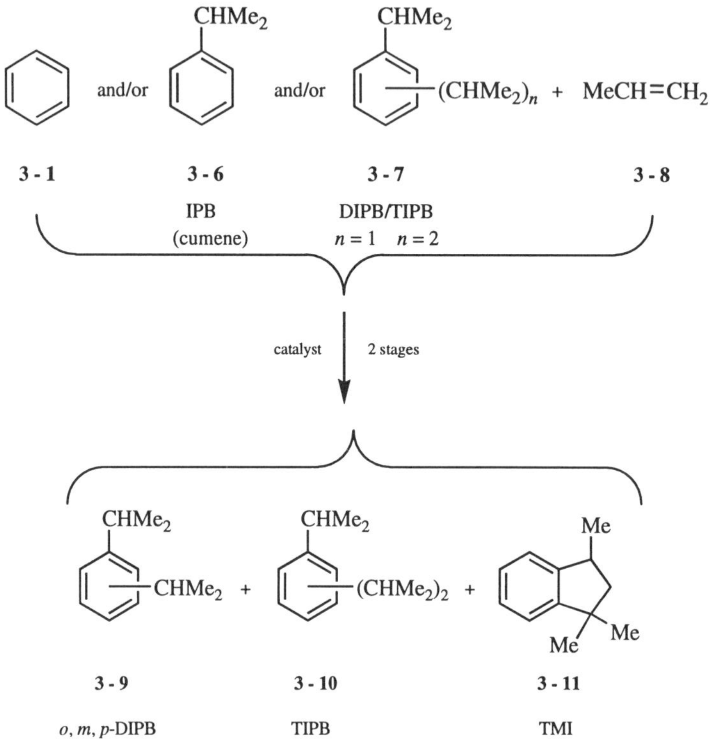

IPB = isopropylbenzene (cumene); DIPB = diisopropylbenzene;
TIPB = trisispropylbenzene; TMI = 1,1,3-trimethylindane

A good listing and analysis of the literature on the process given in Eqs. (3.4)–(3.6) until about 1970, including a choice of the best data for design and cost estimates, was presented in 1972 in Stanford Research Institute's unique PEP reports.[17]

A good description of the pioneering work on this process at the Distillers Chemicals & Plastics, Ltd., working with Hercules Inc. in the 1950s and 1960s has also been given, showing all the pitfalls and problems; this paper also mentions the contributions by Russian workers, and by Phenolchemie, Hibernia, and Scholven-Chemie in Germany.[18] Although it was said that the yield of resorcinol from 99.5%-pure *m*-diisopropylbenzene "can approach 70% of the theoretical," it is likely that considerable processing was required to obtain sufficiently pure dihydroperoxide, **3-13**, for cleavage, and that very careful recycle of intermediate products was required. These difficulties with the process at the time were a good part of the reason it was not commercialized; likely, another reason was that the market for resorcinol at the time was small.

A scheme for the production of crude diisopropylbenzene (DIPB) was shown above in Eq. (3.4).The boiling points of *o*-, *m*-, and *p*-DIPB are 203.8 °C, 203.2 °C, and 210.4 °C, respectively, at atmospheric pressure. Therefore, the *o*-/*m*-DIPB are

(3.5)

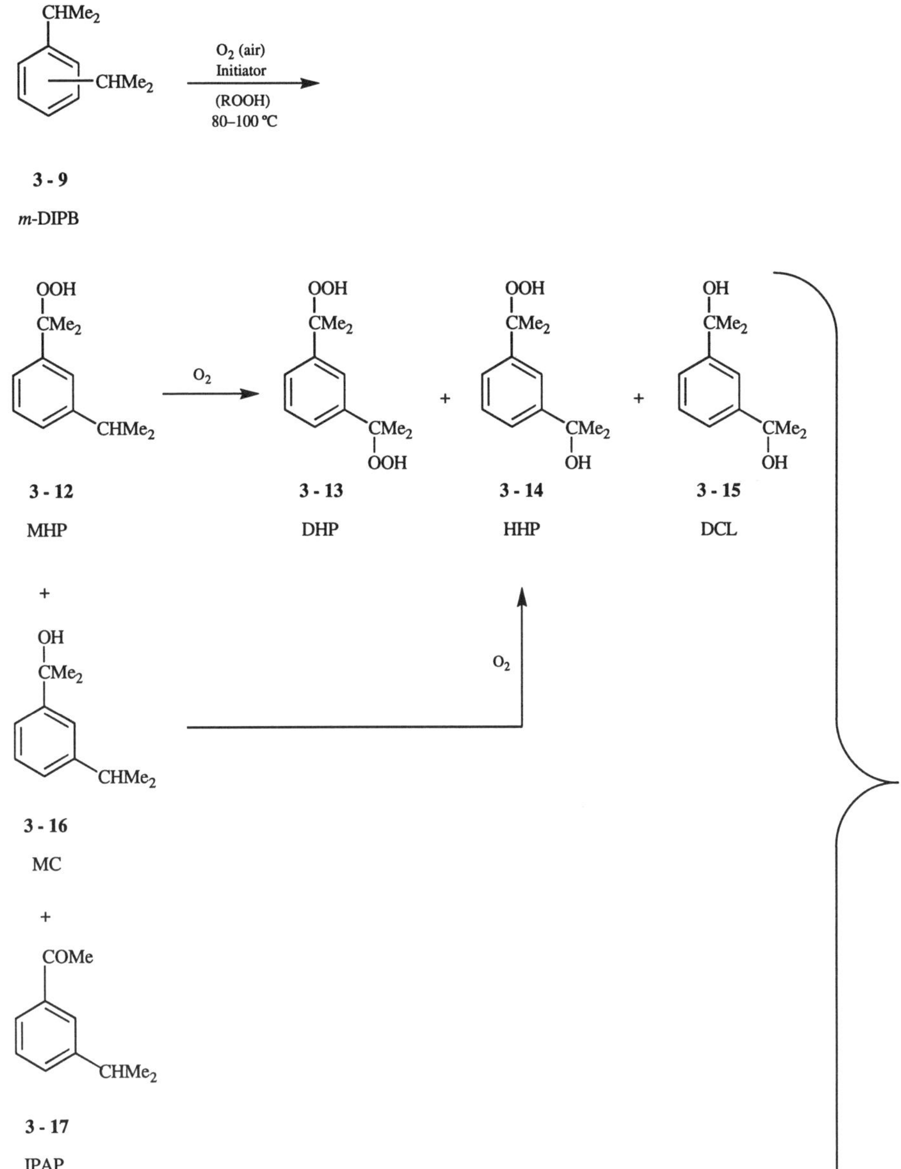

m-DIPB = *m*-diisopropylbenzene; MHP = monohydroperoxide;
MC = monocarbinol; IPAP = isopropylacetophenone;
DHP = dihydroperoxide; HHP = hydroxyhydroperoxide;
DCL = dicarbinol

(3.6)

(3.6a)

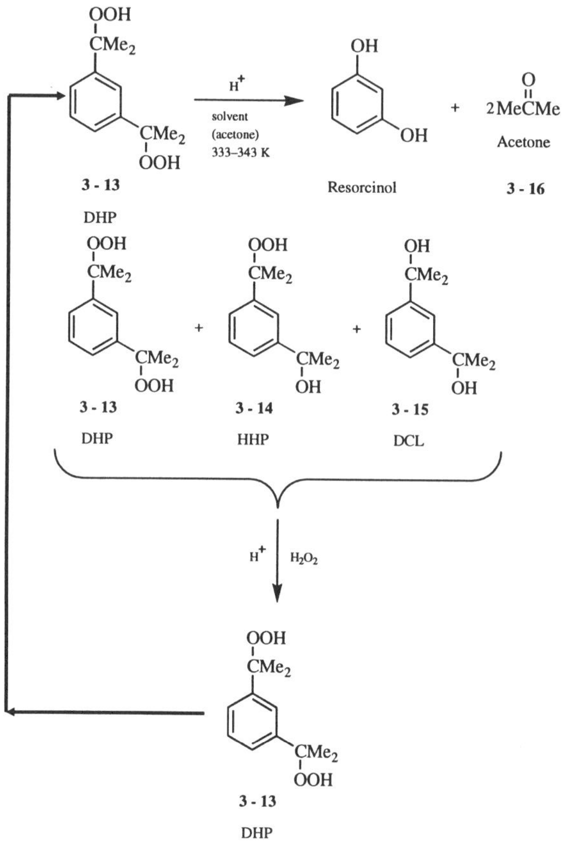

separated from the *p*-isomer in a very efficient (50 theoretical plates) distillation column(s). To obtain *m*-DIPB, ***m*-3-9**, of sufficient purity, low in *o*-DIPB, the alkylation/transalkylation/isomerization process must provide a product of very low *o*-isomer content. Not only does the *o*-DIPB not oxidize in the process and will require a bleed on the recycle stream to oxidation, but the *o*-DIPB can also be converted to the trimethylindane (TMI), **3-11**, especially if aluminum chloride is the alkylation catalyst. And TMI and its unwanted oxidation products also have to be bled from the recycle stream. The older literature[17] indicates that a good process for *m*-DIPB, as described in several Shell Development Co. patents, would be the alkylation of cumene (IPB), **3-6**, with propylene over a silica-alumina catalyst at 150–200 °C in a first reactor, the alkylate from this reactor going to a transalkylation/isomerization step with recycled TMI and *p*-DIPB, together with benzene, using a silica alumina catalyst at 250–300 °C. The alkylate produced contains DIPB with 70/30 *m*-/*p*-isomer ratio and containing < 1% *o*-DIPB.

An interesting way to upgrade the *m*-isomer content of the DIPB fraction was found by scientists at Mobil Oil Corp., based on the selective cracking of the *p*-isomer.

Using a feed containing 68.9% *m*-DIPB/23% *p*-DIPB and a ZSM-5 zeolite catalyst[19] or a HZSM-12 catalyst[20] at 350 °C or 400 °C, a DIPB fraction containing 92.5% *m*-DIPB or 97.5% *m*-DIPB, respectively, was obtained; benzene was the major cracking product.

Another patented way to obtain DIPB with a high *m*-isomer content involved either the isomerization of a 60/40 *p*-DIPB/*m*-DIPB mixture, the transalkylation of cumene with triisopropylbenzene, or the alkylation of cumene with propylene, all in anhydrous hydrogen fluoride/boron trifluoride at –20 °C to 0 °C, to produce a DIPB fraction containing 99+% of the *m*-isomer.[21] Likely, these processes are not commercially attractive for safety and health concerns.

The separation of *m*-/*p*-DIPB by selective absorption has also been patented. Passing a 5% solution in heptane of a 50/50 *m*-/*p*-DIPB feed over a Ca-exchanged Zeolite L at room temperature gave an absorbate that contained 2.54 times more *m*-DIPB than *p*-DIPB. High-purity *m*-DIPB could then be obtained in several passes.[22]

Finally, cumene is produced commercially in very large volume, by alkylation of benzene with propylene, for use in making phenol by the hydroperoxidation/cleavage route. The byproduct of this process is a higher-boiling distillation residue. These cumene bottoms contain some cumene and small amounts of butyl-,amyl-,hexyl-, and other alkylbenzenes, with *o,m,p*-DIPB as the largest component; pure DIPB cannot be obtained economically from this residue by distillation because of the close boiling points of the various impurities. By reacting cumene bottoms with propylene and benzene in the presence of catalysts such as $AlCl_3$/HCl, a DIPB fraction can be obtained which contains the *m*-/*p*-isomers in 2/1 ratio.[23] Another patent for upgrading cumene bottoms containing 12.7 wt% *o*-DIPB, 26.6% *m*-DIPB, and 35.3% *p*-DIPB consisted of contacting them with a rare earth-exchanged zeolite, SK-500 (Union Carbide Corp.), at 150 °C to give an isomerized product containing no *o*-DIPB, 39.2% *m*-DIPB, and 31.2% *p*-DIPB.[24] In the same way, a cumene bottom similar to the preceding composition was treated with an acid clay at 159 °C for 1.5 hr to give a product containing 37.5% *m*-DIPB, 29.3% *p*-DIPB, and no *o*-DIPB.[25] Potentially, the cumene bottoms are the cheapest, plentiful source of *m*-DIPB.

Regarding the hydroperoxidation/cleavage of *m*-DIPB, the status of this process up to about 1970 was discussed earlier. Worldwide interest in this process continued at a steady pace, but intensified development work was done in Japan which resulted in the optimization of process steps and successful commercialization. Mitsui Petrochemical Industries, Ltd., which had already operated a *m*-/*p*-cresol plant from isopropyltoluene and had pioneered the commercial production of hydroquinone from *p*-diisopropylbenzene, brought a 3000 t/yr resorcinol from *m*-DIPB plant on stream in April, 1980; Sumitomo Chemical Co., which had also pioneered the production of *m*-/*p*-cresol from isopropyltoluenes by hydroperoxidation/cleavage, brought a 5000 t/yr resorcinol from *m*-DIPB plant on stream in early 1981 (see also Chapter 13). An important part of their process improvements was the optimization of the step shown above as Eq. (6a), i.e., the extraction of a DHP(**3-13**)/HHP(**3-14**) fraction from the oxidate and the acid-catalyzed oxidation of the HHP to DHP with hydrogen peroxide.

This increased the overall yield of resorcinol from *m*-DIPB over the prior art processes (70% yield max) to the 80% yield range. Both Sumitomo and Mitsui obtained extensive patent coverage on their process improvements in Japan and worldwide, covering the overall new process as well. Sumitomo also patented the hydroperoxidation/cleavage of mixed *m*-/*p*-DIPB, with subsequent separation of resorcinol and hydroquinone. For simplicity and brevity, the following paragraphs largely cover Sumitomo's and Mitsui's U.S. patents in a little more detail.

It was found at Sumitomo that DIPB-hydroperoxides could be isolated from the crude oxidation products with strongly basic anion-exchange resins and desorbed by elution with aqueous alkali, ketones, alcohols, or ethers.[26]

Another Sumitomo patent disclosed that a mixture of *m,p*-DIPB-dihydroperoxide, containing byproducts such as HHP, **3-14**, could be cleaved in acetone solution with sulfuric acid as the catalyst to give an 80% yield of resorcinol from *m*-DHP, **3-13**; the *p*-DHP gave hydroquinone in high yield.[27]

The continuous process for the air oxidation of DIPB in the presence of caustic was improved at Sumitomo by adding an emulsifying agent and by intensive mixing; this enhanced the reaction rate in a safe manner.[28]

Byproducts such as carbinols, ketones, and styrenes, remaining in the DIPB-oxidation products solution after the DHP had been extracted with dilute alkali, were extracted with water/methanol containing a small amount of sodium hydroxide; this treatment of this stream effectively bled off undesirable impurities.[29]

Workers at Sumitomo improved the continuous oxidation of *m*- and/or *p*-DIPB at 90–95 °C by keeping the monohydroperoxide in the equilibrated feed in the 20–40% range; under these conditions the rate of dihydroperoxide formation was improved.[30]

Also at Sumitomo, the acid-catalyzed cleavage of *m*- or *p*-dihydroperoxide (DHP) was improved by keeping the carbinol impurity at 0.16 per 1.0 DHP; under such conditions the yield of resorcinol from *m*-DHP, **3-13**, was 92%.[31]

An improved method for the continuous extraction of DHP from aqueous, alkaline solution was invented at Sumitomo, based on a multistage operation with a temperature gradient from 0–85 °C, countercurrent flow of the aqueous, alkaline solution, and using extractants such as 4-methyl-2-pentanone (methyl isobutyl ketone).[32]

Sumitomo patents on the separation of crude mixtures of resorcinol and hydroquinone (from the air-oxidation of mixed *m,p*-DIPB) utilized slurrying the crudes in esters, ethers, alcohols, or acrylonitrile to precipitated hydroquinone, leaving resorcinol in solution.[33] Other examples involved slurrying of the crude product in toluene/water or other mixed solvents to permit a high recovery of the dihydroxybenzenes, based on differences in their solubilities.[34]

In most of the above patents, the inventions were claimed by a group of 10 authors led by Hidekai Suda. The listing of so many inventors on one patent is quite unusual in U.S. patent practice (often, the U.S. Patent Office raises doubts about individual contributions in the case of so many co-inventors). Certainly, it makes clear that a large technical group was involved in developing these processes.

From Mitsui came a method for recovering additional resorcinol from the crude product by thermally cracking the high-boiling resorcinol distillation residue in the presence of an aromatic ketone.[35] Another patent from Mitsui claimed an improved process for resorcinol in which the cleavage of the *m*-DHP was conducted using a silica-alumina catalyst, limiting the water content of the system to 0.3–1.0%; up to 98% yields of resorcinol were obtained.[36]

From work at Mitsui came another improvement in the liquid-phase air oxidation of *m*-DIPB, ***m*-3.9**, in the presence of dilute, aqueous alkali, using an initial oxidation temperature of 100–107 °C, followed by a lower temperature of 82–93 °C after byproducts began to build up, and carrying the oxidation to a stage where 120–140% monohydroperoxide (calculated as such based on iodometric titration) was obtained—this was twice the conversion of DIPB described in prior art processes. Then the byproducts *m*-HHP, **3-14**, and DCL, **3-15**, were converted to *m*-DHP by oxidation with hydrogen peroxide in the presence of sulfuric acid as the catalyst.[37] Likely, this process contemplated the elimination of the recycle of partial oxidation products.

An improved purification of the crude resorcinol obtained by *m*-DHP cleavage was patented by Mitsui, consisting of the recrystallization of the crude from toluene containing modest amounts of at least one compound from the group of C_1-C_4-alkylphenols and C_2-C_4-acylphenols.[38]

Another improvement from Mitsui was in the oxidation of byproduct *m*-HHP, **3-14**, and DCL, **3-15**, with hydrogen peroxide/sulfuric acid, by using a water-immiscible solvent such as toluene at up to 70 °C and removing water simultaneously via the toluene azeotrope; the yield of the dihydroperoxide precursor to resorcinol was increased.[39]

From Mitsui's laboratories came an improved recovery of the resorcinol obtained by the cleavage of *m*-DHP, involving first removal of the (silica/alumina) catalyst, then distilling off the acetone while adding 20–70% water, based on the resorcinol; the bottoms from the distillation column were resorcinol, obtained in up to 99% recovery.[40]

A new method was developed at Mitsui for thermally cracking the high-boiling residue from the resorcinol distillation column, using ammonium sulfate/bisulfate as the cracking catalyst at 250 °C and 8–20 mm Hg; the recovery of resorcinol was improved.[41]

Mitsui also obtained a combination process patent claiming improvements by pretreating with hydrogen peroxide/sulfuric acid a solution containing mostly *m*-DHP and *m*-HHP and/or *m*-DCL, under conditions which did not substantially cause cleavage of *m*-DHP and simultaneously removing byproduced water as its azeotrope with toluene. This is followed, in a second step, by the *m*-DHP cleavage in the substantial absence of H_2O_2.[42]

One more Mitsui patent covered the acid-catalyzed cleavage of *m*-DHP to resorcinol and acetone, using acetone/toluene as the cosolvents, treating the cleavage product with 20% aqueous sodium sulfate and a small amount of 25% aqueous NaOH, and finally recovering resorcinol from the separated organic phase.[43]

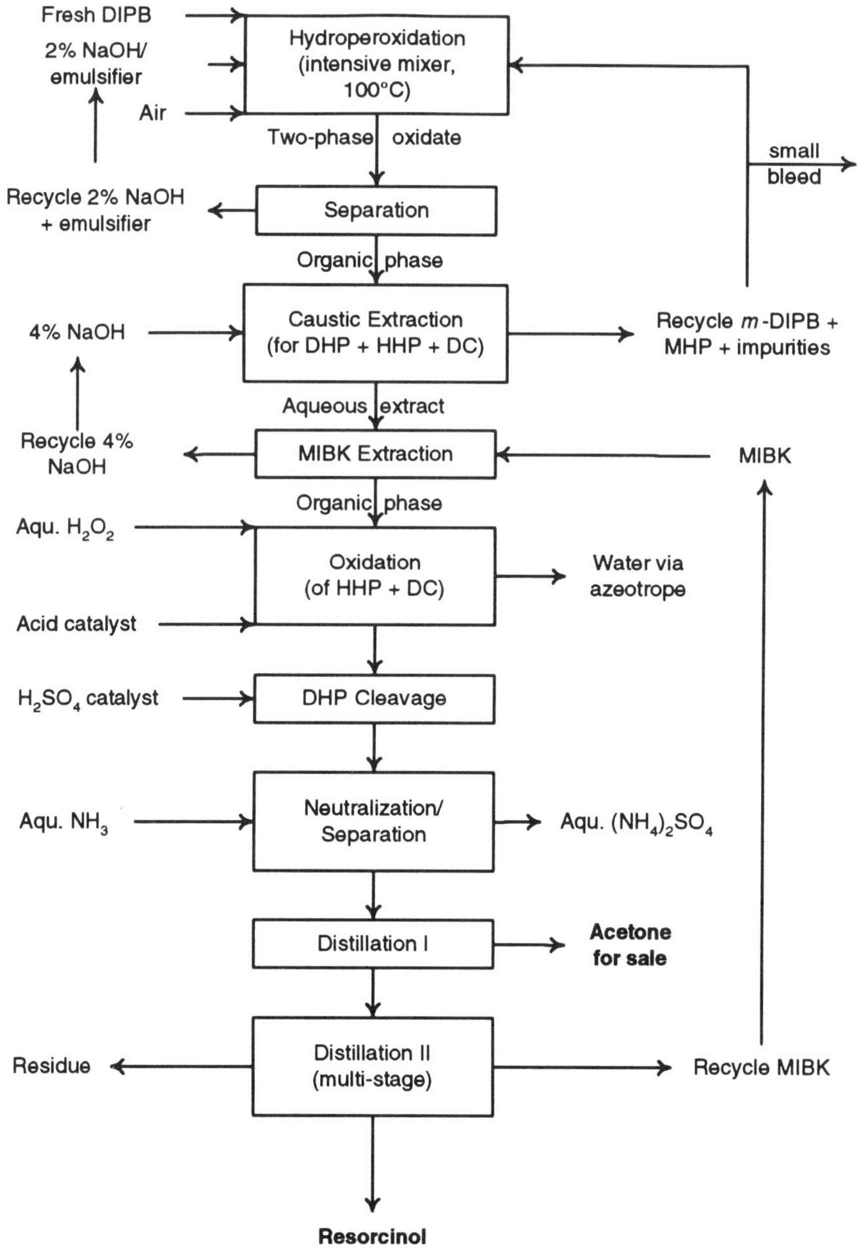

Figure 3.2. Conjectural process—resorcinol from the hydroperoxidation/cleavage of *m*-diisopropylbenzene.

The preceding nine U.S. patents of Mitsui give facets and variations of their DIPB hydroperoxidation/cleavage route to resorcinol in copious detail, but do not really give a definitive clue to their commercial practice. Indeed, neither Sumitomo's nor Mitsui's patents allow certainty about how their resorcinol plants run. A hypothetical version of the modern process is shown in the form of a flow chart in Fig. 3.2, to refocus the operations involved.

As regards the several patents on making resorcinol and hydroquinone by the hydroperoxidation/cleavage of mixed *m*/*p*-DIPB, this is a good idea which would be up against marketing problems with the output. Likely, it is preferable to separate the *m*- and *p*-DIPB and process these separately, as needed.

In this connection, it is interesting to note that UOP reported a separation of catechol, resorcinol, and hydroquinone by selective adsorption/desorption on a Ca, Ba, or Li-exchanged Y-type zeolite; a Koppers patent claimed the separation of the benzenediols by selective complexing with Ca, Mg, or Li halides and diethyl ether.[44]

The staff of Mitsui Petrochemical Industries also developed a process for the simultaneous production of resorcinol and phloroglucinol (1,3,5-benzenetriol) from a mixed feed of *m*-DIPB/1,3,5-triisopropylbenzene by the hydroperoxidation/cleavage route.[45]

Further, Mitsui patented the preparation of 5-methylresorcinol (orcinol) by the hydroperoxidation/cleavage of 3,5-diisopropyltoluene.[46]

There have been many other companies and institutions that worked from the 1970s to the present on aspects of the above route to resorcinol, as evidenced by the many patents issued, e.g., to Kuraray, Chemische Werke Huels, Hoechst, Union Rheinische Braunkohlen Kraftstoff A.G., various technical institutes in the former USSR and Poland, Shell, Signal Chemical, Goodyear, and INDSPEC Chemical Corp. So far, none of these patent holders has gone commercial. Eastman Kodak and Goodyear manufacture hydroquinone from *p*-diisopropylbenzene, each with large captive uses for the product.

While it is not possible to sketch all of these inventions here, a few examples will give their flavor.

An interesting, off-the-beaten-path process for the hydroperoxidation of *m*-DIPB involved first the oxidation of the *m*-DIPB to a low level of conversion (to 10% or more of *m*-MHP, **3-12**), followed by the addition of sulfolane and continuing the oxidation at 125 °C (higher than the usual 80–100 °C); the rate of oxidation was substantially faster than usual and, yet, byproduct formation remained low.[47] Likely, hydrogen bonding of the hydroperoxide(s) to the sulfone retarded decomposition reactions.

INDSPEC's recent patents on this route to resorcinol cover the hydroperoxidation of *m*-DIPB under anhydrous, nonalkaline conditions at 85–95 °C,[48] the decomposition of the dihydroperoxide (*m*-DHP) in the presence of a trace amount of BF_3 as the catalyst,[49] the oxidation of the hydroxyhydroperoxide, *m*-HHP (**3-24**), with hydrogen peroxide to *m*-DHP before cleavage,[50] and an overall process patent incorporating the separate, preceding features.[51] This last patent claims an overall 80–86% yield of resorcinol based on *m*-DIPB.

Resorcinol diacetate is a useful monomer for the preparation of polyesters by transesterification with aromatic dicarboxylic acids, with evolution of acetic acid. An interesting preparation of resorcinol diacetate was patented recently by Eastman Kodak, where the hydroperoxidation of *m*-DIPB and the cleavage of the hydroperoxides was carried out in the presence of acetic anhydride, followed by distillation to give resorcinol diacetate and *m*-isopropylphenol acetate which was recycled to the oxidation step (the actual examples in the patent only detail the preparation of hydroquinone diacetate from *p*-DIPB in this manner).[52]

3.3. RESORCINOL FROM THE ACID HYDROLYSIS OF *m*-PHENYLENEDIAMINE

This process first received industrial attention at I.G. Farbenindustrie A.G. in Germany during World War II.[53] In the 1970s and 1980s it was investigated intensively by Koppers Co. in the U.S. and by Mitsui Toatsu Chemical Co. in Japan. Equations (3.7), (3.8), and (3.9) describe a practical version of this process.

(3.7)

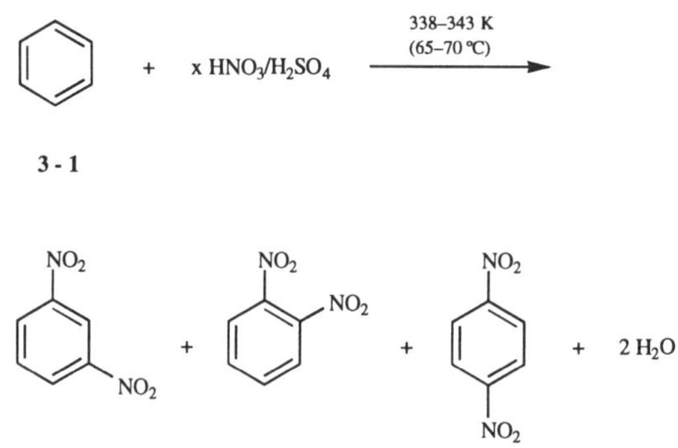

90% 8% 8%

m-DNB *o*-DNB *p*-DNB

3 - 18 3 - 19

Removed by caustic wash

(3.8)

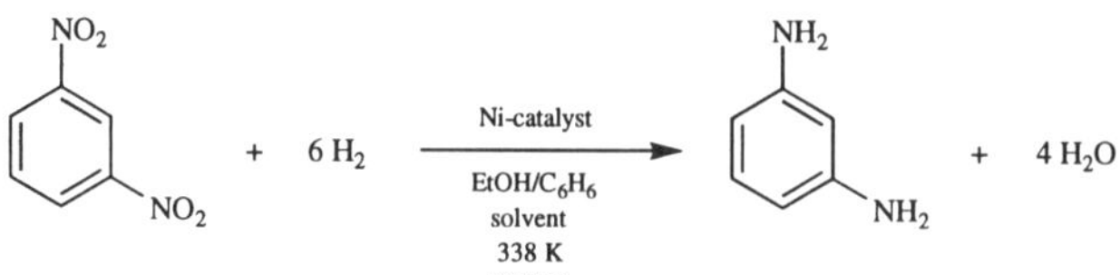

3 - 18 3 - 20

m-DNB MPDA

(3.9) 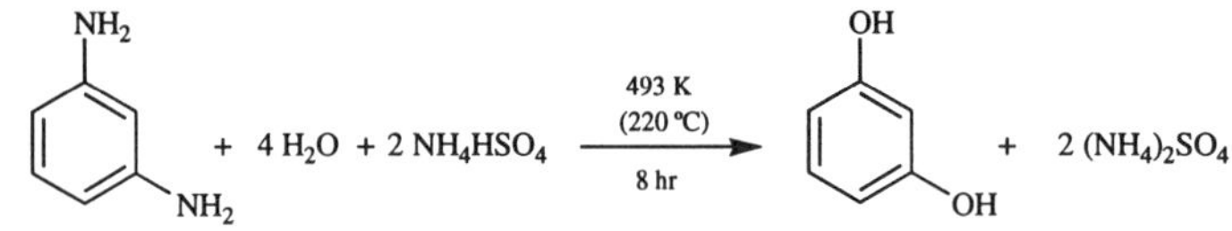

3 - 20 Resorcinol

MPDA DNB = dinitrobenzene; MPDA = *m*-phenylenediamine

The acid hydrolysis of *m*-phenylenediamine (MPDA), **3-20**, requires special construction materials with corrosion resistance at relatively high temperatures. The hydrolysis reaction works well with phosphoric acid,[54] but the disposal of byproduct ammonium phosphate, the purity of the phosphoric acid used, and organic impurities are important side issues. Phosphoric acid is more effective than hydrochloric acid for the hydrolysis of MPDA.[55] Sulfuric acid works well if used at a 1.2–2.2/1.0 mole ratio of H_2SO_4/MPDA.[56] Apparently the best method was to use ammonium bisulfate for the hydrolysis of MPDA. After reaction at *ca* 220 °C/3 hr an 89% yield of resorcinol was obtained; the byproduct ammonium sulfate could be thermally decomposed at *ca* 300–310 °C to recover ammonium bisulfate for recycle to the hydrolysis.[57] Another method of handling the hydrolysis liquor, after the extraction of the organics, was to precipitate pure ammonium sulfate (for sale) from the mixture of ammonium sulfate and bisulfate in water by the addition of methanol.[58]

The aryldiamine hydrolysis route could also be used to make resorcinol derivatives. Thus, the acid hydrolysis of toluenediamine gave methylresorcinol[59] and the hydrolysis of 3,5-diaminobenzoic acid gave 3,5-dihydroxybenzoic acid (alpha-resorcylic acid).[60]

The extensive work by the staff of Mitsui Toatsu Chemicals Co. on this process resulted in many patents covering, e.g., the hydrogenation of *m*-DNB, **3-17**, using a Pd or Pt catalyst and hydrolysis of MPDA, **3-20**, with sulfuric acid or ammonium bisulfate; hydrogenation of *m*-DNB and hydrolysis of MPDA with sodium bisulfate; hydrogenation of *m*-DNB and hydrolysis of MPDA with hydrochloric acid; hydrogenation of crude *m*-DNB/acid hydrolysis of crude MPDA, which gave a 93% yield of resorcinol based on *m*-DNB, showing that *o*- and *p*-phenylenediamine had a slower rate of hydrolysis than MPDA; hydrolysis of MPDA with an organic acid, such as acetic acid; catalytic hydrogenation of *m*-DNB in the presence of sodium or ammonium bisulfate, followed by MPDA hydrolysis at higher temperature; hydrolysis of MPDA hydrochloride, which was said to permit a wider choice of materials of reactor construction, such as Hastelloy, Teflon, lead, and glass; catalytic hydrogenation of *m*-DNB in the presence of hydrochloric acid, followed by the hydrolysis step without first isolating MPDA.[61] Obviously a massive patenting strategy was used, but the processes do not seem to have been commercialized.

A patent to Mitsubishi Chemical Industries claimed the preparation of methyl- and dimethylresorcinols by the hydrolysis of the appropriate diaminotoluene or diaminoxylene with hydrochloric acid in a titanium autoclave at 250 °C.[62]

From Olin Corp. came a patent on the preparation of methylresorcinol by the hydrolysis of toluenediisocyanate (TDI) still residue with aqueous phosphoric acid.[63]

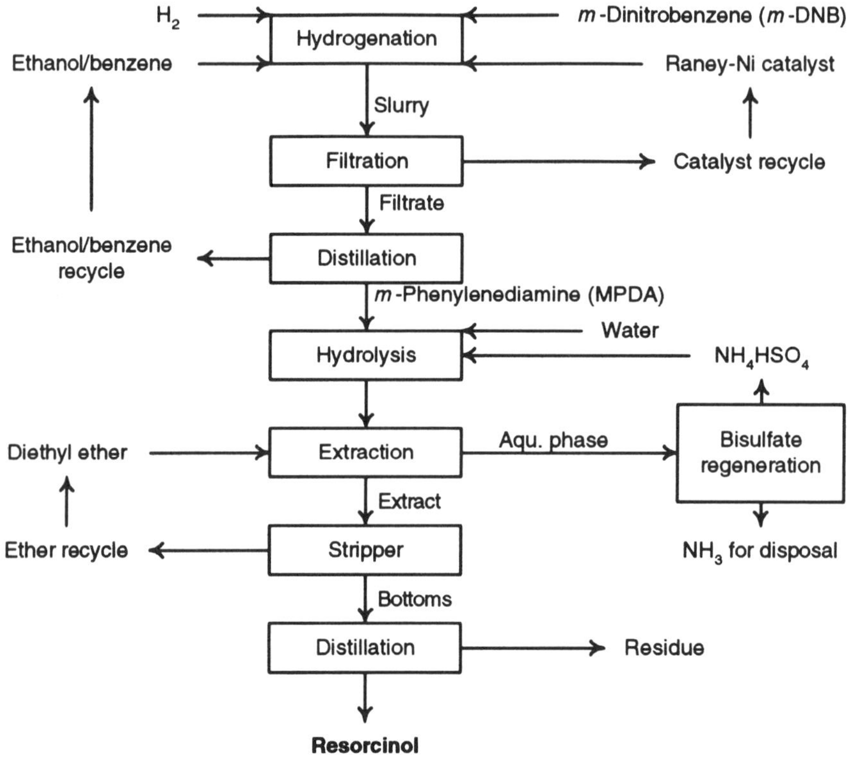

Figure 3.3. Resorcinol from *m*-phenylenediamine (MPDA) by acid hydrolysis (Koppers process).

Once again, a block flow diagram (Fig. 3.3) is given to summarize the preparation of resorcinol by the hydrolysis of MPDA, **3-20**, with ammonium bisulfate, since this is an attractive version of the process.

3.4. RESORCINOL FROM THE ACETONE/ACRYLIC COMPOUND CONDENSATION/CYCLIZATION/ DEHYDROGENATION

This route to resorcinol was under intensive development at Stamicarbon N.V. (part of the DSM organization in the Netherlands) and Hoechst A.G. in Germany in the 1970s and early 1980s. This work was also covered extensively by European, U.S., and Japanese patents. Process improvement patents were obtained by Koppers Co. in the 1980s. Japanese companies did little work on this route to resorcinol.

The general chemistry involved is shown in Eqs. (3.10), (3.10a), (3.10b), (3.11), and (3.12). The following sketches of developments are mostly based on U.S. patents.

(3.10)

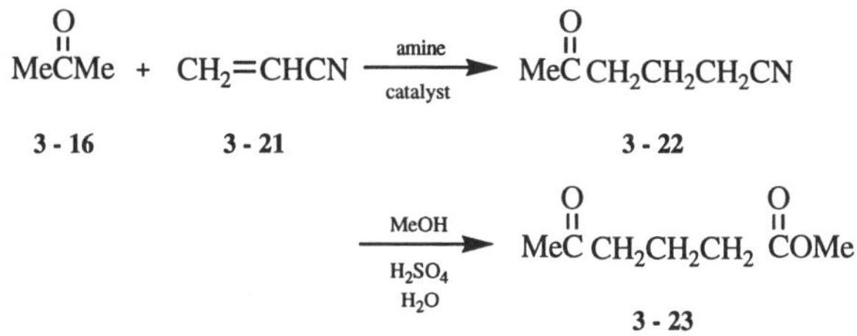

(3.10a)

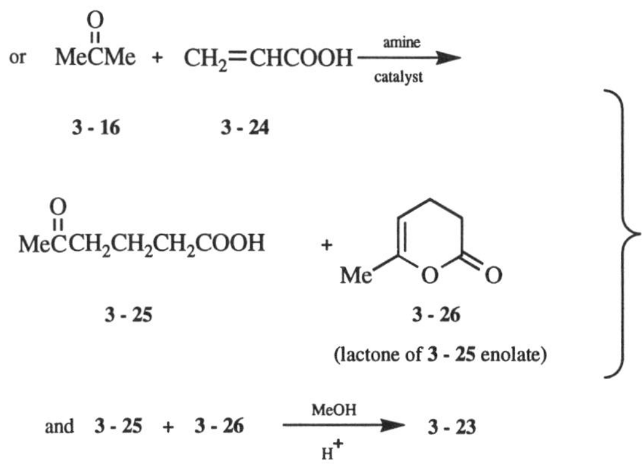

(3.10b)

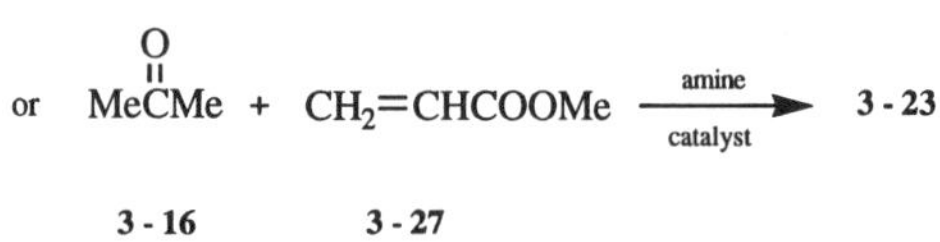

(3.11)

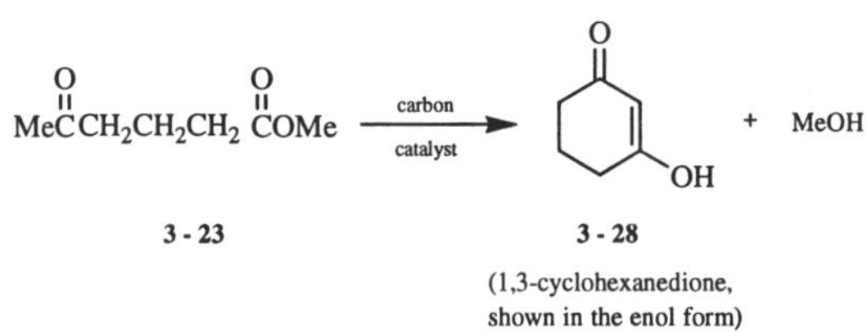

(3.12)

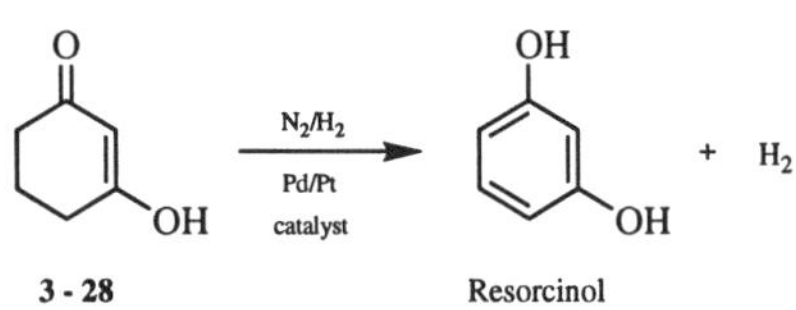

3.4.1. Stamicarbon's Version

Acrylonitrile, **3-21**, was reacted with acetone (large excess), **3-16**, in the presence of small amounts of *N*-isopropylacetone imine, isopropylamine, and benzoic acid; a 91% yield of 5-oxohexanonitrile, **3-22**, was obtained, at 85% conversion of acrylonitrile.[64] (By acid-catalyzed hydrolysis of the nitrile and esterification of the carboxylic acid with methanol, the ester **3-23** can be obtained.)

Methyl 5-oxohexanoate, **3-23**, together with hydrogen was passed over a carbon catalyst at 300–330 °C and *ca* 0.11 LHSV (liquid hourly space velocity) to produce 1,3-cyclohexanedione, **3-28**, in 93+% yield, alas at low conversion of the methyl 5-oxohexanoate.[65]

Methyl 5-oxohexanoate, **3-23**, together with hydrogen and nitrogen (diluent) were passed over various metal-on-carrier catalysts at *ca* 325–345 °C; with a 1.6% Pd-/0.5% Pt-on-carbon catalyst and the feed at 0.15 LHSV, a 4% yield of 1,3-cyclohexanedione plus an 84% yield of resorcinol were obtained at modest per pass conversion of the methyl 5-oxohexanoate.[66]

The preparation of 5-oxohexanonitrile, **3-22**, was improved by keeping the amount of oxygen dissolved in the charge (acetone, acrylonitrile, catalyst) below 20 ppm. After completion of the reaction, the unreacted materials and catalysts were distilled off and the residue was diluted with an inert solvent, such as mesityl oxide (a byproduct of the cyanoethylation reaction), before distillation to recover 5-oxohexanonitrile. The formation of polymer-like deposits in the reactor was minimized and the yield of the desired product was improved.[67]

A two-stage process for resorcinol was disclosed, first converting methyl 5-oxohexanoate, **3-23**, to 1,3-cyclohexanedione, **3-28**, over a carbon catalyst, and then dehydrogenating the dione contained in the methyl 5-oxohexanoate over a 10% Pd-on-C catalyst at *ca* 200 °C to resorcinol.[68]

3.4.2. Hoechst's Version

A high (up to 98%) yield of 1,3-cyclohexanedione was obtained from methyl 5-oxohexanoate, **3-23**, by cyclization in the presence of a stoichiometric amount of a strong base, such as sodium methoxide, in solvents such as *N,N*-dimethylformamide, *N,N*-dimethylacetamide, hexamethylphosphoramide, or dimethyl sulfoxide at *ca* 25 °C for 1 hr, followed by acidification.[69] This was a substantial improvement over the prior art in this Claisen condensation.

In another patent, 5-oxohexanoic acid, **3-25**, and/or its lactone, **3-26**, were cyclized and dehydrogenated in a N_2/H_2 stream at *ca* 300 °C over a 2% Pd-on-alumina catalyst to give up to 73% selectivity to resorcinol at up to 50% conversion of the feed; modest amounts of 1,3-cyclohexanedione and phenol were the main byproducts.[70]

Alkyl-substituted 1,3-cylcohexanediones were obtained in high yield directly from ethyl acrylate and a dialkyl ketone in the presence of stoichiometric amounts of sodium methoxide at *ca* 30 °C/3 hr in diethyleneglycol dimethyl ether, or *N,N*-dimethylformamide and similar solvents; the conversion of the acrylate was nearly 100%.[71]

Another invention covered the conversion of 5-oxohexanoic acid, **3-25** (optionally alkyl substituted), over silica/alumina catalysts at 330–360 °C to mixtures of 1,3-cyclohexanedione (optionally alkyl substituted) and the lactone, **3-26**, at 15–60% conversion of the feed.[72]

Further improvement was obtained by adding methyl 5-oxohexanoate, **3-23**, to a solution of an equimolar amount of sodium methoxide in diethyleneglycol diethyl ether in 1 hr/25 °C. At that point GC analysis of the mixture indicated that the ester had been 100% converted; the mixture was acidified to pH 3.5 with hydrochloric acid and the sodium chloride was filtered off. The filtrate contained 1,3-cyclohexanedione in 94% yield. Without isolating the dione, the solution was heated to 190 °C and dehydrogenated in the presence of an 1% Pd-on-C catalyst, then filtered and distilled to give an 88% yield of resorcinol and a 2.6% yield of phenol.[73]

Separately patented was the liquid-phase dehydrogenation of 1,3-cyclohexandiones in diethyleneglycol dialkyl ether (and similar) solvents at 180–230 °C in the presence of a Pd, Pt, Ru, or Rh catalyst to give resorcinols in up to 94% yield, with a small amount of phenol as a byproduct.[74] If methanol was used as the solvent, a 47% yield of resorcinol monomethyl ether was obtained.

The vapor-phase dehydrogenation of methyl 5-oxohexanoate over catalyst mixtures of 2.4% Pd-on-C/0.6% Pt-on-C/4.5% NaCl-on-C and 2% Th-on-C in a 1/1 H_2/N_2 stream was claimed to give up to a 72% yield of resorcinol at 93% conversion of the ester fed.[75]

5-Oxoalkanoic acids were prepared by the condensation of the appropriate ketone with an alpha,beta-unsaturated acid in the presence of a secondary amine catalyst. For instance, the condensation of acetone with acrylic acid in the presence of a catalytic amount of diethyl amine gave 85% selectivity and 75% conversion of the acrylic acid to 5-oxohexanoic acid, **3-25**.[76] This acid was cyclized in the vapor phase at 380 °C in a stream of nitrogen over an aluminosilicate catalyst. The crude catalyzate was taken up in toluene and the product contained 25% toluene, 30% 1,3-cyclohexanedione, 14% 6-methyl-3,4-dihydro-2-pyranone (the lactone, **3-26**), 8% water, 18% 5-oxohexanoic acid, and 5% byproducts. This solution was dehydrated/stripped, the residue was taken up in toluene again, cooled, filtered, and the cake was washed with toluene and dried to recover 84% of the contained 1,3-cyclohexanedione as a 99% pure product.[77]

3.4.3. Koppers's Version

The vapor-phase cyclization of methyl 5-oxohexanoate, **3-23**, over an activated carbon catalyst with a large surface area was improved by using toluene vapor as the diluent (instead of the large volumes of nitrogen and/or hydrogen used in Stamicarbon's and Hoechst's patents). For example, at 400 °C, a 36% conversion of methyl 5-oxohenoate to 1,3-cyclohexanedione was obtained, and the ultimate yield of the dione was 100%; the dione could be readily isolated in crystalline form.[78]

A combination process patent claimed the cyclization of methyl 5-oxohexanoate to 1,3-cyclohexanedione, as above. In the second step, the dione was dissolved in isopropanol or butyl acetate and dehydrogenated over a 5% Pd-on-C catalyst at 170–180 °C to give up to 98% conversion of the 1,3-cyclohexanedione to 98% pure resorcinol.[79]

The Michael condensation of acetone with methyl acrylate was improved by using a primary amine catalyst and a solid acid catalyst, such as sodium acid phthalate, aluminosilicate, silica/alumina zeolite, or a sulfonated perfluorinated resin like DuPont's Nafion H. The acid catalyst could be recovered by filtration and recycled. The conversion of methyl acrylate was 90% and an 85% per pass yield of methyl 5-oxohexanoate, **3-23**, was obtained.[80]

A block flow diagram for making resorcinol from acetone and methyl acrylate, essentially based on information given in Koppers patents, is shown in Fig. 3.4.

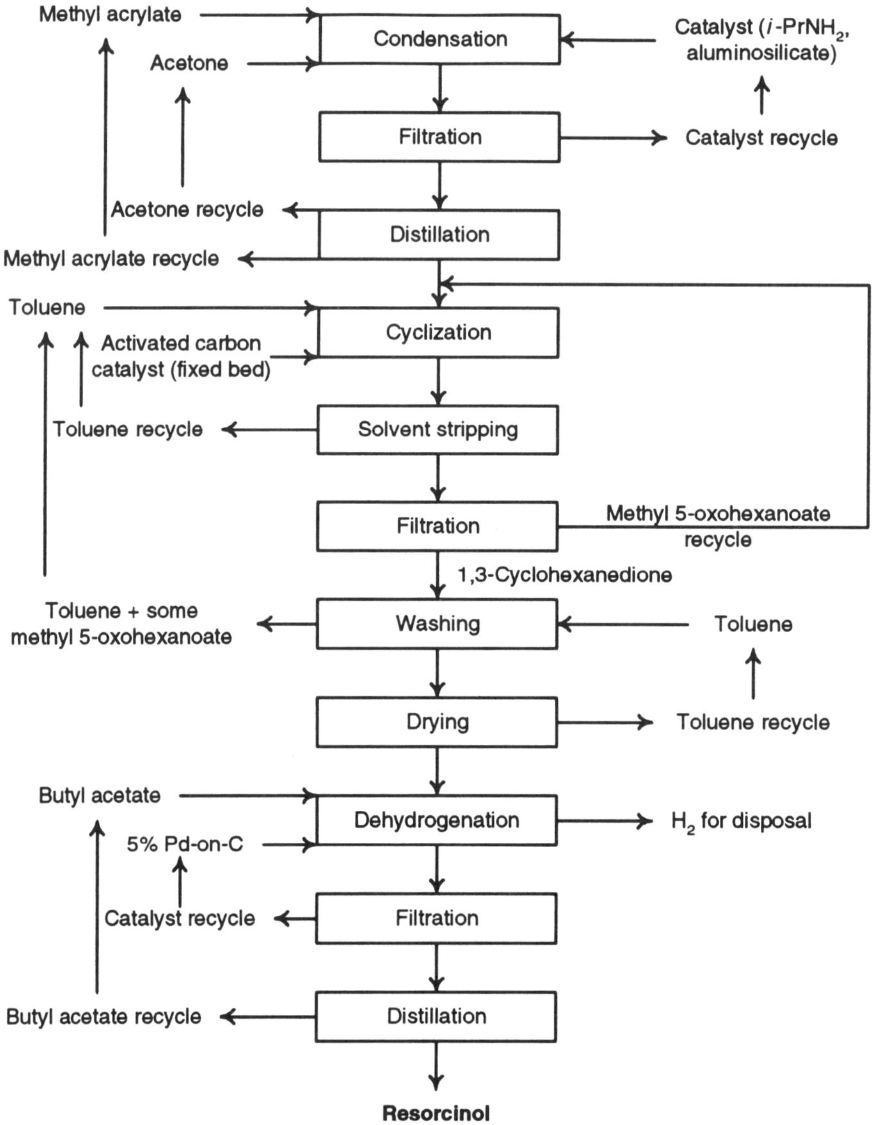

Figure 3.4. Resorcinol by the acetone/methyl acrylate condensation/cyclization/dehydrogenation (Koppers process).

To date, this type of process has not been commercialized, in good part because the chemical cost is high compared to the chemical cost of the benzene disulfonation/caustic fusion and the *m*-diisopropylbenzene hydroperoxidation/cleavage routes to resorcinol.

Stamicarbon/DSM commercialized the first step of their process, condensing acetone with acrylonitrile to make 5-oxohexanonitrile, **3-22**, for another purpose. The 5-oxohexanonitrile was cyclized/dehydrogenated to produce 2-methylpyridne (and 2,3-dimethylpyridine by a similar route) in a plant of 4000 t/yr capacity.

As shown earlier, this route to resorcinol is also valuable for making resorcinol derivatives which are not easily accessible by other routes. For example, starting with 2-butanone (methyl ethyl ketone) and methyl acrylate one can obtain a mixture of 2-methyl- and 4-methylresorcinol, by starting with acetone and methyl crotonate one can get 5-methylresorcinol, from acetone and methyl methacrylate one can make 4-methylresorcinol and, starting with acetone and dimethyl maleate, can lead to 3,5-dihydroxybenzoate.

3.5. RESORCINOL FROM LOW-TEMPERATURE COAL PYROLYZATES, SHALE OIL, AND PLANT SOURCES

2-Methylresorcinol has been produced for many years by Coalite in the United Kingdom from low-temperature coal tar. In the beginning, the low-temperature carbonization of coal, in the 450–700 °C range, was carried out to obtain a low-smoke fuel (coke) for household use as the main product. The pyrolyzate obtained in this process also produces a crude, aqueous liquor byproduct which contains the water-soluble, mono-, and di-hydric phenols. To increase the yield of dihydric phenols, the contact time of the aqueous liquor with the crude oil is extended; the crude oil is recycled to the coal carbonization stage. The enriched aqueous liquor is countercurrently extracted with isobutyl acetate to concentrate the phenols in the extract. This extract is then fractionated by distillation to give cuts containing monohydric phenols, catechol, 4-methylcatechol, 2-methylresorcinol, and a cut rich (*ca* 30%) in resorcinol.[81] Coalite still produces and sells today a *ca* 90% pure 2-methylresorcinol made by this process.

Similarly, the volatiles (kerogen is the oil fraction) obtained by the pyrolysis at *ca* 500–650 °C of certain oil shales are a source of methyl-, dimethyl-, trimethyl-, and ethylresorcinols. Extensive research and development on the pyrolysis of the oil shales, the recovery and upgrading of the resorcinols from the volatiles by extraction of the liquors/fractional distillation/crystallization techniques, and on the utilization of these products has been conducted in Estonia/USSR (and its successor states) since the 1960s, based on the native oil shale (kukserite). A large body of literature derived from these studies, the majority in Russian. Most of the description of this work in English is in the form of abstracts, which contain insufficient detail to be sure about the commercial process practiced in Estonia. The main methylresorcinol which can be obtained in enriched form from this source is 5-methylresorcinol (orcinol). The massive application work on these materials mostly parallelled the uses of resorcinol, e.g., in adhesives and rubber. Some similar application work was done in Japan. The oil-shale kerogen was said to be a low-cost source of resorcinols, but it is hard to tell if the cost-allocation practices in the region were the same as those used by the Western nations.

A paper was published in English which compared the reactivity and spectra of oil shale-derived alkylresorcinol/formaldehyde wood laminating resins with phenolic and resorcinol/formaldehyde resins used commercially for such applications. This paper gave a fair review of the literature and reported that these alkylresorcinol resins had a higher reactivity in cure than the resorcinol resins; the data were insufficient to permit judgment of the relative merits of these resins in performance in the field.[82]

Another paper in English reported on the strength and durability of oil-shale-derived alkylresorcinol/formaldehyde adhesives in laminating wood. These adhesives were found to be largely comparable in strength to the adhesives made from synthetic resorcinol; however, the kerogen-derived adhesives showed low wood failure, while the adhesives made with synthetic resorcinol gave the desirable, high wood failure (i.e., the failure occurs in the wood, not in the adhesive bond).[83]

In Estonia, technical work on the production of alkylresorcinols from oil shale kerogen has continued into the 1990s. For instance, the manufacture of water-soluble alkylresorcinols was improved by pyrolyzing the shale at 560–600 °C, heating the volatiles in the presence of 3.2–3.6% fibrous kaolin catalyst, and condensing the pyrolizate.[84] When Ru-zeolites, aluminosilicates, or iron-ore-supported chromia catalysts were used in the treatment of the volatiles from the semicoking of kukserite, a significant increase in the production of phenols and resorcinols was claimed.[85] Also, a cross-flow scheme was patented for the extraction of alkylphenols and alkylresorcinols with water from a distillation cut, bp 230–250 °C, obtained from oil shale kerogen.[86]

Regarding the production of resorcinol from certain plant materials, work on this source increased in the aftermath of the oil crisis in the 1970s, when fears arose about the availability of petroleum as a feedstock for chemicals. There are some pertinent references to this work in Chapters 5 and 14. One example of work on obtaining resorcinol from plants is the following: In the mid-1970s, an investigation was reported citing earlier work, on using heartwood extracts (methanol solvent) of the black wattle tree (*Acacia Mearnsii*) as a source of resorcinol. The wattle heartwood averaged 45% of timber volume and contained about 8% extractives, including several flavanoids which contained a resorcinol moiety. Alkali fusion of the extractives yielded 8% (based on the weight of the extractives) resorcinol. A modest yield improvement was possible if the extract was partitioned between ethyl acetate and water before fusion. Concentrated waste sulfite liquors obtained in rayon manufacture from wattle wood in South Africa were said to have potential as a source of resorcinol as well.[87] (In most circumstances, synthetic resorcinol seems more accessible and less costly.)

3.6. RESORCINOL FROM HALOAROMATIC PRECURSORS

An older patent described a process starting with the selective sulfonation of the *m*-isomer in an *o,m,p*-dichlorobenzene mixture. The obtained 2,4-dichlorobenzenesulfonic acid was converted to its sodium salt and hydrolyzed with aqueous sodium hydroxide at 180–200 °C for 3 hr, acidified/desulfonated, extracted with ether, and the ether extract was evaporated to give resorcinol of good quality in up to 85% yield.[88]

The direct alkaline hydrolysis of *m*-dichlorobenzene in the presence of a copper compound to benzenediols was claimed to occur under milder conditions, i.e., at 250–300 °C, if carried out in the presence of alkali metal fluorides or bifluorides.[89]

Scientists at the Dow Chemical Co. found that the addition of copper salts of long-chain alkylnaphthalenesulfonic acids improved the alkaline hydrolysis of dihalobenzenes to produce dihydric phenols, including resorcinol (these patents do not give specific examples).[90] There are patents from Dow and others from the 1960s showing that the aqueous or nonaqueous alkaline hydrolysis of *o,p*-chlorophenol, with or without a copper catalyst, gave mixtures of catechol, resorcinol, and hydroquinone plus higher boiling products, their ratio depending in part on the amount of alkali used (see, e.g., Ref. 91). This indicated that direct displacement of halide by hydroxide as well as intermediate benzyne formation were taking place.

Surprisingly, it was claimed that the aqueous hydrolysis of *m*-chlorophenol or *m*-dichlorobenzene with alkali, alkaline earth, or magnesium salts of organic acids gave only resorcinol. For example, the hydrolysis of *m*-chlorophenol with potassium acetate, using a cuprous oxide catalyst, gave a product containing 94% resorcinol.[92]

While some of the above results may seem attractive, nowadays the actual or potential toxicity of the starting materials and impurities therein, of the intermediates and of chlorinated trace byproducts, is likely to make these routes not the first choice.

An interesting variation on the theme deserves mention: *m*-xylene was chlorinated to obtain 2,4-/4,6-dichloro-*m*-xylene and this mixture was oxidized to the dichlorophthalic acids. The latter were hydrolyzed/decarboxylated with aqueous sodium hydroxide in the presence of Cu_2O at 230–250 °C; the dichlorophthalate conversion was 93% and the yield of resorcinol was 97%.[93]

Alkylresorcinols can be made from haloalkylphenols. For instance, 5-methylresorcinol was made by the hydrolysis of 4-chloro-*m*-cresol with aqueous potassium hydroxide at 175 °C/8 hr in 66% yield.[94]

3.7. ROUTES STARTING WITH OXYGENATED CYCLOHEXANES/CYCLOHEXENES

Processes of this type have been of interest to industrial scientists for some time. These routes have proven to be difficult to optimize to a stage where commercialization could be considered. A few examples follow.

In the process sketched in Eq. (3.13), the acid-catalyzed reaction of cyclohexanone, **3-29**, with acetic anhydride gave cyclohexenyl acetate, **3-30**, which was air-oxidized in the presence of a cobalt (naphthenate) catalyst at 80 °C in benzene to

(3.13) 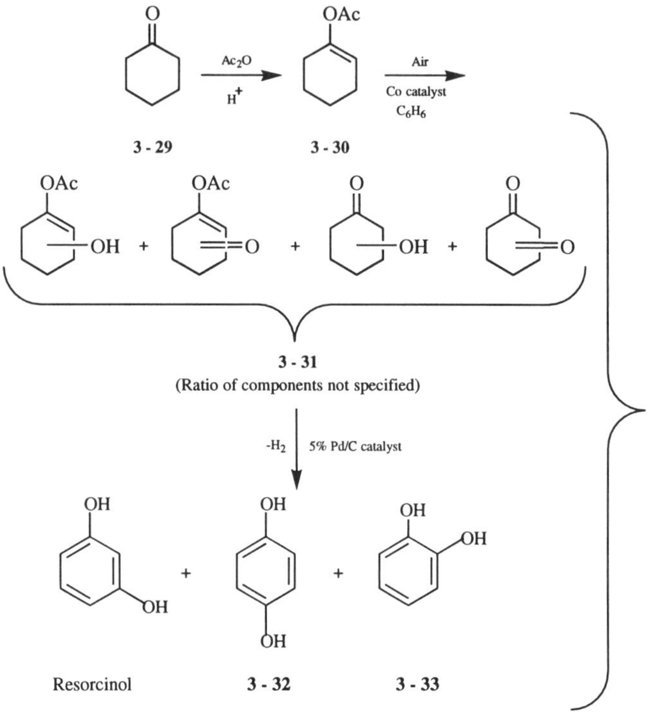

3 - 29 3 - 30

3 - 31
(Ratio of components not specified)

Resorcinol 3 - 32 3 - 33

give a 35% conversion of **3-30** to a mixture of dioxygenated cyclohexenes/cyclohexanes, **3-31**. This mixture was stripped of mono-oxygenated cyclohexanes, the residue was diluted with 1,2,4,-trimethylbenzene and dehydrogenated in the presence of a 5% Pd-on-C catalyst to give a 67% yield of resorcinol, 11% hydroquinone, **3-32**, and 3% catechol, **3-33**.[95]

Another process in this category involved the thermal rearrangement of 2,3-epoxycyclohexanone, **3-34**, to 1,3-cyclohexanedione, which can be dehydrogenated to make resorcinol.[96]

Also, 2-cyclohexenone, **3-35**, was hydrated in the presence of an acid ion-exchange resin at 60 °C to give 3-hydroxycyclohexanone, **3-36**, with 99% selectivity at 34% conversion of the starting material[97]; this compound can be dehydrogenated to produce resorcinol.

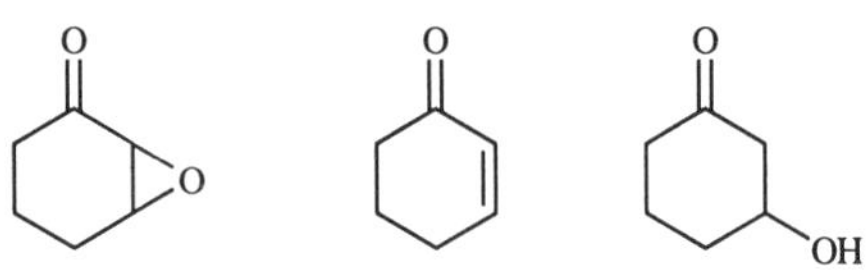

3 - 34 3 - 35 3 - 36

A 3/1 mixture of resorcinol and catechol was obtained by the dehydrogenation of 2,3-epoxycyclohexanone, **3-34**, over a supported Pt-catalyst in an inert solvent at 130–160 °C; phenol was a byproduct.[98]

One patent just claimed the dehydrogenation of 3-hydroxycyclohexanone, **3-36**, over a Pd-on-C catalyst with an oxygen-containing gas at 50–100 °C; the conversion of **3-36** was 95% and the yield of resorcinol was 80%.[99]

A recent patent on this route came from Catalytica Associates. Phenol was hydrogenated to make cyclohexanone, which was dehydrogenated to 2-cyclohexenone, **3-35**. This was hydrated to 3-hydroxycyclohexanone, **3-36**, in the presence of an acid catalyst (ion-exchange resin), and compound **3-36** was oxidatively dehydrogenated, optionally via 1,3-cyclohexandedione, to resorcinol.[100] This was a 4–5 step process, with high to modest conversion of starting materials in each step requiring several different catalysts; catalyst life was not reported. There is a lot of processing/recycling of materials. For these reasons, the process is unlikely to be competitive with current commercial processes for resorcinol.

3.8. RESORCINOL DERIVATIVES FROM NONCYCLIC PRECURSORS OTHER THAN ACETONE/ACRYLIC COMPOUNDS

The dialkyl ketone/acrylic compound condensation/cyclization/dehydrogenation route to resorcinol was outlined in some detail in Section 3.4 above. There are a number of related reactions, starting, e.g., from beta-ketoesters and alpha,beta-unsaturated esters, that have been studied for the production of resorcinol and derivatives. A few examples are given here.

5-Alkylresorcinols and precursors were prepared, as shown in Eq. (3.14), by the condensation of alkyl-substituted propiolates, **3-37**, with diesters of 3-oxoglutaric acid, **3-38**, in the presence of sodium hydride or sodium ethylate to give diesters of resorcinol-2,6-dicarboxylic acid, **3-39**, which were hydrolyzed and decarboxylated to produce the 5-alkylresorcinols. For example, starting from dimethyl 3-oxoglutarate and methyl 2-octinoate, 5-pentylresorcinol (olivetol), **3-40** (R_1 = *n*-pentyl), was obtained in good overall yield.[101]

Another approach to olivetol is shown in Eq. (3.15). 3-Nonene-2-one, **3-41**, and dimethyl malonate, **3-42**, were condensed/cyclized in the presence of sodium methoxide to give the methyl ester of 4-carboxy-5-pentyl-1,3-cyclohexanedione, **3-43** (shown in the enol form). The latter was aromatized/decarbomethoxylated with bromine in *N,N*-dimethylformamide to give olivetol, **3-44**, in 62% overall yield.[102]

A novel preparation of 5-methylresorcinol (orcinol) was patented. In this process, ethyl crotonate was condensed with ethyl acetoacetate in the presence of an equimolar amount of sodium hydride dispersed in mineral oil to give 4-carbomethoxy-5-methyl-1,3-cyclohexanedione in 60% yield. The latter was hydrolyzed/decarboxylated with aqueous KOH, followed by acidification with HCl to produce 5-methyl-1,3-cyclohexanedione in 77% yield, and this intermediate was dehydrogenated, using a 5% Pd-on-C

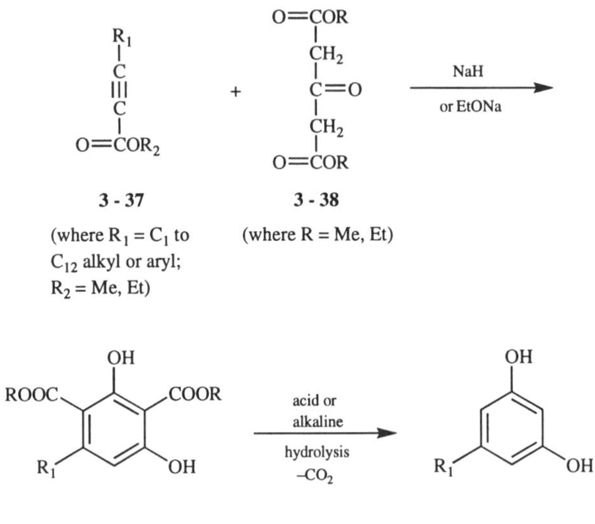

3 - 37 (where $R_1 = C_1$ to C_{12} alkyl or aryl; R_2 = Me, Et)

3 - 38 (where R = Me, Et)

3 - 39

3 - 40

(3.14)

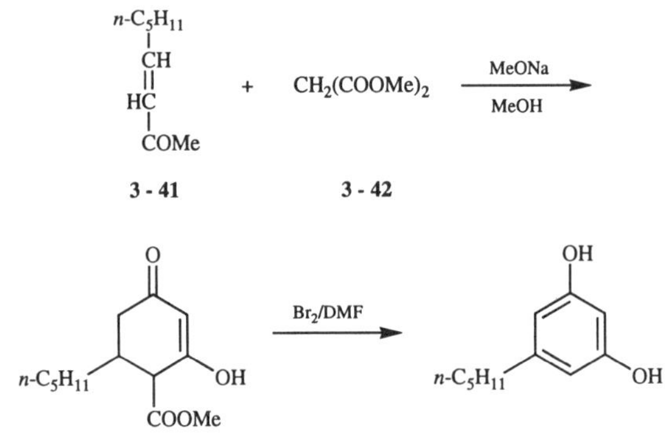

3 - 41

3 - 42

3 - 43

3 - 44

(3.15)

catalyst, to give 5-methylresorcinol in 71%% yield.[103] Related preparations of 5-alkyl-1,3-cyclohexanediones and their aromatization with bromine were studied extensively at the University of Sydney[104]; this paper also gave a good review of the pertinent journal literature to 1974–75.

Another novel route to 5-substituted resorcinols, summarized in Eq. (3.16), involved the addition of phenylsulfinylacetate esters, **3-45**, to alpha,beta-unsaturated ketones, **3-46**, to produce 1,3-cyclohexanedione derivatives, **3-47**, which were aromatized by thermal elimination of benzenesulfinic acid to make a variety of 5-substituted resorcinols, **3-48**, in up to 60% yields.[105]

A synthesis of resorcinols from diketene and ketones was reported, outlined in Eq. (3.17). Thus, the reaction of diketene, **3-49**, and ethyl thallio(I)acetoacetate, **3-50**,

(3.16)

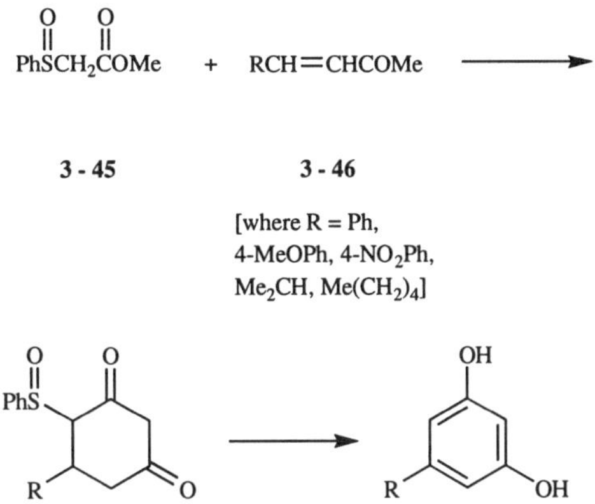

3-45 3-46

[where R = Ph, 4-MeOPh, 4-NO_2Ph, Me_2CH, Me$(CH_2)_4$]

3-47 3-48

(3.17)

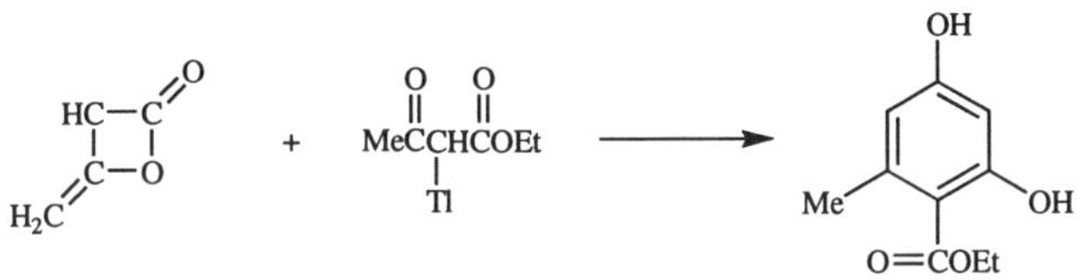

3-49 3-50 3-51

in tetrahydrofuran at 0 °C for 40 hr gave a 49% yield of ethyl 2-methyl-4,6-dihydroxybenzoate (4-carboethoxy-5-methylresorcinol), **3-51**.[106,107]

As an example of the synthesis of resorcinol derivatives useful in fragrances, esters of 3-alkyl-6-methyl-beta-resorcylic acid (2-alkyl-4-carboalkoxy-5-methylresorcinols), **3-53**, were prepared either by the aromatization of the appropriately substituted 1,3-cyclohexanedione with sulfuryl chloride[108] or by treatment of an alpha-pyrone, **3-52**, with sodium methoxide, followed by acidification, as shown in Eq. (3.18).[109]

A final, fragrant note: 2,4-Dihydroxy-3,6-dimethylbenzonitrile (2,5-dimethyl-4-cyanoresorcinol) was prepared from the appropriate 1,3-cyclohexanedions in 60% yield by aromatization with chlorine in acetic acid; this and related products were used to enhance the aroma of tobacco and perfumes.[110]

(3.18)

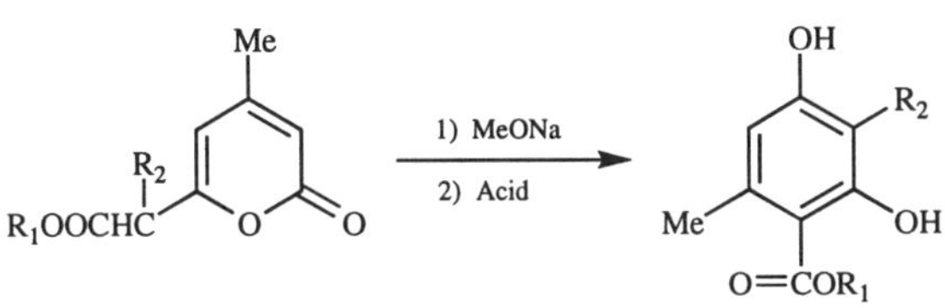

3-52 3-53

(where R_1, R_2 = C_1 to C_{10} alkyl)

3.9. MISCELLANEOUS ROUTES

A few examples of syntheses off the beaten path are given here.

m-Cresol was sulfonated with sulfuric acid at 25–80 °C. The sulfonate was isomerized at 175 °C to maximum content of the 5-sulfo isomer, then desulfonated at 110–140 °C with added water to hydrolyze selectively only the 4- and 6-sulfo-*m*-cresols. The mixture was then neutralized and evaporated, and the residue was fused with KOH to obtain 5-methylresorcinol (orcinol) and/or 3,5-dihydroxybenzoic acid in modest yield, as well as recovered *m*-cresol for recycle.[111]

The fusion of disodium 3,3′-diphenylsulfonedisulfonate with KOH gave resorcinol and a small amount of byproduced *m*-phenolsulfonate.[112]

The acetoxylation of anisole (methyl ether of phenol) with potassium peroxydisulfate in acetic acid in the presence of $Pd(OAc)_2{\cdot}2,2'$-bipyridine as the catalyst gave a 193% yield [based on the $Pd(OAc)_2$] of *o-/m-/p*-acetoxyanisole in 3/61/36 ratio; a small amount of $PhOCH_2OAc$ was a byproduct.[113]

Phthalic anhydride and water vapor were passed at 300 °C over a catalyst containing 4% CuO/3% Zr(IV) oxide/3.6% K_2O/89.4% Al_2O_3 at 3 sec CT (contact time). The catalyzate contained 28 wt% resorcinol; the balance of the organics were phenol, biphenyl, biphenyl ether, benzene, and benzoic acid.[114]

There were several approaches to make resorcinol from *m*-isopropylphenol. For instance, in one patent *m*-isopropylphenyl acetate was oxidized with oxygen in the presence of copper phthalocyanine as the catalyst and *tert*-butylhydroperoxide as the initiator to give a product containing 95.8% *m*-hydroperoxyisopropylphenyl acetate. The hydroperoxide cleavage/acetate hydrolysis needed to make resorcinol was described only in general terms.[115] The starting material, *m*-isopropylphenol, was prepared by alkylating phenol with propylene in the presence of an acid catalyst, isolating a *m,p*-isopropylphenol mixture and selectively dealkylating the *p*-isomer with a high silica zeolite, ZSM-5, catalyst and recovering the *m*-isopropylphenol, with the phenol going to recycle.[116] Another method involved the isomerization of *o-/m-/p*-isopropylphenol with the superacid, HF/SbF_5, whereby substantially pure *m*-isopropylphenol was obtained.[117] Yet another method involved the alkylation of phenol with propylene in HF/BF_3 at 0–100 °C to give a 59% yield of 99% pure *m*-isopropylphenol, plus recovered phenol and some diisopropylphenol, which can be transsalklylated to make more *m*-isopropylphenol.[118]

Another variation on the theme was the hydroperoxidation of *m*-isopropylphenol trifluoromethanesulfonate or *m*-isopropylphenol trimethylsilyl ether, followed by cleavage of the hydroperoxide and hydrolysis of the ester or ether to produce resorcinol.[117–119]

An unusual method for making resorcinol diacetate, **3-55**, in 37% yield involved the oxidation of *m*-cresol acetate, **3-54**, with oxygen at 237 °C in acetic anhydride using sulfuric acid as the catalyst, as shown in Eq. (3.19).[120]

Finally, there were a number of reports on the electrophilic hydroxylation of phenols with H_2O_2 in superacids to give resorcinols. For instance, the reaction of 2,4-xylenol with hydrogen peroxide in HF/SbF_5 at –40 °C gave a 58% yield of

(3.19) 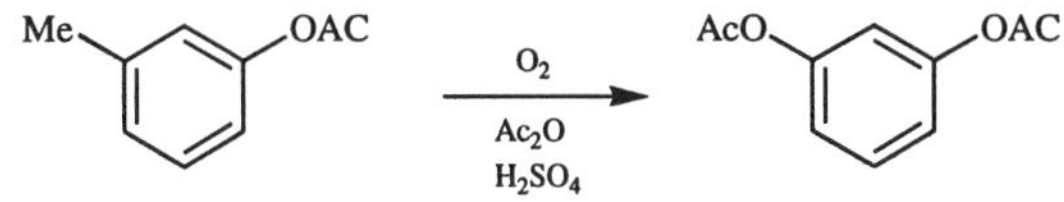

3 - 54 3 - 55

4,6-dimethylresorcinol plus a 32% yield of 2,4-dimethylresorcinol.[121] The reaction of *p*-cresol with H_2O_2 in HF/SbF_5 at –41 °C gave a 48% yield of 4-methylresorcinol.[122] Similarly, the reaction of phenyl acetate with 95% H_2O_2 in HF/SbF_5 gave an 87% yield of hydroxylated phenyl acetate, containing the *o*-/*m*-/*p*-isomers in 6/64/30 ratio.[123]

3.10. REFERENCES

1. X. Bindschedler and Y. Busch, *Chem. Ind., 1*, 370 (1878).
2. B.I.O.S. (British Intelligence Objectives Subcommittee) Final Report No. 1153, 24, p. 366.
3. E. E. Gilbert, *Sulfonation and Related Reactions*, Interscience, New York, 1965.
4. H. Cerfontain, *Mechanistic Aspects in Aromatic Sulfonation and Desulfonation*, Interscience, New York, 1968.
5. C. H. Bamford and C. F. H. Tipper (eds.), *Comprehensive Chemical Kinetics*, Vol. 13, pp. 56–77, Elsevier, New York, 1972.
6. Ger. Offen. DE 1,063,131 (Jan. 21, 1960; G. Otten, O. Scherer, and H. Schild to Hoechst A.G.).
7. Ger. Offen. DE 1,104,500 (Apr. 13, 1961; O. Scherer, R. Huebner, and G. Otten to Hoechst A.G.); *C.A., 56*, 418A (1962).
8. Netherlands Pat. Appl. 6,411,232 (March 29, 1965; to Hooker Chemical Co.); *C.A., 63*, 11434c (1965).
9. Belg. Pat. 698,821 (Nov. 22, 1967; J. O'Brochta to Koppers Co.).
10. U.S. Pat. 4,302,403 (Nov. 24, 1981; K. G. Reabe, H. Dressler, and F. M. Covelli to Koppers Co., Inc.).
11. Jpn. Kokai Tokkyo Koho JP 02,011,530 [90,011,530] (Jan. 16, 1990; A. Matsura, M. Tagaki, and T. Hotta to Kawasaki Steel Corp.); *C.A., 112*, 197839 (1990).
12. U.S. Pat. 4,633,024 (Dec. 30, 1986; R. Ueno, K. Sakota, Y. Naito, and M. Kishimoto to Kabushiki Kaisha Ueno Seiyaku Oyo Kenkyujo); earlier, a similar idea was presented in U.S. Pat. 3,499,939 (March 10, 1970; J. T. Stephan and A. L. Lomax to Stepan Chemical Co.).
13. Japan. Pat. JP 71,041,529 (Dec. 8, 1971; Y. Yamada *et al.* to Ube Industries, Ltd.); *C.A, 76*, 45925 (1972).
14. Indian IN 150,060 (July 10, 1982; to Gharda Chemicals Pvt., Ltd.); *C.A., 98,* 178929 (1983).
15. Jpn. Kokai Tokkyo Koho JP 78,020,493 (June 27, 1978; to Mitsui Toatsu Chemical Co., Inc.); *C.A., 89*, 2818 (12978).
16. Jpn. Kokai Tokkyo Koho JP 80,034,776 (Sept. 9, 1980; to Taoka Chemical Co., Ltd.); *C.A., 94,* 83766 (1981).
17. H. W. Scheeline, *SRI Process Economics Program Report No. 79, "Dihydroxybenzenes"*, by C. Y. Yen, pp. 31–61 (1972).
18. A. R. Graham, *7th World Petroleum Congress Proc., 1967(5)*, 29–40.
19. U.S. Pat. 4,205,189 (May 27, 1980; L. B. Young and G. T. Burress to Mobil Oil Corp.).
20. U.S. Pat. 4,230, 894 (Oct. 28, 1980; L. B. Young to Mobil Oil Corp.).
21. U.S. Pats. 4,547,604 and 4,547,606 (both Oct. 15, 1985; G. A. Olah to PCKU Produits Chimiques Ugine Kuhlmann); these patents are identical in text and differ only in the claims.
22. Jpn. Kokai Tokkyo Koho JP 79,122,229 (Sept. 21, 1979; K. Taniguchi and K. Iwata to Mitsui Petrochemical Industries, Inc.); *C.A., 92*, 94030 (1981).

23. U.S. Pat. 3,927,134 (Dec. 16, 1975; Y. Tadahisa and F. Kentaro to Toyo Soda Manufacturing Co., Ltd.).
24. Jpn. Kokai Tokkyo Koho JP 77,139,030 (Nov. 19, 1977; E. Kudo, T. Saito, and H. Kouchi to Hitachi Chemical Co., Ltd.); *C.A., 88*, 120757 (1977).
25. Jpn. Kokai Tokkyo Koho JP 77,142,019 (Nov. 26, 1977; E. Kudo, T. Saito, and H. Kawachi to Hitachi Chemical Co., Ltd.); *C.A., 88*, 120759 (1977).
26. U.S. Pat. 3,927,114 (Dec. 16, 1975; H. Suda *et al.* to Sumitomo Chemical Co.).
27. U.S. Pat. 3,928,468 (Dec. 23, 1975; H. Suda *et al.* to Sumitomo Chemical Co.).
28. U.S. Pat. 3,933,921 (Jan. 20, 1976; H. Suda *et al.* to Sumitomo Chemical Co.).
29. U.S. Pat. 3,950,431 (Apr. 13, 1976; H. Suda *et al.* to Sumitomo Chemical Co.).
30. U.S. Pat. 3,953,521 (Apr. 27, 1976; H. Suda to Sumitomo Chemical Co.).
31. U.S. Pat. 3,923,908 (Dec. 2, 1975; H. Suda *et al.* to Sumitomo Chemical Co.).
32. U.S. Pat. 4,059,639 (Nov. 22, 1977; H. Hirozaku *et al.* to Sumitomo Chemical Co.).
33. U.S. Pat. 3,911,030 (Oct. 7, 1975; H. Suda *et al.* to Sumitomo Chemical Co.).
34. U.S. Pat. 3,969,420 (July 13, 1976; H. Suda *et al.* to Sumitomo Chemical Co.).
35. U.S. Pat. 4,192,958 (March 10, 1980; I. Hashimoto, H. Nakagawa, and H. Nambu to Mitsui Petrochemical Industries, Inc.).
36. U.S. Pat. 4,229,597 (Oct. 21, 1980; T. Taguchi, T. Ago, and I. Hashimoto to Mitsui Petrochemical Industries).
37. U.S. Pat. 4,237,319 (Dec. 2, 1980; H. Nambu, I. Hashimoto, and I. Imai to Mitsui Petrochemical Industries, Ltd.).
38. U.S. Pat. 4,239,921 (Dec. 16, 1980; I. Hashimoto *et al.* to Mitsui Petrochemical Industries, Ltd.).
39. U.S. Pat. 4,267,387 (May 12, 1981; I. Imai *et al.* to Mitsui Petrochemical Industries, Ltd.).
40. U.S. Pat. 4,273,623 (June 16, 1981; I. Hashimoto *et al.* to Mitsui Petrochemical Industries, Ltd.).
41. U.S. Pat. 4,283,567 (Aug. 11, 1981; H. Namba, H. Nakagawa, and M. Isibasi to Mitsui Petrochemical Industries, Ltd.).
42. U.S. Pat. 4,283,570 (Aug. 11, 1981; H. Nakagawa, N. Hirowatari, and T. Nakamura to Mitsui Petrochemical Industries, Ltd.).
43. U.S. Pat. 4,355,190 (Oct. 19, 1982; H. Nambu, F. Matsunaga, and H. Nakagawa to Mitsui Petrochemical Industries, Ltd.).
44. U.S. Pat. 4,827,049 (May 2, 1989; H. A. Zinnen to UOP Inc.); U.S. Pat. 4,424,381 (Jan. 3, 1984; G. Leston to Koppers Co., Inc.).
45. Jpn. Kokai Tokkyo Koho JP 02,273,637 [90,273,637] (Nov. 8, 1990; T. Ogino to Mitsui Petrochemical Industries, Ltd.); *C.A., 114*, 163723 (1991). Also Jpn. Kokai Tokkyo Koho JP 03,044,344 [91,044,344] (1991; T. Ogino to Mitsui Petrochemical Industries, Ltd.); *C.A., 115*, 71103 (1991).
46. Jpn. Kokai Tokkyo Koho JP 60,089,440 [85,089,440] (May 20, 1985; to Mitsui Petrochemical Industries, Ltd.); *C.A., 103*, 87644 (1985).
47. U.S. Pat. 4,108,907 (Aug. 22, 1978; H. Dressler to Koppers Co.).
48. U.S. Pat. 4,935,551 (June 19, 1990; C.-Y. Wu to INDSPEC Chemical Corp.).
49. U.S. Pat. 4,847,436 (July 11, 1989; C.-Y. Wu to INDSPEC Chemical Corp.).
50. U.S. Pat. 4,847,437 (July 11, 1989; C.-Y. Wu to INDSPEC Chemical Corp.).
51. U.S. Pat. 4,849,549 (July 18, 1989; C.-Y. Wu to INDSPEC Chemical Corp.).
52. World Pat. PCT 88/09324 (Dec. 1, 1988; D. E. Van Sickle and B. W. Overturf to Eastman Kodak Co.); *C.A., 110*, 193563 (1989).
53. U.S. OTS Report PB764, Dr. Fitzky, *The Acid Hydrolysis of Aniline and m-Phenylenediamine*, I. G. Farbenindustrie A.G. (1942).
54. Brit. Pat. GB 1,146,408 (March 26, 1969; N. P. Greco to Koppers Co., Inc.); *C.A., 71*, 49573 (1969).
55. U.S. Pat. 3,462,497 (Aug. 19, 1969; N. P. Greco to Koppers Co., Inc.).
56. U.S. Pat. 3,862,245 (Jan. 21, 1975; N. P. Greco to Koppers Co., Inc.).
57. U.S. Pat. 3,862,246 (Jan. 21, 1975; N. P. Greco to Koppers Co., Inc.).
58. U.S. Pat. 3,902,859 (Sept. 2, 1975; N. P. Greco to Koppers Co., Inc.).
59. U.S. Pat. 3,933,925 (Jan. 20, 1976; N. P. Greco to Koppers Co., Inc.).

60. U.S. Pat. 3,984,466 (Oct. 5, 1976; N. P. Greco to Koppers Co., Inc.).
61. Jpn. Kokai Tokkyo Koho JP 52,106,826 (July 9, 1977); Jpn. Kokai Tokkyo Koho JP 52,106,827 (July 9, 1977); Jpn. Kokai Tokkyo Koho JP 78,002,431 (June 29, 1978); Jpn. Kokai Tokkyo Koho JP 78,007,630 (Jan. 24, 1978); Jpn. Kokai Tokkyo Koho JP 78, 119,838 (Oct. 18, 1978); Japan. Pat. 85,012,333 (1985); Japan. Pat. 85,012,329 (1985); Japan. Pat. 85,023,654 (Aug. 6, 1985); Japan. Pat. 85, 039,055 (Nov. 13, 1985); Japan. Pat. 85,041,057 (Sept. 13, 1985); Japan. Pat. 85,032,614 (July 29, 1985); all to Mitsui Toatsu Chemicals, Inc.
62. Jpn. Kokai Tokkyo Koho JP 73,103,513 (Dec. 25, 1973; H. Obara, J. Onodera, S. Hattori, A. Matsukama, and I. Takagishi to Mitsubishi Chemical Industries, Ltd.); *C.A., 80*, 132824 (1974).
63. U.S. Pat. 4,118,586 (Oct. 3, 1978; S. L. Goldstein and J. S. Babiec to Olin Corp.).
64. U.S. Pat. 3,780,082 (Dec. 18, 1973; J. J. M. Deumens and S. H. Groen to Stamicarbon B.V.).
65. U.S. Pat. 3,932,511 (Jan. 13, 1976; S. E. Schaafsma, J. E. L. Claasens, and E. J. M. Verheijen to Stamicarbon B.V.).
66. U.S. Pat. 3,950,438 (April 13, 1976; S. E. Schaafsma, J. E. L. Claasens, and E. J. M. Verheijen to Stamicarbon, B.V.).
67. U.S. Pat. 4,117,000 (Sept. 26, 1978; T. Balg, C. M. A. Cramers, and H. J. G. Maessen to Stamicarbon B.V.).
68. U.S. Pat. 4,154,965 (May 15, 1979; P. J. N. Meijer, A. M. F. Vanholsaet, and E. T. M. Wolters to Stamicarbon B.V.).
69. U.S. Pat. 3,922,307 (Nov. 25, 1975; W. H. Mueller to Hoechst A.G.).
70. U.S. Pat. 4,018,833 (Apr. 19, 1977; W. H. Mueller and H. Fernholz to Hoechst A.G.).
71. U.S. Pat. 4,028,417 (June 7, 1977; W. H. Mueller and T. Weil to Hoechst A.G.).
72. U.S. Pat. 4,041,049 (Aug. 9, 1977; W. H. Mueller, H. Hey, and H. Meidert to Hoechst A.G.).
73. U.S. Pat. 4,072, 660 (Feb. 7, 1978; W. H. Mueller, H. Hey, and T. Weil to Hoechst A.G.).
74. U.S. Pat. 4,160,113 (July 3, 1979; W. H. Mueller, K. Riedel, and H. Krekeler to Hoechst A.G.).
75. U.S. Pat. 4,250,336 (Feb. 10, 1981; W. H. Mueller, K. E. Mack, and H. Hey to Hoechst A.G.).
76. U.S. Pat. 4,377,709 (March 11, 1983; W. H. Mueller to Hoechst A.G.).
77. U.S. Pat. 4,487,971 (Dec. 11, 1984; H. Litterer and H. Meidert to Hoechst A.G.).
78. U.S. Pat. 4,399,310 (Aug. 16, 1983; N. P. Greco to Koppers Co., Inc.).
79. U.S. Pat. 4,431,848 (Feb. 14, 1984; N. P. Greco to Koppers Co., Inc.).
80. Ger. Offen. DE 3,302,547 (Nov. 3, 1983; N. P. Greco to Koppers Co., Inc.); *C.A., 100*, 5882 (1984).
81. J. Gumb, *Chem. Brit., 4 (7)*, 291 (1968).
82. A. W. Christiansen, *Int. J. Adhesion Adhesives, 4 (3)*, 109 (1984).
83. B. H. River, *Forest Prod. J., 36 (4)*, 25 (1986).
84. USSR Pat. SU 1,698,247 (Dec. 15, 1991; N. V. Koroleva and I. V. Solodovnikova/Moscow Chemical-Technological Institute and G. M. Krzhizhanovskii/Power Engineering Institute); *C.A., 116*, 237797 (1992).
85. N. V. Koroleva and I. V. Solodovnikova, *Koks Khim., 1992 (3)*, 33; *C.A., 118*, 149781 (1993).
86. L. Tiikma and L. Melder, *Khim. Tverd. Topl. (Moscow), 1992 (4)*, 83; *C.A., 117*, 174755 (1992).
87. D. G. Roux, D. Ferreira, J. J. Botha, and D. C. F. Garbutt, *Appl. Polymer Symp. No. 28*, 1365 (1976).
88. U.S. Pat. 2,835,708 (May 20, 1958; J. Kamlet to Goodyear Tire and Rubber Co.).
89. USSR Pat. SU 144,489 (Feb. 15, 1962; to N. N. Vorozhtsov, Jr., T. D. Rubina, and G. G. Yakobson); *C.A., 57*, 12384 (1963).
90. U.S. Pat. 3.247,234, (Apr. 19, 1966) and U.S. Pat. 3,301,908 (Jan. 31, 1967), both by E. A. Ebach, E. P. Merica, and J. C. Valenta to Dow Chemical Co.
91. Brit. Pat. GB 834,254 (May 4, 1960; to Farbenfabriken Bayer A.G.); *C.A., 54*, 20986 (1960). Brit. Pat. GB 1,075,138 (July 12, 1967; to Dow Chemical Co.). Czech. Pat. 182,598 (Apr. 15, 1980; to J. Cerny and Z. Kvapil).
92. U.S. Pat. 3,413,341 (Nov. 26, 1968; K. F. Bursack, H. J. Moltzan, and E. L. Johnston to Frontier Chemical Co.).
93. U.S. Pat. 3,903,177 (Sept. 2, 1975; S. Suzuki to Chevron Research Co.).

94. Brit. Pat. GB 1,140,447 (Apr. 23, 1969; H. Pflugk to Union Rheinische Braunkohlen Kraftstoff A.G.); *C.A., 71,* 70296r (1969).
95. Ger. Offen. DE 2,033,656 (Jan. 21, 1971; M. B. Sherwin and G. M. Sugerman to Chem Systems, Inc.); *C.A., 74,* 87614 (1971).
96. U.S. Pat. 3,627,833 (Dec. 14, 1971; to M. A. Tobias).
97. Ger. Offen. DE 2,205,225 (Aug. 9, 1973; K. Riedel and H. Krekeler to Farbwerke Hoechst A.G.); *C.A., 79,* 115215 (1976).
98. Jpn. Kokai Tokkyo Koho JP 52, 065,235 (Jan. 11, 1977; S. Unemura *et al.* to Ube Industries K.K.); *C.A., 88,* 120775 (1978).
99. Jpn. Kokai Tokkyo Koho JP 76,091,215 (Aug. 10, 1976; K. Tadamitsu to Mitsui Toatsu Co., Ltd.); *C.A., 85,* 159664 (1976).
100. Ger. Offen. DE 3,708,851 (Sept. 24, 1987; J. E. Fellsmann, R. J. Saxton, and P. Tung to Catalytica Associates); *C.A., 108,* 55649 (1987).
101. Ger. Offen. DE 2,359,410 (June 12, 1974; J. Gosteli to Ciba-Geigy A.G.); equivalent is U.S. Pat. 4,020,098 (Apr. 26, 1977); *C.A., 84,* 91205 (1981).
102. A. Focella, S. Teitel, and A. Brossi, *J. Org. Chem., 42,* 3456 (1977); also Ger. Offen. DE 2,002,815 (July 30, 1970; A. Brossi, A. Focella, and S. Teitel to Hoffmann-LaRoche and Co.); *C.A., 73,* 66258 (1970).
103. U.S. Pat. 3,865,884 (Feb. 11, 1975; R. G. Agusto, F. A. Taylor, and H. L. Briggs to Canadian Thermo Images, Ltd.).
104. L. Cleaver, J. A. Croft, E. Ritchie, and W. C. Taylor, *Aust. J. Chem., 29,* 1989 (1976).
105. A. A. Jaxa-Chamiec, P. G. Sammes, and P. D. Kennewell, *J. Chem. Soc., Chem. Commun., 1978(3),* 118; also same authors, *J. Chem. Soc., Perkin Trans. 1, 1980(1),* 170.
106. T. A. Hase, E. Suokas, and K. McCoy, *Acta Chem. Scand., Ser. B, 32,* 701 (1978).
107. T. Kato, M. Sato, and H. Kimura, *J. Chem. Soc., Perkin Trans. 1, 1979(2),* 529.
108. U.S. Pat. 4,339,593 (July 13, 1982; B. J. Willis, D. I. Lerner, and D. H. R. Barton to Fritzsche, Dodge and Olcott Co.).
109. Eur. Pat. Appl. EP 74,497 (March 23, 1983; H.-G. Schmidt to Dynamit Nobel A.G.); *C.A., 99,* 22134 (1983).
110. U.S. Pat. 4,115,431 (Sept. 19, 1978; J. B. Hall, M. A. Sprecker, E. J. Shuster, F. L. Schmitt, and J. F. Vinals to International Flavors and Fragrances).
111. U.S. Pat. 4,009,212 (Feb. 22, 1977; G. Leston to Koppers Co., Inc.).
112. Jpn. Kokai Tokkyo Koho JP 47,038,937 (Dec. 6, 1972; K. Yamada, M. Nakagishi, S. Nagai, H. Yamane, A. Takaoka, K. Hareda, H. Yoshida, and T. Konishi to Ube Industries K.K.); *C.A., 78,* 58030 (1973).
113. L. Eberson and L. Joensson, *J. Chem. Soc., Chem. Commun., 1974,* 885.
114. Ger. Offen. DE 3,224,156 (Dec. 29, 1983; R. Schalenbach *et al.* to Bayer A.G.); *C.A., 100,* 174426 (1984).
115. U.S. Pat. 4,200,763 (Apr. 29, 1980; A. D. Sheng-Hong to The Upjohn Co.).
116. U.S. Pat. 4,405,818 (Sept. 20, 1983; G. E. Stead and L. B. Young to Mobil Oil Corp.).
117. U.S. Pat. 4,424,382 (Jan. 3, 1984; G. A. Olah to PCUK Produits Chimiques Ugine Kuhlmann); also Brit. Pat. GB 2,137,195 (June 5, 1985) and Brit. Pat. 2,136,419 (May 22, 1985), both to ATO Chimie/Ugine Kuhlmann.
118. U.S. Pat. 4,423,254 (Dec. 27, 1983; G. A. Olah to PCUK Produits Chimique Ugine Kuhlmann).
119. U.S. Pat. 4,339,614 (July 13, 1982; G. A. Olah to PCUK Produits Chimique Ugine Kuhlmann).
120. V. Ya. Kugel, G. A. Klinger, I. V. Vygodskaya, F. I. Novak, and A. N. Bashkirov, *Neftekhimiya, 21(3),* 454 (1981); *C.A., 95,* 114963 (1981).
121. J.-P. Gessen, J. C. Jacquesy, and M. P. Jouannetaud, *J. Chem. Soc., Chem. Commun., 1980(23),* 1128.
122. Eur. Pat. Appl. EP 41,441 (Dec. 9, 1981; M. P. Jouannetaud, J.-P. Gessen, and J.-C. Jacquesy to Produits Chimiques Ugine Kuhlmann); *C.A., 97,* 72065 (1982).
123. Eur. Pat. Appl. EP 97,564 (Jan. 4, 1984; G. Morellet, J.-C. Jacquesy, and M.-P. Jouannetaud to PCUK Produits Chimiques Ugine Kuhlmann); *C.A., 101,* 6815 (1984).

4

The Use of Resorcinol in Rubber Compositions

4.1. INTRODUCTION AND OVERVIEW

The use of resorcinol and resorcinol resins to improve the adhesion of organic and inorganic reinforcing materials to natural/synthetic rubber compositions is the major commercial application of resorcinol. For this reason, this use is a separate topic here. Other uses of resorcinol resins are discussed in Chapter 5, and the many uses of resorcinol/derivatives as additives for other polymeric compositions and organic materials are presented in Chapter 11.

The largest use for rubber is in the manufacture of tires for automobiles and trucks (world production of these tires was *ca* 700 million units in 1986[1]), off-the-road vehicles, aircraft, racing cars, motorcycles, and bicycles. Correspondingly, the major application of resorcinol at present is as a bonding/stiffening agent in rubber compounds used for building tires, as practiced by many, if not most, tire manufacturers. These tires are complex composite materials made up of rubber, a variety of chemicals, and organic and/or inorganic reinforcements in the form of yarn, fabric, or cord. Reference 1 provides an excellent review of how tires are constructed for performance. To give some idea of the complexity, a picture of a cross-section of a steel-belted radial tire is given in Fig. 4.1. Again, the main reason for the use of resorcinol/resorcinol resins in tires is to improve the adhesion of the rubber to the reinforcing material under actual service conditions, e.g., moisture and speed which cause heat buildup. The performance of the composite is tested first in the laboratory, and then in extensive road tests of the tires—these are lengthy and costly procedures.

Resorcinol/resins also can contribute to interlayer adhesion and to improving the mechanical properties of the rubber compound. Thus, resorcinol can be used in areas other than the rubber/reinforcement interface.

Tire construction and performance have come a long way from the early solid tire to the inflatable tire. For instance, before 1940 the reinforcement mainly used was cotton (although Michelin had introduced steel-cord reinforced tires in 1936), followed by rayon in the 1940s, nylon 6.6 and nylon 6 in the 1950s, polyester and glass

Figure 4. Goodyear high-performance tire cross-section. Copyright of Goodyear Tire and Rubber Co., published here with permission.

cords in the 1960s, and aramid and steel cord in the 1970s. Today, the steel-belted radial tire construction predominates in the U.S.A. and in Europe in automobile and truck tires because of their overall excellent performance, ranging from long life, safety, good handling and ride to good price/value; the public has recognized these features.

Resorcinol was first used in the form of aqueous resorcinol/formaldehyde/latex (RFL) dip systems applied to rayon cord. Excess RFL is removed from the dipped cord and the coated fiber is dried and baked in an oven to develop the best performance. All subsequently used organic and glass fibers/fabrics required an RFL dip to improve their adhesion to rubber compounds. The optimum rubber latex for this use was found to be a styrene/butadiene/vinyl pyridine latex. For best performance, polyester fibers were found to require a pretreatment with an epoxy resin or a blocked di- or polyisocyanate before the RFL dip application. ICI developed a one-dip system, called H-7, which was based on a resorcinol/*p*-chlorophenol/formaldehyde resin and rubber latex formulation. Aramid fibers were also found to require a pretreatment with an epoxy resin before application of the RFL dip system.

When the RFL dips were first developed, resorcinol was condensed with formaldehyde at pH 9.0–9.5 and the resin solution was added to the diluted latex; the dip was aged before use. However, some users had problems with the reproducibility of the dip; the interaction of unreacted formaldehyde with the ammonia present in the dip

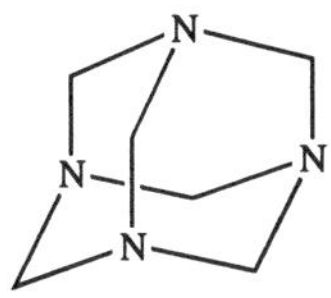

4-1

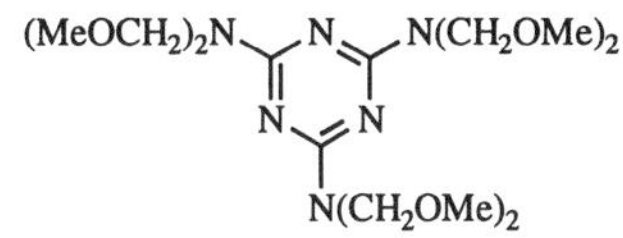

4-2

formulation also caused problems. Therefore, preformed resorcinol/formaldehyde resin solutions were developed for use in the RFL dip systems (e.g., INDSPEC Chemicals' resins R-2170, R-2200, and R-50). These resins still required the addition of some formaldehyde to the dip formulations, but the results were reproducible and side reactions were minimized.

Generally, RFL dip pickup is in the 4–6 wt% (dry basis) range, but fibers used in V-belts and glass-fiber tire cords require much higher (up to 20 wt%) dip pickup. Because exposure of the RFL-treated fabrics to high temperature/humidity and UV light degrades the properties of the adhesive film and the fabrics, the treated fabrics are stored in a black polyethylene film wrap at < 38 °C and 50% RH (relative humidity) before use.

Further improvements in the rubber-to-fabric adhesion were found to be attainable by adding resorcinol or resorcinol resins (named the bonding agent or "methylene acceptor" in the trade) to the rubber compound together with hexamethylene tetramine (HMTA), **4-1**, or hexamethoxymethyl melamine (HMMM), **4-2**, as the "methylene donor" which, together, formed a resorcinol resin network in the rubber compound during processing at up to *ca* 150 °C. The addition of a high-surface-area silica powder was found to give additional improvements—this was the so-called HRH system. Resorcinol itself was found to be fuming, i.e., volatile to a degree, during the processing of these compounds, requiring good ventilation and, with increasing environmental awareness, good scrubbing systems on the exhaust ducts in the plants. The resorcinol resins were less fuming and began to be tailored to minimize fuming.

When steel cord emerged as a reinforcement in the steel-belted radial tires, it was found that brass plating (or zinc plating, less used) was required for corrosion protection of the cord; the brass plating also interacted with the sulfur in the vulcanization systems used in the rubber compounds. RFL dips were found to be useless for improving the adhesion of the steel cord to the rubber. Usually, small amounts of cobalt (rarely nickel) salts of organic acids were added to improve adhesion. The combination of both cobalt salt and resorcinol (resin)/methylene donor—usually hexamethoxymethylmelamine, HMMM—gave further improvement with the steel-cord-to-rubber adhesion. In this case the resorcinol/resin is added at the 2–4 phr level. Hexamethylenetetramine, HMTA, was found to be undesirable as the methylene donor in these systems, because it was capable of generating amines/ammonia which could corrode the plated steel cord.

Uniroyal Chemical Co. developed a resorcinol/acetaldehyde resin (Resorcinol Donor R-6) in combination with TCBQ (2,3,5,6-tetrachlorobenzoquinone) as an adhesion promoter system for brass-plated steel cord, which does not require cobalt salt addition to the rubber compound.

Similar adhesion-promoting systems are also used in many nontire products with fabric or wire reinforcements, such as drive belts, material transport belts, specialty hose, and others; compared to tires these are smaller volume markets.

A massive effort has gone into the improvement of tire construction, composition, and performance by the tire companies and by their suppliers worldwide, and this competitive effort continues. The results are most visible in the steadily better tires which become available. The results of these industrial efforts are also reflected in the massive, pertinent patent literature and in articles in journals. It is apparent that each facet of the total rubber compound, i.e., not only the type of reinforcing materials but also the type of rubber, sulfur and carbon black levels, type of accelerator, antioxidants, and other additives, can influence adhesion and the mechanical properties. The rubber compounder's role is crucial. The compounds actually used by each tire company are valuable trade secrets.

As far as the use of resorcinol and resorcinol resins in rubber compounds for tire construction are concerned, a few of the highlighting references discuss the early use of RFL dips for textile reinforcements,[2] the introduction of vinylpyridine-containing terpolymers in the latex portion,[3] reviews of patents on the pretreatment of polyester cord with a blocked isocyanate, with or without an epoxy resin application, before the RFL dip,[4] the introduction of the direct-bonding agents in rubber compounds with added silica (called the HRH system in the U.S.),[5] and a general review of RFL dip technology.[5a]

There have been good reviews/papers published on the influence of curing agents on rubber-to-textile and rubber-to-steel cord adhesion;[6] on the adhesion of rubber to (brass-plated) steel cords in tires;[7] on systems for the adhesion of aramid, polyester, and glass cords/fabrics to rubber;[8] on the polymer interfaces in tire technology;[9] on the combination of cobalt and resorcinol (resin) bonding agents with methylene donors for adhering rubber compounds to brass-plated steel cord;[10] and on the use of phenolic resins in the rubber industry, including tackifier resins and reinforcing resins as well as resorcinol resins.[11]

A good review of various bonding agents as adhesion promoters both in rubber-to-metal and in rubber-to-textile applications, including mechanisms of action of the bonding agents, was given in Oct. 1993 by R. F. Seibert.[11]

These reviews will help those readers who wish to, or need to, dig deeper into these technologies.

4.2. EXAMPLES OF ACTUAL APPLICATION RESULTS

As already indicated above, the development of a competitive or superior reinforced rubber compound for tires is a complex business. Some of the factors to be considered are raw materials of a considerable variety (rubber and fiber/cord), a large variety and amount/proportion of additives of all kinds, the material prices/cost and availability. The high-priced cobalt, for instance, has gone through large price/availability swings for many years, doubtless requiring the help of commodity futures/hedging techniques by the suppliers and/or the tire companies to dampen the turmoil caused by these price gyrations. Other factors to be considered are the processing conditions for the rubber compound and the large variety of performance tests. The formulations must be tailored to the application and the location in the complex composite. A specific balance of properties is needed for each application/area.

The evaluation results are usually available from the patent literature only. Unfortunately, each patent gives only limited data, compositions, type and range of tests, test conditions, and/or measurement units. Therefore, a judgment of the merits of one patented composition vs another is usually based on experience and intuition. Gaining certainty in appraising an invention may require in-house tests.

In order to give a feel for the value of resorcinol in reinforced rubber compounds, one test compound has been chosen and a set of basic test data for a steel-cord-reinforced compound is given. Table 4.1 gives the composition of the rubber masterbatch used. Table 4.2 shows the performance of this compound with 0.1 phr (part per hundred resin) cobalt alone, cobalt with resorcinol at two levels, and cobalt with three resorcinol resins as the bonding agent, with hexamethoxymethylmelamine, HMMM, as the methylene donor, for improving the steel wire-to-rubber adhesion, with sufficient mechanical performance data given to provide some insight.

All the data in Tables 4.1 and 4.2 were obtained from, and published by permission of, INDSPEC Chemical Corp. An explanation of the terms and abbreviations used is given in the footnotes to Table 4.2.

The masterbatch of the composition given in Table 4.1 is a fairly standard composition for testing and is easily related to by the experts in the industry.

The selected evaluation data given in Table 4.2 show that with cobalt alone (the control) a reasonably good balance of properties was obtained, except for the dramatic drop-off in wire-to-rubber adhesion after 16 hr/120 °C steam aging of the test

Table 4.1. Rubber Compound Formulation

Masterbatch	Phr
Natural rubber SMRCV60	100
N-326 carbon black	55
Zinc oxide	8.0
Stearic acid	1.2
Santoflex 13 (antioxidant/antiozonant)	2.0
Flectol H (antioxidant)	1.0
Santoguard PVI (prevulcanization inhibitor/retarder)	0.2
Total	167.4
Cobalt metal	0.10
Insoluble sulfur	3.75
Santocure NS (accelerator)	0.6

specimens. The use of cobalt with resorcinol or a resorcinol resin (Penacolite[R] resin of INDSPEC Chemical Corp.) showed a dramatic improvement of the wire-to-rubber adhesion after 16 hr/120 °C steam aging. Also, M_H is improved while M_L, an indicator of the processability of the compound, is little changed. The scorch time, T_{S2}, is little changed, while the T'_{25} and T'_{90}, which relate to cure and torque development, are reasonably comparable. The G' (dynamic elastic modulus) is improved—better stiffness indicates a reinforcing effect; the tan delta is lower (for most applications this is desirable) while the tensile strength properties are good and the hardness of the compound is improved.

It should also be noted that resorcinol in this compound, as indicated in Table 4.2, was quite fuming during processing (in this laboratory test on a 2-roll mill, up to 5

Table 4.2. Comparison of Resorcinol and Penacolite Resin as Bonding Agents to Improve the Adhesion of Steel Cord to Rubber

Bonding system	Control	Resorcinol	Resorcinol	Penacolite resin
Cobalt metal,[a] phr	0.1	0.1	0.1	0.1
Bonding agent, HMMM[b]	—	50/50	50/50	60/40
Loading, phr	—	3	4	4
Fuming (on 2-roll mill)	—	very high (3+)	very high (3+)	low (0–1)
Rheometer Cure/150 °C[c]				
M_H, dN-m	50.7	60.6	63.5	60.0
M_L, dN-m	6.3	5.8	6.0	6.8
T_{S2}, min	4.3	4.0	3.7	4.4
T'_{25}, min	6.4	5.7	5.4	6.3
T'_{90}, min	13.0	16.0	16.4	15.4
Wire Adhesion, N-%RC[d]				
Initial	1245-90	1268-90	1250-85	1312-90
After 16 hr steam aging (120 °C)	329-10	1063-80	1180-85	1174-90
Dynamic Mechanical Properties[e]				
G', MPa	14.7	26.8	32.2	26.7
Tan delta	0.232	0.213	0.211	0.207
Tensile Properties[f]				
50% Modulus, MPa	2.2	3.6	4.4	3.8
Tensile strength, MPa	27.7	24.6	24.0	26.2
Elongation, %	468	403	377	429
Shore A Hardness	75	81	83	82

[a]Added as a 12.9% cobalt naphthenate.

[b]Hexamethoxymethylmelamine.

[c]Using the Monsanto Oscillating Disc Rheometer 100; M_H = high torque; M_L = minimum torque before cure (proportional to the viscosity of the compound); T_{S2} = scorch time (time to incipient cure); T'_{25} = time to 25% cure development; T'_{90} = time to 90% cure development.

[d]Test method ASTM D2229; 3/4" wire pullout; 3×0.2+6×0.35 mm wire construction; wire coating 63.5% Cu in brass; N = force for 3/4" wire pullout; %RC= %rubber coverage on pulled wire.

[e]Rheometrics instrument RM-800; test specimen 18/12.5/4 mm; G' @ 0.2% strain, tan delta @ 2.0% strain, 1 Hz, room temperature, where G' = dynamic elastic modulus (proportional to the stiffness) and tan delta = G''/G' (related to hysteresis; usually, small values are desirable).

[f]Instron instrument data.

wt% of the resorcinol charged may be lost during processing at up to 150 °C). The use of the resorcinol resins showed that they minimized this fuming problem while maintaining a good balance of properties.

Basic data collections of this type then provide guidance to the expert compounder about the value of a particular bonding agent and, if satisfied, these data can be used to begin development/optimization of their own formulation/application.

4.3. SELECTIONS FROM THE PATENT LITERATURE, 1983–1993

The references[2–11] given in the preceding overview reviewed the field, in relation to the use of resorcinol, up to the early 1980s. This section is intended to give an insight for industrial developments worldwide since then. The given outlines of selected patents focus on the modification of materials and recipes, all using resorcinol or resorcinol resins. Actual test data are not quoted because, even if available, the test measurement units, reference standards, and rubber compounds varied widely so that absolute comparisons cannot be made. Even the given, limited selection of references shows that, in spite of the turmoil in the U.S. in this industry in the 1980s (massive takeovers, restructuring, etc.), development work continued and much attention was focussed on the use of resorcinol/derivatives in optimizing the performance package.

Vulcanizable rubber composites were developed with improved adhesion to reinforcement materials of all types. These compositions contained a phenol/resorcinol or *m*-aminophenol/aldehyde coresin and a phthalimide compound, or a phenol/resorcinol or *m*-aminophenol/trimethyl-trihydroxyflavan (a condensate of resorcinol with acetone)/aldehyde resin, or similar phenol/resorcinol/aldehyde resins and an arylsulfonamide as the bonding agent together with a methylene donor.[12]

Another patent proposed a homogeneous mixture of 95–30% carbon black and 5–70% comminuted resorcinol as the "methylene acceptor" in compositions for bonding rubber compounds to the reinforcement. The carbon black/resorcinol blend gave better flowability, nondusting properties, and reduced sublimation of resorcinol at compounding temperatures.[13]

Cresol or xylenol/resorcinol/formaldehyde resins were said to give improved adhesive dips together with rubber latexes for polyester fiber used in rubber reinforcement.[14]

Reinforcing fibers, such as nylon, polyester, or polyvinyl alcohol fibers, were treated with a RFL dip and an aging inhibitor for rubber (e.g., 2-mercaptobenzimidazole) before bonding the fiber and the rubber; the degradation of bonding on aging was minimized.[15]

A sulfur-vulcanizable rubber composition was developed which was nontoxic, gave good mechanical properties and very good adhesion to brass-plated steel cord, and which had good retention of adhesion after humidity aging. It contained an active silica as part of the reinforcing filler (the other part being carbon black), a phenol/resorcinol (not more than 30% of the total phenols)/aldehyde resin, a cobalt salt of an organic acid, and a conventional "methylene donor."[16]

Carbon fibers were treated with a resorcinol/allyl chloride reaction product and a RFL dip to improve the adhesion of this fiber to rubber.[17]

A polyester-reinforced rubber tire, claimed to have better durability than a steel-cord reinforced tire, was constructed with a polyester fabric embedded in an adhesion-promoting composition containing neoprene, butadiene rubber, resorcinol resin, formaldehyde/silica, and other additives.[18]

Improved fiber (e.g., nylon) cord to sulfur-vucanizable rubber (e.g., EPDM) adhesion was obtained by dipping the fiber in a dispersion containing a resorcinol/formaldehyde resin, chlorosulfonated polyethylene, and a natural rubber latex.[19]

A one-step dip for polyester or aramid cords and fabrics was invented which contained RFL, an acrylic resin, and a reversibly blocked polyisocyanate. The adhesion of the rubber to these cords was said to be as good as the adhesion of cords treated with the conventional 2-dip systems.[20]

Resorcinol esters, such as the mono-benzoate or the mono-abietate, in conjunction with a methylene donor in a vulcanizable rubber compound, were reported to give enhanced adhesion of rubber to organic fiber or metallic reinforcing materials. The resorcinol esters have lower volatility than resorcinol which is a plus in processing.[21]

Vulcanizable, compounded natural rubber containing 4 phr resorcinol as a (high silica) zeolite adduct showed good mechanical properties and improved adhesion to reinforcements vs a similar composition using resorcinol and ordinary silica.[22]

The adhesion of carbon fibers to rubber compounds was improved by treating the fibers first with a plasma at 150 kHz and 40 kW, then coating them with a RFL dip.[23]

Reinforcing fibers (e.g., polyamide) for rubber were treated with a RFL dip containing a pentaerythritol ester such as, e.g., pentaerythrityl-tetrakis-3-(3,5-di-*tert*-butyl-4-hydroxyphenyl)propionate; the adhesion of the fiber to the rubber compound was improved, long-lasting, and had superior heat resistance.[24]

Carbon fibers useful as reinforcements for vulcanizable rubber compounds were precoated with an epoxy resin, then coated with a RFL dip to improve the adhesion of the fiber to the rubber.[25]

Heavy-duty radial tires were built using, as the reinforcement aramid, cords coated with a natural rubber/resorcinol-hexamethymenetetramine/silica composition; the tires had improved durability.[26]

Improved adhesion of aramid fiber to vulcanizable rubber was obtained by coating the fiber with a RFL dip which contained neoprene rubber and 2,6-bis-(2′,4′-dihydroxyphenylmethyl)-4-chlorophenol in the dip formulation.[27]

A vulcanizable rubber composition with improved adhesion to steel cord was developed which contained Co-resinate, a resorcinol/substituted phenol/formaldehyde resin and/or beta-naphthol, hexamethoxymethylmelamine, and silica.[28]

The adhesion of aramid fibers to a vulcanizable rubber composition was substantially improved by treating the fibers first with sodium hydride in dimethylsulfoxide, then with an epoxy resin, and, finally, with a RFL dip.[29]

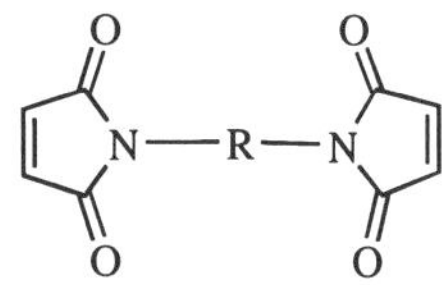

4-3

(where R = arylene or alkylene)

The adhesion of aramid (Teijin's HP-50) and polyester fibers to a vulcanizble rubber composition was greatly improved by treating the fibers with a cold plasma, then ion-plating with nylon-6 vapor, and finally coating the fibers with a RFL dip.[30]

A tire with good durability against mechanical fatigue and thermal deterioration was made from a vulcanizable rubber compound containing conventional ingredients, reinforcements, resorcinol/formaldehyde resin, hexamethoxymethylmelamine, and an organic peroxide.[31]

Nylon cord with better retention of strength after dipping was prepared by twisting the cord and heating for 10–20 sec/160 °C for nylon-6, or 190 °C for nylon-66, before treatment with a RFL dip.[32]

Radial, brass-plated steel-belted tires were constructed with a rubber sheet containing a resorcinol/formaldehyde resin between the tread and the steel cord to obtain good adhesion between cord and rubber even at elevated temperatures and exposure to high humidity.[33]

A vulcanizable rubber compound with improved adhesion to glass and bare or brass-plated steel cord was invented which contained filler, activators, plasticizer, vulcanizing agent, accelerator, resorcinol/hexamethylenetetramine, and calcium hydroxide (inorganic base).[34]

Organic fiber materials, used for reinforcing rubber, were first treated with a RFL dip, then coated with a bis-maleimide compound, **4-3**; the treated fibers had excellent adhesion to rubber with improved durability of the bond strength.[35]

In another patented process, glass or organic fibers were first treated with an aniline/resorcinol/formaldehyde coresin, then with a RFL dip to improve the adhesion of the fibers to the rubber; the bonds had good performance at high temperature and humidity, and mechanical failure was reduced.[36]

A novel, adhesion-improving dip for tire cords consisted of preparing a phenol/formaldehyde resole, then adding resorcinol or a resorcinol resin as a curing agent, and finally adding a vinylpyridine-copolymer latex; good adhesion between cord and rubber was attained.[37]

An adhesive dip for glass fibers was reported which improved the adhesion of fiber to rubber and maintained the mechanical properties of the composite in use, consisting of a conventional RFL dip and a fluorocarbon resin (e.g., PTFE) latex.[38]

An improved tire tread compound was developed which contained carbon black (with I_2 absorption 20–50 mg/g), added Sumikanol 620 (a resorcinol resin), and hexamethoxymethylmelamine. This compound showed 10% higher impact resilience and 16% higher cut resistance vs a similar compound without these additives.[39]

Brass-plated steel cords were coated with hexamethoxymethylmelamine, then embedded in a vulcanizable rubber compound containing resorcinol or resorcinol resin as the methylene acceptor. The rubber-to-cord adhesion was much improved over similar compounds where the cord was not precoated with hexamethoxymethylmelamine.[40]

A natural rubber composition with good adhesion to aramid/steel cord bilayered belts was patented. The compound contained 0.05–0.5 part Co-naphthenate, 0.5–3.0 parts resorcinol resin, and 1–5 parts hexamethoxymethylmelamine. The composite showed good durability at high speed and resistance to aging at high humidity.[41]

Alkyl-substituted novolak resins were prepared by first reacting 1.0 m resorcinol with 0.1–2.0 m of an unsaturated hydrocarbon (such as piperylene or dipentene) and the with 0.1–1.0 m formaldehyde, as methyl formcel (formaldehyde in methanol), at 60–100 °C using an acid catalyst. This resin was added to a vulcanizable rubber compound containing rubber, carbon black, activators, antidegradent, and cobalt naphthenate in a Banbury mixer at 150 °C, then compounded with sulfur accelerator and methylene donor at 100 °C. The cured compound showed excellent adhesion to brass-plated steel cord, with good rubber coverage of the cord, even after steam/humidity aging, as well as improved hardness and stiffness vs the same compound containing Co-naphthenate without a resorcinol resin or containing a nonalkylated resorcinol novolak.[42]

Aramid cords or fabrics with improved adhesion to vulcanizable rubber compositions were prepared by degreasing the fiber/fabric, treating it with a plasma, coating it with a dip containing resorcinol/formaldehyde resin, an acrylonitrile/butadiene/vinylpyridine copolymer latex, Vulcabond E, a blocked isocyanate, and an ethylene urea. The adhesion of this RFL-dipped aramid to a vulcanizable rubber compound was substantially improved over a conventionally treated aramid with an epoxy sub-coat and a RFL top-coat.[43]

Glass-fiber cords for rubber reinforcement were coated with RFL dips containing lignin; the treated cords had good resistance to thermal aging and flex fatigue.[44]

Polyester fiber cords were dipped in a 40% aqueous resorcinol solution, heated to 200 °C for 1 minute, then dipped in a RFL dip containing Pyratex, SBR latex, and a nitrile rubber latex. The adhesive strength of the fiber to rubber was greatly improved over the older types of RFL-dipped fiber.[45]

The adhesion of compounded natural rubber to the reinforcement and the mechanical properties of the composite were improved when the compound contained 30 phr zeolite 13X, which incorporated 20 wt% resorcinol and 20 wt% hexamethylenetetramine, vs a compound which contained the same amount of free resorcinol/hexamethylenetetramine without the zeolite.[46]

Brass-plated steel cords were coated with an acetone solution of 6-diallyamino-1,3,5-triazine-2,4-dithiol to improve their adhesion to a vulcanizable rubber composition which contained conventional additives, with a resorcinol/formaldehyde resin as the "methylene acceptor" and hexamethoxymethylmelamine as the "methylene donor".[47]

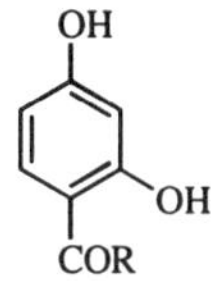

4-4
(where R = alkyl or aryl)

A resorcinol resin which is low-fuming (low in free resorcinol content) and nonhygroscopic, useful for improving the adhesion of rubber to tire cord, was prepared by alkylating resorcinol with styrene, then reaction with 37% aqueous formaldehyde using toluenesulfonic acid as the catalyst, neutralization and devolatilization under vacuum at 160 °C.[48]

A vulcanizable rubber composition containing a methylene donor and a 4-acylresorcinol, **4-4**, especially 4-benzoylresorcinol (**4-4**, R = phenyl), as the methylene acceptor showed low fuming (compared to resorcinol) during compounding, and gave cure and reinforcing properties similar to resorcinol itself in place of 4-benzoylresorcinol.[49]

Organic fibers, especially aramid cord, were treated with a sub-coat of RFL latex and a (second) top-coat of chlorosulfonated polyethylene rubber/polyisocyanate. The treated cords showed excellent adhesion to vulcanizable hydrogenated nitrile rubber compositions.[50]

An excellent heat-resistant adhesive composition for bonding a vulcanizable rubber compound to the reinforcing fiber, especially polyester cords, was developed. The fibers were treated with 5–50 parts of latexes of copolymerized butadiene 25–65/ethylenically unsaturated carboxylic acid(s) 0.5–15/acrylonitrile 2–30% and 50–95 parts of latexes of copolymerized butadiene 40–80/acrylonitrile 2–30/vinylpyridine 5–35/ethylenically unsaturated carboxylic acid(s) 0–15, and 7–30 parts of a resorcinol/formaldehyde resin.[51]

Textile cord was dipped in an emulsion of a carboxylated butylene rubber latex and/or a butadiene/vinylpyridine latex containing sucrose, a copper or zinc salt of an inorganic acid, and a resorcinol/formaldehyde resin. The coated cords showed increased adhesion to vulcanizable poly-*cis*-isoprene rubber compounds.[52]

Polyester fibers with good adhesion to vulcanizable rubber compounds were prepared by drawing the fibers in the presence of spinning oils, heat-setting, then sub-coating with a polyepoxide, curing at 30–80 °C, and finally treating the fibers with a RFL dip containing a blocked poly-isocyanate and/or aziridines, then curing by heating.[53]

Glass fibers were treated with an emulsion containing ammonia, nitrile rubber, chlorosulfonated polyethylene, and a resorcinol/formaldehyde resin, then dried at 280 °C for 22 sec, then treated with a dispersion containing polyisocyanate, hydrogenated nitrile rubber, and chlorinated rubber, then dried at 155 °C for 48 sec. The treated cords had good adhesion to vulcanizable rubber compounds and the mechanical properties of the cured composites were resistant to degradation by heat and water.[54]

Aliphatic polyamide tire cords were treated with an RFL dip and a halogenated phenol/resorcinol/formaldehyde precondensate/SBR or styrene/butadiene/vinylpyridine copolymer latex. The treated cords showed good adhesion to vulcanizable ethylene/propylene copolymer rubber compositions.[55]

Tire cords, especially nylon-6 cords, were coated with a RFL dip containing silica to give improved adhesion to vulcanizable rubber compounds, with good retention of tensile and bonding strength compared to a similar dip without silica.[56]

Polyester fibers with improved adhesion (over prior-art systems) to vulcanizable rubber compounds were prepared by coating the spun fiber with an emulsion containing a glycerol ether having two propylene chlorohydrin units, and with potassium carbonate, drawing the fibers, heat-setting at 240 °C, and making cords; the cords were treated with a RFL dip. The treated cords had substantially improved adhesion to vulcanizable rubber compounds and showed good steam-aging performance compared to cords treated by prior-art methods.[57]

The adhesion of aramid fiber cords to vulcanizable natural rubber compositions was much improved by treating the fibers with a RF/poly(vinyl chloride) latex dip.[58]

An improved treatment for organic fibers, especially poly(ethylene terephthalate) tire cord, consisted of a first dip containing a reaction product of isopropyl isocyanate with a polyoxyalkylene ether of an aralkylated phenol, ethyl methyl ketoxime-blocked MDI, sorbitol polyglycidyl ether, and a RFL composition. The treated fibers were heated at 90 °C (reduced foaming of this dip vs the prior-art dip was observed) and then coated with a second dip containing the blocked polyisocyanate and the RFL; the double-dipped cord gave enhanced adhesion to vulcanizable rubber compounds.[59]

The adhesion of aramid films to vulcanizable rubber compounds was improved by exposing the aramid to a corona discharge before treatment with a RF/chloroprene latex dip.[60] Fatigue-resistant poly(vinyl alcohol) fibers for tire construction, with good adhesion to vulcanizable rubber compounds, were made by first dipping the cords in an aqueous composition containing terephthalaldehyde as a crosslinking agent, drying the coating, and then immersing the cord in a RFL dip, followed by heat-treating the cord at 150–205 °C.[61]

A modified resorcinol novolak resin, less hygroscopic than conventional resins of this type, was prepared by condensing nonylphenol with methyl formcel (formaldehyde in methanol) under alkaline conditions, optionally also adding cashew nutshell liquid, and condensing this resole with resorcinol. This modified resorcinol resin was compounded with a black natural rubber composition; the vulcanized compound showed improved mechanical properties.[62]

In a U.S. patent it was reported that compounding a black natural rubber master batch with 3.5 phr aralkylated resorcinol resin (which was low-fuming in processing) and 2.5 phr hexamethoxymethylmelamine, sheeting out the compound, cooling, then mixing with sulfur, a vulcanization accelerator, and curing at 149 °C gave enhanced adhesion to brass-plated steel wire, 95% rubber coverage, and better mechanical properties compared to a rubber compound containing a conventional resorcinol resin.[63] A similar German patent, using a somewhat different preparation of the styrenated resorcinol resin, also claimed improved adhesion of the vulcanizable rubber compound to the plated steel wire and other advantages.[64] There was another U.S. patent issued in which resorcinol/formaldehyde resins styrenated to a lesser degree were also low in hygroscopicity, low-fuming in processing, and valuable bonding agents in vulcanizable rubber compounds.[65]

Dehydrated, low molecular weight resorcinol/formaldehyde condensation products useful, e.g., as adhesion promoters in vulcanizable rubber compounds, were prepared by condensing excess resorcinol with *s*-trioxane (cyclic formaldehyde trimer) in the presence of an acid catalyst and an inert, nonpolar, water-immiscible organic solvent, such as toluene, with azeotropic removal of water during the reaction. The resulting condensate was insoluble in the solvent, was free of water, and had molecular weight 287.[66]

Granulated resorcinol-silica mixtures were patented for use in improving the adhesion of vulcanizable rubber to organic fibers or steel cord.[67]

Aramid cords (Teijin's HP 50) were treated with a plasma, then with a RFL dip containing Nipol 2619 FS and 2,6-bis-[(2,4-dihydroxyphenyl)methyl]-4-chlorophenol, then heated at 150–240 °C; thus-treated cords had improved adhesion to vulcanizable rubber compounds.[68]

Glass fiber cords were coated with a RFL dip also containing a butadiene rubber latex, chlorosulfonated polyethylene, and a polyisocyanate. The coated fibers were dried and then coated with a suspension of ethylene/propylene rubber and carbon black. The double-dipped cords gave better adhesion to vulcanizable rubber compounds.[69]

Polyester filaments were spun, dipped in a RFL dip, dried, drawn, then relaxed at 160 °C in a specified time/temperature regimen. These prebonded filaments, useful as tire cords, had favorable adhesion to vulcanized rubber compounds.[70]

Dimensionally stable organic fiber cords were prepared by twisting conjugate fibers of 30–90% poly(ethylene-2,6-naphthalenedicarboxylate) as the core and a polyamide sheath (e.g., nylon-66), then treating with a RFL dip. The treated fibers showed excellent retention of tensile strength in a fatigue test and good adhesion to vulcanizable rubber compounds.[71]

Polyester or aramid fibers suitable for reinforcing chlorosulfonated polyethylene rubber or chloroprene rubber were prepared by first treating the fibers with polyepoxies, blocked polyisocyanates, and nitrile or chloroprene rubber latexes, followed by a RFL dip containing a 2-chlorophenol/resorcinol/formaldehyde condensation product.[72]

A novel preparation of fatigue-resistant, multifilament tire cords with good adhesion to rubber was described. Sixty-four PET [poly(ethylene terephthalate)] filaments were coated with a lubricant containing glycerol-epichlorohydrin copolymer and twisted with 129 nonfinished PET filaments; the bundle was treated with a RFL dip.[73]

A vulcanizable natural rubber composition, for use in steel-belted tires, with good mechanical strength and durability was compounded as follows: 100 parts natural rubber, 30 parts silica, 35 parts carbon black, 2.5 parts bis-[3-(triethoxysilyl)propyl]tetrasulfide (a silane coupling agent), 10 parts zinc oxide, 2 parts cobalt stearate, 1 part resorcinol, 1 part hexamethylenetetramine, 1 part vulcanization accelerator, and 5 parts sulfur were used in the usual manner.[74]

A patented method of making aramid or polyester fiber cords with improved flexural fatigue endurance and adhesion to a chloroprene rubber compound consisted of impregnating the cords with an ethyl acetate solution of a polyisocyanate (crude MDI) and then with a resorcinol formaldehyde resin/neoprene latex dip.[75]

High-strength glass fibers were coated with a RFL dip containing also a polybutadiene latex and an antioxidant, then dried at 250 °C for 1 min, twisted and coated with a toluene solution of zinc methacrylate, a halogenated polymer (Hypalon-40-6), a polyisocyanate, and benzoquinone oxime dibenzoate. Vulcanized, hydrogenated nitrile rubber compounds reinforced with these fibers showed high heat (120 °C) and flex stress resistance.[76]

Fatigue-resistant carbon fibers were prepared by treating the fibers with a RFL dip, then twisting the fibers, and heat treating at 220 °C for 1 minute. When embedded in a vulcanized rubber compound, these carbon fibers showed good adhesion to the rubber and good tensile strength retention.[77]

Vulcanizable diene (e.g., high *cis*-butadiene) rubber compositions, optionally containing a methylene donor to form a resin *in situ*, were modified by adding 12 phr alpha-(2,4-dihydroxy)phenyl-*N*-phenyl nitrone, **4-5** [made by reacting 2,4-dihydroxybenzaldehyde (beta-resorcylaldehyde) with nitrobenzene and hydrogen in the presence of a 5% Pt-on-C catalyst in dimethylsulfoxide/ethanol as the solvent]; 75 parts of this modified *cis*-butadiene rubber was compounded with 25 parts isoprene rubber and other compounding ingredients, and then vulcanized. The finished composite showed improved tear resistance as measured by the Strebler adhesion values.[78]

To improve the adhesion of polyester fibers to vulcanized natural rubber compositions, the fibers were first irradiated with a pulsed laser beam (380 nm), then treated with a RFL dip, dried at 140 °C/120 sec, and heated at 230 °C/90 sec before embedment in the rubber compound.[79]

The manufacture of dimensionally stable polyester fibers useful in vulcanizable rubber compositions was patented. The process involved melt spinning of the polyester (mostly PET), passing the strands through a column of a gas at 365 °C, solidifying the fibers by air-cooling in a specified manner (blowing air from the center to the surface of the fibers), treating with a RFL dip, drying, heat-treating at 240 °C/60 sec, and forming cords. These cords had high tenacity and low heat-shrinkage.[80]

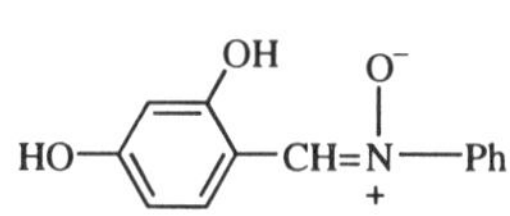

4-5

An improved treatment for aramid fibers was developed, consisting of treating the fibers first with a polyepoxide, liquid rubber, and a blocked polyisocyanate, drying and heat-treatment at 235 °C/60 sec, then coating with a RFL dip (using a chloroprene rubber latex), drying, and heating at 200 °C/ 60 sec, then embedding the treated cords in a vulcanizable (chloroprene) rubber compound. The cured composite showed high peel strength and 80% retention of peel strength in a flex test.[81]

A composition for tire treads was developed based on a vulcanizable compound of 20/80 natural/SB rubber containing 5 phr of Sumikanol 610 (R/F resin) and 5 phr of a rosin gum resin. The cured compound showed good resistance to cuts, heat-aging, and heat-buildup.[82]

A treatment for fibers to improve their bonding to hydrogenated NBR was developed. Thus, polyester cords were treated with an epoxy-terminated polyurethane and a polyamine (from methyl ethyl ketone solution), heat-treated at 200 °C/1 min, then treated with an alkaline, aqueous RFL dip, again heat-treated at 200 °C/1 min, then embedded in unvulcanized, hydrogenated NBR-containing zinc methacrylate, Zetpol 2020, silica, and an organic peroxide (Peroximon F 40). After vulcanization at 170 °C/30 min, the composition showed excellent peel strength.[83] A similar patent claimed the same vulcanizable NBR compound, but the polyester cords were treated with epoxy-modified polyols, polyisocyanates, and lastly with a RFL dip, then dried before embedment in the rubber.[84]

Addition products of *ca* 2/1 mole ratio of resorcinol/di-epoxide (e.g., the diglycidyl ether of bisphenol A) were prepared and found to be good bonding and reinforcing agents, when used in conjunction with either hexamethylenetetramine or hexamethoxymethylmelamine as the methylene donors, for (steel) reinforced, vulcanizable rubber compounds. These novel resorcinol/epoxy compound adducts also had low volatility, gave low fuming in compounding, improved the physical and mechanical properties of the cured composites, and showed improved unaged and steam-aged adhesion of the cords to the rubber compound.[85]

4.4. NONTIRE APPLICATIONS OF RESORCINOL IN RUBBER COMPOUNDS

The following, selected examples will illustrate the scope of these uses.

Rubber compositions for tennis balls consisted of, e.g., 90 parts natural rubber, 10 parts high-styrene resin, 10 parts resorcinol, 6 parts hexamethylenetetramine, and compounding additives. The mixed, kneaded, and vulcanized compound gave a material with high resilience and a good balance of properties for this use.[86]

An adhesive composition for plastics to rubber was prepared consisting of a RF/rubber latex (JSR 0561) with a small amount of *N*-bromosuccinimide. This adhesive was used to laminate an EPDM rubber film with a nylon-66 film and the bond showed high strength.[87]

Core compositions for golf balls contained 100 parts butadiene rubber, 20 parts zinc acrylate, 15 parts zinc oxide, 20 parts silica, 15 parts resorcinol, and 4 parts antioxidant, mixed at 120–150 °C; 5 parts Percumyl D and 4 parts hexamethylenetetramine were then blended in at 60–70 °C and the compound was vulcanized at 140–160 °C to form the cores. The cores were molded with ionomer covers to make golf balls with high durability and impact resilience.[88]

Glass or organic fibers were coated with an RF/hydrogenated nitrile rubber latex. These fibers had improved adhesion to vulcanizable rubber compounds containing saturated nitrile rubbers. The latter compositions were useful for making belts and hoses.[89]

Endless polyamide belts were made by cutting the ends to tapered, toothy shapes, applying resorcinol, cresol, and/or formic acid to the sections to be adhered, joining the ends, clamping and heating at 100 °C/5–20 min. These belts were working well at 150 °C for a much longer time than belts with conventionally bonded ends.[90]

Composite fiber cords were prepared from PET fibers as the core and nylon-66 as the sheath, spinning, drawing, and treating with a RFL dip. The treated fibers were twisted into cords used for embedding in vulcanizable rubber compositions and formed into power-transmission belts with good fatigue resistance.[91]

Poly(vinyl alcohol) short fiber-reinforced rubber compositions were prepared. Vinyon short fibers were treated with a RFL dip and dried at 150 °C. For instance, 17 parts treated short fiber, 100 parts EPDM rubber, and appropriate additives were blended, sheeted, and vulcanized to give a product with a good balance of mechanical properties, substantially better than the same compound without the short Vinyon fibers.[92]

A crosslinkable rubber composition consisting of 5–15 parts ethylene/10–25 parts vinyl acetate/10–25 parts ethyl acrylate/50–70 parts butyl acrylate/1–8 parts methoxymethyl acrylate copolymer was useful for oil-resistant products. For instance, 100 parts copolymer, 100 parts carbon black, 30 parts plasticizer, 3 parts resorcinol, 2.5 parts sodium laurate, and 2 parts 2-undecylimidazole were vulcanized to give a composite with good tensile properties and good performance at elevated and cold temperatures, useful for the fabrication of lubricating seals, packings, and hoses.[93]

Flooring materials were prepared from rubber scrap. Thus, nylon-66 cloth was treated with a RF/SBR/chloroprene rubber latex, dried, coated with chloroprene rubber (Denka A-90), and bonded to chloroprene rubber-treated scrap tire strips. The composite had good adhesion of fabric to rubber and low shrinkage at 120 °C.[94]

Glass fiber cords were treated with a conventional RFL dip, which also contained isobutylene/isoprene copolymer and, optionally, a dicarboxylated butadiene/styrene copolymer. The treated cords were used as the reinforcement in compositions for timing belts for automobile engines.[95]

A mixture of 90±5 wt% shredded rubber tires (scrap) and 10±5 wt% of a resorcinol/formaldehyde resin, optionally with added methylene donor, sulfur vulcanization package, and/or a fireproofing resin, was said to be useful as a road paving material or could be molded to make guardrails or barrier walls.[96]

High-strength glass fiber cords were treated with an ammoniated RFL dip containing an antioxidant, dried at 250 °C/1 min, then coated with a toluene solution of zinc-methacrylate, Hypalon-40, a polyisocyanate, and benzoquinone oxime dibenzoate, dried again at 130 °C/1 min and embedded in a vulcanizable hydrogenated NBR compound. After cure at 120 °C/1 hr, this composite showed high strength retention in a flex test. The latter composition was useful for automotive timing belts.[97]

Another patented finish for glass fiber strands consisted of a RFL dip and fluorine-containing water- and oil-proofing agents. After heating, the strands were treated with a halogen-containing polymer adhesive, dried again, and used to reinforce vulcanizable hydrogenated nitrile rubber compounds. Belts made from these compositions had good mechanical properties and were water-resistant.[98]

A new variation for finishing glass fibers involved the standard RFL dip and a nitrile rubber latex dip, then coating the fibers with a hydrogenated nitrile rubber solution and drying. These coated fibers were used to reinforce a vulcanizable hydrogenated nitrile rubber compound; after cure, these compositions showed very good tensile strength/retention and were suitable for use in making belts.[99]

A dip for poly(vinyl alcohol) fabric was prepared, consisting of a standard RFL mix with added butadiene rubber latex, SBR latex, and oleyl poly(oxyethylene ether). After drying, the fabric was used to reinforce a vulcanizable EPDM compound. The cured composite had good peel strength.[100]

4.5. RESOTROPIN

This compound is an adduct (acid–base type) of resorcinol and hexamethylenetetramine. Considerable work on resotropin, **4-6**, as an additive for rubber was done in the former Soviet Union.

Resotropin used to be prepared by mixing equimolar amounts of resorcinol and hexamethylenetetramine in water and filtering and drying the solids; the storage stability of this product is not good. For example, a Russian manual on additives for polymers gave a shelf life of three months for resotropin.[101] Therefore, work continued on process improvements. In the U.S., for instance, a process was developed for reacting resorcinol and hexamethylenetetramine in the presence of a small amount of water or other solvent in a dough mixer (the shelf life of this product was not described).[102]

A particularly high-purity grade of resotropin, storage stable for 5+ years at ambient conditions, was obtained in high yield by reacting resorcinol and hexamethylenetetramine in methanol solution.[103]

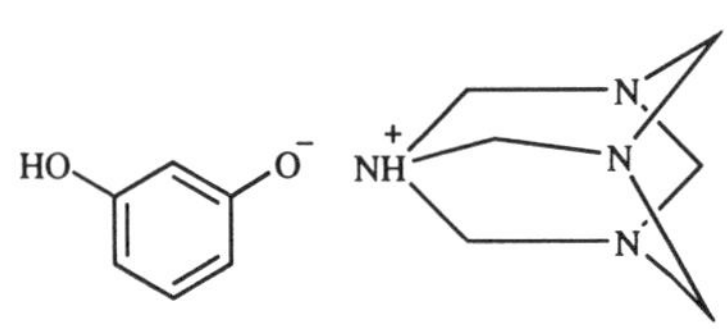

4-6

A few examples of reports from the USSR on the use of resotropin in rubber follow.

The addition of resotropin together with silica to vulcanizable rubber compounds substantially improved the adhesion of the compound to rayon and nylon-6 cord, as well as giving improved tear strength and heat-aging resistance.[104]

A vulcanizable synthetic/natural rubber compound with high adhesion to brass-plated steel cords was obtained by adding 0.5–2 parts resotropin, 0.3–2 parts *p*-nitrosophenol or diphenylamine, and 0.3–1 part cobalt naphthenate, in addition to the usual compounding ingredients, to the rubber.[105]

The addition of a mixed monomethyl-, dimethyl-, and ethylresorcinol (presumably obtained from shale pyrolyzates)/formaldehyde condensate and resotropin to a vulcanizable rubber compound improved the adhesion to organic textile and brass-plated steel cord reinforcements.[106]

The adhesion of vulcanizable butyl rubber or EPD compounds to steel (cords) was reported to be improved by the addition of up to 5 parts resotropin (modifier RU-1) or a 5-methylresorcinol/hexamethylenetetramine adduct (modifier ARU).[107]

The addition of 1–5 parts resotropin to a vulcanizable diene rubber compound was claimed to improve the tack of the stock and the mechanical properties of the vulcanizate.[108]

Dust-free resotropin was prepared by using a mineral oil or a dialkyl phthalate as the dust suppressant (*ca* 3 wt% on the resotropin).[109]

Vulcanizable unsaturated rubber compounds with sulfonamide accelerators, Altax, *N*-phenyl-*N'*-isopropyl-*p*-phenylenediamine, and resotropin showed low heat buildup.[110]

An adhesive for tire cords (polyamide) to improve their adhesion to rubber contained 2–4 parts dibutyl phthalate, 40–60 parts colloidal silica, 40–60 parts resotropin, 100 parts neoprene, 4–6 parts zinc oxide, 4–6 parts rosin, 285–345 parts naphtha, and 285–345 parts ethyl acetate.[111]

In the U.S., resotropin was evaluated by the United States Pipe and Foundry Co. and recommended as an adhesion promotor for vulcanizable natural rubber compounds to brass-plated steel tire cords in the 1970s.[112] However, to date there seems to be little, if any, actual use of resotropin in this application in the U.S.

4.6. THIO- AND DITHIO-bis-RESORCINOL

The few, chosen examples from the patent literature which follow indicate a sustained interest in this type of resorcinol derivative for various uses in vulcanizable rubber compounds. Thio- and dithio-bis-resorcinol are also named (2,4-dihydroxyphenyl)sulfide and disulfide, respectively; sometimes, the names di-(2,4-dihydroxyphenyl) sulfide and disulfide appear. Often, the product under investigation was unlikely to have been a pure compound.

Polyester fiber reinforcement was treated with a RFL-type dip in which the phenolic resin portion was a resorcinol/(2,4-dihydroxyphenyl)sulfide or disulfide/for-

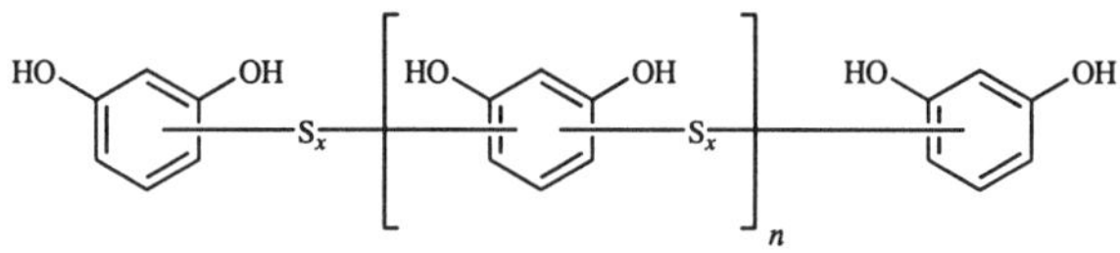

4-7
(where $x = 2$–8 ; $n = 0$–15)

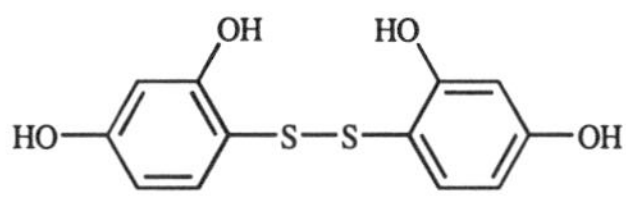

4-8

maldehyde resin. The treated cord showed improved adhesion to the vulcanized rubber at high temperature and dynamic stress.[113]

Sulfur-curable butyl or EPDM rubbers loaded with a kaolin clay filler and 4-mercaptoresorcinol/(2,4-dihydroxyphenyl)disulfide showed good compounding behavior; the vulcanized compound had a desirable balance of properties.[114]

Polyester (PET) fiber was treated with a RFL-type dip containing (2,4-dihydroxyphenyl) disulfide; vulcanizable rubber compounds adhered well to the treated fiber.[115] A similar RFL-type dip containing a polymeric resorcinol-sulfur monochloride condensation product, **4-7**, was used to treat PET, nylon-6, and rayon fibers. The treated fibers showed improved adhesion to vulcanizable rubber compounds.[116] Resorcinol-sulfur monochloride condensation products were reported to be useful heat stabilizers for neoprene rubber.[117]

Di-(2,4-dihyroxyphenyl)disulfide, **4-8**, was a useful vulcanization accelerator in vulcanizable chloroprene compounds, giving good scorch stability and a good balance of physical/mechanical properties.[118]

Polyester (PET) fibers with improved, heat-stable adhesion to vulcanizable rubber compounds were prepared by first treating the fiber with a polyepoxide and then with a RFL-type dip containing a polymeric resorcinol/sulfur condensation product.[119]

4.7. CONCLUDING REMARKS

The above material amply demonstrates that resorcinol, certain resorcinol derivatives, or appropriately chosen resorcinol/(form)aldehyde resins are "methylene acceptors" which, together with "methylene donors" such as HMTA or HMMM, improve the properties of vulcanizable rubber compounds because they can act as reinforcing agents, likely due to the formation of interpenetrating networks (IPN) in the rubber, and because they improve the adhesion of the fibers, textile, or cord to the rubber compound. Resorcinol has a desirable melting point, 110 °C, for processing. The commercial, solid resorcinol resins for this use are designed for appropriate softening points. They have good solubility properties (e.g., in the preparation of latex dips), they improve the processability of the rubber compounds, they have a high reactivity with methylene donors, they improve the mechanical properties, such as the tear resistance of the vulcanizate, and they have a stabilizing (antioxidant) effect. Particularly important is the improvement of the adhesion of reinforcing fiber to rubber by the resorcinol/resorcinol resin/methylene donor addition; likely, this is due to an interlocking effect and hydrogen bonding. A chemical linkage of phenolic to rubber has been proposed via *ortho* (to the phenolic hydroxyl group) methylol groups, which can form quinone methides capable of adding to double bonds or undergoing allylic substitution reactions in the rubber. Indeed, with more than one methylol group, crosslinking (vulcanization) of rubber can occur. Such mechanisms were first proposed by Hultzsch,[120] van der Meer,[121] and Giller.[122] In the case of resorcinol and resorcinol resins, the *o*-methylolphenol/quinone methide mechanism for their interaction with rubber is less likely because RFL dips are cured, with loss of methylol groups, on the fiber before contact with the rubber compound. When resorcinol/resins are used as rubber additives they are advanced by reaction with hexamethylenetetramine (HMTA) or hexamethoxymethylmelamine (HMMM) and, in either case, methylol groups are not formed. This is not to say that a small amount of quinone methide, and reaction with rubber, may not occur due to, e.g., thermal splitting of a Mannich base intermediate or by an acid-catalyzed mechanism.

In this chapter only the use of resorcinol/derivatives as reinforcing agents and bonding agents in vulcanizable rubber compounds was presented. There has been, and there will always be, much work on substitutes for this use of resorcinol. As one example, the use of 2-naphthol in place of resorcinol was proposed by American Cyanamid Co. in the 1970s. To date, however, resorcinol/resorcinol resins have retained their position in this application because they produce a favorable balance of properties, including long-proven cost effectiveness.

4.8. REFERENCES

1. R. S. Bhakuni, S. K. Mowdood, W. H. Wadell, I. S. Rai, and D. L. Knight, *Encyclopedia of Polymer Science and Engineering, 2nd ed., Vol. 16*, 834 (1989), Wiley, New York.
2. Ger. Pat. 725,499 (Feb. 24, 1936; P. Baumann to Continental Gummiwereke A.G.); U.S. Pat. 2,128,229 (Aug. 30, 1938; W. H. Charch and D. B. Maney to E. I. duPont de Nemours and Co.); Brit. Pat. 477,380 (Dec. 23, 1937; to DuPont Rayon Co.).
3. U.S. Pat. 2,561,215 (July 17, 1951; C. J. Mighton to DuPont Co.).
4. U.S. Pat. 2,994,671 (Oct. 17, 1956; W. L. Thompson to DuPont Co.); U.S. Pat. 3,307,966 (March 7, 1967; C. J. Shoaf to DuPont Co.).
5. Neth. Pat. Appl. 6,613,883 (Apr. 3, 1967; to Farbenfabriken Bayer A.G.).

5a. N. K. Porter, *J. Coated Fabrics, 21*, 230 (1992).

6. K.-D. Albrecht, *Rubber Chem. Technol., 46*, 981 (1973).
7. W. J. van Ooij, *Rubber Chem. Technol., 52*, 605 (1979)—this covers the literature from 1970–1978; W. J. van Ooij, *Rubber Chem. Technol., 57*, 421 (1984)—this covers the literature from 1979–1983; W. J. van Ooij, J. Giridhar, and J. A. Ahn, *Kautschuk, Gummi, Kunstst., 44*, 348 (1991)—this covers the literature from 1984–1989.
8. T. S. Salomon, *Rubber Chem. Technol., 58*, 562 (1985).
9. A. Lechtenboehmer, H. G. Moneypenny, and F. Mersch, *Brit. Polymer J., 22*, 265 (1990).

10. G. R. Hamed and J. Huang, *Rubber Chem. Technol., 64*, 285 (1991).
11. R. Sattelmeyer, *Kautschuk, Gummi, Kunstst., 44*, 131 (1991).
11a. R. F. Seibert (Uniroyal Chemical Co., Inc.), Paper No. 52, "A Review of Bonding Agents as Adhesion Promoters in Rubber to Metal and Rubber to Textile Applications," presented at the 144th Meeting of the Rubber Division, American Chemical Society, Orlando, Florida, Oct. 1993.
12. Jpn. Kokai Tokkyo Koho JP 58,160,330 (Sept. 22, 1983), *C.A., 101*, 112248 (1984); Jpn. Kokai Tokkyo Koho Jpn. 58,160,329 (Sept. 22, 1983), *C.A., 100*, 193356 (1984); and Jpn. Kokai Tokkyo Koho JP 58,160,331 (Sept. 22, 1983), *C.A., 100*, 193355 (1984); all to Sumitomo Chemical K.K.
13. U.S. Pat. 4,381,949 (May 3, 1983; G. Kuehner, S. Wolff, and L. Rothbuehr to Degussa A.G.).
14. U.S. Pat. 4,376,854 (March 15, 1983; T. Yamaguchi, K. Kashima, and N. Kaifu to Hdogaya Chemical Co., Ltd.).
15. Jpn. Kokai Tokkyo Koho JP 59,011,384 (Jan. 20, 1984; to Mitsubishi Belting K.K.); *C.A., 101*, 8349 (1984).
16. Eur. Pat. 52,286 (May 30, 1984; to Hoechst A.G.); *C.A., 97*, 73746 (1982).
17. Jpn. Kokai Tokkyo Koho JP 58,109,684 (June 30, 1983; to Yokohama Rubber K.K.); *C.A., 100*, 8266 (1984).
18. Ger. Offen. DE 3,318,059 (Nov. 22, 1984; J. Ippen and F. Stuettgen to Bayer A.G.); *C.A., 102*, 114972 (1985).
19. Jpn. Kokai Tokkyo Koho JP 60,115,679 [85,115,679] (June 22, 1985; to Seitetsu Chemical Industry Co. and Sumitomo Chemical Co., Ltd.); *C.A., 104*, 6137 (1986).
20. U.S. Pat. 4,448,813 (May 15, 1984; T. S. Salomon to B. F. Goodrich Co.).
21. U.S. Pat. 4,605,696 (Aug. 12, 1986; D. A. Benko, S. K. Mowdood, P. H. Sandstrom, W. H. Waddell, III, and L. G. Wideman to The Goodyear Tire and Rubber Co.).
22. Ger. (East) DD 237,516 (July 16, 1986; G. Merkmann, H. Merkmann, W. Timm, C. Geissman, H. Knoll, S. Borgmann, and R. Thiel to VEB Gummiwerke Thueringen); *C.A., 106*, 215361 (1986).
23. Jpn. Kokai Tokkyo Koho JP 62,090,377 [87,090,377] (Apr. 24, 1987; S.Takahashi, Y. Suzuki, S. Ueno, and S. Yoshida to Yokohama Rubber Co., Ltd. and Shin-Etsu Chemical Industry Co., Ltd.); *C.A., 107*, 219006 (1987).
24. Japan. Pat. JP 87, 003,863 (Jan. 27, 1987; Y. Komatsu, Y. Takagi, I. Yoshida, and H. Shoji to Toray Industries, Inc.).
25. Ger. Offen. DE 3,642,930 (June 19, 1987; S. Takahashi, Y. Suzuki, H. Ogawa, and K. Ichimaru to Toho Rayon Co., Ltd. and Yokohama Rubber Co., Ltd.); *C.A., 107*, 328397 (1987).
26. Fr. Demande FR 2,587,277 (March 20, 1987; S. Yamada, M. Taniguchi, M. Takahashi, and K. Kato to Bridgestone Corp.); *C.A., 107*, 238420 (1987).
27. Jpn. Kokai Tokkyo Koho JP 62,141,180 [87,141,180] (June 24, 1987; Y. Igawa and Y. Yoshida to Toray Industries, Inc.); *C.A., 108*, 23202 (1987).
28. Jpn. Kokai Tokkyo Koho JP 62,232,445 (Oct. 12, 1987; to Bridgestone Tire K.K.); *C.A., 108*, 76940 (1988).
29. Jpn. Kokai Tokkyo Koho JP 62,039,633 [87,039,633] (Feb. 20, 1987; M. Tayanagi, S. Ueda, and H. Nagasawa to Asahi Chemical Industry Co.).
30. Eur. Pat. Appl. EP 239,915 (Oct. 7, 1987; M. Furukawa and T. Takata to Teijin Ltd.); *C.A., 108*, 95970 (1987).
31. Jpn. Kokai Tokkyo Koho JP 63,247,102 (Oct. 13, 1988; M. Tomio, H. Takino, and S. Inada to Toyo Rubber Industries K.K.); *C.A., 110*, 174967 (1989).
32. Jpn. Kokai Tokkyo Koho JP 63,203,841 (Aug. 23, 1988; K. Oshima and T. Nishikawa to Bridgestone Tire K.K.); *C.A., 110*, 59308 (1989).
33. Jpn. Kokai Tokkyo Koho JP 63,049,501 [88,049,501] (March 2, 1988; S. Tsukada and H. Kaido to Yokohama Rubber Co.); *C.A., 108*, 222963 (1988).
34. U.S. Pat. 4,782,106 (Nov. 1, 1988; G. Fricke, H. Hoffmann, and H. Umland to Continental A.G.).
35. Jpn. Kokai Tokkyo Koho JP 63,243,380 (Oct. 11, 1988; M. Ito, Y. Iseda, and H. Tagaki to Bridgestone Tire K.K.); *C.A., 110*, 214570 (1989).

36. Jpn. Kokai Tokkyo Koho JP 63,270,873 (Nov. 11, 1988; T. Yotsumoto to Bridgestone Tire K.K.); *C.A., 110*, 194475 (1989).
37. Ger. Pat. DE 2,717,767 (Feb. 18, 1988; E. F. Kalafua and S. C. Sharma to General Tire and Rubber Co.); equivalent is U.S. Pat. 4,040,999 (Aug. 9, 1987).
38. Jpn. Kokai Tokkyo Koho JP 63,126,974 (May 30, 1988); A. Kozuka to Asahi Fiber Glass K.K.).
39. Jpn. Kokai Tokkyo Koho JP 01,122,706 [89,122,706] (May 16, 1989; H. Tomokuni, A. Okamure, and Y. Nomura to Toyo Rubber Industry Co., Ltd.); *C.A., 112*, 58051 (1990).
40. Jpn. Kokai Tokkyo Koho JP 01,129,037 [89,129,037] (May 22,1989; to Yokohama Rubber Co., Ltd.); *C.A., 112*, 8605 (1990).
41. Jpn. Kokai Tokkyo Koho JP 01,168,505 [89,168,505] (July 4, 1989; Y. Ishikawa, H. Kaido, T. Takei, and S. Takahashi to Yokohama Rubber Co., Ltd.); *C.A., 112*, 8621 (1990).
42. U.S. Pat. 4,889,891 (Dec. 26, 1989; B. Durairaj, Alex Peterson, Jr., and G. Salee to INDSPEC Chemical Corp.).
43. Jpn. Kokai Tokkyo Koho JP 01,174,676 [89,174,676] (July 11, 1989; Y. Igawa, I. Yoshida, I. Ikeda, and H. Okamoto to Toray Industries, Inc.); *C.A., 112*, 120420 (1990).
44. Jpn. Kokai Tokkyo Koho JP 01,284,532 [89,284,532] (Nov. 15, 1989; A. Okamura and M. Sekiguchi to Nippon Glass Fiber Co.); *C.A., 112*, 141,495 (1990).
45. Jpn. Kokai Tokkyo Koho JP 01,318,047 [89,318,047] (Dec. 22, 1989; T. Komatsu to Asahi Chemical Industry Co.); *C.A., 112*, 200312 (1990).
46. Ger. Pat. (East) DD 276,295 (Feb. 21, 1990; H. Knoll, S. Bergmann, G. Merkmann, H. Merkmann, and K. Giessmann to VEB Chemiekombinat Bitterfeld); *C.A., 114*, 8119 (1991).
47. Jpn. Kokai Tokkyo Koho JP 03,000,741 [91,000,741] (Jan. 7, 1991; T. Uchama, H. Kaido, and Y. Ishikawa to Yokohama Rubber Co., Ltd.); *C.A., 114*, 249113 (1991).
48. Eur. Pat. Appl. EP 419,741 (Apr. 3, 1991; R. T. Hood and R. M. Lamars to INDSPEC Chemical Corp.); *C.A., 115*, 51667 (1991).
49. U.S. Pat. 4,892,908 (Jan. 9, 1990; B. Durairaj, A. Peterson, Jr., and G. Salee to INDSPEC Chemical Corp.).
50. Eur. Pat. Appl. EP 353,473 (Feb. 7, 1990; A. Fujiwara, H. Matsuoka, S. Onoe, and K. Kusaka to Bando Chemical Industries, Ltd.); *C.A., 113*, 99181 (1990).
51. Jpn. Kokai Tokkyo Koho JP 02,263,883 [90,263,883] (Oct. 26, 1990; Y. Toyoda and S. Mitsushiba to Sumitomo Naugatuck Co.); *C.A., 114*, 165973 (1991).
52. USSR Pat. SU 1,535,870 (Jan. 15, 1990; A. B. Filimonov *et al.* to Kazan Chemical-Technological Institute); *C.A., 112*, 236738 (1990).
53. Jpn. Kokai Tokkyo Koho JP 01,132,883 [89,132,883] (May 25, 1989; T. Matsuda, T. Okada, and J. Kuwata to Toray Industries, Inc.); *C.A., 112*, 22176 (1990).
54. Jpn. Kokai Tokkyo Koho JP 02,141,441 [90,141,441] (May 30, 1990; H. Murakami, T. Fujinaga, H. Momotake, and T. Morimoto to Central Glass Co., Ltd.); *C.A., 114*, 124350 (1991).
55. Jpn. Kokai Tokkyo Koho JP 02,167,346 [90,167,346] (June 27, 1990; D. Takahashi to Yokohama Rubber Co., Ltd.); *C.A., 114*, 104093 (1991).
56. Jpn. Kokai Tokkyo Koho JP 02,243,326 [90,243,326] (Sept. 27, 1990; K. Takahira and E. Komura to Toyo Rubber Industry Co., Ltd.); C.A., 114, 64049 (1991).
57. Eur. Pat. Appl. EP 395,435 (Oct. 31, 1990; N. S. Anderson and A. L. Promislow to Hoechst Celanese Corp.); *C.A., 114*, 64050 (1991).
58. Eur. Pat. Appl. EP 376,623 (July 4, 1990; A. Fujiwara and K. Hata to Bando Chemical Industries, Ltd.); *C.A., 113*, 154149 (1990).
59. Jpn. Kokai Tokkyo Koho JP 02,160,979 [90,160,979] (June 20, 1990; T. Masuda, T. Okada, and T. Sasaki to Toray Industries Inc.); *C.A., 114*, 25660 (1991).
60. Jpn. Kokai Tokkyo Koho JP 02,173,126 [90,173,126] (July 4, 1990; H. Kasatani and T. Azuma to Asahi Chemical Industry Co., Ltd.); *C.A., 113*, 213553 (1990).
61. Jpn. Kokai Tokkyo Koho JP 02,084,587 [90,084,587] (May 26, 1990; M. Sato and S. Iwasaki, to Bridgestone Corp.); *C.A., 113*, 193266 (1990).
62. U.S. Pat. 5,030,692 (June 4, 1991; B. Durairaj to INDSPEC Chemical Corp.).

63. U.S. Pat. 5,021,522 (June 4, 1991; B. Durairaj, A. Peterson, Jr., R. M. Lamars, and R. T. Hood to INDSPEC Chemical Corp.).
64. Ger. Offen. DE 4,001,606 (July 25, 1991; W. Hesse, D. Leicht, and R. Sattelmeyer to Hoechst A.G.); *C.A., 115*, 185158 (1991).
65. U.S. Pat. 5,049,641 (Sept. 17, 1991; R. T. Hood and R. M. Lamars to INDSPEC Chemical Corp.).
66. Ger. Offen. DE 4,002,417 (Aug. 1, 1991; K. Altenbach to Hoechst A.G.); *C.A., 115*, 233709 (1991).
67. Ger. Offen. DE 4,013,258 (Oct. 31, 1991; S. Wolff, U. Goerl, and K. Meier to Degussa A.G.); *C.A., 116*, 42956 (1992).
68. Jpn. Kokai Tokkyo Koho JP 03,220,372 [91,220,372] (Sept. 27, 1991; M. Furukawa and T. Takada to Teijin, Ltd.); *C.A., 116*, 85634 (1991).
69. Jpn. Kokai Tokkyo Koho JP 03,269,177 [91,269,177] (Nov. 29, 1991; K. Hattori, C. Ito, and H. Kojima to Asahi Fiber Glass Co., Ltd.); *C.A., 116*, 153651 (1992).
70. Patentschrift (Switz.) CH 677,499 (May 31, 1991; P. Reinthaler and A. Mueller to Rhone-Poulenc Viscosuisse S.A.); *C.A., 115*, 138165 (1991).
71. Jpn. Kokai Tokkyo Koho JP 03,074,203 [91,074,203] (March 28, 1991; N. Inada and M. Kikuchi to Bridgestone Corp.); *C.A., 115*, 185171 (1991).
72. Jpn. Kokai Tokkyo Koho JP 03,167,367 [91,167,367] (July 19, 1991; H. Watanabe and T. Takada to Teijin, Ltd.); *C.A., 116*, 85607 (1992).
73. Jpn. Kokai Tokkyo Koho JP 03,045,738 [91,045,738] (Feb. 27, 1991; K. Yabuki to Toyobo Co., Ltd.); *C.A., 114*, 230485 (1991).
74. Jpn. Kokai Tokkyo Koho JP 03,007,743 [91,007,743] (Jan. 14, 1991; H. Imura, H. Matsumoto, H. Ogami, and K. Sueyoshi to Toyo Rubber Industry Co., Ltd.); *C.A., 115*, 51668 (1991).
75. Jpn. Kokai Tokkyo Koho JP 04,033, 975 [92,033,975] (Feb. 5, 1992; K. Hata to Bando Chemical Industries, Ltd.); *C.A., 117*, 28604 (1992).
76. Jpn. Kokai Tokkyo Koho JP 04,050,237 [92,050,237] (Feb. 19, 1992; K. Hatsutori and C. Ito to Asahi Fiberglass K.K.); *C.A., 117*, 28592 (1992).
77. Jpn. Kokai Tokkyo Koho JP 04,002,833 [92,002,833] (Jan. 7, 1992; A. Iiyama and M. Kumada to Honda Motor Co., Ltd.); *C.A., 116*, 175897 (1992).
78. U.S. Pat. 5,091,449 (Feb. 25, 1992; J. Cantillo, R. W. Strozier, and W. P. Francik to The Goodyear Tire and Rubber Co.).
79. Jpn. Kokai Tokkyo Koho JP 04,136,267 [92,136,267] (May 11, 1992; H. Watanabe and T. Takada to Teijin Ltd.); *C.A., 117*, 113492 (1992).
80. Jpn. Kokai Tokkyo Koho JP 04,163,314 [92,163,314] (June 8, 1992; S. Murase, K. Ueda, K. Nishikawa, and H. Yokoyama to Unitika Ltd.); *C.A., 117*, 173217 (1992).
81. Jpn. Kokai Tokkyo Koho JP 04, 180,937 [92,180,937] (June 29, 1992; A. Azuma to Bando Chemical Industries, Ltd.); *C.A., 117*, 235637 (1992).
82. Jpn. Kokai Tokkyo Koho JP 04,189,844 [92,189,844] (July 8, 1992; Y. Suzuki, T. Enomoto, and M. Tsuji to Yokohama Rubber Co., Ltd.); *C.A., 117*, 235646 (1992).
83. Jpn. Kokai Tokkyo Koho JP 04,198,325 [92,198,325] (July 17, 1992; A. Azuma to Bando Chemical Industries, Ltd.); *C.A., 118*, 104592 (1993).
84. Jpn. Kokai Tokkyo Koho JP 04,198,326 [92,198,326] (July 17, 1992; A. Azuma to Bando Chemical Industries, Ltd.); *C.A., 118*, 126283 (1993).
85. U.S. Pat. 5,244,725 (Sept. 14, 1993; H. Dressler and A. Peterson, Jr. to INDSPEC Chemical Corp.).
86. Jpn. Kokai Tokkyo Koho JP 61,228,039 [86,228,039] (Oct. 11, 1986; K. Ono, T. Saito, M. Kitayama, and Y. Fukura to Bridgestone Corp.); *C.A., 108*, 113979 (1987).
87. Jpn. Kokai Tokkyo Koho JP 62,179,581 [87,179,581] (Aug. 6, 1987; T. Murachi and S. Nakane to Toyoda Gosei Co., Ltd.); *C.A., 108*, 151756 (1987).
88. Jpn. Kokai Tokkyo Koho JP 62,014,869 [87,014,869] (Jan. 23, 1987; K. Kageyama, M. Machida, and Y. Inomata to Yokohama Rubber Co., Ltd.); *C.A., 107*, 116796 (1987).
89. Eur. Pat. Appl. EP 285,094 (Oct. 5, 1988; to Nippon Zeon K.K.); *C.A., 110*, 25155 (1989).
90. Jpn. Kokai Tokkyo Koho JP 02,186,145 [90,186,145] (July 20, 1990; to Habasit A.G.); *C.A., 113*, 192959 (1990).

91. Jpn. Kokai Tokkyo Koho JP 02,150,528 [90,150,528] (June 8, 1990; A. Ogura, T. Sato, and M. Sato to Toray Industries, Inc.); *C.A., 113*, 158178 (1990).
92. Jpn. Kokai Tokkyo Koho JP 03,177,443 [91,177,443] (Aug. 1, 1991; H. Araki, M. Nakamura, T. Koyama, and H. Iwatsubo to Sumitomo Seika Chemicals Co., Ltd.); *C.A., 115*, 281854 (1991).
93. Jpn. Kokai Tokkyo Koho JP 03,045,607 [91,045,607] (Feb. 27, 1991; T. Ueda and S. Ueki to Kurashiki Kako K.K.); *C.A., 115*, 116160 (1991).
94. Jpn. Kokai Tokkyo Koho JP 03,294,587 [91,294,587] (Dec. 25, 1991; S. Takahashi and S. Takasugi to Yokohama Rubber Co., Ltd.); *C.A., 116*, 216171 (1992).
95. Jpn. Kokai Tokkyo Koho JP 03,065,536 [91,065,536] (March 20, 1991; M. Sekiguchi and A. Okamura to Nippon Glass Fiber Co., Ltd.); *C.A., 115*, 164872 (1991).
96. Can. Pat. Appl. CA 2,024,973 (March 11, 1992; to A. Fehr); *C.A., 117*, 152595 (1992).
97. Jpn. Kokai Tokkyo Koho JP 04,050,237 [92,050,237] (Feb. 19, 1992; K. Hatsutori and C. Ito to Asahi Fiberglass K.K.); *C.A., 117*, 28591 (1992).
98. Jpn. Kokai Tokkyo Koho JP 04,057,833 [92,057,833] (Feb. 25, 1992; A. Okamura and M. Sekiguchi to Nippon Garasu Sen-i K.K.); *C.A., 117*, 50515 (1992); see also JP 04,057,839 [92,057,839], same authors and assignee, for a similar composition.
99. Jpn. Kokai Tokkyo Koho JP 04,103,634 [92,103,634] (Apr. 6, 1992; N. Saito, H. Tezuka, T. Usui, N. Achinami, and K. Hattori to Honda Motor Co., Ltd; Yamashita Gomu K.K.); *C.A., 117*, 113356 (1992).
100. Jpn. Kokai Tokkyo Koho JP 04,168,181 [92,168,181] (June 16, 1992; K. Terajima and M. Ichikawa to Toyoda Gosei Co., Ltd.); *C.A., 117*, 193371 (1992).
101. K. B. Piotrovskii and K. Yu. Salnis, (eds.), *Auxiliary Substances For Polymeric Materials*, Vol. 2 (translated by R. J. Moseley), Rubber and Plastic Research Association of Great Britain, 1967.
102. U.S. Pat. 4,069,220 (Jan. 17, 1978; H. P. Orem, D. R. Hart, and J. E. Hill to Jim Walter Resources, Inc.).
103. Eur. Pat. Appl. EP 210,791 (Feb. 4, 1987; H. Dressler to Koppers Co., Inc.); *C.A., 106*, 119447 (1987).
104. Z. V. Shchichko, D. B. Boguslavskii, Yu. Y. Pivovarova, N. M. Litvin, G. A. Blokh, and Kh. N. Borodushkina, *Kautch. Rezina (Engl. Transl.), 25(1)*, 18 (1966).
105. USSR Pat. SU 478,836 (Apr. 15, 1974; to D. B. Boguslavskii).
106. Brit. Pat. GB 1,527,992 (Oct. 11, 1978; to I. K. Alexeeva and 13 other inventors).
107. A. S. Yarmolenko, A. G. Shvarts, and N. A. Zhovner, *Int. Polymer Sci. Technol., 10 (12)*, T/94 (1983).
108. USSR Pat. SU 981,324 (Dec. 15, 1982; A. A. Deletorskii *et al.* to Moscow Institute of Fine Chemical Technology; Scientific-Research Institute of the Rubber Industry, Scientifc-Research Institute of the Tire Industry); *C.A., 98*, 180888 (1983).
109. USSR Pat. SU 1,182,038 (Sept. 16, 1983; A. A. Orchinnikov, L. A. Protiven, V. I. Khlybov, and L. A. Klimenkov to Research Institute of Chemistry for Polymer Materials); *C.A., 105*, 208597 (1986).
110. USSR SU 1,564,158 (May 15, 1990; A. V. Steblev, L. D. Eremina, Z. K. Nikitina, and A. N. Radaev to the Scientific-Research Institute of the Tire Industry); *C.A., 113*, 154147 (1990).
111. USSR Pat. SU 1,645,274 (Apr. 30, 1991; to L. S. Petrovskaya and N. N. Shatunova); *C.A., 116*, 108105 (1991).
112. W. K. Cunningham and D. R. Hart, *J. Elastomers Plastics, 7*, 380 (1975).
113. U.S. Pat. 3,538,972 (Nov. 10, 1970; P. A. Yorick and C. T. Bills to Ashland Oil and Refining Co.).
114. Fr. Demande 2,011,598 (March 6, 1970; E. F. Cluff and J. J. Verbanc to D. I. DuPont DeNemours and Co.); *C.A., 73*, 99837 (1970).
115. U.S. Pat. 3,705,868 (Dec. 12, 1972; T. Honda, E. Kubota, Y. Fukuura, and Y. Kobiyashi to Bridgestone Tire Co., Ltd.).
116. U.S. Pat. 3,723,239 (March 27, 1973; S. Takemura, E. Yoshino, T. Honda, E. Kubota, and Y. Fukuura to Bridgestone Tire Co., Ltd.); similar is Jpn. Kokai Tokkyo Koho JP 88,012,503 (March 19, 1988; to Bridgestone Tire Co.).
117. Jpn. Kokai Tokkyo Koho JP 076,112,857 (Oct. 5, 1976; K. Hirosha and H. Fujita to Dainichi-Nippon Cables, Ltd.); *C.A., 86*, 56576 (1977).

118. Jpn. Kokai Tokkyo Koho JP 058,122,941 (July 21, 1983; to Sumitomo Chemical K.K.); *C.A., 99*, 196454 (1983).
119. Jpn. Kokai Tokkyo Koho JP 59,228,080 [84,228,080] (Dec. 21, 1984; to Teijin Ltd.); *C.A., 102*, 168205 (1985).
120. K. Hultzsch, *Chemie der Phenolharze,* Springer-Verlag, Berlin, 1956.
121. S. van der Meer, *Recl. Trav. Chim. Pays-Bas, 61*, 898 (1942).
122. A. Giller, *Kautschuk, Gummi, Kunstst., 19*, 188 (1966).

5

Resorcinol/Formaldehyde Resins—Adhesives for Wood, and other Nonrubber Applications

5.1. INTRODUCTION

This chapter reviews and presents a perspective of the special qualities of the resorcinol/formaldehyde resins as adhesives for wood and in a variety of other, commercial or potential uses, with the exception of their use as reinforcing resins and bonding agents in vulcanizable rubber compounds which was dealt with separately in Chapter 4.

Good prior overviews of this subject matter were given in the following references: The production and general uses of phenolic resins up to the mid-1960s was presented in detail, in German, in *Kunststoff-Handbuch.*[1] Phenolic resins as adhesives for wood were discussed, with references up to about 1980, in Pizzi's treatise.[2] A valuable discussion of the chemistry and applications of phenolic resins, with references up to about 1984, was given in a book by Knop and Pilato.[3] The use of resorcinol/formaldehyde resins as adhesives was well reviewed, with references up to 1969, by Moult.[4]

5.2. ADHESIVES FOR WOOD

5.2.1. Overview

Although about 1.7 billion pounds of phenol/formaldehyde (P/F) resins were used in the U.S. in 1986 as adhesives (as a part of all adhesives for wood) mainly in the manufacture of plywood, oriented strand board, and wafer board, only about 8 million pounds of phenol/resorcinol/formaldehyde (P/R/F) resins and resorcinol/formaldehyde (R/F) resins were used that year for the manufacture of laminated lumber beams and finger-jointed products.[5] The reason for the relatively modest use of resorcinol-containing resins as adhesives for wood is their much higher cost (5–15 times higher) than the P/F resins. It is the unique performance of the P/R/F and R/F resins in specialized applications that accounts for their continued use as adhesives for wood and their use/potential in many other applications.

Resorcinol reacts with formaldehyde in aqueous solution about 12 times faster than phenol. The reaction, at a ratio of 0.67 m F/1.0 m R, occurs at a good rate even for neutral pH, is accelerated at alkaline conditions, shows a minimum at pH 3.5, with the reaction rate accelerating again at lower pH. The reaction at the 4- and 6-position of the resorcinol ring with formaldehyde has been shown to be about 5 times faster than the reaction at the 2-position on the resorcinol ring.[6] The reaction of resorcinol with formaldehyde in water takes place at a relatively fast rate at temperatures as low as 10–20 °C. These features—fast reaction of resorcinol (and precondensates) with formaldehyde even at slightly alkaline pH and at relatively low temperature—are part of the reasons for the commercial use of resorcinol/formaldehyde resins as adhesives for wood.

The actual adhesives for wood consist of two parts: the novolak-type resin solution and the powdery hardener.

The resorcinol/formaldehyde resins are prepared at a 1.0 R/0.5–0.7 F mole ratio to make a water-soluble product; if more formaldehyde were used, gelatinous or insoluble resin would begin to form before all the resorcinol had reacted. The larger use of resorcinol for wood adhesives is in making phenol/resorcinol/formaldehyde coresins to lower the cost. A low molecular weight resole is first prepared by reacting phenol with excess formaldehyde at alkaline pH, and the methylol groups contained in the resole are then reacted with resorcinol. To retain the most desirable features of the R/F resins, the P/R/F coresins usually contain at least 50 wt% resorcinol. These resins, after blending with the hardener, require intermediate temperatures (around 50 °C) for full cure. The high-solids resin solutions are usually prepared/modified by the addition of alcohols (such as ethanol and/or ethylene glycol) to temper the exothermic resinification and to modify the solubility, dilutability, viscosity, solids content, rate of reaction with the hardener, pot life of the glue mix, and the wettability and penetration of the surface of the wood. Other additives may be used, e.g., to modify the thixotropic behavior of the liquid resin.

The hardener part of the adhesive is a blend of paraformaldehyde (polyoxymethylene of controllable molecular weight to attain a desirable, selected reactivity/gel time) of suitable particle size with a filler/extender such as walnut shell, coconut shell, or tree bark flour which also provides gap-filling properties to the adhesive after the resin solution and the hardener are mixed.

In addition to their high reactivity at neutral or alkaline pH (at acid pH the wood can be degraded) even at low temperatures, the attractiveness of the R/F or P/R/F resin-hardener blended adhesives for wood includes high bond strength, (i.e., high wood failure is observed in the tests—the wood fails before the adhesive bond), creep resistance, resistance of the bond to high humidity and water, the ability to maintain bond strength under highly cyclic conditions of low and high temperatures/humidity/loads, and the excellent long-term durability of the bonds.

In contrast, the P/F resin adhesives (without added resorcinol) require an acid pH and elevated temperatures (*ca* 90 °C)/longer time for full cure. Thus, the resorcinol resins and coresins not only show superior adhesive quality and performance, they also permit the bonding of larger parts at desirably lower temperatures, as well as time (faster cure) and energy savings.

The resorcinol resin adhesives first rose to importance in the 1940s during World War II. In that period laminated wood panels were produced for use in airplane construction, laminated wood propellers were used on airplanes and boats, and wooden-hulled ships were constructed for use as submarine hunters and mine sweepers. Indeed, wooden minesweepers, with their nonmagnetic hulls, are still being constructed in the U.S.—see Figs. 5.1 and 5.2—and in other countries, as well as pleasure craft with wood structures of various kinds. After World War II, the resorcinol adhesives gained prominence in the construction of laminated wooden beams, arches, and other structural elements. These laminates were initially preferred for exterior

Figure 5.1. Construction of a wooden hull for a minesweeper. Copyright Peterson Builders, Inc., published here with permission.

Figure 5.2. The finished, wooden-hulled minesweeper. Copyright Peterson Builders, Inc., published here with permission.

construction, but became widely used in interior applications because of their proven reliability and durability. Architects produced a variety of attractive designs that took advantage not only of the physical properties of these materials but also expressed the warmth, esthetic appeal, and contemporary feeling made possible with these laminated wood products. Recent examples of applications of R/F resin laminated wood construction are shown in Fig. 5.3 (I-beams), Fig. 5.4 (bridge), Fig. 5.5. (church ceiling).

Wood is a complex material, differing in chemical makeup and structure/properties from species to species and from tree to tree within the same species. As a consequence, in developing an adhesive for a particular application, the moisture content of the wood (optimal about 8–12% when resorcinol adhesives are used), hardwood or softwood type, heartwood or summerwood, preparation of the cut surface, and many other factors have to be considered. Preservative-treated wood requires its own type of adhesive. Many of these complexities have been well reviewed in a recent publication by River *et al.*[7]

For the development of an adhesive for a particular type of wood and type of operation, the product has to be tailored for the application process: hand spread by

Figure 5.3. Laminated, wooden I-beams. Copyright Trust Joist Corp., published here with permission.

Figure 5.4. Glen Echo Park Bridge, Glen Echo, Md. Copyright Sentinel Structures, Inc., published here with permission.

Figure 5.5. Roof of St. James Lutheran Church, Lake Forest, IL. Copyright Sentinel Structures, Inc., published here with permission.

trowels or brushes, automatic pumping or metering, extrusion or roller application. For curing by heat (steam, hot air, or RF = radio frequency), the open assembly life and clamped assembly time must be optimized. The strength and durability of the bond have to be tested. For instance, a cyclic delamination test for water resistance is done in a pressure vessel. Environmental and health considerations must be taken into account for the use of both the adhesive and the cleanup.

Resorcinol adhesives have been developed for the do-it-yourself market; e.g., Weldwood Waterproof® glue is a two-component adhesive of this type.[8]

Less common applications may arise and require additional modification of the adhesives, such as the development of adhesives for insulation panels used in LNG (liquid natural gas) carrier ships, or the tailoring of an adhesive for bonding a fire-retardant, plastic laminate to fire-retardant particle board.

Indeed, the same type of resorcinol adhesive can be used to bond cotton, rayon, nylon, phenolic resins, melamine resins, urea/formaldehyde resins, epoxy resins, and polyurethanes to each other; polyesters and metals have to be primed before bonding with a resorcinol adhesive. Plastics with a lower polarity of their surface, such as polystyrene, polyolefins, polyvinyl chloride, and Teflon, are not readily adhered with R/F adhesives.

The glue line of resorcinol adhesives is of dark color, an undesirable feature in some applications. Resin spills may stain the wood. Thus, care is necessary in using R/F-type adhesives.

5.2.2. Selections from the Literature

The following brief outline of selected references, mostly patents, from the 1970s to 1993 brings into focus the attempts that were and are being made to utilize resorcinol for developing improved products. Included are some references on the improvement of low-cost adhesives (Urea/F and P/F resins) by the addition of resorcinol or resorcinol resins, on ways to speed up the cure of phenolic adhesives by the addition of *m*-aminophenol, and on the use of polyphenolic plant extracts as coreactive extenders, with potential to lower the cost of the adhesive.

Lower ski plates/edges were prepared by impregnating/laminating wood with an aqueous ethanol solution of resorcinol and dimethylolcarbamide in 1.0/0.4–1.5 mole ratio in the presence of ≤1.5 mole *N*-methylolcaprolactam, drying at below 60 °C, and curing at 80–120 °C with clamping of the assembly. The laminates had high wear and water resistance, and were flexible.[9]

A paper published in 1973 by Tiedeman and co-workers reported the modification of P/F resoles with *m*-aminophenol to produce water-soluble resins. These modified resins reacted rapidly with formaldehyde, e.g., with 55% methyl formcel (55% formaldehyde, 35% methanol, 10% water) at room temperature. The gel time was *ca* 30 sec. This resin system performed well in bonding wood.[10]

Much work was undertaken at Weyerhauser Co. in the late 1960s/early 1970s on very fast curing *m*-aminophenol (MAP; several other arylamines were studied as well) modified adhesive systems and practical application methods for these systems. Many of the neat adhesive formulations gelled so quickly at ambient temperature that wood penetration was poor and pot life was too short. Their several patents for the bonding of wood using these systems involved, e.g., applying a P/R/F resin with excess aldehyde to one of the surfaces to be bonded and an adhesive made by reacting a phenolic resole (containing reactive methylol groups) with MAP to the other surface, then clamping the pieces together; or, the phenolic resole/MAP reaction product was blended with thickened 55% methyl formcel, applied to both surfaces, and the pieces were pressed together. MAP (arylamine) modified amino resins and ketone-aldehyde resins were also covered.After cure for several minutes, strong and waterproof bonds were obtained.[11] These attractive systems appear not to have been commercialized, doubtless one of the reasons being that the price-conscious wood product industry could not accommodate the higher cost of these modified adhesives.

Returning to developments with resorcinol resins, novel P/R/F adhesives of low free resorcinol content were prepared by reacting 60–70 m of phenol and 60–70 m of aqueous formaldehyde at about 60–100 °C and pH 8–10, then with 30–40 m of resorcinol, followed by devolatilization of the mixture to remove water and unreacted phenol, finally adding water back, adding sufficient formaldehyde to react all the free resorcinol, and adjusting to pH 9 to obtain a clear resin solution.[12]

Cellulosic pulp sludges, obtained as a waste material in the chemical pulping of wood, were used as an extender for P/R/F resins and used to produce nonsagging adhesives for wood.[13]

R/F resins for use as wood adhesives were modified with methyl-, ethyl-, or carboxymethylcellulose to reduce internal stresses and shrinkage of the final adhesive mix.[14]

A P/F resin used as a binder for composition (particle) board was modified by adding 2–4% R/F resin to accelerate the cure.[15]

A rapidly curing adhesive was prepared from an alkylresorcinol (shale-oil derived)/formaldehyde resin mixed with Portland cement and paraformaldehyde.[16]

An adhesive was prepared which contained 100 parts urea/formaldehyde resin, 10 parts 2,4,6-tris-(2,4-dihydroxybenzyl)resorcinol, 14 parts wood flour, and 1 part ammonium chloride; this adhesive showed good resistance to boiling water while the same adhesive made without the resorcinol derivative failed this test.[17]

A phenolic resole containing at least 65% monomeric methylolphenols and free formaldehyde was reacted with resorcinol to eliminate essentially all free formaldehyde; this liquid, water-dilutable resin was used to make adhesives for finger-jointing or laminating wood.[18]

An acetone/formaldehyde resin containing free formaldehyde was reacted with resorcinol in water until all the free formaldehyde was consumed. A liquid resin was obtained which had good storage stability and reacted quickly with hardener at room temperature to form adhesives for wood.[19]

R/F resins or P/R/F resins were modified with low-cost, water-soluble proteins such as casein, soybean powder, blood powder, or gelatin to give resins suitable for gluing wood articles such as athletic goods, boats, or the outside walls of wood houses.[20]

Urea/formaldehyde (U/F) resins were reacted with resorcinol to make resins useful as wood adhesives which were curable at neutral or alkaline pH with formaldehyde.[21]

A 1/2 P/F mole ratio resin was mixed with 1–5 parts resorcinol, 1–5 parts formaldehyde, and 1–5 parts sodium carbonate per 100 parts resin.This mix cured rapidly (11 min/110 °C) and was highly extendible with inexpensive fillers without impairing the cure rate and bond strength. This adhesive was used, e.g., in making plywood.[22]

A P/F resole resin was modified with polyvinyl alcohol (PVA), thermally reacted at alkaline pH, and then blended with a R/F resin. This modified resin had the same water resistance, weathering resistance, and durability as a conventional R/F resin. It was used to prepare adhesives which had improved substrate coating properties, good penetration of the wood surface, and showed flexibility of the adhesive layer.[23]

A R/F resin adhesive for wood was prepared by blending the resin with a hardener comprising paraformaldehyde and a ketone (e.g., acetone)/formaldehyde precondensate as a suspending agent; there was no dusting problem. The precondensate resin participated in the cure of the system and strong bonds were formed.[24]

Aqueous R/F or P/R/F resins were blended with 5–60 wt% (based on the resin) of dried tea (fresh or waste) powder and mixed with paraformaldehyde, optionally containing extenders such as talc or walnut shell powder, to prepare an adhesive suitable for bonding wood with a high water content (>20%).[25]

A mixture of 1/1 (by weight) of resorcinol and a polyphenolic extract from the bark of Siberian larch was blended with 20 phr paraformaldehyde and 12 phr coconut shell flour, mixed with water to give a mixture containing 65% solids (pH 6.0) which was coated on lauan boards and cured with clamping for 5 min/80 °C to give a laminate with high water resistance and high wood failure.[26]

Plywood was spread with a mixture of 100 parts melamine/urea/formaldehyde resin, 10 parts resorcinol, 10 parts wheat flour, and 0.5 part ammonium chloride, covered with polyester fabric and press-cured for 1 min/115 °C to give a laminate with improved surface hardness and weatherability.[27]

Birch board was bonded by applying a urea resin contact adhesive (Insta-Bond R-101-A) to one board and a hardener containing 6 parts resorcinol/100 parts 1 N HCl to another board. After press-cure for 10 min/25 °C, the laminate was water resistant and durable.[28]

Formaldehyde emissions during laminating wood with amino resins were reduced by first treating the wood with ammonium borate solution, joining the boards with a resorcinol-modified adhesive, then press-curing for 30 min/120 °C. This laminate had a formaldehyde emission of 0.16 ppm vs 14.2 ppm for laminate made with the unmodified adhesive.[29]

Fast-curing laminating adhesives for wood were prepared by making a P/F resole using a sodium sulfite catalyst (pH 8.5–9.2), advancing the resin to a specified viscosity, reacting the resole with resorcinol, diluting the resin with ethylene glycol/ethanol/water to make a resin solution of 40–60 wt% solids content and 2.0–3.5 poise viscosity, and adding 4–8 wt% potassium carbonate and fumed silica (Aerosil 200). This resin was blended with paraformaldehyde/walnut shell flour to prepare the adhesive for laminating wood.[30]

A fast-curing adhesive was obtained by adding 38 wt% sodium nitrate to a P/R/F resin (56% solids content) which was applied to one board, while a hardener mix containing paraformaldehyde, aqueous formaldehyde, a filler, and potassium carbonate was applied to the other board. The joined boards were press-cured at room temperature and could not be separated after 5 min vs after 40 min when the same mix without the sodium nitrate was used. The function of the nitrate was said to be water removal via hydrate formation.[31]

Fire-retardant plywood was made by adhering fire-retardant treated veneers (lauan) with a mixture of 100 parts melamine/phenol/formaldehyde copolymer, 15 parts wheat flour, 5 parts water, and 5 parts R/F resin. The assembly was press-cured for 3 min/120 °C to give strong adhesive bonds, much stronger than when a M/P/F resin alone was used.[32]

An odorless, room-temperature-curing adhesive was prepared by blending 100 parts partially hydrolyzed/acetoacetylated poly(vinyl acetate) emulsion, 10–50 parts R/F resin or P/R/F coresin, and 20–30 parts hexamethylenetetramine. The blend gave excellent initial adhesion and was resistant to water and boiling water.[33]

An adhesive having good bond strength (100% wood failure) and water resistance was prepared by refluxing a solution of 60.5 g phenol, 24.1 g water, 6.2 g methanol, and 25.3 g formaldehyde for 2 hr/90–92 °C, then adding 3.9 g resorcinol, refluxing 2 hr, followed by adding 24.3 g resorcinol and refluxing for 1 hr. This resin, which had a reduced free resorcinol content vs the P/R/F resins of the prior art, was diluted to 37.7% solids content, adjusted to pH 8.2, and mixed with 2.7 g paraformaldehyde/1.6 g wood flour.This adhesive was used to bond beechwood strips and cured at 90 °C; very good bonding and resistance to boiling water was attained.[34]

An adhesive was developed for bonding high-density eucalyptus wood. To a P/F resole (Chemplex 775) was added 3 parts novolak resin (GC 1499), 0.4 part *m*-aminophenol, fillers, alkali, and water. This adhesive had a fast cure time and controlled penetration suitable for making plywood from veneers.[35]

A sprayable adhesive was prepared by reacting 440 parts resorcinol, 170 parts 37% formaldehyde, and 160 parts ethanol for 6 hr/65–70 °C, adding 2 parts powdered silica, then mixing with a hardener consisting of 10 parts paraformaldehyde and 5 parts wood flour.This adhesive was used for making 5-ply laminated wooden bowls which had good bond strength and water resistance.[36]

An adhesive for application by extrusion was prepared by reacting a solution of 440 parts resorcinol, 170 parts 37% formaldehyde, and 160 parts ethanol, with an adjustment to pH 5.6–6.0, for 6 hr at 65–70 °C. To 100 parts of this resin solution was added 2 parts powdered silica and 10 parts paraformaldehyde to give a composition with 40 min gel time at 40 °C; mixing with another 10 parts paraformaldehyde/5 parts wood flour gave an adhesive with a viscosity suitable for application by an extruder. Hemlock boards (5-ply) were bonded with this adhesive with 60 min open assembly time at 30 °C and press-cured at 40 °C for 24 hr to give a laminate which showed 93% wood failure, good water resistance, and good performance in the boil test, claimed to be much better than a commercial R/F resin.[37]

A two-component adhesive for automatic mixing/application consisted of a R/F resin prepared with added ethanol/glycerine using toluenesulfonic acid as the catalyst and blended with a P/F resole containing a small amount of hydroxyethylcellulose and coconut shell flour. Douglas fir bonded with this adhesive showed better bond strength and resistance to boiling water than when powdered paraformaldehyde/coconut shell flower was used as the hardener.[38]

Fire-resistant wooden doors were manufactured by vacuum-impregnating boards of wood with an ammonium polyphosphate solution, then laminating the boards in a staggered pattern with a R/F resin adhesive, and assembling the doors with a decorative frame. These doors did not burn through when contacted for >60 min with a flame.[39]

Unseasoned wood pieces could be bonded with a relatively high viscosity (7000 cP) R/F resin; the moisture in the wood did not inhibit the formation of strong bonds after press-cure/drying.[40]

A modified U/F resin adhesive was prepared consisting of 100 parts U/F resin, 10 parts tris-(2,4,6-dihydroxybenzyl) resorcinol, 14 parts wheat flour, and 1 part ammonium chloride. Plywood was made by spreading the plies with this adhesive and press-curing for 80 sec at 125 °C. The specimen had good bond strength and was resistant to boiling water. Plywood made with the same adhesive but without the added resorcinol derivative failed in these tests.[41]

Due to the "polyphenolic" (flavanoids, tannins) nature of certain plant extracts available from, e.g., mangrove, pine, and wattle (mimosa in South Africa, acacia in Australia) trees, there has been longstanding interest in using such materials in adhesives. Indeed, a commercial use of such adhesives appears to have occurred.[42] During the oil crisis in the 1970s, interest in such plant extracts by the institutional and industrial communities increased because they were available from renewable resources and believed to be low cost. Research on the use of such polyphenolics has extended to the present. Often, resorcinol or resorcinol resins are added to make viable products. A few pertinent references follow.

Useful adhesives for wood were prepared from the condensed flavanoid tannins obtained from the bark extract of the black wattle; modifications with P/R/F resins were described.[43] Similar types of adhesives were developed for particle board binders.[44,45]

Cold-set tannin/R/F adhesives for use with structural wood were prepared, with the added resorcinol at levels as low as 5%.[46]

Laminated, wooden beams were prepared using wattle tannin adhesives modified with small amounts of phenol and resorcinol. These beams had to be heated during the curing stage and proper humidity had to be maintained to keep the timber from drying out, in order to obtain good bonding. The cost of laminating at higher temperature was said to be compensated for by the lower cost of the adhesive.[47]

In Chile, commercially produced pine-tannin-extract adhesives were modified with P/R/F coresins to make fast-setting ("honeymoon") adhesives for the fingerjointing of timbers.[48] The use of apitong (*Dipterocarpus spp.*) extractives modified with R/F resins as adhesives for resinous and nonresinous timbers was investigated.[49]

5.3. FIRE-RETARDANT, FIBER-REINFORCED COMPOSITES

5.3.1. Introduction and Selections from the Literature

Filled and unfilled phenolic molding compounds, laminates, and composites have been known and used for a long time. Their use was based on their excellent dimensional stability, high thermal resistance, good physical properties, and, often, the low cost of finished parts. A high-technology example of the use of phenolic resin composites was as an ablative material for the nose cones of space vehicles.[1,3] In recent decades the use of fiber-reinforced phenolic composites began to be aimed at applications where their fire retardance and low smoke generation in fires was of prime importance, driven by increasingly stringent governmental regulations of organic materials/composites used in transportation and building applications. Also, such resin systems were developed for use by a variety of application techniques, including hand layup, filament winding, pultrusion, prepreg, resin transfer molding (RTM), and liquid injection molding (LIM) techniques. The use of resorcinol began to be considered for systems of coresins with P/F resoles, and/or with furfural or furfuryl alcohol, enabling cures with base or acid catalysts at lower temperatures and/or shorter cure time. The following selections from the literature concern only the advances made with resorcinol-containing systems.

A tough, fire-resistant resin for use in making glass-fiber-reinforced composites comprised a mixture of resorcinol with a P/F resole or a partially cured P/R/F resin and paraformaldehyde plus furfural or furfuryl alcohol. These resins cured at room temperature, had good resistance to embrittlement and cracking, low flame spread, low smoke evolution, good heat release, and a long shelf life. The fiber glass composites could be used to make fire doors, aircraft or building panels, electrical conduits, piping, and coatings.[50]

Fire-retardant resins which were nonpunking (nonsmoldering, nondripping) when exposed to flame consisted of a mixture of two phenolic polymer solutions with at least a boric acid catalyst, one or both of the prepolymer solutions being an acid-catalyzed R/furfural (excess) resin, or one of the prepolymers being a mixture of mostly unreacted resorcinol/furfural (slight excess).[51] Similar fire-retardant materials were based on cresol/xylenol and/or resorcinol resins with furfural or mixtures of furfural and paraformaldehyde, using an acid catalyst or a polyamine (Versamid) curing agent reactive with either one or both the aldehyde and the phenolic component. These materials could be reinforced with carbon fibers, graphite, silica, metal oxides, or silicates.[52]

Resins for making fiber-reinforced composites consisted of a P/F resole to which resorcinol was added at 80 °C and, after cooling, nonaethylene glycol diacrylate was added. A resin of this type had improved mechanical strength and contained ("only") 2.9% free phenol and 0.8% free formaldehyde.[53]

A flame-retardant material was prepared by first reacting one mole of resorcinol with less than one mole of furfural at alkaline pH and mixing the resin with a P/F resole and zinc borate as a curing catalyst. These resins could be reinforced with fiber glass, carbon fiber, or organic fibers to prepare flame-retardant composites.[54]

Thermosetting resins, suitable for liquid injection molding techniques such as LIM, RIM (reaction injection molding), RTM, and SMC (sheet molding compounds) and for pultrusion, consisted of a high-solids P/F resole or a R/F resole and/or a P/R/F novolak, and resorcinol as a "compatibilizer," phenolsulfonic acid as a reactive catalyst, and a polyalkylene glycol. These resins had reduced viscosity, improved clarity and strength, formed less bubbles and voids, and were suitable for the preparation of (glass) fiber-reinforced composites.[55]

A resin suitable for making copper-clad prepreg laminates consisted of a varnish A [made by reacting 60 g 3-HOC_6H_5-$OP(OPh)_2$, 100 g phenol, 40 g nonylphenol, 140 g 37% formaldehyde, and 5 g triethylamine at reflux for 2 hr, then diluting with methanol to 50% solids] and a varnish B (a resin made from 200 g phenol reacted with 100 g tung oil in the presence of *p*-toluenesulfonic acid catalyst, then refluxed for 2 hr with 250 g benzoguanamine, 120 g phenol, 460 g 37% formaldehyde, and 12 g dimethylamine, and finally diluted with triphenyl phosphate to 50% solids). Kraft papers were impregnated with solutions A, coated with solution B, devolatilized, layered, and pressed with copper foil for 1.25 hr at 170 °C to give laminates with excellent flame retardancy (UL-94 test rating V-0), good moisture and heat resistance, good punchability, and dimensional stability.[56]

Improved resorcinol-modified phenolic resin binders for fiber-reinforced plastics, excluding paraformaldehyde as a curing agent and requiring little or no nonreactive solvent to achieve appropriate viscosity, consisted of blends of a P/F resole resin, furfural (reactive solvent), resorcinol, and/or a R/F novolak resin, fillers such as aluminum trihydrate (ATH), kaolin, or talc, if desired, alkaline catalyst (magnesium hydroxide, sodium hydroxide), and a silane bonding agent. This mix was applied to glass fiber or fabric (e.g., chopped strand mat) and molded/press-cured at 25 °C to elevated temperatures (e.g., 85 °C/10 min) to give a fire-resistant composite with excellent mechanical properties even at high temperatures, with low smoke/toxic emission in fire tests.[57]

Resins with good pot life and impact strength for use in sheet molding compounds were prepared, e.g., by blending 60 parts P/F resole resin containing omega-amino-caprolactam and a methoxymethylated polyamide, 30 parts R/F novolak resin, 5 parts propylene carbonate, 10 parts talc, 1 part stearic acid, and 4 parts 65% phenolsulfonic acid. This blend was coated on a polyethylene film, overlaid with a glass fiber mat, coated with the resin blend, topped with polyethylene film, and press-rolled to give a sheet molding compound which was stable for 15 days.[58]

Finally, a polyester fiber-reinforced, low-temperature formable, structural material was developed. A porous material of polyester fiber was saturated with a solution of a P/F resole and resorcinol in 9/1 ratio, dried at 80 °C/20 min, and hot-pressed at 150 °C/1 min to give a fully cured, shaped composite.[59]

Table 5.1. FIRE $PRF_2^{®}$ 1000 Series Resins

Resin	Pot life (minutes @ 75 F)	Mix viscosity (Poise @ 23 °C)	Uses
1000 FM	9–12	15	Hand lay-up; spray-up; RTM
1030 FM	25–30	10	Filament winding; impregnation; RTM; complex hand lay-up
1060 FM	30–45	7	Impregnation; large RTM; filament winding; postcure recommended

Table 5.2. Specifications for the Components of FIRE PRF_2®

Specifications	Component A	Component B
Total solids, Abderhalden	N/A	72–78
Viscosity, cps @ 73 F (23 °C)	50–350	500–3000
pH	3.0–5.0	8.0–10
Mix ratio	By wt%	By volume
FIRE PRF_2 1000, Comp. A	46	4 parts
FIRE PRF_2 1000, Comp. B	34	3 parts
Filler, $BaSO_4$	20	1 part
or alumina hydrate	20	2 parts

Table 5.3. Examples of Release and Cure Schedule of FIRE PRF_2 Laminates

Temperature	Release[a]	Cure
61 F (16 °C)	20–24 hr	1 month
73 F (23 °C)	16–20 hr	2 weeks
90 F (32 °C)	9–12 hr	11 days
131 F (55 °C)	1–3 hr	16–24 hr
170 F (77 °C)	0.5–2 hr	12–16 hr
194 F (90 °C)	0.5 h	4–6 hr
266 F (130 °C)	0.25 hr	1 hr

[a]Release times listed are for 3-mm-thick, flat sheets. Thicker or complex parts may require longer times for release.

Table 5.4. Typical Physical Properties of FIRE PRF_2 1000 Laminates[a]

	Glass content	
Test procedure	25 wt%	35 wt%
Flexural strength, psi-ASTM D790	23,800	30,000
Flexural modulus, psi-ASTM D 790	8.6×10^5	1.1×10^6
Tensile strength, psi-ASTM 638	10,700	16,000
Tensile modulus, psi-ASTM D 638	1.0×10^6	1.1×10^6
Tensile elongation, %-ASTM D 638	1.4	1.0–2.2
Shore D hardness-ASTM D 2240	75–85	75–85
Specific gravity-ASTM D 369	1.7	1.8
Limiting oxygen index-ASTM D 2863	52–65	52–55
Smoke density-ASTM E-84 Tunnel Test	10	10
Flame spread-ASTM E-84 Tunnel Test	10–20	5–15
Maximum specific optical density, corrected flaming mode-ASTM E 662	20	27
Coefficient of linear expansion	22×10^{-6}	—
Izod impact, notched, ft. lb./in./in.	13.6	16.2
Heat deflection temperature 2264 psi, -ASTM D 648	396 F (202 °C)	—

[a]The properties listed are for glass reinforcement made of 1 ply C-glass veil, 2 ply 2 oz. mat, and 1 ply 6 oz. woven roving. Reinforcement content and type, methods of fabrication, and degree of cure may affect properties.

Table 5.5. FIRE PRF_2 Laminates—Fire/Smoke/Toxicity Test Results and Approvals

Factory Mutual 4922 All FIRE PRF2 1000 series resins are "Identified Component Resin/Systems for Use in the Manufacturing of Fume and/or Smoke Exhaust Ducts Without Sprinklers and Dampers"	AFNOR NF-X-70-100 (French Standard for Smoke Emissions and Toxicity) Rated F-1 (smokeless)
ASTM E-84: Class 1 Flame spread 10 Smoke development 10	AFNOR NF P 92-501 (French Standard for Fire Protection) Rated M-1 (fireproof)
Urban Mass Transit Authority Guidelines *ASTM E-162 Flame Spread Index* Specification: <25. Test result: 1 ASTM E-662 Flaming Mode *Specific Optical Density* (of smoke) Specs.: Ds 1.5 Ds 4 <100 <200 Test results: 0 6	*Federal Aviation Regulation* 14 CFR Part 25 25.853 Max. Rate of Heat Release, Spec.: <65 / Rate of Heat Release @ 2 min, Spec.: <65 Test results: 1/16 in. 45.3 36.7 1/ 8 in. 71.3 4.3 3/16 in. 13.6 5.3
New York State Combustion Toxicity Test LC 50 by the UPITT test—86.5 g (better than red oak—20 g)	*Smoke Emissions (ASTM 662)* Specific Optical Density @ 4 min. Spec.: <200 Result: 6

Table 5.6. Typical Characteristic of Resorciphen® 2074-A/2026-B

Specifications	2074-A	2026-B
Total solids, Abderhalden	N/A	78–81
Viscosity, cps @ 73 F (23 °C)	2500–6000	500–3000
pH	3.0–5.0	0.5–2.0
Mix ratio	**Parts by weight**	
Resorciphen 2074-A	74	
Resorciphen 2026-B	26	
Internal release agent	1	
Antifoam DF-210	0.5	
Filler	As required	
General characteristics (typical)	**2074-A**	**2026-B**
Appearance @73 F (23 °C)	Amber liquid	Red liquid
Specific gravity, ASTM D 369, @71.6 F (22 °C)	1.20 (approx.)	1.26 (approx.)
Flash point, Pensky-Martens Closed Cup, ASTM D 93	190 F (87.7 °C)	>212 F (>100 °C)
Total solids in mix	94% (before reaction)	
% Water in mix	6% (before reaction)	
Mix viscosity @ 73 F (23 °C), cps[a]	2500–9000	
pH of mix	3.54.–4.0	

[a]The properties of the mix will depend on the type and amount of filler.

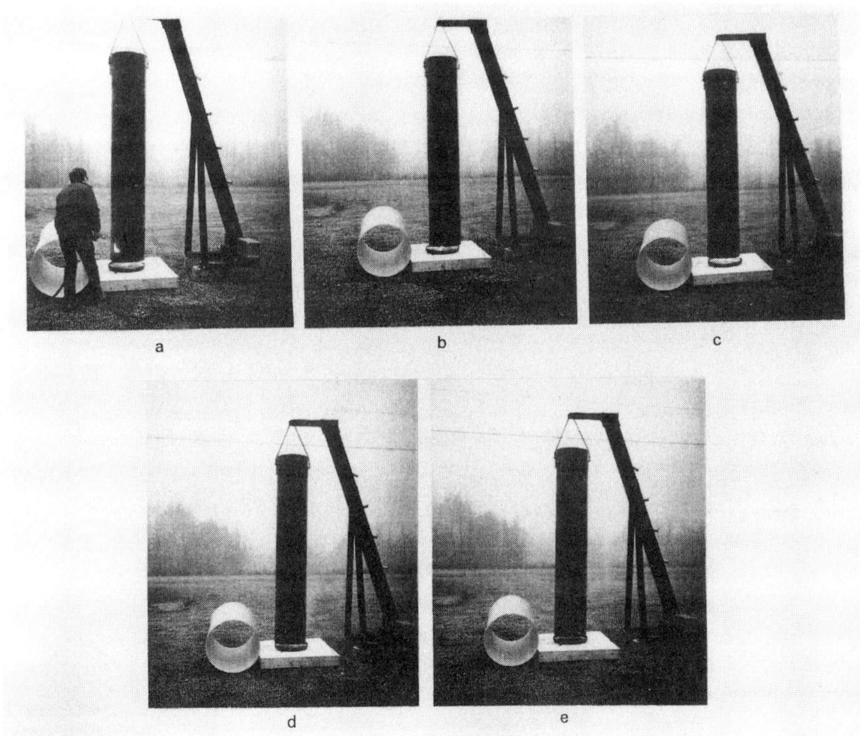

Figure 5.6. Factory Mutual vertical fire test of ductwork made with Fire PRF^2 1000™ resin. a) Start of ignition; b) 30 seconds into the test; c) one minute into the test (the duct is self-extinguished, no smoke emission); d) same duct, after one minute in repeat test; e) after two fire tests, the duct is structurally intact. Copyright of INDSPEC Chemical Corp., published here with permission.

5.3.2. Specific Examples of Applications

The details given here are intended not only to show the excellent performance of the composites but also to give a glimpse of the extensive tailoring of resins and the extensive testing required to support commercial development. All the data in the following tables and figures were obtained from INDSPEC Chemical Corp. and are published here with INDSPEC's permission.

INDSPEC Chemical is offering commercially resorcinol-containing phenolic resin systems for general application in fire-retardant composites with low flame-spread and smoke emission. Georgia Pacific Corp. sells a resorcinol-containing resin for duct-work applications, only. INDSPEC's Fire PRF_2 1000™ series of two-part (phenolic resole and resorcinol and/or resorcinol resin) systems have a six-month shelf life. The three versions of precatalyzed resin available to date differ in pot life and viscosity, for use with various application techniques, as shown in Table 5.1. Further data, including specifications and mix ratios for components A and B, and fillers such

Figure 5.7. Factory Mutual vertical fire test of ductwork made with brominated polyester resin containing 5% Sb_2O_3. a) Start of ignition; b) 30 seconds into the test; c) one minute into the test; d) still burning three minutes into the test; e) four minutes into the test, the duct collapsed. Copyright of INDSPEC Chemical Corp., published here with permission.

Table 5.7. Typical Mechanical Properties of Pultruded Profiles Made With Glass Roving/Resorciphen™ ResinA/B Mix

Roving type	PPG 788
Glass content (wt%)	76–77
Flexural strength (psi)	162,000–175,000
Flexural modulus (psi)	6,500,000–7,000,000
Short beam shear (psi)	6,600
Compressive strength (psi)	93,000
In-plane shear strength (psi)	5,700

Table 5.8. Typical Physical Properties of Pultruded Profiles Made With Glass Mat/Roving and Filled Resorciphen® Resin A/B Mix

	Filler[a]	
Test procedure	Clay	Hydrated alumina
Longitudinal		
Flexural strength ASTM D 790 (psi)	78,500	78,700
Flexural modulus ASTM D 790 (psi)	3,490,000	3,330,000
Transverse		
Flexural strength ASTM D 790 (psi)	13,300	12,700
Flexural modulus ASTM D 790 (psi)	1,300,000	1,170,000
Longitudinal		
Tensile strength ASTM D 638 (psi)	78,000	81,800
Tensile modulus ASTM D 638 (psi)	5,040,000	5,100,000
Transverse		
Tensile strength ASTM D 638 (psi)	4,100	3,950
Tensile modulus ASTM D 638 (psi)	779,000	630,000
Short beam shear ASTM D 2344 (psi)	4,000	4,500
Notched izod impact ASTM D 256 (ft. lb./in.)		
Longitudinal	72	82
Transverse	4.1	4.7
Water absorption ASTM D 570 (%)[b]		
24 hr @ 75 F	2.6	2.5
2 hr at boil	6.0	3.9

[a]62–67 wt% glass, 20 phr filler in resin mix.
[b]Improved water absorption is available with alternate fillers.

as barium sulfate or aluminum trihydrate, ATH, in the Fire PRF_2 1000 system are given in Table 5.2. A typical cure and release schedule for the laminates/composites at various temperatures is presented in Table 5.3, a list of typical physical properties of the composites is given in Table 5.4, and the results of various tests aimed at meeting and/or exceeding various industry and governmental standards for flame resistance, smoke development, and toxicity are grouped together in Table 5.5.

Finally, a pictorial comparison of a fire test at Factory Mutual Laboratories of a glass-reinforced duct made with Fire PRF_2 1000 resin and a similar duct made with a fire-resistant brominated polyester resin containing 5 wt% antimony trioxide is shown in Figs. 5.6 and 5.7. The duct was suspended vertically over a pan containing a standard amount of heptane which was ignited. The duct made with Fire PRF_2 1000 resin self-extinguished 1 minute into the test with no smoke emission; the test was repeated with the same test specimen with the same result and structural integrity was maintained. The duct made with the 5% Sb_2O_3-containing brominated polyester resin emitted heavy black smoke 30 seconds into the test, began to burn 1 minute into the test, and continued to burn with collapse in flames 4 minutes into the test.

Table 5.9. Fire, Smoke, and Smoke Toxicity Test Data and Electrical Properties of Profiles Made With Glass Mat/Roving and Resorciphen® Resin A/B

		Filler	
Test procedure		Clay	Hydrated alumina
Limiting Oxygen Index ASTM D 2863 (%)		>55	>55
NBS Smoke Chamber ASTM E 662			
Flaming mode			
D 1.5		0	0
D 4		2	1
D max. corrected		21–44	19–45
Flame Spread ASTM E 84		5	5
Smoke Developed ASTM E 84		5	5
Smoke Toxicity Mil M-14G			
Gas	NYC transit spec. maximum (ppm)	Level detected (ppm)	
Hydrogen chloride	10	0	—
Hydrogen cyanide	10	1	—
Ammonia	500	0	—
Aldehydes	30	6	—
Nitrogen oxides	100	24	—
Carbon monoxide	1000	310	—
Carbon dioxide	10000	1525	—
New York State Combustion Toxicity (U-Pitt) LC50, grams		91	—
Arc Resistance ASTM D 495 (sec)		183	182
Dielectric Strength ASTM D 149 (V/mil)		110	115
Dielectric Breakdown ASTM D 149 (kV)[a]		28	25

[a]Dielectric breakdown ASTM D 149 (kV) = 67–70 with alternate fillers.

Another fire-resistant, resorcinol-containing two-part phenolic resin system for composites processed by pultrusion was developed by INDSPEC Chemical Corp. and named Resorciphen™ 2074-A/2026-B. This resin system provides articles with very good mechanical strength, low flame spread, and very low smoke emission in fire tests. In pultrusion, the glass fiber (roving) is first passed through a resin bath and then the impregnated reinforcement material goes through heated dies, with finished product (rods, bars, ducts, etc.) being pulled continuously from the last die. Resorciphen™ 2074-A/2026-B is a two-component matrix resin system; its specifications, pt A/pt B mix ratio, and typical characteristics are listed in Table 5.6. Typical mechanical properties of the pultruded profiles are given in Table 5.7, typical mechanical properties of filled resin/mat and roving profiles are given in Table 5.8, and fire, smoke, and

Table 5.10. Airborne Emissions of Regulated Monomers During Pultrusion Operations Using Resorciphen® Resin A/B

	OSHA limit (ppm)	Measured results (ppm)
Formaldehyde, TWA—Full Shift, 29 CFR 1910.1048		
Operators (averaged)	0.75	0.12
Area samples		
@ bath	0.75	0.15
@ die	0.75	0.27
Formaldehyde, Short Term—15 Min, 29 CFR 1910.1048		
Operators (averaged)	2.0	0.45
Area samples		
@ bath	2.0	0.50
@ die	2.0	0.18
Phenol, TWA—Full Shift, 20 CFR 1910.1000		
Operators (averaged)	5.0	0.037
Area samples		
@ bath	5.0	0.033
@ die	5.0	0.094

Figure 5.8. Pultruded GRP shapes, produced with Resorciphen™ resin system. Copyright of INDSPEC Chemical Corp., published here with permission.

smoke toxicity test data and electrical properties of test specimens of pultruded profiles are given in Table 5.9. Test data relating to OSHA (Occupational Safety and Health Administration) regulations on airborne emissions of formaldehyde and phenol during pultrusion operations with the Resorciphen A/B resin system are shown in Table 5.10, the results being well below OSHA limits.

Figure 5.8 shows examples of pultruded profiles prepared with the glass-reinforced Resorciphen A/B resin system.

Finally, it should be noted that resin waste disposal is regulated as well. In the Toxicity Characteristics Leaching Procedure, TCLP, Resorciphen™ resin waste did not exhibit TCLP toxic characteristics (EPA Test Methods SW 846/8240, SW 846/8270); resin waste did not exhibit characteristic waste codes D001, D002, or D003.

5.4. ION-EXCHANGE RESINS

The first synthetic, organic ion-exchange resins were sulfonated P/F resins, developed in the mid-1930s in Britain by Adams and Holmes.[60] This invention was licensed by I.G. Farbenindustrie A.G. Wolfen and commercialized/further developed in the late 1930s to 1940s under the trade name Wofatit. This included the resorcinol-containing resins listed in Table 5.11. These cation-exchange resins found their largest use in boiler water treatment.[61]

In the 1950s, these resins were largely replaced in the U.S. and overseas by the sulfonated styrene/divinylbenzene copolymers which had improved chemical stability. Nonetheless, interest in resorcinol-containing, phenolic ion exchangers and similar products has continued to this date, as evidenced by the following, selected references.

A chelating coresin prepared by condensation of 8-hydroxyquinoline, resorcinol, and formaldehyde showed high selectivity for the separation of copper from cobalt, nickel, calcium, and magnesium.[62] Another copper-selective resin with high sorption capacity was a coresin made from 4-acetylresorcinol (also called resacetophenone or 2,4-dihydroxyacetophenone) oxime with resorcinol and formaldehyde.[63]

An ion-exchange coresin with good capacity for molybdate and tungstate ion was prepared by reacting resorcinol and polyethylene polyamine with hexamethylenetetramine.[64]

A phenolic chelating resin was prepared by the co-condensation of resorcinol, iminodiacetic acid, phenol, and formaldehyde. This type of resin removed ferric and/or cupric ions from water and could be regenerated by treatment with acid.[65] A straight R/iminodiacetic acid/F coresin removed both cesium and strontium from aqueous nuclear wastes at alkaline pH.[66]

Benzo-13-crown-4/F and benzo-9-crown-3/F resins were crosslinked with R/F; these coresins showed good selectivity for the extraction of lithium from saline solutions at alkaline pH.[67]

Table 5.11. Resorcinol-Containing Ion-Exchange Resins

Resin	Composition
Wofatit A	Sulfomethylated R/F resin
Wofatit K	R/Benzaldehydedisulfonic acid condensation product
Wofatit C	R/Resorcylic acid/P/F coresin

A dibenzo-18-crown-6/R/F coresin showed an absorption capacity of 19 mg/g for gold; thus, the resin was valuable for the separation and recovery of gold from solution.[68]

A sulfonated shellac/R/F cation-exchange coresin had an ion-exchange capacity of 4.85 meq/g; the resin lost <2% of its capacity after heating in water at 95 °C/10 hr.[69]

A R/F resin, prepared at alkaline conditions and heat-cured, was highly selective for cesium, even in the presence of sodium ions in high concentration. The resin was useful for the removal of cesium from aqueous, radioactive waste.[70]

The coordinating behavior and ion-exchange properties of a 4-acetylresorcinol/urea/F resin was studied. Polymeric chelates of the $(ML_2)_n$-type were prepared containing Cu(II), Ni(II), Co(II), Mn(II), Zn(II), Vo(II), and UO_2(II) ions.[71]

5.5. SOIL STABILIZERS AND DRILLING AIDS

Soil stabilizers are needed to control seepage of water through the soil as well as to increase the load-bearing capacity of the soil, particularly of shifting, loose, sandy, or granular soil. The drilling chemicals of interest here mostly are aids in post-primary oil production and became of most interest during the oil crisis of the 1970s, when increased recovery of oil from abandoned wells by steam or gas injection seemed both necessary and profitable. The following, selected references illustrate this use of resorcinol/resins.

Soils of low permeability, such as finely grained silts or slimy sands, were stabilized and strengthened by injecting, at ambient temperature and the low pressure necessary to displace entrained air, aqueous solutions of resorcinol, formaldehyde, and, as a catalyst, a mixture of sodium bicarbonate and ammonia or ammonium persulfate. The *in situ* formation of R/F resin produced the strengthening of the soil.[72]

Thermally stable, foamable, gelled compositions consisting of a high molecular weight polyacrylamide solution and resorcinol, sodium bicarbonate, formaldehyde, and a surfactant for crosslinking were suitable for selectively plugging oil-bearing subterranean formations to increase oil production by steam or gas injection.[73]

The recovery of oil from subterranean reservoirs was improved by selective modification of the strata with solutions of resorcinol and an aldehyde (formaldehyde, furfural, glyoxal, etc.) and with a controlled pH sequence on the alkaline side which allowed the emplacement of the gel-forming composition at the desired distance away from the bore hole and permitted the production of a uniform gel throughout the reservoir. The permeability of the treated strata could be restored by treatment with aqueous sodium hypochlorite or a mineral acid.[74]

An oil well was injected with an aqueous gel composition formed from xanthan gum crosslinked with ions of a transition metal such as Cr(III), resorcinol, and formaldehyde in a brine solution. The initial gel selectively entered the pores in a more permeable zone where the gel rehealed and formed a shear and thermally stable gel. The treated formation yielded improved recovery of oil by steam or water flooding or by carbon dioxide injection.[75]

5.6. BINDERS FOR FOUNDRY CORES, MOLDS, REFRACTORIES, AND FRICTION MATERIALS

Here, too, the cost of raw materials is a major consideration for commercial use. The binders made with resorcinol contain only a small amount of it to improve the cold setting, rate of cure, etc. In some cases the low-cost resorcinol still residue obtained from the benzene disulfonation/caustic fusion process is employed to good effect.These applications will be better understood by use of the following examples from the patent literature.

A mixture of 44–77 wt% reducing sugar, 5–22% urea, 4–19% formaldehyde, and up to 1% resorcinol (or other polyhydric phenol) was added to an aqueous binder for sand molds used in metal casting. The molds produced had improved mechanical properties, separability from the casting, and did not emit noxious fumes.[76]

Composite crosslinking agents/resin binders for foundry cores and sand molds, which were both cold and hot setting, consisted of 25 parts reducing sugar (e.g., dextrose or glucose), 4 parts water, 9 parts resorcinol, 62 parts 85% phosphoric acid and furfuryl alcohol. The sand molds made with this binder had good mechanical properties, strippability from the core box, and gave an excellent finish to the metal castings.[77]

A solution containing 1 part water, 0.4–99 parts resorcinol still residue, and optionally 0.25 part sodium hydroxide was added to molding sands and clays. The molded sand had better green strength and the additive acted as a plasticizer for the clay; the castings made in these sand molds had a smoother surface finish.[78]

Casting cores or molds were made from a mix containing 5–8 wt% aluminum cement or its mixture with porzolana cement, 80–88% sand, 0.05–0.4% surfactant, and 3–4% P/F or P/R/F resin as well as 0.1–1.0% LiCl as a setting catalyst.[79]

A sand mix containing 1–8% P/F novolak resin (based on the sand), 5–22% curing agent (e.g., hexamethylenetetramine), and 1.0–10% resorcinol (based on the novolak) gave castings with reduced cycle time and temperature in the shell molding process.[80]

A resinous binder for abrasive materials consisted of an aqueous solution of a P/F resole and either a P/R/F coresin or a mixture of resorcinol and formaldehyde at specified ratios. This mix cured faster than without the resorcinol component and less environmental pollution due to free phenol or formaldehyde was incurred.[81]

Sand (1 kg) was mixed with 20 g of P/R/F precondensate (methanol added) and 20 g of 40% glyoxal, molded and cured to give a mold with excellent compressive strength (15 kg/cm^2).[82]

A useful binder for abrasive materials was prepared by reacting 400 parts resorcinol still residue at 125 °C with 81 parts 37% aq. formaldehyde, then refluxing and distilling to remove 67 parts volatiles; hexamethylenetetramine was a suitable curing agent.[83]

A nonaqueous tap-hole mix for blast furnaces consisted of a binder made from resorcinol still residue, cellosolve, hexamethylenetetramine, polyethylene glycol, and a refractory filler (such as coke breeze, pyrophillite, calcined fire clay, fine silicone carbide, ball clay, or ferrosilicon). This binder was noncarcinogenic (conventionally, coal tar had been used as the binder) and did not emit noxious fumes on firing. The binder was also useful in preparing ramming mixes, gunning mixes, and mortars which formed refractory material on firing.[84] Similar binders were useful in making carbon-containing refractory brick or other shapes.[85]

Another tap-hole mixture for blast furnaces was prepared by mixing 40 parts of a R/F novolak resin solution in ethylene glycol with 5% added silicon powder and 60% refractory solids. This mix had good extrudability and, after firing, had high strength at high temperatures.[86]

A high-viscosity slurry useful for manufacturing molded brake pads was made by adding a liquid double salt (ammonium or potassium) of a dibasic acid and monobasic phosphoric acid to a dry mixture of reinforcing fiber (steel, glass, mineral, organic, or calcium-sodium metaphosphate), fillers (carbon and graphite), and a P/F/R resin. Thereafter, magnesium oxide (a catalyst) and paraformaldehyde were added and heated molds were filled with this slurry. The pads were then removed from the molds and postcured in a heated chamber.[87]

A binder composition for enhancing the green strength of SO_2-hardening foundry molding compounds was patented which consisted of a furfuryl alcohol/formaldehyde/resorcinol resin.[88]

A cold-setting ("no-bake") foundry binder composition was developed consisting of a P/F resole resin, silica sand, benzenesulfonic acid in methanol as the catalyst, and a 20% solution of resorcinol in furfuryl alcohol (55% resorcinol in water or 60% resorcinol in 25% aqueous methanol could be used instead). The molded forms had good compressive strength.[89]

A binder for shell molding was prepared from a P/F resole resin, which was mixed with resorcinol and dehydrated at 110 °C/700 mm Hg. The resulting resin was blended with silica sand and small amounts of calcium stearate and water. The molded shapes were fast curing and had improved mechanical properties (vs binders made without resorcinol).[90]

5.7. PHENOLIC RESIN FOAMS, AEROGELS, AND RELATED POLYMER FOAMS

In these applications, the use of resorcinol as an additive or of R/F resins as such is said to result in advantages such as faster foam preparation, corrosion protection, reduction in the amount of catalyst used, protective coatings, reduction in free formaldehyde content, and the preparation of aerogels (very light foams).

Usually, phenolic foams are made by mixing a blowing agent with a soft or fluid resin and an acid catalyst; the exothermic resinification reaction produces sufficient heat to vaporize part or all of the blowing agent to attain the foam structure. In the 1970s, Freons were the most-used blowing agents, but in the 1980s the ozone-depleting potential of the Freons released to the atmosphere was recognized and work began on environmentally nonhazardous blowing agents; likely, this work is still going on.

The following collection of selected reference material gives a glimpse of the use of resorcinol in (phenolic) foam products.

A phenolic foam was made from 100 parts P/F (resole) resin, 7.5 parts 80% aqueous resorcinol, 1.5 parts silicone oil, 5.5 parts mineral pigments, 4 parts Freon, and 5.5 parts 35% HCl; the foam had little or no corrosive action on metals.[91] When a similar mixture was blended with a dicyandiamide/formaldehyde resin, a fire-resistant, self-extinguishing foam with good insulating and mechanical properties was obtained.[92]

A microporous sheet of polyvinyl chloride (PVC) was prepared by adding 5 g 35% aq. formaldehyde and 3 g resorcinol dissolved in 5 g water to 20 g of a 50% solids PVC latex. To this blend was added 1 g mineral acid and the mixture was poured onto a flat mold to give a 0.5-mm-thick layer. The mold was covered with a glass plate to prevent the evaporation of water, precured at 70–90 °C, and the wet, solidified resin was removed and heated for a few minutes to 180 °C to cure the resin and to remove volatiles. A microporous sheet of resin was obtained which could be used as a battery separator.[93]

A solution of 19.5 parts resorcinol in 2.76 parts water was heated to 60 °C, mixed with 0.4 part 65% aqueous *p*-toluenesulfonic acid, treated with 11.1 parts 38% aq. formaldehyde over 2 hr/95 °C, and cooled to 50 °C. An additional 66.6 parts aq. 65% *p*-toluenesulfonic acid was added and the blend was concentrated in vacuo at 40 °C to 20% residual water content. To this R/F novolak solution was added 100 parts of a P/F resole, 15 parts 1,1,1-trichloro-2,2,2-trifluoroethane as the blowing agent, and 1 part polyethylene glycol nonylphenyl ether (surfactant).The mixture was stirred and began to foam after 25 sec. The foam product density was 34.4 kg/m^3 and an average cell size <0.003 mm^3 was obtained.[94]

A resinous, solid foam was prepared from a solution of 138.3 g diisobutylnaphthalenesulfonate sodium salt, 44 g resorcinol, and 118 g 85% phosphoric acid in 440 ml water by adding 40 ml 35% aq. formaldehyde. After 1 hr/105 °C this foam showed a 10% weight loss and after 2 yr/25 °C the total weight loss was 33% (the amount of water in the sample).[95]

Amino-resin precondensates or caprolactam-modified amino-resins were foamed with pentane or hexane in the presence of 1–30 phr resorcinol as a foam-controlling agent to give strong, stable foams.[96]

A closed-cell phenolic resin foam, with an insulation value similar to polyurethane foams, was prepared from a P/F resole which was neutralized, concentrated, and acidified, then reacted with an amount of urea and resorcinol in a mole ratio equal to the free formaldehyde content of the resole, adding a surfactant as a cell stabilizer and a blowing agent (such as Freon 113), and finally initiating foam formation by adding an acid catalyst to reduce the pH of the mix to 4. The heat of reaction raised the temperature to 70–75 °C/4–8 min. The foam was cured by heating to 120 °C. This product had good insulation value (initial K value <0.12; after 8 days aging, K value 0.18), had 85% closed-cell content (after 10 days/50 °C), <0.2% retained blowing agent, density 1.75–2.75 pcf, low moisture uptake and friability, and did not generate fumes as toxic as those of polyurethane in a fire.[97] A similar preparation of a closed-cell phenolic foam was patented in which only resorcinol (in place of urea and resorcinol) was added to the P/F resole; a higher peak exotherm (85–95 °C) was reached faster (3–6 min).[98]

A nonflammable thermal insulator was prepared from 100 parts P/F resole, 15 parts 1,1,1-trichloro-2,2,2-trifluoroethane, 1 part surfactant (SH-193), and 20 parts 65% aq. phenolsulfonic acid. This mix was foamed in a mold at 70 °C/10 min, cooled to 25 °C, and coated with a R/F resin to give a water-resistant, hard foam material.[99]

Corrosion-resistant foams were prepared from 100 parts P/F resole resin and 5–30 parts resorcinol (or alkylresorcinols), acid catalyst, Freon 11, butane, or hexane as the blowing agent, and stabilizers such as polysiloxane or alkylphenols.These foams not only had improved corrosion resistance but were also less hygroscopic, and showed less surface embrittlement than similar materials made without resorcinol. These foams were intended for use as ceiling and insulation materials, cores for fireproof doors, etc.[100]

A recent development, pioneered by R. W. Pekala, is the R/F resin aerogel technology (melamine/F aerogels were made also). Thus, very low density R/F resin open-cell foams, later called aerogels, were prepared by reacting, in dilute aqueous solution, resorcinol and formaldehyde in the presence of a base (sodium carbonate) in sealed vials at 85 °C/7 days to form a polymeric gel. The deep-red gels were acid-washed and the water was displaced with acetone (repeated exchange with fresh solvent), then submitted to supercritical extraction with CO_2. The resulting foam had densities from 50–100 mg/ml. The transparency of the foam indicated an ultrafine cell size, below the wavelength of visible light. These materials were said to be suitable for high-energy physics applications, such as inertial confinement fusion[101]; a U.S. patent application was made.[102] The particle and cell size, surface area, and density of these aerogels were said to depend largely on the catalyst concentration in the preparation.[103] Carbonized R/F resin aerogels were also prepared and their microstructure was studied.[104] In a patent issued on the R/F aerogel process, the addition of chloroplatinic acid or palladium chloride in *N,N*-dimethylformamide to the aerogels prior to solvent exchange was included; the finished foams were pyrolized to give carbonized foams with uniform distribution of Pt or Pd and high catalytic function.[105] In 1989, a small startup company in California, Thermolux, L.P., appeared to be looking for commercial applications of such aerogels in insulation for refrigerators and containers, and in air filters. (In mid-1993, Thermolux was no longer in business.)[106]

Several review-type papers and publications on studies of specific properties of these R/F resin aerogels have appeared. For example, a study of the thermal conductivity of monolithic R/F aerogels, was reported,[107] a paper on the thermal and electrical conductivity of the carbonized aerogels was published,[108] work on the photoconductivity of the carbon aerogels was reported,[109] and two broader reviews of these (carbonized) aerogels appeared.[110,111] Obviously, commercial applications of these products will be aimed at areas where cost is secondary. For instance, there is said to be commercial potential for carbon aerogels in supercapacitors.[106]

Low-density carbonized composite foams were made by preparing an inverse, aqueous emulsion of styrene/divinylbenzene, polymerizing/crosslinking the oil using sodium persulfate as the catalyst and heating to attain a foam, filling the foam cells in vacuo with a 2% aqueous R/F resin solution containing sodium carbonate as the catalyst. The foam was then dried and pyrolized at 1050 °C/12 hr in argon to make a carbonized foam with density 46 mg/cm^3 and very small cell diameter. The product was said to be useful as a laser target in controlled fusion reactions, as catalyst supports, and for structural supports.[112]

Finally, a sprayed-on cellular coating was produced from a mix of a P/F resole resin, Toresin T 30 (a methoxymethylated polyamide), resorcinol, and caprolactam as a thinner; 100 parts of this low-viscosity mix was blended with 3 parts foam stabilizer, 15 parts Freon 113, and 15 parts 65% aq. phenolsulfonic acid. The blended mix was sprayed on a veneer to give a foamed, well-adhering coating with bulk density 0.052.[114]

5.8. ADDITIVES FOR CONCRETE

The possibilities for resorcinol in this area are shown by the following selected reports.

R/F resins were added at the 0.05–0.2 wt% level to concrete mixtures (Portland cement, aggregate, water). The modified concrete was said to have increased water resistance.[115]

Polymer concrete with a compressive strength of >2000 psi (>13.78 MPa) after 1 hr/24 °C was obtained by mixing a P/R/F resin with Portland cement and an aggregate containing 1–10% water. The addition of *ca* 14.7–17.5% type III Portland cement to the polymer concrete gave a faster rate of phenolic resin formation and improved the properties of the composite.[116]

An additive, added at 0.05–5.0 wt% to the cement (in a concrete mix), consisting of a water-soluble R/F resin or catechol/F resin, improved the slump and the compressive strength of the concrete.[117]

A superplasticizer with improved water-reducing properties for cement slurries, mortars, grout, or concrete was developed, consisting of neat, sulfomethylated (treated with formaldehyde and sodium sulfite) or formaldehyde-reacted resorcinol still residue (as obtained from the benzene disulfonation/caustic fusion process); alternatively, the still residue could be co-condensed with ligninsulfonates (a byproduct of wood pulping) or with sodium naphthalenesulfonate and formaldehyde to make a suitable product. Blends of neat or modified resorcinol still residue with sodium naphthalenesulfonate/formaldehyde condensation products also worked well as superplasticizers for improving the workability and compressive strength of cementitious compositions.[118] Similarly, a plasticizing additive for mortar and concrete mixes was developed which was made up of 83.9% water, 9.5% resorcinol still residue, 5.8% sodium hydroxide, and 0.8% formaldehyde.[119]

Polyelectrolyte complexes as plasticizers for concrete were patented; they were made by reacting polyethylene polyamine with formaldehyde and a R/F prepolymer.[120]

5.9. TANNING AGENTS FOR ANIMAL HIDES

Tanning agents combine with the collagen portion of animal skins to produce leather. Phenol/aldehyde resins with an appropriate degree of sulfonation have long been known to be useful in this application. In the older literature, resorcinol-containing tanning agents were described as obtained by condensing R/NH_3/F (these were acid-soluble resins), R/acrolein with post-treatment by sodium bisulfite (anionic tanning agents), or R/crotonaldehyde/$NaHSO_3$ and similar modifications. A few examples from the literature follow.

Animal hides were tanned by a solution of chromium salts, formaldehyde, and 0.5–5.0 wt% (based on the hides) of resorcinol; leather of improved quality was obtained.[121]

In a new version, animal skins were tanned by impregnating and drumming with an aqueous solution, buffered to pH 4.5–4.7, of tris-(hydroxymethyl) nitromethane and a polyhydric phenol such as resorcinol for 1 hr, then readjusting the pH to 6.5–7.0 and drumming until the skins were tanned. The hides could then be treated with a second tanning agent, e.g., basic chromium sulfate, glutaraldehyde, or a blend of vegetable tannins consisting of chestnut, sulfonated quebracho, and wattle, to produce leather of high quality.[122]

An improved method for dyeing/tanning leather (and other materials) consisted of treatment with a reactive dye (e.g., the Procion dyes of ICI or the Drimarines of Sandoz), chitosan, and a synthetic tanning agent such as a R/F precondensate resin.[123]

5.10. MICROCAPSULES

These products are polymeric, hollow spheres of small diameter, filled, in most commercial applications, with an oil or solution of chemicals; release of the internal phase occurs by breaking the capsule. The formation of the filled microcapsules is generally induced from aqueous suspensions or dispersions of polymer and filler, e.g., by coacervation of a hydrophilic colloid. These products are used for dispensing dyes, drugs, flavors, adhesives, perfumes, agricultural chemicals, and other ingredients. The largest use is in carbonless copy paper; capsules containing leuco dye (colorless precursor) are deposited on paper and withheld from contact with the color developer until the capsules are broken by pressure applied by pencil or a stylus. More information on such technologies is given in Chapter 10, in the description of Mead's Cycolor process.

The function of resorcinol in this art can be as a reagent in forming the walls of the microcapsules, as a R/F resin, as a hardening accelerator, and/or to improve the heat resistance of the capsules. The following selections from the patent literature sketch the ideas involved.

A pressure-sensitive recording paper was prepared. A leuco dye (e.g., crystal violet lactone) was dissolved in tetrachloroethylene or toluene and encapsulated with mixtures of gum arabic, gelatin, and methyl vinyl ether/maleic anhydride copolymer, hardened with glutaric dialdehyde or uranyl nitrate and modified with a R/F resin. The encapsulated dye was deposited on paper.[124] (Development occurred when the capsules were broken and the dye contacted an acid developer, such as a clay.)

In another invention, a phenolic compound such as resorcinol or a water-soluble R/F resin was added during the formation of oil-filled microcapsules from an aqueous suspension of gelatin, followed by coacervation of the colloid by pH adjustment and hardening with formaldehyde in the presence of carboxymethyl cellulose as a viscosity controller. The addition of resorcinol increased the wall thickness and density of the capsules and the microcapsules had better heat resistance.[125]

Small polymer capsules containing solvent were also prepared as follows. A dispersion of 1–10 μm droplets of 150 ml xylene in 200 ml water which contained 160 ml of 5% aq. polyvinyl alcohol (PVA), 12 g resorcinol, and 3 g urea (used to inhibit the agglomeration of the capsules) was mixed at 95 °C, acidified with 10 ml 10% sulfuric acid, and reacted with 30 ml 37% aq. formaldehyde added over 2 hr, then held at 55 °C/1.5–3 hr. The suspension was neutralized to pH 4.5 to give xylene-filled microcapsules whose walls were made up of a R/F resin containing 2–110% PVA.[126]

A novel way of preparing a dispersion of microcapsules containing leuco dyes was by high-speed mixing of 70 parts isopropylnaphthalenes, in which 2 parts crystal violet lactone and 0.5 part benzoyl leucomethylene blue were dissolved, with 150 parts 5% aq. Gohsenol T-350 (presumed to be PVA), 5 parts urea, 0.5 part resorcinol, and enough 10% HCl to reach pH 3.4, then mixing with 12.5 parts 37% aq. formaldehyde, holding at 55 °C/2 hr, and neutralizing at this point with 20% NaOH. The resulting dispersion had a strong odor of formaldehyde which was eliminated by the addition of 25 parts 25% aq. hydroxylamine hydrochloride.[127] Another method for removing free formaldehyde from microcapsules made by a similar formulation (where the main polymer ingredient was an isobutene/maleic acid copolymer sodium salt) was to place the capsules in a cellulose acetate membrane bag and dialyze with running water for

3 hr. The microcapsules were odorless and had good resistance to heat, solvents, and moisture.[128]

In another patented process, a mixture of 100 g 1/4 melamine/F resin, 50 g 1/1.8 urea/F resin, 158 g aq. urea/F/tetraethylenepentamine resin hydrochloride, 62 g water, 1 g triethanolamine, and 0.3 g sodium dodecylbenzenesulfonate was intensively agitated with 200 ml xylene to form an emulsion which was held at 30 °C/pH 3.6 for 1 hr, acidified to pH 3.0, mixed with 20 ml 10% aq. resorcinol solution, and mixed for 18 hr to obtain microcapsules. To a solution of 100 parts neoprene, 1 part P/F resin, and 1 part coumarone/indene resin in 500 parts toluene were added 30 parts of the microcapsules. The coated capsules were deposited on PVC sheets and dried 2 hr/40 °C. When the coated sheet was pressed at 10 kg/cm^2 with another PVC sheet and left at room temperature for 24 hr, good adhesion between sheets was obtained; when the resorcinol was left out of the formulation, adhesion between sheets was poor.[129]

A process was developed for forming microcapsules by using urea or melamine and formaldehyde, with a small amount of resorcinol added (to reduce the permeability of the wall). The reaction was carried out in the presence of a polyol as a modifier; sodium bisulfite was used to remove free formaldehyde. In this patent, the encapsulated oil was trimethylolpropane triacrylate containing a leuco dye and a photopolymerization initiator (a system used in Mead's Cycolor color copying process).[130]

Another patented way to remove residual formaldehyde from dispersions of microcapsules with walls made of amino resins was by reaction with 2-imidazolidinone/urea and/or (alkyl)resorcinol.[131]

Microcapsules were also prepared by the *in situ* condensation of U/F. Thus, a 1/1 isobutylene/maleic anhydride copolymer was treated with poly(ethylene oxide) monoethyl ether and aqueous NaOH, then mixed with urea and resorcinol (10 wt% of the urea) and an organic solvent solution of a leuco dye. The resultant emulsion was treated with 37% aq. formaldehyde to give a slurry of dye-filled capsules. Copy paper coated with these capsules showed better color-forming ability and had better storage stability.[132]

5.11. COATINGS FOR POLYMERIZATION REACTORS

Reactors for the suspension or emulsion polymerization, e.g., of vinyl chloride to make PVC, suffer from fouling of the walls of the reactor due to encrustation with polymer. Coating the reactor internals with a variety of polar polymers, including resorcinol-containing polymers, can reduce or eliminate this undesirable scale formation. Pertinent, selected references, all patents of B.F. Goodrich Co., to the use of resorcinol in this application are presented below.

The formation of polymer crusts on the internal surfaces of polymerization reactors was eliminated by coating with a *p*-chlorophenol/R/F coresin.[133]

The interior surfaces of polymerization reactors were coated with an aqueous solution containing 5% R/F prepolymer, 3.3% NaOH, and 2% NaOCl. When the coating was applied before each of three successive runs were made, negligible fouling occurred.[134]

Polymerization vessels were coated with an aqueous alkaline or ammoniacal solution of a R/acetaldehyde water-soluble resin. The encrustation of the reactor internal surfaces with polymers of vinyl or vinylidene halides, acrylate esters, acrylonitrile, and other vinylidene monomers was substantially eliminated. Prior, cumbersome techniques for removing polymer buildup from the reactor surfaces, such as solvent cleaning or hydraulic or mechanical cleaning, were overcome.[135]

5.12. LITHOGRAPHY, PRINTED CIRCUIT BOARDS, MAGNETIC TAPES, AND ENCAPSULATED SEMICONDUCTORS

These are examples of the use of resorcinol in high technology. Additional information on this area is presented in Chapter 10.

A sublimable adhesive was prepared for temporarily joining articles (such as semiconductor materials), for cutting, grinding, or polishing. The preparation involved melting biphenyl, heating to 105–115 °C, adding 1–10 wt% resorcinol, and cooling. In use, the surfaces to be adhered were heated to 100–110 °C and the solid (stick) adhesive was drawn along one of the surfaces; the two surfaces were then joined and cooled.[136]

An amount of 60–90 parts of metal powder (Cu, Ni, Co) was dispersed in 10–40 parts of a binder resin such as a R/F resin containing *m*-aminophenol (or the *o*- or *p*-isomer), as a reducing agent and 3–8 parts (based on the metal) of a phenylenediamine derivative, e.g., *N*-(2-hydroxyethyl)-*p*-phenylenediamine. This material had excellent electroconductivity and could be used to make a circuit on a printed board.[137]

An alkali-soluble resin was prepared from 5 parts resorcinol mono-trimethylsilyl ether, 2.1 parts 2-methylresorcinol, and 3.7 g 35% aq. formaldehyde. This novolak resin was blended with a reaction product of 5-chlorosulfonyl-1,2-naphthoquinone-2-diazide/*p*-cresol dissolved in ethyleneglycol diacetate, coated on Si (wafer), and dried. The coating was useful as a mask in making printed circuit boards by ion etching.[138]

A *p*-*tert*-butylphenol/*p*-phenylphenol/F resin (presumably containing methylol groups) was reacted with resorcinol to give a resin with softg. point 120 °C. This resin (100 parts) was blended with 200 parts acrylonitrile/butadiene rubber, 60 parts silica, and 0.3 part palladium chloride in a 1/1 cellosolve acetate/methyl ethyl ketone solvent sufficient to produce a 23% solids content. This adhesive was coated on both sides of a phenolic or epoxy paper-base laminate. The coated substrate was then roughened in a chromic acid bath, washed with water, and finally coated with an electroless copper plating solution. This composition had excellent heat resistance in soldering and was used in making printed circuit boards.[139]

A photosensitive composition was prepared comprising 10–50 wt% *o*-naphthoquinone diazide and 1–80% straight or modified R/F resin. This composition had excellent chemical resistance (not against alkali), a long press life, and had excellent solubility in alkali; it was especially suitable for the preparation of lithographic printing plates.[140]

A resin was prepared by reacting a novolak epoxy resin, a phenolic novolak such as a R/F resin low in prepolymer, and an organic phosphine (e.g., triphenylphosphine) as a cure accelerator which also improved the moisture resistance and electrical characteristics at high temperature. These latter properties were enhanced by the addition of an organic phosphorous acid compound (such as the calcium salt of octadecylphosphate). These resins were useful in the preparation of highly reliable resin-encapsulated semiconductors.[141]

A coating composition was prepared from iron oxide powder, water, a sulfosuccinate dispersant, an ethoxylated fatty acid amide, a *N*-hydroxymethyl-acrylamide/resorcinol condensation product, an acrylic acid/acrylamide copolymer with resorcinol and formaldehyde, and a polyacrylate latex. This dispersion was coated on a suitable substrate to prepare magnetic recording tapes.[142]

A positive-type photosensitive composition was prepared from *o*-naphthoquinone diazide and an alkali-soluble novolak resin, especially an alkylresorcinol/F resin; this composition was useful for preparing printed circuit boards.[143]

A P/R/F coresin (100 parts) was reacted with 30 parts glycidyl methacrylate using dimethylaminomethyl-phenol as the catalyst. The adduct contained 15% acrylate groups; 40 parts of this resin was mixed with 60 parts 2-hydroxyethyl methacrylate, 10 parts talc, and 5 parts photopolymerization initiator (Irgacure 651 of Ciba-Geigy). This UV light-curable resin composition was useful as an etching resist for printed circuit boards.[144]

m-Aminophenol/F resins were reacted with monoisocyanates which contained olefinic unsaturation, e.g., an addition product of toluenedisocyanate and 2-hydroxyethyl acrylate. The products were storage-stable photopolymers. These polymers were relatively low cost because they were simple to prepare; they could be used to make protective coatings, lithographic printing plates, resists, and circuit boards.[145]

A *m*-aminophenol/*m*-cresol/*p*-cresol/F coresin (10 parts) and 3.5 parts 5-chlorosulfonyl-1,2-naphthoquinone-diazide reaction product with 2,3,4-trihydroxybenzophenone was coated on a silicon wafer, dried, prebaked, exposed to a UV pattern, and processed to give a well defined image. This was a positive-working photoresist composition.[146]

5.13. MISCELLANY

This last section of R/F resins and coresins shows additional varieties of their applications as binders and adhesives by means of selective examples from the literature.

A stable pyrotechnic composition was prepared which contained a liquid P/R/F resin and an oxidizing agent such as ammonium nitrate or perchlorate.[147]

A surgical adhesive with negligible irritability of local parenchymateous tissue contained 60% aq. gelatin solution and resorcinol in 3/1 ratio, with formaldehyde being added at the point of application.[148]

R/F resins (and others, such as U/F resins) could be used for making corrugated paperboard. The resin solution or dispersion was applied to the tops of the flutes of a corrugated paperboard web and the curing agent (formaldehyde) was applied to a liner; web and liner were contacted and bonded instantaneously.[149]

Another adhesive for corrugated board manufacture consisted of an aqueous dispersion of cooked amylose containing suspended starch solids and modified with a R/F resin. These adhesives were waterproof after cure.[150]

A fireworks composition resistant to decomposition under dampness was made by mixing a 60% alcoholic solution of R/F prepolymer, oxidizer (e.g., a nitrate or perchlorate), a color former, and 37% aq. formaldehyde in a combustible or noncombustible adsorbent as the curing agent. The mix was extruded as a strand, or casings were filled with this mix, with curing at *ca* 35 °C.[151]

A 3 mil thick polyethylene terephthalate (PET) film was coated on its right edge with a 20% solution of resorcinol in methyl ethyl ketone; after the solvent evaporated a thin layer of resorcinol remained. The left edge and the right edge of the PET film were then contacted and hot-pressed to form a tubular laminate with good bond strength (catechol or hydroquinone could be used instead of resorcinol).[152] In another version of this invention, a stick adhesive was prepared by dissolving 2 wt% polyester (Mylar) in resorcinol and shaping the melt by casting/cooling. The adhesive was then used to bond PET film by rubbing the stick over the film.[153]

Enzymes (not specified) could be fixed on a carrier made by suspension polymerization of R/1° or 2° amine/F in water. This resin suspension was contacted with an aqueous solution of the enzyme (5–50 mg/ml as the protein) at pH 3–9 and 0–50 °C/20 min–2 hr. Optionally, the enzyme could be further fixed by treatment with a dialdehyde such as glutaraldehyde. The retention of enzymatic activity was excellent. Inactivated enzyme could be washed off the carrier by treatment with 3 N NaOH at 80–100 °C/10 min–2 hr. The resin could be reused as an enzyme carrier.[154]

Polyvinyl chloride (PVC) plastisols useful as sealants, adhesives, and corrosion-protective coatings in the auto industry were prepared by modifying PVC/phthalate plasticizer compositions, prepared in a known manner, with pigments and other additives such as 1–2 wt% epoxidized vegetable oil (stabilizer) and/or a metal stearate, and 2–6 wt% MgO, 6–18 wt% hexamethylenetetramine, 3–9 wt% melamin, and/or 1–12 wt% resorcinol. This mix was applied to the substrate at 130–140 °C and gave good adhesion in 40–50 min.[155]

A modified gypsum, useful for plastering and for making high-strength building products, consisted of 60–62 parts gypsum hemihydrate, 3–12 parts formaldehyde, 9–18 parts resorcinol, 2–3 parts hexamethylenetetramine, and the balance water. After curing, this mix had a compressive strength of 28.0–42.4 MPa (27.5–42.0 MPa after 30 cycles of alternate wetting and drying) and a bending strength of 18.0–26.6 MPa (17.6–26.1 MPa after the same 30 cycles of wetting and drying).[156]

Resins were developed which were useful for removing (waste) products such as urea, ammonia, and/or formaldehyde from aqueous liquids. These resins were useful in the hemodialysis of blood or for purifying water in swimming pools or aquariums.

The resinous materials were incompletely condensed R (or phloroglucinol)/(di-)aldehyde condensation products having free methylol groups; they could be crushed and used to fill cartridges optionally also containing charcoal. For example, a suitable resin was prepared by dissolving 1.0 m resorcinol and 0.33 m ferric chloride (catalyst) in 250 ml water and adjusting to pH 8 with sodium carbonate; 1.0 m formaldehyde (37% in water containing 10% methanol) was then added, keeping the temperature below 60 °C, holding at 60 °C/4 hr, and subsequently adjusting to pH 2 with HCl and washing the precipitate free of iron. The slurry was then again made alkaline with sodium carbonate to pH < 10, another 2.0 m formaldehyde was added, and the mix was held 4 hr/70 °C. The resin formed was neutralized, washed free of salts and formaldehyde, crushed to 0.05–1.0 mm particle size, conditioned sequentially with $K_2S_2O_5$, $NaBH_4$, and $Ca(HCO_3)_2$, washed again, and freeze-dried to give a resin with an absorption capacity per gram of resin of 15 mg urea, 0.9 mg PO_4^{3-}, 0.7 mmol K^+, and 2.5 μmol phenol out of a human hemofiltrate.[157]

An improved ballistic material for flexible body armor was developed, consisting of multiple plies of cloth woven with an aramid, e.g., Kevlar thread, one or more plies of which were coated with a R/F resin latex which also filled the interstices between the threads of a treated ply.[158]

A process for making corrugated paper board with improved wet strength was developed which used a starch adhesive and a small amount of R/F adhesive mixed just before use; the mix was ripened for 15 min before application.[159]

A gellable, solvent-free sealant and sound insulating composition was prepared which was useful as an adhesive and corrosion-protective material in foundries and auto manufacture. This material consisted of 100 parts PVC plastisol, 30.3 parts dibutyl phthalate, 36.4 parts dioctyl phthalate, 30 parts epoxidized vegetable oil, 30.3 parts epoxidized fatty acid monohydric alcohol ester, 9.1 parts melamine, 18.2 parts hexamethylenetetramine, 3.0 parts resorcinol, and 60.7 parts fillers. This mix was spread on primed sheet metal and cured 1 hr/137 °C to give a well-adhered elastic coating which could be readily topcoated.[160]

Sheets of thermoplastics, such as PET or polyarylates, were coated with a P/R/F coresin (e.g., Plyophen 5015) to prepare heat-resistant films or sheets useful as electrical insulators.[161]

Improved pressing fabrics for paper making were prepared by treating nylon-66 staple fibers (terminal NH_2-conc. ≥6 mequiv/kg) with 0.5% ethylenediamine and 2.0% formaldehyde at 54.4 °C/0.5 hr, rinsing and treating with 1.5% resorcinol and 1.0% formaldehyde at 54.4 °C/1 hr, rinsing again, drying, and curing at 149 °C/10 min. This treatment increased the intrinsic viscosity of the fibers by >4% and increased the durability of the fabric in use.[162]

Spheres of macroporous phenolic resins were prepared by mixing 0.5 m resorcinol, 1.0 m 37% aq. formaldehyde, 0.1 m $MgCl_2 \cdot 6H_2O$ (catalyst), and 0.33 m Me Cerosoft (filler?) in 150 ml water containing 10 g 24% aq. NaOH for 30 min/30 °C, then adding 400 ml tetrachloroethylene, and removing the water with stirring at 100

°C. Resin spheres were obtained with a specific surface area 200 m^2/g, pore volume 0.5 ml/g, and average pore diameter 88 Å.[163]

As a final invention in this category, a gelatin/resorcinol/glutaraldehyde resin was prepared which was useful in the adhesion of a prosthesis to biological tissue.[164]

5.14. REFERENCES

1. R. Vieweg and E. Becker, eds., *Kunststoff-Handbuch*, Vol. X, "Duroplaste," p. 19ff., Carl Hanser Verlag, Munich, 1968.
2. A. Pizzi, in *Wood Adhesives—Chemistry and Technology* (A. Pizzi, ed.), p. 105, Marcel Dekker, New York and Basel, 1983.
3. A. Knop and L. A. Pilato, *Phenolic Resins*, Springer-Verlag, Berlin, Heidelberg, New York, Tokyo, 1985.
4. R. H. Moult, in *Handbook of Adhesives*, 2nd edition (I. Skeist, ed.), p. 417, Van Nostrand Reinhold, New York, 1977.
5. G. E. Meyers, *Adhesives Age, 31(10)*, 32 (1988).
6. A. Pizzi, R. M. Horak, D. Ferreira, and R. D. Roux, *Cellul. Chem. Technol, 1979 (13)*, 753.
7. B. H. River, C. B. Vick, and R. H. Gillespie, in *Treatise on Adhesion and Adhesives*, Vol. 7 (J. D. Minford, ed.), p. 1, Marcel Dekker, New York, Basel, Hong Kong, 1991.
8. *Consumer Reports, Jan. 1988*, 49.
9. Ger. Offen. DE 2,200,909 (Sept. 28, 1972; A. Y. Aarna *et al.* to Tallinski Politechnitscheski Institute; Derewojobratywajuschtschij Kombinat Wijsnurk); equiv. is U. S. Pat. 3,829,528 (Aug. 13, 1974; same assignation).
10. G. T. Tiedeman and M. R. Sanclemente, *J. Appl. Polymer Sci., 17*, 1813 and, with H. A. Smith, *ibid.*, 1819 (1973).
11. U.S. Pat. 3,518,159 (June 30, 1970); U.S. Pat. 3,784,525 (Jan. 8, 1974); U.S. Pat. 3,880,694 (Apr. 29, 1975), all by H. G. Freeman, G. F. Baxter, and G. Graham to Weyerhauser Co.
12. U.S. Pat. 3,919,151 (Nov. 11, 1975; R. H. Moult and J. J. Colligan to Koppers Co., Inc.).
13. U.S. Pat. 3,907,728 (Sept. 23, 1975; A. L. Lambuth and W. M. Hearon to Boise Cascade Corp.).
14. USSR Pat. SU 467,914 (May 28, 1975; to A. S. Freidin, Sh.-G. Klauzner, A. B. Sholokova, *et al.*); *C.A., 83*, 18508 (1975).
15. U.S. Pat. 3,903,041 (Sept. 2, 1975; L. F. Bornstein to Georgia-Pacific Corp.).
16. USSR Pat. SU 483,419 (Apr. 2, 1976; V. M. Khrulev, R. I. Rukov, and V. A. Zaburunov to Novos Civil Engineering); *C.A., 84*, 46388 (1977).
17. Jpn. Kokai Tokkyo Koho JP 50,109,932 [75,109,932] (Aug. 29, 1975; Y. Yamamoto, K. Higuchi, and K. Okada to Eidai K. K.); *C.A., 84*, 46388 (1977).
18. U.S. Pat. 4,061,620 (Dec. 6, 1977; M. F. Gillern to Weyerhauser Co.).
19. Ger. Offen. DE 2,606,103 (Oct. 14, 1976; M. F. Gillern to Weyerhauser Co.); *C.A., 86*, 17588 (1977).
20. Jpn. Kokai Tokkyo Koho JP 51,098,740 [76,098,740] (Aug. 31, 1976; H. Yamada and T. Ishikawa to Aica Kogyo K. K.); *C.A., 85*, 178522 (1976).
21. U.S. Pat. 4,032,575 (June 28, 1977; E. A. Blommers and R. H. Moult to Koppers Co., Inc.).
22. Can. Pat. CA 1,009,780 (May 3, 1977; to Sumitomo Durez Co.).
23. Jpn. Kokai Tokkyo Koho JP 52, 006,742 [77,006,742] (Jan. 19, 1977; M. Yamao, K. Tanaka, and T. Yamada to Hohnen Oil K. K.; Mitsubishi Gas Chemical Co.; Inc., Yukata Horumarine K.K.); *C.A., 87*, 136980 (1977).
24. Belg. Pat. Appl. 867,051 (Nov. 13, 1978; to Casco AB); similar is U.S. Pat. 4,175,005 (Nov. 20, 1979; to S.-E. Anderson).
25. Jpn. Kokai Tokkyo Koho JP 53,143,636 [78,143,636] (Dec. 14, 1978; T. Nogi to Taoka Kagaku Kogyo K.K.); *C.A., 98*, 138787 (1979).
26. Jpn. Kokai Tokkyo Koho JP 57,159,867 [82,159,867] (Oct. 2, 1982; to Toyama Prefecture); *C.A., 98*, 127335 (1983).

27. Jpn. Kokai Tokkyo Koho JP 58,036,451 [83,036,451] (March 3, 1983; to Sumitomo Bakelite Co.); *C.A., 99*, 124347 (1983).
28. Jpn. Kokai Tokkyo Koho JP 60, 028,479 [85,028,479] (Feb. 13, 1985; to Shoei Kagaku Kogyo K. K.); *C.A., 102*, 221863 (1985).
29. Jpn. Kokai Tokkyo Koho JP 61,043,501 [86,043,501] (March 3, 1986; I. Kosho, H. Hirai, N. Asaumi, and I. Fujiwara to Otsuka Chemical Co., Ltd.); *C.A., 106*, 35004 (1986).
30. U.S. Pat. 4,608,408 (Aug. 26, 1986; R. T. Hood and R. L. Bender to Koppers Co., Inc.).
31. PCT Int. Appl. WO 87,5041 (Aug. 27, 1987; I. Karlsson to Casco Nobel AB); *C.A., 108*, 23077 (1987).
32. Jpn. Kokai Tokkyo Koho JP 62, 103,101 [87,103,101] (May 13, 1987; Y. Daimaru, Y. Takigawa,T. Takebe, M. Tekebe, H. Egami, and Y. Yokoyama to Dainippon Ink and Chemicals, Inc.; Daiken Kogyo Co., Ltd.); *C.A., 107*, 60895 (1987).
33. Jpn. Kokai Tokkyo Koho JP 61,296,078 [86,296,078] (Dec. 26, 1986; T. Kamikaseda and M. Yamamoto to Kanebo NSC K.K.); *C.A., 107*, 41363 (1987).
34. S. African ZA 89,000,216 (Oct. 25, 1989; A. Pizzi and S. Vermeulen to Council for Scientific and Industrial Research); *C.A., 113*, 7507 (1990).
35. PCT Int. Appl. WO 90,006,347 (June 14, 1990; P. J. Collins, Y. Yazaki, and M. J. Delaney to Commonwealth Scientific and Industrial Research Organization; Chemplex Australia Ltd.); *C.A., 113*, 193787 (1990).
36. Jpn. Kokai Tokkyo Koho JP 02,133,485 [90,133,485] (May 22, 1990; M. Ogawa and S. Hayashi to Oshika Shinko K.K.); *C.A., 113*, 133835 (1991).
37. Jpn. Kokai Tokkyo Koho JP 02,127,489 [90,127,489] (May 16, 1990; M. Ogawa and S. Hayashi to Oshika Shinko K.K.); *C.A., 113*, 116769 (1990).
38. Jpn. Kokai Tokkyo Koho JP 03,031,384 [91,031,384] (Feb. 12, 1991; Y. Kato to Aica Kogyo Co., Ltd.); *C.A., 115*, 10496 (1991).
39. Jpn. Kokai Tokkyo Koho JP 03,053,901 [91,053,901] (March 7, 1991; N. Hirai, H. Yoshida, and Y. Matahira to Shizuokaken Mokuzai Kyodo Kumiai Rengokai); *C.A., 115*, 73845 (1991).
40. Eur. Pat. Appl. EP 493,010 (July 1, 1992; J. R. Parker, J. B. M. Taylor, D. V. Plackett, and T. D. Lomax to New Zealand Secretary of Forestry); *C.A., 117*, 153038 (1992).
41. Jpn. Kokai Tokkyo Koho JP 50, 109,932 [75,109,932] (Aug. 29, 1975; Y. Yamamoto, K. Higuchi, and K. Okada to Eidai K.K.); *C.A., 84*, 46388 (1976).
42. K. F. Plomley, *Tannin-Formaldehyde Adhesives for Wood. II. Wattle Tannin Adhesives*, Div. Forest Prod. Techn. Paper No. 39, Commonwealth Scientific and Industrial Research Organization, Australia, Melbourne, 1966.
43. H. M. Saayman and J. A. Oatley, *Forest Prod. J. , 26(12)*, 27 (1976).
44. J. M. Dinwoodie, *J. Inst. Wood Sci., 8(2)*, 59 (1978).
45. A. Pizzi, *Adhesives Age, 21(9)*, 32 (1978).
46. A. Pizzi *et al., Holz Roh-Werkst., 46(2)*, 67 (1988).
47. H. Scharfetter, *Forest Prod. J., 25(3)*, 30 (1975).
48. E. Von Leyser and A. Pizzi, *Holz Roh-Werkst., 48(1)*, 25 (1990).
49. C. C. Moredo, Jr. and T. Sakuno, *Int. J. Adhes. Adhes., 11(4)*, 228 (1991).
50. U.S. Pat. 4,053,047 (Oct. 11, 1977); U.S. Pat. 4,076,873 (Feb. 28, 1978); U.S. Pat. 4,107,127 (Aug. 15, 1978), all to L. E. Shea.
51. U.S. Pat. 4,275,170 (June 23, 1981) and U.S. Pat. 4,309,527 (Jan. 5, 1982), both by L. E. McAllister and H. Dietrich to Fiber Materials, Inc.
52. U.S. Pat. 4,409,361 (Oct. 11, 1983) and U.S. Pat. 4,471,089 (Sept. 11, 1984), both by L. E. McAllister and H. Dietrich to Fiber Materials, Inc.).
53. Jpn. Kokai Tokkyo Koho JP 63,159,465 [88,159,465] (July 2, 1988; M. Inoue, K. Kono, Y. Yamashita, and K. Ueda to Mitsui Toatsu Chemicals, Inc.); *C.A., 110*, 59104 (1989).
54. U.S. Pat. 4,820,576 (Apr. 11, 1989; H. E. Duryea to Fiber Materials, Inc.).
55. U.S. Pat. 4,880,893 (Nov. 14, 1989; P. A. Waitkus to Plastics Engineering Co.).

56. Jpn. Kokai Tokkyo Koho JP 01,318,046 [89,318,046] (Dec. 22, 1989; N. Inmaki and N. Honda to Toshiba Chemical Corp.); *C.A., 112*, 218392 (1990).
57. U.S. Pats. 5,075,413 and 5,075,414 (both Dec. 24, 1991; T. H. Daily, Jr. to INDSPEC Chemical Corp.).
58. Jpn. Kokai Tokkyo Koho JP 04,126,759 [92,126,759] (Apr. 27, 1992; I. Ide, T. Ise, and T. Korogi to Lignyte Co., Ltd.); *C.A., 117*, 172744 (1992).
59. PCT Int. Appl. WO 92,016,572 (Oct. 1, 1992; S. Horiki and K. Ito to Nagoya Oil Chemical Co., Ltd.); *C.A., 118*, 192968 (1993).
60. B. A. Adams and E. L. Holmes, *J. Soc. Chem. Ind., 54*, 1-GT (1935).
61. G. Schwachula, H. Hoffmann, and A. Eser, *Z. Chem., 30*, 270 (1990).
62. L. D. Pennington and M. B. Williams, *Ind. Eng. Chem., 51(6)*, 759 (1959).
63. V. Sykora and F. Dubsky, *Collect. Czech. Chem. Commun., 32(9)*, 3342 (1967).
64. USSR Pat. SU 231,800 (March 9, 1970; to V. A. Ilin, I. V. Samborski, L. L. Grachev, A. B. Pashkov, and L. A. Domina); *C.A., 73*, 26340 (1970).
65. Jpn. Kokai Tokkyo Koho JP 55,040,744 [78,040,744] (Sept. 18, 1978; M. Hirai and Y. Iwatani to Unitika K.K.); *C.A., 93*, 135804 (1980).
66. U.S. Pat. 4,423,159 (1983; A. Ebra and R. M. Wallace to U.S. Department of Energy).
67. J. Bartulin, M. Parra, A. Ramirez, and H. Zunga, *Polym. Bull. (Berlin), 24(2)*, 129 (1990).
68. D. Shen, *Lizi Jiaohuan Xu Xifu, 7(4)*, 281 (1991); *C.A., 117*, 28198 (1992).
69. K.K. Sharma, P. C. Gupta, and R. Prasad, *Res. Ind., 36(4)*, 246 (1991); *C.A., 117*, 71118 (1992).
70. PCT Int. Appl WO 91,009,891 (July 11, 1991; J. P. Bibler and R. M. Wallace to U.S. Department of Energy); *C.A., 115*, 234200 (1991).
71. K. D. Patel and M. M. Patel, *Synth. React. Inorg. Met.-Org. Chem., 23(2)*, 299 (1993); *C.A., 118*, 192796 (1993).
72. U.S. Pat. 3,332,245 (July 25, 1967; C. Caron to Soletanche).
73. U.S. Pat. 4,300,634 (Nov. 17, 1981; R. L. Clampitt to Phillips Petroleum Co.).
74. Eur. Pat. Appl. EP 177,324 (Apr. 9, 1986; P. W. Chang, C. N. Meltz, G. D. Gruetzmacher, and R. A. Totino to Pfizer Inc.); *C.A., 105*, 155982 (1986).
75. U.S. Pat. 4,903,766 (Feb. 27, 1990; P. Shu to Mobil Oil Corp.).
76. Ger. Offen. DE 2,117,774 (Oct. 28, 1971; to R. Matalon); *C.A., 76*, 36773 (1972).
77. U.S. Pat. 3,551,365 (Dec. 29, 1970; to R. Matalon).
78. Ger. Offen. DE 2,100,232 (Oct. 28, 1971; R. E. Melcher to Whitehead Bros., Inc.); *C.A., 76*, 36775 (1972).
79. Neth. Pat. Appl. 143,444 (Oct. 15, 1974; to Centro Sperimentale Metalico).
80. U.S. Pat. 4,089,837 (May 16, 1978; M. Luttinger, G. E. Cremeans, and L. J. Nowacki to Manley Bros., Inc.; Borden, Inc.).
81. Ger. Offen. DE 2,738,267 (March 1, 1979; W. Hesse, R. Sattelmeyer, and E. Teschner to Hoechst A.G.); *C.A., 90*, 187968 (1979).
82. Jpn. Kokai Tokkyo Koho JP 81,091,956 (July 25, 1981; to Kao Soap Co., Ltd.); *C.A., 96*, 39469 (1982).
83. Eur. Pat. Appl. EP 47,992 (March 24, 1982; W. Hesse, F. Landauer, and K. Schmiedel to Hoechst A.G.); *C.A., 96*, 218757 (1982).
84. U.S. Pat. 4,381,355 (Apr. 26, 1983; F. W. Henry, Jr., H. E. Anthony, III, and S. Banerjee to General Refractories Co.).
85. U.S. Pat. 4,387,173 (June 7, 1983; F. W. Henry, Jr., M. Wood, Jr., and L. Rankovic to General Refractories Co.).
86. Ger. Offen. DE 3,437,385 (May 15, 1985; T. F. Vezza to Dresser Industries, Inc.); *C.A., 103*, 41503 (1985).
87. U.S. Pat. 4,617,165 (Oct. 14, 1986; to P. H. S. Tsang, J. P. Coyle, and S. K. Rhea).
88. Jpn. Kokai Tokkyo Koho JP 62,016,839 [87,016,839] (Jan. 26, 1987; S. Takahasi, M. Sawa, and K. Kiuchi to Kao-Quaker Co., Ltd.); *C.A., 106*, 218132 (1987).
89. U.S. Pat. 4,644,022 (Feb. 17, 1987; R. Iyer to Acme Resin Corp.).

90. Jpn. Kokai Tokkyo Kho JP 01,315,411 [89,315,411] (Dec. 20, 1989; K. Mori, H. Matsushina, H. Kunimoto, and A. Ishigaki to Dainippon Ink and Chemicals, Inc.); *C.A.*, *113*, 7481 (1990).
91. Fr. Pat. 1,578,941 (Aug. 22, 1969; to Compagnie de Saint-Gobain); similar is U.S. Pat. 3,907,723 (Sept. 23, 1975; M. Pretot to Certainteed Products Corp.).
92. Fr. Pat. 1,588,803 (Apr. 24, 1970; D. Hanton to Compagnie de Saint-Gobain); *C.A.*, *73*, 121245 (1970).
93. U.S. Pat. 3,551,362 (Dec. 29, 1970; E. K. H. Decker to W. R. Grace & Co.).
94. Ger. Offen. DE 2,549,441 (May 13, 1976; R. Francombe to Ciba-Geigy A.G.); *C.A.*, *85*, 47682 (1976).
95. Ger. Offen. DE 2,834,794 (Feb. 14, 1980; H. Baumann to Schaum Chemie W. Bauer G.m.b.H.); *C.A.*, *93*, 96117 (1980).
96. Jpn. Kokai Tokkyo Koho JP 81,072,027 and 81,072,028 (both dated June 16, 1981; to Matsushita Electric Works, Ltd.); *C.A.*, *95*, 151699 (1981).
97. U.S. Pat. 4,546,119 (Oct. 8, 1985; J. Lunt, E. J. MacPhearson, and P. J. Meunier to Fiberglass Canada, Inc.).
98. Eur. Pat. Appl. EP 242,620 (Oct. 28, 1987; P. J. Meunier, J. Lunt, and E. J. MacPhearson to Fiberglass Canada, Inc.); *C.A.*, *108*, 39115 (1987).
99. Jpn. Kokai Tokkyo Koho JP 62,044,434 [87,044,434] (Feb. 26, 1987; C. Tsukamoto, H. Okuyama, R. Miyagaki, and T. Yasui to Toyo Rubber Industry Co., Ltd.); *C.A.*, *107*, 60799 (1987).
100. Jpn. Kokai Tokkyo Koho JP 59,124,940 (1984; to Asahi Organic Chemical Industry Co., Ltd.); *C.A.*, *101*, 172457 (1984).
101. R. W. Pekala and R. E. Stone, *Polymer Prepr.*, *29(1)*, 204 (1988).
102. U.S. Pat. Appl. U.S. 199,404 (Sept. 1, 1989; R. W. Pekala to U.S. Department of Energy; University of California, Livermore; Lawrence Livermore National Laboratory).
103. R. W. Pekala, *Mater. Res. Soc. Symp. Proc.*, *1990*, 171 (Polym. Based Mol. Compds.), 285.
104. R. W. Pekala, C. T. Alviso, and J. D. LeMay, *J. Non-Cryst. Solids*, *125(1–2)*, 67 (1990).
105. U.S. Pat. 4,997,804 (March 5, 1991; R. W. Pekala to U.S. Department of Energy).
106. Information kindly supplied by Dr. R. W. Pekala, Chem. Mater. Sci. Dept., Lawrence Livermore National Laboratory, Livermore, CA 94550.
107. X. Lu, M. C. Arduini-Schuster, J. Kuhn, O. Nilsson, J. Fricke, and R. W. Pekala, *Science (Washington)*, *255(5047)*, 971 (1992).
108. X. Lu, O. Nilsson, and J. Fricke, *J. Appl. Phys.*, *73(2)*, 581 (1993).
109. M. Hosoya, G. Reynolds, M. S. Dresselhaus, and R. W. Pekala, *J. Mater. Res.*, *8*, 811 (1993).
110. R. W. Pekala, C. T. Alviso, F. M. Fong, and S. S. Hulsey, *J. Non-Cryst. Solids*, *145(1–3)*, 90 (1992).
111. R. W. Pekala, C. T. Alviso, and J. D. Lemay in *Chemical Processing of Advanced Materials* (L. L. Hench and J. K. West, eds.), p. 671ff., Wiley, New York, 1992.
112. U.S. Pat. 4,992,254 (Feb 12, 1991; F. M. Kong to U.S. Department of Energy).
113. Jpn. Kokai Tokkyo Koho JP 03,014,843 [91,014,843] (Jan. 23, 1991; Y. Ueda and T. Hashiba to Showa High Polymer Co., Ltd.); *C.A.*, *115*, 10213 (1991).
114. Jpn. Kokai Tokkyo Koho JP 04,106,136 [92,106,136] (Apr. 8, 1992; I. Ide, T. Ise, and S. Suzukawa to Lignite K.K.); *C.A.*, *117*, 92393 (1992).
115. USSR Pat. SU 309,924 (Sept. 20, 1971; A. V. Satalkin, V. A. Solntseva, and E. I. Ivanov to Leningrad Railway Transport Engineering Institute); *C.A.*, *76*, 17423 (1972).
116. T. Sugama, L. E. Kukacka, and W. Horn, *J. Appl. Polymer Sci.*, *26*, 1453 (1981).
117. U.S. Pat. 4,414,034 (Nov. 8, 1983; S. A. Lipowski to Diamond Shamrock Corp.).
118. U.S. Pat. 4,440,577 (Apr. 3, 1984; A. W. Marcellis and G. S. Johnson to Diamond Shamrock Corp.).
119. USSR Pat. SU 1,047,863 (Aug. 8, 1984; K. F. Paus *et al.*, to Belgorod Construction Materials); *C.A.*, *100*, 11796 (1984).
120. USSR Pat. SU 1,645,267 (Apr. 30, 1991; G. A. Rezchikov, L. I. Simonanko, and V. I. Stambulko to All-Union Scientific Research Institute of Natural Gas; Moscow Engineering-Construction Institute); *C.A.*, *116*, 129954 (1992).

121. USSR Pat. SU 174,753 (Sept. 27, 1965; A. N. Mikhailov, S. M. Bresler, F. V. Miranov, and V. P. Suvorova to State Leather Factory, Ostashkov); *C.A., 64*, 2296 (1966).
122. U.S. Pat. 3,617,186 (Nov. 2, 1971; W. Windua, M. L. Happich, and G. S. Showell to U.S. Secretary of Agriculture).
123. Eur. Pat. Appl. EP 209,330 (Jan. 1, 1987; B. C. Burdett and S. M. Burkinshaw to Cogent Ltd.); *C.A., 106*, 157912 (1987).
124. Fr. Demande 2,012,741 (March 20, 1970; R. G. Bayless and D. J. Striley to National Cash Register Co.); *C.A., 74*, 32785 (1971).
125. Belg. Pat. 770,566 (Dec. 1, 1971; to Fuji Photo Film Co., Ltd.).
126. Ger. Offen. DE 2,253,050 (May 24, 1973; R. L. Hart, D. D. Emrick, and R. G. Bayless to National Cash Register Co.); *C.A., 77*, 54542 (1973).
127. Jpn. Kokai Tokkyo Koho JP 80, 145,524 (Nov. 13, 1980; to Kanzaki Paper Manufacturing Co.); *C.A., 94*, 210574 (1981).
128. Jpn. Kokai Tokkyo Koho JP 81, 021,639 (Feb. 28, 1981; to Kanzaki Paper Manufacturing Co.); *C.A., 95*, 63346 (1981).
129. Ger. Offen. DE 3,401,056 (July 19, 1984; Y. Igarashi and M. Takahashi to Kureha Chemical Industry Co., Ltd.); *C.A., 101*, 212342 (1984).
130. U.S. Pat. 4,594,370 (June 10, 1986; D. G. Adkins to The Mead Corp.).
131. Jpn. Kokai Tokkyo Koho JP 88,232,838 (Sept. 28, 1988; to Kanzaki Paper Mfg. Co.); *C.A., 110*, 145052 (1989).
132. Jpn. Kokai Tokkyo Koho JP 04,004,034 [92,004,034] (Jan. 8, 1992; H. Nozawa, M. Hirashima, and T. Kawame to Kuraray Co., Ltd.); *C.A., 117*, 28420 (1992).
133. Belg. Pat. 879,419 (Feb. 1, 1980; L. Cohen to B. F. Goodrich Co.); *C.A., 93*, 72884 (1980).
134. Belg. Pat. 879,573 (Feb. 5, 1980; L. Cohen to B. F. Goodrich Co.); *C.A., 93*, 27281 (1980).
135. PCT Int. Appl. WO 82,002,555 (Aug. 5, 1982; J. L. Dorsch and D. J. Smith to the B. F. Goodrich Co.); *C.A., 98*, 91156 (1983).
136. USSR Pat. SU 744,017 (June 30, 1980; to A. E. Bogatyrev, V. A. D'yakonov, *et al.*); *C.A., 93*, 169191 (1980).
137. Jpn. Kokai Tokkyo Koho JP 58,225,168[83,225,168] (Dec. 27, 1983; H. Takahashi, N. Morozumi, S. Takanezawa, and K. Nakoa to Tokyo Shibaura Denki K.K.); *C.A., 101*, 25093 (1984).
138. Ger. Offen. DE 3,515,210 (Oct. 31, 1985; Y. Saotome, H. Gokan, K. Saigo, M. Suzuki, and Y. Ohnishi to NEC Corp.); *C.A., 104*, 177742 (1985).
139. Eur. Pat. Appl. EP 192,365 (Aug. 27, 1986; H. Takahashi, N. Morozumi, S. Takanezawa, and K. Nakao to Hitachi Chemical Co.); *C.A., 106*, 6129 (1982).
140. U.S. Pat. 4,477,553 (Oct. 16, 1984; T. Yamamoto, K. Goto, and M. Uhara to Konishiroku Photo Industry Co., Ltd.).
141. U.S. Pat. 4,617,584 (Oct. 14, 1986; H. Ikeya and A. Hatanaka to Tokyo Shibaura Denki Kabushiki Kaisha).
142. U.S. Pat. 4,652,388 (March 24, 1987; to M. H. Gold).
143. Eur. Pat. Appl. EP 227,487 (July 1, 1987; Y. Hosaka, I. Nozue, M. Takatori, Y. Harita, and I. Nozue to Japan Synthetic Rubber Co.); *C.A., 108*, 140771 (1988).
144. Jpn. Kokai Tokkyo Koho JP 62, 096,513 [87,096,513] (May 6, 1987; M. Ashida and M. Zama to Showa Highpolymer Co., Ltd.); *C.A., 107*, 238707 (1987).
145. Ger. Offen. DE 3,630,996 (March 17, 1988; H. Ahne and W. Plundrich to Siemens A.G.); *C.A., 109*, 38861 (1988).
146. Jpn. Kokai Tokkyo Koho JP 03,001,142 [91,001,142] (Jan. 7, 1991; Y. Nakano, M. Kada, and S. Ito to Mitsubishi Petrochemical Co., Ltd.); *C.A., 114*, 256981 (1991).
147. Brit. Pat. GB 1,202,390 (Oct. 24, 1968; M. J. Armstrong to Astra Fireworks, Ltd.); *C.A., 73*, 89731 (1970).
148. C. W. Cooper and R. D. Falb, *Ann. N.Y. Acad. Sci., 146(1)*, 214 (1968).
149. U.S. Pat. 3,536,561 (Oct. 27, 1970; D. J. L. Griffiths to Westvaco Corp.).
150. U.S. Pat. 3, 532,648 (Oct. 6, 1970; D. L. Wilhelm to A. E. Staley Manufacturing Co.).

151. Ger. Offen. DE 1,446,913 (Dec. 16, 1971; to Comet-Apparatebau G.m.b.H.)
152. U.S. Pat. 3,558,424 (Jan. 26, 1971; E. P. Weiterer to American Cyanamid Co.).
153. U.S. Pat. 3,558,548 (Jan. 26, 1971; E. P. Weiterer to American Cyanamid Co.).
154. Jpn. Kokai Tokkyo Koho JP 53, 113,083 [78,113,083] (Oct. 3, 1978; Y. Hashimoto, H. Kaide, and K. Kimura to Kyowas Fermentation K.K.); *C.A., 90*, 50621 (1979).
155. Ger. (East) Pat. DD 145,175 (Nov. 26, 1980; E. Born, H. Hesse, and G. Henke to Kombinat VEB Chemische Werke Buna); *C.A., 94*, 193859 (1981).
156. USSR Pat. SU 979,290 (Dec. 7, 1982; F. P. Spiridonov and T. I. Termysheva to Chuvach State University); *C.A., 98*, 203475 (1983).
157. U.S. Pat. 4,392,963 (July 12, 1983; H. Perl, D. Nussbaumer, H. Kluever, and H. Beer to Sartorius G.m.b.H.).
158. U.S. Pat. 4,522,871 (June 11, 1985; to R. A. Armellino, Jr. and S. E. Armellino).
159. Ger. Offen. DE 3,615,233 (May 6, 1986; R. B. Neale and K. Linke to N. B. Love Industries Pty., Ltd., New South Wales, Australia); *C.A., 107*, 9179 (1987).
160. Ger. (East) Pat. DD 244,559 (Apr. 8, 1987; E. Born, M. Goritzka, G. Bageritz, G. Mehl, U. Schumann, W. Taenzer, and K. Hanemann to VEB Chemisch- und Lackfabrik Doellnitz); *C.A., 107*, 200618 (1987).
161. Jpn. Kokai Tokkyo Koho JP 62,238,742 [87,238,742] (Oct. 19, 1987; K. Iida, H. Yamato, and Y. Sumino to Takiron Co., Ltd.); *C.A., 108*, 95822 (1987).
162. Eur. Pat. Appl. EP 392,682 (Oct. 17, 1990; L. Normandin to Albany International Corp.); *C.A., 114*, 166635 (1991).
163. Jpn. Kokai Tokkyo Koho JP 04,059,806 [92,059,806] (Feb. 26, 1992; T. Mori, K. Kondo, T. Nakano, and K. Mukai to Yunichika K.K.); *C.A., 117*, 27938 (1992).
164. PCT Int. Appl. WO 93,004,711 (March 18, 1993; R. G. Johnson to Beth Israel Hospital); *C.A., 118*, 175888 (1993).

6

m-Aminophenol

6.1. INTRODUCTION

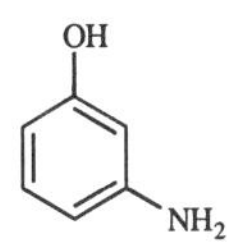

6-1

m-Aminophenol (MAP), **6-1**, is an important starting material for dyes, including a variety of leuco (or latent) dyes used in imaging technology, optical bleaches and fluorescent agents (see Chapter 10), drugs (see Chapter 8), agricultural chemicals (see Chapter 7), and high-performance polymers (see Chapter 9). Although the synthesis of *m*-aminophenol(s) by the reaction of resorcinol with ammonia or (di)alkylamines has been known for about 100 years,[1,2] the preferred manufacturing process had been for many years a route involving the sulfonation of nitrobenzene to *m*-nitrobenzenesulfonic acid, the reduction of the nitro group to give *m*-aminobenzenesulfonic acid, followed by the caustic fusion of sodium *m*-aminobenzenesulfonate to yield sodium *m*-aminophenolate. The *m*-aminophenol was isolated/purified by neutralization, filtration, and recrystallization; the overall yield of *m*-aminophenol from nitrobenzene was estimated by SRI International to be 58%.[3] (Doubtless, process improvements were made by the producers using this process.) The nitrobenzene/sulfonation/reduction/caustic fusion route is still being used commercially, notably by ACNA in Italy, the largest producer of MAP, as well as a few smaller manufacturers. Sumitomo Chemical Co. used this process as well into the early 1980s and had licensed their knowhow to Hindustani Organic Chemicals, Ltd. of India which started up a plant on this basis with a capacity for MAP of 1.5 mill. lb /yr in 1970.[4] The output of the Indian plant was said to be devoted to making sodium *p*-aminosalycilate (PAS; see Chapter 8), a tuberculostatic agent.

Sumitomo Chemical Co., however, developed the process for making MAP from resorcinol and commercialized it in a plant in Japan with a 2.2 mill.lb /yr capacity in

1987.[5] Subsequently, Mitsui Chemical Co. in Japan also opened a resorcinol-based MAP facility. Both Sumitomo and Mitsui had earlier pioneered the commercial production of resorcinol by the *m*-diisopropylbenzene hydroperoxidation/cleavage route. It is almost certain that the resorcinol amination route for MAP is the most economical process, at least for a producer of resorcinol.

The following sections provide selected references, from *ca* 1970 on, to processes for making MAP from resorcinol (a flow diagram for this process is given), 1,3-cyclohexanedione, and related compounds. The literature on reacting resorcinol with primary or secondary amines, and the N-alkylation of MAP as routes *N,N*-dialkylaminophenols are covered also because these compounds are starting materials of the synthesis of leuco dyes used in fast-growing imaging technologies (see Chapter 10). The given information demonstrates substantial development efforts on these processes/products, leading to the realization that there were only a few winners from all this activity.

6.2. *m*-AMINOPHENOL FROM RESORCINOL

The simplified equation (6.1) outlines this process.

(6.1)

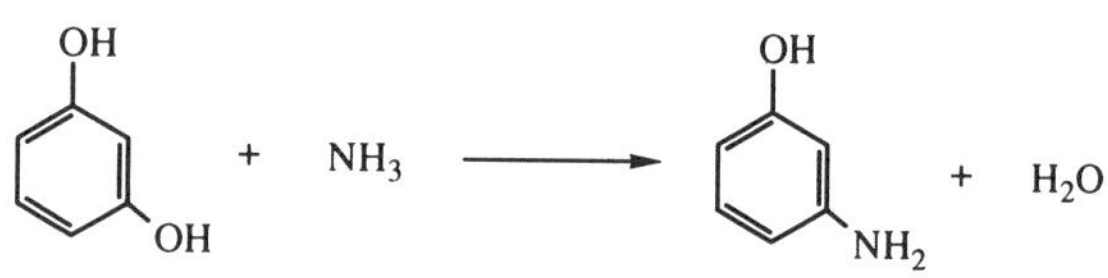

Before going into details, a comment on yield data given in the (patent) literature is helpful. Often, especially in literature abstracts (such as *C.A.* and Derwent reports), yield (or selectivity) data for MAP from resorcinol are given without giving the per-pass conversion of resorcinol. Even where conversion and yield numbers are given, these are analytical data; in-hand yields of high-purity MAP can be expected to be lower than the yields by analysis due to losses of product in the isolation/separation of the crude products and in the purification step(s).

In the older literature, catalysts such as ammonium sulfite, ammonium chloride, ammonium phosphate, ammonium arsenate, boric acid, or zinc chloride were used to promote the slow amination of resorcinol with aqueous ammonia at *ca* 100–200 °C and up to 100 atm (1470 psig) pressure; the MAP was isolated by crystallization/filtration or by extraction/distillation/crystallization. In these older processes, large amounts of inorganic salts had to be handled, there were problems with the recycle of mother liquors containing resorcinol and the easily oxidized MAP, and large amounts of waste water were produced. In some cases the conversion of resorcinol was kept low to minimize byproducts such as *m*-phenylenediamine and/or higher molecular weight materials. There was also a color stability problem with MAP, since MAP darkens on storage unless the purity is very high.

For all these reasons, and also because of the increasing markets for MAP and its derivatives, industrial development work continued from the 1970s to today.

Thus, a nonaqueous process was developed involving the passage of anhydrous ammonia through molten resorcinol at 150–180 °C. The crude product was distilled and the distillate was extracted with diisopropyl ether or water/diisopropyl ether (presumably, MAP was then separated from resorcinol by extraction with aqueous mineral acid, etc.); the conversion of resorcinol was low. In a continuous process using this procedure, employing a two-stage cascade apparatus, 125 g/hr resorcinol and a total of 34 g/hr ammonia were fed at 160 °C to the first and second stage, with an 8-hr residence time; the conversion of resorcinol was 31% and the yield of (selectivity for) MAP was 85%.[6]

A new process variation was described in which resorcinol was reacted with ammonia or primary or secondary amines and an alkanoic (e.g., propionic) acid in xylene solution at 200 °C in a pressure vessel to give a *ca* 65% yield of the appropriate *m*-(alkyl)aminophenol.[7]

A vapor-phase amination of resorcinol was developed using oxides of P, Zn, V, Mo, or W as the catalyst. For example, a 60% aqueous solution of resorcinol and ammonia was passed over a P_2O_5-on-silica catalyst at 320 °C for 5 hr to obtain a 31% conversion of resorcinol and 99% selectivity to MAP.[8] (Presumably, the lifetime of this catalyst was low due to the water in the feed.)

A highly selective process for aminating resorcinol with ammonia or primary or secondary amines was based on the use of Sb, V, or Fe compounds and ammonium compounds as the catalysts. For instance, 45 mmol resorcinol was heated for 2 hr/230 °C with 112 mmol 20% ammonia, 4.4 mmol $Fe_3(PO_4)_2{\cdot}8H_2O$, and 18.7 mmol $(NH_4)_3PO_4{\cdot}3H_2O$ under nitrogen in an autoclave to give a 52% conversion of resorcinol and 100% selectivity to MAP.[9]

A high conversion of resorcinol (or hydroquinone) was achieved on amination with ammonia or ethylamine in the presence of $Ni_3(PO_4)_2$ or $Zn_3(PO_4)_2$ and ammonium phosphates as the catalysts. For instance, with resorcinol and ammonia at 270 °C, the resorcinol conversion was 93% and a 73% yield of MAP plus an 18% yield of *m*-phenylenediamine were obtained.[10] (Likely, this crude product would be difficult to purify.)

A novel, vapor-phase amination of resorcinol was developed in which a Ga-oxide-on-silica catalyst was used. Passing 60% aqueous resorcinol and ammonia at 1/10 mole ratio over this catalyst at 320 °C and GHSV 1200 hr^{-1} for 5 hr, a 26% conversion of resorcinol and 96% selectivity to MAP were obtained.[11]

High conversion of resorcinol and high selectivity to MAP were claimed when resorcinol was reacted with aqueous ammonia in the presence of ammonium molybdate. Thus, heating 110 parts resorcinol with 194 parts 27.5 wt% NH_4OH, and 35 parts $(NH_4)_6Mo_7O_{24}{\cdot}H_2O$ at 200 °C in an autoclave gave 68.9 parts MAP (said to be 90.5% selectivity) at 93.8% conversion of resorcinol.[12]

Another vapor-phase process was patented in which resorcinol was aminated in phenol solution with ammonia or primary or secondary amines in the presence of a

Mo-containing catalyst [prepared by heating a mixture of ammonium paramolybdate and $Al(NO_3)_3{\cdot}9H_2O$ for 24 hr/100 °C and 5 hr/500 °C]. For example, when resorcinol and ammonia in phenol were heated for 6 hr/210 °C in an autoclave in the presence of this catalyst, a 95% conversion of resorcinol and a 91.3% yield of MAP plus a 4.2% yield of *m*-phenylenediamine were obtained.[13]

Another patent claimed improvements by reacting resorcinol with ammonia in water in the absence of a catalyst. Thus, a solution of 2.0 m resorcinol in 2.5 m ammonia (50% aqueous solution) was heated for 2 hr/230 °C at 32–36 kg/cm^2 pressure to give a 60% conversion of resorcinol and 85% selectivity to MAP.[14] A follow-up patent described carrying out this process, with or without a catalyst, and included the purification of the product by extraction under an inert gas (such as nitrogen) of resorcinol with an aliphatic, dialkyl ether in which MAP is insoluble to give MAP of high quality. For instance, 1.0 m resorcinol was aminated with 1.5 m ammonia (28% aqueous solution) in a stirred autoclave, which had been flushed with nitrogen, for 3 hr/240 °C; the excess ammonia was then purged and GC analysis of the product indicated a 60% conversion of resorcinol and an 84% yield of MAP were obtained. The crude was then distilled (water at 100 mm Hg, the organics at 5 mm Hg) and the molten distillate, *ca* 95 g, was contacted with 200 g di-*n*-butyl ether under a nitrogen atmosphere at 120 °C, then cooled to 30 °C to crystallize the MAP which was filtered. This product was treated similarly with 200 g fresh di-*n*-butyl ether (optionally in the presence of water containing a surfactant to keep the crystals from sticking to the walls of the reactor), filtered again, and vacuum-dried at 60 °C to give 45.3 g (68.6% in-hand yield based on the resorcinol converted) of MAP of 99.3% purity. The two mother liquors were combined, stripped of di-*n*-butyl ether, and the residue was recycled.[15]

An alternate, preferably vapor-phase, process consisted of the amination of 1.0 m resorcinol with 5.0–30 m ammonia at 230–500 °C over a gallium oxide catalyst.[16]

A new, aqueous process involved the amination of resorcinol in the presence of a water-soluble catalyst. For instance, 110 g (1.0 m) resorcinol, 35 g $(NH_4)_6Mo_7O_{24}{\cdot}4H_2O$, and 194 g (3.1 m) 27.5% aq. ammonia were reacted at 200 °C and 22–33 kg/cm^2 pressure to give a 93.8% conversion of resorcinol and 90.5% selectivity to MAP (the in-hand yield of recrystallized MAP was close to 70%).[17]

Another patented variation was the reaction of resorcinol with ammonia in the absence of a catalyst in a monohydric phenol as the solvent at 180–260 °C in a stirred autoclave. For example, resorcinol and ammonia in 1/2 mole ratio in phenol solvent were reacted under nitrogen at 230 °C/5 hr to give a 65% conversion of resorcinol and 84% selectivity to MAP.[18]

Finally, the reaction of resorcinol with ammonia in aromatic hydrocarbons, such as toluene or xylene, or in a monohydric phenol as the solvent in the presence of an aluminosilicate catalyst, such as an activated montmorillonite clay or various high-silica zeolites, at *ca* 220 °C/5 hr gave, e.g., a 91% conversion of resorcinol and an 88.6% ultimate, in-hand yield (based on recycling unreacted resorcinol) of

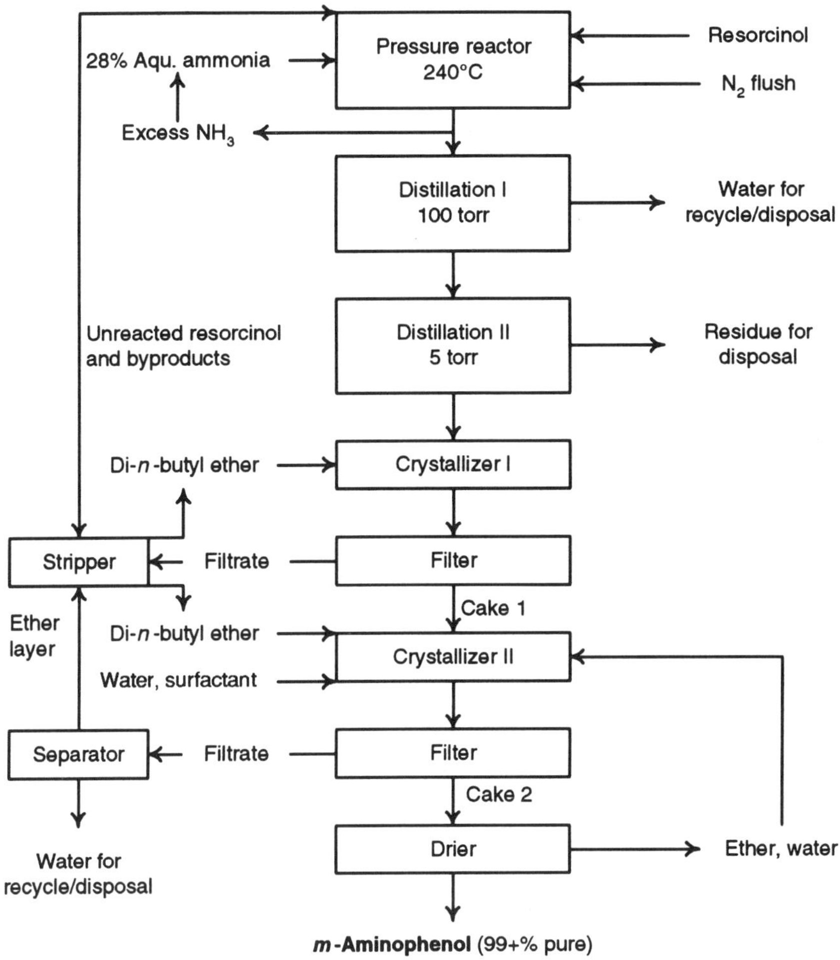

Figure 6.1. Process flow diagram for *m*-aminophenol from resorcinol.

MAP. The crude product was filtered to remove the catalyst, the filtrate was distilled, and the crude MAP was recrystallized from alcohol (methanol, butanol) to obtain high-purity MAP; the used catalyst could be recycled many times.[19]

The actual processes used by Sumitomo and Mitsui in their MAP-from-resorcinol plants are, of course, not known in any detail. A hypothetical process flow diagram for such a process is given in Figure 6.1, based on the description contained in U.S. Pat. 4,585,896 which issued to Sumitomo Chemical Co. in 1986.[15]

6.3. *N*-ALKYL- AND *N,N*-DIALKYL-*m*-AMINOPHENOLS FROM RESORCINOL OR *m*-AMINOPHENOL

All of the above processes for MAP itself from resorcinol can be used here by substituting alkylamines, dialkylamines, or cycloalkylamines for ammonia. The same problems of incomplete conversion of resorcinol, formation of byproducts, and separation and purification of products apply here as well. The stability of the products can be worse (e.g., reagent chemicals were received from supply houses as black tars) than with MAP. The melting points of the N-alkylated products are often lower than that of MAP (mp 122–123 °C) and their solubilities in organic solvents are higher than that of MAP, which can be a problem in recovering recrystallized material in high yield from crudes. In part because of these problems, an often-used way to make the N-alkylated MAPs had been the N-alkylation of MAP with alkyl halides in the presence of a base, or with alcohols in the presence of a strong acid.

Because of the continuously growing demand for the *N,N*-dialkyl-MAP-derived leuco dyes in imaging technology such as fax transmission, much process development work for these starting materials was done in the 1980s and continues today. Separate process patents for making N-alkylated MAPs were applied for and issued. Selected references follow.

Thus, resorcinol was aminated with heterocyclic amines in the presence of phosphoric acid or phosphite esters. For example, 770 parts resorcinol, 653 parts morpholine, and 41 parts 50% H_3PO_4 were reacted for 12hr/210 °C and 6 hr/230 °C to give the N-substituted MAP, **6-2**, in 86% yield; substituting pyrrolidine or piperidine (for the morpholine) gave compounds **6-3** and **6-4**, respectively, in good yields.[20]

The amination of resorcinol (and other dihydric phenols) with ammonia or a secondary or primary alkylamine in phenol as the solvent worked reasonably well. For instance, the reaction of 36.4 mmol resorcinol with 72.8 mmol Et_2NH in 182 mmol phenol at 250 °C/5 hr under nitrogen in an autoclave gave a 73% conversion of resorcinol and 70% selectivity to *m*-(diethylamino)phenol.[21]

m-(Cycloalkylamino)phenols were also produced by the alkylation of MAP with a cycloalkanone in the presence of a reducing agent such as Zn, Fe, Sn, or Pd-on-C with HCl, H_2SO_4, or CH_3COOH in water or in organic solvents at 5–50 °C; the yields of the *N*-cycloalkyl-MAPs were in the 50–95% range.[22]

N-Monoalkyl- or *N,N*-dialkyl-MAPs were prepared by reacting MAP with alkyl halides in water under heat and pressure, feeding ammonia (deoxidizing agent, base) continuously. This process was said to improve the selectivity to *N,N*-dialkyl-MAP, to

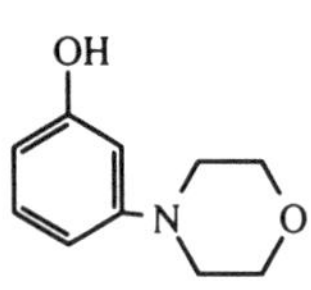

6-2

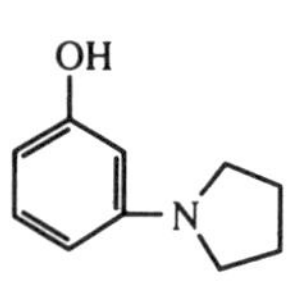

6-3

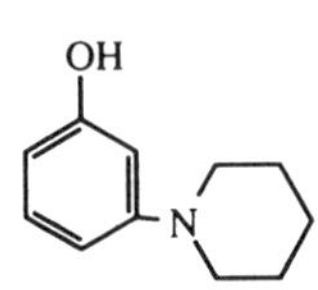

6-4

facilitate the separation of products, to produce less waste water and sludge, and to prevent the deterioration of reactants.[23]

A noncorrosive, reductive N-alkylation method for MAP was patented in which MAP was reacted with aldehydes in the presence of hydrogen and Pt- or Pd-on-C catalysts. Thus, MAP was charged to a stirred autoclave with methanol as the solvent; an equimolar amount of acetaldehyde was fed successively over 2 hr at 40 °C and 10 kg/cm^2 H_2 to give *m*-(ethylamino)phenol with 88.1% selectivity and 96.9% conversion of MAP.[24] When a 3/1 mole ratio of acetaldehyde/MAP was used under the same conditions, *m*-(diethylamino)phenol was obtained with 96.2% selectivity and 100% conversion of MAP.[25]

The N-dibutylation of MAP with butyl iodide and sodium carbonate as the acid-trapping agent in aqueous isopropanol at reflux for 6 hr gave a 100% conversion of MAP and 96.5% selectivity to the desired product.[26]

An alternate process for *N,N*-(dibutylamino)phenol involved the reaction of MAP with an excess (5–35% over the stoichiometric requirement) of butyraldehyde in methanol in the presence of 0.5–5.0 wt% (based on MAP) of acetic acid or butyric acid, 5% Pd-on-C catalyst, and 10 kg/cm^2 hydrogen pressure at 42 °C (max). The charge was worked up by filtration and stripping the filtrate of solvent and unreacted butyraldehyde, then vacuum-distilling the residual, crude product to give an 84% yield of *m*-(dibutylamino)phenol of 98.5% purity.[27]

A preparation of *N*-monoalkyl-*m*-aminophenol by the direct amination of resorcinol was patented. For instance, a mixture of resorcinol and methylamine was reacted at 160 °C and 10 kg/cm^2 pressure for 2 hr to give a 49.2% conversion of resorcinol and 92.9% selectivity to *m*-(methylamino)phenol. The crude product was purified by extractive distillation with tetraethylene glycol, the distillate containing 30.6 wt% resorcinol, 58.3% *m*-(methylamino)phenol, and 2% *m*-$C_6H_4(NHMe)_2$; this distillate was treated countercurrently with ethylene dichloride to obtain *m*-(methylamino)phenol of 96.6% purity.[28]

A new wrinkle in the direct amination of resorcinol with alkylamines consisted of the removal of water as it was generated in the reaction. For instance, a charge of resorcinol, dimethylamine, and aqueous phosphoric acid in phenol was refluxed at 200 °C for 5 hr with continuous removal of water. A 61.2% yield (based on the resorcinol charged) of *m*-(dimethylamino)phenol was obtained (without removal of water the yield was 51.4%).[29]

N-Cyclohexyl-*N*-methyl-*m*-aminophenol was prepared by first reacting resorcinol with cyclohexylamine at 170 °C for 12 hr to give *m*-(cyclohexylamino)phenol. To the crude product was added xylene and methyl *p*-toluenesulfonate and the mixture was refluxed for 10 hr to obtain an 85% yield (based on the resorcinol converted) of *N*-cyclohexyl-*N*-methyl-*m*-aminophenol.[30]

A preparation of *N*-alkoxyalkyl-*N*-alkyl-*m*-aminophenol was patented. For example, resorcinol was first reacted with 2-methoxyethylamine for 12 hr/130–135 °C, then the excess alkoxyalkylamine was distilled off, and the residue was treated with isobutylbromide for 8 hr/80–90 °C; *m*-[*N*-(2-methoxyethyl)-*N*-isobutylamino]phenol was obtained in 85% yield.[31]

Another two-step N-alkylation process for MAP was patented. For instance, 660 parts resorcinol, 483 parts $BuNH_2$, and 33 parts H_3PO_4 were heated in an autoclave for 8 hr/200 °C. The monoalkylation product was then treated with 863 parts BuBr in water in the presence of KI and a quaternary ammonium compound to give 850 parts *m*-(dibutylamino)phenol.[32]

6.4. *m*-AMINOPHENOL(S) FROM 1,3-CYCLOHEXANEDIONE(S)

The general reaction scheme is shown in Eq. (6.2). 1,3-Cyclohexanedione (shown in the enol form), **6-5**, is aminated to produce the 3-amino-2-cyclohexen-1-one, **6-6**, which is dehydrogenated to the corresponding MAP, **6-7**.

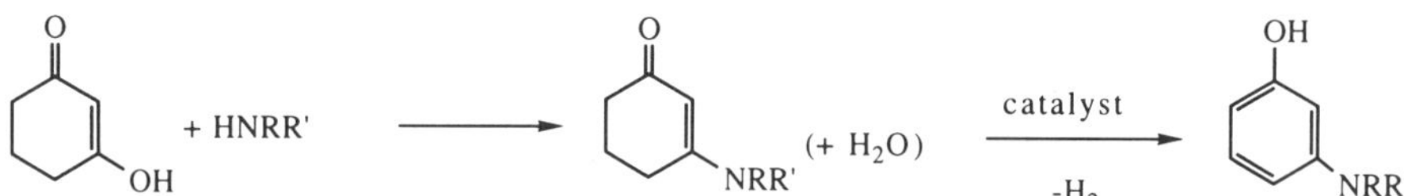

(6.2) (where R, R' = H or alkyl)

6 - 5 **6 - 6** **6 - 7**

This process is also applicable to the preparation of *N*-aryl-*m*-aminophenols by using an arylamine as the aminating agent; these products are not discussed further here. This 1,3-cyclohexanedione route to *m*-aminophenols is not important industrially for economic reasons at this time, but may be useful in the case of certain specialties, e.g., in the preparation of ring-alkylated *m*-aminophenols via alkyl-1,3-cyclohexanediones.

Three examples of this approach from the patent literature follow.

The amination of 1,3-cyclohexanedione with diethylamine in benzene at reflux for 5 hr, with azeotropic removal of water, followed by cooling or concentrating the solution gave 3-diethylamino-2-cyclohexen-1-one (**6-6**, R,R′ = Et) in 66% yield. This enaminoketone was dehydrogenated in the presence of a 10% Pd-on-C catalyst in triethyleneglycol at 193–196 °C until the evolution of hydrogen stopped; the yield of *N,N*-diethyl-*m*-aminophenol was 85% of the theory. MAP and *m*-(isopropylamino)phenol were prepared similarly from the appropriate enaminoketone precursor in 92–95% yield.[33] A vapor-phase version of the dehydrogenation step, using a fixed-bed catalyst , at *ca* 340 °C was patented also.[34]

A recent patent to duPont claimed the dehydrogenation of 3-amino-2-cyclohexen-1-one, using a Pd-on-C or Pd/Pt-on-C catalyst, in a solvent in the presence of a base. The MAP thus produced could be used without isolation/purification to prepare 3,4′-diaminodiphenyl ether, a monomer for making aramids.[35] (Interestingly, the 3-amino-2-cyclohexen-1-one was obtained by the catalytic hydrogenation of *m*-

phenylenediamine to 3-amino-2-cyclohexen-1-imine, which was hydrolyzed to the precursor to MAP.)[36]

6.5. STABILIZATION OF *m*-AMINOPHENOL

High-purity *m*-aminophenol is colorless and color-stable for a long time, at least if stored in a closed container. Nonetheless, the following two patents claim stabilizers to prevent the discoloration of MAP (the patent abstract did not give the purity of the starting MAP).

Thus, it was claimed that MAP to which 0.0005 wt% $NaHSO_3$ was added was color-stable when left in the dark at room temperature for two months.[37]

Another way to stabilize the low color of MAP was reported to be the addition of 500 ppm triphenylphosphine; the product was color-stable for a long time even when stored under moist air.[38]

6.6. REFERENCES

1. Ger. Pat. 49,060 (Aug. 21, 1889; to A. Leonhardt & Co.).
2. Ger. Pat. 121,683 (May 13, 1901; to Badische Anilin- Und Soda-Fabrik).
3. SRI (Stanford Research Institue)-PEP (Process Economics Program) Report No. 171B, *Pesticides and Intermediates—Supplement B*, by Y.-C. Yen and D. J. Lin, with contributions by C.-W. Lee, Sept. 1986, p. 169.
4. *Jap. Chem. Week, Sept. 17, 1970*, p. 3.
5. *Chem. Week, Aug. 20, 1986*, p. 73.
6. Ger. Offen. DE 2,140,786 (Feb. 22, 1973; R. Berthold, A. Topp, and G. Schaeffer to Farbwerke Hoechst A.-G.); *C.A., 78*, 124270 (1973).
7. Ger. Offen. DE 2,426,186 (Dec. 18, 1975; H. J. Quadbeck-Seeher and H. Hoch to BASF A.-G.); *C.A., 84*, 89816 (1976).
8. Jpn. Kokai Tokkyo Koho JP 55,053,250 (Apr. 18, 1980; to Mitsui Petrochemical Industries, Ltd.); *C.A., 93*, 114135 (1980).
9. Jpn. Kokai Tokkyo Koho JP 80,004,338 (Jan. 12, 1980; T. Nagaoka, S. Shimizu, and H. Hayashi to Mitsui Petrochemical Industries, Ltd.); *C.A., 93*, 7837 (1980).
10. Jpn. Kokai Tokkyo Koho JP 80,015,412 (Feb. 2, 1980; T. Nagaoka, S. Shimizu, and H. Hayashi to Mitsui Petrochemical Industries, Ltd.); *C.A., 93*, 167890 (1980).
11. Jpn. Kokai Tokkyo Koho JP 80,053,246 (Apr. 18, 1980; to Mitsui Petrochemical Industries, Ltd.); *C.A., 94*, 15389 (1981).
12. Jpn. Kokai Tokkyo Koho JP 61,007,239 [86,007,239] (Jan. 13, 1986; F. Matsunaga, E. Kato, T. Kimura, and Y. Isoda to Mitsui Petrochemical Industries, Ltd.; Honshu Chemical Industry Co., Ltd.); *C.A., 104*, 186128 (1986).
13. Jpn. Kokai Tokkyo Koho JP 60,215,654 [85,215,654] (Oct. 29, 1985; M. Tanaka, K. Taniguchi, and K. Saeki to Mitsui Petrochemical Industries, Ltd.); *C.A., 106*, 49767 (1987).
14. Eur. Pat. Appl. EP 197,633 (Oct. 15, 1986; H. Harada, H. Maki, and S. Sasaki to Sumitomo Chemical Co., Ltd.); *C.A., 106*, 66891 (1987).
15. U.S. Pat. 4,585,896 (Apr. 29, 1986; H. Harada, H. Maki, and S. Sasaki to Sumitomo Chemical Co., Ltd.).
16. Jpn. Pat. JP 86,051,577 (Nov. 10, 1986; to Mitsui Petrochemical Industries, K.K.).
17. Jpn. Pat. JP 86,007,239 (Jan. 13, 1986; F. Matsunaga, E. Kato, T. Kimura, and Y. Isoda to Mitsui Petrochemical Industries, Ltd.; Honshu Chemical Industry Co., Ltd.).

18. Jpn. Kokai Tokkyo Koho JP 81,073,048 (June 17, 1981; to Mitsui Petrochemical Co., Ltd.); *C.A.*, *95*, 150185 (1981).
19. Eur. Pat. Appl. EP 449,546 (Oct. 2, 1991; H. Dressler to INDSPEC Chemical Corp.); equiv. U.S. Pat. 5,130,489 (July 14, 1992; same assignment).
20. Ger. Offen. DE 2,900,193 (July 24, 1980; H. J. Kollinger and H. Scheuermann to BASF A.-G.); *C.A.*, *93*, 186187 (1980).
21. Jpn. Kokai Tokkyo Koho JP 81,073,048 (June 17, 1981; to Mitsui Petrochemical Industries, Ltd.); *C.A.*, *95*, 150185 (1981).
22. Eur. Pat. Appl. EP 181,505 (May 21, 1986; N. Yonese and M. Kondo to Kanzaki Paper Manufacturing Co., Ltd.); also, U.S. Pat. 4,713,488 (Dec. 15, 1987; same assignment).
23. Jpn. Kokai Tokkyo Koho JP 62,048,654 [87,048,654] (March 3, 1987; H. Harada, H. Maki, and S. Sasaki to Sumitomo Chemical Industries, K.K.); *C.A.*, *107*, 58645 (1987).
24. Jpn. Kokai Tokkyo Koho JP 62,258,345 [87,258,345] (Nov. 10, 1987; H. Harada, H. Maki, and S. Sasaki to Sumitomo Chemical Co., Ltd.); *C.A.*, *109*, 149059 (1988).
25. Jpn. Kokai Tokkyo Koho JP 62,258,346 [87,258,346] (Nov. 10, 1987; H. Harada. H. Maki, and S. Sasaki to Sumitomo Chemical Co., Ltd.); *C.A.*, *109*, 149060 (1988).
26. Jpn. Kokai Tokkyo Koho JP 02,101,053 [90,101,053] (Apr. 12, 1990; T. Deu, K. Tsuzukibashi, Y.Suzuki, S. Sanpei, K. Hatano, and T. Kurihara to Koriyama Kasei Co., Ltd.); *C.A.*, *113*, 40149 (1990).
27. Jpn. Kokai Tokkyo Koho JP 03,063,250 [91,063,250] (March 19, 1991; H. Maki, M. Kawasaki, H.Shimizu, and T. Ito to Sumitomo Chemical Co., Ltd); *C.A.*, *115*, 73626 (1991).
28. Jpn. Kokai Tokkyo Koho JP 03,020,251 [91,020,251] (Jan. 29, 1991; M. Kondo, Y. Kurano, S. Yamamoto, and M. Tanaka to Mitsui Petrochemical Industries, Ltd.); *C.A.*, *114*, 228523 (1991).
29. Jpn. Kokai Tokkyo Koho JP 03,099,043 [91,099,043] (Apr. 24, 1991; M. Kondo and N. Hirowatari to Mitsui Petrochemical Industries, Ltd.); *C.A.*, *115*, 182822 (1991).
30. Jpn. Kokai Tokkyo Koho JP 04,077,460 [92,077,460] (March 11, 1992; M. Nakatsuka, A. Otsuji, and T. Yamaguchi to Mitsui Toatsu Kagaku K.K.); *C.A.*, *117*, 89957 (1992).
31. Jpn. Kokai Tokkyo Koho JP 04,300,856 [92,300,856] (Oct. 23, 1992; M. Nakatsuka, A. Otsuji, and T. Yamaguchi to Mitsui Toatsu Chemicals, Inc.); *C.A.*, *118*, 128198 (1992).
32. Eur. Pat. Appl. EP 514,687 (Nov. 25, 1992; M. Hauptreif and H. Reichelt to BASF A.G.); *C.A.*, *118*, 193632 (1992).
33. U.S. Pat. 4,212,823 (July 15, 1980; W. H. Mueller to Hoechst A.-G.).
34. Ger. Offen. DE 3,120,245 (Dec. 9, 1982; W. Reimann and W. H. Mueller to Hoechst A.-G.); *C.A.*, *98*, 143117 (1983).
35. PCT Int. Appl. WO 92,018,445 (Oct. 29, 1992; S. E. Jacobson to E. I. duPont de Nemours and Co.); *C.A.*, *118*, 24107 (1992).
36. U.S. Pat. 5,149,874 (Sept. 22, 1992; S. E. Jacobson to E. I. duPont de Nemours and Co.).
37. Jpn. Kokai Tokkyo Koho JP 62,061,956 [87,061,956] (March 18, 1987; H. Harada, H.Maki, and S. Sasaki to Sumitomo Chemical Co., Ltd.); *C.A.*, *107*, 115359 (1987).
38. Jpn. Kokai Tokkyo Koho JP 02,209,851 [90,209,851] (Aug. 21, 1990; T. Ogino and H. Ishida to Mitsui Petrochemical Industries, Ltd.); *C.A.*, *114*, 6015 (1991).

7

Agricultural Chemicals, Including Veterinary Products

7.1. INTRODUCTION

A wide variety of derivatives of resorcinol and *m*-aminophenol, ranging from simple compounds to complex structures and compounds modeled on natural products, have been reported to be useful as agricultural chemicals. Often, certain types of compounds received intensive attention from the agrochemical research community for a decade or more before interest waned. For example, work on carbamates of phenolic compounds was most active in the 1950s through the early 1970s; diphenyl ethers and phenoxy-diphenyl ethers were pursued vigorously in the 1970s and 1,3-cyclohexanediones have been worked on by many companies from the mid-1970s to the present. Other examples of items of prolonged interest are the 7-hydroxycoumarins, -chromans, -chromanones, and -chromenes (particularly their phosphates and thio-phosphates), benzofurans, 7-alkoxyisoflavones, phenyl ureas, derivatives of 2,6-dihydroxybenzoic acid (gamma-resorcylic acid), analogs of the foregoing types but based on *m*-aminophenol, zearalenones/zearalenols, and others.

Some of these compounds were not made by synthesis but by extraction from natural or fermentation products. This variety of structures useful as herbicides, insecticides, fungicides, acaricides (miticides), sugar-enhancers, plant and animal growth regulators, or veterinary products appear both in the commercial entities and in the experimental compounds.

There were many more participants in the agricultural chemical industry in the 1970s, and earlier, than in the 1990s. As environmental regulations, testing requirements, potential liabilities, and the cost of research/development escalated rapidly, there was a great restructuring in this industry. The accident (sabotage?) in Union Carbide's Sevin carbamate insecticide plant in Bhopal, India, which cost close to 4000 lives and injured many more, resulted in litigation which is still ongoing in 1993; this accident was not only a terrible tragedy but it also fed the fires of the opponents of the industry and its products. Not only accidents but the toxicity of these chemicals and their residues in the ground, in water, and in the food chain became major issues. The agricultural chemical industry, which helped to increase agricultural productivity a great deal, which helped to make food abundant in many regions as well as affordable and safe, had little time to enjoy the success. Marginal producers and companies who thought their resources would be better employed elsewhere sold their agrochemical businesses to those U.S. and foreign operators who had larger volumes and profits here, who committed themselves to stay in and do whatever it takes, to search for more effective (smaller application) and less toxic compounds, even to use biotechnology to come up with safer products. At the same time, "organic farming" (no chemicals) became fashionable but, likely, it became apparent that this meant lower supply and higher prices, and this wave diminished.

As far as the chemistry is concerned, all of the above factors were reflected in a tremendous volume of agrichemical patent literature post-World War II into the 1970s and 1980s, and a decided drop-off thereafter. Nonetheless, there were pockets of concentrated effort in the past ten years, as will become apparent in the selection from the literature later on.

7.2. COMMERCIAL PRODUCTS

Coumaphos, **7-1**, the diethyl thiophosphate of 3-chloro-4-methyl-7-hydroxycoumarin, is an anthelmintic and antiparasitic agent sold by Bayer (Bayvet/Miles in the U.S.), primarily in cattle dips for fleas, ticks and other insects, mites, and scabies. Certain formulations of this insecticide were tradenamed Co-Ral, Baymix, Meldane, Asuntol, and Perizin.

Haloxon, **7-2**, is a coumarin-type anthelmintic (treatment for worms) for use in ruminants, pigs, and horses; it was developed in the 1960s and sold by Wellcome as Galloxon® and by Cooper as Halox®.

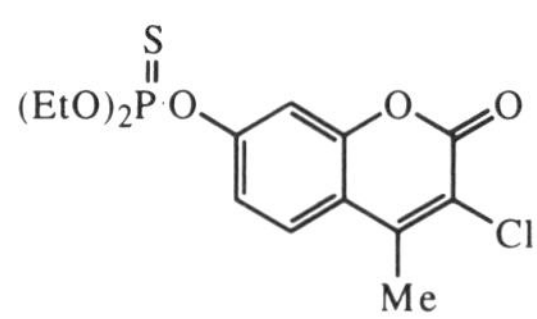

7-1

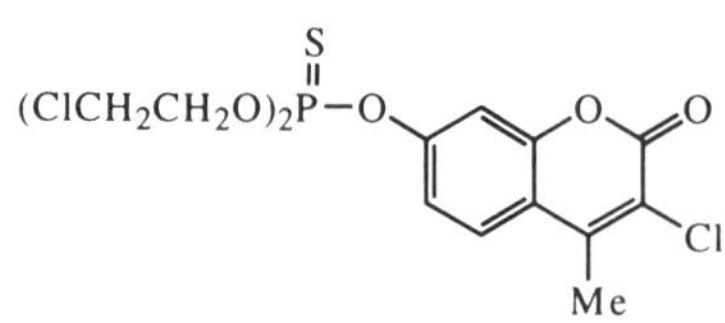

7-2

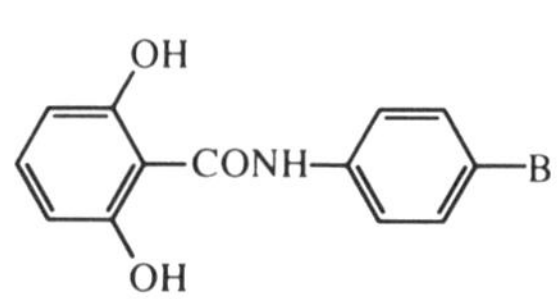

7-3

Resorantel, **7-3**, 4′-bromo-2,6-dihydroxybenzanilid, was developed by Hoechst in the 1960s and is used as an anthelmintic for sheep and other ruminants.

Benzoximate, **7-4**, ethyl *O*-benzoyl-3-chloro-2,6-dimethoxybenzhydroximate, is an acaricaide (miticide) developed by Nippon Soda Co. in the early 1970s; it is used to control spider mites on fruit trees and grapes in Japan and Europe.

Isoxaben, **7-5**, *N*-[3-(1-ethyl-1-methylpropyl)-5-isooxazolyl]-2,6-dimethoxybenzamide, evolved from research at Elanco Products Co. (Eli Lilly) in the early 1980s; it is a selective, pre-emergence herbicide for use in cereals, vinyards, on fruits, turf, and ornamentals against broadleaf weeds. Tradenames are Flexidor, EL-107, Gallery, X-Pand, and Cent-7 (Dow-Elanco).

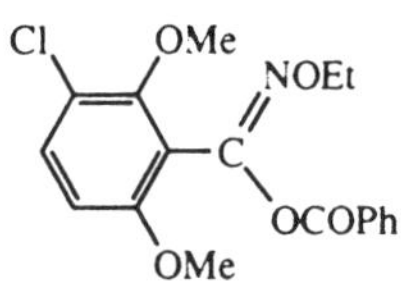

7-4

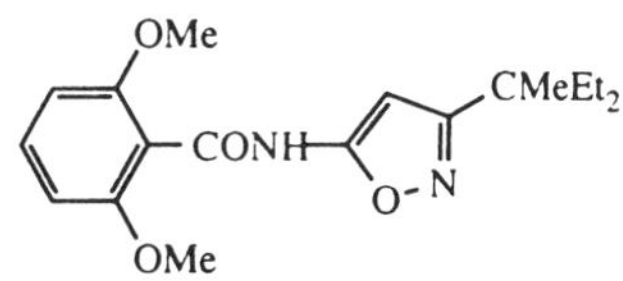

7-5

Chlomethoxynil, **7-6**, 2,4-dichlorophenyl-3′-methoxy-4′-nitrophenyl ether, also called X-52, is a selective herbicide used in rice fields; it was brought to market in the early 1970s by Nihon Nohyaku Co.

Goal®, **7-7**, 2-chloro-4-trifluoromethyl-3′-ethoxy-4′-nitrophenyl ether, was developed by Rohm & Haas Co. in the mid-1970s; this phenyl ether compound (also named Oxyfluorfen and Rh-2915) is used as a selective, pre- and post-emergent herbicide for control of witchweed in corn and weeds in soybeans, fruit trees, almonds, peanuts, grapes, and other crops.

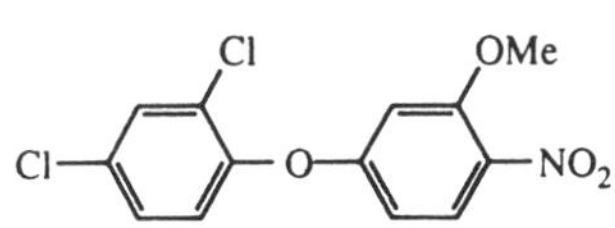

7-6

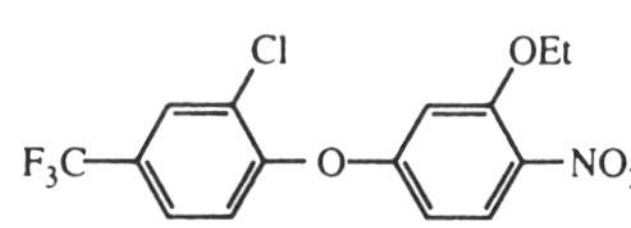

7-7

In the 1970s, Nippon Soda developed and began to market the herbicide alloxydim, **7-8**, a 1,3-cyclohexanedione-type compound which gives post-emergence control of grassy weeds in broadleaf crops. Certain formulations of alloxydim are tradenamed Kusaguard, Fervin, Clout, and Tritex (Nippon Soda).

Sethoxydim, **7-9**, another 1,3-cyclohexanedione derivative with the chemical name 2-[1-(ethoxyimino)butyl]-3-hydroxy-5-(2-ethylthio)propyl-2-cyclohexen-1-one, was invented by Nippon Soda in the late 1970s and licensed for marketing in the U.S. and Europe by BASF under the tradename Poast. It is several times more potent than alloxydim as a post-emergence herbicide for the control of grasses and other

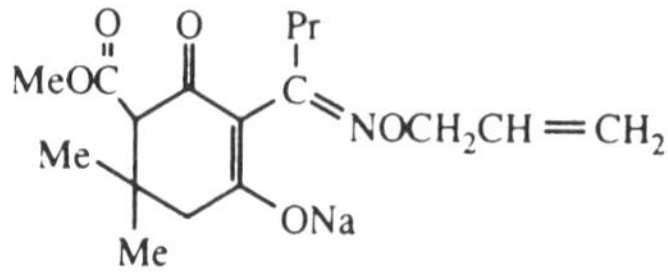

7-8

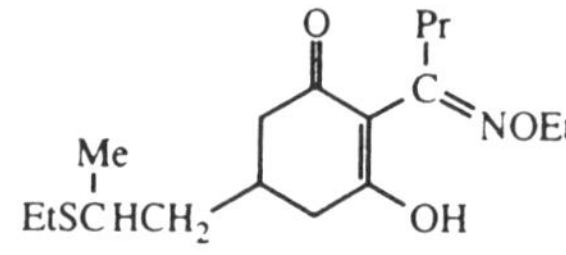

7-9

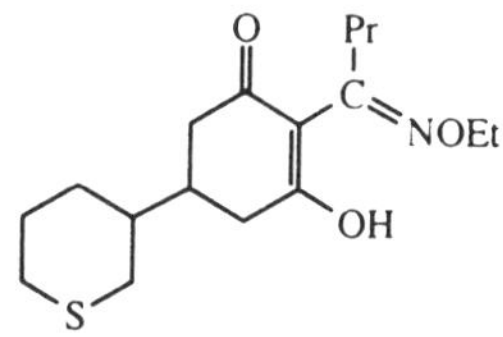

7-10

annual and perennial weeds in broadleaf crops, such as soybeans and cotton. Other tradenames for Sethoxydim products are Checkmate, Fervinal, Sertin, Grasidim, and Nabu (Nippon Soda).

In 1984, sales of alloxydim and sethoxydim were reported to be $40 mill. The developers of these herbicides won the 31st Okochi Grand Prize For Technology in Japan.[1] (This paper also shows the synthesis of alloxydim and sethoxydim.)

Another cyclohexanedione-type compound, Cycloxydim, **7-10**, 2-[1-(ethoxyimino)butyl]-3-hydroxy-5-(2*H*-tetrahydrothiopyran-3-yl)-2-cyclohexen-1-one, was discovered by BASF in the 1980s and also is a selective, post-emergence herbicide for controlling grasses and other weeds in broadleaf crops such as vegetables, alfalfa, cocoa, coffee, potatoes, and soybeans. Other tradenames for Cycloxydim formulations are BASF 517H, Focus, Laser, and Stratos.

Prohexadione, **7-11**, 2-propionyl-5-carboxycyclohexane-1,3-dione, was developed as KUH-833 by Kumiai as a growth regulator for wheat, barley, rice, and turf.

Tralkoxydim, **7-12**, another cyclohexanedione-type, is marketed in several formulations by ICI under the tradenames Achieve, Grasp, and Splendor, as a selective, systemic herbicide to control grassy weeds (e.g., *Setaria viridis*) in cereals.

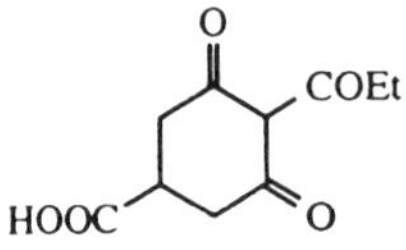

7-11

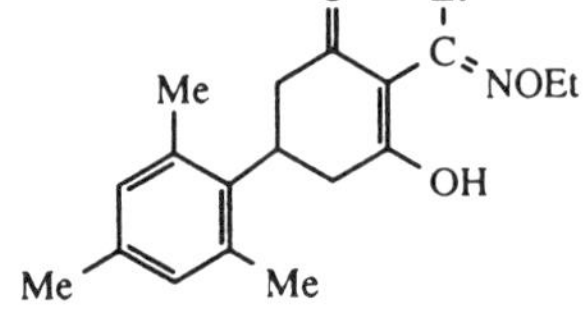

7-12

Ciba-Geigy's CGA 163935, **7-13**, ethyl 4-cyclopropyl(hydroxy)methylene-3,5-dioxocyclohexanecarboxylate, was developed as a growth regulator for the prevention of lodging (to reduce stem growth) in cereals and winter oilseed rape. Certain formulations of this product are tradenamed Modus and Vision (Ciba-Geigy).

Cloproxydim, **7-14**, is also a cyclohexanedione derivative developed by Chevron in the mid-1980s for the post-emergence control of annual and perennial grasses in broadleaf crops (e.g., soybean, cotton, flax, potatoes, sugar beets), vegetable crops, and trees. A formulation of cloproxydim was sold under the tradename Select (Chevron).

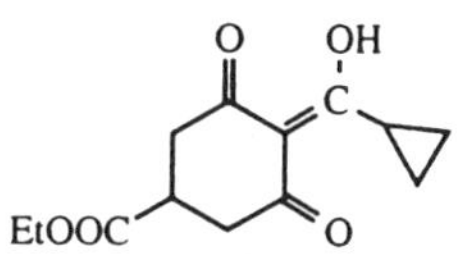

7-13

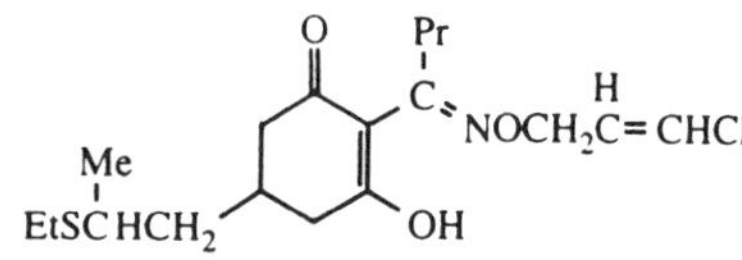

7-14

There appear to be several new 1,3-cyclohexanedione derivatives about to be commercialized as herbicides or pesticides, developed by ICI (and Stauffer, acquired by ICI), Nippon Soda, and others. These materials are among the selections from the literature given later on.

A number of important agricultural chemicals derived from *m*-aminophenol were developed in the 1960s through the 1980s.

Phenmedipham, **7-15**, 3-methoxycarbonylaminophenyl *N*-(3-methylphenyl)carbamate, was developed and marketed under the tradename Betanal by Schering A.G. as a post-emergence herbicide for sugar and fodder beets; it became very successful and was licensed to several other companies (e.g., in Finland, Czechoslovakia, and the former USSR). The formulated phenmedipham was sold under the tradenames Betanal (Schering), Alegro (Rhone-Poulenc), and Kemifam (Kemira). In 1986, the price of 100% phenmedipham was given as \$49/lb. and Schering was said to have a plant capacity of 5 mill. lb./yr.[2]

Schering A.G. also commercialized a related carbamate compound, desmedipham, **7-16**, 3-ethoxycarbonylaminophenyl *N*-phenylcarbamate, for use as a herbicide for beet crops; the formulated product was tradenamed Betanal AM and Betanex (Schering).

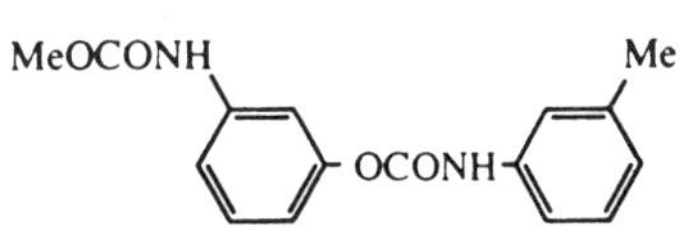

7-15

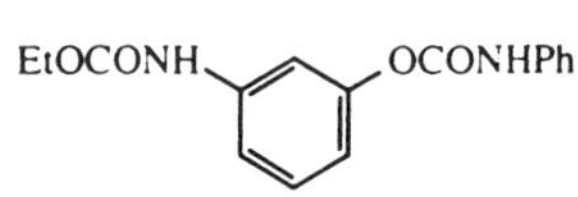

7-16

Karbutilate, **7-17**, 3-(3,3′-dimethylureido)phenyl *tert*-butylcarbamate, was developed at FMC Corp. and marketed under the tradename Tandex as a total soil herbicide

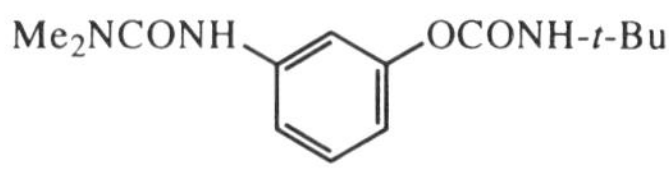

7-17

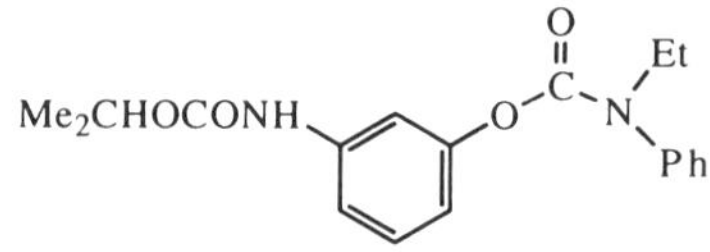

7-18

for pre- and post-emergence use on grasses, broadleafed weeds, brush, and vines in noncrop applications; the rights to this product were later sold to Uniroyal Chemical Corp.

Phenisopham, **7-18**, 3-isopropoxycarbonylaminophenyl *N*-ethyl-*N*-phenylcarbamate, was developed by Schering A.G. (sold under the tradename Diconal) as a weed killer for cotton fields.

Tetrafluoron, **7-19**, *N,N*-dimethyl-*N*-[3-(1′,1′,2′,2′-tetrafluoroethoxy)phenyl]-urea, was patented by Hoechst in 1971 and marketed under the tradenames Tomilon and Fletoxuron for the control of grasses and many other weeds in crops such as cotton, alfalfa, and sugarcane.

Mepronil, **7-20**, 3′-isopropoxy-2-methylbenzanilid, was introduced in 1981 by Kumiai Chemical Industries as a foliar fungicide for fruit trees, ornamentals, rice, and vegetables; one formulation is tradenamed Basitac.

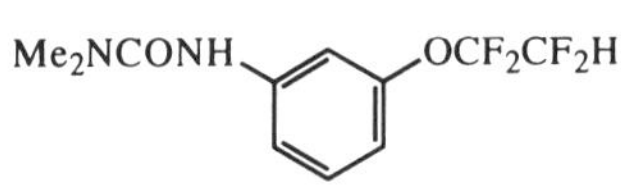

7-19

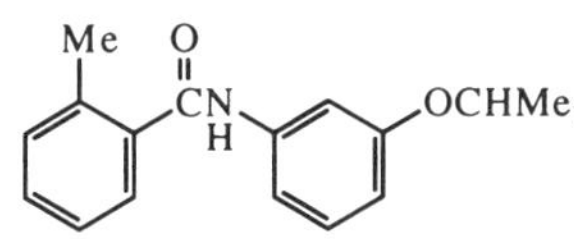

7-20

Moncut (NNF-136), **7-21**, alpha, alpha, alpha-trifluoro-3′-isopropoxy-*o*-toluanilide, is a fungicide introduced by Nihon Nohyaku Co. in the early 1980s for use on rice, cereals, fruits, potatoes, and others.

Formetanate (tradenamed Dicarzol and Carzol [the hydrochloride]), **7-22**, 3-dimethylaminomethyleneiminophenyl methylcarbamate, was invented by Schering A.G. and targeted for use as an acaricide and insecticide (against *Diptera, Hemiptera* and *Lepidoptera*) on ornamentals, fruits, vines, vegetables, beets, and lucerne.

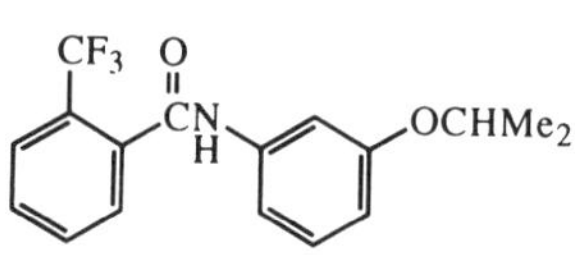

7-21

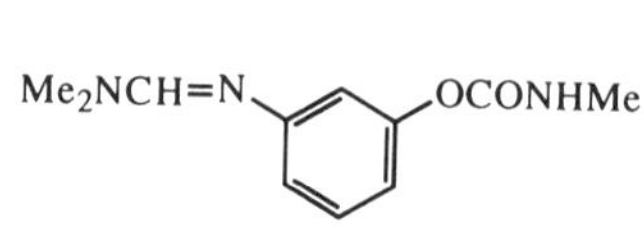

7-22

The group of compounds known as resorcylic acid lactones (RALs) embraces two commercial products. Ralgro (Zeranol, zearanol), **7-23**, an anabolic (growth-promoting) agent, was used in the form of an (ear) implant in beef cattle and feedlot lambs. Ralgro was reported by International Mineral and Chemicals (IMC) to have had sales of about \$30 mill in 1981.[3] Ralgro is prepared by reduction of the carbonyl group in zearalenone, **7-24**, itself an anabolic agent, obtained from the mycelia of the fungus *Giberella zeae* (*Fusarium graminearum*).

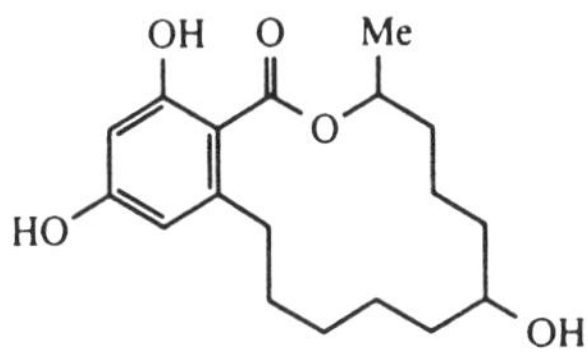

7-23

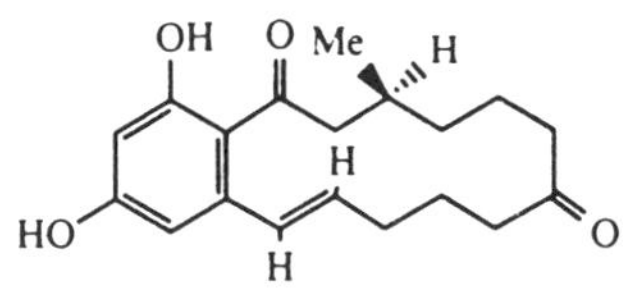

7-24

Finally, rotenone, **7-25**, contains a resorcinol moiety and is an example of an agrochemical isolated from a natural product, the derris root, and from cubé; it is used as an insecticide and grubicide.

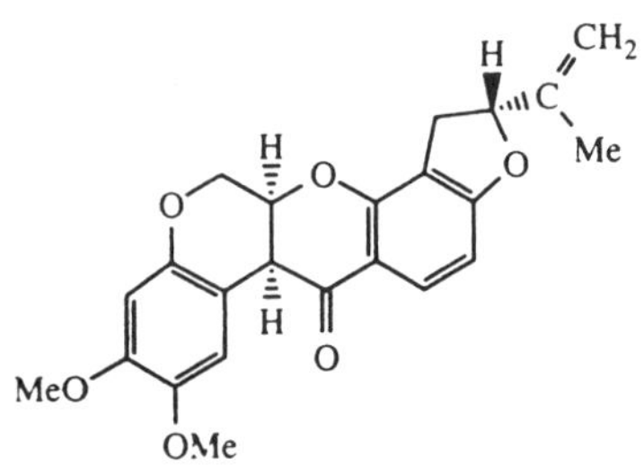

7-25

7.3. A SAMPLING OF DEVELOPMENTS FROM 1980–1992

On reading the patent literature on resorcinol- or *m*-aminophenol-related agricultural chemicals, it becomes clear that a tremendous effort was required to obtain broad patent coverage—almost every patent covers many, often dozens, of compounds. However, in the following selections, only one compound, representative of the invention, is shown for each patent. It will be seen that differences between patented products often appear modest, but modest structural changes often produce significant differences in action and selectivity. It also becomes apparent that considerable skill and ingenuity in synthesis is involved, but the synthesis is only a start in developing a useful chemical. For example, in the case of herbicides there is extensive testing needed to define the scope of the use, the optimum application rate, the formulation of appropriate media for application, such as solutions, emulsified concentrates, suspensions, dusts, wettable powders, granules, and the like; also needed are the greenhouse tests pre- and post-emergence of weeds, the field tests, and, obviously and very importantly, even though not always described in the patents, the toxicity studies, the tests of persistence and of toxic residues. If all that looks favorable, the work on the governmental permitting processes, the EPA registration, the packaging, labeling, and so on, has to be started, modified, and finished. This is a monumental and very costly task which only the strongest companies are able and willing to continue nowadays.

The following, limited selections from the literature do indicate the broad scope of the development work.

7.3.1. Cyclohexanediones

From 1980 on, this is by far the largest category of research and development on agricultural chemicals related to resorcinol. It will be seen that there were a lot of competitive products worked on by most of the majors in this field. The following selections illustrate the intensity of this work.

The cyclohexane-1,3-dione derivative, **7-26**, was found to control *Setaria* weeds in soybeans at 20 g/10 a.[4] Dimedone (5,5-dimethyl-1,3-cyclohexanedione) derivatives, like compound **7-27**, were also patented by Nippon Soda Co. as herbicides.[5]

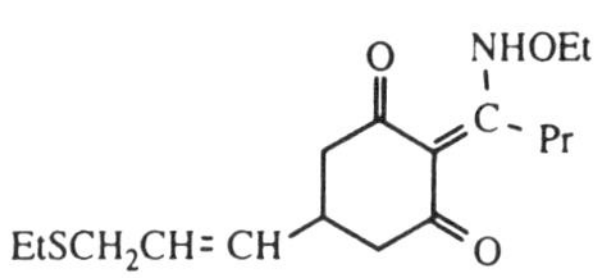

7-26

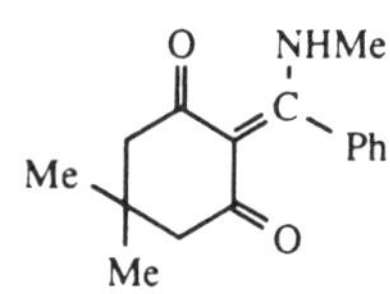

7-27

Cyclohexandiones, such as **7-28**, were found to be plant growth regulators; applied at 50 g/10 a to soybeans at the 1.5-leaf stage, the compound gave a substantial increase in axiallary buds in three weeks.[6]

Metal salts of 1,3-cyclohexanedions, e.g., the Co(II) salt of compound **7-29**, prepared in the presence of NaOH (to form the enolate) in acetone, were found to be effective herbicides at 15 g/10 a.[7]

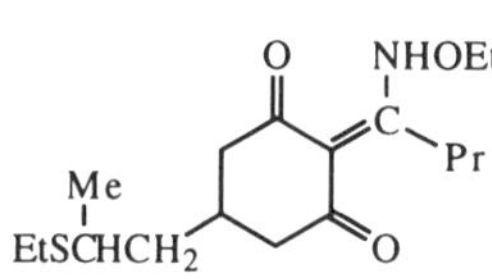

7-28

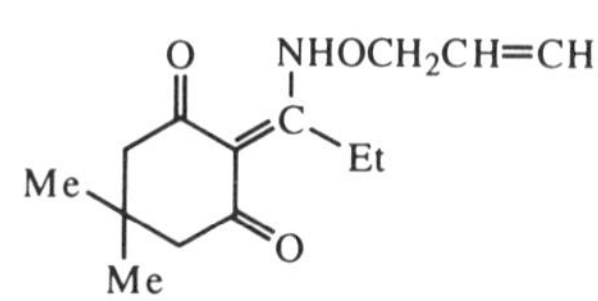

7-29

A synergistic herbicide mixture against grasses was invented, consisting of 0.125% solutions each of 2-chloro-2′,6′-dimethyl-*N*-(pyrazol-1-ylmethyl)acetanilide and 2-(1-ethoxyaminobutylidene)-5-(2-ethylthiopropyl)cyclohexane-1,3-dione.[8]

In further variations on the theme, Nippon Soda Co. discovered that 2-alkylidene-4-carboethoxycyclohexane-1,3-diones, e.g., **7-30** at 31 g/10 a, were efficient herbicides.[9] 5-Alkylthioalkyl-2-alkoxyiminoalkyl-3-hydroxycyclohex-2-ene-1-ones, such as **7-31**, were also effective herbicides.[10]

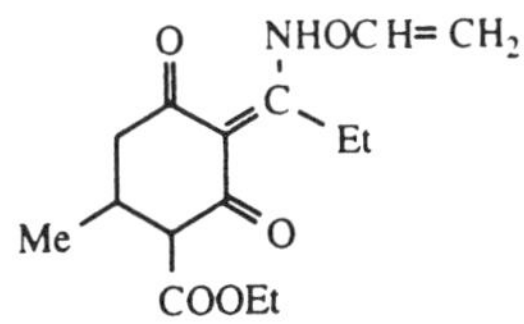

7-30

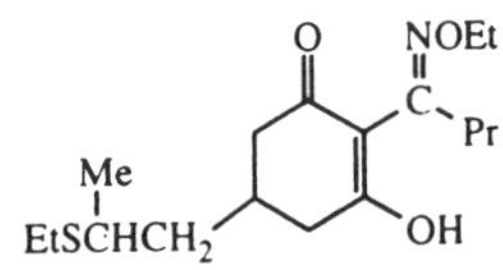

7-31

Thiazolylurea derivatives such as, e.g., **7-32**, prepared by reacting 2-bromo-5,5-dimethyl-1,3-cyclohexanedione with *N*-methylthiourea, followed by reaction of the intermediate thiazolylamine with methyl isocyanate, were found to be broad-spectrum herbicides at 2–4 kg/ha in pre- and post-emergence application.[11]

Esters of 2-(alkoxyiminoalkyl)-3-hydroxy-2-cyclohexen-1-ones, such as **7-33** (at 5 g/10 a), were found to be powerful herbicides.[12]

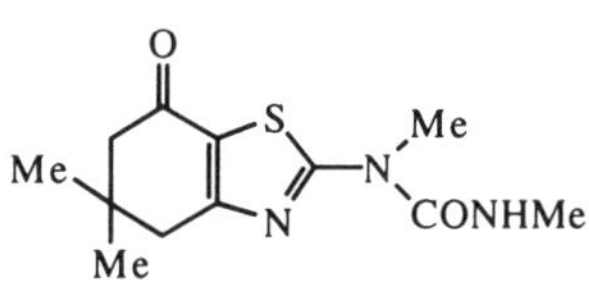

7-32

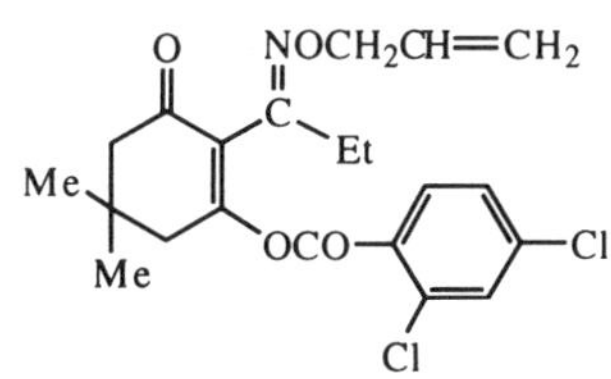

7-33

Complex carbamates of 2-aryl-3-hydroxy-5,5-dimethyl-2-cyclohex-1-one, exemplified by compound **7-34**, showed both excellent acaricidal and insecticidal

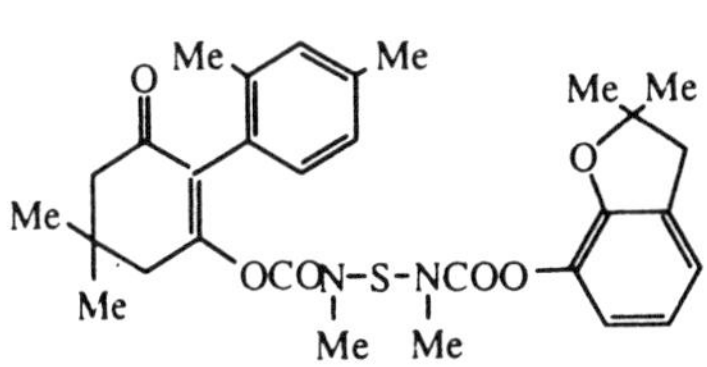

7-34

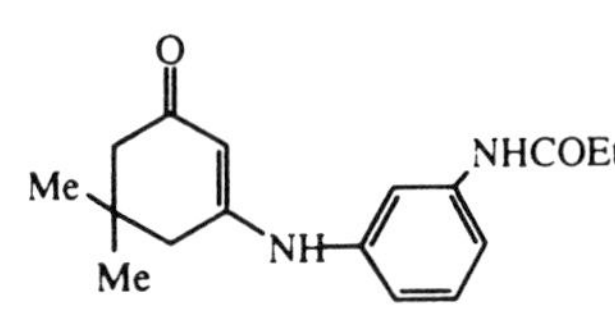

7-35

activity; the chosen example, **7-34**, gave excellent control of bean aphids, 2-spotted mite eggs, Southern army worm, Mexican bean beetle, and house flies.[13]

N-(Oxocyclohexenyl)-*N'*-acyl-*m*-phenylenediamines, prepared by reacting dimedone with a N-acylated *m*-phenylenediamine and typified, e.g., by structure **7-35**, had herbicidal activity.[14]

Novel 3-(aminoaryloxy)-2-cyclohexen-1-ones, such as, e.g., **7-36**, prepared by reacting 3-chloro-5,5-dimethyl-2-cyclohexene-1-one (obtained from 5,5-dimethyl-1,3-cyclohexanedione) with *m*-hydroxyphenyl *N,N*-dimethylurea, showed fair pre-emergence and excellent post-emergence control of certain grasses and broadleaf weeds.[15]

2-(1-Oximinoalkyl)-3-alkoxy-5,5-dimethyl-2-cyclohexen-1-ones such as, e.g., **7-37**, were demonstrated to have herbicidal activity in the laboratories of Chevron Research Co.[16]

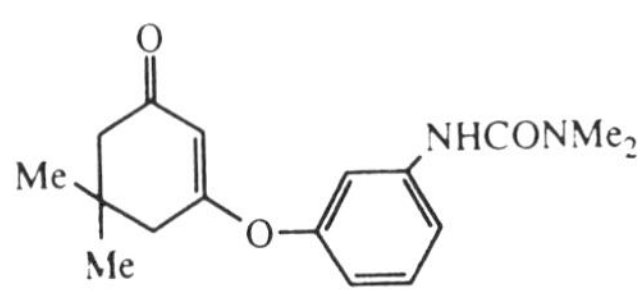

7-36

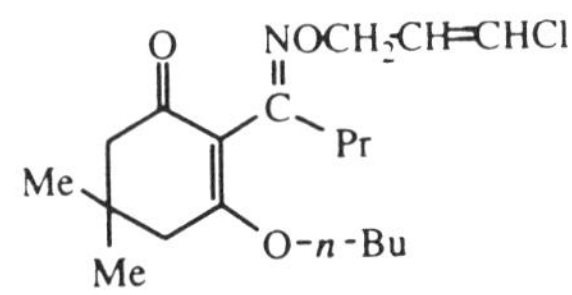

7-37

2-(1-Oximinoalkylidene)-5-aryl-1,3-cyclohexanediones such as, e.g., **7-38**, controlled grassy weeds in grain with less injury to the crop than similar compounds of the prior art.[17]

The phenylacetate ester of the 2-(1-oximinoalkylidene)-3-hydroxy-5,5-dimethyl-2-cyclohexen-1-one, **7-39**, showed good herbicidal activity against *Echinochloa crus-galli, Digitaria adscendens and Chenopodium album*; similar compounds also had herbicidal activity.[18]

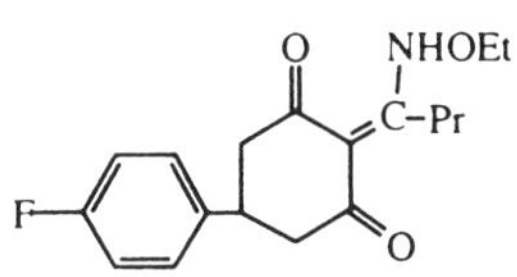

7-38

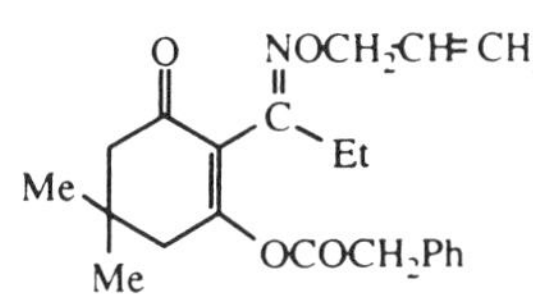

7-39

The terephthalate diester of the 2-(1-oximinoalkylidene)-3-hydroxy-4-methoxycarbonyl-5,5-dimethyl-2-cyclohexen-1-one, **7-40**, had herbicidal activity comparable to that of dimethyl tetrachloroterephthalate.[19]

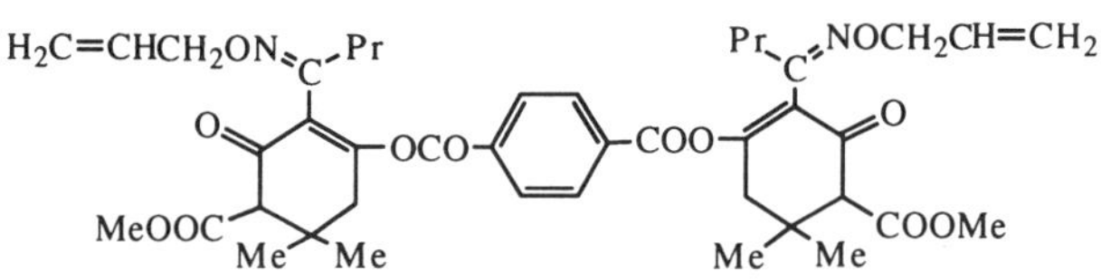

7-40

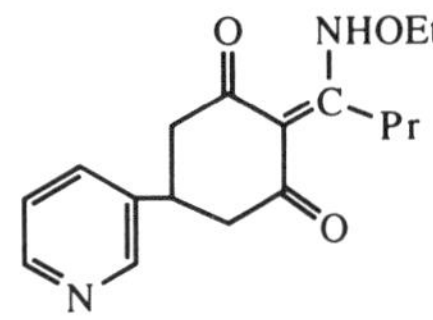

7-41

2-(1-Ethoxyiminobutylidine)-5-(3-pyridyl)-1,3-cyclohexanedione, **7-41**, was shown to be an effective herbicide for soybean crops.[20]

Esters of 2-aryl-3-hydroxy-5,5-dimethyl-2-cyclohexen-1-one were found to enhance the sugar content of plants. Thus, compound **7-42** was spray-applied to sorgum at 8 lb./acre and gave a 396% increase of the sugar content of the sap in comparison to the control without this agent.[21]

2-(1-Ethoxyiminoalkylidene)-5-aryloxymethyl-1,3-cyclohexanediones were effective herbicides. For example, compound **7-43** was a selective herbicide useful for wheat and soybean crops.[22]

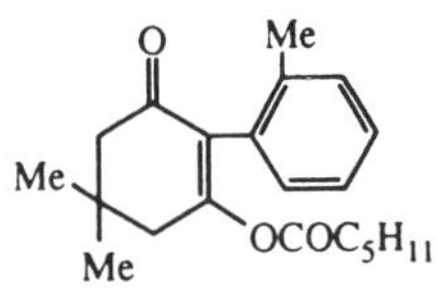

7-42

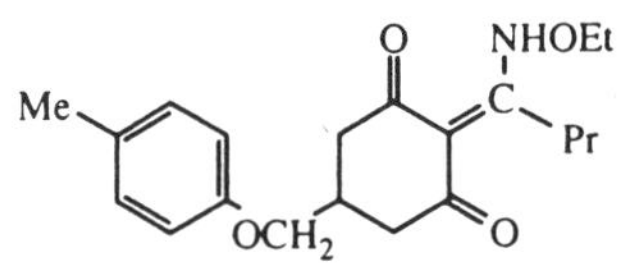

7-43

2-(Ethoxyiminoalkylidene)-3-hydroxy-5-alkylthioaryl-2-cyclohexen-1-ones, such as **7-44**, were developed by ICI Australia as herbicides.[23] In the same laboratories, compound **7-45** was developed and gave 100% control of *Digitaria sanguinalis* at a 0.025 kg/ha application rate.[24]

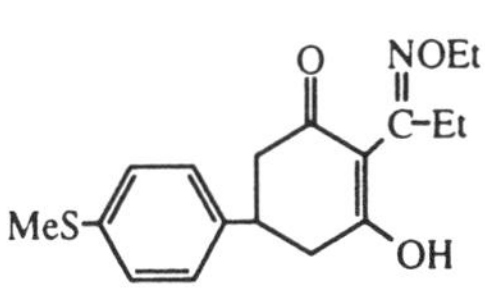

7-44

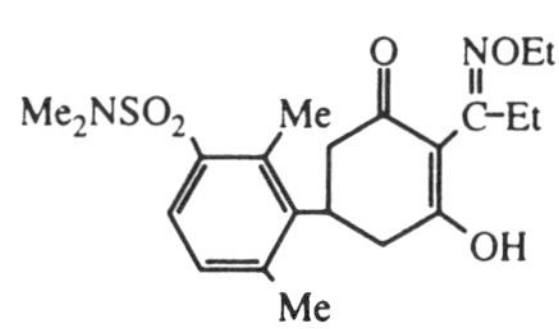

7-45

2-Anilinomethylidene-5-(2-ethylthiopropyl)-1,3-cyclohexanedione, **7-46**, was patented by Teijin as a herbicide useful against graminaceous and broadleafed weeds.[25]

The synthesis and herbicidal activities of sulfur-containing derivatives of 1,3-cyclohexanedione were reviewed in 1982.[26]

2-(1-Alkoxyiminoalkylidene)-5-(4-chlorobenzyloxyethylidene)-1,3-cyclohexanedione, such as, e.g., **7-47**, was found to be a selective herbicide for undesirable grasses in grain fields.[27]

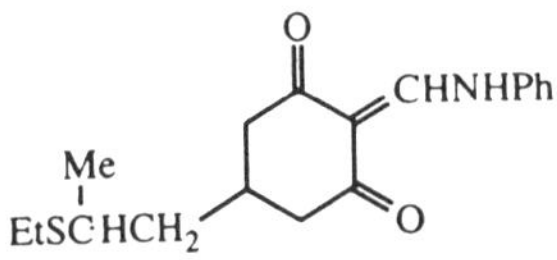

7-46

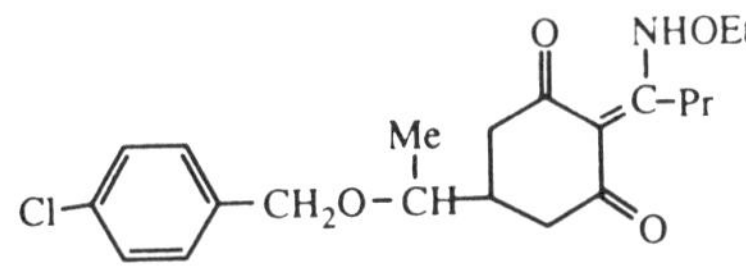

7-47

2-(1-Alkoxyiminoalkyl)-3-hydroxy-5-(4-indanyl)-2-cyclohexen-1one, **7-48**, was a selective post-emergent herbicide for *Digitaria sanguinalis* in rice at the 0.025 kg/ha application rate.[28] A similar, 5-mesityl-1,3-cyclohexanedione derivative, **7-49**, gave 100% control of *Avena fatua* and *Alopercurus myosuroides*, with little or no damage to winter wheat and spring barley at the 0.04 kg/ha application rate.[29]

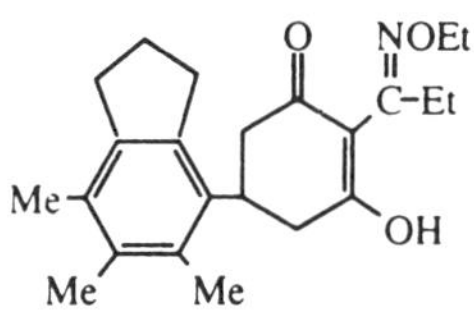

7-48

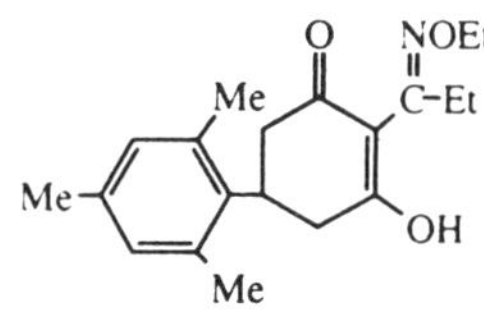

7-49

2-(1-Allyloxyaminobutylidene)-5-cyclohexylidenomethyl-1,3-cyclohexanedione, **7-50**, was seen to have good herbicidal activity with less propensity to damage wheat than the N-ethyl- or N-vinyl analogs.[30]

2-Aroyl-1,3-cyclohexanediones such as, e.g., **7-51**, were found to be excellent pre- and post-emergence herbicides for a variety of undesirable plants at the 4.48 kg/ha application rate.[31] (Note the relative simplicity of this compound vs structures of the prior art.)

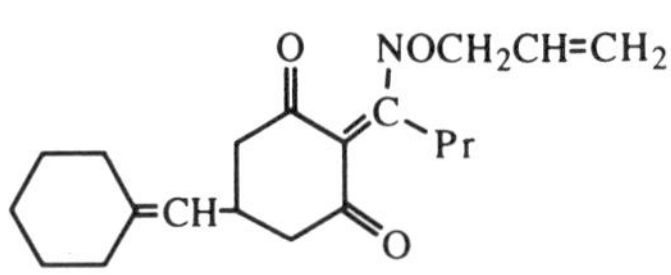

7-50

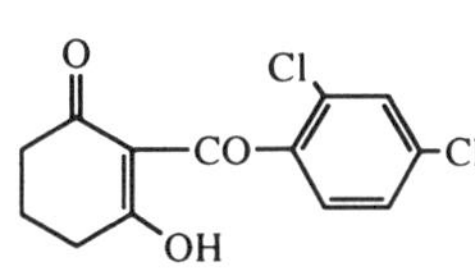

7-51

(Thio)phosphates of 2-aryl-3-hydroxy-2-cyclohexen-1-ones were developed which had both insecticidal and herbicidal properties. For instance, compound **7-52** controlled bean aphids, mites (adults and eggs), Southern army worms, Mexican bean beetles, and house flies, and was a pre-emergence herbicide for rye grass.[32]

Certain condensation products of 1,3-cyclohexanediones with substituted phenoxypropionic acids were post-emergence herbicides against annual and perennial grasses and other weeds. For instance, compound **7-53** gave excellent control of green foxtail (*Setaria virides*), wild oat (*Avena* fatua), watergrass (*Echinochloa crus-galli*), and curly dock (*Rumex crispus*).[33]

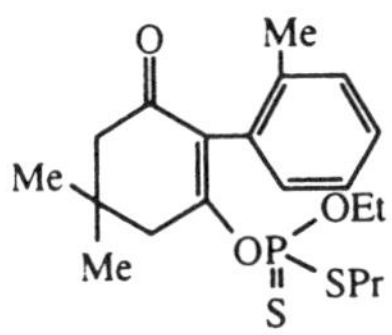

7-52

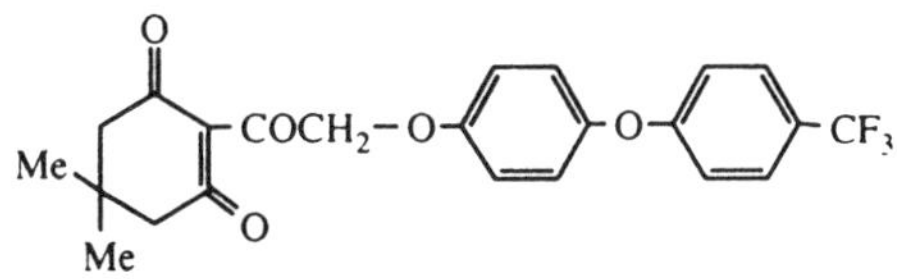

7-53

Carbonate esters of 2-aryl-3-hydroxy-2-cyclohexen-1-one had biological activity. For example, the ester **7-54** had acaricidal and herbicidal activity.[34]

2-(1-Ethoxyiminopropyl)-3-hydroxy-5-(2,4,6-trimethylpyrimidin-5-yl)-2-cyclohexen-1-one, **7-55**, had post-emergence herbicidal activity; at an application rate of 1 kg/ha, this compound totally controlled wild oats with no or little damage to wheat.[35]

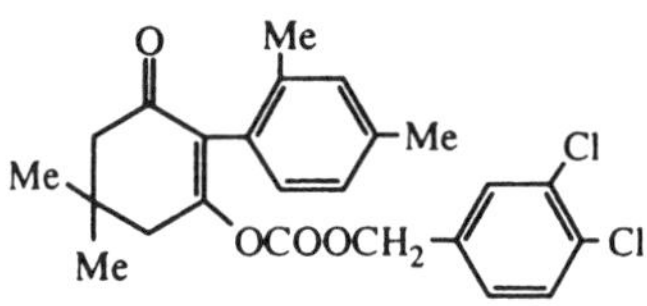

7-54

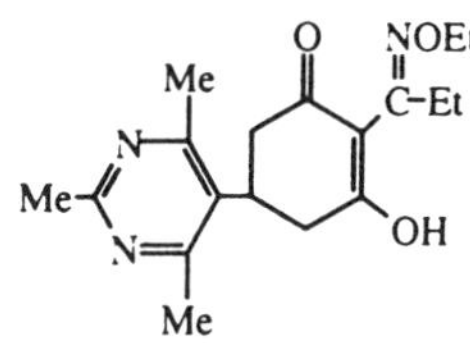

7-55

In a further play on the theme of 2-(1-alkoxyiminoalkyl)-1,3-cyclohexanediones, compound **7-56** was found to give 100% pre-emergence control of weeds at 0.5 kg/ha and to have no harmful effect on wheat or corn.[36]

2-[1-(3-chloroallyloxyamino)alkylidene]-5-alkylthioalkyl-1,3-cyclohexanediones were patented as herbicides for controlling grasses in broadfleaf crops. Thus, compound **7-57**, in post-emergence tests at 0.7 $\mu g/cm^2$, gave 100% control of *Digitaria*, wild oat, and other grasses, and showed little effect on soybeans.[37]

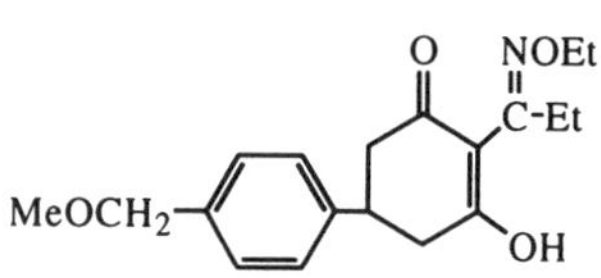

7-56

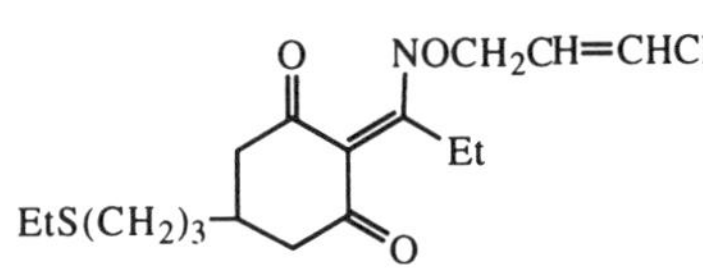

7-57

Enol esters of *ortho*-unsubstituted 2-aryl-1,3-cyclohexanediones were claimed as effective herbicides and acaricides. For instance, compound **7-58** gave complete pre-emergence control of crabgrass (*Digitaria sanguinalis*) without any harm to tomatoes; this compound also controlled 2-spotted mites (adults or eggs).[38]

2-(1-Alkoxyiminoalkyl)-5-heterocyclyl(alkyl)-1,3-cyclohexanediones were developed as selective pre-emergence herbicides. As an example, compound **7-59** was effective against grasses, such as *Alopecurus myosuroides*, at an application rate of 0.06 kg/ha, without any harm to wheat.[39]

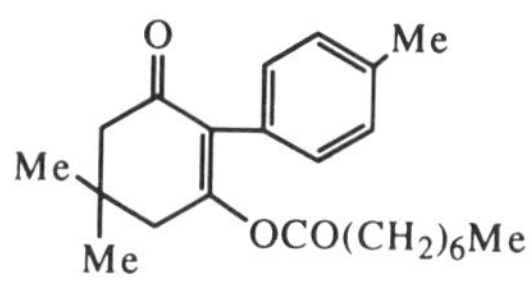

7-58

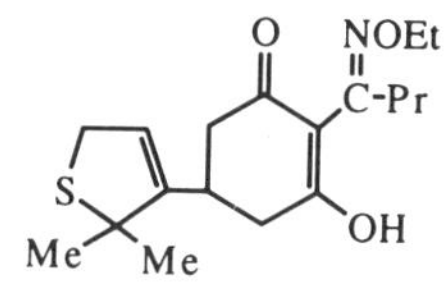

7-59

2-(2-Chlorophenylacetyl)-1,3-cyclohexanedione, **7-60**, and similar compounds were shown to be effective pre- and post-emergence weed killers at 4.48 kg/ha.[40]

2-(1-Ethoxyaminobutylidene)-5-(1-phenylpyrrol-3-yl)-1,3-cyclohexanedione, **7-61**, and similar compounds were found to be pre-emergence herbicides against grasses at an application rate of 3.0 kg/ha, and these compounds did not damage crops.[41]

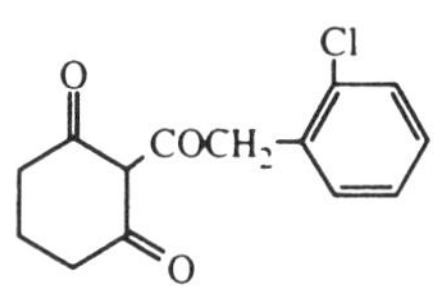

7-60

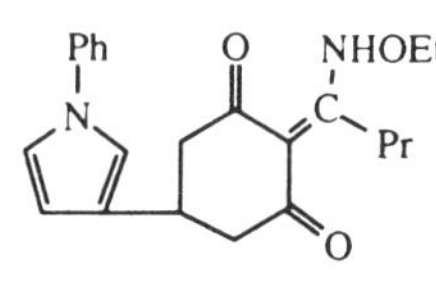

7-61

High herbicidal acticity and selectivity was obtained with 2-(1-allyloxyiminobutyl)-3-hydroxy-5-(4-aminophenyl)-2-cyclohexen-1-one, **7-62**, and related materials. Thus, **7-62** killed foxtail at 0.25 kg/ha and did little or no harm to wheat.[42]

5-Benzofuranyl-1,3-cyclohexanediones were prepared, tested, and found to be excellent herbicides. For instance, compound **7-63** at 0.02 kg/ha gave good control of *Echinochloa crus-galli* without any damage to winter wheat and rice.[43]

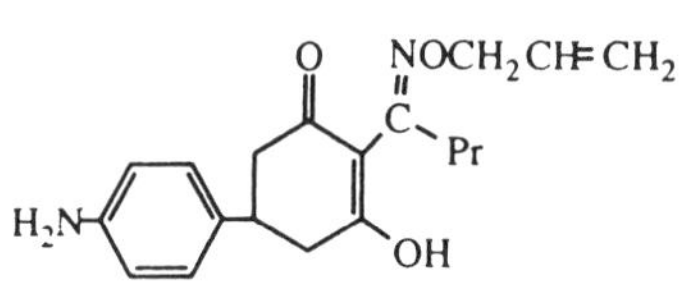

7-62

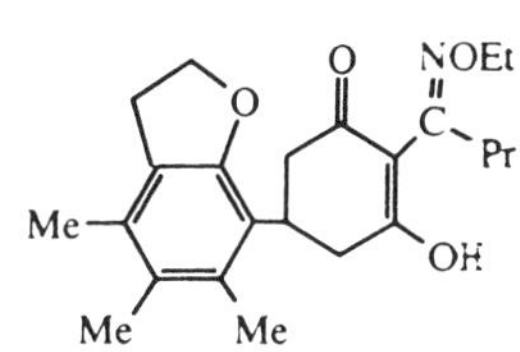

7-63

Esters and amides of acylated 1,3-cyclohexanedione-5-carboxylic acids were synthesized as a further extension in this class of herbicides and plant growth regulators. Thus, compound **7-64** gave 90–100% control of grasses such as *Lolium perenne* at 0.1–2.5 kg/ha.[44]

2-(1-Alkoxyaminoalkyl)-3-hydroxy-5-(1,3-dioxepanyl)-2-cyclohexen-1-ones and related compounds were prepared and tested for herbicidal activity. For instance, compound **7-65**, at 0.25 kg/ha in post-emergence application, gave better inhibition of the growth of grasses than 1,3-cyclohexanediones of the prior art, and they did not injure broadleaf crops such as soybeans.[45]

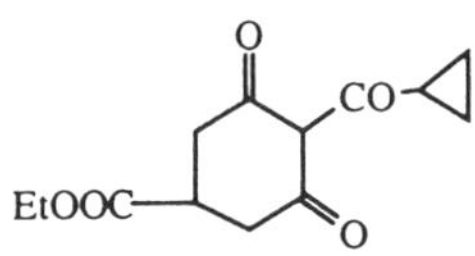

7-64

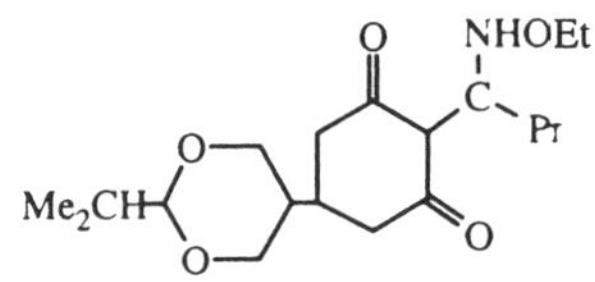

7-65

Heterocyclic 2-substituted 1,3-cyclohexanediones were found to be selective herbicides. As an example, compound **7-65a** at 1 lb /acre gave 100% control of velvetleaf and mustard.[46]

5-Tetrahydronaphthyl-substituted 1,3-cyclohexanediones were found to be good pre-emergence herbicides. For example, compound **7-66** at 1.0 kg/ha gave 100% control of ryegrass with <10% injury to, e.g., peas.[47]

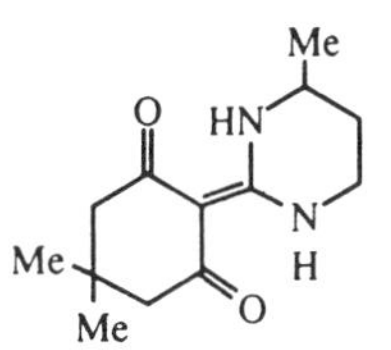

7-65a

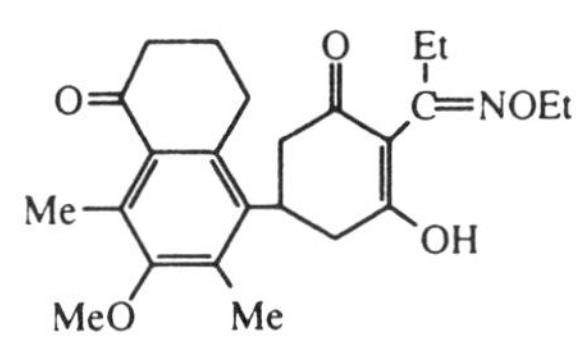

7-66

Esters or amides of 2-(1-hydroxyalkylidene or 1-alkoxyaminoalkylidene)-1,3-cyclohexanedione-5-carboxylic acid were prepared and shown to have pre- and post-emergence herbicidal and plant growth regulating activity. For example, compound **7-67**, sprayed on soybeans ("Hark" variety) at 0.1 kg a.i./ha, markedly increased the number and weight of the harvested siliques (pods) on the leading shoot.[48]

Herbicidal *trans*-2-[(3-chloroallyloxyimino)alkyl]-5-(substituted sulfinylalkyl)-1,3-cyclohexanediones were developed. Thus, compound **7-68** showed 100% control of grasses such as crabgrass, with 0% injury to broadleaf plants such as mustard, in both pre- and post-emergence application.[49]

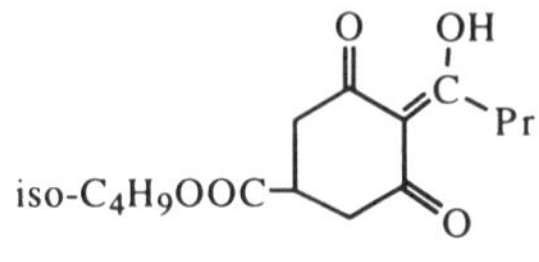

7-67

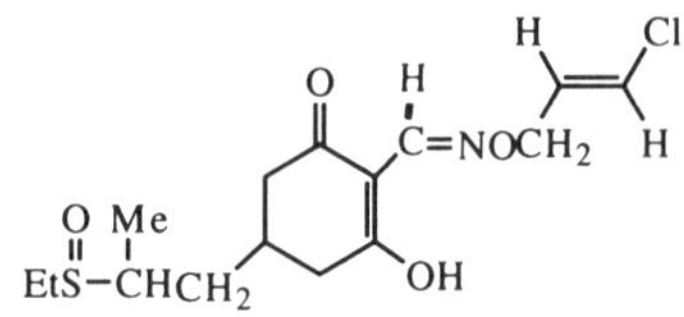

7-68

Certain novel 2-(2-substituted benzoyl)-1,3-cyclohexanediones, e.g., compound **7-69**, were synthesized and found to have herbicidal activity. At 4.48 kg/ha in pre-emergence application, **7-69** gave 100% control of *Setaria viridis, Rumex crispus*, and other weeds.[50] Similarly, certain 2-(2-nitrobenzoyl)-1,3- cyclohexanediones were selective herbicides; e.g., compound **7-70**, applied pre-emergence at 4.48 kg/ha, gave 100% control of *Setaria viridis, Brassica juncea*, and other weeds.[51]

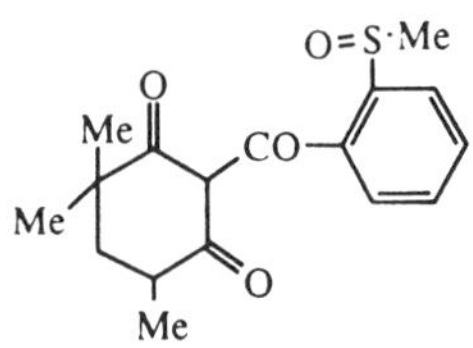

7-69

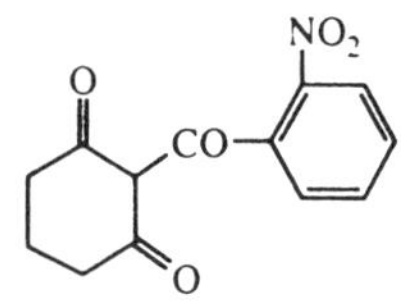

7-70

Herbicidal/biocidal esters of 2-aryl-1,3-cyclohexanediones were patented. For instance, compound **7-71** showed outstanding miticidal activity against *Tetranychus urticae* (adult and egg) and also gave 100% control of crabgrass with no injury to bean, corn, tomato, cotton, or soybean.[52]

A process for making herbicidal 2-acyl-1,3-cyclohexanediones was patented, involving the rearrangement of the corresponding 3-acyloxy-2-cyclohexen-1-ones in the presence of a catalytic amount of a cyanide source and a molar excess of trialkylamine, potassium carbonate, or trisodium phosphate. In an alternate process, the rearrangement was performed in the presence of a stoichiometric amount of KCN or LiCN and a catalytic amount of a crown ether (e.g., 18-crown-6).[53]

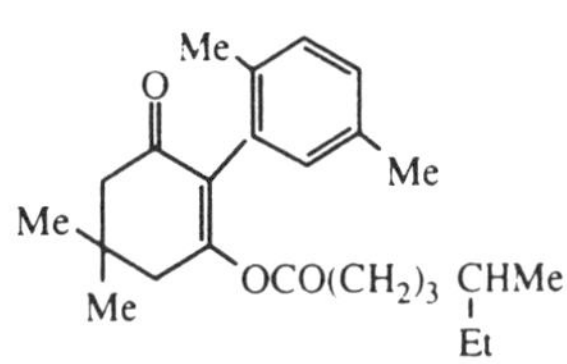

7-71

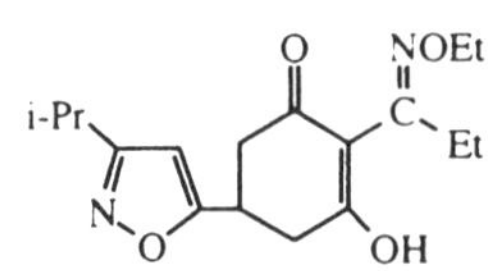

7-72

2-(Oximinoalkyl)-5-isoxazolyl-1,3-cyclohexanediones were shown to be excellent herbicides. For instance, compound **7-72**, at 3 kg/ha, gave 100% control of *Graminaceae* weeds.[54]

New 2-[1-(3-chloroallyloxyamino)alkylidene]-5-alkylthioalkyl-1,3-cyclohexanediones were good, selective post-emergence herbicides; e.g., compound **7-73**, at 0.28 g/cm^2, 100% controlled *Echinochloa crus-galli*, Johnson grass, and gave 93–99% kill of wild oats and yellow foxtail with no harm to soybeans.[55] The sodium salt of a similar compound, 2-[1-*trans*-(3-chloroallyloxyamino)ethylidene]-5-(2-ethylthiopropyl)-1,3-cyclohexanedione, sprayed at 200 mg/L on tobacco, gave 100% tobacco sucker (auxiliary shoot) control.[56]

2-(Pyrimidinecarbonyl)-1,3-cyclohexanediones were useful as herbicidal agents. As an example, compound **7-74** at 4.48 kg/ha gave 100% control of foxtail.[57]

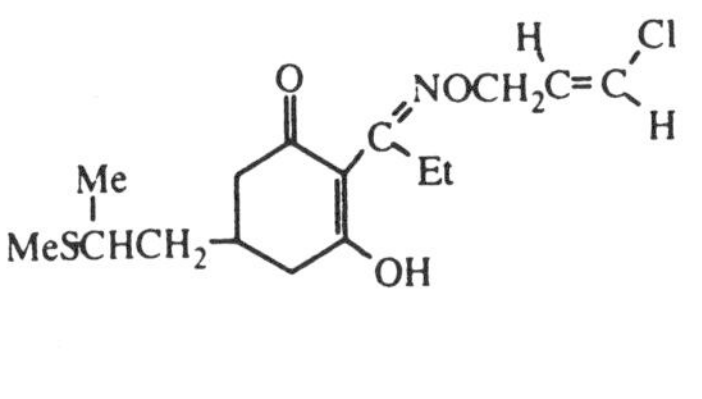

7-73

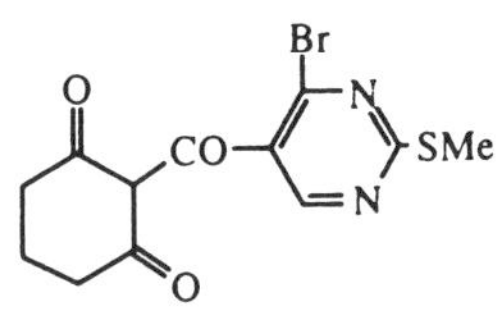

7-74

2-(1-Ethoxyiminobutyl)-3-hydroxy-5-ethyl-5-methoxycarbonyl-2-cyclohexen-1-one, **7-75**, and similar compounds were demonstrated to be herbicides and plant growth regulators. Thus, compound **7-75** gave 100% control of *Panicum spp.* and *Lolium rigidum* at 5 kg/ha in both pre- and post-emergence application.[58]

2-(1-Alkoxyiminoalkyl)-3-hydroxy-5-[2-(fluorophenylthio)ethyl]-2-cyclohexen-1-ones were excellent selective herbicides. For example, compound **7-76**, in pre-emergence application at 2.5 g/a, gave 100% inhibition of *Echinochloa crus-galli, Sorghum halepense,* and *Setaria viridis* without harming soybeans or cotton.[59]

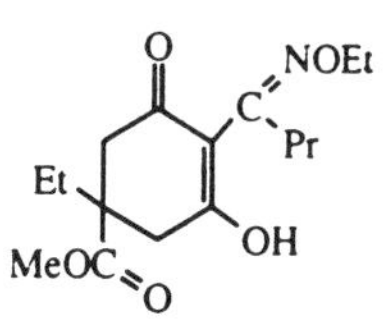

7-75

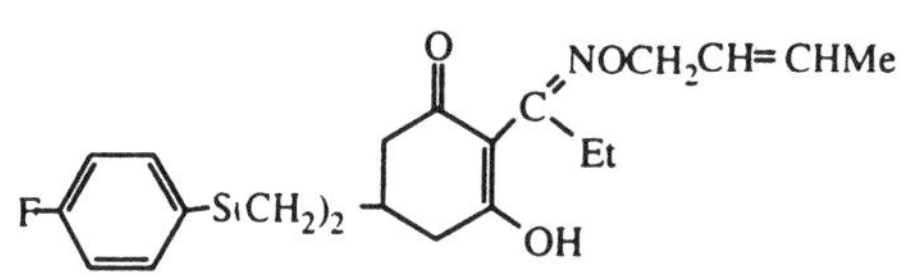

7-76

2-(1-Alkoxyiminoalkyl)-3-hydroxy-5-(2-pyridylthio)alkyl-2-cyclohexen-1-ones were studied and found to be selective herbicides. For instance, compound **7-77** in post-emergence application at 40 g/a gave 100% control of barnyard grass in rice paddies.[60]

2-Benzoyl-5-phenyl-1,3-cyclohexanediones were powerful herbicides. For instance, compound **7-78**, applied pre-emergence at 4.48 kg/ha, gave 100% control of 6 (e.g., *Setaria viridis*) out of 7 weeds tested.[61]

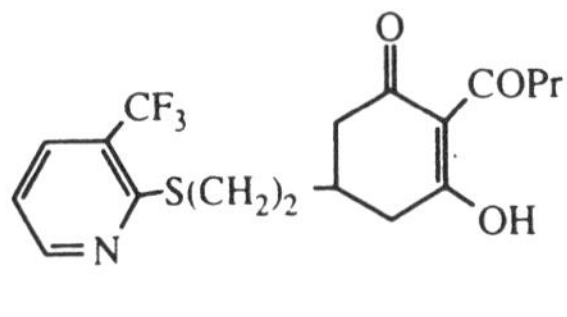

7-77

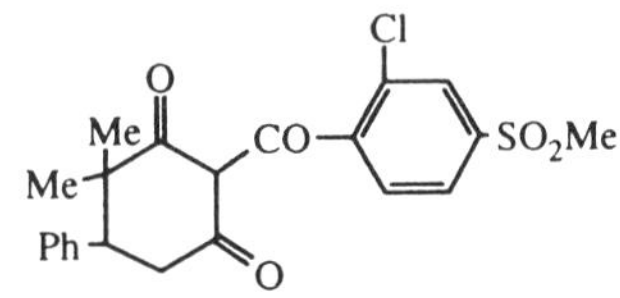

7-78

2-(1-Allyloxyiminoalkyl)-3-hydroxy-6-aryl-2-cyclohexen-1-ones were found to be very good herbicides as well. For example, compound **7-79**, used at 5 kg/ha, gave complete control of *Echinochloa crus-galli* and *Avena sativa*.[62]

2-Heterocyclecarbonyl-1,3-cyclohexanedions and -cyclohexanetriones were good herbicides. Thus, compound **7-80**, applied pre-emergence at 4 lb /acre, gave 100% control of *Setaria virides*.[63]

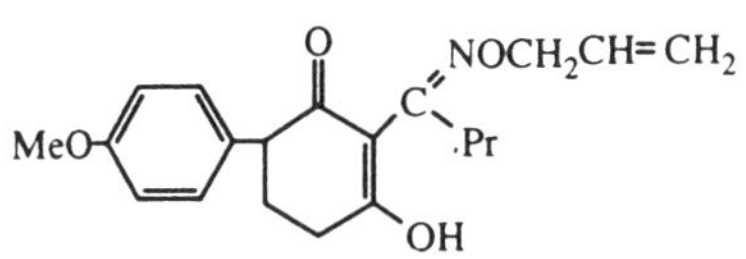

7-79

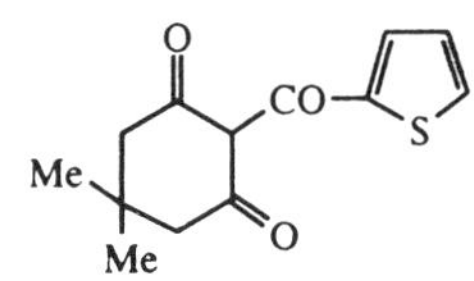

7-80

2-Aroyl-4-hydroxy (or mercapto)-1,3-cyclohexanediones were prepared and recognized to be excellent herbicides. Thus, compound **7-81** was applied post-emergence at 4.48 kg/ha and gave 100% kill of *Setaria viridis, Echinochloa crus-galli, Ipomea lacunosa, Abutilon theophrasti,* and *Brassica juncea*.[64]

2-[2-(Aryl)acetyl]-1,3-cyclohexanediones were synthesized, tested, and proven to be good herbicides. Thus, compound **7-82** in pre-emergence application at 4.48 kg/ha gave 90–100% control of green foxtail, watergrass, morning glory, velvetleaf, Indian mustard, and other weeds.[65]

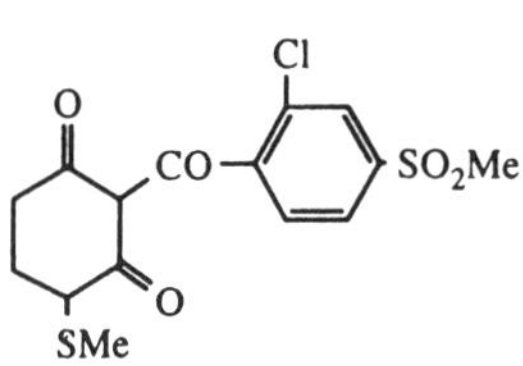

7-81

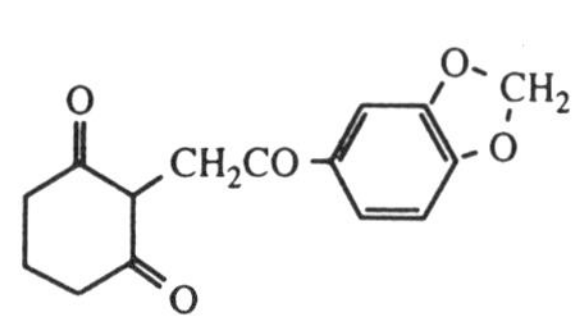

7-82

2-(Alkylaminomethylene)-4-phenylalkyl-5,5-dimethyl-1,3-cyclohexanediones were recognized as useful herbicides against annual and perennial weeds. For example, compound **7-83**, applied post-emergence at 50 g/acre, gave 100% control of *Abutilon theophrasti* and *Datura*, without harming soybean seedlings.[66]

2-Aroyl-1,3-cyclohexanediones were shown to be antidotes for sugarcane herbicides. Thus, post-emergence use of 2,4-D (as the dimethylamine salt) at 2.16 kg/ha damaged the sugarcane, but when 2-(2-chloro-4-methanesufonyl)phenyl]-1,3-cyclohexanedione was applied simultaneously at 1 kg/ha, the sugarcane was unharmed.[67]

2-(1-Alkoxyiminoalkyl)-3-hydroxy-5-[2-(2-fluoro-4-trifluoromethylphenylthio)-ethyl]-2-cyclohexen-1-ones were superb herbicides. For instance, compound **7-84** at 40 g/are gave 100% control of *Echinochloa crus-galli* and wild oats.[68]

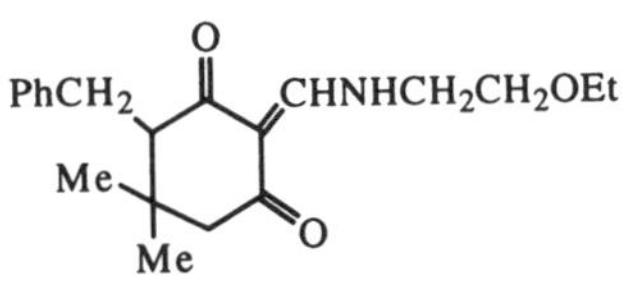

7-83

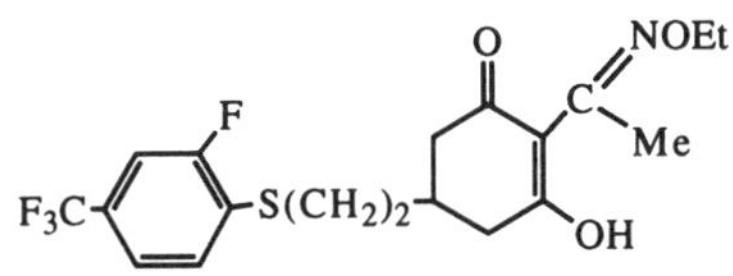

7-84

2-(1-Alkoxyiminoalkyl)-3-hydroxy-5-(4-ethylphenyl)-2-cyclohexen-1-ones were discovered to be herbicides. For instance, compound **7-85**, applied pre-emergence at 0.5 kg/ha, was strongly herbicidal against grassy plants.[69]

2-(2-nitro-4-methylsulfonyl)benzoyl-3-hydroxy-5-(3-methoxypropyl)-2-cyclohexen-1-one, **7-86**, and similar compounds were useful as herbicides. Compound **7-86**, applied pre-emergence at 100 g/10 are, gave 100% inhibition of barnyard grass.[70]

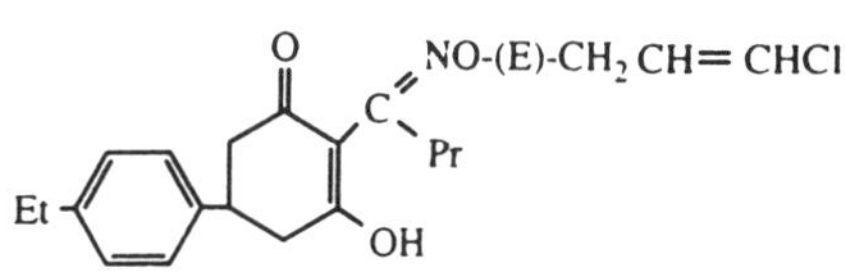

7-85

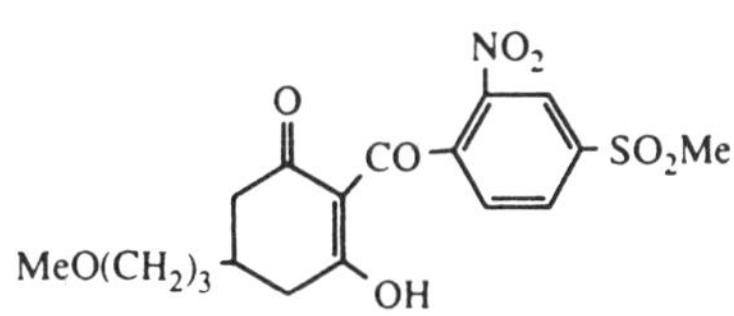

7-86

2-(2-Methyl-3-methoxycarbonyl-4-methanesulfonyl)benzoyl-1, 3-cyclohexanedione, **7-87**, and related materials were more powerful herbicides than atrazine (a commercial herbicide). Compound **7-87** at 0.5 g/are gave >90% control of *Eleusine indica, Digitaria adscendens,* and *Panicum dichotomiflorum* without harming corn; the application of atrazine at 4 g/are gave only 40–70%, 40–70%, and 5–20% control, respectively, of these weeds.[71]

2-(1-Alkoxyiminoalkyl)-3-hydroxy-5-phenoxyphenyl-2-cyclohexen-1-ones were found to be efficient and selective post-emergence herbicides. As an example, compound **7-88**, at 1 kg/ha foliar application, controlled *Echinochloa crus-galli* without injuring linseed, sugar beet, or soybean.[72]

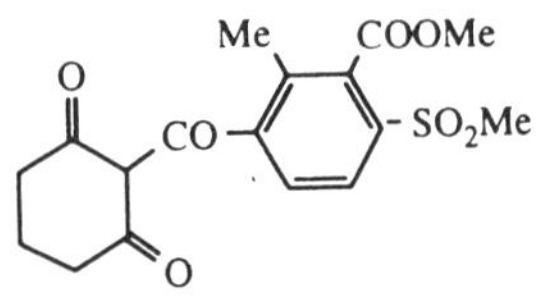

7-87

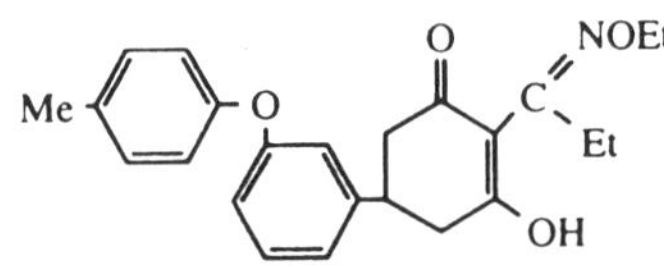

7-88

Lithium salts of 2-(1-alkoxyiminoalkyl)-3-hydroxy-5-alkylthiocycloalkyl-2-cyclohexen-1-ones were prepared and found to be storage-stable herbicides and plant growth regulators. Thus, compound **7-89**, applied post-emergence at 250 g/ha, gave 100% kill of most monocotyledonous plants without harming soybeans and sunflower; such compounds inhibited the growth of cover crops, grasses, and grains, and modified soybean growth.[73]

Alkyl 4-aryl-2-haloimidazole-5-carboxylates were safeners for 2-(1-alkoxyiminoalkyl)-1,3-cyclohexanediones used as herbicides. For example, ethyl 2-chloro-4-phenylimidazole-5-carboxylate, at an application rate of 0.06 kg/ha, eliminated injury to wheat caused by 2-(1-ethoxyiminobutyl)-3-hydroxy-5-ethylthioisopropyl-2-cyclohexen-1-one (Sethoxydim, **7-9**) used at 0.06 kg/ha.[74]

Novel 2-aroyl-1,3-cyclohexanediones were prepared for use as herbicides. For instance, compound **7-90**, in pre-emergence application at 0.25 lb./acre, gave 100% control of *Setaria viridis* and velvetleaf.[75]

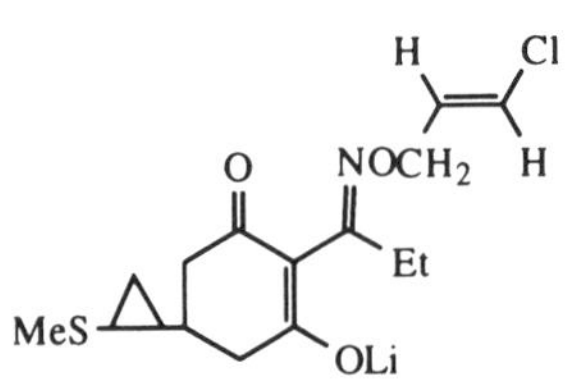

7-89

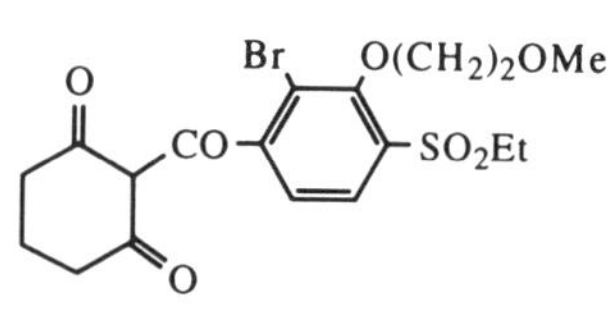

7-90

2-Iminobenzyl-1,3-cyclohexanediones were found to be broad spectrum herbicides which did not damage crops. For instance, compound **7-91**, at 0.4 kg/ha application rate, gave 100% control of *Setaria viridis, Digitaria adscendens, Chenopodium album, Abutilon avicenae,* and other weeds, without harming corn, barley, wheat, or soybeans.[76]

2-Pyridylcarbonyl-3-hydroxy-2-cyclohexen-1-ones were synthesized and proven to be effective herbicides. Thus, compound **7-92**, in pre-emergence application, gave almost complete control of *Solanum nigrum* and other weeds, without damaging corn and other crops.[77]

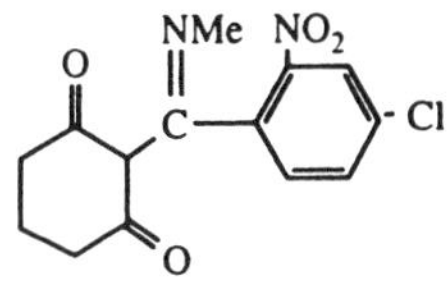

7-91

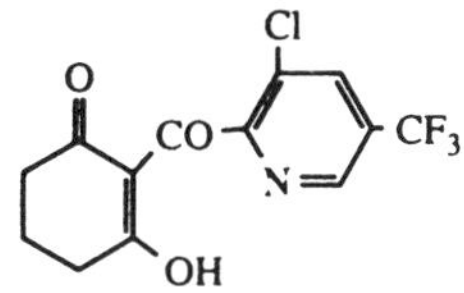

7-92

2-Acyl-5-pyrazolyl-1,3-cyclohexanediones and related compounds were found to be insecticides. For instance, compound **7-93**, applied at 200 ppm, gave 90% control of *Nilaparvata lugens* larvae in rice seedlings.[78]

It was shown that 4,4,5-trimethyl-2-(2-chloro-3-methoxy-4-methylsulfonylbenzoyl)-1,3-cyclohexanedione, **7-94**, applied at 6.3 g/10 are, gave 100% control of *Digitaria adscendens, Amaranthus blitum,* and other weeds, while the isomer, 4,4,6-trimethyl-2-(2-chloro-3-methoxy-4-methylsulfonylbenzoyl)-1,3-cyclohexanedione, gave only 20–79% control of the same weeds.[79]

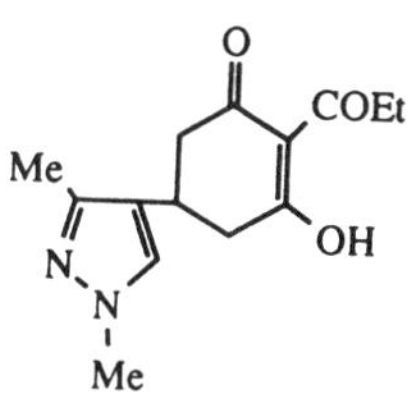

7-93

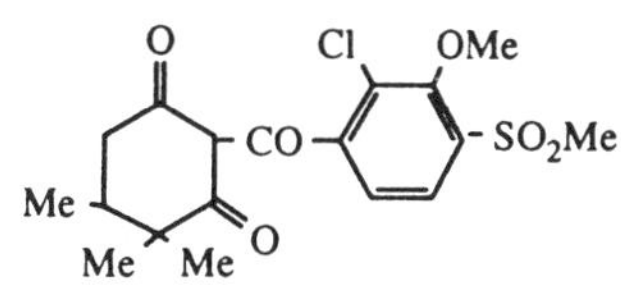

7-94

The substituted 2-aroyl-1,3-cyclohexanedione, **7-95**, and similar products were discovered to give, at 100 g/are, 100% control of crabgrass, foxtail, Indian mallow, and other weeds.[80]

New varieties of 2-aroyl-1,3-cyclohexanediones were shown to be selective herbicides for rice. Thus, compound **7-96**, in pre-emergence application at 0.06 kg/ha,

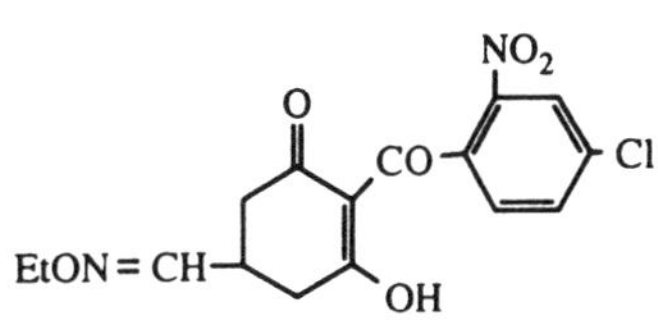

7-95

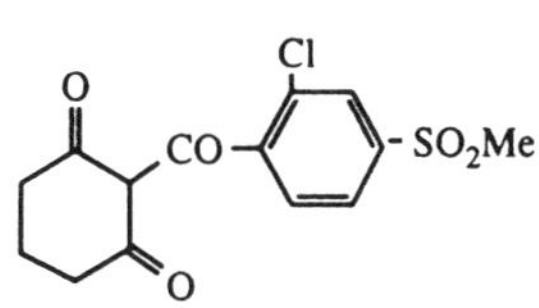

7-96

gave 100% control of *Cyperus montis, Scirpus maritimus*, and other weeds, without harming rice plants.[81]

Substituted 2-(1-alkoxyiminoalkyl)-3-hydroxy-5-pyridyl-2-cyclohexen-1-ones were useful pre-emergence herbicides. For instance, compound **7-97** at 0.5 lb /acre gave 100% control of yellow foxtail, Johnson grass, barnyard grass, and wild oat.[82]

2-Acyl-5-cyano-1,3-cyclohexanediones were both herbicides and plant growth regulators. Thus, compound **7-98** inhibited the growth of barley and wheat more effectively than (2-chloroethyl)trimethylammonium chloride.[83]

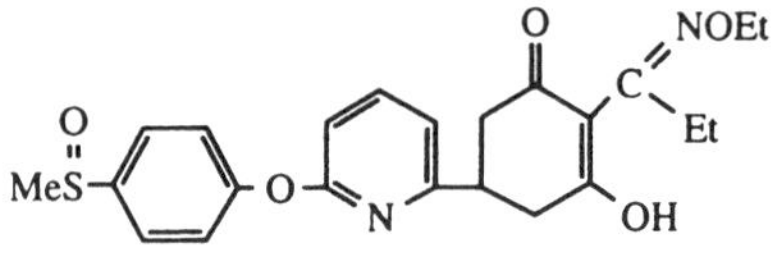

7-97

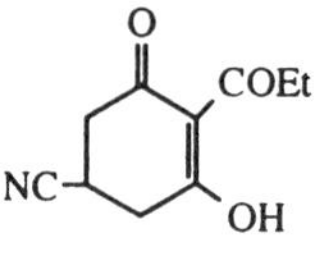

7-98

2-(1-Oximinoalkyl)-3-hydroxy-5-ethylsulfonamidoalkyl-2-cyclohexen-1-ones were prepared and demonstrated to be useful herbicides. As an example, compound **7-99**, applied post-emergence at 0.4 kg/ha, gave 100% control of wild oats without harming wheat.[84]

3-(Pyridylmethylamino)-2-cyclohexen-1-ones, prepared from 1,3-cyclohexanedione, were discovered at Nippon Soda to be fine insecticides. Thus, compound **7-100** at 125 ppm killed 100% aphids and green rice leaf hoppers.[85]

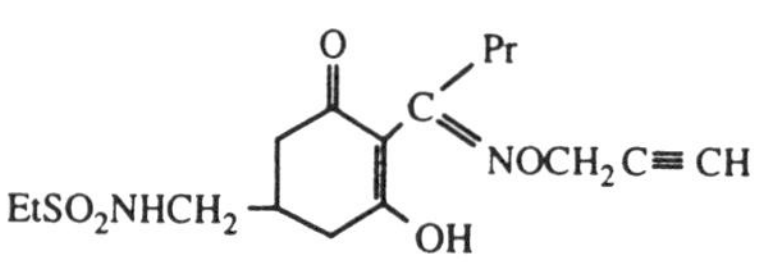

7-99

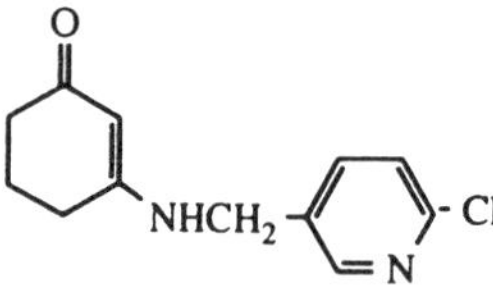

7-100

2-(Alkynylalkenyloxyiminoalkyl)-hydroxy-2-cyclohexen-1-ones were shown to be selective herbicides. For example, compound **7-101** was an excellent post-emergence weed killer against grasses in oat crops.[86]

Novel 2-aroyl-1,3-cyclohexanediones on an inert carrier were patented as selective weed control agents. For instance, compound **7-102**, applied pre-emergence at 25 g/10 a, gave 100% control of barnyard grass, small-flower umbrella plant, arrowhead, and Japanese bulrush but did not injure rice plants.[87]

2-(3-Alkoxybenzoyl)-1,3-cyclohexanediones were discovered which were excellent herbicides. For instance, compound **7-103**, applied pre-emergence at 0.25 lb /acre

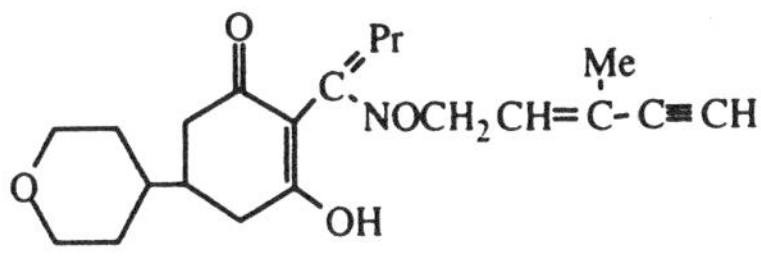

7-101

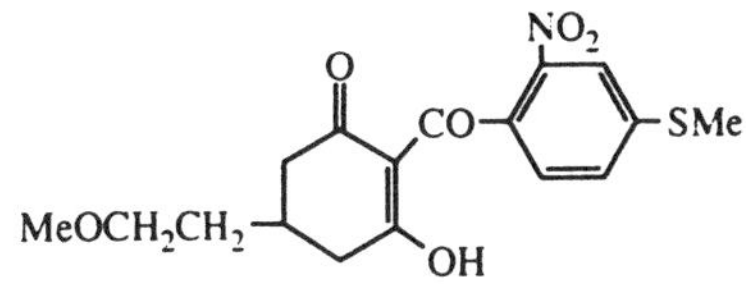

7-102

and post-emergence at 0.28 lb./ha, 100% controlled *Echinochloa crus-galli, Brachiaria platyphylla, Abutilon theophrasti,* and *Xanthium.*[88]

A review of the activities and properties of the 1,3-cyclohexanedione-type herbicides was published in 1992.[89]

A synergistic mixture of the commercial herbicide, bromoxynil, with 2-(2-chloro-4-methylsulfonyl)benzoyl-1,3-cyclohexanedione (or similar compounds) at 16 and 18 g/h, respectively, gave close to 100% control of *Amaranthus retroflexus* and *Brassica Kaber*; each of the components alone was less effective than the mixture.[90]

5-(Pyridin-4-yl)-2-cyclopropylcarbonyl-3-hydroxy-2-cyclohexen-1-one (and related compounds), **7-104**, was found at 200 ppm to give >90% control of the insects, *Niloparvata lugens* and *Nephotettix cincticeps.*[91]

Haloalkoxy-substituted 2-benzoyl-1,3-cyclohexanediones were prepared for application as herbicides and plant growth regulators. For example, compound **7-105** at 1.25 kg/ha gave 80–100% pre- and post-emergence control of *Stellaria media.*[92]

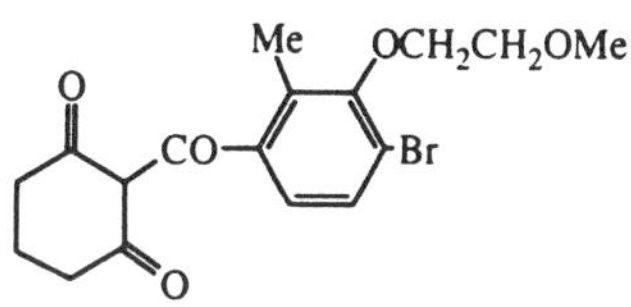

7-103

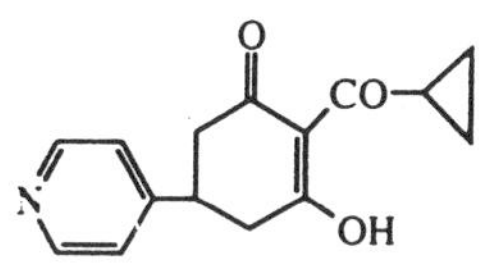

7-104

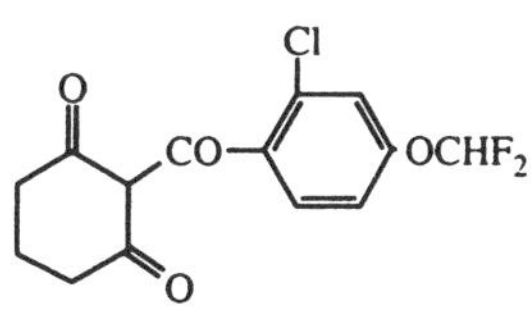

7-105

7.3.2. Phenyl Ethers

This is another class of commercial agricultural chemicals which includes several resorcinol-derived products (see Section 7.2). The substantial developmental activity in the 1980–1992 period is demonstrated by the following selections from the literature, and includes herbicides, fungicides, and an insecticide.

Sulfonates of hydroxyphenyl ethers were patented as effective herbicides, applied at 125–500 g/10 are. An example of these materials is compound **7-106**.[93]

Tetrahydrofuranyl ethers of hydroxy-nitrophenyl ethers were found to be practical herbicides. Thus, compound **7-107**, applied post-emergence at 1 g/are, gave total control of, e.g., *Echinochloa crus-galli* in rice.[94]

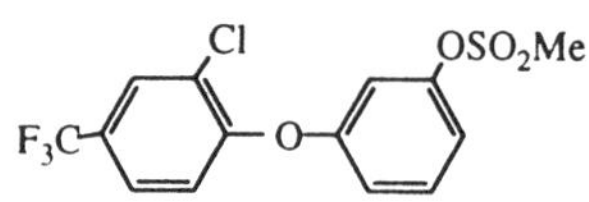

7-106

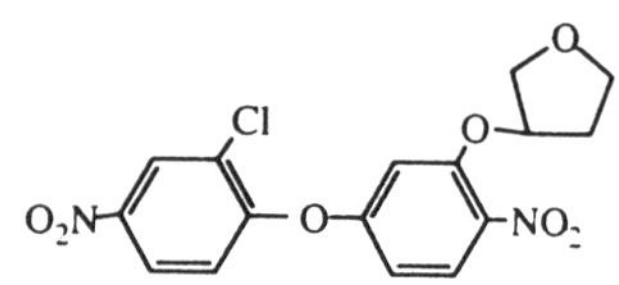

7-107

A series of esters of carboxyalkyl-thioalkyl ethers of hydroxyphenyl ethers were prepared and found to have good herbicidal and plant growth regulating activity. One example from this patent is compound **7-108**.[95]

1,3-Dioxanylalkyl ethers of hydroxy-nitrophenyl ethers also had good herbicidal activity. For example, compound **7-109**, applied at 5 g/are, gave 100% control of barnyard grass in rice.[96]

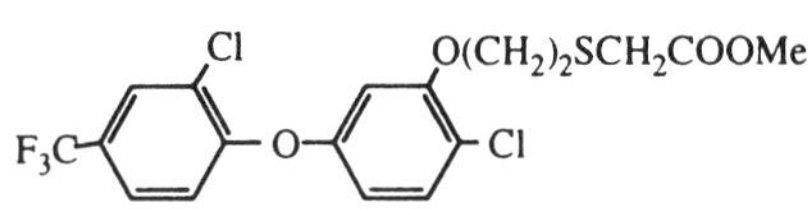

7-108

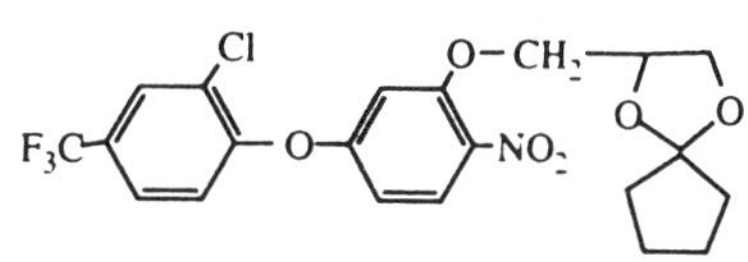

7-109

Carbamates of hydroxyphenyl ethers were found to be useful herbicides. Thus, compound **7-110** showed good activity against *Echinochloa crus-galli, Scirpus juncondes*, and other weeds.[97]

A series of fluoroalkyl ethers of 3-hydroxy-4-nitrophenyl ethers were synthesized and found to have excellent herbicidal activity and extremely low toxicity to useful crops. Compound **7-111**, for instance, applied at 20 g/are pre-emergence and at 2 g/are post-emergence, gave 100% suppression of barnyard grass (*Echinochloa crus-galli*)

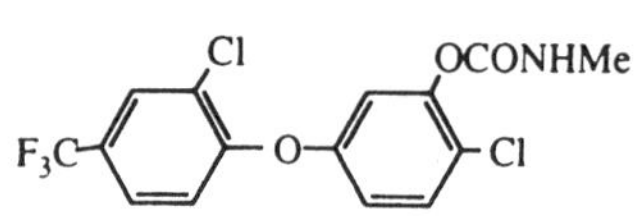

7-110

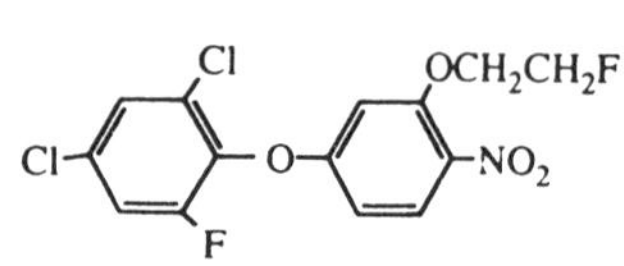

7-111

and other broadleafed and narrow-leafed common weeds, with no damage to rice plants.[98]

Oxyfluorfen (Goal®), **7-7**, and related compounds were patented by Rohm & Haas Co. as selective herbicides.[99]

Substituted resorcinol di-(4-trifluoromethylphenyl) ethers were also found to be effective herbicides. Thus, compound **7-112**, applied at 10 lb /acre, gave 99–100% control monocot and dicot weeds.[100]

Furfuryl amides of (phenoxyphenoxy)alkanoic acids were useful in controlling undesirable vegetation. For instance, compound **7-113**, at a 0.5 lb /acre pre-emergence application rate and 0.25 lb /acre post-emergence application, gave 100% suppression of wild mustard, pigweed, velvetleaf, morningglory, and other weeds with slight or no damage to wheat, rice, and corn.[101]

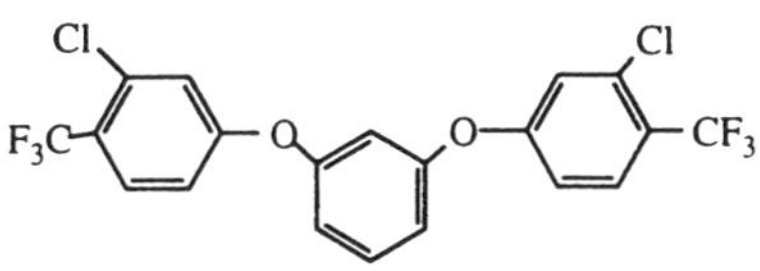

7-112

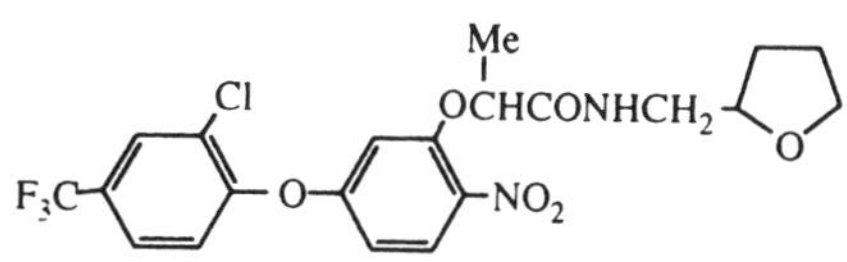

7-113

Esters of (phenoxyphenoxy)alkanecarboxylic acids were prepared and developed as effective and selective herbicides. For example, compound **7-114** at 1 kg/ha application gave 100% kill of *Sesbania exaltata, Amaranthus tetroflexus,* and other weeds, without harming wheat and with little injury to barley, corn, sorghum, and rice.[102]

Heterocyclic amides of phenoxyphenoxyalkanoic acids were patented as useful pre-emergence herbicides. Thus, compound **7-115**, applied at 0.125 lb /acre, gave total control or was severely phytotoxic to wild mustard, pigweed, velvetleaf, and sugar beet, but was little or not at all toxic to soybean, cotton, wheat, rice, and oats.[103]

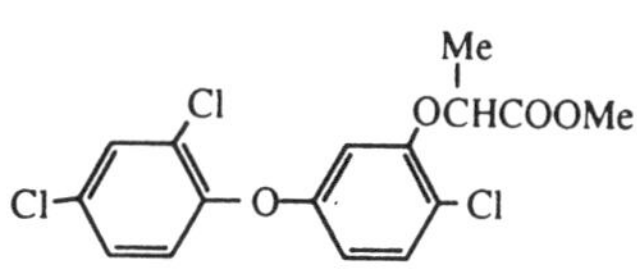

7-114

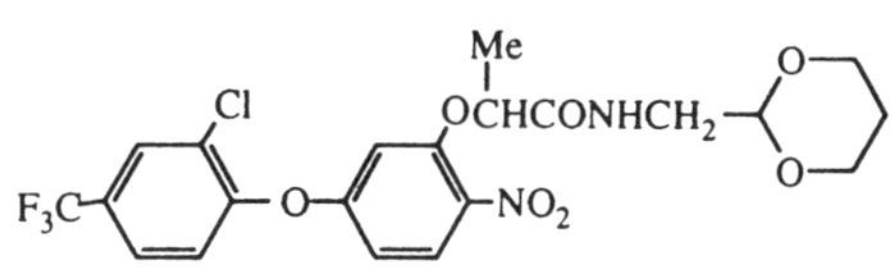

7-115

A large number of novel 2-chloro-4-trifluoromethyl-3′-hydroxy(or 3′-amino)-4-nitro-diphenyl ethers were prepared and established to be highly effective herbicides at 0.25–4.0 lb /acre. For instance, compound **7-116**, used at 2 lb /acre in pre-emergence application, gave 98–100% control of monocot and dicot weeds without much, if any, harm to peanuts, soybeans, or wheat.[104]

(Phenoxyphenoxy)alkyl phosphinates, phosphonates, and related compounds were synthesized and shown to be useful for weed control. For instance, compound **7-117**, dosed at 10 lb /acre in pre-emergence application, gave 95–99% control of grasses and broadleaf weeds, and at 10 lb /acre in post-emergence applications gave 81% control of grasses and 95% control of broadleaf weeds.[105]

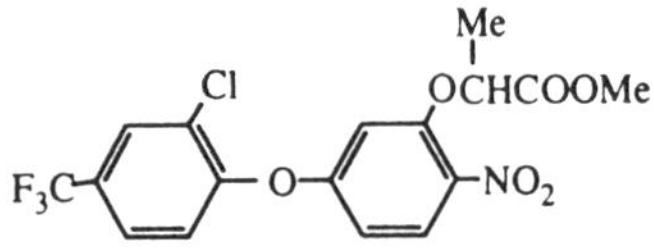

7-116

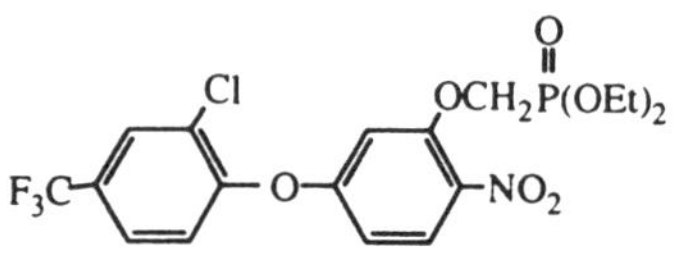

7-117

Certain arylsulfonylated phenyl ethers were synthesized and found to be good herbicides. Thus, compound **7-118**, in post-emergence application at 1 g/are, gave 100% control of cocklebur (*Xanthium pensylvanicum*) and did not harm soybean.[106]

The (*R*)-enantiomer of the diphenyl ether derivative, **7-119**, was prepared and was seen to be a more powerful herbicide than its known racemate.[107]

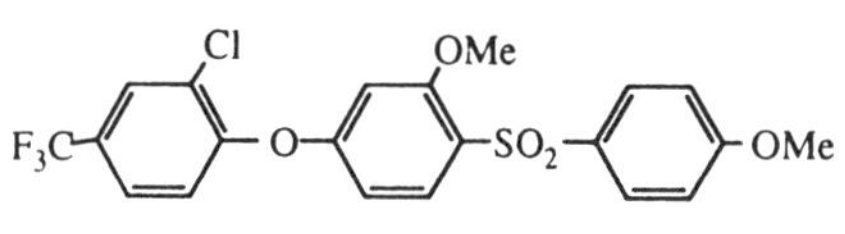

7-118

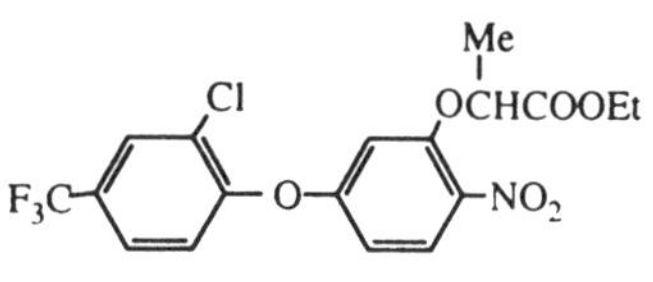

[(R)-enantiomer]

7-119

The (phenoxyphenoxy)propionyl-thiourea compound, **7-120**, was made and shown to give excellent control of *Avena fatua, Echinochloa crus-galli*, and *Setaria* in sugar beet, cotton, and soybean plantings.[108]

Derivatives of ([pyridinyloxy]-2-nitro-5-phenoxy)propionamides were synthesized, screened, and found to be selective herbicides useful for controlling weeds in valuable crops. Thus, compound **7-121**, applied at 62.5 ppm, gave 100% post-emer-

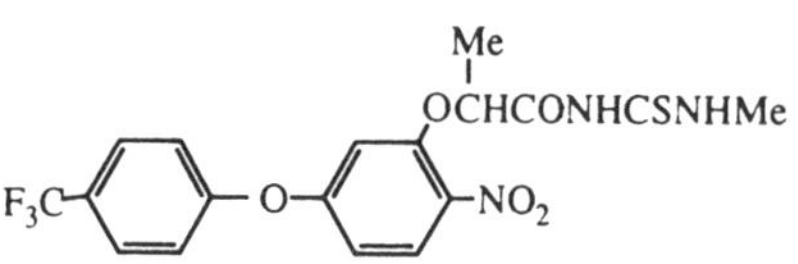

7-120

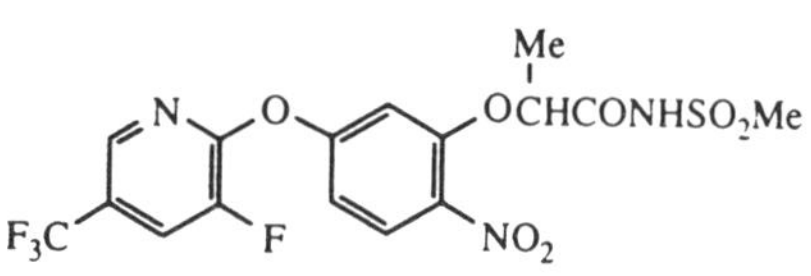

7-121

gence control of morning glory, velvetleaf, and jimson weed, with little damage to corn and soybeans.[109]

Alkylideneaminooxymethyl phenoxy(or pyridyl)phenoxypropionates were prepared for use as the active ingredients in herbicidal compositions. For instance, compound **7-122** was useful as an excellent pre- and post-emergence herbicide, used at 0.01–1.0 kg/ha, against monocotyledonous and dicotylodenous weed grasses in rice and cereal crops.[110]

Alkoxycarbonyl-substituted (phenoxyphenoxy)propionic acid esters were made and shown to be effective herbicides. For instance, compound **7-123**, used at 0.125–0.5 kg/ha, gave excellent control of *Ipomea purpurea* and many other weeds.[111]

Alkyl (phenoxyphenoxy)propenoates were prepared and were useful as fungicides. As an example, compound **7-124**, applied at 40 ppm as a 0.05% spray, gave 100% control of *Erysiphe graminis.*[112]

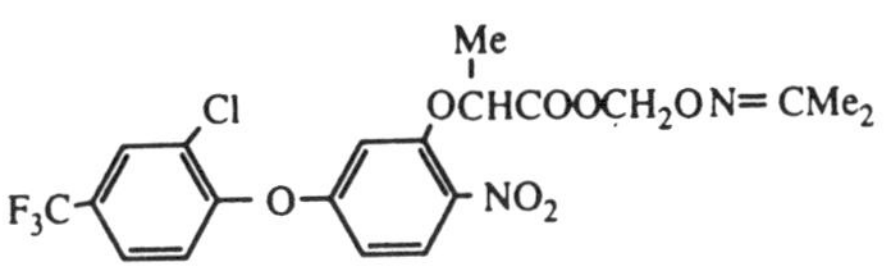

7-122

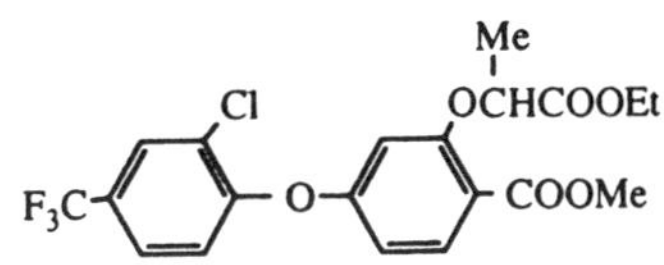

7-123

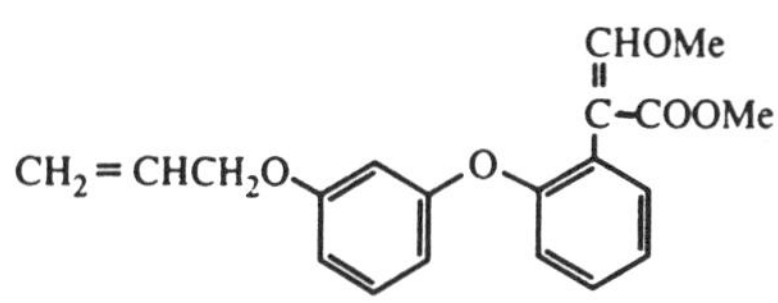

7-124

Novel fluorinated phenyl ethers were made and found to be excellent, selective herbicides. Thus, compound **7-125**, at 1 kg/ha, gave 96–100% control of *Setaria viridis, Abutilon theophrasti*, and other weeds without injury to cotton or rice.[113]

Oximes of carboxaldehyde-substituted phenyl ethers were made and were quick-acting herbicides against broad- and narrow-leafed weeds. For example, compound **7-126**, used at a 0.5 kg/ha rate, gave 100% control of *Abutilon theophrasti, Aranthus mangostanus, Astralagus sinicus, Solanum nigrum,* and other weeds.[114]

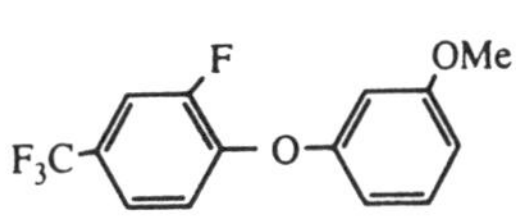

7-125

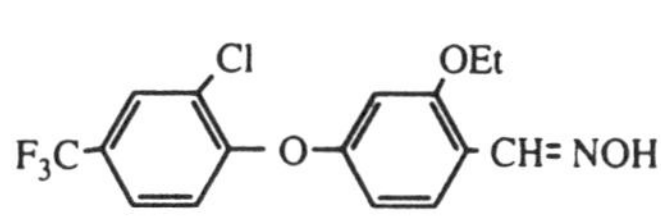

7-126

The (*R*)-isomer of [phenoxy(or pyridyloxy)phenoxy]propionic esters was found to be much higher in herbicidal activity than the corresponding (*S*)-isomer or the racemate. Thus, compound **7-127** (*R*)-isomer, applied at 0.031 kg/ha, was five times more active than the (*S*)-isomer and 2.5 times better than the racemate against *Sorghum halepense*.[115]

Certain substituted phenyl ethers were prepared which proved to be good insecticides. For instance, compound **7-128** was tested at 50 ppm for metamorphosis inhibitory activity against brown rice plant hopper nymphs and gave 100% inhibition.[116]

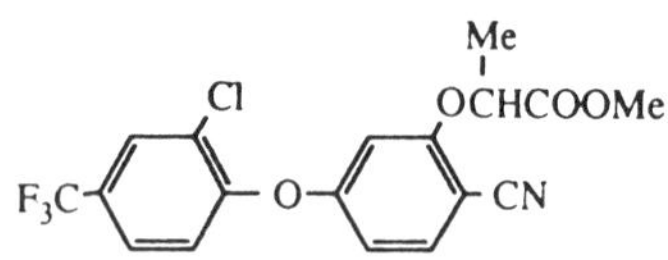

[(R)-isomer]

7-127

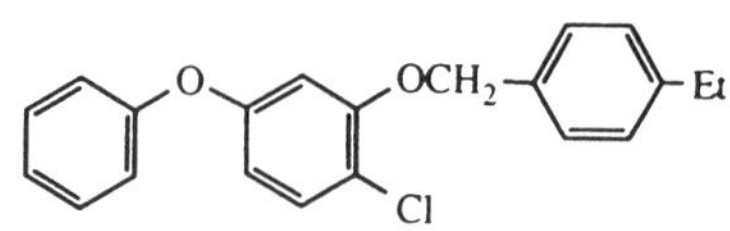

7-128

The (*R*)-isomer of 3-(ethoxycarbonylethoxy)-4-methoxycarbonyl-3′-chloro-4′-trifluoromethyldiphenyl ether, **7-123**, was synthesized and found to give 100% control of pigweed in Indian mallow at 0.3 kg/ha application, while the (*S*)-isomer gave little or no control of pigweed.[117]

A large number (110) of phenoxy(or pyridoxy)-benzoxazines was prepared and found to be good herbicides. Thus, compound **7-129**, used at 5 g/10 are, produced 100% control of large crabgrass and barnyard grass.[118]

2-[(Phenoxyprimidinyloxy)phenyl]-2-(methoxyimino)acetamides were developed which had excellent herbicidal properties. Thus, compound **7-130** (*E*-isomer), used in the form of a spray at 100 ppm, gave total control of *Puccinia recondita, Erysiphe graminis tritici, Septorium nodorum, Pyricularia oryzae*, and other weeds.[119]

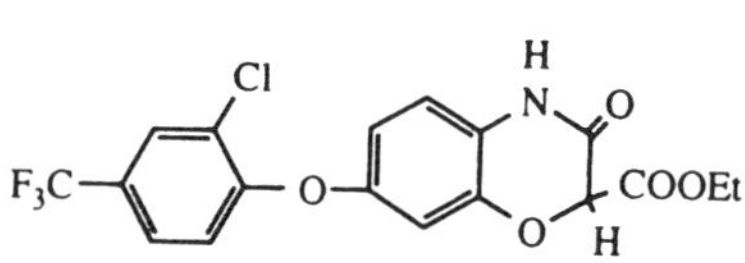

7-129

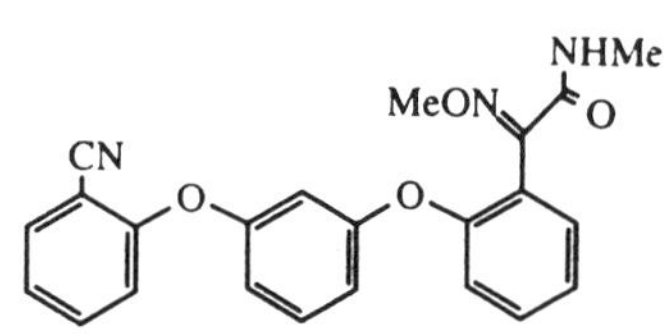

7-130

7.3.3. Resorcylic Acid Derivatives

The compounds covered here all have a beta-resorcylic acid (2,4-dihydroxybenzoic acid) or gamma-resorcylic acid (2,6-dihydroxybenzoic acid) moiety in their structure.

N-(1,3,4-Thiadiazol-2-yl)-2,6-dialkoxybenzamides were prepared and demonstrated to be useful insecticides. For instance, compound **7-131** at 2.5 ppm gave 100% control of *Spodoptera eridania* larvae.[120]

Amides of 2,6-diphenoxybenzoic acid were made and shown to be pre- and post-emergence herbicides. Thus, *N,N*-dimethyl-2,6-diphenoxybenzamide, **7-132**, post-emergence applied at 1 kg/ha, gave total control of barnyard grass with no harm to wheat or rice.[121]

N-Benzoyl-*N'*-phenylureas were developed for pesticide use; e.g., compound **7-133** showed insecticidal activity at 800 ppm.[122]

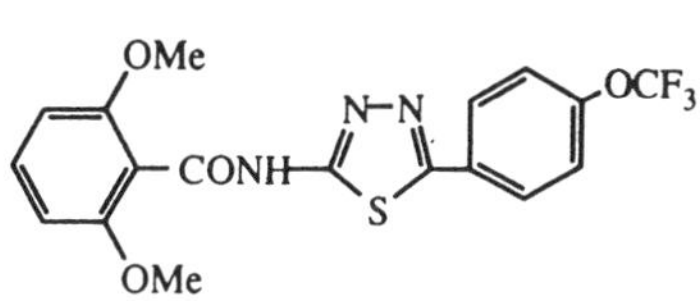

7-131

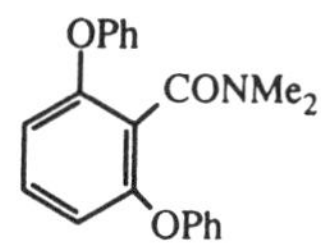

7-132

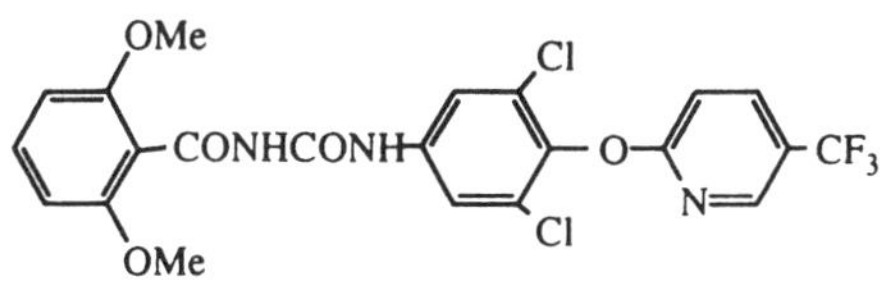

7-133

New (1,3,4-thiadiazin-2yl)-2,6-dialkoxybenzamides were studied and shown, as the sodium salts, to be effective herbicides. For example, compound **7-134**, in pre-emergence application at 0.5 lb/acre, gave 100% control of lambsquarter and mustard weed with no injury to corn, cotton, soybean, wheat, rice, and other crops.[123]

N-(2,6-Dimethoxybenzoyl)-*N'*-arylureas were synthesized and demonstrated to be herbicides. Thus, compound **7-135**, in pre-emergence application at 1lb/acre, gave total control of lambsquarter and pigweed.[124]

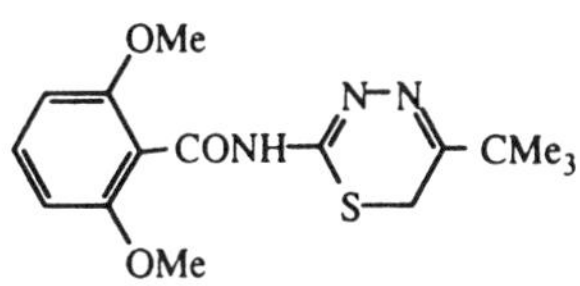

7-134

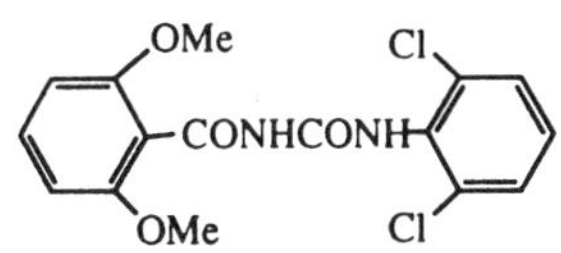

7-135

A number of derivatives of resorcylic acid lactones (RALs) were prepared and shown to be useful agricultural chemicals. Thus, alcoholysis of alpha-zearanol gave compound **7-136** which, in a diet of 1/1 sugar/low-fat dried milk, gave a 68% decrease in the reproduction of adult houseflies.[125] Carbonate esters of zearanol were patented as improved animal growth promoters.[126] Hexahydro-16-deoxy-alpha-zearanol-7-acetate was prepared and treated with Bu_4N^+ HSO_4^- and Chlorox to give compound **7-137**, which showed anabolic activity, e.g., in sheep.[127]

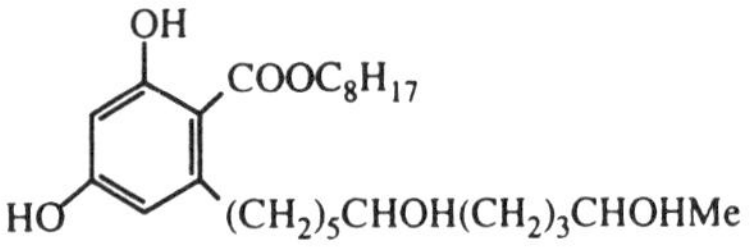

7-136

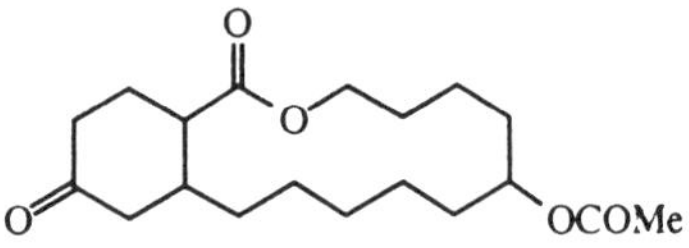

7-137

Zearanol 6′-carbonates were prepared and patented as growth promoters in ruminants.[128] Ring-opened RAL derivatives, such as compound **7-138**, were prepared and claimed as animal growth promoters.[129]

Monorden, **7-139**, was found to be an inhibitor of chitin synthesis and was used as an insecticide; at 25 ppm it controlled *Manduca sexta* and *Mamestra.*[130]

2,6-Bis(pyrimidinyloxy)benzoates, such as compound **7-140**, were demonstrated to be effective herbicides.[131]

In forestry practice, outside of conventional agrichemicals, a Russian patent reported that sodium *p*-aminosalicylate, **7-141**, applied as a 0.1–0.2% aqueous solution to conifers during the tapping stages, increased oleoresin production to 176% (versus 100% yield without this stimulator).[132]

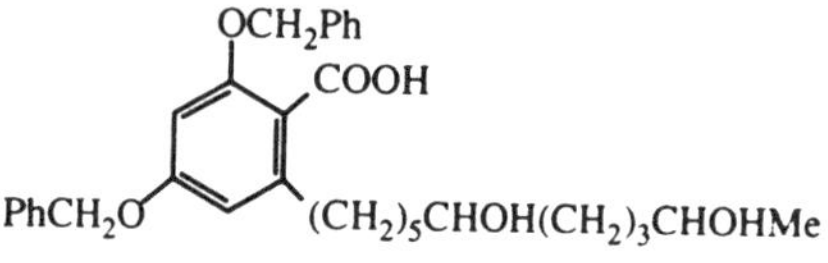

7-138

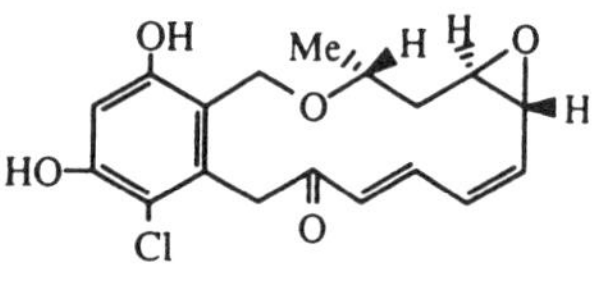

7-139

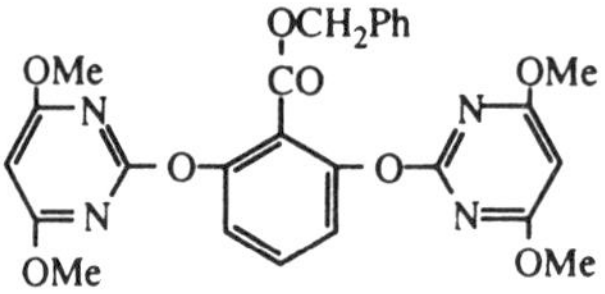

7-140

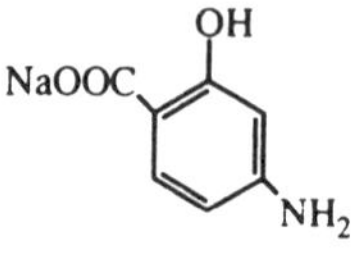

7-141

7.3.4. Heterocyclic Derivatives of Resorcinol

In this category can be found herbicides, insecticides, fungicides, and antifeedants.

Many bis-(pyridyloxy)benzenes and related compounds were prepared which were useful as herbicides. Included in this series were resorcinol-derived materials such as compound **7-142** which, applied pre-emergence at 1 kg/ha, gave 100% control of wild oats and ryegrass in wheat.[133]

4-Chromanone-7-phosphates and -phosphonates, such as compound **7-143**, were shown to be insecticidal to mosquito larvae and two-spotted spider mite.[134]

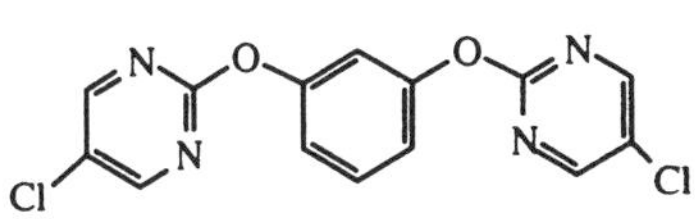

7-142

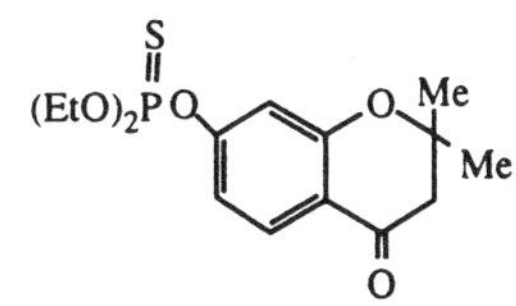

7-143

Polyalkyl-7-hydroxychromenes were prepared at Mitsui Petrochemical Industries and these materials were useful as herbicides, fungicides, and intermediates for insecticides. Thus, compound **7-144**, in pre-emergence application at 5 kg/ha, gave 80% control of *Cyperus difformis* and 90% control of lambsquarter. The *O*-ethyl-*S*-ethyl phosphorthioate of compound **7-144** was said to be effective for the control of the diamond-back moth (*Plutella maculipennsis*).[135] Indeed, *O*-ethyl-*S*-ethyl phosphates of compound **7-144** were claimed in a patent of Kumiai Chemical Co. as insecticides (equal to or better than Dipterex, Sumithion, Orthene), miticides, and nematocides.[136]

Heterocyclic patalenes, prepared by the treatment of 1,3-cyclohexanedione dioximes with S_2Cl_2, were found to be herbicides. Thus, compound **7-145** was reported to have herbicidal activity at 5 kg/ha.[137]

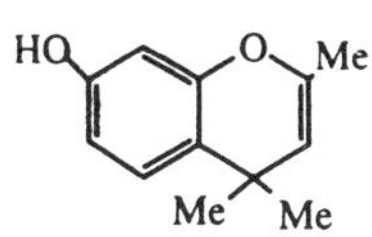

7-144

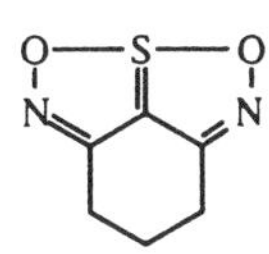

7-145

Polyalkyl-5-hydroxychromenes, e.g., compound **7-146**, were prepared and were intermediates for agricultural chemicals as well as bactericides (at 1000 ppm) and germicides (at 100 ppm).[138]

2-Alkoxy-2,4,4-trimethyl-7-hydroxychromans and their phosphorothiolates, such as compound **7-147**, were prepared and found to have insecticidal, acaricidal, and fungicidal activity at 100–500 ppm.[139] Certain stilbene derivatives were found to be

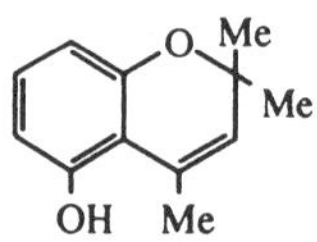

7-146

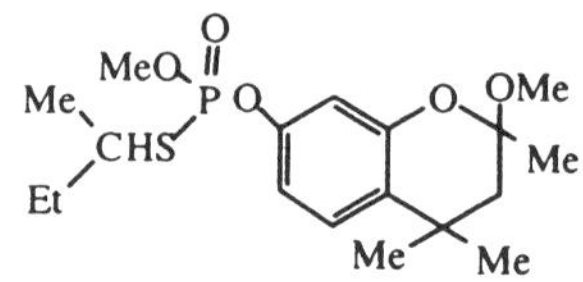

7-147

useful fungicides. Thus, compound **7-148** was active against *Trichophyton mentagropytes* and *Mucor racemosus*, the LD_{50} being 547.9 mg/kg (i.p.) in mice.[140]

Anti-juvenile hormones were prepared which were effective in the control of insects by inhibiting the actions of juvenile hormone; they induced precocious maturation of immature insects, resulting in death either before or after the molting to an incompetent precocious adult. Sterilization of mature insects was also observed. An example is compound **7-149**; at a dose of 1.5 μg/cm^2 it inducted precocious metamorphosis in 90% of immature milkweed bugs (*Oncopetus fasciatus*) and at 7.5 μg/cm^2 it caused sterilization in 100% of adult milkweed bugs.[141]

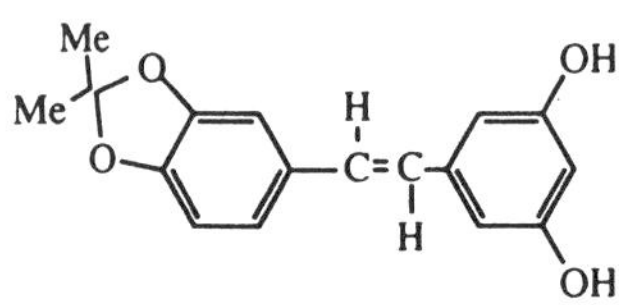

7-148

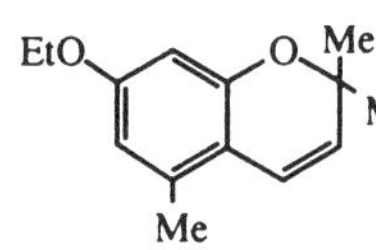

7-149

4-Methyl-7-hydroxy[2*H*]benzopyran-2-one, **7-150** (4-methyl-7-hydroxycoumarin), and its esters or ethers, at 50–2000 g/ha together with Cu, Zn, Mg, or B-chelates, were plant growth regulators, suitable for increasing the yield of cereals, soybean, sunflower, and corn, the latter by 10–17%.[142]

7-Alkoxy-isoflavone derivatives were prepared and were useful as fodder additives. For instance, compound **7-151** was used as a metabolic stimulant.[143]

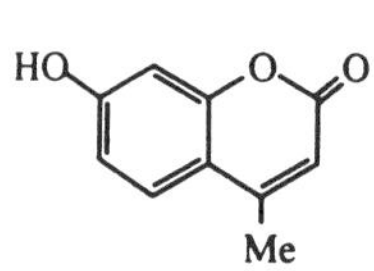

7-150

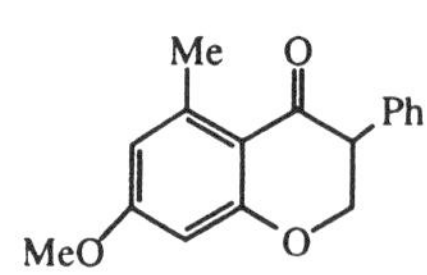

7-151

3-Chloro-4-methyl-7-hydroxycoumarin, **7-152**, an agrochemical intermediate, was prepared by condensing resorcinol with methyl acetoacetate in octane in the presence of sulfuric acid with removal of the distillate, holding for 1 hr/120–125 °C, cooling, adding acetic acid and sulfuryl chloride, and heating at 75–80 °C/2 hr to give the target compound in 98% yield and 88% purity.[144] (This is an illustration of the difficulties involved in making some of the chemicals described here.)

Pyrimidinyloxy- and triazinyloxy-phenoxypropionic acids and esters were synthesized and shown to be plant growth regulators. Thus, compound **7-153** in spray application inhibited by 80% the elongation of kidney bean nodes and enhanced the formation of lateral buds.[145]

Derivatives of 3,4-dihydro-2*H*-benzopyran were prepared for use as pesticides. For example, an 0.2 wt% solution of compound **7-154** together with 0.1 wt% methoprene was used against the termite *Prorhinotermes simplex* and gave 10% white soldiers and 10% intercastes after 14 days.[146]

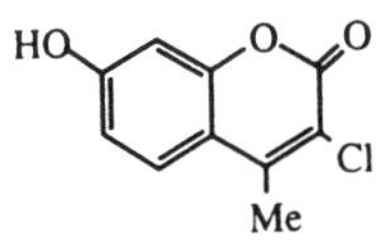

7-152

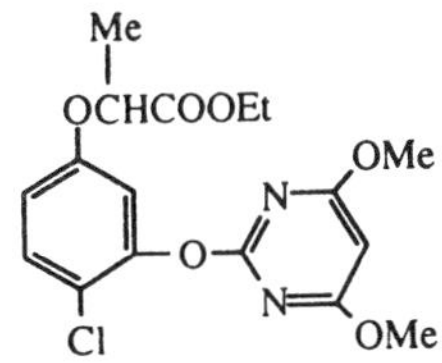

7-153

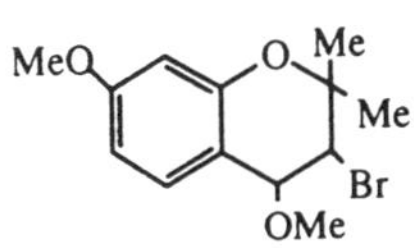

7-154

7.3.5. Miscellaneous Resorcinol Derivatives

Certain novel benzophenones were prepared and found to be broad spectrum fungicides suitable for use on food crops since, under normal conditions, they did not leave a toxic residue. For instance, 2,2′-dihydroxy-4,4′-diacryloxybenzophenone, **7-155**, used at 25 lb /acre, gave 90% control of powdery mildew on beans after 14 days; at lower concentration this compound gave 100% control of bean rust (*Uromyces phaseoli*) and cucumber anthracnose (*Colletotrichum lagenarium*).[147]

A large number (89) of phosphoramidothioates of nitroresorcinols were prepared and demonstrated to be fungicides. For example, compound **7-156**, at a 200 ppm

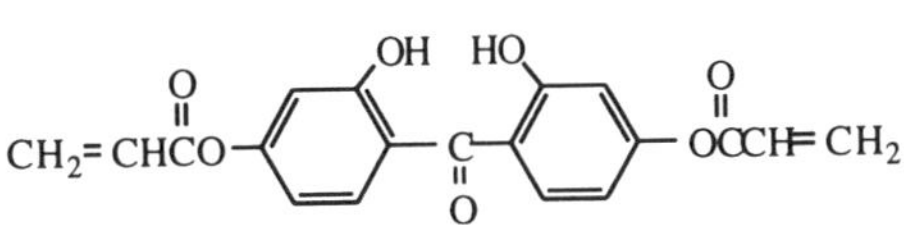

7-155

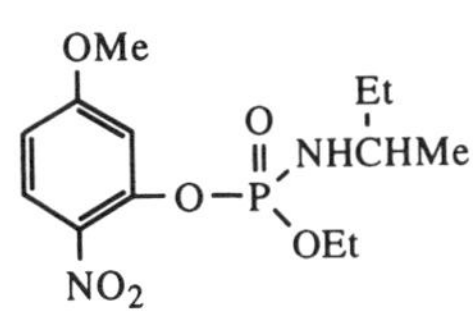

7-156

application rate, was more effective on cucumbers against *Pseudoperonospora cubensis* than Daconil.[148]

In tests of several compounds on the growth and yield of pearl millet, five cultivars of this plant (*Pennisetum typhoides*) were sprayed with 0.5 ml/L resorcinol. This treatment increased grain yield by 25%, hastened the emergence of ears on the main tiller, and increased the leaf area.[149]

Several di-(propargyloxy)benzenes were prepared and found to be fungicides. As an example, compound **7-157**, applied in the form of a 0.03–0.1 g/L solution, gave good control of *Botrytis cineres, Pseudocercosporella herpotrichoides, Rhizozctonia solani*, and other fungi.[150]

Clavamine, **7-158**, isolated from the poison glands of the Joro spider, *Niphila clavata*, and also synthesized, had insecticidal activity. Thus, compound **7-158** at 3×16^{-6} M killed 70–90% of Culex mosquito larvae after 3 hr.[151]

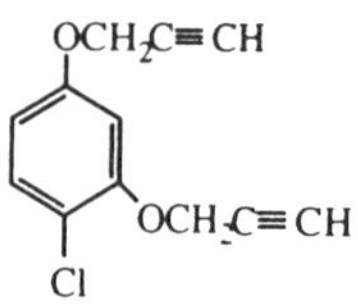

7-157

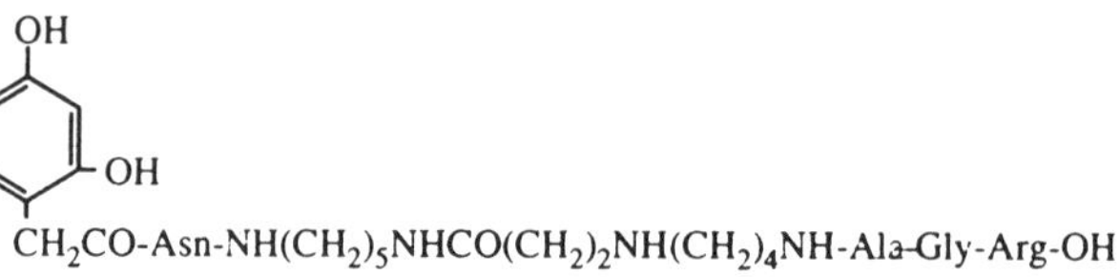

7-158

4,6-Di-*tert*-butylresorcinol, its alkali metal salts, or a phosphite derivative were reported to be biocides, particularly useful as fungicides against wood-destroying fungi. Thus, these compounds were useful for wood preservation.[152]

7.3.6. Derivatives of *m*-Aminophenol

In the 1980s, research on the use of *m*-aminophenol as a starting material for agricultural chemicals continued at a good pace. Just a few examples follow.

In the carbamate category, new carbanilic acid [3-(alkylcarbonylamino)-phenyl]esters were prepared as herbicide candidates; e.g., compound **7-159** at 5 kg/ha gave 100% control of *Brassica* and *Solanum plants*.[153]

Mixtures of biscarbamates and acifluorfen were found to be synergistic herbicides; thus, post-emergence application of 1 kg/ha phenmedipham (Betanal, **7-15**) with 0.05 kg/ha acifluorfen gave 100% control of *Abutilon* and *Amaranthus retroflexus*, with only slight damage to soybeans; each herbicide by itself was less effective than the mixture.[154]

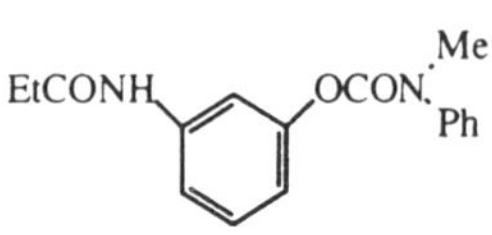

7-159

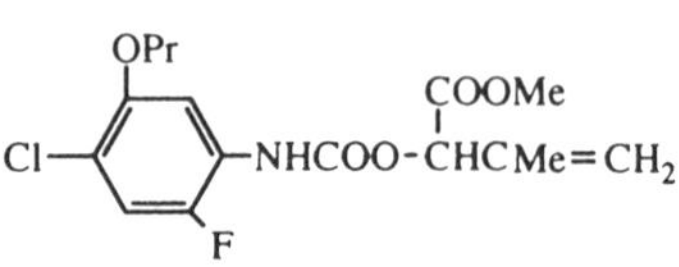

7-160

Novel *N*-phenylcarbamate herbicides were prepared; e.g., compound **7-160**, applied to the soil at 80 g/are as an emulsion, was effective in controlling *Panicum crus-galli, Pharbitis purpurea,* and *Abutilon avicennae.*[155]

In the phenylurea category, substituted [(cyclopropylmethoxy)phenyl]ureas were synthesized and shown to be herbicides. For instance, compound **7-161** at 2 lb/acre gave 100% post-emergence control of broadleaf weeds.[156]

m-(Phenylalkoxy)phenyl-*N*-methoxy-*N*-methylureas were prepared and tests showed that they were useful for the selective control of dicotyledonous and certain monocotyledonous weeds in agronomic crops. Compound **7-162** at 1.0 kg/ha, for example, gave 100% control of cocklebur and jimsonweed, with no harm to barley or sorghum.[157]

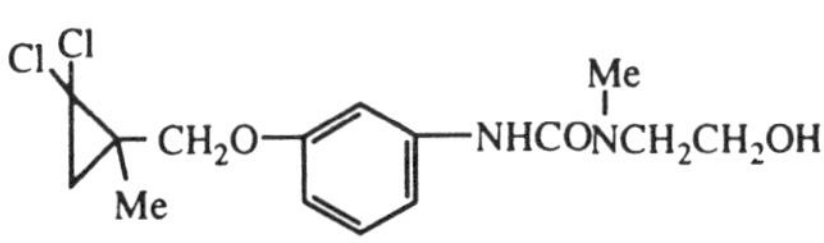

7-161

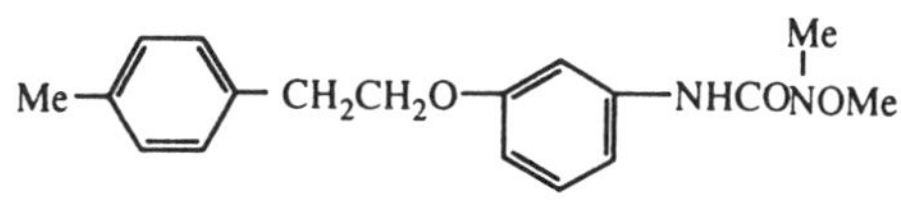

7-162

Carbanilic acid 3-ureidophenyl esters were shown to be useful herbicides; thus, compound **7-163**, in both pre- and post-emergence application at 5 kg/ha, completely controlled *Sinapis weed.*[158]

m-Cyanoalkoxyphenylureas were prepared. For instance, compound **7-164** showed good herbicidal activity against mono- and di-cotyledonous plants, with little or no damage to corn, cereals, cotton, and soybean.[159]

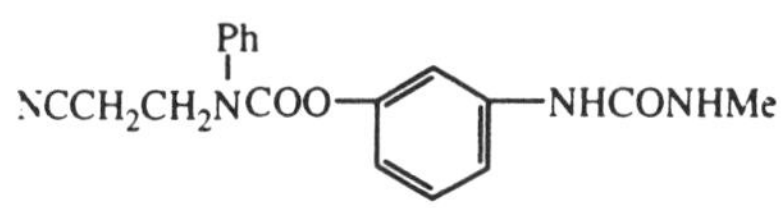

7-163

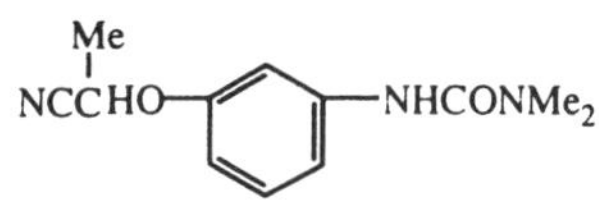

7-164

A series of novel phenylureas were prepared and shown to be useful as herbicides. For instance, compound **7-165** gave excellent control of *Echinochloa crus-galli.*[160]

Certain benzoylphenylureas were patented as insecticides; e.g., compound **7-166**, applied as a 5% emulsion, killed >90% of *Plutella xylostella* larvae.[161]

Nitro-phenylaminoureas were prepared and proved to be pre- and post-emergence herbicides; an example of these products is compound **7-167**.[162]

Certain substituted benzanilide ethers, e.g., compound **7-168**, were prepared which were useful agricultural fungicides.[163]

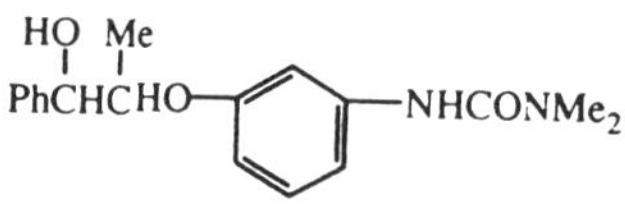

7-165

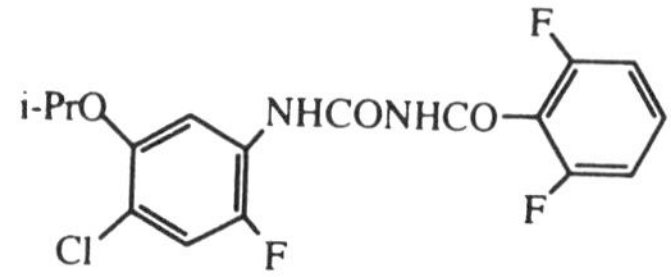

7-166

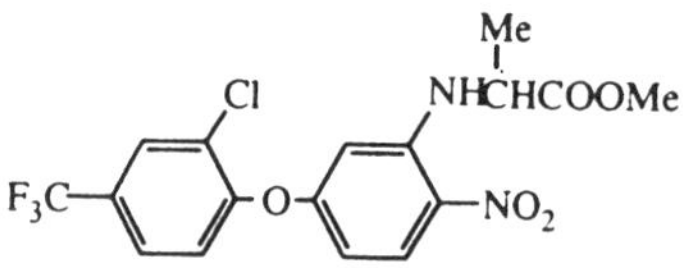

7-167

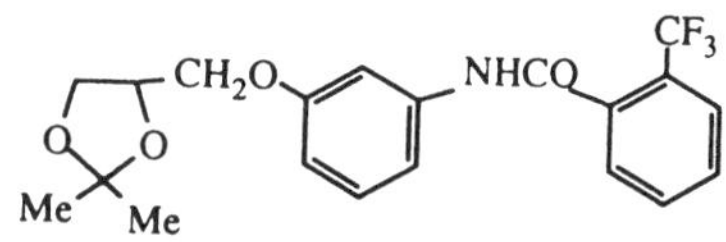

7-168

Heteroaromatic ether herbicides were invented. Thus, compound **7-169** at 5 kg/ha gave 90% control of *Echinochloa crus-galli* and other weeds.[164]

Diphenyl ether (and pyridyl phenyl ether) derivatives were made which were useful for combatting weeds. For instance, compound **7-170**, at 0.56 kg/ha pre-emergence application, gave total control of *Setaria viridis* (foxtail).[165]

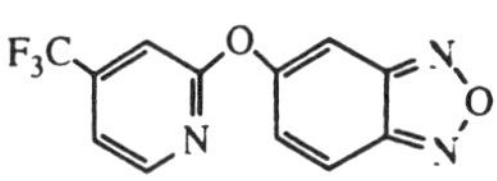

7-169

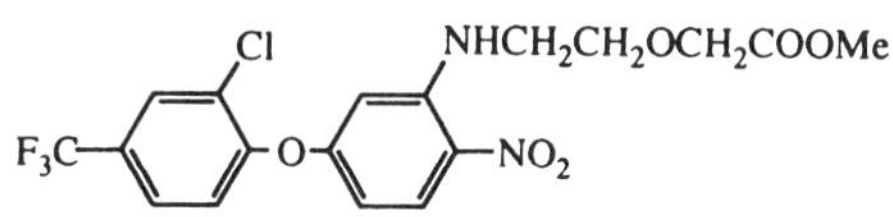

7-170

N-Arylformamide oximes, exemplified by compound **7-171**, were synthesized and shown to be effective fungicides at 10–100 ppm, and miticides at 125–500 ppm.[166]

Novel fungicidal carbanilates were developed; e.g., compound **7-172** at 30 ppm prevented the infection of rose petals with *Botrytis cinerea*.[167]

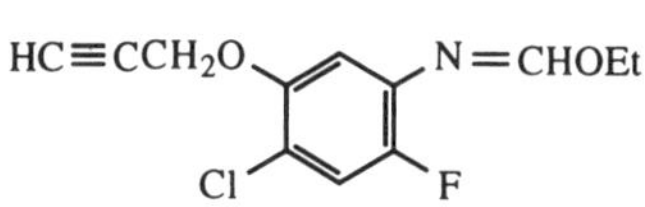

7-171

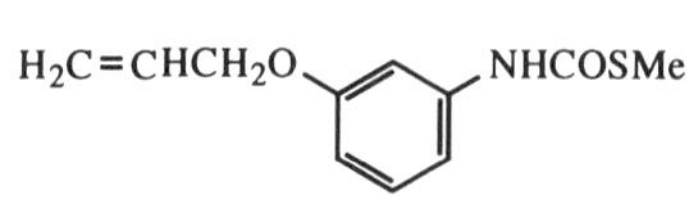

7-172

Hexahydrophthalanilic acid derivatives were prepared and found to be good herbicides. Thus, compound **7-173**, applied at 60 g/10 are, gave 90–100% control of *Amaranthus retroflexus* and Indian mallow, with no damage to soybeans.[168]

Bis-(aminophenyl)phosphates and analogs were demonstrated to be pesticides. For example, compound **7-174**, in phosphate buffer at pH 7 at 5.5×10^{-5} M concentration, reduced the activity of eel acetylcholinesterase to 85% within 30 seconds.[169]

Certain cinnoline derivatives were prepared as male sterilants for *Liliopsida* in hybrid seed production. For instance, compound **7-175**, when used at 1.4 g/L in the treatment of onion, gave 100% male sterility in this pest.[170]

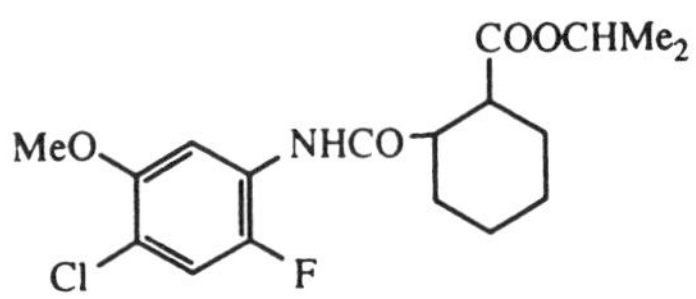

7-173

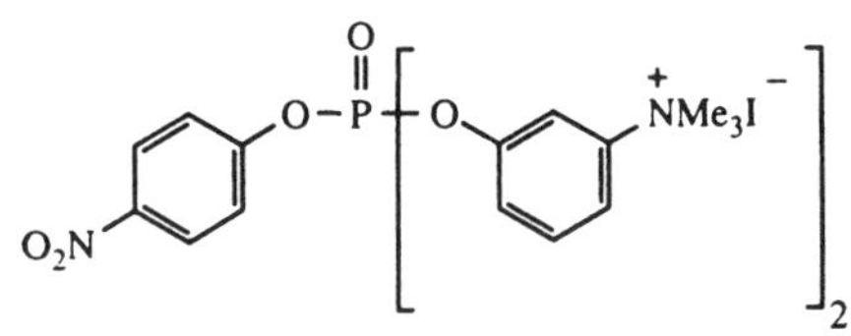

7-174

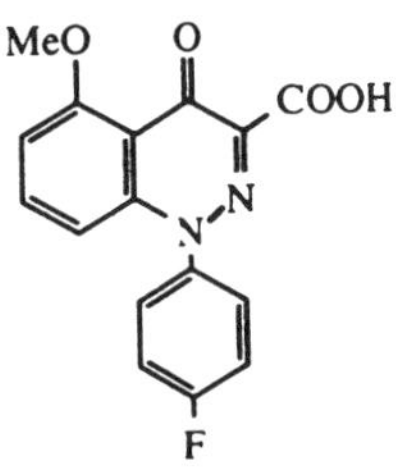

7-175

7.4. REFERENCES

1. N. Sato, Y. Uchiyama, M. Asada, I. Iwataki, and T. Takematsu, *Chemtech, 1988*, 430.
2. Y.-C. Yen and D.-J. Lin, with contributions by C.-W. Lee, *Pesticides and Intermediates*, Process Economics Program (PEP) Report No. 171B, SRI International, Sept. 1986, Menlo Park, CA.
3. *Chemical Week, March 17, 1982.*
4. Jpn. Kokai Tokkyo Koho JP 80,089,283 (July 5, 1980; to Nippon Soda Co.); *C.A., 93*, 162729 (1980).
5. Jpn. Kokai Tokkyo Koho JP 80,019,209 (May 24, 1980; to Nippon Soda Co.); *C.A., 93*, 220422 (1980).
6. Jpn. Kokai Tokkyo Koho JP 80,115,802 (Sept. 6, 1980; to Nippon Soda Co.); *C.A., 94*, 11639 (1981).
7. Jpn. Kokai Tokkyo Koho JP 80,025,178 (July 4, 1980; to Nippon Soda Co.); *C.A., 94*, 65205 (1981).
8. Ger. Offen. DE 2,920,300 (Nov. 20, 1980; K. Eicken and B. Wuerzer to BASF A.-G.); *C.A., 94*, 78441 (1981).
9. Jpn. Kokai Tokkyo Koho JP 80,023,817 (June 25, 1980; to Nippon Soda Co.); *C.A., 94*, 120973 (1981).
10. Jpn. Kokai Tokkyo Koho JP 81,068,660 (June 9, 1981; to Nippon Soda Co.); *C.A., 95*, 168641 (1984).

11. Eur. Pat. Appl. EP 32,879 (July 29, 1981; H. G. Brunner and W. Fory to Ciba-Geigy A.-G.); *C.A., 96*, 20091 (1982).
12. Jpn. Kokai Tokkyo Koho JP 81,164,161 (Dec. 17, 1981; to Nippon Soda Co.); *C.A., 96*, 180861 (1982).
13. U.S. Pat. 4,303,669 (Dec. 1, 1981; T. D. J. D'Silva to Union Carbide Corp.).
14. U.S. Pat. 4,265,658 (May 5, 1981; R. A. Felix to Stauffer Chemical Co.).
15. U.S. Pat. 4,336,062 (June 22, 1982; R. A. Felix to Stauffer Chemical Co.).
16. Belg. BE 891,190 (March 16, 1982; T. Luo to Chevron Research Co.); *C.A., 97*, 38561 (1982).
17. Eur. Pat. Appl. EP 54,682 (June 30, 1982; D. Jahn, W. Rohr, R. Becker, and B. Wuerzer to BASF A.-G.); *C.A., 97*, 181972 (1982).
18. Jpn. Kokai Tokkyo Koho JP 82,112,364 (July 13, 1982; to Nippon Soda Co.); *C.A., 97*, 197890 (1982).
19. Jpn. Kokai Tokkyo Koho JP 57,140,757 [82,140,757] (Aug. 31, 1982; to Nippon Soda Co.); *C.A., 98*, 53420 (1983).
20. Eur. Pat. Appl. EP 66,195 (Dec. 8, 1982; D. Jahn, R. Becker, N. Goetz, and B. Wuerzer to BASF A.-G.); *C.A., 98*, 160705 (1983).
21. U.S. Pat. 4,351,666 (Sept. 28, 1982; J. F. Koerwer to Union Carbide Corp.).
22. Jpn. Kokai Tokkyo Koho JP 58,052,248 [83,052,248] (March 28, 1983; to Ihara Chemical Industry Co., Ltd. and Kumiai Chemical Industry Co., Ltd.); *C.A., 99*, 70385 (1983).
23. Eur. Pat. Appl. EP 82,694 (June 29, 1983; A. Serban, K. G. Watson, G. J. Bird, G. J. Farquharson, and T. L. Houston to ICI Australia Ltd.); *C.A., 99*, 157894 (1983).
24. Eur. Pat. Appl. EP 85,529 (Aug. 10, 1983; K. G. Watson, G. J. Bird, L. E. Cross, G. J. Farquharson, and L. A. Garson to ICI Australia Ltd.); *C.A., 100*, 34275 (1984).
25. Jpn. Kokai Tokkyo Koho JP 58,110,561 [83,110,561] (July 1, 1983; to Teijin Ltd.); *C.A., 99*, 157993 (1983).
26. I. Iwataki, M. Shibuya, and H. Ishikawa (Nippon Soda Co.), in *Pestic. Chem.: Hum. Welfare Envir., Proc. Int. Congr. Pestic. Chem., 5th, 1982 (Publ. 1983)*, 1, 151; *C.A., 99*, 104819 (1983).
27. Eur. Pat. Appl. EP 70,495 (Jan. 26, 1983; J. Dieter, R. Becker, N. Goetz, and B. Wuerzer to BASF A.-G.); *C.A., 99*, 122000 (1983).
28. Eur. Pat. Appl. EP 86,588 (Aug. 24, 1983; A. Serban, K. G. Watson, G. J. Bird, L. E. Cross, and G. J. Farquharson to ICI Australia Ltd.); *C.A., 100*, 156380 (1984).
29. Eur. Pat. Appl. EP 80,301 (June 1, 1983; R. B. Warner, A. Serban, K. G. Watson, G. J. Bird, L. E. Cross, and G. J. Farquharson to ICI Australia Ltd.); *C.A., 100*, 102943 (1984).
30. Ger. Offen. DE 3,219,315 (Nov. 24, 1983; D. Jahn, R. Becker, M. Keil, W. Himmele, and B. Wuerzer to BASF A.-G.); *C.A., 100*, 120579 (1984).
31. Eur. Pat. Appl. EP 90,262 (Oct. 5, 1983; W. J. Michaely and G. W. Kraatz to Stauffer Chemical Co.); *C.A., 100*, 51151 (1984).
32. U.S. Pat. 4,409,153 (Oct. 11, 1983; L. E. Hodakowski to Union Carbide Corp.).
33. U.S. Pat. 4,482,727 (Nov. 13, 1984; D. L. Lee to Stauffer Chemical Co.).
34. U.S. Pat. 4,436,666 (March 13, 1984; T. N. Wheeler to Union Carbide Corp.).
35. Eur. Pat. Appl. EP 102, 823 (March 14, 1984; G. K. Watson and G. J. Farquharson to ICI Australia Ltd.); *C.A., 101*, 38468 (1984).
36. Brit. Pat. Appl. GB 2,137,200 (Oct. 3, 1984; G. J. Farquharson to ICI Australia Ltd.); *C.A., 102*, 131585 (1985).
37. Belg. BE 903,349 (Apr. 1, 1986; to Chevron Research Corp.); *C.A., 105*, 129472 (1986).
38. U.S. Pat. 4,526,723 (July 2, 1985; T. N. Wheeler to Union Carbide Corp.).
39. Ger. Offen. DE 3,429,437 (June 20, 1985; M. Keil, D. Jahn, R. Becker, N. Goetz, B. Wuerzer, and N. Meyer to BASF A.-G.); *C.A., 103*, 160381 (1985).
40. Eur. Pat. Appl. EP 162,336 (Nov. 27, 1985; D. L. Lee, D. R. James, and W. J. Michaely to Stauffer Chemical Co.); *C.A., 104*, 168139 (1986).
41. Ger. Offen. DE 3,430,229 (Oct. 10, 1985; R. Becker, D. Jahn, M. Keil, H. Theobald, W. Spiegler, and B. Wuerzer to BASF A.-G.); *C.A., 104*, 109455 (1986).

42. Eur. Pat. Appl. EP 136,703 (Apr. 10, 1985; M. Keil, R. Becker, D. Jahn, N. Goetz, U. Schirmer, and B. Wuerzer to BASF A.-G.); *C.A., 103*, 178042 (1985).
43. U.S. Pat. 4,511,391 (Apr. 16, 1985; A. Serban, K. G. Watson, G. J. Bird, and G. J. Farquharson to ICI Australia Ltd.).
44. Eur. Pat. Appl. EP 177,450 (Apr. 9, 1986; H. G. Brunner and H. Mueller to Ciba-Geigy A.-G.); *C.A., 105*, 133410 (1986).
45. Ger. Offen. DE 3,340,265 (May 15, 1985; D. Jahn, R. Becker, W. Spiegler, and B. Wuerzer to BASF A.-G.); *C.A., 103*, 215326 (1985).
46. U.S. Pat. 4,537,616 (Aug. 27, 1985; D. T. Manning to Union Carbide Corp.).
47. Eur. Pat. Appl. EP 176,294 (Apr. 2, 1986; G. J. Farquharson, K. G. Watson, and G. J. Bird to ICI Australia Ltd.); *C.A., 105*, 42385 (1986).
48. U.S. Pat. 4,618,360 (Oct. 21, 1986; H.-G. Brunner to Ciba-Geigy Corp.).
49. S. African ZA 85,007,678 (Nov. 26, 1986; T. Luo to Chevron Research Co.).
50. Eur. Pat. Appl. EP 186,119 (July 2, 1986; C. G. Carter, L. David, and W. J. Michaely to Stauffer Chemical Co.); *C.A., 105*, 152711 (1986).
51. Eur. Pat. Appl. EP 186,118 (July 2, 1986; C. G. Carter to Stauffer Chemical Co.); *C.A., 105*, 133519 (1986).
52. U.S. Pat. 4,659,372 (Apr. 21, 1987; T. N. Wheeler to Union Carbide Corp.).
53. U.S. Pat. 4,695,673 (Sept. 22, 1987; J. B. Heather and P. D. Milano to Stauffer Chemical Co.).
54. Ger. Offen. DE 3,609,181 (Sept. 24, 1987; D. Kolassa, M. Keil, U. Schirmer, H. Theobald, R. Becker, B. Wuerzer, and N. Meyer to BASF A.-G.); *C.A., 108*, 33624 (1988).
55. Ger. Offen DE 3,627,410 (Feb. 26, 1987; T. Luo and W. Loh to Chevron Research Co.); *C.A., 106*, 213492 (1987).
56. U.S. Pat. 4,701,205 (Oct. 20, 1987; A. Omid and P. K. Bassi to Chevron Research Co.).
57. U.S. Pat. 4,708,732 (Nov. 24, 1987; C. G. Carter to Stauffer Chemical Co.).
58. Eur. Pat. Appl. EP 219,343 (Apr. 22, 1987; A. J. Liepa and K. N. Winzenberg to Dunlena Pty., Ltd.); *C.A., 107*, 23012 (1987).
59. Eur Pat. Appl. EP 254,514 (Jan. 27, 1988; K. Arai, N. Mito, K. Morita, and N. Hirata to Sumitomo Chemical Co., Ltd.); *C.A., 109*, 37607 (1988).
60. Eur. Pat. Appl. EP 253,537 (Jan. 20, 1988; K. Arai, N. Mito, K. Morita, and N. Hirata to Sumitomo Chemical Co., Ltd.); *C.A., 108*, 200233 (1988).
61. U.S. Pat. 4,767,447 (Aug. 30, 1988; D. L. Lee and W. J. Michaely to Stauffer Chemical Co.).
62. Eur. Pat. Appl. EP 291,114 (Nov. 17, 1988; T. Gilkerson, D. C. Jennens, and W. R. Shaw to Shell Internationale Research Maatschappij B. V.); *C.A., 110*, 212209 (1989).
63. Eur. Pat. Appl. EP 283,261 (Sept. 21, 1988; J. E. D. Barton, D. Cartwright, C. G. Carter, J. M. Cox, D. L. Lee, G. Mitchell, F. H. Walker, and F. X. Woolard to Imperial Chemical Industries PLC; ICI Americas, Inc.); *C.A., 110*, 95273 1989).
64. Eur. Pat. Appl. EP 268,795 (June 1, 1988; D. L. Lee to Stauffer Chemical Co.); *C.A., 110*, 38743 (1989).
65. U.S. Pat. 4,822,905 (Apr. 18,1989; D. L. Lee, D. R. James, and W. J. Michaely to Stauffer Chemical Co.).
66. Jpn. Kokai Tokkyo Koho JP 01,272,561 [89,272,561] (Oct. 31, 1989; M. Kishida, T. Asami, S. Yoshida, Y. Inoue, and N. Takahashi to Nihon Nohyaku Co., Ltd.); *C.A., 112*, 217309 (1990).
67. Braz. Pedido PI BR 88,004,709 (Apr. 18, 1989; M. K. Alterman, A. Do, and L. Fernando to ICI Americas, Inc.); *C.A., 112*, 50600 (1990).
68. Jpn. Kokai Tokkyo Koho JP 01,180,871 [89,180,871] (July 18, 1989; K. Arai, K. Morita, N. Mito, and N. Hirata to Sumitomo Chemical Co., Ltd.); *C.A., 112*, 76606 (1990).
69. Eur. Pat. Appl. EP 332,076 (Sept. 13, 1989; D. Kolassa, J. Kast, M. Keil, U. Schirmer, N. Meyer, and K. O. Westphalen to BASF A.-G.); *C.A., 112*, 118452 (1990).
70. Jpn. Kokai Tokkyo Koho JP 01,068,351 [89,068,351] (March 14, 1989; N. Ito, S. Muraki, and Y. Kawakita to Nisso Petrochemical Industries Co., Ltd.); *C.A., 111*, 173758 (1989).

71. Jpn. Kokai Tokkyo Koho JP 01,143,851 [89,143,851] (June 6, 1989; T. Tsunoda, N. Yanaka, E. Ooya, M. Baba, K. Suzuki, T. Nawamaki, and S. Watanabe to Nissan Chemical Industries, Ltd.); *C.A., 111*, 189597 (1989).
72. Eur. Pat. Appl. EP 381,292 (Aug. 8, 1990; T. Gilkerson, C. J. Foster, and R. Stocker to Shell Internationale Research Maatschappij B.V.); *C.A., 114*, 81092 (1991).
73. Ger. Offen. DE 3,941,160 (June 21, 1990; H. Tobler and M. Van Holzen to Ciba-Geigy A.-G.); *C.A., 114*, 61593 (1991).
74. Ger. Offen. DE 3,902,772 (Aug. 2, 1990; U. Wriede, G. Hamprecht, H. Koehler, T. Kuekenhoehner, B. Wuerzer, and K. O. Westphalen to BASF A.-G.); *C.A., 114*, 6508 (1991).
75. PCT Int. Appl. WO 90,005,712 (May 31, 1990; W. J. Michaely to ICI Americas, Inc.); *C.A., 113*, 171675 (1990).
76. Jpn. Kokai Tokkyo Koho JP 02,003,649 [90,003,649] (Jan. 9, 1990; Y. Inayoshi, M. Yamazaki, and S. Aizawa to Hodayaga Chemical Co., Ltd.); *C.A., 112*, 231320 (1990).
77. Eur. Pat. Appl. EP 353,187 (Jan. 31, 1990; H. G. Brunner to Ciba-Geigy A.-G.); *C.A., 113*, 40462 (1990).
78. Eur. Pat. Appl. EP 395,422 (Oct. 31, 1990; H. Hamaguchi, T. Hino, E. Kohno, N. Ando, T. Nishimatu, and H. Kodama to Nihon Nohyaku Soda Co., Ltd.); *C.A., 114*, 247265 (1991).
79. Jpn. Kokai Tokkyo Koho JP 03,209,361 [91,209,361] (Sept. 12, 1991; H. Adachi, A. Ueda, and H. Hosaka to Nippon Soda Co., Ltd.); *C.A., 116*, 78591 (1992).
80. PCT Int. Appl. WO 91,001,289 (Feb. 7, 1991; A. Ueda, S. Suga, Y. Miyazawa, T. Aihara, and K. Tomida to Nippon Soda Co., Ltd.); *C.A., 114*, 206633 (1991).
81. PCT Int. Appl. WO 91,005,470 (May 2, 1991; L. Willms, H. J. Zeiss, K. Bauer, and H. Bieringer to Hoechst A.-G.); *C.A., 115*, 87503 (1991).
82. U.S. Pat. 5,013,352 (May 7, 1991; L. D. Markley, C. T. Hamilton, B. A. Swisher, and J. Secor to Dow-Elanco).
83. Eur. Pat. Appl. EP 422,537 (Apr. 17, 1991; J. Kast, N. Meyer, U. Misslitz, J. Schubert, J. Jung, W. Rademacher, K. O. Westphalen, and B. Wuerzer to BASF A.-G.); *C.A., 115*, 70985 (1991).
84. PCT Int. Appl. WO 92,008,696 (May 29, 1992; K. N. Winzenberg and A. J. Liepa to Dunlena Pty., Ltd.); *C.A., 117*, 191383 (1992).
85. PCT Int. Appl. WO 92,000,964 (Jan. 23, 1992; H. Ohishi, T. Iihama, K. Ishimitsu, and T. Yamada to Nippon Soda Co., Ltd.); *C.A., 117*, 7806 (1992).
86. Eur. Pat. Appl. EP 481,354 (Apr. 22, 1992; J. Kast, N. Meyer, U. Misslitz, A. Harreus, H. Rang, M. Gerber, H. Walter, and K. O. Westphalen to BASF A.-G.); *C.A., 117*, 48339 (1992).
87. U.S. Pat. 5,110,343 (May 5, 1992; U. Akiyoshi, O. Haruhito, T. Ihara, H. Ishikawa, K. Tomida, and H. Hosaka to Nippon Soda Co., Ltd.).
88. U.S. Pat. 5,092,919 (March 3, 1992; N. H. Nguyen to Imperial Chemical Industries PLC).
89. I. Iwataki, *Ration. Approaches Struct., Act., Ecotoxicol. Agrochem, 1992*, 397; *C.A., 118*, 54190 (1993).
90. Fr. Demande FR 2,675,340 (Oct. 23, 1992; R. H. Hewett and B. M. Luscombe to Rhone Poulenc Agriculture Ltd.); *C.A., 118*, 185837 (1993).
91. Jpn. Kokai Tokkyo Koho JP 04,230,660 [92,230,660] (Aug. 19, 1992; E. Kono, K. Ando, H. Hamaguchi, and A. Kaneoka to Nihon Nohyaku Co., Ltd.); *C.A., 117*, 247159 (1992).
92. Eur. Pat. Appl. EP 502,492 (Sept. 9, 1992; H. Stark, K. Bauer, and H. Bieringer to Hoechst A.-G.); *C.A., 117*, 233600 (1992).
93. Jpn. Kokai Tokkyo Koho JP 80,154,953 (Dec. 2, 1980; to Mitsui Toatsu Chemicals, Inc.); *C.A., 94*, 156552 (1981).
94. Fr. Demande FR 2,471,979 (June 26, 1981; T. Yoshimoto, T. Toyama, K. Igarishi, M. Ura, Y. Enomoto, Y. Funakoshi, and Y. Hojo to Mitsui Toatsu Chemicals, Inc.); *C.A., 96*, 6554 (1981).
95. Eur. Pat. Appl. EP 25,773 (March 25,1981; H. Szczepanski, O. Rohr, and H. Rempfler to Ciba-Geigy A.-G.); *C.A., 95*, 115077 (1981).
96. Eur. Pat. Appl. EP 23,725 (Feb. 11, 1981; H. Sekioka, M. Takenaka, Y. Kawaguchi, S. Takamura, M. Nishimura, and M. Watanabe to Ube Industries, Ltd.); *C.A., 95*, 24525 (1981).

97. Jpn. Kokai Tokkyo Koho JP 81,025,143 (March 10, 1981; to Mitsui Toatsu Chemicals, Inc.); *C.A.*, *95*, 61819 (1981).
98. U.S. Pat. 4,323,389 (Apr. 6, 1982; T. Yoshimoto, T. Inoue, H. Michiyama, T. Harayama, O. Morikama, Y. Hojo, T. Baba, T. Toyama, M. E. Ura, and Y. Takasawa to Mitsui Toatsu Chemicals, Inc.).
99. U.S. Pat. 4,419,122 (Dec. 6, 1983; C. Swithenbank to Rohm and Haas Co.).
100. U.S. Pat. 4,419,123 (Dec. 6, 1983; C. Swithenbank to Rohm and Haas Co.)
101. U.S. Pat. 4,404,018 (Sept. 13, 1983; L. J. Stach and T. Hokama to Velsicol Chemical Corp.).
102. Patentschrift (Switz.) CH 637,361 (July 29,1983; O. Rohr, B. Boehner, K. Burdeska, and G. Pissiotas to Ciba-Geigy A.-G.); *C.A.*, *100*, 6071 (1984).
103. U.S. Pat. 4,402,729 (Sept. 6, 1983; T. Hokama to Velsicol Chemical Corp.).
104. U.S. Pat. 4,419,123 (Dec. 6, 1983; C. Swithenbank to Rohm and Haas Co.).
105. U.S. Pat. 4,456,464 (June 26, 1984; S.-F. Lee and C. A. Heurick to Zeocon Corp.).
106. U.S. Pat. 4,486,223 (Dec. 4, 1984; H. Konishi, N. Hino, H. Matsumoto, and R. Yoshida to Sumitomo Chemical Co., Ltd.).
107. Ger. Offen. DE 3,319,287 (Nov. 29, 1984; H. Foerster, B. Gallenkamp, U. Priesnitz, H. J. Riebel, L. Eue, and R. R. Schmidt to Bayer A.-G.); *C.A.*, *102*, 113056 (1985).
108. Ger. Offen. DE 3,407,493 (Sept. 5, 1985; U. Kraatz, W. Fuehrer, L. Eue, R. R. Schmidt, and H. J. Santel to Bayer A.-G.); *C.A.*, *104*, 148559 (1986).
109. U.S. Pat. 4,599,105 (July 8, 1986; S. K. Malhotra and B. C. Gerwick, III, to The Dow Chemical Co.).
110. U.S. Pat. 4,792,616 (Dec. 20, 1988; P. Winternitz and R. Zurflueh to Hoffmann-LaRoche, Inc.).
111. Eur. Pat. Appl. EP 281,103 (Sept. 7, 1988; S. Azuma, T. Hiramatsu, K. Nakagawa, and Y. Ichikawa to Teijin Ltd.); *C.A.*, *110*, 38747 (1989).
112. Eur. Pat. Appl. EP 341,845 (Nov. 15, 1989; J. M. Clough, C. R. A. Godfrey, P. J. DeFraine, M. G. Hutchings, and I. Ferguson to Imperial Chemical Industries PLC); *C.A.*, *112*, 197854 (1990).
113. Brit. UK Pat. Appl. GB 2,225,014 (May 23, 1990; D. Cartwright to Imperial Chemical Industries PLC); *C.A.*, *113*, 190898 (1990).
114. PCT Int. Appl. WO 90,001,874 (March 8, 1990; S. Azuma, K. Nakagawa, T. Hiramatsu, and Y. Ishikawa to Teijin Ltd.); *C.A.*, *113*, 36398 (1990).
115. PCT Int. Appl. WO 91, 019,694 (Dec. 26, 1991; S. Azuma, T. Hiramatsu, and Y. Ishikawa to Teijin Ltd.); *C.A.*, *116*, 193933 (1992).
116. Eur. Pat. Appl. EP 484,688 (May 13, 1992; A. Shuto, N. Sakamoto, H. Kisada, H. Fujimoto, and K. Umeda to Sumitomo Chemical Co., Ltd.); *C.A.*, *117*, 89952 (1992).
117. Jpn. Kokai Tokkyo Koho JP 04,041,463 [92,041,463] (Feb. 12, 1992; S. Azuma, T. Hiramatsu, and Y. Ichikawa to Teijin Ltd.); *C.A.*, *117*, 7660 (1992).
118. Eur. Pat. Appl. EP 479,420 (Apr. 8, 1992; N. Kawaguchi, H. Fukami, R. Sago, and K. Ikai to Suntory Ltd.); *C.A.*, *117*, 26581 (1992).
119. Brit. UK Pat. Appl. GB 2,253,624 (Sept. 16, 1992; J. M. Clough, C. R. A. Godfrey, and P. J. DeFraine to Imperial Chemical Industries PLC); *C.A.*, *118*, 38945 (1993).
120. Israeli Pat. IL 55,355 (Aug. 31, 1982; to Eli Lilly and Co.); *C.A.*, *99*, 70741 (1983).
121. U.S. Pat. 4,191,554 (March 4, 1980; W. A. Gregory to E. I. DuPont de Nemours and Co.).
122. Jpn. Kokai Tokkyo Koho JP 58,039,657 [83,039,657] (March 8, 1983; to Ishihara Sangyo Kaisha, Ltd.); *C.A.*, *99*, 5520 (1983).
123. U.S. Pat. 4,436,549 (March 13, 1984; T. D. Thibault to Eli Lilly and Co.).
124. U.S. Pat. 4,452,631 (June 5, 1984; K. W. Burow, Jr. and G. W. Johnson to Eli Lilly and Co.).
125. U.S. Pat. 4,886,899 (Dec. 12, 1989; M. J. Jacobs to Pitman-Moore, Inc.).
126. U.S. Pat. 4,751,239 (June 14, 1988; E. B. Hodge to International Minerals and Chemical Corp.).
127. Eur. Pat. Appl. EP 312,353 (Apr. 19, 1989; M. J. Jacobs to Pitman-Moore, Inc.); *C.A.*, *112*, 7397 (1989).
128. Eur. Pat. Appl. EP 325,428 (July 26, 1989; E. B. Hodge to Pitman-Moore, Inc.); *C.A.*, *112*, 76975 (1989).

129. U.S. Pat. 5,041,610 (Aug. 20, 1991; K. A. Drengler to International Minerals and Chemical Corp.).
130. Jpn. Kokai Tokkyo Koho JP 04,066,507 (March 2, 1992; M. Taniguchi to Santory K. K.); *C.A., 117*, 21977 (1992).
131. Eur. Pat. Appl. EP 472,113 (Feb. 26, 1992; N. Wada and R. Yoshida to Kumiai Chemical Industry Co., Ltd.; Ihara Chemical Industry Co., Ltd.); *C.A., 93*, 114562 (1980).
132. USSR Pat. SU 1,277,929 (Dec. 23, 1986; to Urals Forestry Institute); *C.A., 107*, 156809 (1987).
133. Eur. Pat. Appl. EP 8,192 (Feb. 20, 1980; A. Serban, K. G. Watson, and R. B. Warner to ICI Australia Ltd.); *C.A., 93*, 114562 (1980).
134. U.S. Pat. 4,194,029 (March 18, 1980; H.-L. Lam to Stauffer Chemical Co.).
135. Eur. Pat. Appl. EP 21,827 (July 1, 1981; I. Hashimoto and H. Namba to Mitsui Petrochemical Industries, Ltd.); *C.A., 95*, 80730 (1981).
136. Eur. Pat. Appl. EP 54,589 (June 30, 1982; I. Chiyomaru, H. Sugiyama, K. Nitta, K. Fujimori, T. Hirano, and O. Tada to Kumiai Chemical Industry Co., Ltd.); *C.A., 97*, 212043 (1982).
137. Jpn. Kokai Tokkyo Koho JP 58,000,987 [83,000,987] (Jan. 6, 1983; to Shell Internationale Research Maatschappij B.V.); *C.A., 98*, 179392 (1983).
138. Jpn. Kokai Tokkyo Koho JP 57,109,779 [82,109,779] (July 8, 1982; to Kumiai Chemical Industry Co., Ltd.; Mitsui Petrochemical Industries, Ltd.); *C.A., 97*, 198111 (1982).
139. Jpn. Kokai Tokkyo Koho JP 59,088,479 [84,088,479] (May 22, 1984; to Mitsui Petrochemical Industries, Ltd.); *C.A., 101*, 171099 (1984).
140. Jpn. Kokai Tokkyo Koho JP 60,120,875 [85,120,875] (June 28, 1985; to Snow Brands Milk Products Co., Ltd.); *C.A., 104*, 68839 (1986).
141. U.S. Pat. 4,542,150 (Sept. 17, 1985; W. S. Bowers to Cornell Research Foundation, Inc.).
142. PCT Int. Appl. WO 86,005,069 (Sept. 12, 1986; A. Balint, J. Benczik, F. Fodor, A. Horvath, B. Karaczonyi, E. Toemoerdi, L. Zbisko, S. Gaal, I. Gardi, *et al.* to Nitrokemia Ipartelepek); *C.A., 106*, 14723 (1987).
143. Ger. Offen. DE 2,432,799 (Jan. 2, 1986; L. Feuer, L. Farkas, M. Nogradi, B. Vermes, A. Gottsegen, and A. Wolfner to Chinoin Gyogyszer es Vegyeszeti Termelek Gyara Rt.).
144. U.S. Pat. 4,788,298 (Nov. 29, 1988; R. H. S. Wang to Eastman Kodak Co.).
145. Jpn. Kokai Tokkyo Koho JP 03,287,579 [91,287,579] (Dec. 18, 1991; N. Sasaki, S. Kudo, M. Yoshikawa, K. Endo, and T. Kaji to Mitsubishi Petrochemical Co., Ltd); *C.A., 116*, 194359 (1992).
146. Czech. Pat. CS 269,384 (Jan. 16, 1991; to J. Kahovcova); *C.A., 115*, 29120 (1991).
147. U.S. Pat. 4,329,360 (May 11, 1982; B. M. Resnick to GAF Corp.).
148. Eur. Pat. Appl. EP 89,191 (Sept. 21, 1983; H. Kohsaka, M. Sasaki, and Y. Ishiguri to Sumitomo Chemical Co., Ltd.); *C.A., 100*, 22801 (1984).
149. K. S. Datta, P. Sagar, R. C. Hasija, and R. L. Kapoor, *Ann. Biol. (Ludhiana, India), 2(1),* 9 (1986); *C.A., 106*, 45661 (1987).
150. Eur. Pat. Appl. EP 249,567 (May 7, 1987; P. Place, R. Pepin, and F. Chretien to Rhone-Poulenc Agrochemie BP).
151. Jpn. Kokai Tokkyo Koho JP 02,069,497 [90,069,497] (March 8, 1990; M. Yoshioka and T. Shiba to Takeda Chemical Industries, Ltd.); *C.A., 113*, 132535 (1990).
152. U.S. Pat. 4,645,780 (Feb. 24, 1987; H. Dressler and H. A. Ward to Koppers Co., Inc.).
153. Ger. Offen. DE 2,844,806 (Apr. 24, 1980; G. Boroschewski, L. Nuesslein, and F. Arnst to Schering A.-G.); *C.A., 93,* 167982 (1980).
154. Ger. Offen. DE 3,600,995 (July 16, 1987; H.-J. Kraehmer to Schering A.-G.); *C.A., 107*, 170603 (1988).
155. Jpn. Kokai Tokkyo Koho JP 01,316,306 [89,316,306] (Dec. 21, 1989; J. Takahashi, M. Enomoto, T. Haga, M. Sasaki, and M. Sato to Sumitomo Chemical Co., Ltd.); *C.A., 112*, 212489 (1990).
156. U.S. Pat. 4,361,438 (Nov. 30, 1982; R. A. Felix to Stauffer Chemical Co.).
157. U.S. Pat. 4,308,213 (Dec. 29, 1981; D. M. Spatz and B. Cross to American Cyanamid Co.).
158. Israeli IL 55,039 (July 30, 1982; to Schering A.-G.); *C.A., 99*, 87842 (1983).
159. U.S. Pat. 4,400,205 (Aug. 23, 1983; G. Pissiotas and O. Rohr to Ciba-Geigy Corp.).

160. Jpn. Kokai Tokkyo Koho JP 58,150,558 [83,150,558] (Sept. 7, 1983; to Kanagawa Kagaku Kenkyusho K. K.); *C.A., 100*, 51303 (1984).
161. Jpn. Kokai Tokkyo Koho JP 61,148,158 [86,148,158] (July 5, 1986; A. Kurozomi, T. Hayaoka, and A. Masui to Nippon Kayaku Co., Ltd.); *C.A., 106*, 18157 (1987).
162. Eur. Pat. Appl. EP 80,746 (June 8, 1983; G. B. Paerels and C. W. Raven to Duphar International Research B. V.); *C.A., 99*, 212909 (1983).
163. Ger. Offen. DE 3,139,457 (Apr. 21, 1983; W. Daum, V. Paul, and P. Schiller to Bayer A.-G.); *C.A., 99*, 53770 (1983).
164. Brit. UK Pat. Appl. GB 2,157,679 (Oct. 30, 1985; P. Camilleri and D. Munro to Shell Internationale Research Maatschappij B.V.); *C.A., 105*, 78940 (1986).
165. Eur. Pat. Appl. EP 130,041 (Jan. 2, 1985; A. N. Beaulieu and R. Y. Yih to Rohm and Haas Co.); *C.A., 103*, 6017 (1985).
166. Jpn. Kokai Tokkyo Koho JP 61,053,260 [86,053,260] (March 17, 1986; K. Hayakawa, H. Nishikawa, and A. Hashimoto to Nippon Soda Co., Ltd.); *C.A., 105*, 190659 (1986).
167. Eur. Pat. Appl. EP 274,271 (July 14, 1988; D. R. Baker, K. H. Brownell, and C. Kazerian to Stauffer Agricultural Chemical Co., Inc.); *C.A., 109*, 165714 (1988).
168. Jpn. Kokai Tokkyo Koho JP 01,175,961 [89,175,961] (July 12, 1989; T. Tokunaga, H. Watanabe, K. Tsuzuki, S. Someya, and S. Kora to Tosoh Corp.; Agro-Kanesho Co., Ltd.); *C.A., 112*, 35456 (1990).
169. U.S. Pat. 4,948,908 (Aug. 14, 1990; M. A. Frommer, Y. Segall, and E. Shirin to Ramot Purotech Ltd.).
170. U.S. Pat. 5,129,940 [July 14, 1992; to T. G. Patterson (Orsan)].

8

Pharmaceuticals, Over-the-Counter Medications, and Diagnostic Aids

8.1. COMMERCIAL PRODUCTS

The above index indicates that the resorcinol moiety appears in a considerable variety of pharmaceuticals and related materials, both older and relatively new products. The sources of these products are either synthesis from resorcinol, 1,3-cyclohexanedione (dihydroresorcinol), or *m*-aminophenol, or extraction of natural or fermentation products; many modifications of these techniques are employed. Indeed, the development of some of these commercial compounds was based on the modification of natural products to minimize undesirable side effects and to shape and maximize the therapeutic properties. It is not possible to describe the details of the manufacturing processes of the commercial materials here; these details are available in the primary literature.

8.1.1. Bronchospasmolytics, Antiasthmatics, and Respiratory Stimulants

Orciprenalin, **8-1**, also called metaproterenol, and tradenamed Alupent (Boehringer) and Metapril (Sandoz), is sold as the sulfate salt; it is a beta-sympathomimetic agent used for the treatment of bronchial spasms, asthma, emphysema, and coughs.

Because of a degree of undesirable side effects of such drugs on the heart, beta$_2$-sympathomimetica were developed in the 1960s which primarily cause relaxation of the bronchial musculature. In the latter category are the bronchodilators, fenoterol (available as Berotec® from Boehringer and Warner-Lambert), **8-2**, and terbutalin, **8-3**, sold under the tradenames Bricanyl (Marion Merrel Dow, with povidone) and Brethine (Ciba-Geigy). Reproterol, **8-4**, sold, e.g., as Bronchodil® (Homburg, Schering) and Bronchospasmin® (Asta-Degussa), is a related compound but contains a purine-2,6-dione segment; it is used as an antiasthmatic agent.

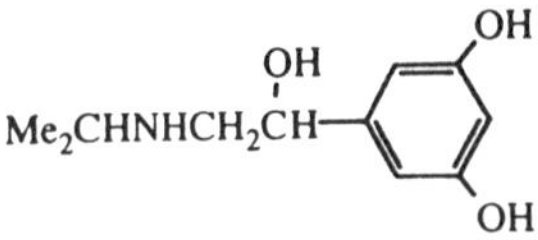

8 - 1

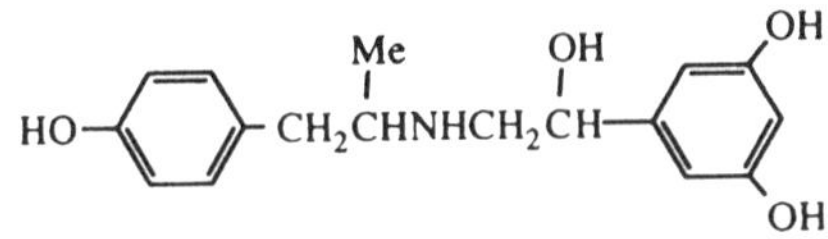

8 - 2

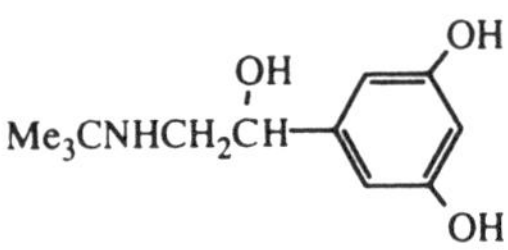

8 - 3

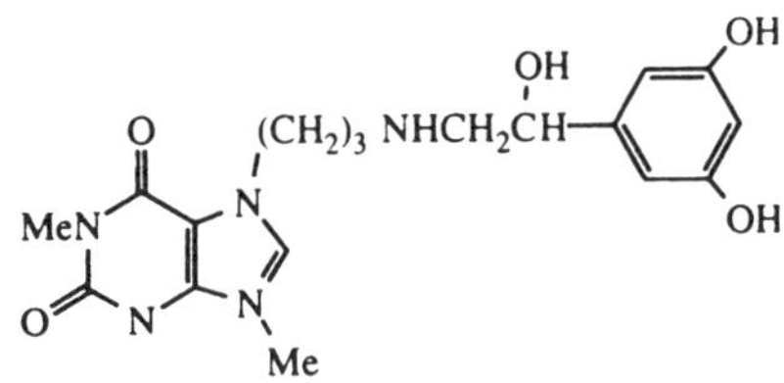

8 - 4

Also in the 1960s, the important antiasthmatic agent, sodium chromoglycolate (Fison's Intal®), **8-5**, was introduced. This drug, delivered via an aerosol inhalator, is particularly effective against allergic reaction-caused asthma and as a prophylactic against asthma attacks. Children with asthma are said to respond well to this drug. Sales of Intal in 1991 were close to \$300 million.[1]

A respiratory stimulant with a chromone structure, dimefline (available from Wallace as Remeflin®), **8-6**, also stems from the 1960s. A more recently developed prophylactic anti-allergenic agent is nedocril (Fison's Tilade®), **8-7**; its sales reached \$30 million in 1991.[1]

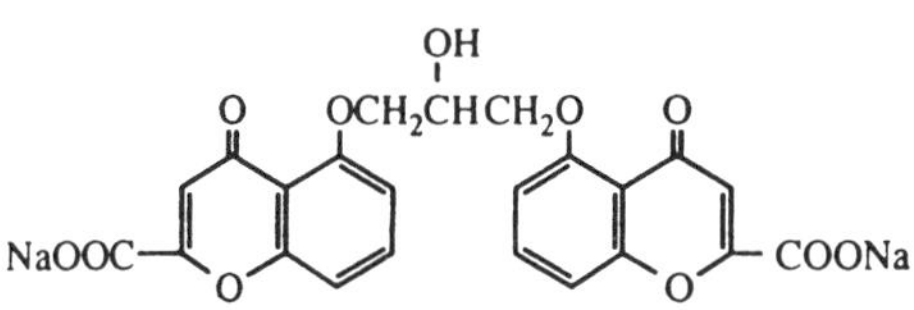

8 - 5

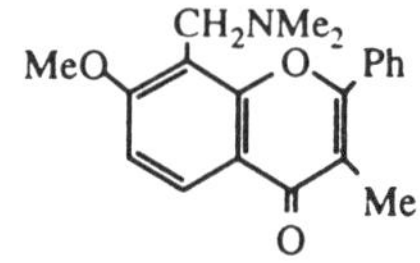

8 - 6

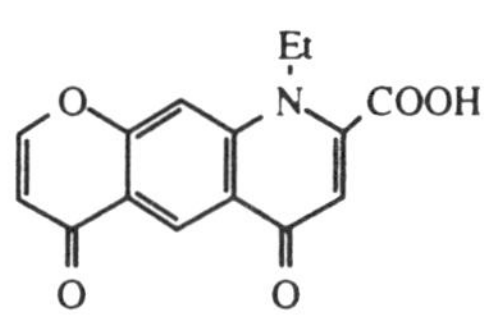

8 - 7

8.1.2. Coronary Vasodilators and Betablockers

Carbochromen (chromonar, sold as Intensain® by Cassella-Abbott), **8-8**, is a vasodilator with a coumarin structure, developed in the early 1960s.

The coronary vasodilator, efloxate (made by Recordati), **8-9**, stems from the 1950s. Also developed in this period was the natural coumarin, visnadin (Carduben® of Madaus and Vibeline® of Roger Bellon), **8-10**, extracted from the *Ammi visnaga* L plant; it was also used as a coronary vasodilator.

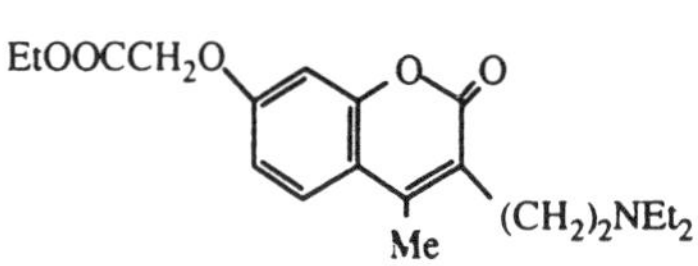

8-8

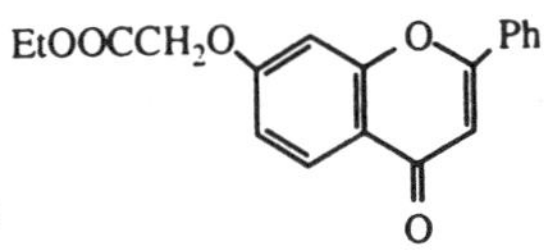

8-9

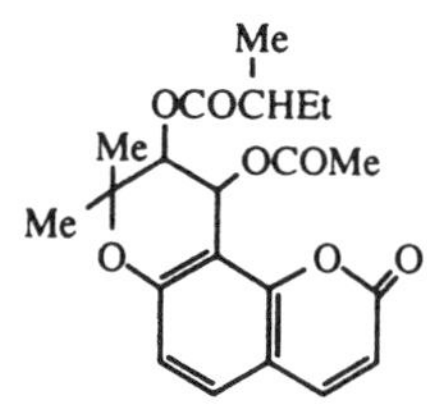

8-10

Betablockers are adrenergic blocking agents, capable of reducing the heart beat and the force of heart contractions. One such betablocker, carteolol (Endak® of Madaus, Mikelan® of Otsuka), **8-11**, was developed in the 1970s; its synthesis starts from 1,3-cyclohenxanedione and its amination product, 3-amino-2-cyclohexen-1-one. Pindolol (Visken® of Sandoz), **8-12**, is another beta-adrenergic receptor blocker with intrinsic sympathomimetic activity; one synthesis route starts from 1,3-cyclohexanedione. Pindolol is an effective antihypertensive agent; it attenuates increases in the heart rate, systolic blood pressure, and cardiac output resulting from exercise.

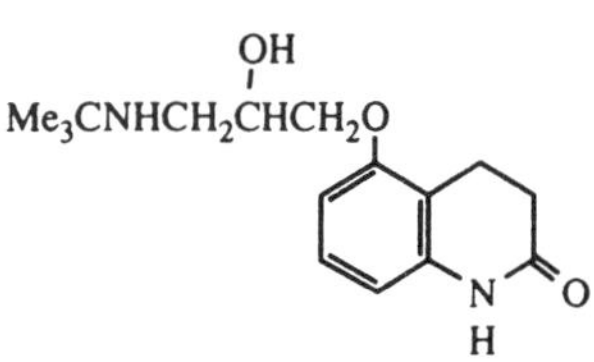

8-11

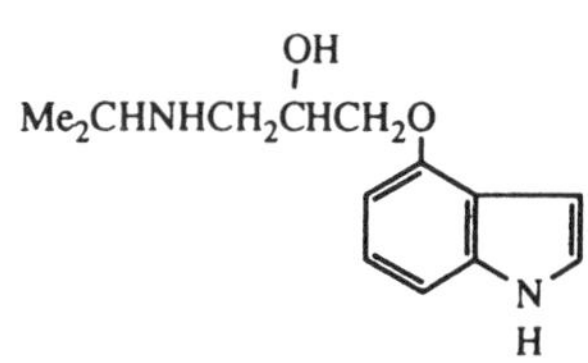

8-12

8.1.3. An Ulcer Medication

Sofalcon (Taisho's Solon®), **8-13**, was developed in the 1980s in Japan as an ulcer therapeutic agent. The drug evolved from studies of the old Chinese herbal medicine, guangdougen, obtained from *Sophora subprostrata*, and is used for the treatment of digestive problems; a large number of chemical modifications of the active principle of guangdougen were made and tested, which led to sofalcon.[2]

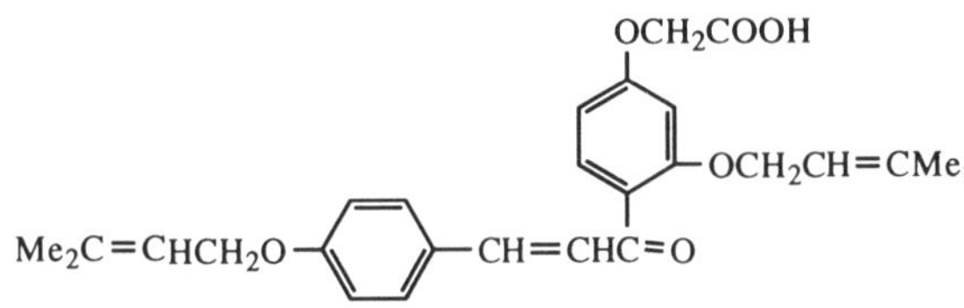

8-13

8.1.4. An Antiseptic and Anthelminthic Agent

4-*n*-Hexylresorcinol, **8.14**, stems from the 1920s. It is an antiseptic compound with a phenol coefficient 45 and has been used in over-the-counter medications such as Sucrets (throat lozenges), also in wound-healing preparations. Hexylresorcinol has also been used since the 1940s as an anthelminthic in human and/or veterinary applications, e.g., against intestinal nematodes, the dwarf tapeworm, the large intestinal fluke, *Fasciolopis buski*, and the hookworm of dogs; it has also been prescribed as a urinary antiseptic.

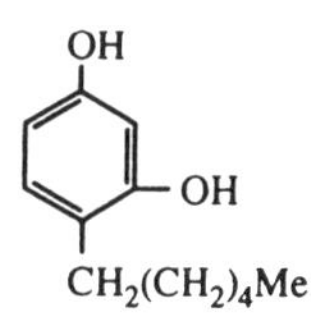

8-14

8.1.5. Choleretic Agents

Hymecromon (Lipha's Mendiaxon®, Roussel's Crodimon®), **8-15**, has been prescribed as a choleretic, anticholinergic, and spasmolytic agent. It is useful in the treatment of bile duct spasms.

Metochalcone (Beta's Chemicol®, Glaxo's Lesidrin®), **8-16**, is a choleretic and a diuretic made from 2,4-dihydroxybenzaldehyde (beta-resorcaldehyde).

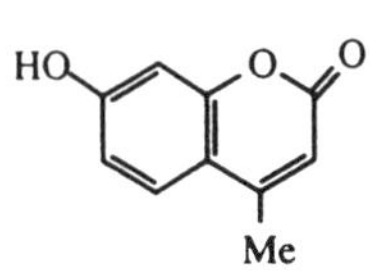

8-15

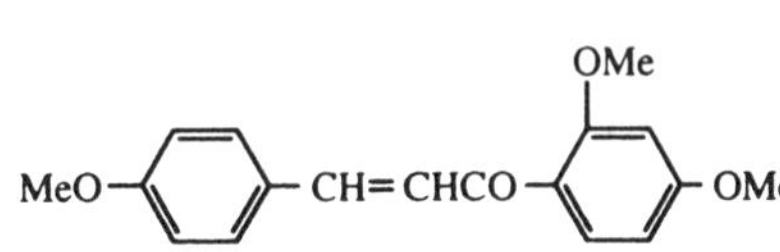

8-16

8.1.6. An Antiadrenergic Agent

Phentolamine, **8-17**, sold by Ciba-Geigy under the tradename Regitine, was developed in 1950 for prescription as an antiadrenergic and sympatholytic agent.

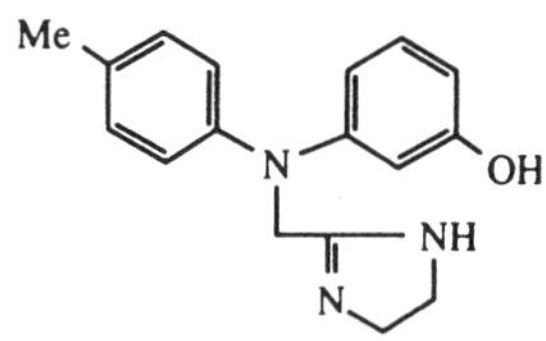

8-17

8.1.7. A Tuberculostatic Drug

p-Aminosalicylic acid (PAS), **8-18**, and its sodium salt, sold by Lilly as Enseals® and by Parke-Davis as Pamisyl® (there are many other tradenames worldwide), was indicated as an antibacterial in the 1940s for the arrest of tuberculosis and, for a time, was the preeminently used drug until the arrival of isonicotinic acid hydrazide and other medications. While the use of PAS has declined, if not ceased, in the U.S., it is still widely used overseas.

Pasiniazid (Hoffmann-LaRoche) is a mixture of PAS and isonicotinic hydrazide, sold as an antitubercular medication.

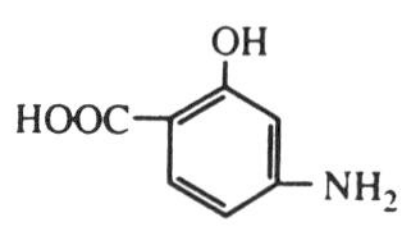

8-18

8.1.8. Antidepressants and Antipsychotics

Molindone hydrochloride (Abbott's Lidone®, DuPont's Mobane), **8-19**, is a neuroleptic and antidepressant, developed in the 1960s, which is prescribed as an antipsychotic. In man it was said to be three times as potent as chloropromazine. Molindone's synthesis starts from 1,3-cyclohexanedione.

Ramoxipride (Astra), **8-20**, and Raclopride (Astra), **8-21**, are antipsychotic agents developed during the 1980s in Sweden. Raclopride has also been investigated for use in the treatment of Alzheimer's disease.

8.1.9. Reversible Cholinesterase Inhibitors

There are three indirect parasympathomimetics derived from *m*-aminophenol which have been known since the 1930s. Demecarium bromide (sold as Humorsol® by Merck and as Tosmilen® by Lentia/Austria), **8-22**, shows prolonged activity and is used, e.g., in the treatment of glaucoma and as a vagotonic agent.

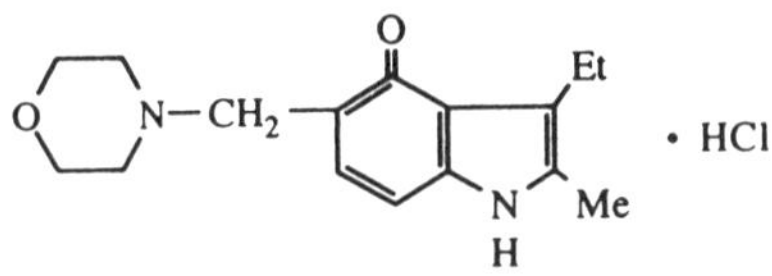

8-19

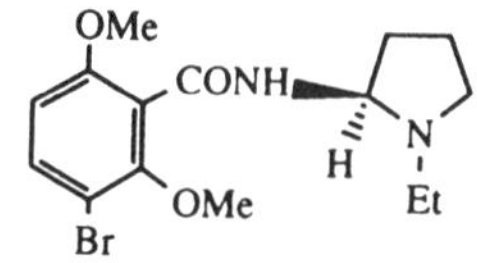

8-20

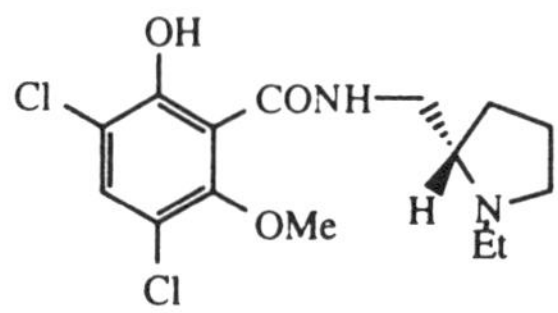

8-21

Edrophonium chloride, **8-23**, sold by Roche as Tensilon®, and neostigmine bromide, **8-24**, have been used as curare (arrow poison) antagonists. Neostigmine was also used in the symptomatic treatment of myastenia gravis to improve muscle strength (also used for dogs with this disease), and after surgery to reverse the effects of certain neuromuscular blocking agents. Neostigmine methylsulfate is used to prevent and treat postoperative distention and urinary retention.

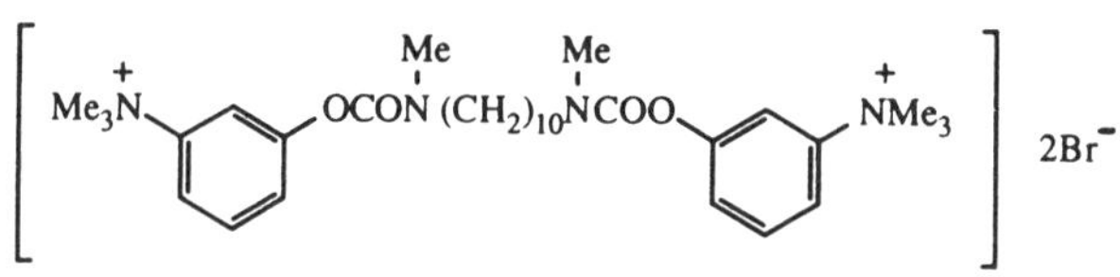

8-22

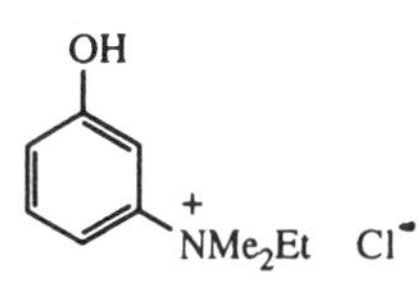

8-23

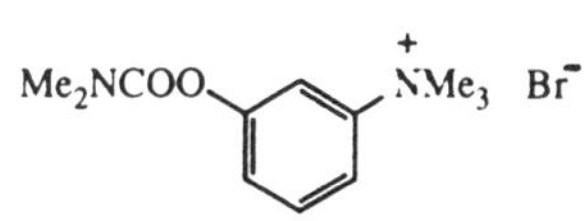

8-24

8.1.10. Antibiotics, Antibacterials, and Antifungal Agents

The semisynthetic antibiotic, methicillin sodium (Bristol Meyers' Staphcillin®, Pfizer's Azapen®; many other tradenames), **8-25**, was developed in the 1950s. It is used as an antibacterial agent which is hydrolysis/deactivation resistant (hydrolysis being catalyzed by penicillinase), but not stable in acid medium—therefore it has to be administered parenterally. In veterinary medicine it is used also as an antimicrobial.

There are several drugs produced by fermentation which contain methylresorcinol moieties in their structure. For example, the antifungal compound, siccanin, **8-26**, was obtained from the cultured broth of *Helminthosporium siccans* Drechsler. The antibacterial, novobiocin (Upjohn's Albamycin®, Hoechst's Inamycin®, and others), **8-27**, is produced by *streptomyces spheroides*; it is used as the sodium or calcium salt.

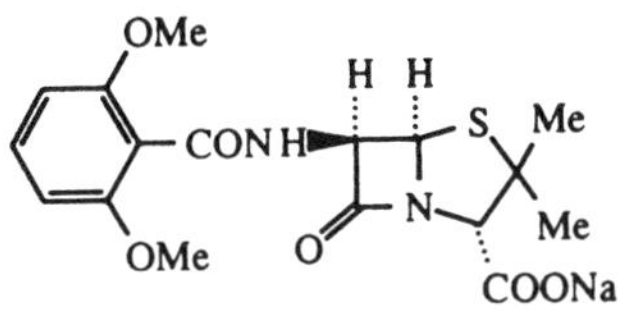

8-25

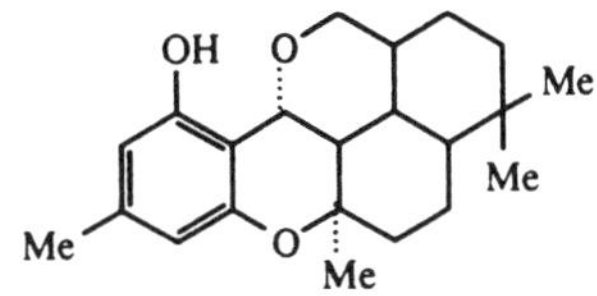

8-26

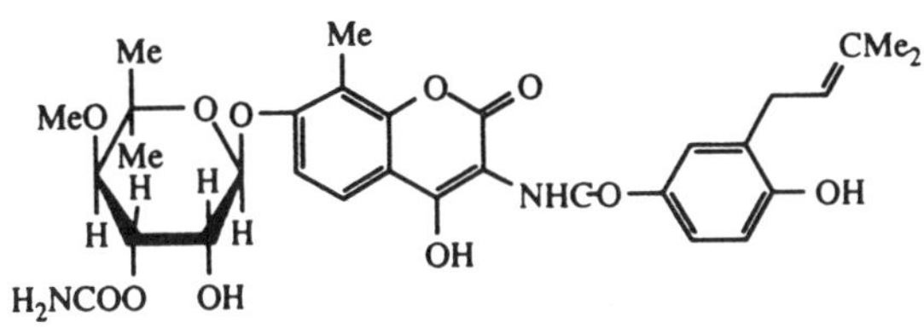

8-27

8.1.11. Antiemetics

Nabilon, **8-28**, sold by Lilly as Cesamet®, was developed in the 1970s and is a potent antiemetic drug. Its major use is in controlling the severe nausea caused by anticancer agents; it has also been considered for treating motion sickness and morning sickness.

Metoclopramide (Robins' Reglan®), **8-29**, has also been suggested as an antiemetic agent. Naboctate (H.G. Pars Pharm.), **8-30**, is an antiemetic and antiglaucoma agent. A related antiemetic is nonabine (Beecham), **8-31**.

8.1.12. Miscellaneous Pharmaceuticals

Ipriflavone (Chinoin), **8-32**, was suggested in the 1980s for the treatment of osteoporosis. Nabazenil (H.G. Pars Pharm.), **8-33**, is another product resulting from studies of cannabis-related compounds; it is used as an anticonvulsant.

Fenocinol (Labs.Lafon), **8-34**, has antispasmodic action. Meptazinol (Wyeth's Meptid®, WY 22811), **8-35**, is an injectable analgesic which can be made starting from 1,3-cyclohexandedione.

8.1.13. Acne and Antiseborrheic Preparations, Dandruff Control

Resorcinol itself is an ingredient often used for its antiseptic and keratolytic action (at least 2% to be effective), along with sulfur, salicylic acid, and/or benzoyl peroxide, in over-the counter remedies for acne, dandruff, and scabies. Examples of these

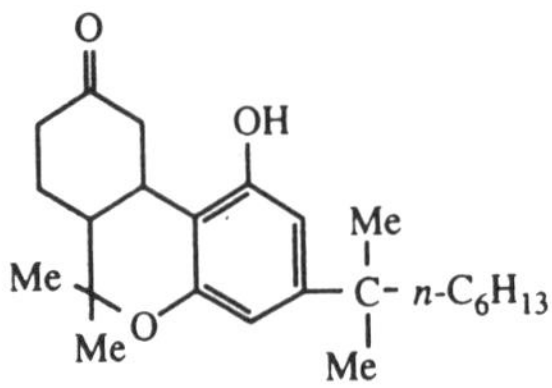

8-28

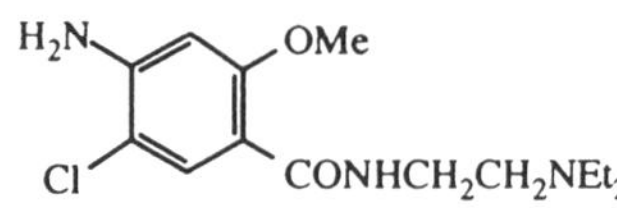

8-29

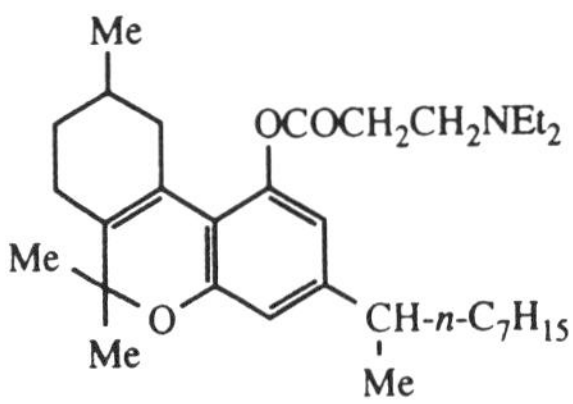

8-30

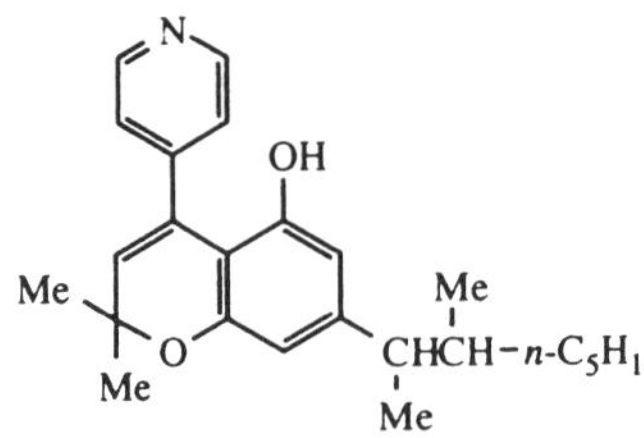

8-31

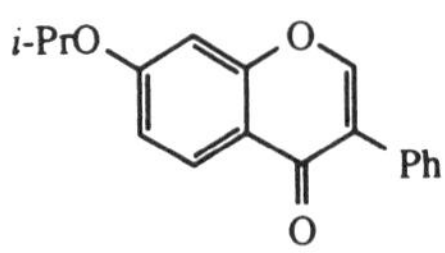

8-32

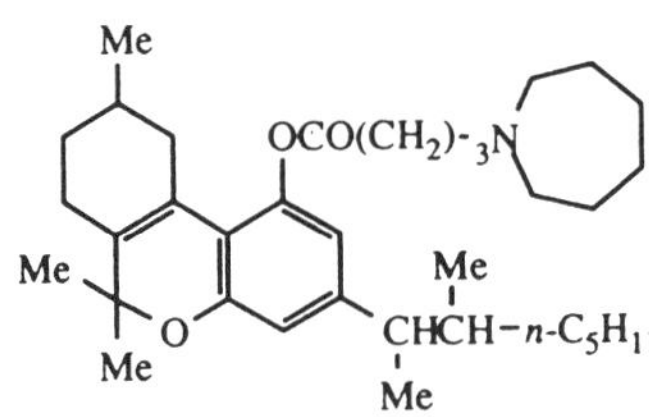

8-33

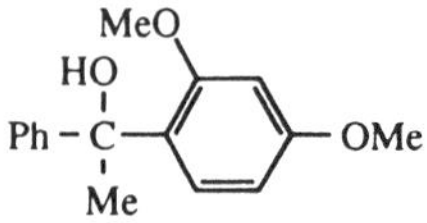

8-34

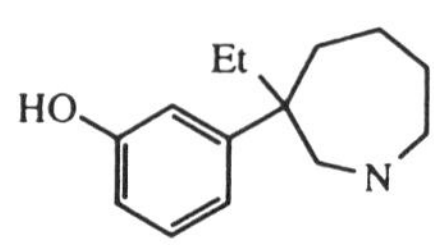

8-35

medications are Acnomel, Clearasil Adult Care, Fostex, Oxy-5, PHisoAc, Sulforcin, Thera-Blem, and Vanoxide (these are all tradenames).

Resorcinol monoacetate, **8-36**, is used similarly to resorcinol as an antiseborrheic (dandruff) and keratolytic agent, and as an antipruritic (against itches of the skin).

In a review, "Topical Acne Drug Products for Over-the-Counter Human Use; Final Monograph," the FDA concluded that over-the-counter acne drug products containing resorcinol 2%, resorcinol monoacetate 3%, salicylic acid 0.5–2%, or sulfur 3–10%, or combinations of resorcinol and sulfur, are generally recognized as safe (GRAS) and effective, under the Federal Food, Drug, and Cosmetics Act.[3]

Thioxolone (Boehringer-Ingelheim's Camyna®, Polfa's Acnosan, and others), **8-37**, is also an antiseborrheic dermatological compound derived from resorcinol. Trioxsalen (ICN's Trisoralen), **8-38**, is a melanizing dermatological agent.

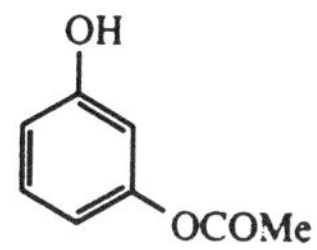

8-36

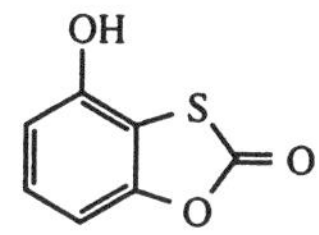

8-37

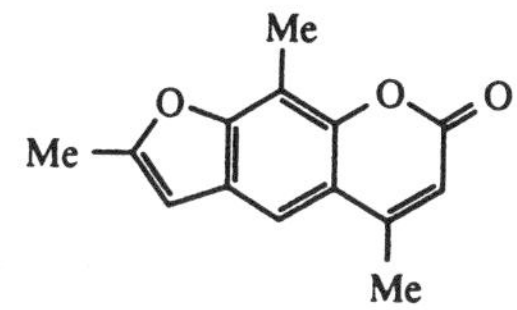

8-38

8.1.14. Ultraviolet Light Screens for Skin Protection

The hydroxybenzophenone-type compounds used to protect plastics from deterioration by UV light (see Chapter 11) have also been used to protect the skin in sun-screen preparations. Examples of these compounds are oxybenzone (2,4-dihydroxybenzophenone), **8-39**, dioxybenzone, **8-40**, mexenon (American Cyanamid; also WB Pharmaceutical's Uvistar®), **8-41**, and sulibenzone (GAF), **8-42**, the latter for use where water solubility is helpful.

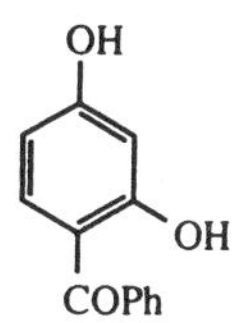

8-39

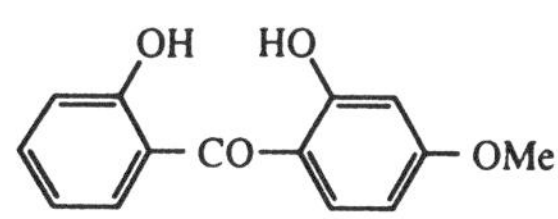

8-40

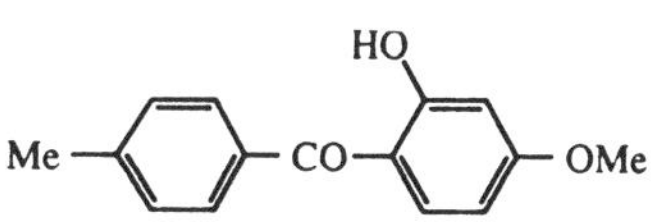

8-41

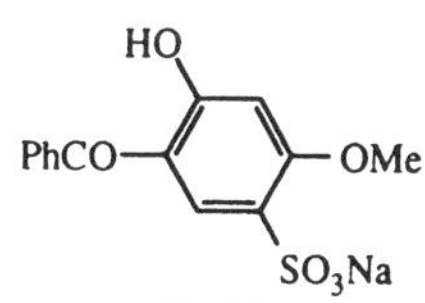

8-42

8.1.15. Diagnostic Aids and Reagents

Fluoresceine sodium (uranine, C.I. Acid Yellow; see also Chapter 10), **8-43**, is used as a diagnostic aid in corneal injuries; in veterinary medicine it is used in the diagnosis of periodic ophthalmia in horses. Erythrosine, **8-44**, has been used as a dental diagnostic agent.

In recent years there have been many uses made of fluorescent resorcinol derivatives as labeling agents in biochemical and medical research applications, such as in the HPLC separation and analysis of hydroxyl group-containing compounds by precolumn derivatization with fluorescers, for cellular pH measurements via fluorescent probes, and others. A particularly interesting and valuable example is found in the automated DuPont Genesis 2000 DNA sequence analyzer, in which four chemically similar but spectroscopically differing succinylfluoresceins are coupled to a different propargylamino-linked dideoxynucleotide triphosphate, e.g., **8-45**. A set of tagged reagents is obtained which are potent chain-terminating substrates for a number of DNA polymerases.[4]

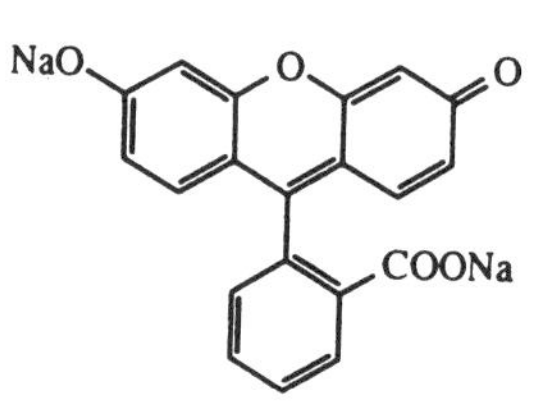

8-43

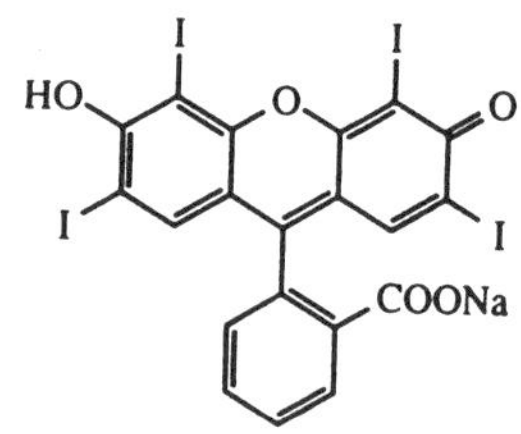

8-44

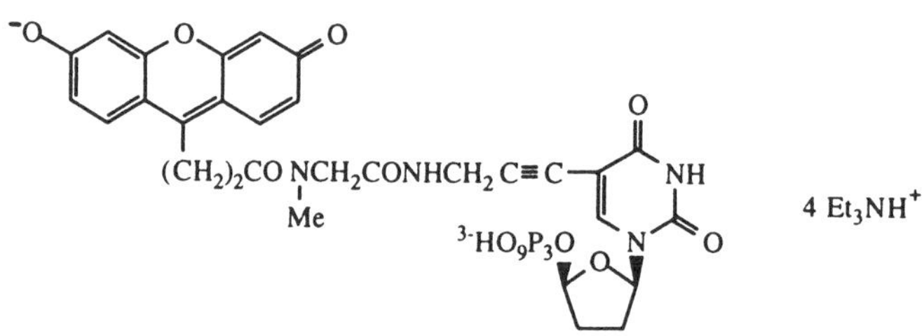

8-45

8.1.16. Outlook

Resorcinol, 1,3-cyclohexanedione, and *m*-aminophenol continue to be often-used starting materials for the synthesis of new drugs for the treatment of a wide range of diseases. Products previously used as racemates are being improved by resolution and/or synthesis to obtain the therapeutically most effective chiral drug. Fluorescence-analytical methods (in which resorcinol derivatives are used) for biologically and medically interesting entities are still in a growth phase.

Perhaps the most intensive development of drugs derived from resorcinol derivatives in the last 15 years or so were the leukotriene inhibitors. Research on allergic reactions of the lung showed that arachidonic acid metabolites formed by the action of lipoxygenases were related to various diseases. Some of the metabolites were found to be members of a group of eicosatetrenoic acids called leukotrienes (some of the compounds previously had been called the SRS-A = slow reacting substances of anaphylaxis). Leukotriene inhibitors were developed; many of these were ethers of 2-propyl-4-acetylresorcinol which showed great promise as antiinflammatories and antiallergic drugs. Many publications described the breadth of these developments, e.g., Reference 5. Two examples of compounds which progressed to clinical trials were LY-171883, **8-46**, and SC-41930, **8-47**; the latter was reported to be in phase-II testing (involving thousands of symptomatic patients) for psoriasis in the U.S. and for ulcerative colitis in Japan.[6] None of the resorcinol-derived leukotriene inhibitors appear to have made it to market to date, an example of the very lengthy trials and tribulations of new drug development.

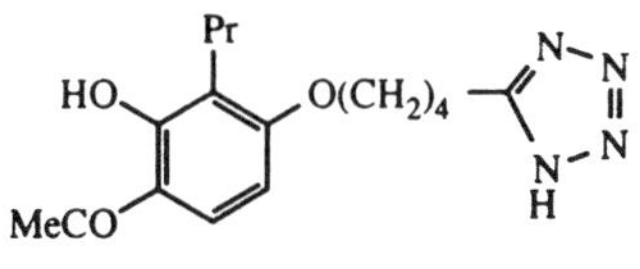

8-46

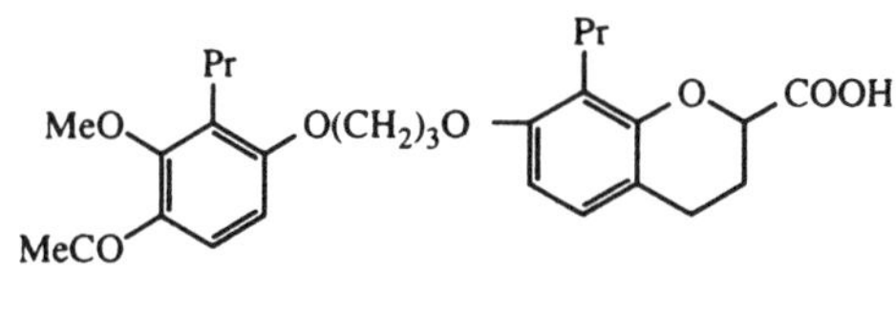

8-47

The following Section 8.2 presents a small selection of pertinent references (165) from the 1980–1992 period, with a few 1993 references, showing clearly the diverse drugs and reagents under development which have a resorcinol moiety in their structure. Surely, a few of these will prove to be valuable additions worthy of commercialization.

8.2. SELECTION OF DEVELOPMENTS SINCE 1980

Research on new drugs is exciting because intricate chemistries, often novel, can be used since the drug unit cost can be high and the doses can be very small. Also, an important new drug is good for humankind, it can generate considerable income for the innovating company, and recognition of many types for the inventor. There is the necessity of interdisciplinary work. The cost and effort required for development is very high—the product must be proven to be safe and effective; side effects must be minimized and the risk of side effects must be balanced against the benefit to be gained. It is said that many thousands of compounds are prepared, tested, and many are patented, for every medication that is commercialized. Penalties for mistakes or oversights can be severe once the side effects become subject to litigation. Every effective new drug prompts massive work by competitors to come up with alternatives outside the original's patent claims if not with improvements on the theme. In some cases, the effort fails totally. At any rate, there is a massive literature generated in this field which would require a massive treatise by itself to report in detail. The following selections from this literature give a feel for many facets of this fascinating endeavor from the miniperspective of resorcinol-related compounds. The indications are there why it takes about 10 years or more from the concept to the marketing of a new drug.

(Methoxyphenyl)glutamines such as compound **8-48** were found to be effective immunosuppressants i.p. or orally in mice.[7]

Amides of *p*-aminosalicylic acid were prepared which were potent central-nervous-system (CNS) depressants and antipsychotics. Thus, compound **8-49** was effective at a clinical dose for adults of 1–200 mg/day taken orally and 1–150 mg/day parenterally.[8]

Substituted perfluoroacylresorcinols were made and determined to be useful active ingredients in pharmaceutical, cosmetic, and pesticidal compositions. For instance, compound **8-50** showed excellent inhibition (MIC-values of 0.31 to 250 μg/ml against gram-positive bacteria such as *staphylococcus aureus* SG 511, *corynebacterium acnes*, yeast, dermatophytes, and molds.[9]

2-Substituted 1,3-cyclohexanediones, e.g., compound **8-51**, were prepared and showed antihypertensive activity.[10]

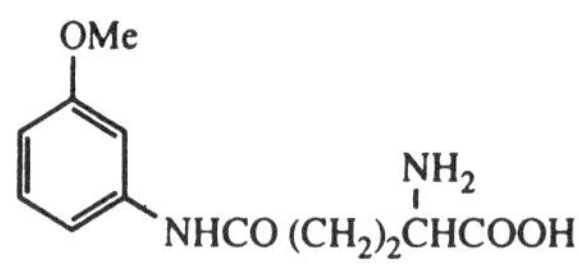

8-48

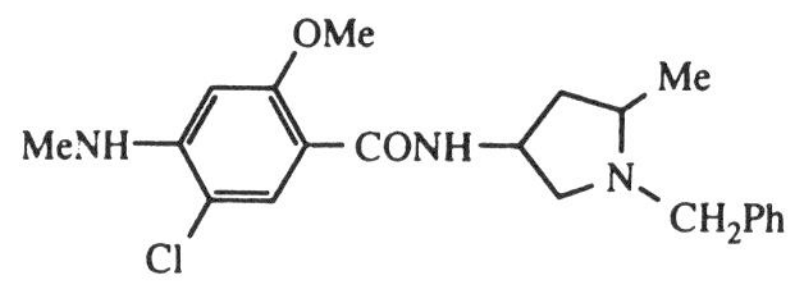

8-49

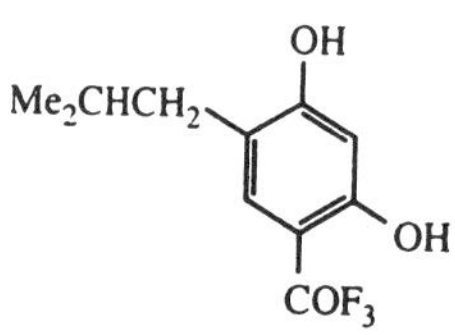

8-50

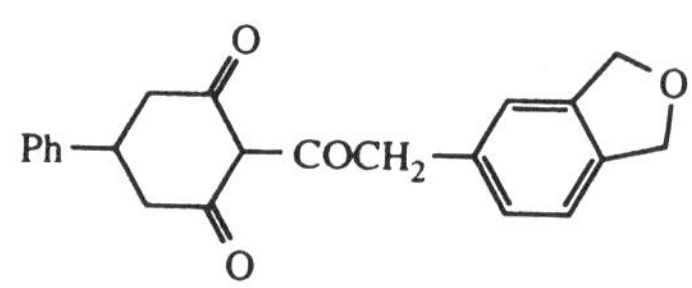

8-51

Psoralen derivatives were synthesized for use in pharmaceutical preparations. Thus, compound **8-52**, at 1% concentration in an ointment, was useful in treating psoriasis without causing erythmia.[11]

1-Phenyl-3-(substituted phenoxy)propylamines were found to be antidepressants. For instance, compound **8-53** showed psychotropic activity and inhibited serotonin and norepinephrine uptake.[12]

Dialkoxy-monorden derivatives were prepared and evaluated. Thus, compound **8-54** was found to have an ED_{50} 0.8 mg/mL against tumorous human nasopharynx cells; this compound had an LC_{50} 0.8 mg/mL against nematodes.[13]

Certain sesquiterpene compounds (containing a resorcinol segment) were prepared and were useful in treating hepatitis. For instance, compound **8-55** was effective in treating induced hepatitis in rats at 5 mg/kg.[14]

N-(Hydroxyalkyl)-7-hydroxycoumarin-3-carboxamides were synthesized and found to be useful in clinical analysis. Thus, compound **8-56** was useful as an instrument calibrator for fluorescent immunoassay methods.[15]

Substituted iso-psoralens were prepared and were useful as photosensitizing agents. For instance, compound **8-57** had low toxicity, produced minimal erythemal irritation, and was useful for the photochemotherapy of psoriasis, as an inhibitor of DNA replication, and for inactivating viruses.[16]

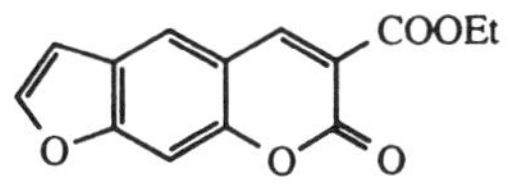

8-52

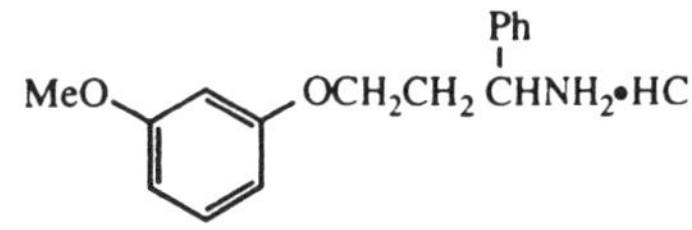

8-53

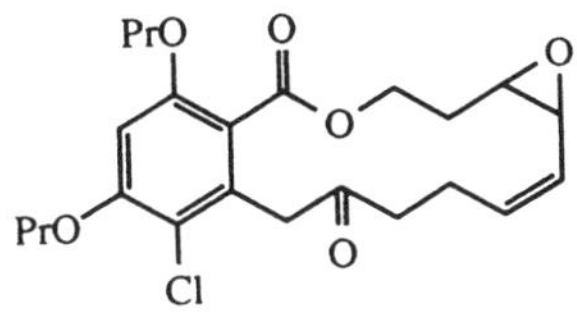

8-54

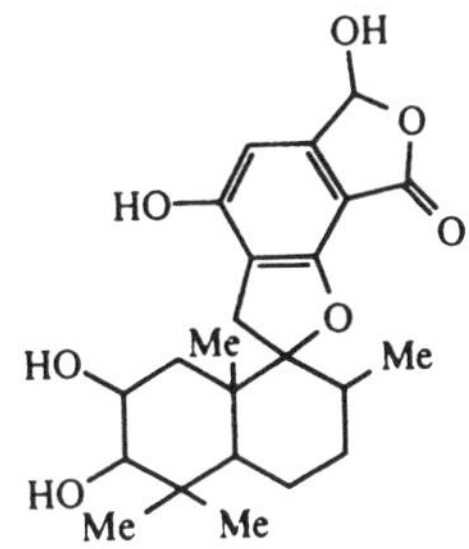

8-55

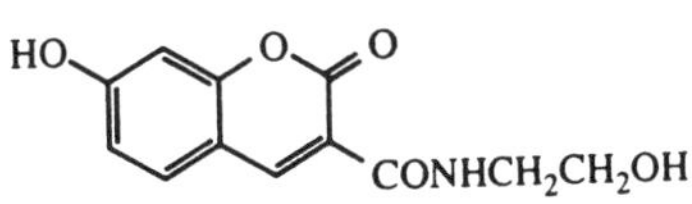

8-56

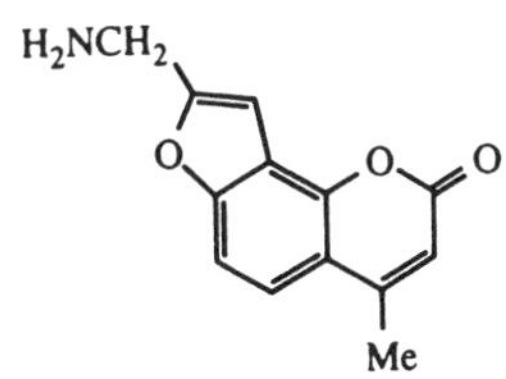

8-57

[(Hydroxyphenoxy)alkoxy]benzoic acids and related compounds were prepared and screened for therapeutic properties. For example, compound **8-58** showed antiallergic activity.[17]

Benzopyranopyridine derivatives containing a resorcinol segment (azacannabinoids) were prepared and demonstrated to be potent antihypertensives. An example in this group is compound **8-59**.[18]

Substituted hydroxy-benzopyranone ethers were synthesized and tested for use as pharmaceuticals. As an example, compound **8-60** was found to be useful in the treatment of allergic asthma and as an antiinflammatory agent.[19]

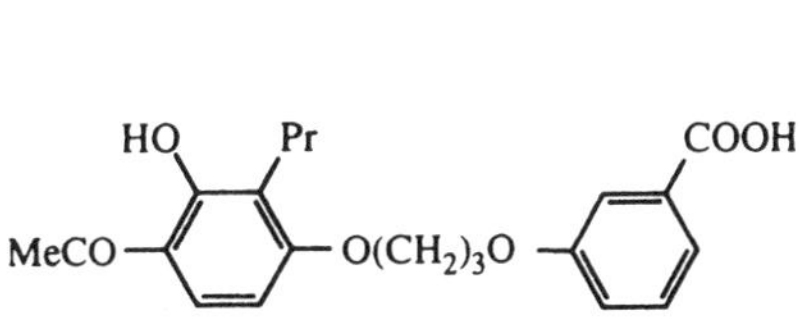

8-58

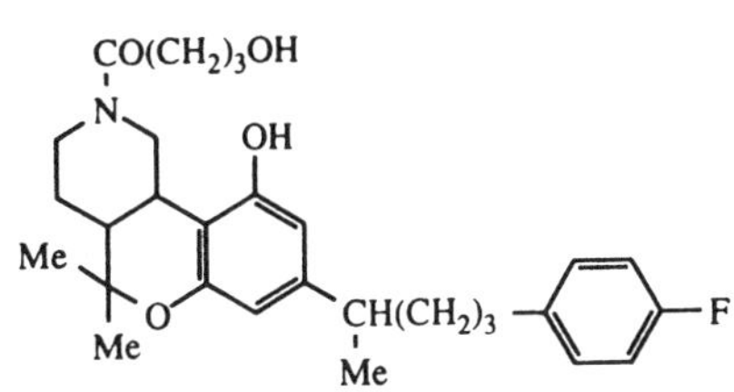

8-59

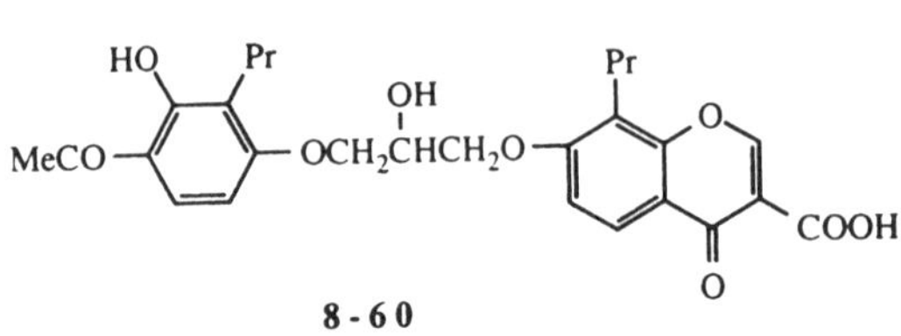

8-60

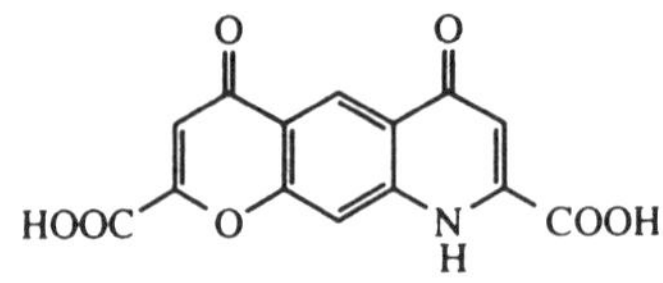

8-61

Pyranoquinoline derivatives (and their salts), such as compound **8-61**, were prepared and found to be useful as antibody-antigen inhibitors.[20]

Thiobenzanilides were prepared from *m*-aminophenol ethers and evaluated as antibacterials. Thus, compound **8-62** was effective against *Pellicularia sasakii* and *Gymnosporangium hareanum.*[21]

3-(*N*-Piperazinyl)-2-cyclohexen-1-ones were prepared. An example is compound **8-63**, which had analgesic properties.[22]

A tripeptide amide of 4-methyl-7-aminocoumarin, **8-64**, was prepared for use as a substrate for the determination of trypsin activity.[23]

N-(Azabicyclooctyl)-*p*-aminosalicylamides were synthesized and evaluated for therapeutic properties. For example, compound **8-65** had both antiemetic properties and CNS activity.[24]

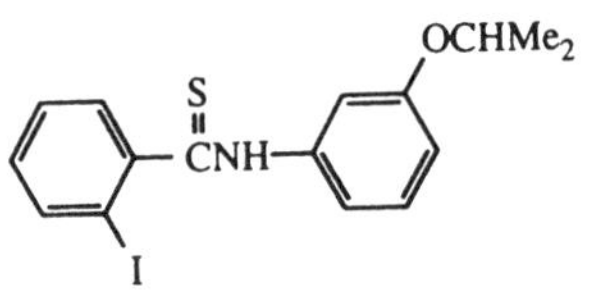

8-62

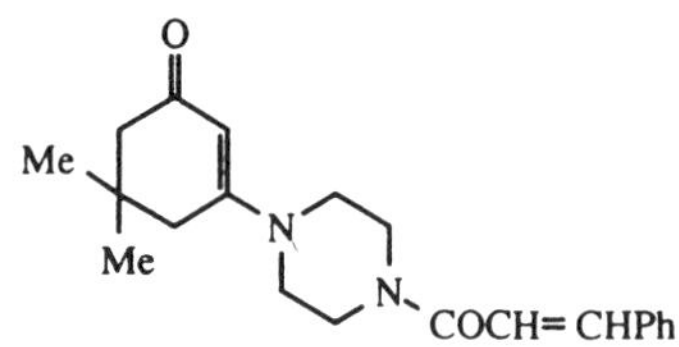

8-63

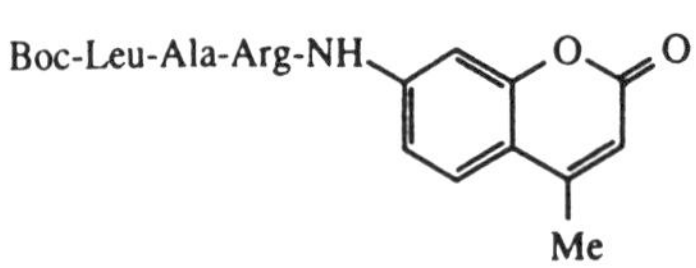

8-64
(Boc = Me_3COOC)

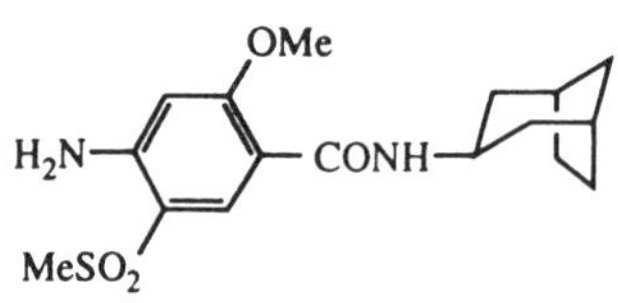

8-65

Δ^9-Tetrahydrocannabinol bis-(2-chloroethyl)carbamate, **8-66**, was found to have antitumor activity with mood-elevating side effects.[25]

Certain phenylpiperidines were prepared as muscle contractants. For instance, compound **8-67** had an *in vitro* ED_{50} 19.7 μM for contraction of rabbit papillary muscle.[26]

A series of benzoxazine derivatives, such as compound **8-68**, was prepared which were useful as platelet aggregation inhibitors, antiinflammatories, antithrombotics, and phosphodiesterase inhibitors.[27]

Bis-esters of resorcinol with ampicillin or amoxycillin and penicillanic acid 1,1-dioxide or 6-beta-(hydroxymethyl)penicillanic acid 1,1-dioxide were patented as useful antibacterials, administered orally or parenterally.[28]

Certain hydroxychromone ethers were prepared and patented as antihypertensives. An example of such products is compound **8-69**.[29]

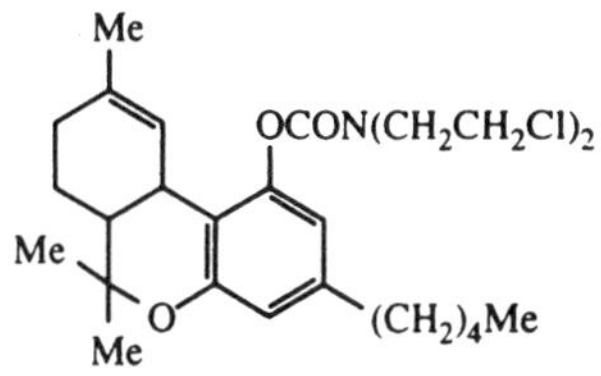

8-66

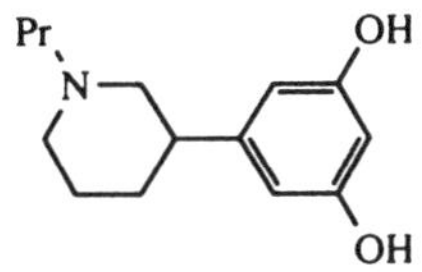

8-67

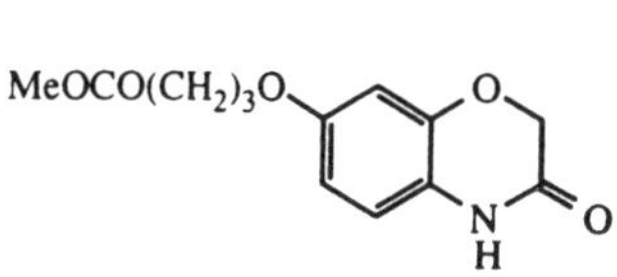

8-68

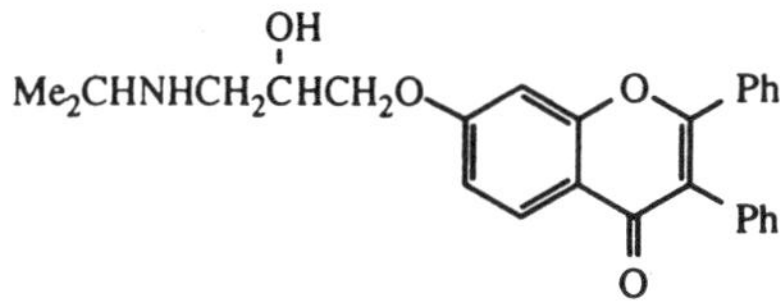

8-69

Diphenoxypropane derivatives were synthesized and tested for antiasthmatic properties, Among the active materials was compound **8-70** which, at 5 mg/kg i.v., gave 93% inhibition of leukotriene D_4-induced bronchoconstriction in guinea pigs; the compound also showed antiinflammatory action in rats.[30]

Certain esters of phenoxyacetic acids were prepared and patented as antianaphylactic agents. An example of the products in this series is compound **8-71**.[31]

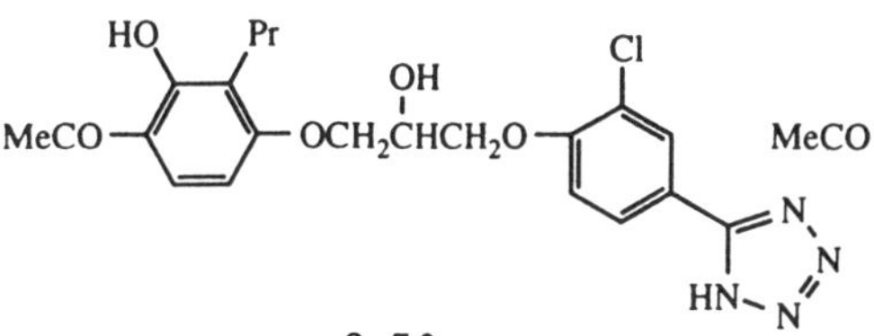

8-70

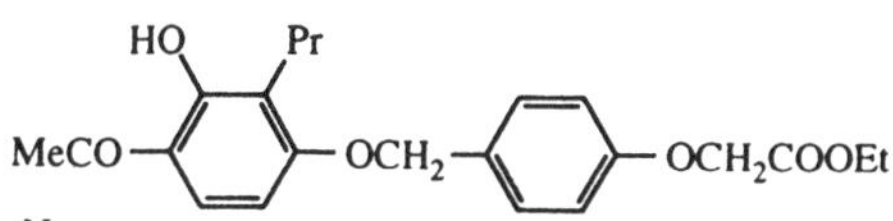

8-71

Novel phenoxyalkylcarboxylic acid derivatives were prepared and tested as leukotriene inhibitors. Compound **8-72**, for example, was tested in guinea pigs at 10 mg/kg and gave 95% inhibition of bronchoconstriction caused by leukotriene E_4.[32]

The 4-phenoxy-*p*-aminosalycylamide derivative, **8-73**, was evaluated and claimed to be a useful psychotropic agent.[33]

Triiodoresorcinol (called Riodoxol in the USSR), **8-74**, was said to be an effective virucide against *Herpes simplex* in chinchillas, and was also a fungicidal and bacte-

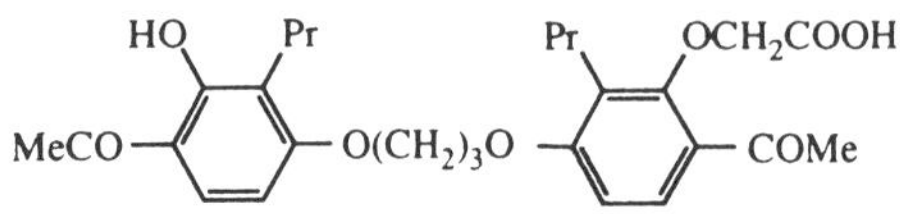

8-72

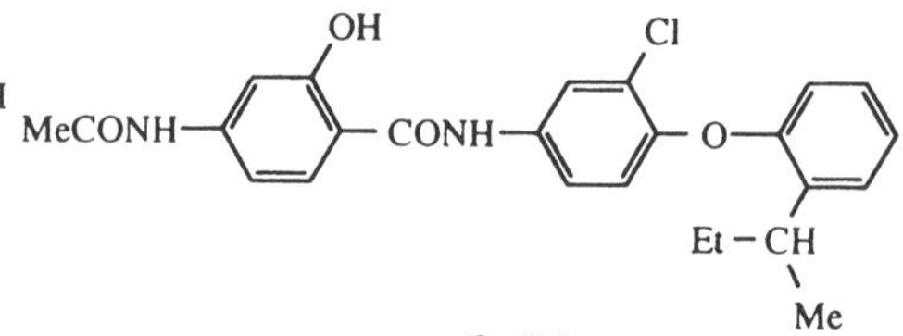

8-73

riostatic agent; it had low toxicity and no allergenic, carcinogenic, mutanogenic, or teratogenic properties. Triiodoresorcinol was also patented as an effective antifungal and antibacterial agent.[34]

A series of novel resorcinol (and catechol) ethers, such as compound **8-75**, was prepared and shown to be useful as antithrombotics (prevent blood clot formation), as inhibitors of thrombane-synthetase, and as thrombin inhibitors.[35]

2-Propargyloxy-4-aminobenzamides were made and demonstrated to have antiemetic activity. For instance, compound **8-76** at 1 mg/kg in dogs gave 50% protection against emesis (vomiting).[36]

New phenoxyisopropanolamine derivatives, e.g., compound **8-77**, were prepared and shown to have useful antihypertensive and alpha-adrenergic blocker activity.[37]

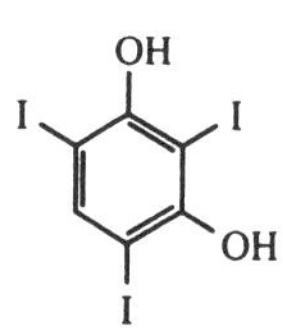

8-74

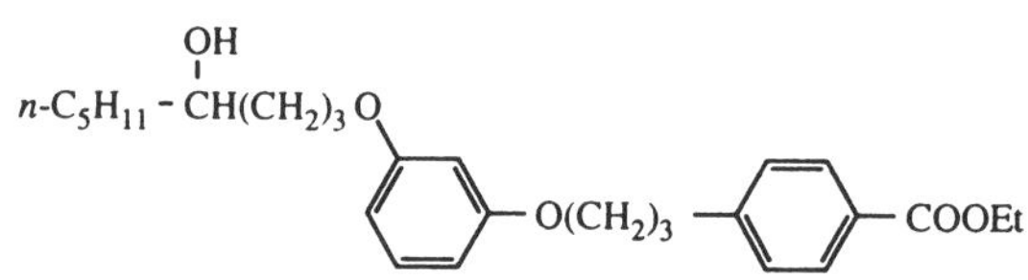

8-75

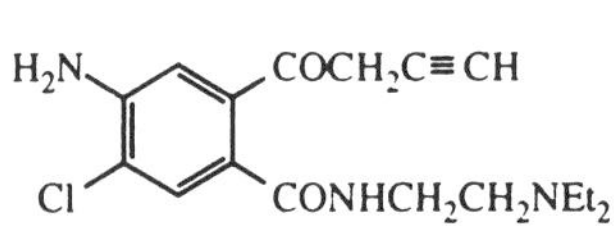

8-76

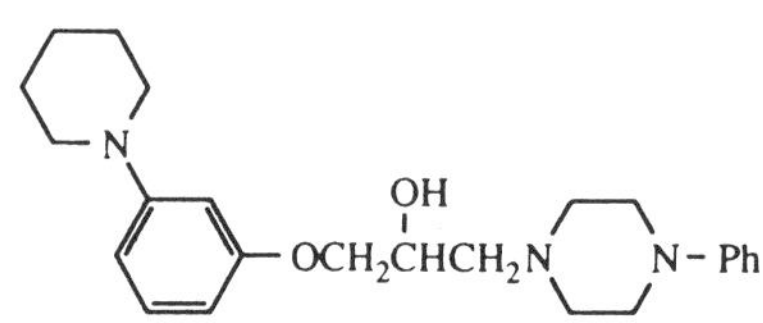

8-77

A series of naphthyloxyalkanecarboxylic acid derivatives was prepared and patented as antiallergenic agents. An example of these products is compound **8-78**.[38]

The *trans*-bis-(resorcylaldoximato)-copper II complex, **8-79**, showed antitumor activity against mouse leukemia L1210 and Ehrlich ascites carcinoma.[39]

Benzisoxazolecarboxylic acids were prepared and screened for medically useful activity. Thus, compound **8-80** (and its pharmaceutically acceptable salts) was found to have a diuretic ED_2 of 27.3 mg/kg.[40]

A series of substituted-alkyl ethers of resorcinol were prepared as leukotriene antagonists. For instance, compound **8-81** at 3×10^{-6} M in vitro gave 100% inhibition

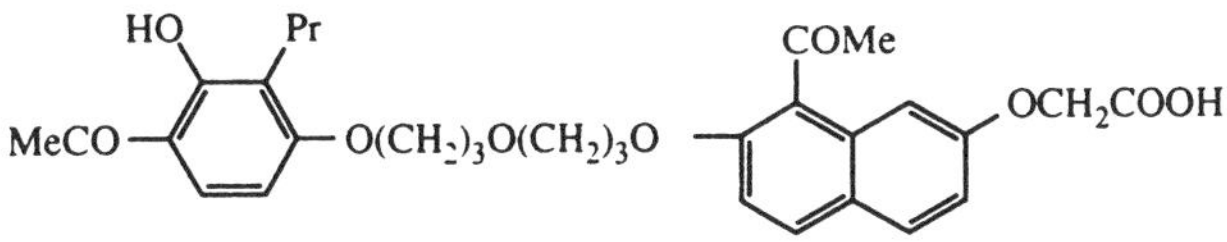

8-78

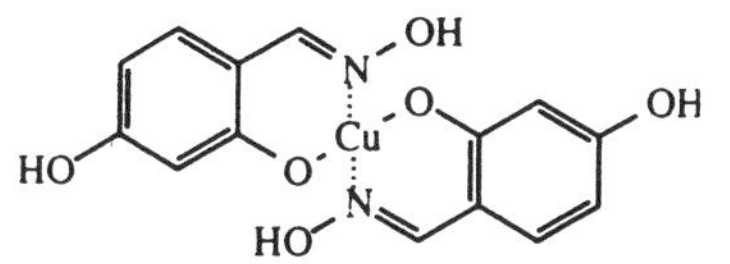

8-79

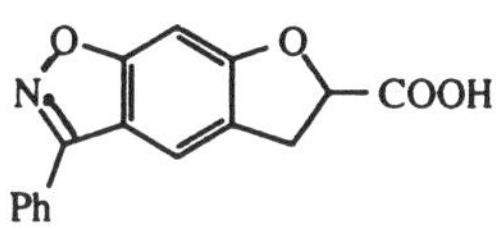

8-80

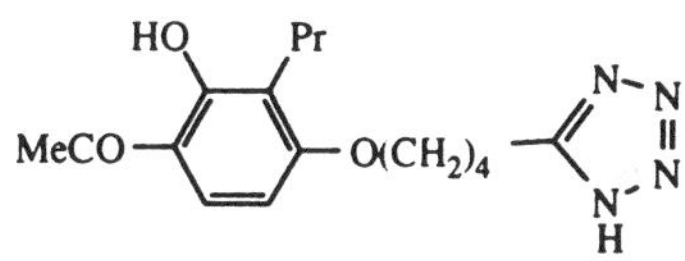

8-81

of leukotriene D_4-caused ileal contractions in guinea pig ilea (the third and lowest division of the small intestine).[41]

Derivatives of 4-[1-(phenyl)-1-hydroxy-prop-1-yl]resorcinol were made and shown to be useful as alcohol intoxication antagonists. Thus, compound **8-82** reduced the alcohol-induced time of unconsciousness in rats as much or more than did L-cysteine.[42]

3-Alkyl-8-chloro-5,6-dihydrofuro-[3,2-*f*]-1,2-benzisoxazole-6-carboxylic acids were found to be useful for the control of sodium in body fluids. Thus, compound **8-83**, at 40 mg/kg given orally in rats, caused the elimination of 2 mequiv. Na^+ in 4 hr.[43]

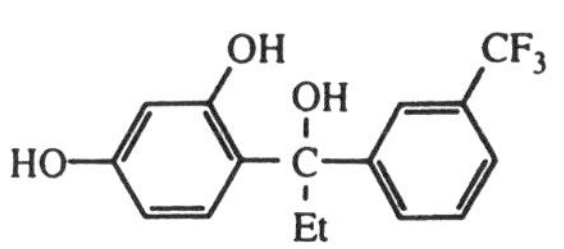

8-82

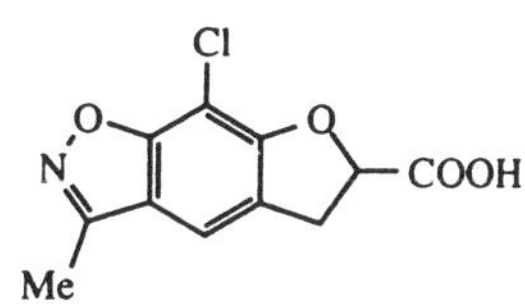

8-83

4-Oxo-4,5,6,7-tetrahydroindole derivatives, such as **8-84**, were synthesized starting from 1,3-cyclohexanedione and were useful intermediates for drugs (e.g., pindolol).[44]

4′-Aminomethylfluorescein derivatives, such as compound **8-85**, were prepared and patented for use in fluorescence polarization immunoassay analyses.[45]

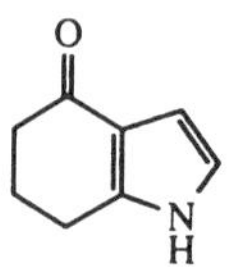

8-84

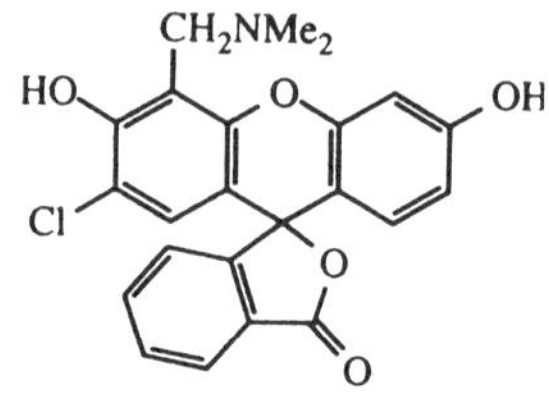

8-85

Novel heterocyclic compounds, such as *N*-[3-(2-benzothiazolyl)methoxy]phenyl-pyrrolidine-2,5-dione, **8-86**, and related compounds were made and claimed to be useful for the treatment of leukotriene-mediated nasobronchial obstructive air passageway conditions, such as allergic rhinitis, allergic bronchial asthma, and the like.[46]

Phenoxyalkoxy-3,4-dihydro-2*H*-1-benzopyran derivatives were made and their salts with pharmaceutically tolerable bases were useful as antiallergenics. One example is compound **8-87**, which had IC_{50} 1×10^{-7} M in the guinea pig ileum test of SRS-A antagonism.[47]

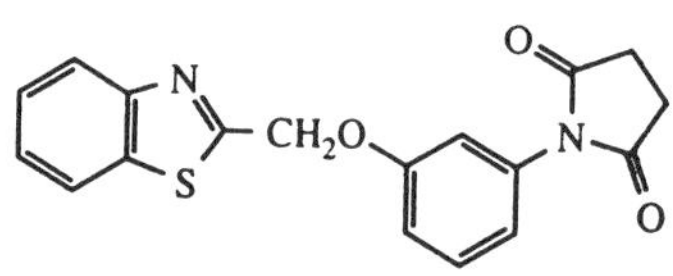

8-86

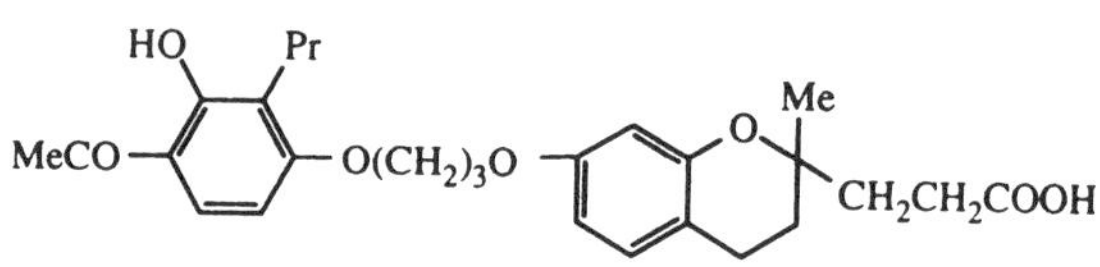

8-87

Rhodamine derivatives, e.g., compound **8-88**, were synthesized for use as chromophoric substrates for proteinases. The kinetic constants for the hydrolysis of these compounds by human plasmin, bovine trypsin, and human thrombin were determined.[48]

5-Substituted chromones, such as compound **8-89**, and their salts were prepared; **8-89** (sodium salt) inhibited passive cutaneous anaphylaxis in rats with ED_{50} 10.4 µmol/kg given intravenously and it also inhibited histamine release in vitro by mast cells of rats.[49]

3,4-Dihydro-10-hydroxy-1-(2*H*)-phenanthrones, such as compound **8-90**, were made starting from 2-arylacetyl-1,3-cyclohexanediones, for use as intermediates in the synthesis of steroids or their analogs.[50]

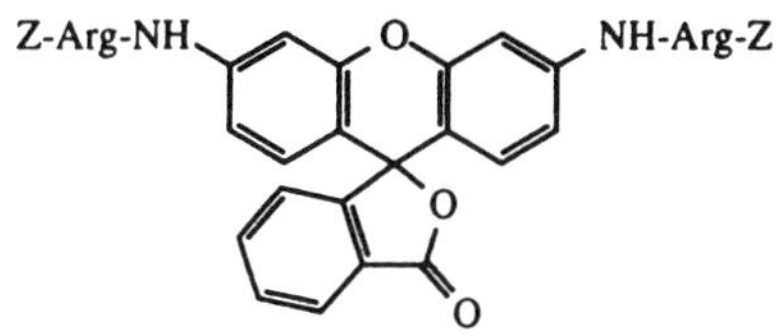

8-88

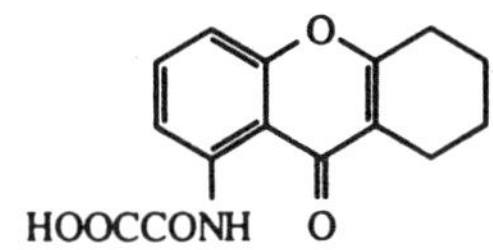

8-89

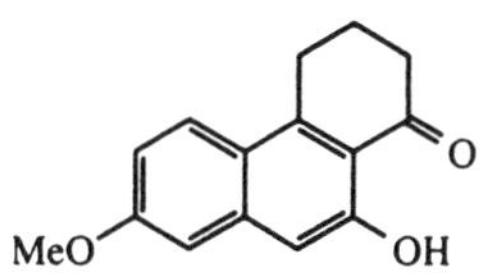

8-90

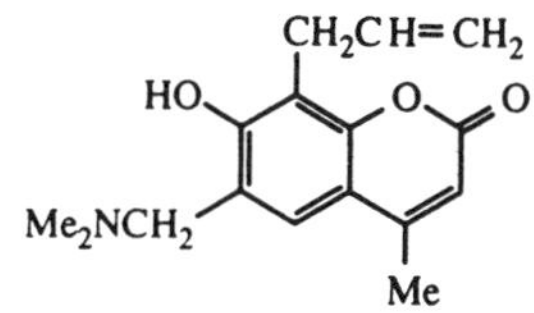

8-91

6-(Aminomethyl)-7-hydroxycoumarins were prepared which inhibited blood platelet aggregation and showed antitumor activity. One example from this series is compound **8-91**.[51]

The thioalkyl ether of resorcinol, **8-92**, was prepared and patented as an allergy inhibitor/antiasthmatic agent.[52]

Certain isoflavones were made and found to be useful as antiinflammatory agents. Thus, compound **8-93** at 50 mg/kg oral dose gave 45% inhibition in the rat paw edema test.[53]

Antagonists of slow reacting substances of anaphylaxis (SRS-A) and leukotrienes C_4, D_4, and E_4, which contained a thiophenoxyalkyl ether of a resorcinol segment, were synthesized and recognized as antiasthmatic, antiallergic, antiinflammatory, and cytoprotective agents. For example, compound **8-94** was useful for the treatment of allergies.[54]

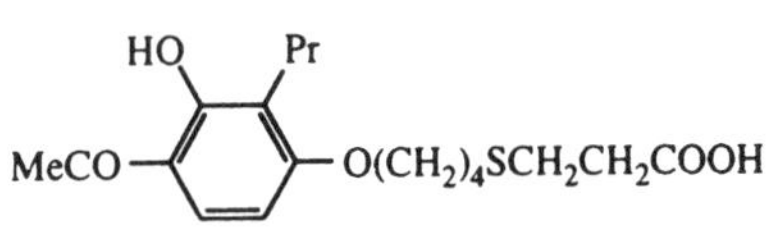

8-92

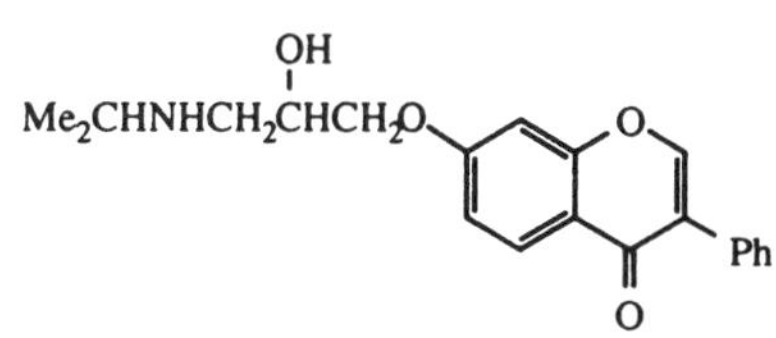

8-93

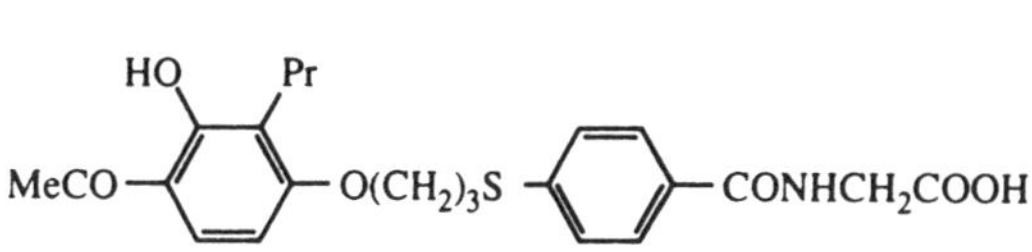

8-94

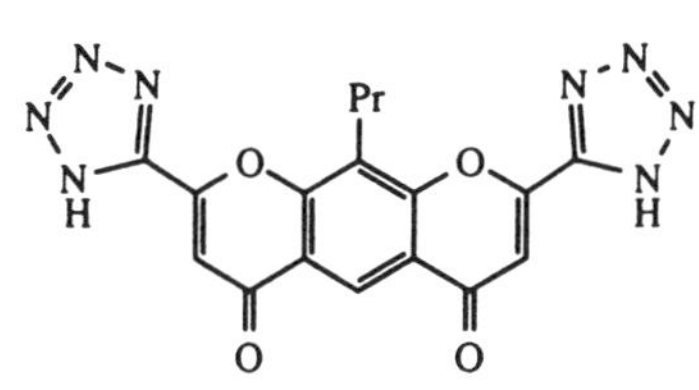

8-95

Benzopyran derivatives such as compound **8-95** were prepared and determined to be useful antiasthma drugs.[55]

4-[3(*R*)-({2-hydroxy-3-(hydroxyphenoxy)propyl}amino)butyl]benzamides and their salts were made and found to be useful as inotropic (affecting the contractility of muscular tissue) agents. For example, compound **8-96** (as the hydrochloride) at 0.5 mg/kg oral dose in conscious, instrumented dogs raised cardiac contractility to 123% of the zero time value after 2.5 hr, while a placebo gave a reduction to 96% of cardiac contractility; **8-96** did not affect the heart rate or arterial pressure.[56]

Pharmaceutical preparations containing flavene (and thioflavene) derivatives such as, e.g., compound **8-97** and their salts, were prepared and claimed to be useful mucolytics and immunostimulants, applicable for the treatment of liver diseases.[57]

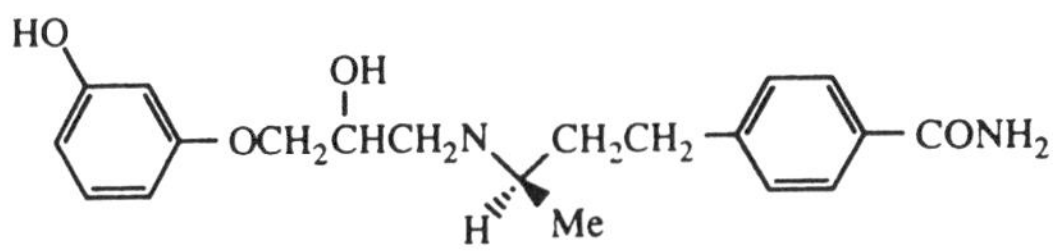

8-96

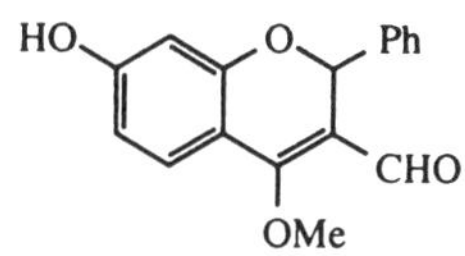

8-97

Benzopyran-4-one derivatives were made which were bone resorption inhibitors useful in treating osteoporosis. Thus, compound **8-98** was shown to be effective as a bone resorption inhibitor in rat fetal long bone culture.[58]

A series of substituted dihydrobenzopyran-2-carboxylates, typified by compound **8-99**, were prepared and said to be useful antiallergy and antiinflammatory agents.[59]

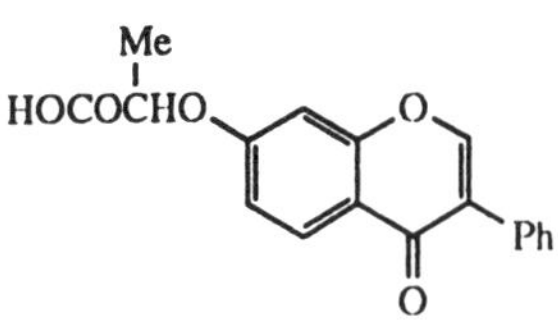

8-98

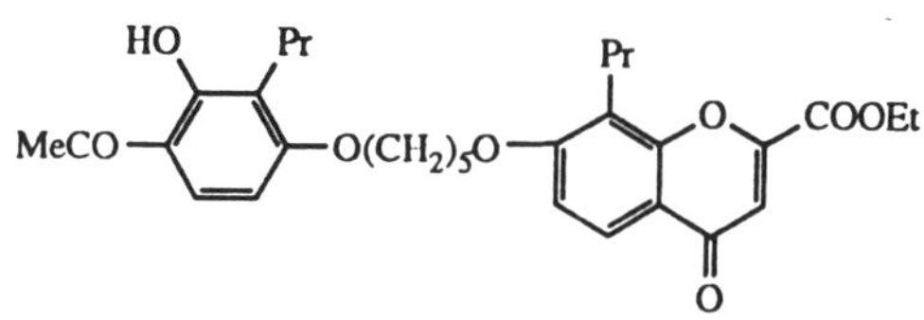

8-99

Hydroxystilbene compounds, such as **8-100**, were extracted from smartweed (e.g., *Polygonum cospidatum S*, *Polygonum orientale L*, etc.). These compounds were proposed as food additives to enhance liver function.[60]

Certain resorcinol ethers, exemplified by compound **8-101**, were synthesized and found to be antiallergy drugs with leukotriene D_4-antagonizing and phospholipase A_2

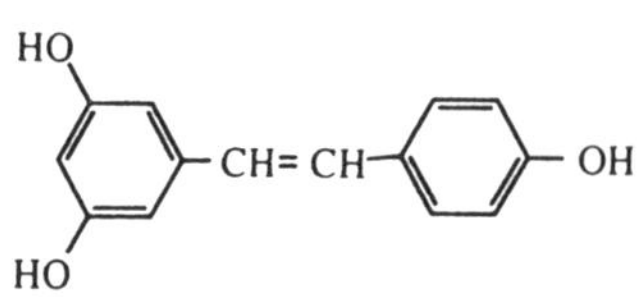

8-100

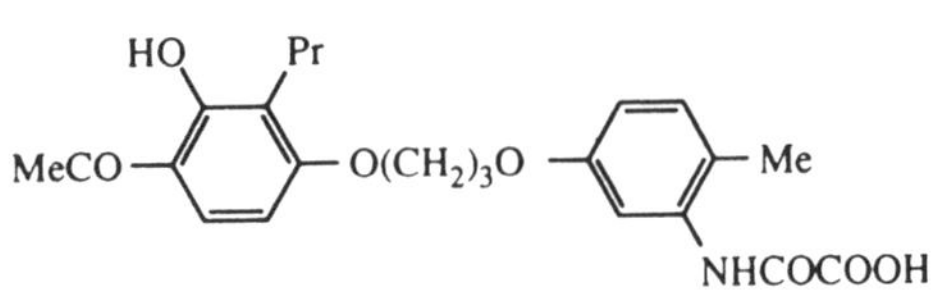

8-101

and C-inhibiting ability. For example, the tris-(2-hydroxyethyl)ammonium salt of compound **8-101** at 0.06 μM gave 50% inhibition of leukotriene D_4-induced contractions of guinea pig ileum; formulations for aerosol inhalator application of this compound were developed.[61]

New ethers of 2-propyl-4-acetylresorcinol were prepared for use as antiallergy drugs. Thus, compound **8-102** was an inhibitor of leukotriene-caused contractions in guinea pig ileum, with $IC_{50} < 10^{-4}$ M.[62]

Substituted chromanon-2-yl alkanols, of the type exemplified by compound **8-103** (monoacetate), and derivatives prepared from these alkanols, were made and patented as leukotriene D_4 inhibitors useful in the treatment of allergies, inflammatory conditions, and coronary vasoconstrictions.[63] This patent contains many references related to the concept and development of leukotriene inhibitors as drugs.

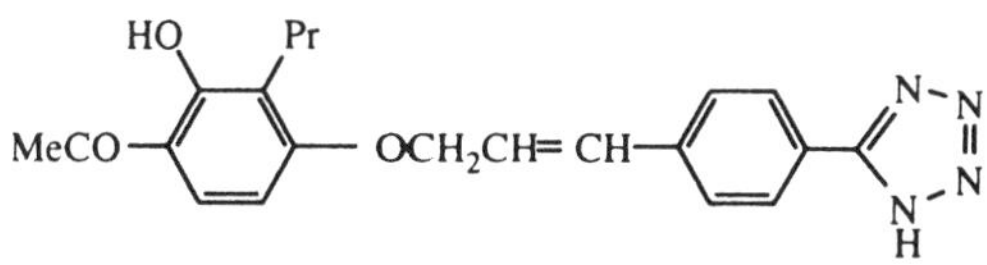

8-102

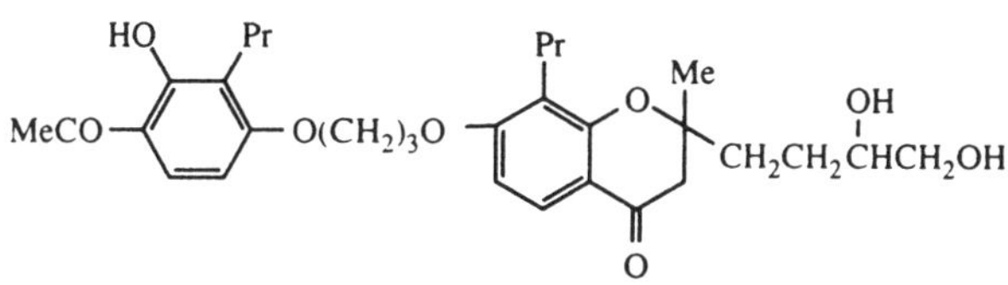

8-103

A thiazolyl-hexahydroquinoline compound, **8-104**, was prepared, starting from 1,3-cyclohexanedione, and was a calcium antagonist reported to be a useful antihypertensive agent; at 50 mg/kg in rats, it reduced blood pressure by 47 mm Hg after 1.5 hr.[64]

Pyrazolopyridine-cycloalkanone derivatives, typified by compound **8-105**, were synthesized and found to have good anxiolytic activity in rats.[65]

Chromogenic (highly colored) and fluorogenic (strongly fluorescing) derivatives of 7-hydroxycoumarins, such as compound **8-106**, were prepared and claimed to be useful for detecting and determining hydrolases.[66]

Schiff bases of 1,4-bis-[(omega-aminoalkyl)amino]-9,10-anthracenediones, and leuco bases thereof, were synthesized; among the aldehydes used for the preparations of the Schiff compounds were 2,4- and 3,5-dimethoxybenzaldehyde and 2-hydroxy-4-methoxybenzaldehyde. These compounds, e.g., **8-107**, were useful as chelating agents and for inducing the regression of leukemia and/or inhibiting tumor growth in mammals.[67]

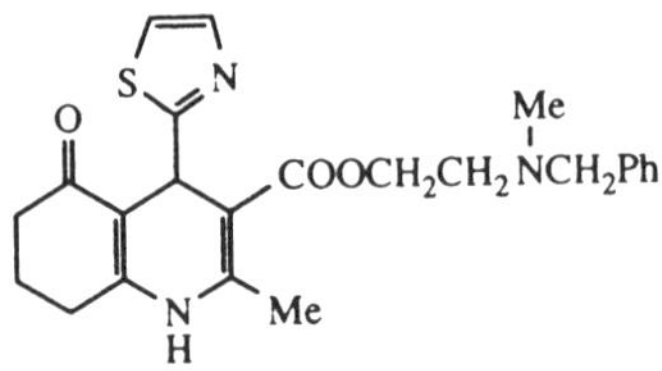

8-104

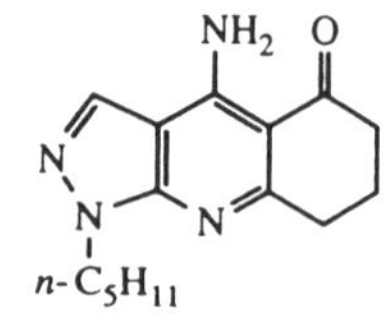

8-105

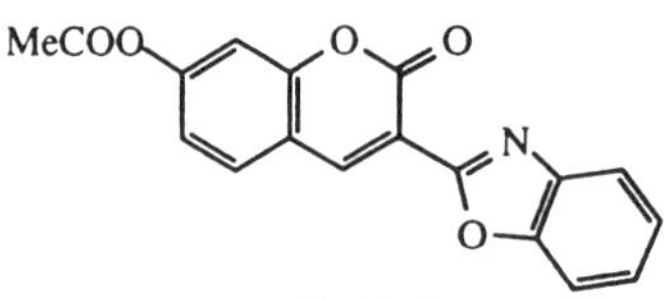

8-106

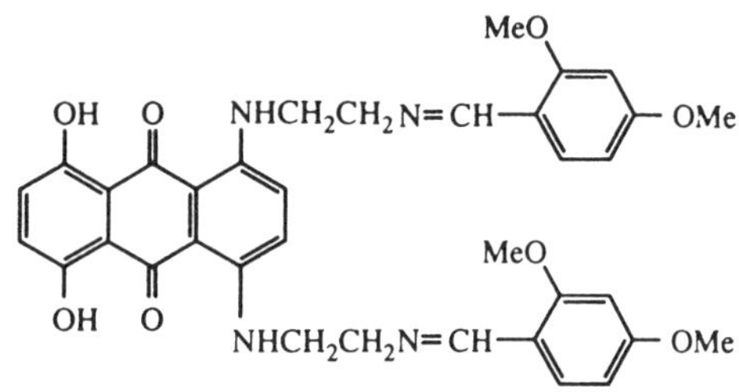

8-107

7-Alkoxy-isoflavones were made and patented for use as nonsteroidal anabolics, in osteoporosis, and as weight-reducing agents and anticholesteraemic agents. An example of these products is compound **8-108**.[68]

Novel phenethanolamines, typified by compound **8-109**, were synthesized and were claimed to increase the force of the heart contraction and to decrease the peripheral vascular resistance.[69]

Certain phenoxyalkyl ethers of 2-propyl-4-acetylresorcinol were prepared as antiasthmatic and antiinflammatory drugs. Thus, compound **8-110** had an ED_{50} of 47 μmol/kg oral dose for inhibition of leukotriene D_4-induced bronchoconstriction in guinea pigs, which was a lower dose than required with similar compounds of the prior art.[70]

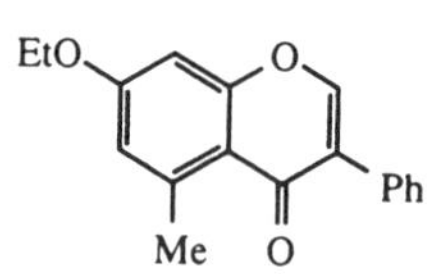

8-108

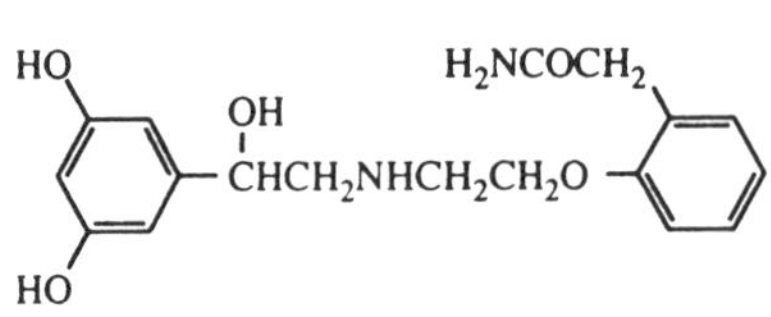

8-109

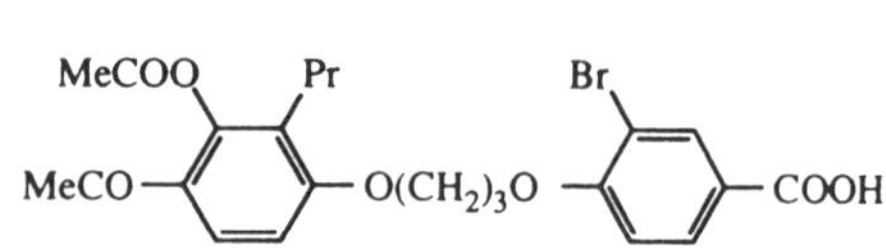

8-110

7-Aminoalkoxybenzopyranones were prepared which had effective antipsychotic and anxiolytic properties. For instance, compound **8-111** was tested for antipsychotic action by the [3*H*] haloperidol binding assay and had an IC_{50} of 1.5×10^{-8} M.[71]

7,8,9,10-Tetrahydro-6*H*-dibenzo[*b,d*]-pyranyloxyaminopropanols, such as compound **8-112**, were synthesized and claimed to be useful antihypoxia agents, i.e., they protected warm-blooded animals from the adverse effects of oxygen deprivation.[72]

Certain 2-(3-hydroxycyclohexyl)-5-alkyl resorcinols were made and found to be effective analgesics. Thus, compound **8-113** was a good analgesic with no side effects when given to rats and mice at up to 5 mg/kg.[73]

Benzofurancarboxylates were made and patented as lipoxygenase inhibitors. These compounds are exemplified by structure **8-114**.[74]

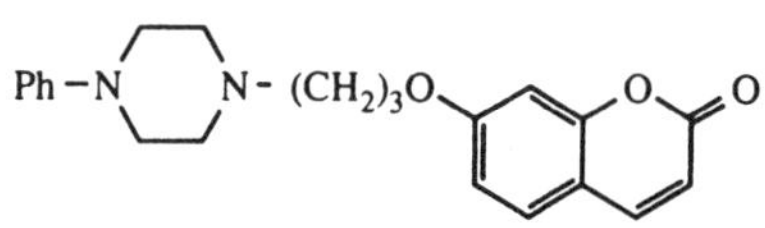

8-111

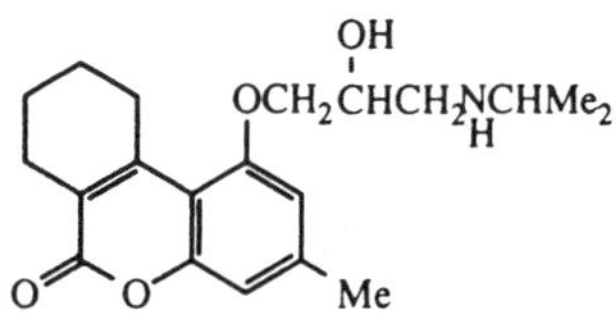

8-112

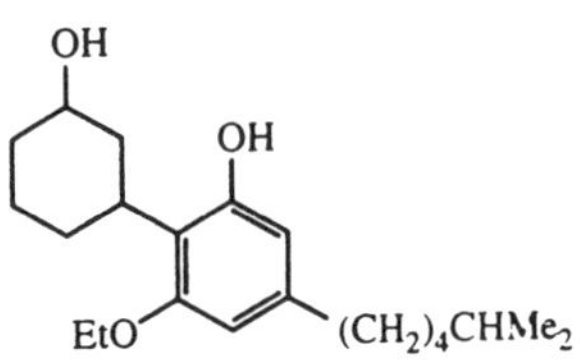

8-113

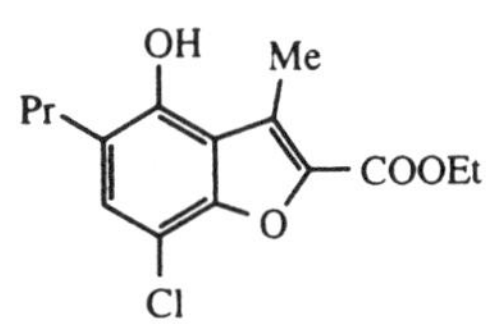

8-114

4-Aminosalicylic acid derivatives and their salts (e.g., the imidazolium salt) were prepared and patented for use as antiinflammatory, antiplatelet aggregating and antithrombotic agents. An example of such compounds is **8-115**.[75]

Heterocyclic-substituted ethers of *m*-aminophenol were made, such as compound **8-116**, which inhibited in vivo the synthesis of 5-hydroxy-eicosatetranoic acid and leukotriene B_4 by arachidonic acid lipoxygenase in rat glycogen-elicited polymorphonuclear leukocytes, and also inhibited effectively leukotriene-caused bronchospasms in guinea pigs.[76]

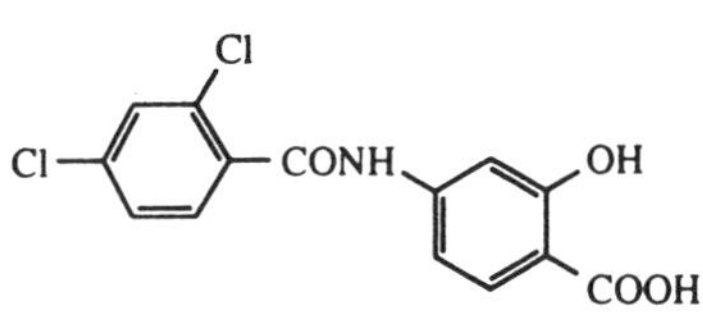

8-115

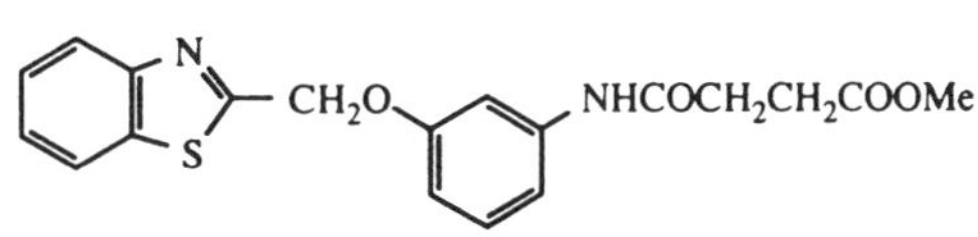

8-116

A skin care composition, consisting of 3–7 parts salicylic acid, about 8–12 parts resorcinol, 8–12 parts lactic acid, and about 60–90 parts ethyl alcohol by weight, was patented. Applied to the face, and after approximately one week, these compositions caused light peeling of the dead surface skin and removal of some blemishes, imparting a youthful and healthy complexion.[77]

New 7-hydroxyflavone derivatives were prepared and shown to be effective antiulcer agents at 50 mg/kg i.p. in mice. An example of these compounds is structure **8-117**.[78]

Xanthone derivatives were prepared which were useful as diuretics and uricosurics. Thus, compound **8-118** given orally to rats at 100 mg/kg increased urine volume by 99.5% and increased the excretion of uric acid by 48.6%.[79]

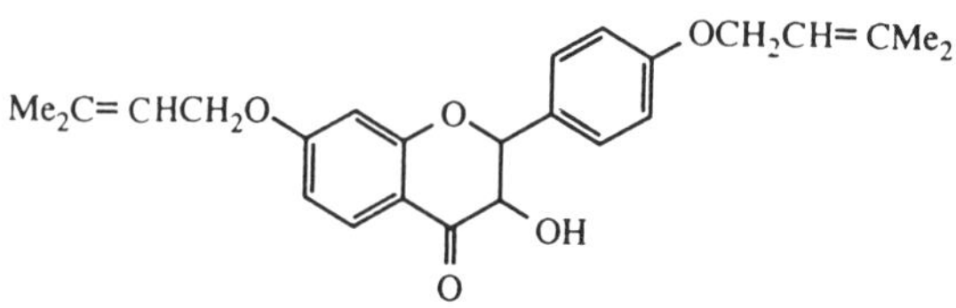

8-117

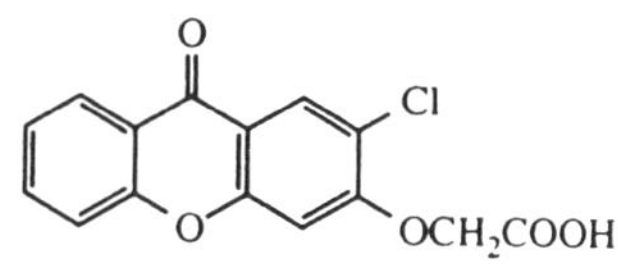

8-118

Esters of 2-fluoroalkyl-4-acetylresorcinol were made which were useful as antiallergy and antiinflammatory agents. For example, compound **8-119**, in tests of leukotriene D_4-induced contraction of guinea pig ileum, had an IC_{50} 0.012 μM/L.[80]

Certain (furanyl)pentenylchromene derivatives were prepared and shown to be anticancer agents. For example, compound **8-120** in vitro inhibited mouse lymphatic leukemic cells with an IC_{50} of 4.0×10^{-3} μmol/mL.[81]

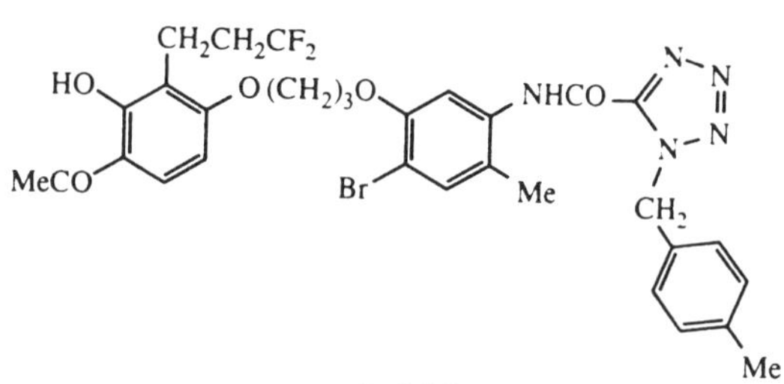

8-119

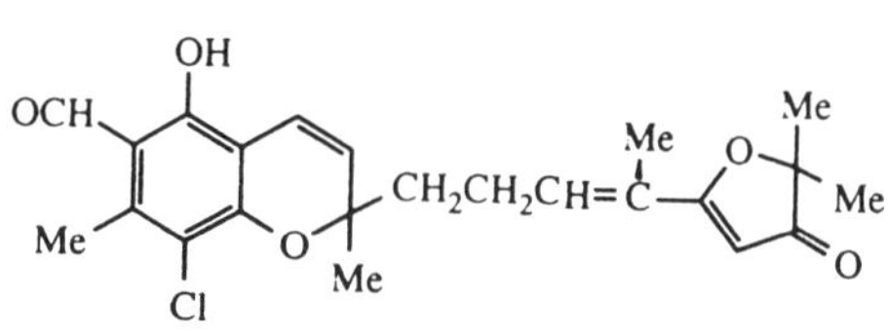

8-120

Novel amides of mycophenolic acid were synthesized and formulated for use in the treatment of inflammatory diseases such as rheumatoid arthritis. An example of such agents is compound **8-121**.[82]

A new synthesis of the analgesic drug, meptazinol **8-122**, was developed which started from 1,3-cyclohexanedione.[83]

2-Arylimidazole derivatives were made and shown to be useful cardiovascular agents. Thus, compound **8-123** given at 0.6 mg/kg i.v. in cats reduced blood pressure by 40 mm Hg.[84]

Isoliquiridigenine, **8-124**, was patented in pharmaceutical formulations for the treatment and prevention of liver diseases and renal failure.[85]

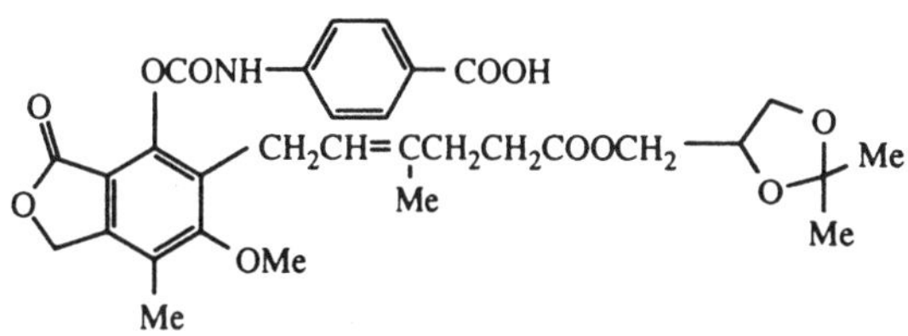

8-121

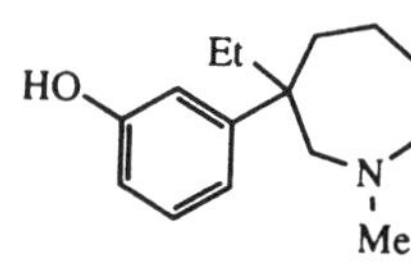

8-122

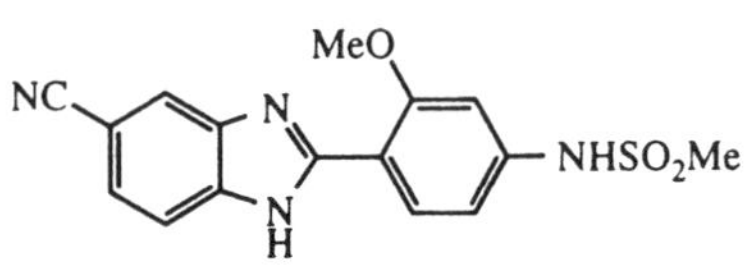

8-123

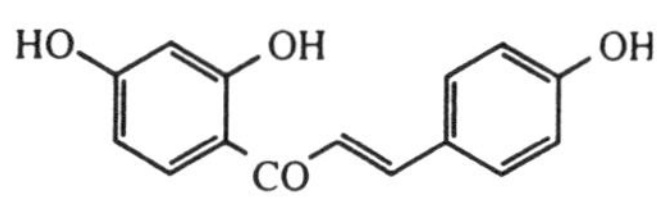

8-124

Yet another diverse and large series (the patent has 93 examples and 43 claims) of ethers of 2-propyl-4-acetylresorcinol was developed and reported to be useful for the treatment of allergic conditions such as asthma, and cardiovascular diseases such as angina and arrhythmias. Just one example of this series is compound **8-125**.[86]

Novel cyclohexanealkyl ethers of 2-propyl-4-acetylresorcinol were prepared which were antagonists of leukotriene D4, C4, and E4, useful as antiasthmatic, antiallergic, antiinflammatory, and cytoprotective drugs. For instance, compound **8-126** at 10 μg/mL inhibited by 62% the ileum contraction in guinea pigs caused by leukotriene D_4.[87]

The biotynilated psoralen derivative, **8-127**, was a neoplasm inhibitor; given in vitro at 1 μg/mL, followed by irradiation with near-UV light, this compound inhibited lymphocyte proliferation by 99%.[88]

Lilly's compound AR-L100, **8-128**, was synthesized and evaluated; it was a potent, nonglycoside inhibitor of Na^+, K^+ ATPase (it is an important pharmacological tool).[89]

Leukotriene-inhibiting benzoyl- and alkanoylresorcinol derivatives were prepared and shown to be useful for the treatment of inflammation. For instance,

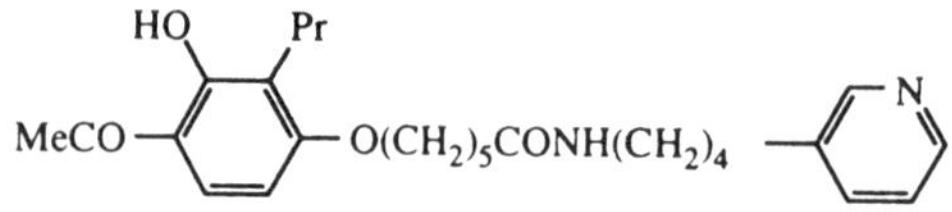

8-125

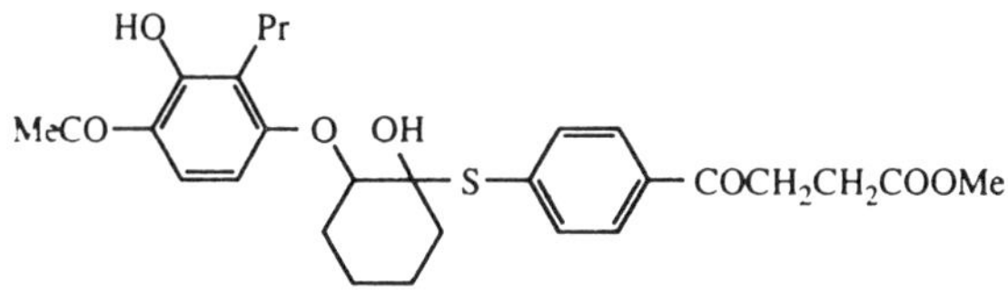

8-126

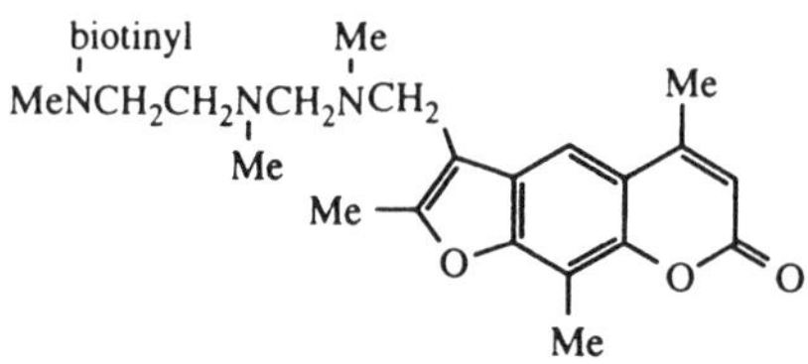

8-127

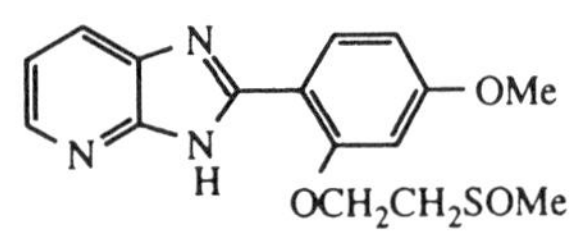

8-128

compound **8-129**, applied topically at 1%, gave 60% inhibition of arachidonic acid-induced ear edema in mice. Precursors of compound **8-129** inhibited the binding of leukotriene B_4 to peripheral human neutrophils by 96% at 10^{-6} M.[90]

Alkoxy-substituted dihydrobenzopyran-2-carboxylates and analogs were prepared and shown to be highly effective antiinflammatory agents. Compound **8-130**, for instance, was five times more potent as a leukotriene B4 antagonist than the hydroxy analog of **8-130** known in the prior art.[91]

5-Hydroxy-3-aminochromans were made, and their enantiomers and salts had a therapeutic effect on the central nervous system and a selective effect on the 5-hydroxytryptamine receptor in mammals. Thus, they were useful in treating depression, anxiety, anorexia, senile dementia, Alzheimer's disease, hypertension, thermoregulatory, and sexual disturbances. An example of these compounds is structure **8-131**.[92]

A process for the preparation of trioxsalen, **8-38**, starting from 7-allyloxy-4,8-dimethylcoumarin, was patented.[93] Trioxsalen is a photosensitizer/pigmentation agent (see the earlier section on commercial products).

Complex ethers of 2-propyl-4-acetylresorcinol were patented as antagonists of leukotrienes and SRS-A (slow-reacting substance of anaphylaxis), which were therefore useful as antiasthmatics, antiinflammatories, and allergy inhibitors. Compound **8-132** is an example of the compounds in this series.[94]

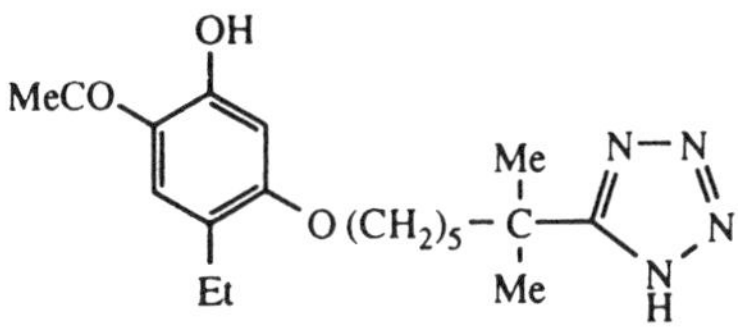

8-129

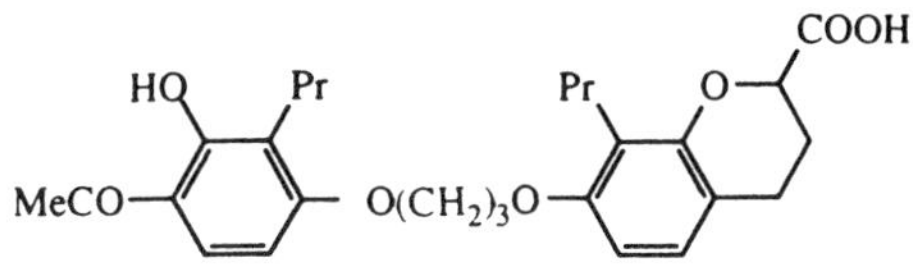

8-130

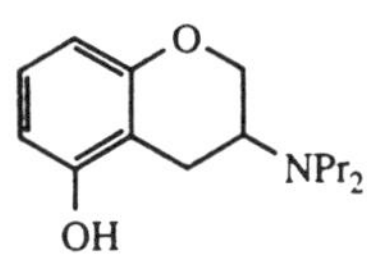

8-131

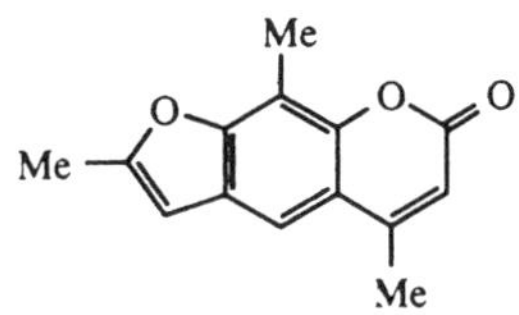

8-38

A new type of cannabinoid compounds, named carenadiols, was prepared and said to be useful in the treatment of patients with various neurological disorders. The compounds were said to have potential application in treating epilepsy, hyperkinetic movement disorders, skeletal muscle spasticity (e.g., in multiple sclerosis and spinal cord injuries), and action tremors. Compound **8-133** is one example of these new cannabinoids.[95]

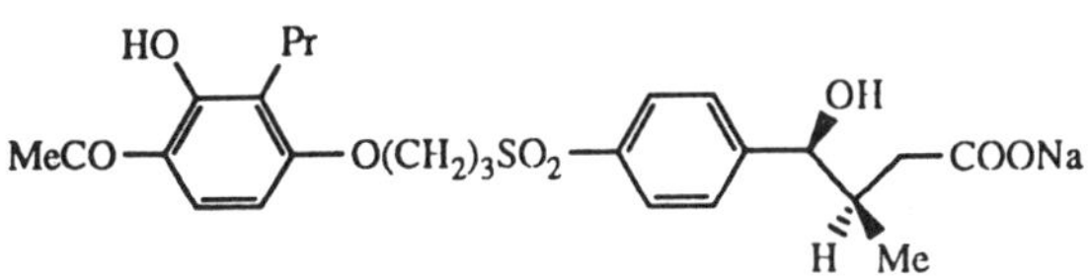

8-132

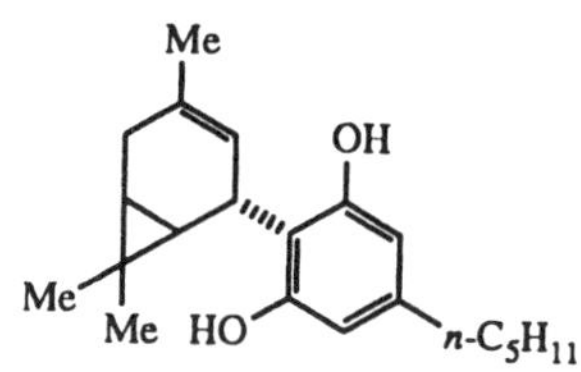

8-133

In another variation on the cannabinoid theme, compounds such as **8-134** were synthesized, tested in mice, and found to be potentially useful in man as analgesics and antiemetics.[96]

Benzofurano-[3,2-*c*]quinolines were made and proven to have bone resorption-inhibiting properties, aimed at treating osteoporosis. Thus, compound **8-135** was given at 300 mg/kg daily for two weeks to rats raised on a low-Ca diet. At the end of the test the rats had 8.1–0.5 mg Ca and 5.1–0.3 mg P/femur, while the unmedicated controls had 4.2–0.6 mg Ca and 3.2–0.4 mg P/femur.[97]

A series of 9-amino-3,4-dihydroacridines and related compounds were prepared, starting with 1,3-cyclohexanedione, and tested in cholinesterase inhibition assays and

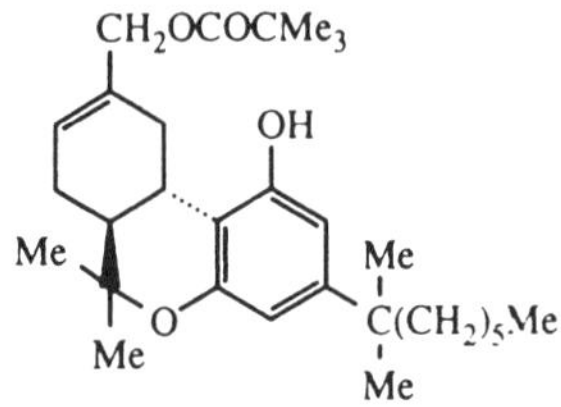

8-134

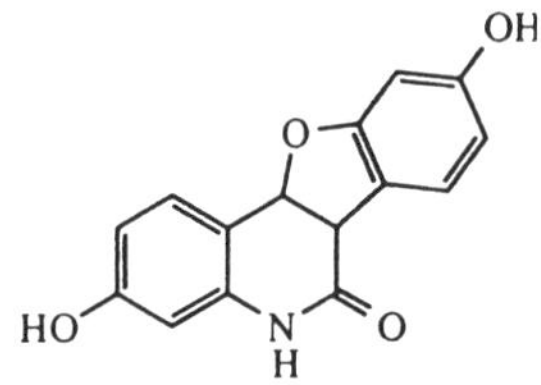

8-135

dark avoidance assays of mice. As a result of these tests, the novel compounds were indicated to be useful for the treatment of memory dysfunctions characterized by decreased cholinergic function, such as Alzheimer's disease. One simple example of this series is compound **8-136**.[98]

3-Bromo-2,6-dimethoxy-(*N*-nortropenyl)benzamide derivatives were made and shown to be tranquilizers with reduced extrapyramidal side effects. For instance, compound **8-137**, given orally to mice, inhibited apomorphine-induced climbing behavior with an ED_{50} 0.33 mg/kg.[99]

Quinoline-2,5-diones were prepared, starting with 1,3-cyclohexanedione, and found to be useful as analgesics, antiinflammatories, and antipyretics. For instance, compound **8-138**, in a heat-induced pain test in mice, doubled reaction times when given orally at 59–155 mg/kg.[100]

Certain 2-cyclohexylidenedithiolanes were synthesized and shown to be liver function protectants. For example, mice treated with 30 mg/kg of compound **8-139** 6 hr before 0.05 mL/kg of carbon tetrachloride was administered showed no liver damage, while significant liver disorder was found when **8-139** was not given.[101]

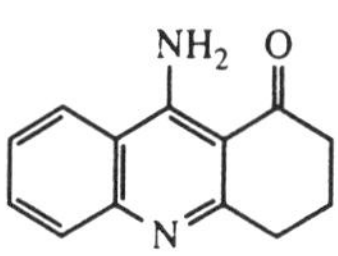

8-136

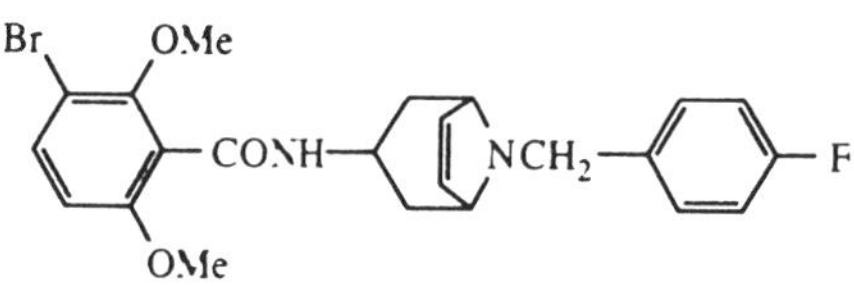

8-137

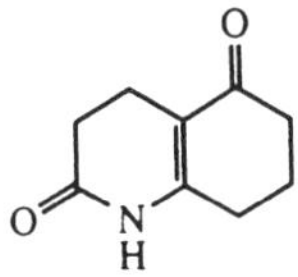

8-138

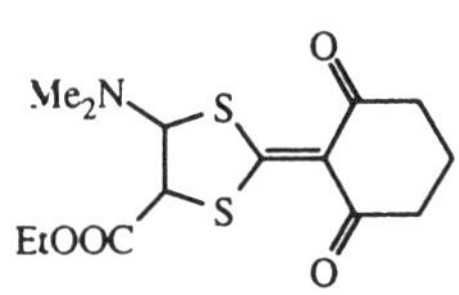

8-139

Surprisingly, simple derivatives of 2-propyl-4-acetylresorcinol were found to be useful as antirheumatics. Thus, compound **8-140**, given at 100 mg/kg to rats, gave 39% inhibition of adjuvant-induced arthritis.[102]

(Propylaminopropoxy)aurones were prepared and indicated to be allergy inhibitors. Thus, compound **8-141** at 1 μM gave 16% inhibition of leukotriene C_4-induced contractions of guinea pig ilial strips.[103]

Phenoxyalkoxy ethers of di- and tetrahydronaphthols were prepared and demonstrated to be antagonists of leukotriene D, therefore useful as antiallergy and antiinflammatory drugs. Compound **8-142** exemplifies these new products.[104]

The 2-allyl-4-acetylresorcinol derivative, **8-143**, and related compounds were made and evaluated as antirheumatics. Thus, compound **8-143**, at 100 mg/kg p.o., gave 35.5% inhibition against edema of adjuvant arthritis in rats. The toxicity of these compounds was relatively low, with an LD_{50} > 1000 mg/kg p.o. in mice.[105]

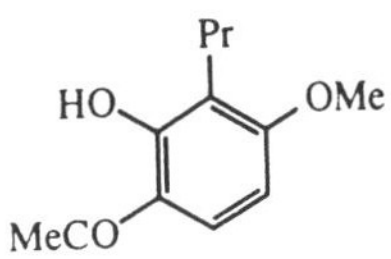

8-140

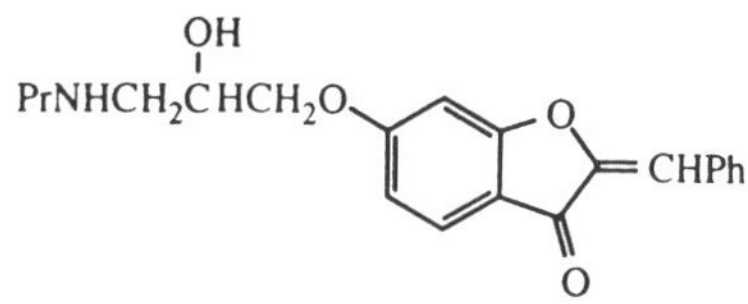

8-141

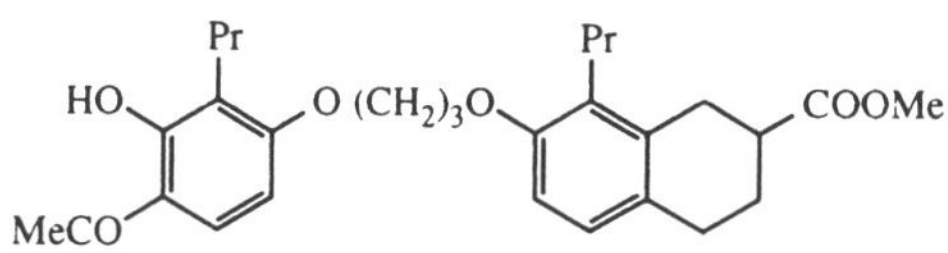

8-142

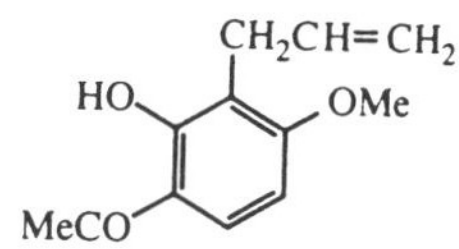

8-143

Benzofuranyloxy-alkylamines and related compounds were prepared and found to be useful for treating arrythmia, myocardial infarction, angina pectoris, and cardiac failure in animals; these compounds were also dopamine and serotonine antagonists. For instance, compound **8-144** had an APD_{75} (action duration potential) of 31 μg/mL.[106]

3-Cyano-7-alkoxycoumarins, e.g., **8-145**, were prepared for use in fluorometric assays for the determination of cytochrome P-450 mixed function oxidases. Thus, compound **8-145** was contacted with an enzyme sample and the fluorescence intensity, caused by ether cleavage, was measured.[107]

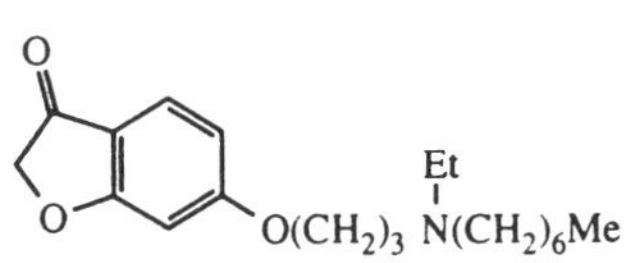

8-144

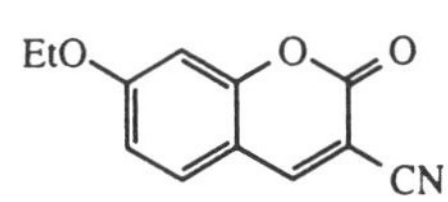

8-145

Oxaloamino-substituted fluorinated resorcinol ethers were synthesized and claimed to be useful as antiallergic and antiinflammatory agents; formulations for application by aerosol were given. Compound **8-146** is an example of drugs in this series.[108]

Resorufin (and resazurin)-aminoglycosides, such as **8-147**, were made and patented for use in the determination of *N*-acetylglucoamidinase. Reagent kits were prepared.[109]

[(Phenoxyalkoxy)phenoxy]alkanoates were demonstrated to be highly effective antiallergenics. Thus, compound **8-148**, given orally at 50 mg/kg, gave 94.8% inhibition of leukotriene D_4-induced bronchoconstriction in guinea pigs.[110]

Poly(prenyl)phenyl sulfates were prepared and were said to be useful as reverse transcriptase inhibitors for the treatment of Aids, as antiinflammatories, antiallergic agents, and as drugs to treat acute pancreatitis and gastric diseases. For instance, compound **8-149** in vitro had an IC_{50} 1.5×10^{-8} M against reverse transcriptase of avian myeloblastomatosis virus.[111]

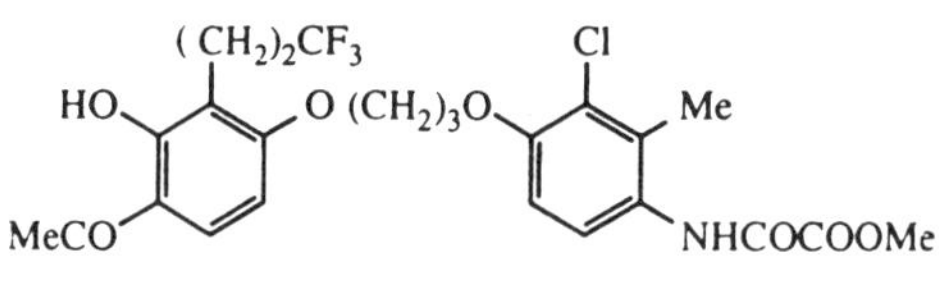

8-146

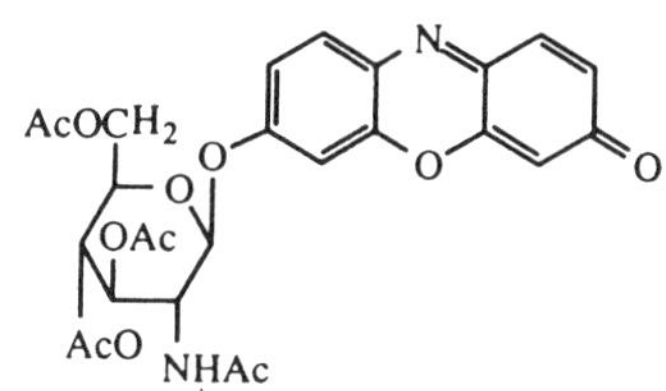

8-147

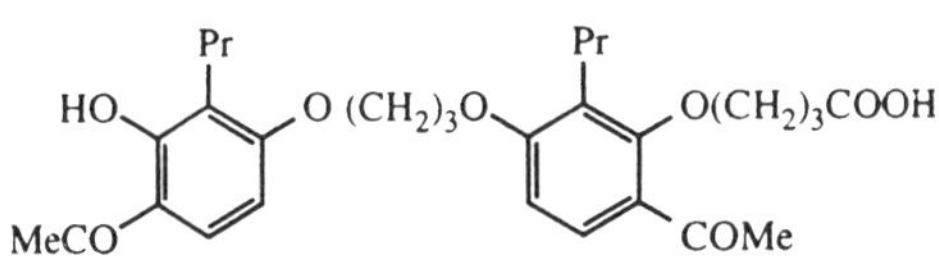

8-148

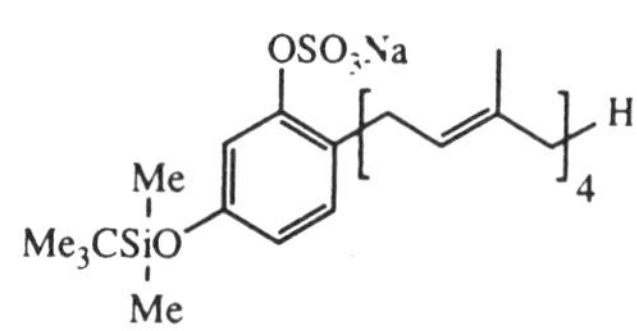

8-149

New annelated indole derivatives were prepared, starting from 1,3-cyclohexanedione, and converted to their pharmaceutically acceptable salts which were useful as selective antagonists of neuronal 5-HT receptors. An example of this group is compound **8-150**.[112]

A stereoselective synthesis of benzenebutanoic acid derivatives, such as **8-151**, was developed; these compounds were leukotriene antagonists.[113]

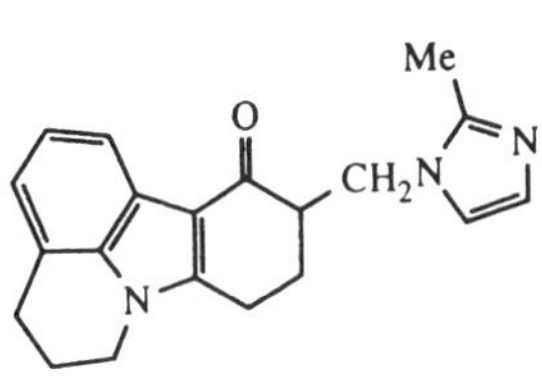

8-150

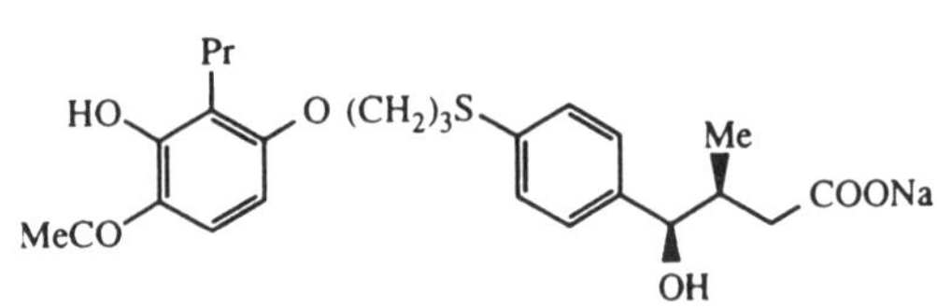

8-151

Physiologically active substance TAN-931 was isolated from cultures of *Penicillium funiculosum*. This substance and its derivatives were aromatase inhibitors which inhibit estrogen synthesis and can lower the estrogen level in the blood of mammals. These compounds were intended for the control of conception, the treatment and prevention of breast cancer, cancer of the uterus and ovaries, and of endometriosis. For example, compound **8-152** in vitro inhibited the aromatase from a homogenized human placenta microsome with IC_{50} 4.3 μg/mL.[114]

Benzoxazolones were synthesized and inhibited both lipoxygenase and cyclooxygenase. These compounds, such as **8-153**, were useful antiallergy and antiinflammatory drugs.[115]

N-[(1-Ethylpyrrolidinyl)methyl]-2,6-dihydroxybenzamide derivatives were prepared and were suggested to be useful for treating emesis, psychiatric disorders, and psychosomatic conditions. Representative of this group is structure **8-154**.[116]

A large number (154 examples in the patent) of ethers of 2-alkyl- or alkenyl-4-acetylresorcinols were prepared and demonstrated to have effective activity in vivo in test systems designed to detect antiinflammatory agents. These tests included carrageenin assays, arachidonic acid-induced ear edema assays, thymocyte assays, mouse endotoxin shock assays, and inhibition of binding ^{3}H-leukotriene B_4 to peripheral human neutrophils tests. Compound **8-155** is an example of an effective antiinflammation agent in this series.[117]

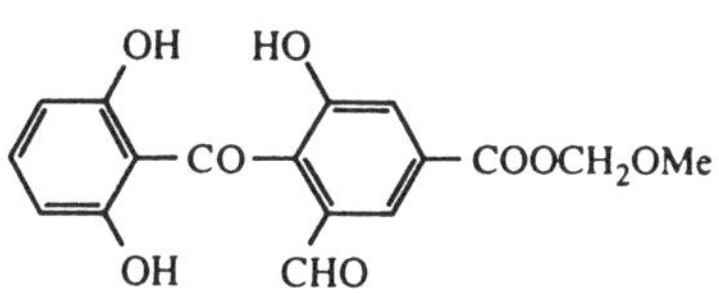

8-152

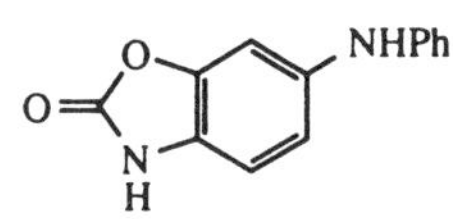

8-153

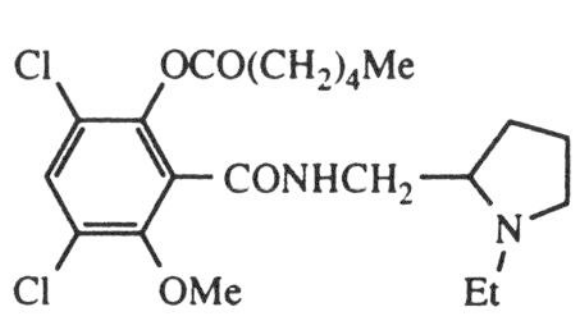

8-154

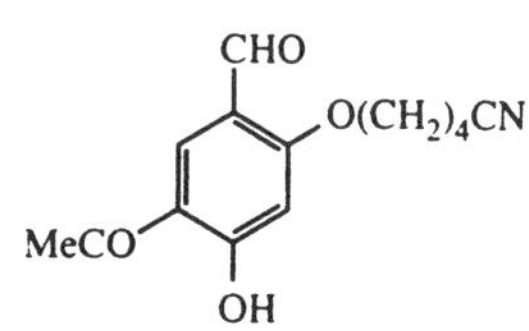

8-155

Quinindolines were prepared and indicated as antidepressants and anxiolytics; e.g., compound **8-156** was shown to have anxiolytic activity at 20 mg/kg p.o. in rats.[118]

7-Carboxy-4-phenylcoumarins were prepared and demonstrated to be aldose reductase inhibitors; such agents are useful for the treatment of diabetic retinopathy, cataracts, nervous disorders, and kidney disorders. Compound **8-157** is an example of these drugs.[119]

Novel bis-aryl phosphate esters were made which were antagonists of PAF (platelet activating factor) induced problems. Compound **8-158**, for instance, inhibited

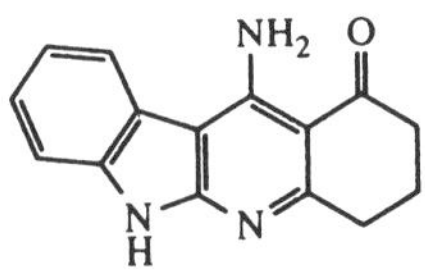

8-156

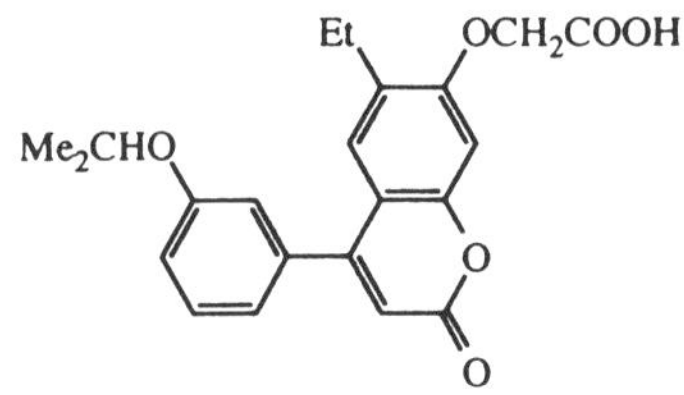

8-157

PAF-induced aggregation in platelet-rich plasma of rabbits, with an IC_{50} 2.13×10^{-6} M; this compound also protected against PAF-caused hypotensive shock in mice and reduced endotoxin-induced hypertension in rats.[120]

Alkylresorcinols, such as 4-isoamylresorcinol, were patented as skin depigmentation agents; they were said to inhibit oxidase activity which contributes to the formation of melanin.[121]

A composition containing 0.1–0.3 wt% salicylic acid, 1.0–2.0 resorcinol, 2.0–2.5 camphor, 0.2–0.5 menthol, and 1.0–2.0 thymol, the rest ethyl alcohol, was patented as an agent to treat gingivitis (inflammation of the gums).[122]

Substituted benzothiazoles were synthesized for use as antitumor agents. Thus, compound **8-159** (as the perchlorate), at 10 mg/kg/day in mice transplanted with P388 tumor cells, gave a 128% increase in the survival rate.[123]

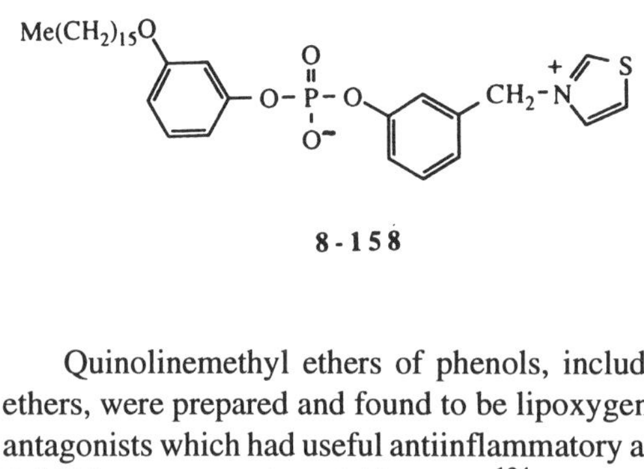

8-158

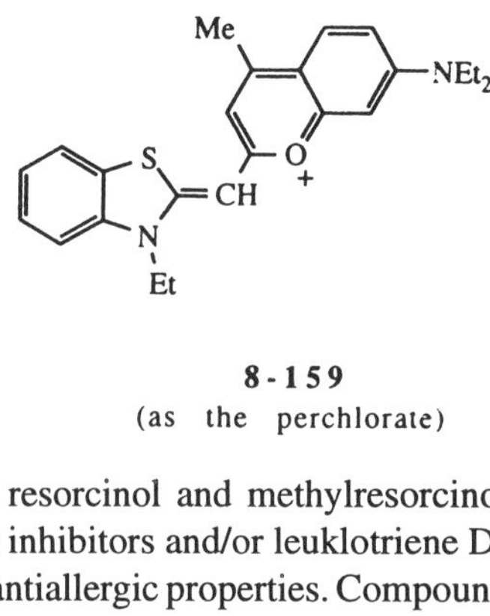

8-159
(as the perchlorate)

Quinolinemethyl ethers of phenols, including resorcinol and methylresorcinol ethers, were prepared and found to be lipoxygenase inhibitors and/or leuklotriene D_4 antagonists which had useful antiinflammatory and antiallergic properties. Compound **8-160** is representative of this group.[124]

New ethers, containing a dihydroxybenzofuran segment, of 2-propyl-4-acetylresorcinol were prepared and shown to be useful as allergy inhibitors. For instance, compound **8-161** was tested in vitro for SRS-A antagonism, using guinea pig ileum, and found to have an IC_{50} 3×10^{-7} M.[125]

Xanthene derivatives were made and found to be leukotriene antagonists. For example, compound **8-162** gave 25% inhibition of the binding of [^{3}H]-leukotriene B_4 to human peripheral neutrophils at 10^{-6} M.[126]

Anticancer agents containing 4-alkylresorcinols were patented. As an example, 4-*n*-hexylresorcinol (0.5 wt% in DMSO), given i.p. at 0.1 mL/10 g body wt for 10 days, inhibited the growth of Cloudman S-91 melanoma cells in mice. The LD_{50} of 4-*n*-hexylresorcinol was determined to be >500 mg/kg p.o. in mice.[127]

7-Aminoalkoxycoumarins were synthesized and recognized as useful for the treatment of cerebrovascular disorders. For instance, compound **8-163** at 100 mg/kg dose given orally to mice gave 100% inhibition of reserpine-induced ptosis (drooping of the upper eyelid).[128]

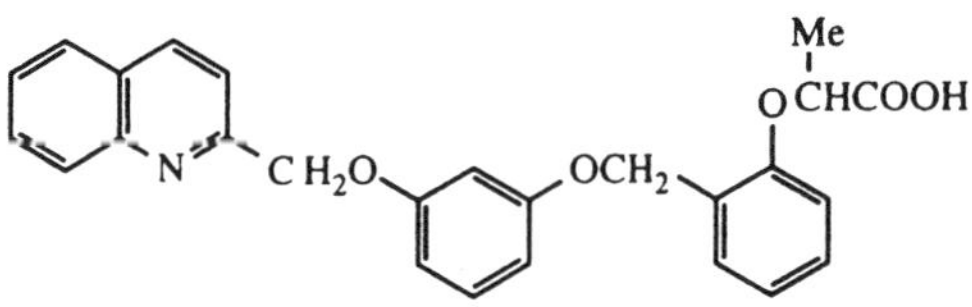

8-160

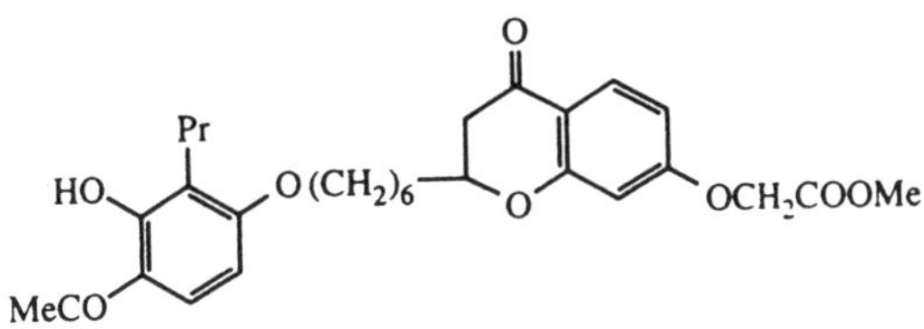

8-161

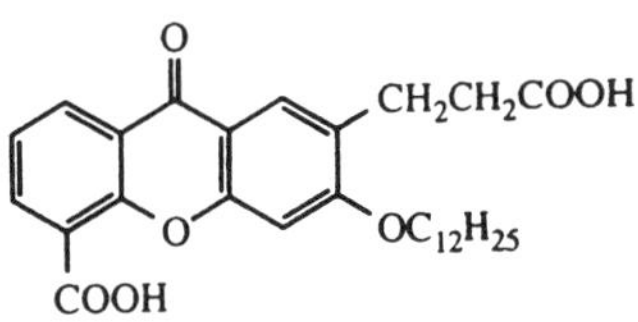

8-162

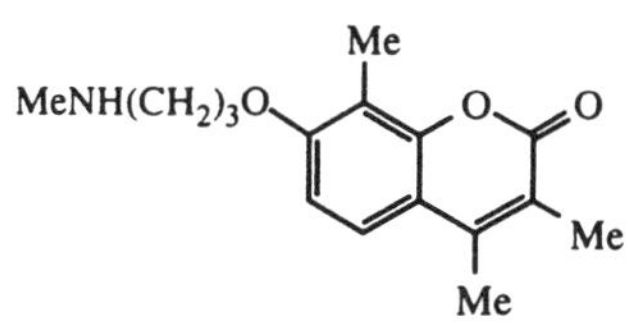

8-163

Fluorescent amino acid derivatives, such as compound **8-164**, were prepared and shown to be useful in the amino acid sequence determination of proteins and peptides.[129]

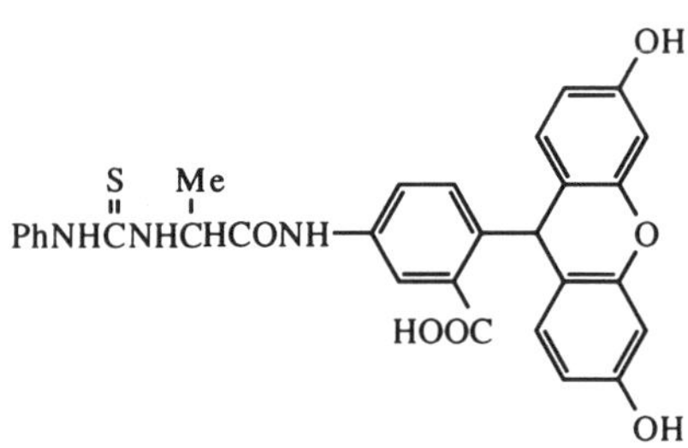

8-164

Many chalcone derivatives containing a resorcinol moiety were prepared and shown to be useful as antiulcer agents, acting as gastric mucosal protectives and gastric secretion inhibitors. Thus, compound **8-165**, given orally at 100 mg/kg, inhibited 93.4% of stress-induced stomach ulcers in rats.[130]

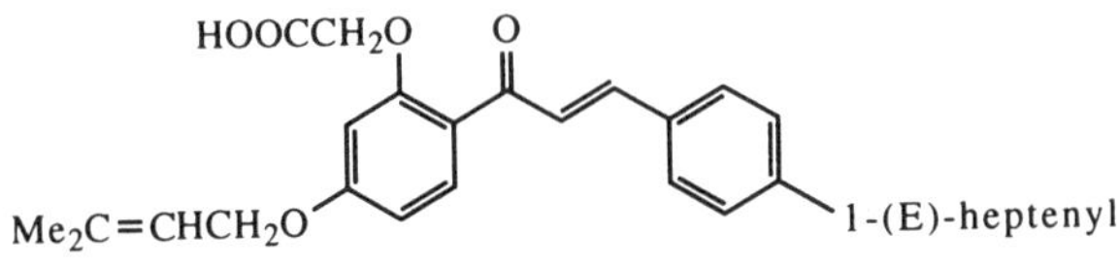

8-165

2-Propylresorcinol ethers containing a 7-alkoxydihydrobenzopyran-2-carboxylic acid portion were developed and were leukotriene B_4 antagonists useful as antiinflammatory agents and in the treatment of LTB4-mediated conditions. An example of this group is compound **8-166**.[131]

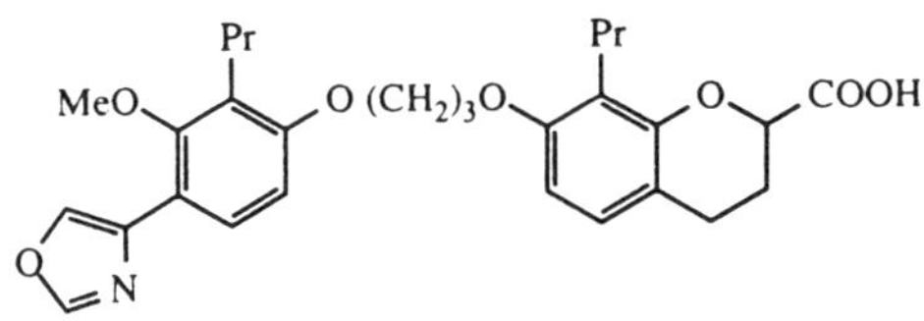

8-166

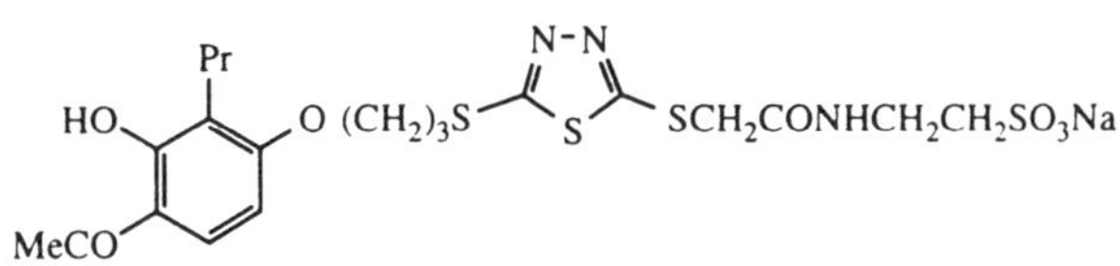

8-167

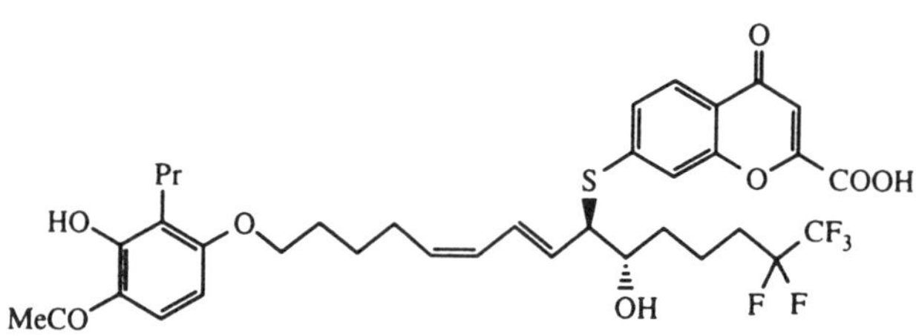

8-168

2-Propyl-4-acetylresorcinol ethers containing taurin amide derivatives were leukotriene antagonists with good water solubility. Thus, compound **8-167** showed an IC_{50} 1.5×10^{-7} M against LTD_4 in guinea pigs in vitro.[132]

Complex substituted, long-chain alkenyl ethers of 2-propyl-4-acetylresorcinol were prepared and shown to be useful as antiinflammatories and antiallergics. For instance, compound **8-168**, in the form of a 0.001–1.0 wt% spray, antagonized leukotriene D_4 in the standard bronchoconstriction test in guinea pigs and also antagonized phospholipase A_2 at 0.5–50 μM.[133]

Ethers of diresorcylalkanols were synthesized and shown to be effective as inhibitors of lipooxygenase, useful in treating allergies and inflammation. For instance, compound **8-169** at 15 μM showed 99% inhibition of 5-hydroxy-eicosatetronic acid and 97% inhibition of leukotriene B_4.[134]

Flavonoids of the basic structure **8-170** were extracted from mulberry or licorice and purified by column chromatography; these flavonoids were found to be inhibitors of sodium, potassium-activated ATPase. One such compound, mulberrofuran A, at 100 μM gave 97% inhibition of sodium, potassium ATPase; even at 1.0 g/kg, given p.o. to mice, no deaths were observed.[135]

Tricyclic benzoxazine derivatives were prepared and shown to be potential anticancer agents. As an example, compound **8-171** was claimed to be effective at 1–100 mg/day.[136]

2-[(3-Aminophenoxy)methyl]morpholine derivatives were developed and indicated to be brain function improvers. Thus, compound **8-172**, at 100 mg/kg p.o. in mice, showed 10% inhibition of reserpine.[137]

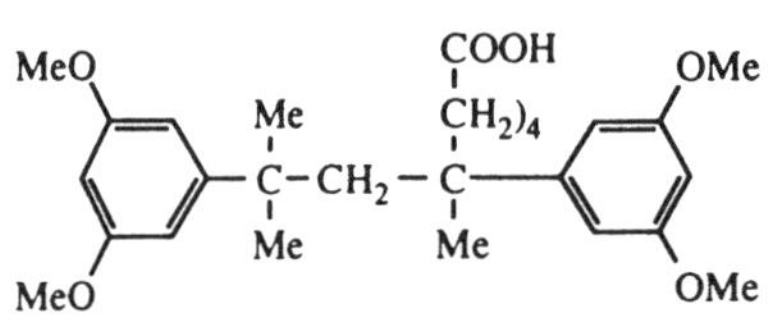

8-169

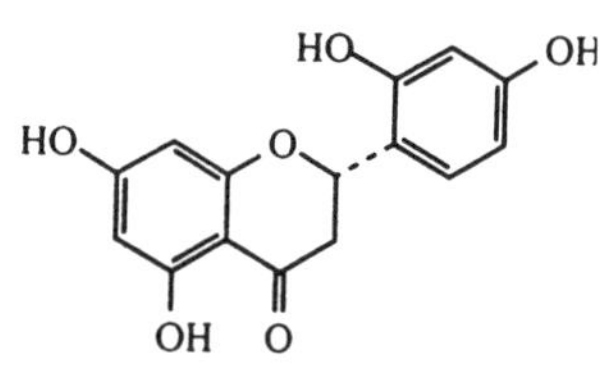

8-170

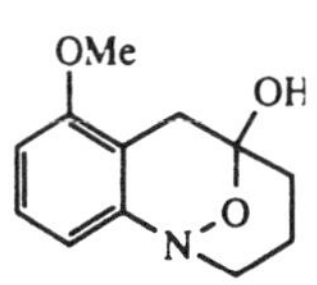

8-171

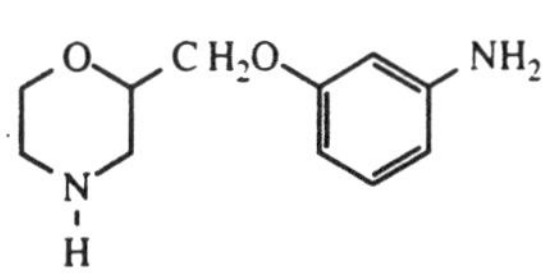

8-172

Benzopyranyl beta-D-thioxyloside analogs were synthesized and claimed as antithrombotics. As an example, compound **8-173**, given to rats at 3 mg/kg p.o. showed 65% inhibition of activated factor X-induced hypercoagulation.[138]

Substituted tricyclic compounds, such as xanthenes and fluorenes, were prepared, tested, and found to be useful leukotriene inhibitors, therefore useful in the treatment of inflammatory diseases and allergic disorders such as asthma. For example, xanthone **8-174** at 10^{-6} M gave 99% inhibition of the binding of ^{3}H-LTB_4 to peripheral human neutrophils. Suitable pharmaceutical forms of these drugs as aerosols, capsules, tablets, etc., were prepared.[139]

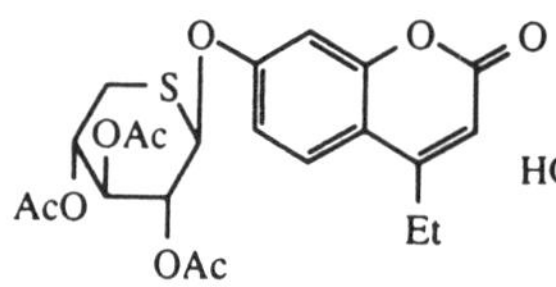

8-173

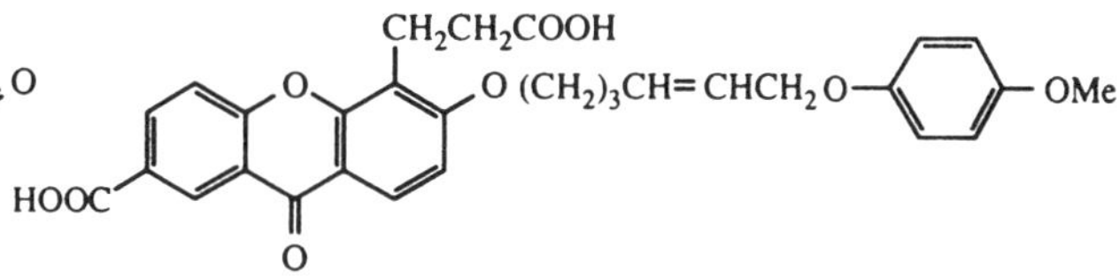

8-174

A hair growth-stimulating preparation was patented; it contained boric acid 8.000, resorcinol 4.250, salicylic acid 2.500, tannin 0.250, lysol 0.050, and phenol 0.025 g/L.[140]

Terpenylresorcinol compounds were made which were tyrosinase inhibitors, antioxidants, and antibacterial agents. For example, a mixture of compounds **8-175** and **8-176** had an ID_{50} 0.18×10^{-4}M as tyrosinase inhibitors, was effective against *Staphylococcus aureus* at 0.1%, and was effective as an antioxidant at 0.01% in the rhodan-iron test.[141]

A silyl-substituted 2,4-dihydroxybenzophenone, **8-177**, was synthesized and had utility in creams, foundations, etc., preventing sunburn at 2%.[142]

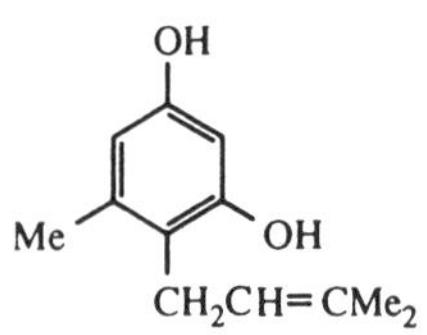

8-175

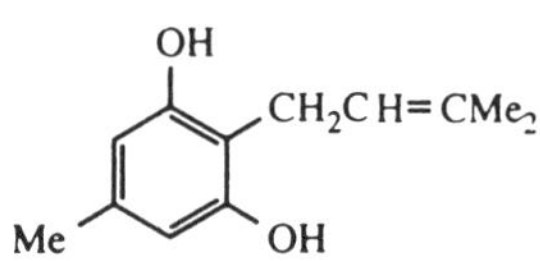

8-176

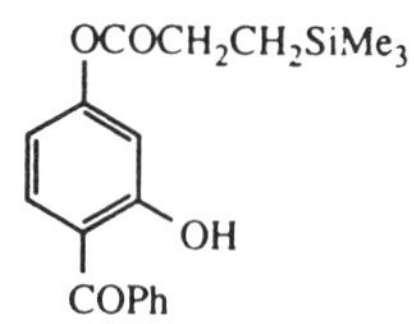

8-177

Pyridylpyrrolothiazole derivatives were prepared and shown to be effective as platelet activating factor (PAF) antagonists. Thus, compound **8-178** in vitro had an IC_{50} 4.5×10^{-8} M for the inhibition of PAF-induced coagulation of rabbit blood platelets.[143]

Topical preparations containing resorcinol glycosides were patented. Thus, a composition containing vaselin 5.0, liquid paraffin 15.0, cetanol 5.0, glycerin monostearate 2.0, polyoxyethylene sorbitan monostearate 2.0, butyl *p*-hydroxybenzoate 0.2, perfumes 0.2, 1,3-butylene glycol 10.0, di-Na edetate 0.1, water 5.75, and *m*-hydroxyphenyl beta-D-glycoside 3.0 parts by weight was mixed to give a nonirritating skin cream, which promoted wound healing and had a skin-lightening effect.[144]

A great many 3,4-dihydro-7-alkoxy-benzopyran-2-alkanoates and analogs were prepared, tested, and found to be leukotriene B_4 antagonists. An example of these products is compound **8-179**, which was 8.9 times more effective than a similar compound of the prior art in inhibiting in vitro the binding of LTB4 at human neutrophils.[145]

A series of aminophenoxyphenoxypropyl-tetrazole derivatives were useful in the treatment of bronchial asthma and allergic rhinitis. For instance, compound **8-180** showed LTD4 binding inhibition with an IC_{50} as low as 3.58×10^{-8} M.[146]

4,5,6,7-Tetrahydroindole derivatives such as compound **8-181** were prepared and found to have antagonistic action against serotoninergic and dopaminergic receptors; thus, they were useful as psychotropic and antihypertensive agents.[147]

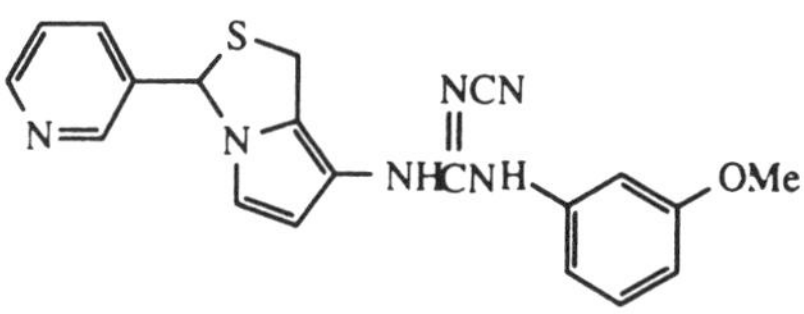

8-178

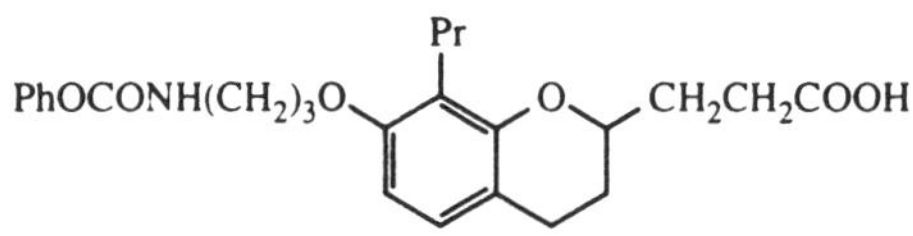

8-179

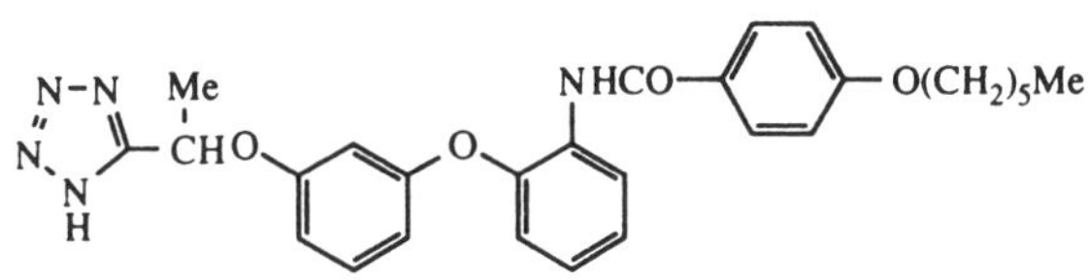

8-180

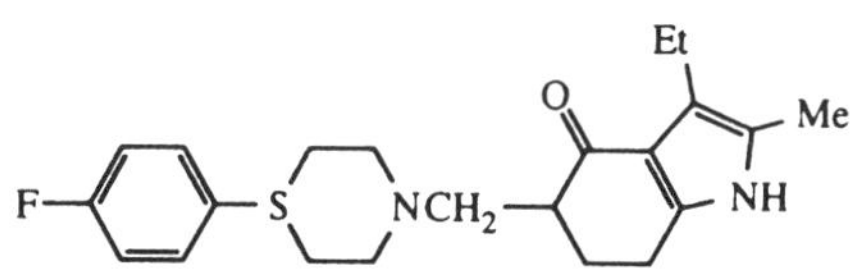

8-181

Certain alkyl-arylureas were synthesized, screened, and found to be useful as antiartherosclerotics. Compound **8-182**, for instance, fed at 50 mg/kg to rats on a cholesterol-cholic acid diet, reduced LDL-cholesterol by 92%.[148]

A novel anticancer agent, resothiomycin **8-183**, was manufactured by cultivation of *Streptomyces* 4511-6 in an oat meal and yeast extract-containing medium at 27 °C for 120 hr; the cells were then removed by filtration and the filtrate was worked up to obtain the active principle. Compound **8-183** had an IC_{50} 15.5 µg/mL against murine leukemia cell L5178Y; the substance also enhanced the anticancer activity of vincristine and actinomycin D.[149]

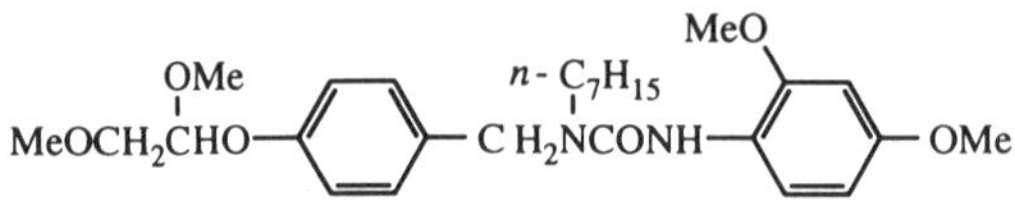

8-182

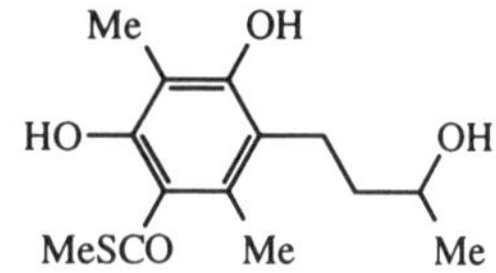

8-183

A substituted aryl ether of 2,6-dihydroxybenzoic acid, compound **8-184** (and similar products), was an endothelin receptor antagonist useful for treating myocardial infarction and renal insufficiency. This substance was obtained by fermentation and showed, at 50 mg/kg i.v. in rats, a 47% inhibition of the pressor effect induced by endothelin-I.[150]

Antibiotic M143-37F11, previously obtained as a fermentation product, is now produced from the synthetic intermediate, **8-185**.[151]

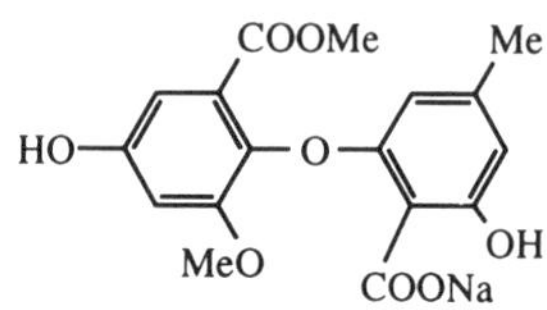

8-184

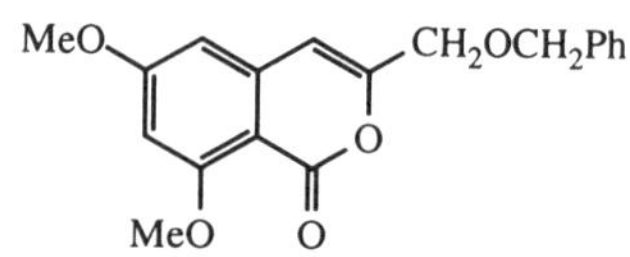

8-185

p-Aminosalicylic acid derivatives of thiazole were synthesized and found to be antiinflammatory agents. An example of these drugs is compound **8-186**.[152]

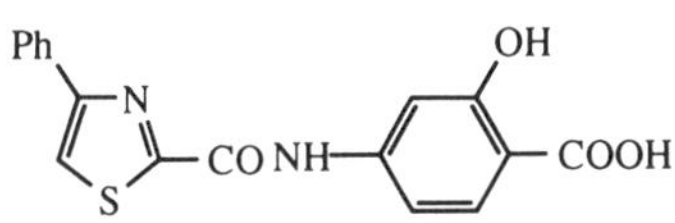

8-186

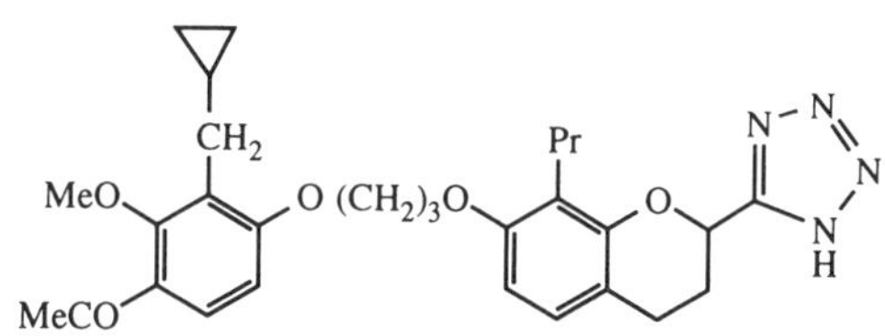

8-187

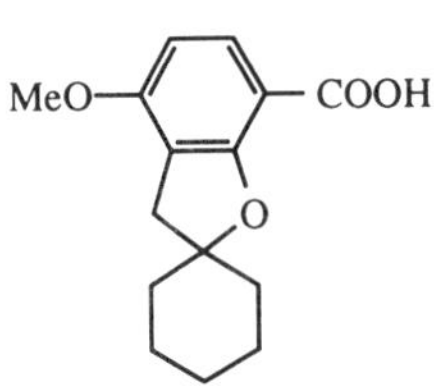

8-188

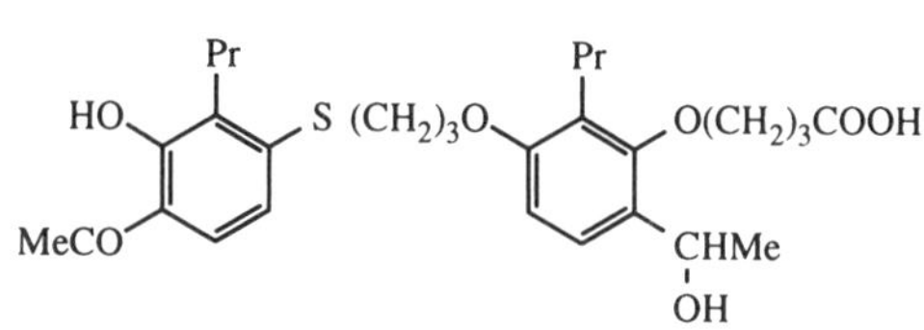

8-189

Tetrazolylbenzopyran derivatives were prepared and found to be useful as antiinflammatory compounds. For instance, compound **8-187** had a relative potency value of LTB4 antagonists of 219.[153]

Spirobenzofuran-cyclohexanes were prepared and reported to be useful as complement inhibitors, immunosuppressants, and antiinflammatories. For instance, compound **8-188** had an IC_{50} 5 mM for inhibition of complement C5a production.[154]

Phenoxyalkylcarboxylic acid derivatives were prepared and shown to be useful as antiallergy agents. Thus, compound **8-189**, given i.v. in guinea pigs at 1.0 mg/kg, showed 71% inhibition of leukotriene D_4-induced bronchoconstriction.[155]

Sesquiterpenes, such as compound **8-190**, useful for treating noninsulin-dependent diabetes, were obtained from *Stachybotrys* sp. K76 grown in cultured media.[156]

Porphyrin-metal complexes, **8-191**, which contained resorcinol segments, were prepared; they were useful as O_2 carriers.[157]

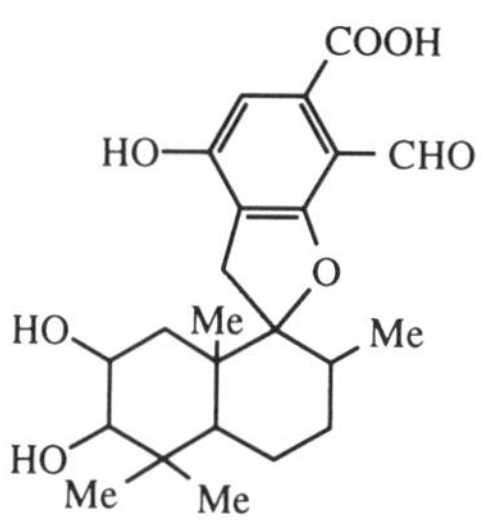

8-190

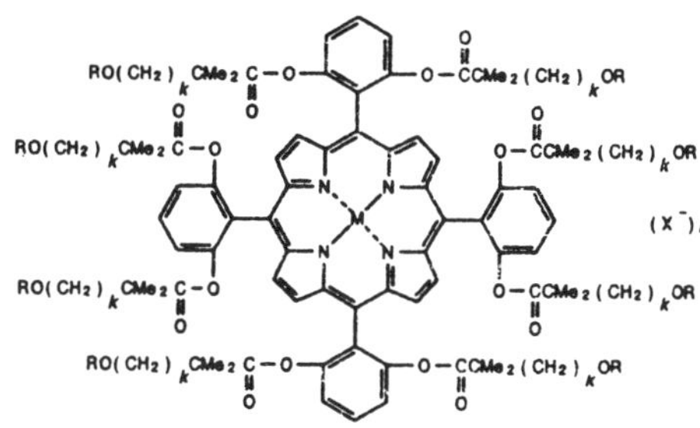

8-191

A stilbene trimer compound, **8-192**, was isolated from the tuber of *scirpus fluviatilis* var. *yagara*; this compound was an actylcholinesterase inhibitor (IC_{50} 2.88 μg/mL) 40 times more effective than physostigmine.[158]

Compositions containing 1% 4-isopropylresorcinol, **8-193**, were acne-preventative; this substance significantly inhibited *Propionobacterium acnes*.[159]

N-Acetyl-beta-D-glucosamine derivatives of 2,7-dichlorofluorescein, such as compound **8-194**, were synthesized as very sensitive substrates for determining *N*-acetyl-beta-D-glucoseaminidase.[160]

The complex *m*-aminophenol derivative, **8-195**, and similar compounds were patented as anticholesteremics, useful in preventing and treating hyperlipidemia and arteriosclerosis.[161]

Benzopyran derivatives, such as compound **8-196**, were prepared and shown to be useful as antiallergic agents.[162]

Derivatives of 5-aminochromone were prepared as potential antitumor agents. For instance, compound **8-197** had an IC_{50} 35 μM in vitro for inhibition of human breast adenocarcinoma MCF7 cells.[163]

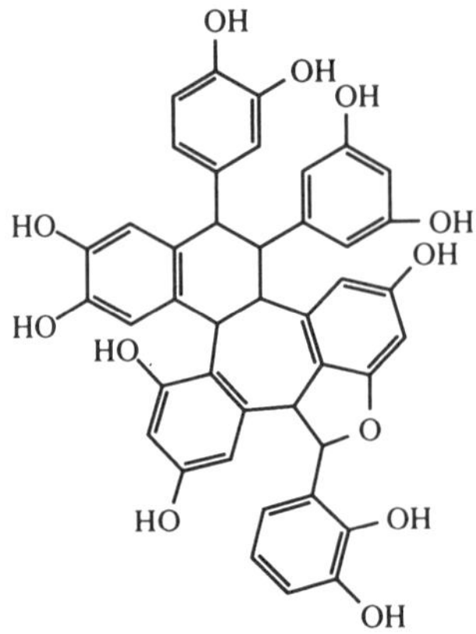

8-192

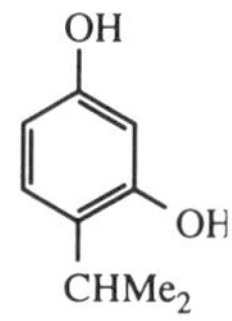

8-193

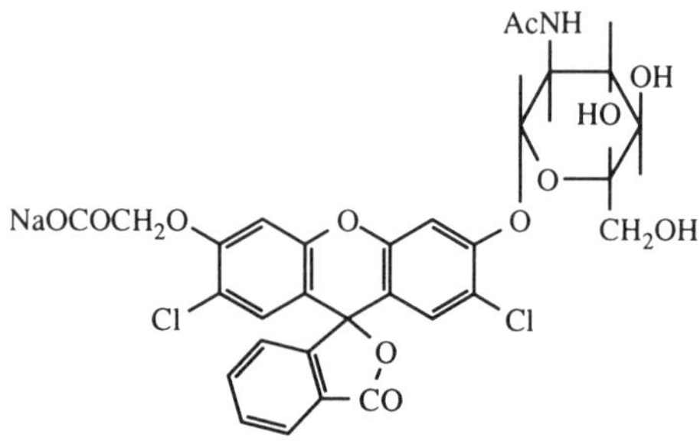

8-194

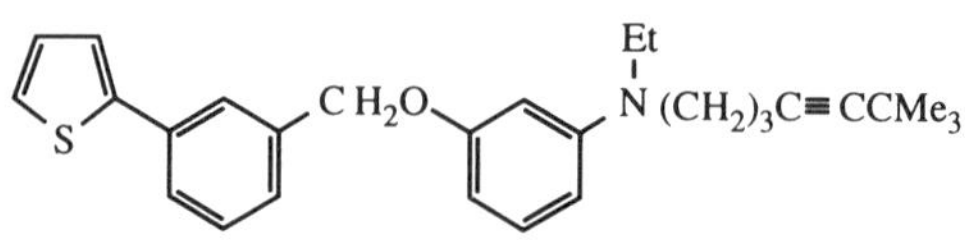

8-195

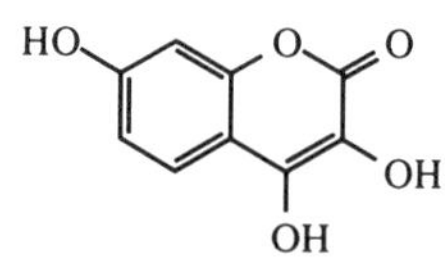

8-196

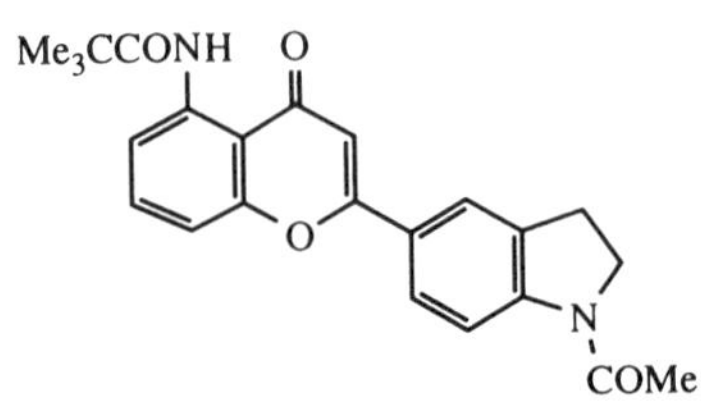

8-197

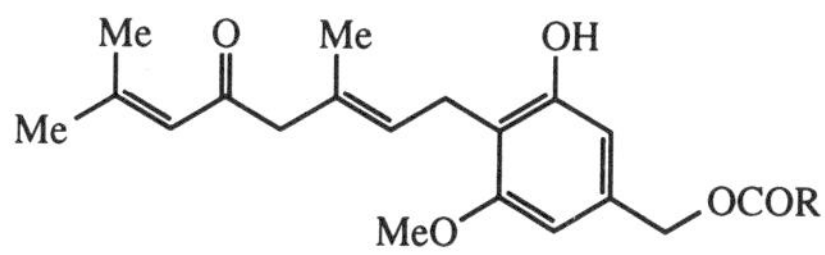

Hericenone C: R=palmityl
D: R=stearyl
E: R=linoleyl

8-198

Hericenone C, D, and E, **8-198**, were extracted from grains of *Hericinium erinaceum*. These compounds were useful for treating inflammation, pain, and senile dementia; the compounds were inhibitors of PGE_3 formation and inducers of nerve-growth factor.[164]

1-(Piperazinylalkyl)carbostyrils were synthesized and shown to be useful drugs. As an example, compound **8-199**, given at 30 mg/kg orally to mice exposed to head trauma, shortened the time required for recovery of spontaneous mobility to 20% of the recovery time of the controls.[165]

Biscoumarins were prepared as cardiovascular agents. Thus, compound **8-200** at 3 mg/kg i.v. in beagles showed high antithrombotic activity.[166]

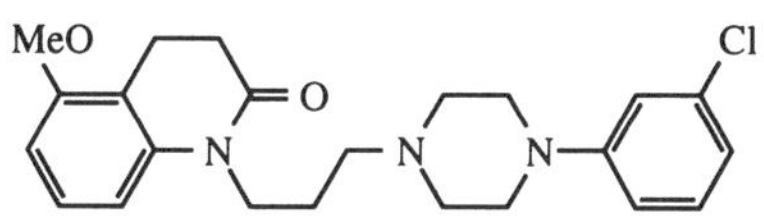

8-199

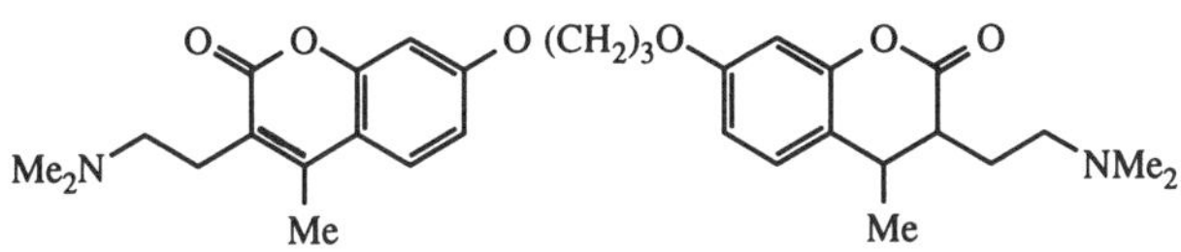

8-200

Mono- and bis-aryl phosphate antagonists of platelet activating factor (PAF) were developed. For instance, compound **8-201** (American Cyanamid CL 184005) was under development as a potential treatment for septic shock in humans.[167]

The 2-benzoyl-1,3-cyclohexanedione derivative, **8-202**, originally prepared for use as an agrochemical, was found to be useful as a drug for the treatment of tyrosinemia. This disease of children involves cirrhosis of the liver, tumors, rickets, and blood and kidney disorders; most patients die before adulthood.[168]

2,6-Dimethoxyxanthone, **8-203**, and similar compounds were synthesized and shown to have potent inhibitory action on arachidonate-induced platelet aggregation.[169]

Finally, it should be mentioned that there were many publications in the 1980–1992 period on antibiotics containing a resorcinol moiety (or moieties); these com-

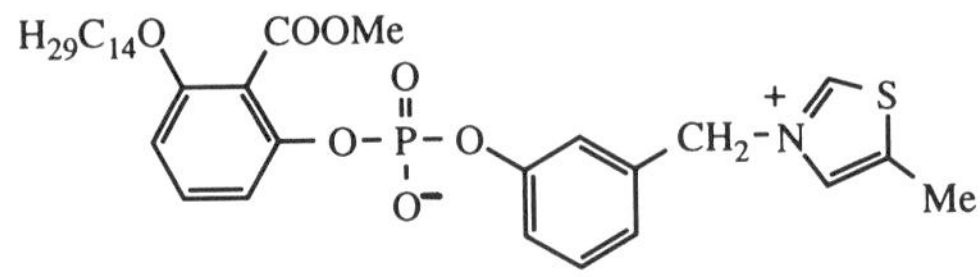

8-201

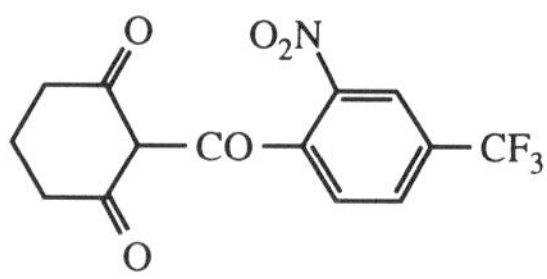

8-202

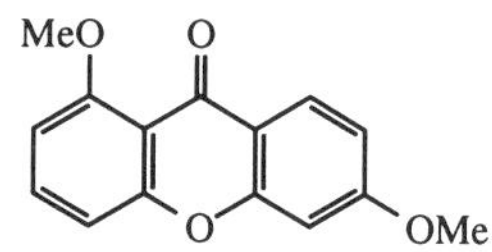

8-203

pounds were extracted from the fermentation broth of certain organisms and were useful as antibacterials or fungicides. Examples are teicoplanin and its derivatives and analogs (many patents by Gruppo Lepetit S.p.A; also by Eli Lilly), derivatives of antibiotic TAN-528A (Takeda), and eremomycin (USSR).

8.3. REFERENCES

1. *Chem. Ind. (London), 1992 (5),* 196.
2. D. A. Lewis, *Chem. Brit., 28,* 143 (1992).
3. *Fed. Regist., 16 Aug. 1991, 56* (159), 41008-20.
4. G. W. Parshall and W. A. Nugent, *Chemtech, 1988 (6),* 382.
5. W. S. Marshall, T.Goodson, G. J. Cullinan, D. Swanson-Bean, K. D. Haisch, L. E. Rinkema, and J. H. Fleisch, *J. Med. Chem., 30,* 682 (1987); R. D. Dillard, F. P. Carr, D. McCullough, K. D. Haisch, L. E. Rinkema, and J. H. Fleisch, *ibid., 30,* 911 (1987); F. J. Brown, P. R. Bernstein, L. A. Cronk, D. L. Dossett, K. C. Heel, T. P. Maduskuie, Jr., H. S. Shapiro, E. P. Vacek, Y. K. Yee, A. K. Willard, R. D. Krell, and D. W. Snyder, *ibid., 32,* 807 (1989).
6. *Chem. Eng. News, 70, Aug. 31, 1992,* 11.
7. Jpn. Kokai Tokkyo Koho JP 80,036,428 (March 14, 1980; H. Munakata, T. Tsutsui, S. Sato, K. Honda, M. Kobayashi, and K. Azuma to Mitsubishi Chemical Industries Co., Ltd.); *C.A., 94,* 66064 (1981).
8. Brit. UK Pat. Appl. GB 2,037,740 (July 16, 1980; M. Takashima, S. Iwanami, and S. Usuda to Yamanouchi Pharmaceutical Co., Ltd.); *C.A., 94,* 15673 (1981).
9. U.S. Pat. 4,225,619 (Sept. 30, 1980; R. Brickl, H. Eberhardt, K.-R. Appel, U. Lechner, and W. Merk to Boehringer Ingelheim G.m.b.H.).
10. Eur. Pat. Appl. EP 17195 (Oct.15, 1980; S. Hamane, S. Kitamura, and Y. Nakajima to Eisai Co., Ltd.); *C.A., 95,* 103336 (1981).
11. Neth. Appl. NL 79,001,466 (Aug. 26, 1980; to Agence Nationale de Valorisation de la Researche); *C.A., 94,* 103336 (1981).
12. U.S. Pat. 4,207,343 (June 10, 1980; E. R. Lavignio, L. J. McShane, and B. B. Molloy to Eli Lilly and Co.).
13. U.S. Pat. 4,228,079 (Oct. 14, 1980; G. J. Calton to W. R. Grace and Co.).
14. Ger. Offen. DE 3,031,788 (March 12, 1981; M. Shinohara, H. Kaise, Y. Nakano, T. Izawa, Y. Oshiro, and W. Miyazaki to Otsuka Pharmaceutical Co., Ltd.); *C.A., 95,* 81275 (1981).

15. U.S. Pat. 4,273,715 (June 16, 1981; R. T. Buckler and J. F. Burd to Miles Laboratories, Inc.).
16. U.S. Pat. 4,294,847 (Oct. 13, 1981; K. D.Kaufmann to T. C. Elder, Inc.).
17. Eur. Pat. Appl. EP 28,063 (May 6, 1981; A. W. Oxford and F. Ellis to Glaxo Group, Ltd.); *C.A., 96*, 34874 (1982).
18. U.S. Pat. 4,305,938 (Dec. 15, 1981; H. E. Zaugg to Abbott Laboratories); see also H. E. Zaugg and J. Kynel, *J. Med. Chem., 26(2)*, 214 (1983).
19. U.S. Pat. 4,252,818 (Feb. 24, 1981; J. Rokach, P. A. Hamel, and R.F. Hirschman to Merck and Co., Inc.).
20. Austrian Pat. 361,483 (March 10, 1981; to Fisons, Ltd.); *C.A., 95*, 7252 (1981).
21. Jpn. Kokai Tokkyo Koho JP 81,095,164 (Aug. 1, 1981; to Kumiai Chemical Industry Co., Ltd.); *C.A., 96*, 6429 (1982).
22. Jpn. Kokai Tokkyo Koho JP 82,032,275 (Feb. 20, 1982; to Nippon Shinyaku Co., Ltd.); *C.A., 96*, 217880 (1982).
23. Jpn. Kokai Tokkyo Koho JP 82,075,956 (May 12, 1982; to Torii and Co., Ltd.); *C.A., 97*, 163507 (1982).
24. Eur. Pat. Appl. EP 55,524 (July 7, 1982; E. A. Watts to Beecham Group PLC); *C.A., 97*, 215785 (1982).
25. U.S. Pat. 4,327,028 (Apr. 27, 1982; W. C. Kaplan to Calcol, Inc.).
26. Eur. Pat. Appl. EP 62,018 (Oct. 6, 1982; J. A. B. Engel, U. A. Hacksell, J. S. M. Hjorth, P. L. Lindberg, A. H. Mattson, J. L. G. Nilsson, D. Sanchez, *et al.* to Aktiebolag Haessle); *C.A., 98*, 71946 (1983).
27. Jpn. Kokai Tokkyo Koho JP 57,209,281 [82,209,281] (Dec. 22, 1982; to Otsuka Pharmaceutical Co. Ltd.); *C.A., 98*, 143437 (1983).
28. U.S. Pat. 4,351,840 (Sept. 28, 1982; D. K. Pirie to Pfizer, Inc.).
29. Eur. Pat. Appl. EP 81,621 (June 22, 1983; E. S. C. Wu to Pennwalt Corp.); *C.A., 99*, 139772 (1983).
30. Eur. Pat. Appl. EP 80,371 (June 1, 1983; A. Nohara and Y. Maki to Takeda Chemical Industries, Ltd.); *C.A., 99*, 194621 (1982).
31. Eur. Pat. Appl. EP 83,228 (July 6, 1983; P. R. Bernstein and A. K. Willard to ICI Americas, Inc.); *C.A., 99*, 194624 (1983).
32. Ger. Offen. DE 3,312,675 (Oct. 13, 1983; M. Carson, R. A. Lemahieu, and W. C. Nason to F. Hoffmann-LaRoche and Co., A.-G.); *C.A., 100*, 51267 (1984).
33. Rom. RO 83,980 (May 30, 1984; E. R. Bozga, F. I. Cunescu, and S. Ciliani-Bibian to Istitutul de Cercetari Chimico-Farmaceutice); *C.A., 102*, 148898 (1985).
34. A. N. Grinev, G. N. Pershin, L. F. Stebaeva, I. S. Nikoleva, V. I. Shvedov, I. A. Kharizomenova, and E. M. Botkina, *Khim.-Farm. Zh., 18(3)*, 374 (1984); *C.A., 101*, 72359 (1984). USSR SU 1,114,422 (Sept. 23, 1984; G. N. Pershin, V. I. Shvedov, E. N. Sytina, A. N. Grinev, T. A. Gusikova, and A. L. Mikerina to S. Ordzhonikidze, All-Union Scientific-Research Chemical-Pharmaceutical Institute [*C.A., 102*, 32248 (1985)]—this patent claims Riodoxol as an antifungal agent; USSR SU 1,029,451 (March 7, 1985; A.P.Khrunova, T.A.Korchevaya, E. V. Vyborova, G. N. Pershin, N. S. Bogdanova, and I. S. Nikolaeva to Moscow Scientific-Research Institute/Chemical/Pharmaceutical Institute) [*C.A., 102*, 215175 (1985)]—this patent claims triiodoresorcinol as an antibacterial agent.
35. Ger. Offen. DE 3,125,059 (Jan. 5, 1983; H. Boeshagen, U. Hoerlein, G. Reinhard, F. Seuter, and E. Perzborn to Bayer A.-G.); *C.A., 98*, 106960 (1983).
36. Span. Pat. ES 516,736 (Oct. 1, 1983; A. F. Foguet, J. A. Ortiz Hernandez, L. Anglada Burniel, and A. Sacristan Munoz to Ferrer International, S.A.); *C.A., 100*, 138775 (1984).
37. Eur. Pat. Appl. EP 89,634 (Sept. 28, 1983; Y. Aoyagi, T. Okubo, T. Tomita, H. Nishida, and H. Enomoto to Nippon Shinyaku Co., Ltd.); *C.A., 100*, 51240 (1984).
38. Ger. Offen. DE 3,416,230 (Nov. 8, 1984; R. A.LeMahieu to F. Hoffmann-LaRoche and Co., A.-G.); *C.A., 102*, 113083 (1985).
39. P. Lumme, H. Elo, and J. Janne, *Inorg. Chim. Acta, 92(4)*, 241 (1984); see also *ibid., 107(2)*, L15–L16 (1985).
40. U.S. Pat. 4,456,612 (June 26, 1984; J. J. Plattner and A. K. Fung to Abbbott Laboratories).

41. Eur. Pat. Appl. EP 108,592 (May 16, 1984; W. S. Marshall and J. P. Verge to Eli Lilly and Co.); *C.A., 101*, 171258 (1984).
42. Eur. Pat. Appl. EP 112,587 (July 4, 1984; E. Toth, J. Torley, G. Fekete, L. Szporny, L. Vereczkey, E. Palosi, L. Kleborich, P. Vittay, S. Gorog, and I. Hajdu to Gideon Richter, Vegyeszeti Gyar Rt.); *C.A., 102*, 5915 (1985).
43. Eur. Pat. Appl. EP 126,342 (Nov. 28, 1984; J. A. Parks and J. J. Plattner to Abbott Laboratories); *C.A., 102*, 149250 (1985).
44. Eur. Pat. Appl. EP 101,004 (Feb. 22, 1984; M. Matsumoto and N. Watanabe to Sagami Chemical Research Center); *C.A., 101*, 23331 (1984). A different process, also starting from 1,3-cyclohexanedione, is described in Jpn. Kokai Tokkyo Koho JP 59,144,758 [84,144,758] (Aug. 8, 1984; to Ube Industries, Ltd.); *C.A., 102*, 62078 (1985).
45. Eur. Pat. Appl. EP 201,751 (Nov. 20, 1986; J. R. Fino and M. T. Shipchandler to Abbott Laboratories); *C.A., 106*, 192343 (1987).
46. U.S. Pat. 4,550,172 (Oct. 29, 1975; J. H. Musser and C. O. Tio to American Home Products Co.).
47. Eur. Pat. Appl. EP 129,906 (Jan. 2, 1985; N. Cohen and G. F. Weber to Hoffmann-LaRoche and Co., A.-G.); *C.A., 103*, 62235 (1985).
48. U.S. Pat. 4,557,862 (Dec. 10, 1985; W. F. Mangel, S. Leytus, and L. L. Melhado to University Patents, Inc.).
49. Eur. Pat. Appl. EP 163,227 (Dec. 4, 1985; T. S. Abram, P. Norman, and B. T. Warren to Miles Laboratories, Inc.); *C.A., 105*, 97312 (1985).
50. USSR Pat. SU 780,435 (Feb. 28, 1985; A. A. Akhrem, F. A. Lakhvich, and T. S. Khlebnikova to Institute of Bioorganic Chemistry, Belorussian Academy of Sciences); *C.A., 103*, 123198 (1985).
51. Belg. Pat. BE 900,225 (Jan. 28, 1985; to Fidia S.p.A.); *C.A., 102*, 220751 (1985).
52. Ger. Offen. DE 3,324,916 (Jan. 17, 1985; R. Umminger, W. G. Friebe, W. Kampe, A. Toesch, and O. H. Wilhelms to Boehringer Mannheim G.m.b.H); *C.A., 103*, 53815 (1985).
53. U.S. Pat. 4,501,755 (Feb. 26, 1985; E. S. Wu to Pennwalt Corp.).
54. Eur. Pat. Appl. EP 150,118 (July 31, 1985; J. Rokach and R. N. Young to Merck Frosst Canada, Inc.); *C.A., 104*, 130266 (1986).
55. Eur. Pat. Appl. EP 150,966 (Aug. 7, 1985; K. J. Gould, J. L. Suschnitzky, and I. D. Dicker to Fisons PLC); *C.A., 104*, 68836 (1986).
56. Eur. Pat. Appl. EP 151,513 (Aug. 14, 1985; K. K. Schmiegel and H. C. Wilson to Eli Lilly and Co.); *C.A., 104*, 50682 (1986).
57. Brit. UK Pat. Appl. GB 2,145,718 (Apr. 3, 1985; C. G. Rimbault to Zyma S.A.); *C.A., 104*, 50731 (1986).
58. Eur. Pat. Appl. EP 146,922 (July 3, 1985; I. Yamazaki and Y. Sawa to Takeda Chemical Industries, Ltd.); *C.A., 104*, 5774 (1986).
59. Eur. Pat. Appl. EP 150,447 (Aug. 7, 1985; M. Miyano and R. L. Shone to G. D. Searle and Co.); *C.A., 104*, 5775 (1986).
60. Jpn. Kokai Tokkyo Koho JP 60,009,455 [85,009,455] (Jan. 18, 1985; to Osaka Yakuhin Kenky); *C.A., 103*, 70127 (1985).
61. Eur. Pat. Appl. EP 165,897 (Dec. 27, 1985; P. Wenk, A. Sallmann, and A. Beck to Ciba-Geigy A.-G.); *C.A., 105*, 60408 (1986).
62. Eur. Pat. Appl. EP 146,333 (June 26, 1985; J. Goldworthy and J. P. Verge to Lilly Industries, Ltd.); *C.A., 103*, 215299 (1985).
63. U.S. Pat. 4,546,194 (Oct. 8, 1985; M. Miyano, C. I. Villamil, and R. L. Shone to G. D. Searle and Co.).
64. U.S. Pat. 4,546,186 (Oct. 8, 1985; M. A. Abou-Gharbia to American Home Products Co.).
65. Eur. Pat. Appl. EP 141,608 (May 15, 1985; J. B. Campbell, Jr. and T. M. Bare to ICI Americas, Inc.); *C.A., 103*, 215280 (1985).
66. Ger. Offen. DE 3,329,394 (Feb. 28, 1985; O. S. Wolfbeis and H. Harnisch to Bayer A.-G.). *C.A., 103*, 22577 (1985).
67. U.S. Pat. 4,540,788 (Sept. 10, 1985; K. C. Murdock to American Cyanamid Co.).

68. Neth. Pat. NL 178,875 (Feb. 1, 1986; to Chinoin Gyogyszer).
69. Ger. Offen. DE 3,528,700 (March 6, 1986; R. F. Newton, D. Middlemiss, A. Naylor, and C. A. Meerholz to Glaxo Group Ltd.); C. A., 105, 42505 (1986).
70. Eur. Pat. Appl. EP 181,125 (May 14, 1986; A. Nohara and Y. Maki to Takeda Chemical Industries, Ltd.); *C.A., 105*, 97161 (1986).
71. Eur. Pat. Appl. EP 175,541 (March 26, 1986; H. A. Dewald, L. D. Wise, and T. G. Heffner to Warner-Lambert Co.); *C.A., 105*, 133752 (1986).
72. U.S. Pat. 4,569,994 (Feb. 11, 1986; R. C. Griffith to Pennwalt Corp.).
73. Ger. Offen. DE 3,619,448 (Dec. 11, 1986; R. Mechoulam and J. J. Feigenbaum to Yissum Research Development Co.; Hebrew University of Jerusalem); *C.A., 106*, 119347 (1987).
74. Eur. Pat. Appl. EP 200,443 (Apr. 19, 1985; J. G. Atkinson, Y. Guindon, and C. K. Lau to Merck Frosst Canada, Inc.); *C.A., 106*, 102081 (1987).
75. PCT Int. Appl. WO 86,003,199 (June 5, 1986; G. Sportoletti to Italfarmaco S.p.A.); *C.A., 105*, 165010 (1986).
76. U.S. Pat. 4,594,425 (June 10, 1986; J. Musser and D. M. Kubraka to American Home Products Corp.).
77. U.S. Pat.4,608,370 (Aug. 26, 1986; to R. B. Aronsohn).
78. Jpn. Kokai Tokkyo Koho JP 61,048,511 [86,048,511] (Oct. 28, 1986; K. Kyogoku, K. Hatayama, S. Yokomori, Y. Kawashima, and J. Sawada to Taisho Pharmaceutical Co., Ltd.); *C.A., 106*, 156148 (1987).
79. Eur. Pat. Appl. EP 175,376 (March 26, 1986; T. Mori, T. Dan, and M. Akima to Chugai Pharmaceutical Co., Ltd.); *C.A., 105*, 152930 (1986).
80. Ger. (East) DD 244,339 (Apr. 1, 1987; A. Beck, A. Sallmann, R. W. Lang, and P. Wenk to Ciba-Geigy A.-G.); *C.A., 108*, 37385 (1988).
81. Jpn. Kokai Tokkyo Koho JP 62,181, 276 [87,181,276] (Aug. 8, 1987; T. Hyama and H. Saimoto to Sagami Chemical Research Center); *C.A., 108*, 37645 (1988).
82. U.S. Pat. 4,686,234 (Aug. 11, 1987; P. H. Nelson, A. C. Allison, and E. M. Engui to Syntex (U.S.A.), Inc.).
83. R. G. Shepherd and A. C. White, *J. Chem. Soc., Perkin Trans. 1, 1987(10)*, 2153.
84. Ger. Offen. DE 3,522,230 (Jan. 2, 1987; E. Mueller, N. Hauel, K. Noll, B. Narr, J. Heider, M. Psiorz, A. Bomhard, V. J. Meel, and W. Diederen to Dr. Karl Thomae, G.m.b.H.); *C.A., 106*, 156466 (1987).
85. PCT Int. Appl. WO 87,007,835 (Dec. 30, 1987; T. Sato, H. Matsumoto, and H. Kakegawa to Dainippon Ink Chemical Industry Co.; Nippon Hypox Laboratories, Inc.); *C.A., 109*, 98856 (1988).
86. U.S. Pat. 4,672,066 (June 9, 1987; M.Carson, L. H. Ru-Jen, and R. A. Mahieu to Hoffmann-LaRoche, Inc.).
87. U.S. Pat. 4,667,055 (May 19, 1987; J. Gillard, J. Rokach, and P. C. Belanger to Merck Frosst Canada, Inc.).
88. PCT Int. Appl. WO 87,005,805 (Oct. 8, 1987; W. A. Saffran, R. L. Edelson, F. P. Gasparro, J. T. Welsh, and C. R. Cantor to Columbia University); *C.A., 109*, 149352 (1988).
89. D. W. Robertson and J. H. Krushinski, *Org. Prep. Proc. Int., 20(3)*, 311 (1988), and references cited therein.
90. Eur. Pat. Appl. EP 276,065 (July 27, 1988; N. G. Bollinger, T. Goodson, Jr., and D. K. Herron to Eli Lilly and Co.); *C.A., 109*, 230544 (1988).
91. Eur. Pat. Appl. EP 292,977 (Nov. 30, 1988; S. W. Djuric, R. L. Shone, and S. S. T. Yu to G. D. Searle and Co.); *C.A., 110*, 174088 (1989).
92. PCT Int. Appl. WO 88,004,654 (June 30, 1988; H. O. J. Hall and S. O. Thorberg to Astra Laekmede AB); *C.A., 109*, 210900 (1988).
93. Indian Pat. IN 163,780 (Nov. 5, 1988; to R. J. Kumar, L. D. K. Gazula, and G. Srimannarayana); *C.A., 112*, 158226 (1990).
94. Eur. Pat. Appl. EP 288,202 (Oct. 26, 1988; H. W. R. Williams and R. N. Young to Merck Frosst Canada, Inc.); *C.A., 110*, 114450 (1989).

95. U.S. Pat. 4,758,597 (July 19, 1988; A. R. Martin, P. F. Conroe, and V. J. Shah to University Patents, Inc.).
96. Ger. Offen. DE 3,735,990 (Apr. 28, 1988; R. Mechoulam, J. Feigenbaum, M. Srebnik, and N. Lander to Yissum Research and Development Co.); *C.A., 110*, 154614 (1989).
97. Eur. Pat. Appl. EP 293,146 (Nov. 30, 1988; Y. Konoshita, S. Ikeguchi, N. Tsutsumi, Y. Ajisawa, and A. Ujiie to Kissei Pharmaceutical Co., Ltd.); *C.A., 110*, 114818 (1984).
98. U.S. Pat. 4,839,364 (June 13, 1989; G. M. Shutske and F. A. Pierrat to Hoechst-Roussel Pharmaceuticals, Inc.).
99. Ger. Offen. DE 3,741,224 (June 22, 1989; R. Sobotta, A. Langbein, H. Merz, and R. Bauer to Boehringer Ingelheim K.G.); *C.A., 112*, 35688 (1990).
100. Ger. Offen. DE 3,808,136 (Sept. 21, 1989; E. Mueller, J. Nikl, A. Heckel, and G. Engelhardt to Dr. Karl Thomae, G. m. b. H.); *C.A., 112*, 138919 (1990).
101. Jpn. Kokai Tokkyo Koho JP 01,242,580 [89,242,580] (Sept. 27, 1989; K. Hiraga, M. Goto, K. Inoue, and M. Ichida to Nippon Nohyaku Co., Ltd.); *C.A., 112*, 158230 (1990).
102. Jpn. Kokai Tokkyo Koho JP 01,121,236 [89,121,236] (May 12, 1989; M. Shimizu, K. Inoe, H. Kohama, H. Oonishi, and M. Sugyama to Terumo Corp.); *C.A., 111*, 194305 (1989).
103. U.S. Pat. 4,806,660 (Feb. 21, 1989; E. S. Wu to Pennwalt Corp.).
104. Eur. Pat. Appl. EP 300,404 (Jan. 25, 1989; M. Miyano, M. A. Stealy, and R. L. Shone to G. D. Searle and Co.); *C.A., 111*, 96867 ((1989).
105. Jpn. Kokai Tokkyo Koho JP 01,197,436 [89,197,436] (Aug. 9, 1989; M. Shimizu, S. Myoka, H. Oonishi, K. Inoe, and M. Sugyama to Terumo Corp., Ltd.); *C.A., 112*, 55238 (1990).
106. PCT Int. Appl. WO 89,005,289 (June 15, 1989; I. Tomino, M. Ishiguro, T. Kitahar, K. Yokoyama, N. Kihara, J. Kamiya, K. Yoshihara, M. Ishii, and A. Mizuchi to Mitsui Petrochemical Industries, Ltd. ; Mitsui Pharmaceuticals, Inc.); *C.A., 112*, 55583 (1990).
107. Brit. UK Pat. Appl. GB 2,211,500 (July 5, 1989; I. N. H. White to Medical Research Council); *C.A., 112*, 55607 (1990).
108. U.S. Pat. 4,820,726 (Apr. 11, 1989; A. Beck, A. Sallmann, R. W. Lang, and P. Wenk to Ciba-Geigy Corp.).
109. Ger. Offen. DE 3,904,731 (Aug. 24, 1989; K. Kasai, S. Tokutake, and N. Yamaji to Kikkoman Corp.); *C.A., 112,* 56578 (1990).
110. Eur. Pat. Appl. EP 306,959 (March 15, 1989; M. Ohashi, K. Awano, T. Tanaka, and T. Kimura to Kyorin Pharmaceutical Co., Ltd.); *C.A., 111*, 133788 (1989).
111. Jpn. Kokai Tokkyo Koho JP 01,294,660 [89,294,660] (Nov. 28, 1989; H. Yamamoto and K. Shiraki to Yamanouchi Pharmaceutical Co., Ltd.); *C.A., 112*, 198843 (1990).
112. Eur. Pat. Appl. EP 297,651 (Jan. 4, 1989; H. H. Haeck, D. Hamminga, I. Van Wijngaarden, and W. Wouters to Duphar International Research, B.V.); *C.A., 110*, 212631 (1989).
113. U.S. Pat. 4,857,659 (Aug. 15, 1989; R. Frenette, J. Y. Gauthier, R. N. Young, R. Zamboni, M. Kakushima, and T. R. Verhoeven to Merck Frosst Canda, Inc.).
114. Eur. Pat. Appl. EP 342,665 (Nov. 23, 1989; T. Kanamuru, T. Hida, and M. Muroi to Takeda Chemical Industries, Ltd.); *C.A., 113*, 5954 (1990).
115. Eur. Pat. Appl. EP 360,566 (March 28, 1990; M. Nakane, T. Kokura, and F. Ito to Pfizer, Inc.); *C.A., 113*, 132163 (1990).
116. U.S. Pat. 4,937,260 (June 26, 1990; T. DePaulis, S. I. Raemsby, and S. O. Oegren to Astra Laekemedal AB).
117. U.S. Pat. 4,945,099 (July 31, 1990; N. G. Bollinger, T. Goodson, Jr., and D. K. Herron to Eli Lilly and Co.).
118. Eur. Pat. Appl. EP 372,962 (June 13, 1990; I. T. Forbes, R. T. Martin, and M. Thompson to Beecham Group PLC); *C.A., 114*, 6486 (1991).
119. Jpn. Kokai Tokkyo Koho JP 02,191,269 [90,191,269] (July 27, 1990; H. Nakayama, M. Ishikura, Y. Ueda, K. Imai, M. Terajima, and A. Suzui to Toyo Pharmaceutical Co., Ltd. ; Daiso Co., Ltd.); *C.A., 113*, 211845 (1990).

120. Eur. Pat. Appl. EP 374,513 (June 27, 1990; A. Wissner, K. Green, and R. E. Schaub to American Cyanamid Co.); *C.A., 113*, 212346 (1990).
121. Eur. Pat. Appl. EP 341,664 (Nov. 15, 1990; S. Torihara, Y. Tamai, M. Shiono, and K. Tasaka to Kuraray Ltd.); *C.A., 113*, 178024 (1990).
122. USSR SU 1,560,215 (Apr. 30, 1990; A. I. Grudnayov to Central Scientific-Research Institute of Stomatology); *C.A., 113*, 178290 (1990).
123. Jpn. Kokai Tokkyo Koho JP 02,172,916 [90,172,916] (July 4, 1990; T. Shishido and H. Okada to Fuji Photo Film Co., Ltd.); *C.A., 114*, 101980 (1991).
124. U.S. Pat. 4,920,131 (Apr. 24, 1990; F.-C. Huang, R. A. Galemmo, Jr., and H. F. Campbell to Rorer Pharmaceutical Corp.).
125. Israeli Pat. IL 80,080 (Sept. 17, 1990; to F. Hoffmann-LaRoche and Co., A.-G.); *C.A., 115*, 92072 (1991).
126. U.S. Pat. 4,996,230 (Feb. 26, 1991; D. M. Gapinski to Eli Lilly and Co.).
127. Jpn. Kokai Tokkyo Koho JP 02,292,213 [90,292,213] (Dec. 3, 1990; H. Tamai, T. Yamane, and K. Tasaka to Kuraray Co., Ltd.); *C.A., 114*, 157192 (1991).
128. Eur. Pat. Appl. EP 411,912 (Feb. 5,1991; H. Yanagisawa, K. Fujimoto, Y. Shimoji, and T. Kanazaki to Sankyo Co., Ltd.); *C.A., 115*, 71396 (1991).
129. Jpn. Kokai Tokkyo Koho JP 03,068,551 [91,068,551] (March 25, 1991; S. Takahashi to Seiko Instruments and Electronics, Ltd.); *C.A., 115*, 72228 (1991).
130. Eur. Pat. Appl. EP 412,803 (Feb. 13, 1991; S. Yakomori, K. Saijo, T. Matsunaga, Y. Nakashima, and K. Hatayama to Taisho Pharmaceutical Co., Ltd.); *C.A., 115*, 8321 (1991).
131. PCT Int. Appl. WO 91,017,160 (Nov. 14, 1991; S. W. Djuric, T. D. Penning, and J. P. Snyder to G. D. Searle and Co.); *C.A., 116*, 83676 (1992).
132. Jpn. Kokai Tokkyo Koho JP 03,068,549 [91,068,549] (March 25, 1991; T. Mase, Y. Ishii, K. Tomioka, and Y. Soeishi to Yamanouchi Pharmaceutical Co., Ltd.); *C.A., 115*, 92274 (1991).
133. Eur. Pat. Appl. EP 419,411 (March 27, 1991; A. Von Sprecher, B. Schaub, and R. W. Lang to Ciba-Geigy A. -G.); *C.A., 115*, 114351 (1991).
134. PCT Int. App. WO 91,005,757 (May 2, 1991; N. Y. Shih and P. Mangiaracina to Schering Corp.); *C.A., 115*, 114147 (1991).
135. Jpn. Kokai Tokkyo Koho JP 03,068,516 [91,068,516] (March 25, 1991; S. Sato M. Chin, H. Mihashi, and T. Nomura to Tsumura and Co., Ltd.); *C.A., 115*, 182941 (1991).
136. Jpn. Kokai Tokkyo Koho JP 03,115,287 [91,115,287] (May 16, 1991; to Fujisawa Pharmaceutical Co, Ltd. ; Colorado State University Research Foundation); *C.A., 115*, 183332 (1991).
137. Jpn. Kokai Tokkyo Koho JP 03,127,783 [91,127,783] (May 30, 1991; H. Yanasigawa, K. Fujimoto, Y. Shimoji, T. Kanezaki, N. Iwata, and Y. Kubo to Sankyo Co., Ltd.); *C.A., 115*, 232261 (1991).
138. Eur. Pat. Appl. EP 421,829 (Apr. 10, 1991; S. Samreth, V. Barberousse, P. Renaut, F. Bellamy, and J. Millet to Fournier Innovation et Synergie); *C.A., 115*, 256557 (1991).
139. U.S. Pat. 4,996,230 (Feb. 26, 1991; D. M. Gapinski to Eli Lilly and Co.).
140. Fr. Demande FR 2,653,996 (May 10, 1991; to Armando Lourenco); *C.A., 115*, 239319 (1991).
141. Jpn. Kokai Tokkyo Koho JP 03,284,612 (Dec. 16, 1991; H. Shibata, Y. Minosasa, K. Matsui, H. Uehara, and H. Tanaka to Naris Cosmetics Co., Ltd.); *C.A., 116*, 214752 (1992).
142. Jpn. Kokai Tokkyo Koho JP 03,287,589 [91,287,589] (Dec. 18, 1991; Y. Yokokawa, M. Yoshida, K. Umishio, and K. Uehara to Shiseido Co., Ltd.); *C.A., 116*, 214698 (1992).
143. Jpn. Kokai Tokkyo Koho JP 03,287,587 (Dec. 18, 1991; T. Mase, Y. Kondo, H. Nagaoka, T. Yamada, and K. Tomioka to Yamanouchi Pharmaceutical Co., Ltd.); *C.A., 116*, 255603 (1992).
144. Jpn. Kokai Tokkyo Koho JP 04,001,116 [92,001116] (Jan. 6, 1992; T. Hamazaki, M. Matsugami, M. Takenochi, K. Utsugi, and H. Nakano to Pola Chemical Industries, Inc., Ltd.); *C.A., 116*, 180944 (1992).
145. PCT Int. Appl. WO 92,000,011 (Jan. 9, 1992; S. W. Djuric, S. H. Docter, and S. S. T. Yu to G. D. Searle and Co.); *C.A., 116*, 214350 (1992).
146. PCT Int. Appl. WO 92,000,285 (Jan. 9, 1992; Y. Yoshimoto, S. Yasufuku, Y. Mikita, K. Inoue, and K. Nakanouchi to Nippon Shinyaku Co., Ltd.); *C.A., 116*, 214505 (1992).

147. Jpn. Kokai Tokkyo Koho JP 04,054,179 [92,054,179] (Feb. 21, 1992; J. Imuda, N. Kihara, A. Mizuchi, K. Horigome, and A. Awaya to Mitsui Sekiyu Kagaku Kogyo K. K.; Mitsui Seiyaku K. K.); *C.A., 117*, 69731 (1992).
148. PCT Int. Appl. WO 92,006,075 (April 16, 1992; K. W. Franzmann to Wellcome Foundation Ltd.); *C.A., 117*, 48124 (1992).
149. Jpn. Kokai Tokkyo Koho JP 04,187,632 [92,187,632] (July 6, 1992; H. Suzuki, T. Nishimura, H. Yamaki, H. Yamaguchi, N. Tanaka, and T. Takeuchi to Microbiochemical Research Foundation); *C.A., 117*, 169501 (1992).
150. Jpn. Kokai Tokkyo Koho JP 04,134,048 [92,134,048] (May 7, 1992; T. Ishimaru, S. Tsuboya, H. Shirafuji, and Z. Terashita to Takeda Chemical Industries, Ltd.); *C.A., 117*, 178329 (1992).
151. Jpn. Kokai Tokkyo Koho JP 04,112,884 [92,112,884] (Apr. 14, 1992; T. Yoshioka, R. Miyata, H. Tone, R. Okamoto, H. Kumagai, M. Ishizuka, and T. Takeuchi to Merican Corp., Ltd.); *C.A., 117*, 171087 (1992).
152. Jpn. Kokai Tokkyo Koho JP 04,154,773 [92,154,773] (May 27, 1992; N. Sugiyama, T. Goto, T. Okano, Y. Yamaura, F. Akaboshi, S. Ono, H. Yoshitomi, Y. Naito, and T. Fukaya to Green Cross Corp.); *C.A., 117*, 212486 (1992).
153. PCT Int. Appl. WO 92,011,252 (July 9, 1992; S. W. Djuric, D. J. Fretland, and S. S. T. Yu to G. D. Searle and Co.); *C.A., 117*, 234018 (1992).
154. PCT Int. Appl. WO 92,010,096 (June 25, 1992; R. D. Sindelar, B. J. Bradbury, T. S. Kaufman, S. H. Ip, H. C. March, Jr., and C. Lee to T Cell Sciences, Inc.; University of Mississippi); *C.A., 117*, 251233 (1992).
155. Eur. Pat. Appl. EP 495,485 (July 22, 1992; M. Ohashi, T. Tanaka, N. Ishikawa, and T. Kishi to Kyorin Pharmaceutical Co., Ltd.); *C.A., 117*, 233602 (1992).
156. Jpn. Kokai Tokkyo Koho JP 03,169,811 [91,169,811] (July 23, 1991; K. Miyasaki and S. Saigo to Otsuka Pharmaceutical Co., Ltd.); *C.A., 116*, 67191 (1992).
157. Jpn. Kokai Tokkyo Koho JP 03,275,687 [91,275,687] (Dec. 6, 1991; H. Tsuchida, T. Komatsu, E. Hasegawa, and H. Nishide to Zaidan Hojin Seisan Kaihatsu Kagaku Kenkyusho; Nippon Oil and Fats Co., Ltd.); *C.A., 117*, 90050 (1992).
158. Jpn. Kokai Tokkyo Koho JP 03,294, 273 [91,294,273 (Dec. 25, 1991; T. Akiyama, K. Ichikawa, Y. Matsui, and H. Haruyama to Sankyo Co., Ltd.); *C.A., 116*, 22211551 (1992).
159. Jpn. Kokai Tokkyo Koho JP 04,169,511 [92,169,511] (June 17, 1992; M. Noda, K. Matsuda, M. Takenochi, N. Satooka, T. Yamane, M. Torihara, and H. Tamai to Pola Chemical Industries, Inc.; Kuraray Co., Ltd.); *C.A., 117*, 220114 (1992).
160. Jpn. Kokai Tokkyo Koho JP 04,221,393 [92,221,393] (Aug. 11, 1992; K. Kasai, K. Okada, and N. Yamatsugu to Kikkoman K. K.); *C.A., 118*, 124963 (1993).
161. Jpn. Kokai Tokkyo Koho JP 04,217,678 [92,217,678] (Aug. 7, 1992; Y. Tsuchiya, H. Masaki, T. Sato, and T. Kamei to Banyu Pharmaceutical Co., Ltd.); *C.A., 118*, 168980 (1993).
162. PCT Int. Appl. WO 92,013,852 (Aug. 20, 1992; H. Takagaki, M. Sakai, H. Takahashi, K. Inazawa, M. Abe, and N. Kimura to Dainippon Ink Chemical Industry Co., Ltd.); *C.A., 119*, 8681 (1993).
163. Jpn. Kokai Tokkyo Koho JP 04,282,380 [92,282,380] (Oct. 7, 1992; Y. Shida, T. Akama, S. Ikeda, M. Kasai, S. Kandori, Y. Hirayama, H. Ishida, and K. Gomi to Kyowa Hakko Kogyo Co., Ltd.); *C.A., 119*, 8593 (1993).
164. Jpn. Kokai Tokkyo Koho JP 04,266,848 [92,266,848] (Sept. 22, 1992; H. Kawagishi, F. Kojima, H. Sakamoto, T. Yoshida, and Y. Ishiguro to Kagome Co., Ltd.); *C.A., 118*, 66842 (1993).
165. Eur. Pat. Appl. EP 512,525 (Nov. 11, 1992; Y. Oshiro, T. Tanaka, T. Kikuchi, and K. Tottori to Otsuka Pharmaceutical Co., Ltd.); *C.A., 118*, 101983 (1993).
166. PCT Int. Appl. WO 92,022,545 (Dec. 23, 1992; A. Romeo and M. Prosdocimi to Fidia S.p.A.); *C.A., 118*, 212 892 (1993).
167. A. Wissner, M. L. Carrol, K. E. Green, S. S. Kerwar, W. C. Pickett, R. E. Schaub, L. W. Torley, S. Wrenn, and C. A. Kohler, *J. Med. Chem., 35(9)*, 1650 (1992).

168. PCT Int. Appl. WO 93,000,080 (Jan. 7, 1993; M. K. Ellis, S. T. Lindstedt, E. A. Lock, M. E. H. Maj, L. C. Mutter, and M. P. Prisbylla to Imperial Chemical Industries PLC); *C.A., 118*, 94329 (1993); see also: *Chem. Brit., 29*, 10 (1993).
169. C. N. Lin, S. S. Liou, F. N. Ko, and C. M. Teng, *J. Pharm. Sci., 82(1)*, 11 (1993); *C.A., 118*, 212814 (1993).

9

The Uses of Resorcinol/Derivatives in Polymers

9.1. INTRODUCTION

Resorcinol, resorcinol derivatives, and resorcinol resins can be used as monomers or as reactive additives which increase the molecular weight of a (pre-)polymer. This section deals with such applications. Indeed, there are several commercial uses, such as, e.g., the use of resorcinol di-(2-hydroxyethyl) ether in polyurethanes and polyesters, 3,4′-diaminodiphenyl ether in aramids, resorcinol diglycidyl ether in high-technology applications, and polyphenylene ethers as high-temperature lubricants. More details on these uses are given under the individual polymers.

Several of the major uses of resorcinol in the phenolic resins area were covered separately. The use of resorcinol, resorcinol derivatives, and resorcinol/(form)aldehyde resins in rubber compounds was presented in Chapter 4, and their use in adhesives for wood and other nonrubber applications was discussed in Chapter 5. The use of resorcinol/derivatives as nonreactive additives for polymers is covered in Chapter 11.

In the polymer applications covered in this chapter, resorcinol and its derivatives serve a special need which can be to make a product with a lower melting point or polymer melt temperature for ease of processing, or an improved balance of physical properties for a particular application, or, often, to make a product with improved high-temperature stability and fire resistance.

Aside from the commercial uses of resorcinol/derivatives, much of the discussion and references given in this chapter refer to raw inventions, to the flow of product development, to competitive activities, to modifications made to tailor the properties of a polymer, and oriented to end uses which are often best described as engineering polymer and high-technology applications. A selection of references is presented which makes it clear that competitive materials abound, often with marginal differences in properties, yet this margin may be the key to commercial success of one over the other. Overall, it becomes clear that massive efforts were made to develop improved polymers using resorcinol or resorcinol derivatives. A large contribution to these efforts has come from Japan. The massive effort alone does not guarantee commercial viability; it is needed to demonstrate a favorable overall cost/performance of a new material, which can gain customers. And an older material may start to lose its market share in the process.

The early 1990s brought the demise of the USSR and the end of the cold war. This has impacted U.S. research on high-technology/military needs for high-performance materials. Such research will remain at a lower level than before for some time.

9.2. MONOMERS BASED ON RESORCINOL OR *m*-AMINOPHENOL

High-purity monomers are required for polymerization reactions. Therefore, much effort is made to come up with good processes and purification procedures for the monomers. Also, new monomers are always looked for to make new or modified, high-performance materials. The following examples illustrate both of these needs as well as the variety of monomers that can be made from resorcinol.

9.2.1. Resorcinol Diacetate

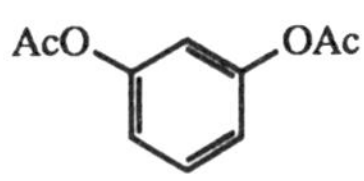

9-1

This monomer, **9-1**, is used often in the preparation of polyarylates by polycondensation with aromatic dicarboxylic acids (the acid interchange method) or their esters (transesterification), with the evolution of acetic acid or acetate esters, respectively. Several examples of such polymerizations appear later in this chapter (see Section 9.5).

It is a sign of the importance of **9-1** that a recent (1992) patent of Sumitomo claimed a process for the preparation of resorcinol diacetate of high purity by the reaction of resorcinol with acetic anhydride, even though this is one of the oldest reactions in organic chemistry. Thus, the reaction of resorcinol with acetic anhydride in 1.00/1.07 mole ratio in the presence of a catalytic amount of pyridine at 100 °C/1 hr was said to give high-purity (100%), colorless resorcinol diacetate.[1]

9.2.2. Resorcinol Di-(2-hydroxyethyl) Ether [HER]

The Chemical Abstracts name for this compound, **9-2**, is 2,2′-(*m*-phenylenedioxy) diethanol; HER is the tradename of INDSPEC Chemical Corp. for **9-2**.

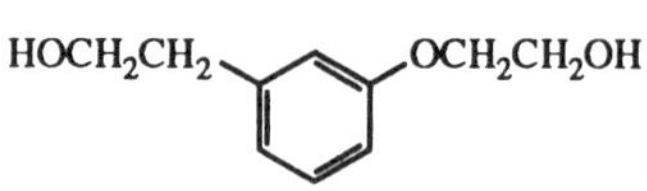

9-2

Resorcinol di-(2-hydroxyethyl) ether is useful as a chain extender for polyurethanes, as a starting material for (*m*-phenylene dioxy) diacetic acid, and as a comonomer for polyester materials with low gas permeability.

A good process for this monomer, **9-2**, is the reaction of resorcinol in aqueous solution with the stoichiometric amount of ethylene oxide in the presence of a catalytic amount of an alkali metal hydroxide, giving **9-2** in 75–85% yield; the crude product is readily recrystallized/purified from water.[2]

A similar process, using triethylamine as the catalyst, was said to give larger granules of **9-2** with lower water content (2.5%) and improved purity.[3]

An excellent process for **9-2** is the reaction of resorcinol with the stoichiometric amount of ethylene carbonate in the presence of a very small amount of a triorganophosphine (e.g., triphenyl phosphine) catalyst at *ca* 150–170 °C; no solvent is needed, the crude product is obtained in quantitative yield and has *ca* 96% purity as well as very low water content, and it can be readily upgraded to 99+% purity by recrystallization from, e.g., alcohols or ketones; environmental problems, compared to the disposal of aqueous liquors with a relatively high BOD (biological oxygen demand), are minimized.[4] A similar process is the reaction of resorcinol with ethylene carbonate in the presence of a small amount of water using an alkali metal carbonate, then adding an alkali metal hydroxide (to dissolve resorcinol and the mono-hydroxyethyl ether), and recovering **9-2** by crystallization/filtration in 94% yield (the purity of the product was not given in the patent).[5]

Finally, the reaction of resorcinol with ethylene carbonate at 160–170 °C in the presence of potassium carbonate as the catalyst was reported in a recent Japanese patent to give a 61% yield of resorcinol mono-(2-hydroxyethyl) ether and a 32% yield of **9-2**.[6]

9.2.3. (*m*-Phenylenedioxy)diacetic Acid

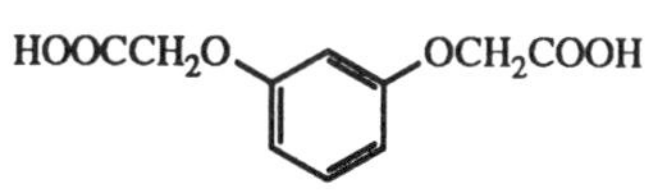

9-3

This dicarboxylic acid, **9-3**, has become of interest recently as a comonomer for the preparation of absorbable, radiation sterilizable polyesters for use, e.g., in making sutures and other surgical devices, and also for the preparation of copolyesters of low gas permeability. The preparation of **9-3** by the old methods [the condensation of alkali resorcinate with sodium chloroacetate or alkyl chloroacetates] has received renewed attention; new methods, involving the oxidation of HER, **9-2**, have been developed, as follows.

The reaction of resorcinol with methyl chloroacetate in 1/2 mole ratio in methanol, with slow addition of the required amount of sodium methoxide, reflux for 17 hr, filtration to remove sodium chloride, and cooling gave crystalline dimethyl (*m*-

phenylenedioxy)diacetate; after two recrystallizations from methanol, a 24% yield of high-purity product was obtained.[7] By the reverse addition of reagents and control of the addition rate, the yield of the dimethyl ester of **9-3** was raised to 90+%.[8]

The preparation of the dicarboxylic acid, **9-3**, was said to be improved by gradually adding an aqueous solution of sodium chloroacetate to an aqueous solution of sodium resorcinate while keeping the pH of the solution at 10; acidification and filtration gave a 72% yield of **9-3**.[9] Similar preparations where the sodium salt of **9-3** was added to aqueous H_2SO_4 at 100 °C, or where sodium sulfite was added in the initial reaction, gave the product, **9-3**, as a low-colored material.[10]

The preparation of **9-3** by the oxidation of HER, **9-2**, with oxygen or air in an aqueous, alkaline medium in the presence of a Pt- or Pd-containing catalyst activated with a Pb or Bi compound was claimed in a patent; no details were given.[11] The oxidation of HER in the same manner at 90 °C in the presence of a Pd/Te-on-C catalyst was reported to give a 97% yield of the dicarboxylic acid, **9-3**.[12]

The preparation of the partial oxidation product, **9-4**, of HER by a similar oxidation process in good yield was described.[13] This product should be homo-polymerizable.

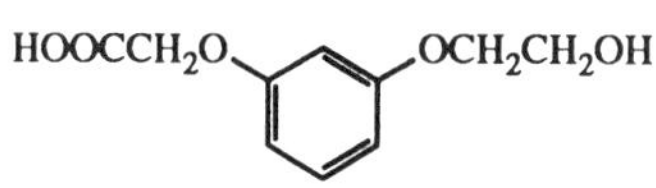

9-4

Finally, the air oxidation of HER, **9-2**, in aqueous, alkaline solution at 100 °C in the presence of a Pd-on-C/Th-on-C catalyst, followed by a standard workup, was said to give the dicarboxylic acid, **9-3**, in 90+% yield and high purity.[14]

9.2.4. 3,4′-Diaminodiphenyl Ether

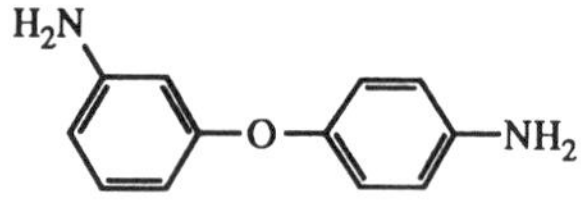

9-5

This arylenediamine, **9-5**, is a comonomer for the preparation of Teijin's aramid, HM-50 (see Section 9.2.6); it can be used as an epoxy curing agent and as a monomer in the preparation of other polyamides and polyimides. For these reasons, extensive work has been done on processes for making high-purity **9-5**. Examples of this work are the following:

3,4′-Diaminodiphenyl ether, **9-5**, was prepared by the hydrogenation of 3-amino-4′-nitrodiphenyl ether (obtained by reaction of alkali *m*-aminophenolate with *p*-nitrochlorobenzene) in the presence of a Pt, Pd, Ru, Raney Ni, Ni-on-diatomaceous earth, or copper chromite catalyst, using alkylene glycol mono-ether/mono-ester as the solvent.[15]

The 3,4′-diaminodiphenyl ether, **9-5**, was also made in 50% yield by reacting *m*-aminophenol with *p*-iodoaniline in dimethylsulfoxide as the solvent in the presence of potassium hydroxide and copper iodide as the catalyst.[16]

The condensation of 3-aminophenol with high-purity 4-chloronitrobenzene in the presence of anhydrous potassium carbonate, followed by catalytic hydrogenation of the obtained 3-amino-4-nitrodiphenyl ether, gave an 82% yield of 3,4′-diaminodiphenyl ether of 99.95% purity.[17]

The intermediate, 3-amino-4′-nitrodiphenyl ether, for **9-5** was obtained in 96% yield by adding powdered sodium hydroxide to a solution of *m*-aminophenol and *p*-chloronitrobenzene in *N*-methyl-2-pyrrolidinone/water; when *N,N*-dimethylformamide/water was used as the solvent, the yield of 3-amino-4′-nitrodiphenyl ether was <40%.[18] A similar condensation of *m*-aminophenol with *p*-chloronitrobenzene in the presence of potassium carbonate, 1,3-dimethyl-2-imidazolidinone (as the solvent), and aqueous trioctyl methylammonium chloride at 130 °C under nitrogen for 4 hr gave 3-amino-4′-nitrodiphenyl ether in 97% yield.[19]

Crude (98.9% pure) 3,4′-diaminodiphenyl ether, **9-5**, was purified by distillation through a 30-plate column at 5/1 reflux ratio and 3 mm Hg top-of-the-column pressure to give **9-5** of 99.99% purity which remained colorless on 10 days storage at room temperature.[20]

9.2.5. 4,6-Diaminoresorcinol

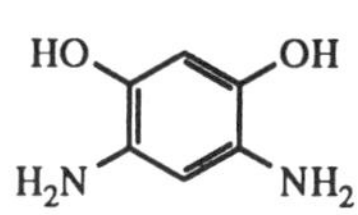

9-6

This monomer, **9-6**, is a starting material for the high-strength, liquid crystalline poly[*p*-phenylenebenzobis-(oxazole)] (PBO; see Section 9.11). 4,6-Diaminoresorcinol is usually made by the reduction of 4,6-dinitroresorcinol.

A preparation of 4,6-dinitroresorcinol was patented which consisted of adding resorcinol to white (free of NO_2) 70–90% nitric acid, with or without sulfuric acid being present, and with added urea (to destroy any nitrous acid that might be present). The procedure minimized or controlled the formation of undesirable byproducts such as 2-nitroresorcinol, 2,4-dinitroresorcinol, and the explosive 2,4,6-trinitroresorcinol, and a 60–65% yield of the desired 4,6-dinitroresorcinol was obtained.[21]

A preparation of the monophosphate salt of 4,6-diaminoresorcinol was patented in which 2-chloro-4,6-dinitroresorcinol was hydrogenated in aqeous phosphoric acid in the presence of a Pd-on-C catalyst, with simultaneous dechlorination and nitro group reduction.[22] The phosphate salt could be used directly in the condensation with terephthalic acid to make a polybenzoxazole.

9.2.6. 1,3-bis-(4-aminophenoxy)benzene (9-7), 1,3-bis-(3-aminophenoxy)-5-chlorobenzene (9-8), and 1,3-bis-(3-aminophenoxy)benzene (9-9)

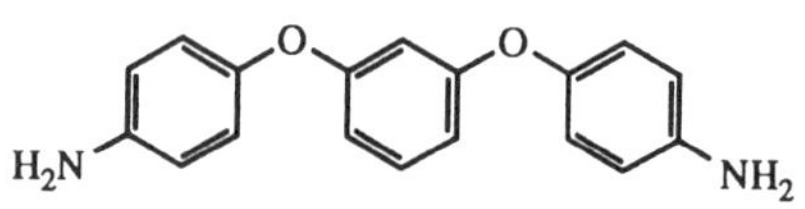

9-7

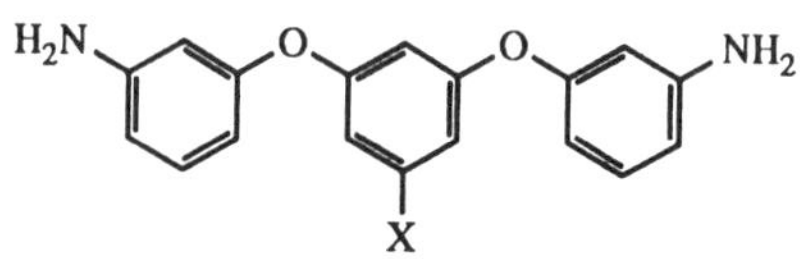

9-8: X=Cl; 9-9: X=H

Again, these aryl ether diamines are useful monomers for making heat-resistant polyamides and polyimides.

1,3-bis-(4-aminophenoxy)benzene, **9-7**, is prepared by the condensation of alkali resorcinate with *p*-chloronitrobenzene, followed by the catalytic hydrogenation of the intermediate, 1,3-bis-(4-nitrophenoxy)benzene, by processes similar to those for making 3,4′-diaminodiphenyl ether described in Section 9.2.4 above.

1,3-bis-(3-aminophenoxy)-5-chlorobenzene, **9-8**, was prepared by heating *m*-aminophenol with aqueous KOH and 1-methyl-2-pyrrolidinone in toluene with azeotropic removal of water, followed by the addition of a solution of 1,3,5-trichlorobenzene in 1-methyl-2-pyrrolidinone at 170 °C and holding the charge for 4 hr/190 °C to obtain the diamine, **9-8**, in 83% yield.[23] A similar process which included the dehalogenation of **9-8** by treatment of the crude, preceding product with hydrogen in the presence of a 5% Pd-on-C catalyst and anhydrous sodium carbonate as the acid acceptor gave an 86% yield of 1,3-bis-(3-aminophenoxy)benzene, **9-9**, of 99.9% purity.[24]

9.2.7. Diamines of Aromatic Ether-Ketones

4,4′-bis-(3-aminophenoxy)benzophenone, **9-10**, was prepared in 87% yield and 99.6% purity by reacting 4,4′-dichlorobenzophenone with *m*-aminophenol in the presence of KOH in 1,3-dimethyl-2-imidazolidone as the solvent for 8 hr/150–160 °C.[25]

1,3-bis[4-(3-aminophenoxy)benzoyl]benzene, **9-11**, was made by reacting 1,3-bis-(4-chlorobenzoyl)benzene with *m*-aminophenol in the presence of hydroxides,

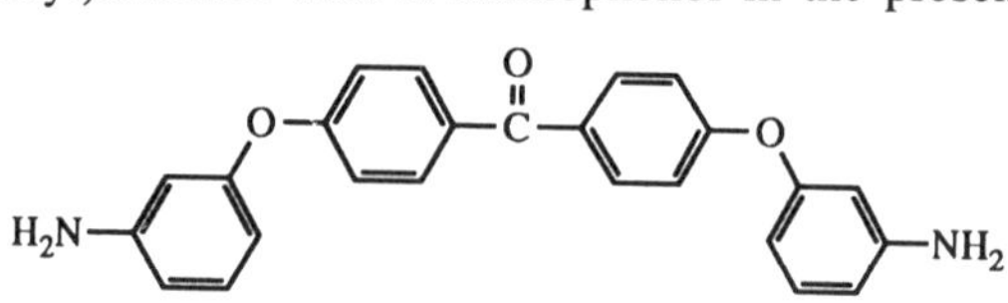

9-10

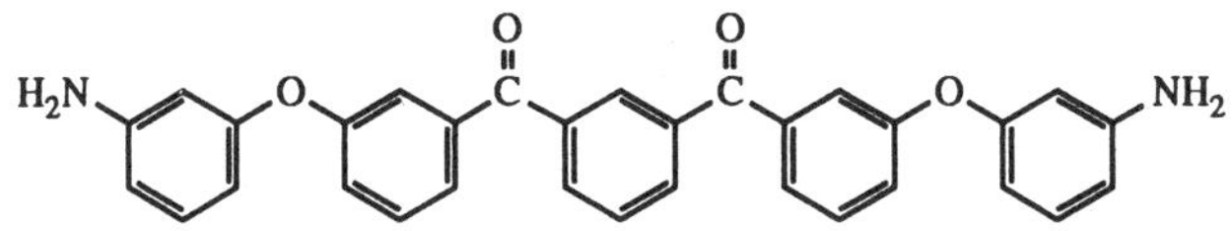

9-11

carbonates, or alkoxides of alkali metals in aprotic solvents (e.g., *N,N*-dimethylformamide) at 120–140 °C.[26]

Another member of this series, 3-amino-4′-(*m*-aminophenoxy)benzophenone, **9-12**, was prepared by the reaction of *m*-aminophenol with 3-amino-4′-chlorobenzophenone in the presence of an alkali base in an aprotic solvent. This diamine, **9-12**, was said to be an improved [vs 4,4′-bis-(*m*-aminophenoxy)benzophenone] curing agent for epoxy resins.[27]

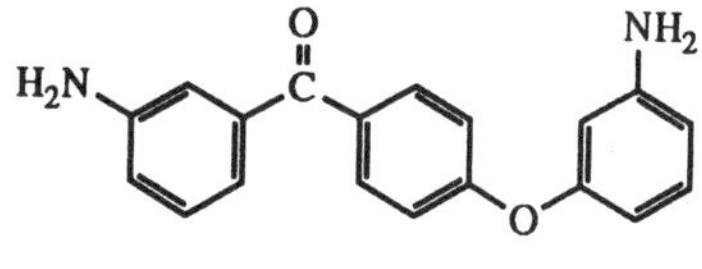

9-12

9.2.8. Diamines from 2,6-Dichloropyridine

These products are made by the reaction of 2,6-dichloropyridine with aminophenols to give mono- and/or di-(aminophenoxy)pyridines. Thus, a solution of *m*-aminophenol in dimethylsulfoxide containing a slight excess of flake KOH (over that needed to make the phenoxide) and toluene was refluxed under nitrogen for the azeotropic removal of water; 2,6-dichloropyridine (0.5 m/1.0 m *m*-aminophenol) was added and the charge was held at 150–160 °C /8 hr, then cooled, and filtered to give a 96% yield of crude 2,6-bis-(3-aminophenoxy)pyridine, **9-13**. The crude was purified by two recrystallizations from isopropanol. This material was said to be a useful precursor of polyimide resin of improved processability and adhesivity.[28]

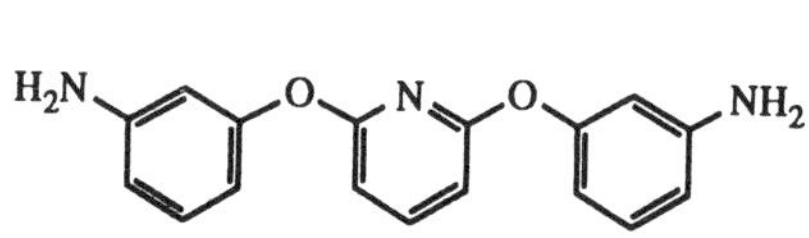

9-13

An alternate process for the diamine, **9-13**, was based on the reaction of 2,6-dichloropyridine with *m*-nitrophenol in the presence of a base to give 2,6-bis-(3-nitrophenoxy)pyridine. This nitro compound was reduced by catalytic hydrogenation or with hydrazine to give 2,6-bis-(3-aminophenoxy)pyridine, **9-13**, in 70–80% yield.[29] The intermediate, 2-(3-aminophenoxy)-6-chloropyridine, was prepared in 95% yield by the reaction of equimolar amounts of *m*-aminophenol and 2,6-dichloropyridine in the presence of potassium carbonate at 120–130 °C for 15 hr; this intermediate product is useful for making unsymmetrical diamines of this type.[30]

9.2.9. Diamines of Aromatic Ether-Sulfones

Examples of these materials are 4,4′-bis-(3-aminophenoxy)diphenylsulfone, **9-14**, and 4-(3-aminophenoxy)-4′-(4-aminophenoxy)diphenylsulfone, **9-15**.

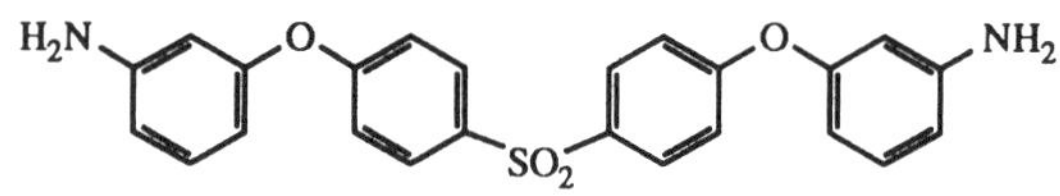

9-14

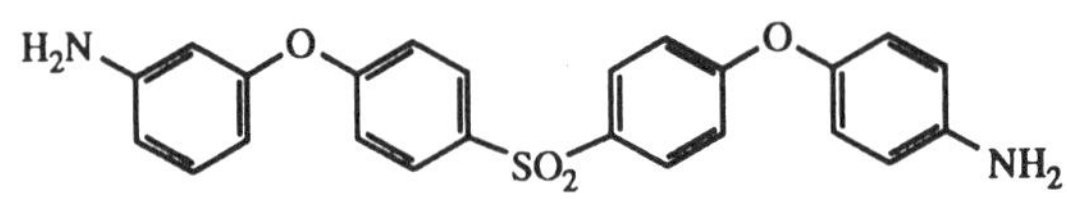

9-15

These materials are made starting with the condensation of *m*-aminophenol with 4,4′-dichlorodiphenylsulfone in the presence of alkali, by the same techniques described in Section 9.2.7 for diamines of aromatic ether-ketones and in Section 9.2.8 for diamines from 2,6-dichloropyridine. These bis-(aminophenoxy)diphenylsulfones are important hardeners for epoxy resins modified with thermoplastic polymers (such as polysulfones or phenoxy resins); the compositions are used with reinforcing fibers to make advanced composites with improved impact resistance and tensile properties plus high compressive strength.[31]

Oligomeric diamines of aromatic ether-sulfones, e.g., **9-16**, were prepared by related techniques and are useful as hardeners for epoxy resins which gave fiber-reinforced composites with excellent physical properties and low moisture absorption.[32]

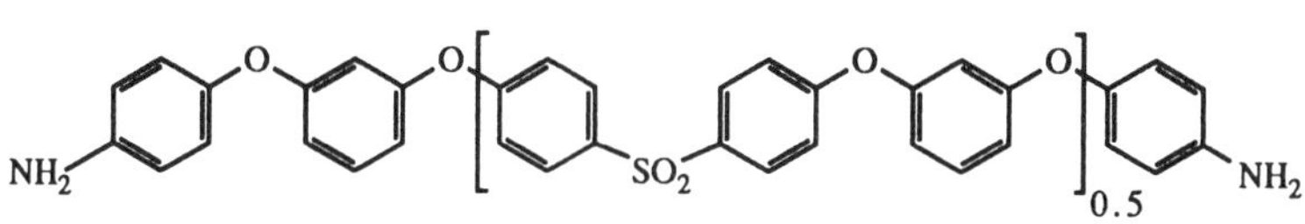

9-16

A further modification is the similar preparation of oligomeric diamines of aromatic sulfide–sulfones; these diamines were also intended as epoxy resin curing agents.[33]

9.2.10. 3,8-Dihydroxy-5*a*,10*b*-diphenylcumarano-2,3,2′,3′-cumaran

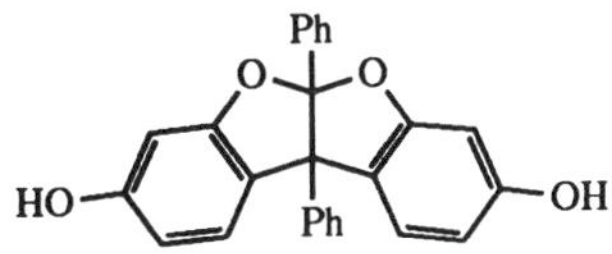

9-17

This unusal monomer, **9-17**. was made by the acid-catalyzed condensation of benzil, Ph–COCO–Ph, with resorcinol.[34] The compound is used in the preparation of high-performance polyesters and polyethers.

9.3. POLYURETHANES

9.3.1. HER as a Chain Extender

The preparation of resorcinol di-(2-hydroxyethyl) ether, **9-2**, tradenamed HER, has been described in Section 9.2.2. The product was test-marketed by Koppers Co. in the 1970s, then licensed to Anderson Development Co. who produced and sold small quantities of HER to makers of polyurethane elastomers.

HER is a useful chain extender for a variety of isocyanate-tipped prepolymers, derived from a variety of polyester-diols or polyether-diols reacted with toluenediisocyanate (TDI) or methylene-bis-(4-phenylisocyanate) (MDI). The polyester/polyether portion forms the soft segment and the chain extender, like HER, the hard segment of the elastomer. HER can be used in liquid casting systems applied, e.g., for making solid tires, and in thermoplastic elastomers and resins used, e.g., for injection molded products such as tubing. Polyurethane elastomers with HER as the chain extender can be formulated to give products with high abrasion resistance, low compression set, high tear resistance, with Shore D hardness in the 55–60 range, and comparable in many other physical properties to similar products made with the commercial chain extender, hydroquinone di-(2-hydroxyethyl) ether (Eastman Chemical's HQEE). The resorcinol-based HER has a lower melting point (89 °C) and, for this reason, offers an advantage in processing over the higher-melting (104 °C) HQEE. In the test-marketing program for HER to the polyurethane industry, favorable comments from the customers included the formation of less cheesy gels, less part shrinkage, better pot life, more workable finished product, the ability to cast larger and better parts, low heat buildup of cast wheels in dynamic tests, the elimination of surface cracking of parts, parts with less creep, and a product with better hydrolytic stability.

HER, **9-2**, can also be used in blends with other chain extenders. A recent example of the use of HER (and other diols) for the cure of isocyanate-tipped prepolymers useful for making fiber-reinforced composites, especially for dynamic applications, has been reported.[35]

A novel polyurethane elastomeric composition, in which HER can be used as the chain extender and which is filled with *ca* 2–8% polymeric microspheres, was intended for acoustic scattering and high reflection loss materials (soundproofing).[36]

9.3.2. Resorcinol Di-(hydroxypropyl) Ether

This product can be made by reacting resorcinol with propylene oxide or propylene carbonate in 1/2 mole ratio in the presence of a base, similar to the reaction of resorcinol with ethylene oxide or ethylene carbonate to make HER, as described in Section 9.2.2. The propoxylated resorcinol is a mixture of isomers, e.g., **9-18** and **9-19**.

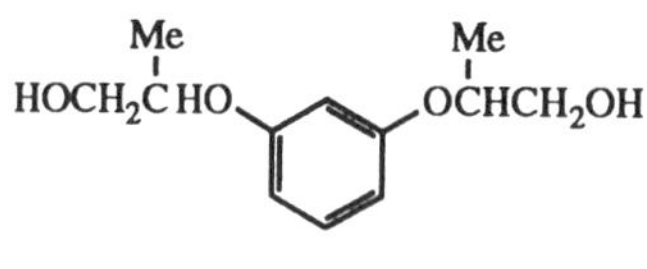

9-18

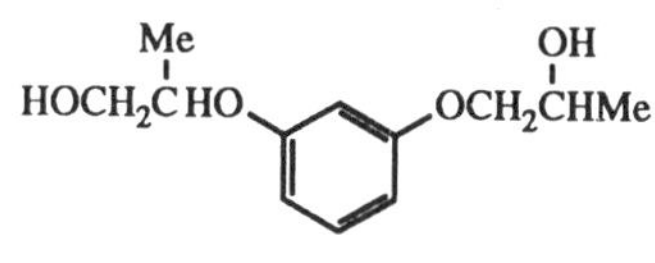

9-19

Recently, the preparation of a liquid MDI [methylene-di-(4-phenylisocyanate)] composition, liquid at 25 °C and storage stable, was patented; it was made by mixing MDI with resorcinol di-(hydroxypropyl) ether in an amount such that the product had an isocyanate group content of 10–30 wt%, and blending with 1–6 wt% tripropyleneglycol.[37] Such compositions were said to be useful for making polyurethane coatings, foams, and elastomers, including reaction injection molded (RIM) products.

9.3.3. Resorcinol/Formaldehyde Resins

An interesting use of resorcinol resins in structural fiber-reinforced materials involved the impregnation of a soft polyurethane sheet with *m*-phenylenediisocyanate, which was hot-pressed with a resorcinol/formaldehyde resin-impregnated glass cloth to prepare a laminate useful, e.g., for automobile interiors.[3]

9.4. EPOXY RESINS

Resorcinol, resorcinol derivatives, and resins made from resorcinol can be used both as the raw material for glycidyl (i.e., 2,3-epoxypropyl) ether-type epoxy resins as well as (partial) curing agents, fluxes, fortifiers, tougheners, accelerators, and adhesion promoters by reaction with epoxy resins. In these applications resorcinol usually competes with bisphenol A (4,4′-isopropylidene-bisphenol), which costs much less on a per pound basis (*ca* \$0.90/lb.) than resorcinol (*ca* \$2.45/lb.). However, on a hydroxyl equivalent weight basis the cost difference becomes much less, and the performance of the resorcinol-containing products is often advantageous in certain applications, particularly where high-temperature resistance is needed.

9.4.1. Resorcinol Diglycidyl Ether

This product, **9-21**, often named RDGE, is made by the reaction of resorcinol with excess epichlorohydrin, with gradual addition of aqueous sodium hydroxide at reflux, shown in simplified fashion by Eq. (9.1).

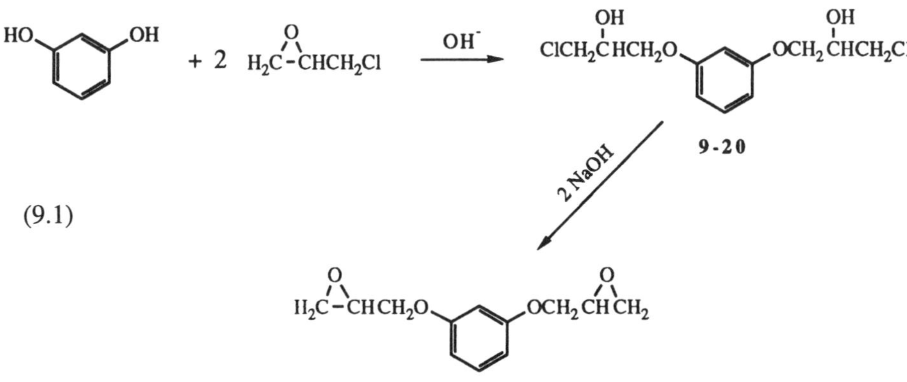

(9.1)

9-21

Water is removed via the epichlorohydrin/water azeotrope, sodium chloride is then removed by filtration, and the excess epichlorohydrin is distilled off. The residue is crude RDGE, which can be upgraded by vacuum distillation. A low-temperature version of this old process, requiring a very long reaction time, was described in a 1965 Japanese patent.[39]

The general process (largely aimed at bisphenol A diglycidyl ether) has been described in many modifications, including two-stage versions with the dichlorohydrin ether, **9-20**, being formed first, e.g., using divalent tin compounds[40,41] or alkylenephosphoranes[42] or tri(*n*-butyl)methylphosphonium bicarbonate[43] as the catalyst. Other process modifications involved the use of solvents in the reaction, e.g., ethers[44] or water-soluble alcohols and/or immiscible aromatic hydrocarbons.[45–47] Generally, these process modifications were said to give products of higher purity (lower epoxy equivalent weight), lower color, and/or lower residual chlorine content. Further, the residual chlorine content of crude epoxy resins was said to be improved by dissolution of the resin in a solvent and post-treatment with more aqueous NaOH.[48,49]

An unusual process involved the reaction of the dihydric phenol with excess epichlorohydrin using a quaternary ammonium or phosphonium compound, or a sulfonium compound, or a basic ion-exchange resin as the catalyst, stripping the mixture to remove excess epichlorohydrin and glycerine dihalohydrin, and repeating these treatments with epichlorohydrin/catalyst until the residual product contained <1000 ppm hydrolyzable chloride. This process did not use NaOH, so that the product was free of sodium chloride; the catalyst could be recovered by washing the product with water.[50]

The useful properties of RDGE, **9-21**, include a very low viscosity (300–500 cs/25 °C), high epoxy content (epoxy equivalent weight 130), low vapor pressure (bp 172 °C/0.8 mm Hg), high flash point (>350 F = 150 °C, open cup), a higher reactivity than the diglycidyl ether of bisphenol A (and many other phenolic glycidyl ethers) with amine or anhydride curing agents, and low shrinkage of the cured resins. These properties give excellent wetting of fillers/reinforcements and thorough penetration of fibers and molds to give faithful reproduction of intricate molds. RDGE can be used to prepare room-temperature curable adhesives, coatings, castings, potting compounds, tooling, and encapsulation resins. The cured resins have good chemical resistance, toughness, dimensional stability, high dielectric strength and electrical resistivity, high surface hardness and heat resistance. RDGE can be used as a reactive diluent for other epoxy resins.

Like some other (low molecular weight) epoxy resins, RDGE poses a skin irritation hazard, particularly after repeated skin exposure (sensitization). Crude RDGE, when fed in corn oil to rats or mice at 25–100 mg/kg doses, was reported to be carcinogenic (caused neoplasms of the forestomach).[51] A study of the mutagenic activity of RDGE reported that it had no clastogenic activity in *in vivo* tests.[52] At any rate, the use of RDGE requires rigorous attention to safe handling procedures.

The following excerpts are examples of the uses of RDGE reported in the literature:

Hard ether waxes were produced by the reaction of $C_{16\text{-}30}$ aliphatic alcohols with RDGE (and other polyepoxides) in the presence of alkaline or amine catalysts,[53] or $SnCl_4$, BF_3, or BF_3 complex (e.g., etherate) catalysts.[54]

For applications in space vehicle (heat shield) components, high-strength, readily processable, char-forming insulating materials were made with RDGE cured with dicarboxylic acid anhydrides; the use of 2-nitroresorcinol diglycidyl ether gave improved char formation.[55]

Strong, cold-vulcanizable polychloroprene adhesives were prepared using an equimolar mixture of RDGE/pentaerythritol in the recipe.[56]

Dental restoration cements with good physical properties (e.g., high compression strength) were prepared using inorganic fillers and the polymerizable binder, 1,3-bis-[2,3-di-(methacryloxy)propoxy]benzene, which was made from RDGE by reaction with methacrylic acid.[57]

Epoxy thermosets for fiber-reinforced composites, having improved toughness for high-performance (e.g., aircraft and aerospace) applications, were prepared using RDGE, 4,4′-diaminodiphenylsulfone curing agent, and amino-terminated aryl ether/sulfone or ketone oligomers as the toughening agent.[58] Another approach to high performance epoxy resins involved a RDGE/4,4′-diaminodiphenylsulfone resin system using amino-terminated polyimide oligomers as the toughening agent.[59]

In the area of imaging technology, a phenoxy resin made from RDGE, resorcinol, and 4,4′-sulfonyldiphenol was described to be suitable for vesicular imaging.[60]

An interesting application is the use of an epoxy resin with RDGE and polyamide-amine hardener to make gasohol-resistant adhesives with good adhesion to nylon, said to be useful in the fabrication of fuel filters and fuel tanks.[61]

Metal oxide- or silica-containing prepolymers and polymers, presumably for coatings, etc., were based on RDGE, using imidazole derivatives as the catalyst.[62]

Anion-exchange materials, such as membranes, were based on a RDGE/polyamine resin which was post-alkylated with alkyl halides, alkyl sulfates, or epichlorohydrin.[63]

An extremely tough, thermosetting bismaleimide reinforced resin system was based on carbon fibers impregnated with bismaleimides, 3,3′-diallylbisphenol A, RDGE, and a polyimide.[64]

Compositions useful for the sealing of semiconductors comprised, e.g., bisphenol A epoxy resin 20, RDGE 80, methyl-hexahydrophthalic anhydride 100, Fuselex RD8 370 (composition unknown), gamma-glycidoxypropyl-trimethoxysilane 1, 2,4-diamino-6-[2-undecylimidazole(1)]ethyl-s-triazine (accelerator) 2, and carbon black 0.2 parts.[65] Another type of composition useful as semiconductor sealants for chip-on-glass mounting contained 30–70% RDGE, 30–70% carboxylates of aliphatic diepoxides, dicarboxylic acid anhydride hardener, powered silica, and tris(dimethylaminomethyl)phenol as the cure accelerator (cure at 120 °C/5 hr).[66]

Porous ion-exchange membranes were prepared from RDGE and polyethylene-polyamine or polyethyleneimine, with cure at 60 °C; the membranes had up to 5.2 mg-equiv/h anion-exchanging capacity.[67]

A radiation-curable composition with improved heat resistance and dielectric properties consisted of 99.0–99.5 wt% acrylated RDGE and 0.5–1.0 wt% of various modifiers.[68]

9.4.2. Glycidyl Ethers of Resorcinol Derivatives and Resins

Estonian shale oil reportedly is a good source of methylresorcinols. Russian patents on epoxy resins based on these raw materials, as well on applications, began to appear in the 1960s. Later on, Japanese companies began to get interested in alkylresorcinol diglycidyl ethers. Thus, an early patent described the preparations of compositions for coatings, sealants, and insulating materials based on mixed bisphenol A diglycidyl ether/methylresorcinol diglycidyl ethers, hardener, plasticizer, and filler.[69] Several Japanese patents appeared on the preparation of methyl- and ethyl-resorcinol diglycidyl ethers and their resinification with amine curing agents; applications were said to be in paints, laminates, etc.[70]

Internally plasticized epoxy resins, giving cured products with good impact resistance, were based on dodecylresorcinol diglycidyl ether.[71]

The diglycidyl ether of 4-methylolresorcinol, prepared by the reduction of 2,4-dihydroxybenzaldehyde (beta-resorcylaldehyde) and reaction with epichlorohydrin and caustic, was reported to be much faster curing at room temperature with diethylenetriamine than RDGE or methylolated biphenol A diglycidyl ether. Such compositions were said to be useful, e.g., for encapsulating heat-sensitive electrical components used in ordnance fuzes.[72]

The acid-catalyzed condensation product of resorcinol with acetone, **9-22**, was converted to the triglycidyl ether; this product was cured with diaminodiphenylmethane to give a heat-resistant resin with good physical properties.[73]

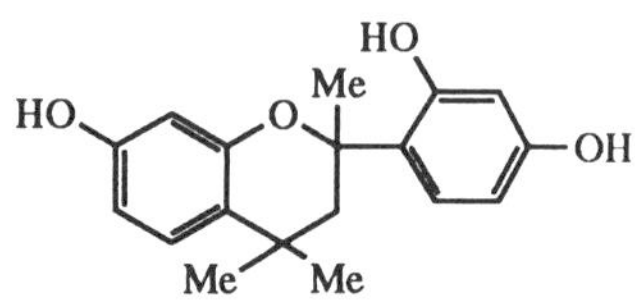

9-22

The still residue (RM 441 of INDSPEC Chemical Corp.) from the distillation of crude resorcinol from the benzene disulfonation/caustic fusion process, which is rich in 2,4,4′-trihydroxybiphenyl, was converted to the polyglycidyl ether; this product was useful, e.g., for making coaltar-based coatings with high resistance to attack by chemicals, solvents, moisture, and to degradation at high temperature.[74]

The glycidyl ethers of resorcinol/ formaldehyde resins, cured with conventional curatives, were shown in 1964 to have better high-temperature performance than similar systems based on bisphenol A diglycidyl ether.[75]

A polyglycidyl ether of a low molecular weight resorcinol/formaldehyde novolak resin (rich in methylene-bis-resorcinol) was found to have low viscosity at casting temperatures, to be faster curing, and the cured resin had better mechanical strength than the corresponding epoxy resins based on phenolic novolaks.[76] This resin was marketed by Dow as QX 3476.

The polyglycidyl ethers of R/P/F resins were reported to have excellent properties in conventional epoxy resin systems/applications, as well as being useful as stabilizers for polyvinyl chloride.[77]

The polyglycidyl ethers of resins made by the acid-catalyzed condensation of 2,6-dimethylolphenol or 2,6-dimethylol-*p*-cresol with resorcinol gave, after cure with 4,4′-diaminodiphenylmethane, resins with excellent flexural strength, flexural modulus and Rockwell hardness, superior glass transition temperatures (as high as 330 °C) and heat distortion temperatures (as high as 289 °C), much better than the bisphenol A-based epoxy resin, a phenol novolak epoxy resin , a cresol novolak epoxy resin or *N,N,N′,N′*-tetraglycidyl-*m*-xylylenediamine cured with 4,4′-diaminodiphenylmethane.[78]

Combustion-resistant polyolefins were prepared by using 3–9 parts of a fire-resistant additive made by successively adding to an alkylresorcinol epoxy resin, **9-23**, first 1 part methylphosphonic acid, then 1–3 parts red phosphorus.[79]

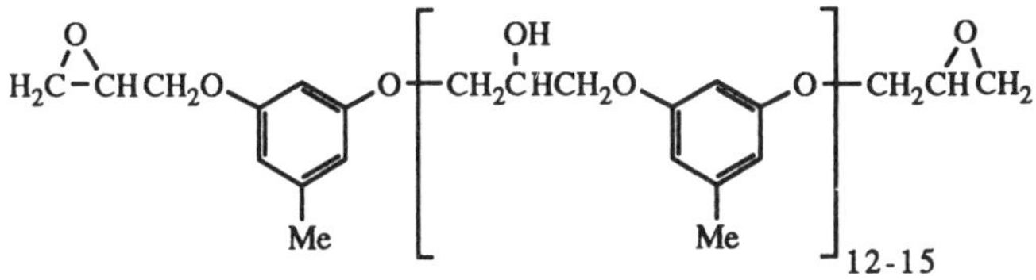

9-23

Urea/formaldehyde resins were modified with 3–25 phr alkylresorcinol novolak epoxy resins and 5–10 phr 5% aqueous hydrazine hydrochloride to give an adhesive with good adhesion, water resistance, and low toxicity in curing.[80]

Novel epoxy resins were made by condensing resorcinol (or other dihydric phenols) with 4,6-dichloromethyl-*m*-xylene, then with epichlorohydrin and alkali; these epoxy resins were said to be curable with conventional curing agents to give resins with high heat distortion temperatures.[81] Similar epoxy resins, based on the condensation product of resorcinol with *p*-xylylene dichloride, were modified with methacrylic acid, butyl acrylate, and methacryloxypropyl silcone oligomer, and the mixture was copolymerized, in the presence of azo-bis-isovaleronitrile as the catalyst, with vinyl siloxane to give an epoxy resin containing dispersed silicone rubber. This product was blended with additional epoxy resin, a phenol novolak, inorganic filler, and other additives, including triphenylphosphine catalyst, to give a potting compound for semiconductors which, after molding and curing at 180–185 °C, was resistant to heat, thermal shock, and solder.[82] Another version of this type of epoxy resin were the polyglycidyl ethers of the resin made by the acid-catalyzed condensation of resorcinol with alpha,alpha-dimethoxy-*p*-xylene; after curing this epoxy resin with 4,4′-diaminodiphenylmethane at 120–170 °C, a heat- and impact-resistant composition was obtained which had a heat distortion temperature 279 °C, glass transition temperature 328 °C, bending strength 9.9 kg/mm^2, flexural modulus 335 kg/mm^2, and elongation 5.6%, generally superior to similar resins based on the diglycidyl ether of bisphenol A.[83]

A resorcinol/*o*-cresol/formaldehyde novolak-based epoxy resin 100 parts, blended with, e.g., RDGE, containing 3–15 parts 2-methylimidazole (or similar compound), and 3–20 parts 2,4-diamino-6-vinyl-*s*-triazine/isocyanuric acid adduct plus a filler gave a composition which gave well-hardening films useful for resist ink for printed circuit boards.[84]

A mixture of bisphenol A epoxy resin 53 parts, resorcinol novolak epoxy resin 13 parts, tetrabromobisphenol A 34 parts and 0.01 part 2-methylimidazole was heated at 150 °C/4 hr, then mixed with 4.4 parts dicyandiamide and 0.2 part *N,N*-dimethylbenzylamine to give a composition suitable for making fire-retardant, glass cloth-reinforced composites.[85]

The epoxy resin derived from a R/F novolak resin or a resorcinol/*p*-hydroxybenzaldehyde resin and tetrabromobisphenol A was dissolved in a 1/1 mixture of 2-butanone (methyl ethyl ketone) and propylene glycol mono-methyl ether, heated at 150

°C for 3 min, and then a catalytic amount of 2-ethyl -4-methylimidazole was added to make a curable mixture for making copper-clad, glass-reinforced composites used for printed circuit boards with improved heat and solder resistance over prior art compositions.[86]

Triglycidyl-*m*-aminophenol, **9-24**, was prepared by reacting *m*-aminophenol with the appropriate amount of epichlorohydrin at 70 °C, heating in the presence of a catalytic amount of $PhCH_2N^+Et_3Cl^-$ at 100 °C to complete the formation of chlorohydrins, then cooling to 50 °C and adding 20% NaOH to produce the epoxy compound, **9-24**.[87]

An epoxy resin based on thio-bis-resorcinol, typified by structure **9-25**, was prepared and had good storage stability; this resin was blended with 4,4′-diaminodiphenylsulfone and cured for 1.5 hr/150 °C and 4 hr/190 °C. After 4 hr in boiling water, the cured resin showed a water absorption 2.2% and a molding tensile strength 10.0 kg/mm^2, tensile modulus 380 kg/mm, elongation 3%, and T_g 250 °C—substantially better than a similarly prepared and cured resin based on a phenol novolak epoxy resin/4,4′-diaminodiphenylsulfone.[88]

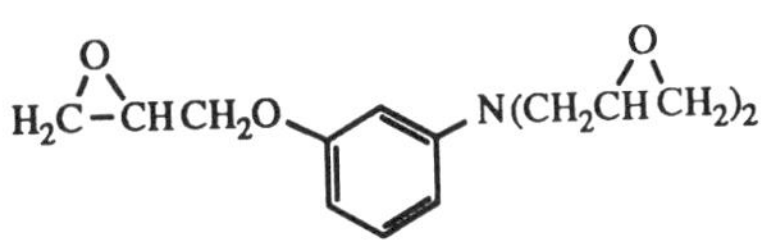

9-24

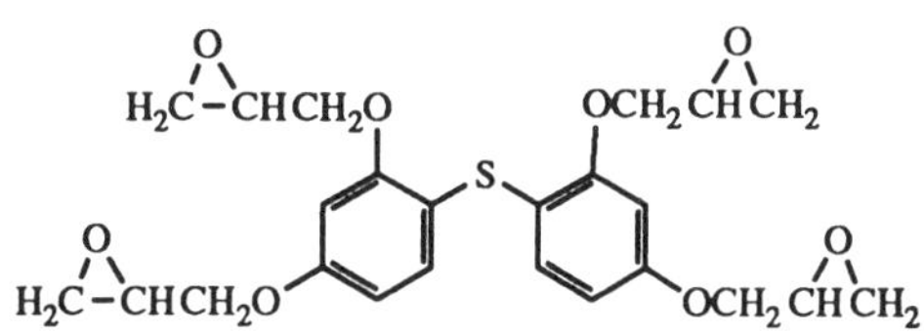

9-25

Calixarene-based epoxy resins, **9-26**, were prepared and were useful for castings, composites, adhesives, and tooling for the aerospace industry. In one example, epoxy resin **9-26** (R = H) was cured with 5-methylbicyclo[2,2,1]hept-5-ene-2,3-dicarboxylic

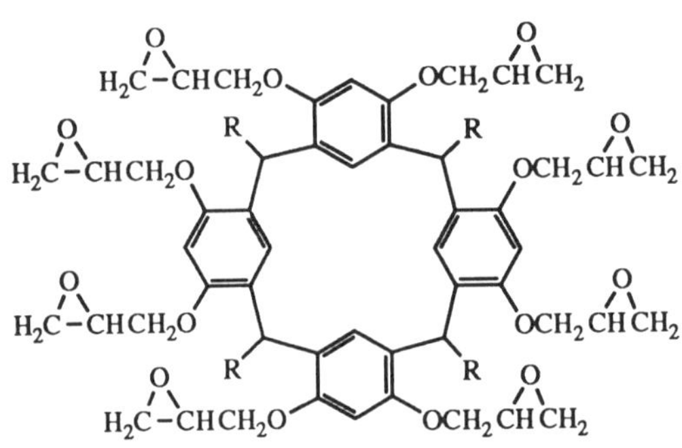

9-26
(R=H, $C_{1\text{-}12}$ alkyl)

acid anhydride (methylnorbornene-2,3-dicarboxylic anhydride) using 1-methylimidazole as the catalyst at 150–250 °C to give a casting with T_g 347 °C.[89]

A similar calixarene-derived epoxy resin, **9-26** (R = Me), was mixed with 4,4′-diaminodiphenylsulfone, modified with Hycar 1300X13 (15%), $BF_3 \cdot EtNH_2$ catalyst was added, and the mixture was poured into a mold and cured at 100–180 °C; the casting had a fracture toughness 0.68 MPa-$m^{0.5}$ and a glass transition temperature T_g 285 °C.[90]

9.4.3. The Use of Resorcinol, Resorcinol Derivatives, or Resins as Curing Agents for Epoxy Resins

The term, curing agent, is meant to include the partial addition to or cure of epoxy resins to give products which retain some epoxy functionality and have higher molecular weight, also called advanced epoxy resins. The cure occurs by the addition of epoxide to the phenolic hydroxy group. A fully cured product would have no epoxy functionality and would be termed a phenoxy resin. The following examples demonstrate the scope of the commercial applications of these processes.

Thus, novolak epoxy resins were reacted with 10–33 mol% of benzenediols, such as resorcinol, to yield solid, fusible products which could be further converted to give hard, infusible resins with high chemical resistance.[91]

An electrical insulating material with excellent tensile strength, hardness, shear strength, bending resistance, and heat deformation temperature was prepared by mixing 25 parts of bis-(2,3-epoxycyclopentyl)ether and 25 parts of a novolak epoxy resin with 50 parts of a resorcinol/formaldehyde resin (Penacolite B-1A of Indespec Chemical Corp.), impregnating glass fiber-reinforced mica paper with this mixture, and curing.[92]

Metal adhesives with high metal-to-metal bond strength (the examples showed the bonding of aluminum) were formulated by mixing a low molecular weight phenol novolak epoxy resin, the triglycidyl ether of 1,1,3-tris-hydroxyphenyl-propane, nylon 11, a 1/1 mixture of resorcinol/catechol as a "fluxing" agent, and melamine as a curing agent to give an adhesive which was cured at 350 F (176 °C).[93]

A liquid mixture suitable as a casting, laminating, or impregnating resin or as an adhesive was prepared, consisting of a liquid phenolic novolak epoxy resin (epoxy equivalent weight 150–300), an approximately equivalent amount of 2-methylresorcinol (2,6-dihydroxytoluene), and a tertiary amine, imidazole, quaternary ammonium halide, or alkali metal alcoholate as the catalyst.This mixture was cured at elevated temperature (e.g., 160 °C). The cured product had high impact strength, heat stability, and resistance to hydrolysis. In view of the high crosslinking density in a resin made up primarily of aryl units, these excellent properties were said to be surprising.[94]

Curable, solventless epoxy resin compositions with long pot life, e.g., one year at room temperature, contained up to 15 wt% of resorcinol or bisphenol A as the accelerator and a small amount of a metal (e.g., Al, Zn, Ti, Zr) acetylacetonate as the curing catalyst. The cured resins had good chemical and hydrolysis resistance, and good electrical and heat aging properties.[95]

An epoxy resin composition of 86 parts diglycidyl ether of bisphenol A, 14 parts novolak epoxy resin, 30 parts xylenol, and 14 parts resorcinol was homogenized, then blended with 37 parts of the 1/1 dipropylenetriamine/formaldehyde reaction product to give an adhesive particularly useful for cementing fireclay or basalt linings in furnaces.[96]

A fire-resistant (UL-94 flammability rating V-0) epoxy resin laminate with high glass transition temperature (235 °C), low coefficient of thermal expansion, and good electrical properties was prepared by dissolving a resorcinol/formaldehyde resin (Penacolite B-19S of INDSPEC Chemical Corp.) 12 parts, tetrabromobisphenol A 5 parts, and *N*-phenyl-*N'*,*N'*-dimethylurea 0.01 part in a mixed methyl amyl ketone/methyl isobutyl ketone solvent at 70 °C, adding bisphenol A epoxy resin (Epikote 1031 of Shell Chemical Co.) 38 parts and tetrabromobisphenol A diglycidyl ether. This solution was used to impregnate glass fibers at 163 °C in order to prepare a prepreg, which was press-cured at 200 °C/200 psi to give the final laminate.[97]

An advanced (with resorcinol) epoxy resin composition was prepared from the triglycidyl ether of 1,1,1-tri-(hydroxyphenyl)methane 100 parts (0.61 epoxy equiv), a dihydric phenol such as resorcinol 3.5 parts (0.064 hydroxyl equiv) and polyoxyethyleneglycol 0.3 part. This mixture was treated with 45% aq. KOH to reduce the residual hydrolyzable chlorine content, taken up in a 75/25 2-butanone/toluene solvent, neutralized with dilute phosphoric acid, washed with water, and stripped of solvent under vacuum to leave a resin with soft. pt. 82 °C and epoxide equivalent weight 197. This resin was a nonsintering solid useful for preparing electrical potting and molding compositions.[98]

A fortified (increased in strength and modulus, yet not brittle) epoxy resin was prepared by compounding a resin-forming polyepoxide, an amine curing agent, and a fortifier of a reaction product of an aromatic polyol such as resorcinol with a diepoxide having one epoxy group less reactive than the other (e.g., vinylcyclohexene dioxide), and heat-curing the mixture. Nonbrittle resins with tensile strength as high as 140 MPa and moduli as high as 4500 MPa were obtained.[99]

Linear phenolic, hydroxyl-terminated resins were prepared by reacting a low molecular weight diepoxy resin with dihydric phenols (bisphenol A preferred, resorcinol was usable); these adducts, mixed with imidazole cure accelerator, were used as curing agents for epoxy powder coatings.[100]

Thermosetting epoxy resin mixtures were formulated consisting of a mixture of trifunctional and difunctional epoxy resins (e.g., epoxy novolak resin and the diglycidyl ether of bisphenol A), dihydric phenol (e.g., bisphenol A or resorcinol), and cure accelerator (e.g., *N*-acylimidazole or imidazole). These mixtures may also contain poly(ether)imides, poly(amid)imides, or poly(ether)sulfones as plasticizers. These products had good storage stability, low viscosity at room temperature, and were useful for moldings, adhesive films, and for making prepregs. The cured resins had good heat stability, impact bending strength, and high glass transition temperature (up to 226 °C).[101]

Phenolic resoles for laminates with excellent heat resistance, dimensional stability, and punchability were made from a brominated epoxy resin (e.g., the diglycidyl ether of tetrabromobisphenol A) reacted with resorcinol in the presence of a catalytic amount of triethylamine, then with a phenol/formaldehyde resole, then triphenylphosphate was added to give a 60% solids varnish used to prepare (Kraft paper) laminates.[102]

Epoxy resin electrical insulators with improved heat resistance were prepared by reacting oligomeric acrylonitrile-allylglycidyl ether copolymer with a low molecular weight R/F resin.[103]

Hydroxyl-terminated azomethines such as compound **9-27**, prepared from terephthaldialdehyde and *m*-aminophenol, were curing agents for epoxy resins to give materials with high T_g values.[104]

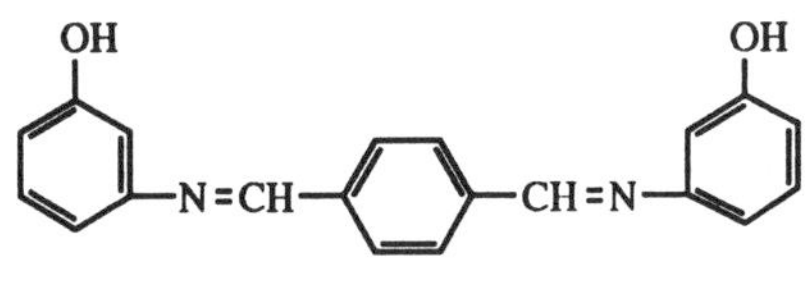

9-27

A primer for improved bonding of polymers to metals, minerals, and plastics was prepared which consisted of, e.g., 40 parts of a 10/90 blend of 3-glycidoxypropyltrimethoxysilane/Cymel 303, 10 parts *m*-aminophenol, and 50 parts methanol. In one test, this primer was applied to glass, air-dried, then bonded with PVC plastisol at 175 °C/7 min; the PVC film adhered strongly at 25 °C, dry or after soaking in water for 1 day/25 °C.[105]

An anticorrosive electrodeposition coating was prepared consisting, e.g., of 3 parts bisphenol A epoxy resin, 0.6 part Hycar CTBN 1300X13 (carboxylated nitrile rubber), 0.6 part resorcinol, diethanolamine curing agent, and butoxyethanol-blocked 3/1 TDI (toluenediisocyanate)/trimethylolpropane adduct. After electrodeposition of this material on steel and cure, a coating was obtained which had good low-temperature bendability and good impact strength.[106] A similar coating composition for primers was patented which contained epoxy resin, CTNBR, resorcinol, dicyandiamide, and 2-phenyl-4-methyl-5-methylolimidazole as a cure accelerator.[107]

1,3-Dihydroxy-4, 6-bis-[alpha-methyl-alpha-(4′-hydroxyphenyl)ethyl]benzene, **9-28**, was prepared by the acid-catalyzed condensation of resorcinol with

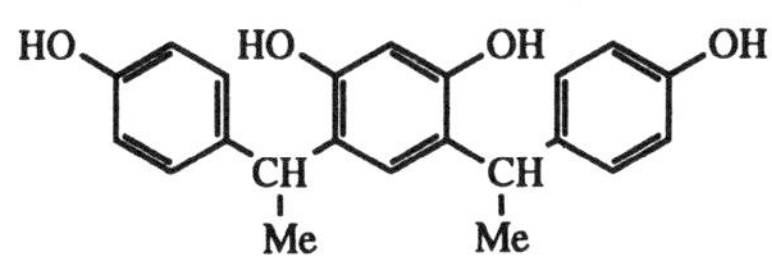

9-28

4-isopropenylphenol. The product was useful as a hardener for epoxy resins or for the preparation of its tetraglycidyl ether, a new epoxy resin.[108]

9.5. POLYESTERS

The materials considered here are partially aromatic polyesters, wholly aromatic polyesters (polyarylates and liquid crystalline aromatic polyesters), polycarbonates, and various copolymers.

There is mounting evidence that the use of resorcinol or resorcinol derivatives, even in relatively small amounts, can contribute very valuable properties to such polymeric compositions. Data on actual commercial products are given where available, but are difficult to discern from the literature because these polymers are sold under tradenames and detailed compositions are not given. The following review makes it clear that this is a very active, current area of research and development.

9.5.1. Partially Aromatic Polyesters

The polycondensation of dimethyl resorcinol-4,6-dicarboxylate (dimethyl 4,6-dihydroxyisophthalate) with various alkyleneglycols in 1-methylnaphthalene at reflux, using 0.1 wt% zinc borate as the catalyst, gave thermoplastic polyesters with high melting points. For example, when ethylene glycol was used, the polyester had mp >340 °C, and when propylene glycol was used, the polyester had mp 328 °C. These polymers could be molded into various shapes or extruded into fibers that could be oriented by drawing.[109]

A polyester block copolymer containing a recurring prepolymer of equimolar parts of 1,12-dodecanediol and alpha,alpha,alpha′,alpha′-tetramethylpimelic acid and a recurring prepolymer made from equimolar parts of resorcinol and 2,3-dichloro-2,3-dimethylsuccinic acid gave a material of high thermal and chemical stability, suitable for making coatings, films, moldings, fibers, filaments, or fabrics.[110]

A polyester, prepared from dimethyl (*m*-phenylenedioxy) diacetate (the dimethyl ester of **9-3**) and ethylene glycol in the presence of dibutyltin oxide as the catalyst, could be shaped and molded to produce useful, high-energy radiation-sterilizable surgical products.[111]

A copolyester was prepared by reacting a base of polyester, made by polycondensation of 1,3-propanediol and dimethyl (*m*-phenylenedioxy)diacetate, with *p*-dioxanone. This copolyester was useful for making radiation sterilizable goods which were also absorbable in the body.[112]

Transparent aromatic polyesters having low oxygen permeability (0.61 cm^3-min/m^2-day-atm for a 200-μm-thick film) was prepared by the polycondensation of isophthalic acid, ethyleneglycol, and (*m*-phenylenedioxy)diacetic acid, **9-3**; these products were useful for making gas-barrier packaging materials.[113]

Polyester resins with low gas permeability were made by the self-condensation of *m*-(2-hydroxyethoxy)phenoxyacetic acid, **9-4**. These resins were useful in making packaging materials for consumable products such as foodstuffs and carbonated beverages.[114]

A well-defined poly(ester crown ether), **9-29**, based on bis-(5-carboxy-1,3-phenylene)-32-crown-10 and bisphenol A, was prepared and described to be of interest in complexation and host–guest chemistry.[115]

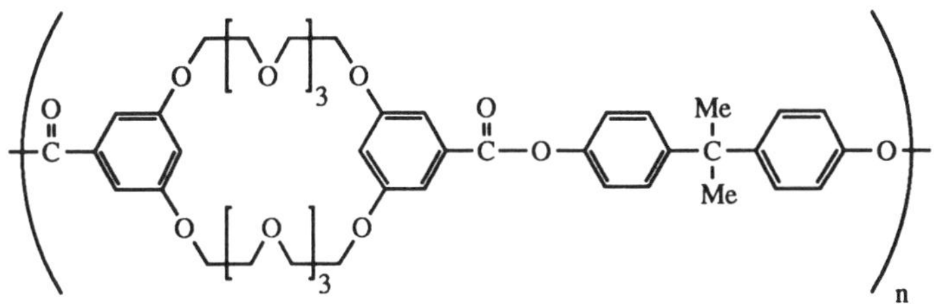

9-29

Copolyesters were made from isophthalic acid/terephthalic acid, ethyleneglycol/resorcinol di-(2-hydroxyethyl) ether (HER, **9-2**), and a small amount of a polyfunctional hydroxy compound. The copolymer was molded into bottles which had good transparency, gas-barrier properties, and attractive economics.[116]

Laminates from polyolefins, polystyrene, and/or polyesters of isophthalic acid and (*m*-phenylenedioxy)diacetic acid, **9-3**, were prepared by coextrusion and were suitable for packaging materials of low oxygen permeability.[117]

9.5.2. Polyarylates

These polymers are made from aromatic diols and aromatic dicarboxylic acids. Much of the early development was done by the school of V.V. Korshak in the 1960s in the USSR. Commercial products are still in the early, growth-from-a-small-base stage of their life cycle (see, e.g., Ref. 118). There seems to be considerable current interest in using resorcinol in such products, as exemplified by the following patents:

Copolyesters having a high bending modulus were prepared from aromatic dicarboxylic acids, especially terephthalic acid, a halo- or C_{1-3}-alkyl-hydroquinone, and resorcinol.[119]

Copolyesters with good melt processability and electrical properties were obtained from halogen-containing dicarboxylic acids, terephthalic acid, hydroxybenzoic acid, and aromatic diols (including resorcinol).[120]

Fiber-forming aromatic copolyesters having high tenacity and modulus were prepared from 55–65 mol% *p*-hydroxybenzoic acid, 15–25 mol% terephthalic acid, 10–23 mol% 2,6-naphthalenediol or 2,6-dihydroxyanthraquinone, and 2–10 mol% resorcinol or *m*-hydroxybenzoic acid.[121]

Heat-resistant polyester molding materials with good gas-barrier (CO_2-barrier) properties, useful for packaging, were made by the polymerization of diphenyl isophthalate with resorcinol.[122]

Fire-resistant polyarylate fibers were made by the polycondensation of equimolar amounts of terephthalic acid with resorcinol diacetate; the polymer was melt spun and drawn to give fibers with a limiting oxygen index 31, compared to LOI 21 for polyethyleneterephthalate (PET) fibers.[123]

Heat-resistant (5 wt% loss at 438 °C), high glass transition temperature (T_g 261 °C), fluorescent polymers were made by the polycondensation of equimolar amounts of fluorescein and 4,4′-dichlorodiphenylsulfone in *N*-methyl-2-pyrrolidinone containing potassium carbonate at 205–215 °C.[124]

A substrate for optical recording disks with small double refraction contained a thermoplastic polyarylate based on resorcinol/iso- or terephthalic acid; this material gave accurate recordings and easy reading.[125]

High-melting, high-tenacity, low-elongation, and low-shrinkage fibers were prepared by the polycondensation of *p*-acetoxybenzoic acid, hydroquinone diacetate, resorcinol diacetate, and isophthalic acid, followed by melt spinning and drawing.[126]

Aromatic polyesters prepared, e.g., from polyethylene terphthalate (PET) and 0.1–30 phr 2-naphthol, triphenylcarbinol, and/or benzenediol (including resorcinol) gave transparent, impact-resistant films with low moisture and oxygen permeability.[127]

Aromatic polyesters (and copolyesters) prepared from 3,8-dihydroxy-5*a*,10*b*-diphenylcumarano-2,3,2′,3′-cumaran, **9-17**, and terephthaloyl or isophthaloyl chloride had high glass transition temperatures (T_g as high as 357 °C), high heat and oxidation resistance, and were flame retardant; they were especially useful for making electrical parts such as insulators, lamp holders, light reflectors, and auto parts such as, e.g., clamps in the motor compartment.[128]

A polyarylate was prepared by the polycondensation of resorcinol with naphthalene-1,6-dicarboxylic acid dichloride; this material had T_g 157 °C, decomposition temperature 450 °C, and tensile modulus 240 kg/mm^2.[129]

A melt-processable polyarylate was prepared by the polycondensation of a 30/10/10 (mol) mixture of 4-hydroxybenzoic acid/terephthalic acid/resorcinol in the presence of acetic anhydride and pyridine at 100–210 °C; this material had a heat distortion temperature 166 °C, flexural modulus 32,000 kg/cm^2, and had good color.[130]

A copolyester prepared from a 96/4 mixture of dimethyl terphthalate/(*m*-phenylenedioxy) diacetic acid and ethylene glycol could be formed into blow-molded bottles and oriented films.[131]

A polyarylate made by the polycondensation of 1.322 parts resorcinol, 0.844 part napthalene-1,3-dicarboxylic acid dichloride, 0.844 part 1,6-naphthalenedicarboxylic acid dichloride, and 0.844 part 1,7-naphthalenedicarboxylic acid dichloride had a T_g of 175 °C, decomposition temperature 465 °C, and tensile modulus 22 kg/mm^2.[132]

9.5.3. Thermotropic Liquid-Crystal Polymers (LCPs)

These high-performance engineering resins are valued for their high mechanical strength and stiffness, and their high temperature and chemical resistance. In the context of this chapter, these polymers consist of aromatic rings linked by ester structures. For high performance, the main monomers have a *para*-orientation of substituents on the aromatic ring and relatively expensive comonomers are being used to maximize performance, including, e.g., *p*-hydroxybenzoic acid, 4,4′-dihydroxy-biphenyl (4,4′-biphenol), 2,6-naphthalenedicarboxylic acid, and 6-hydroxy-2-naphthoic acid. A good, relatively short review of the synthesis, processing, properties, and applications of the LCPs, with references, was given by Yoon *et al.*[133]

Resorcinol has been used in LCPs as a comonomer to aid in reducing the high melt processing temperatures and viscosity while maintaining the high-performance properties. Indeed, the cost of resorcinol is modest, and therefore attractive, compared to the several other, commercially used monomers in this field. The following examples will serve as an illustration of this application:

Liquid crystalline copolyesters with a melt temperature low enough to process in conventional equipment were prepared, e.g., by polycondensation of 33.2 parts terephthalic acid, 64.8 parts 2,6-naphthalenedicarboxylic acid, 38.8 parts resorcinol diacetate, and 66.6 parts hydroquinone dipropionate. This polymer could be melt-spun at 360 °C into fibers with a tenacity >3g/denier.[134]

Thermotropic liquid-crystalline polyester-carbonates were made by polycondensation of hydroquinone, resorcinol, 2,2-dimethyl-1,3-propanediol, methylhydroquinone, diphenyl terephthalate, and diphenyl carbonate in the presence of a catalytic amount of tetrabutyl titanate. These polymers were suitable for making fibers which retained high tenacity and tensile modulus at elevated temperatures.[135]

A series of copolymers of general structure, **9-30**, was prepared with resorcinol as a nonmesogenic unit and a decamethylene spacer; the resorcinol comonomer content was varied from 10–80 mol%. The properties of these materials were studied. With increasing resorcinol content the melt temperature, T_M, of the copolymer and the mesophase range both decreased.[136]

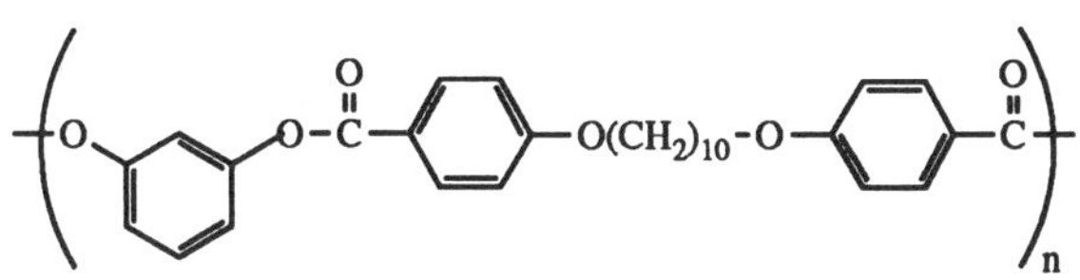

9-30

The polycondensation product of 0.2 m terephthalic acid, 0.1 m *tert*-butylhydroquinone, 0.1 m resorcinol, and 0.86 m acetic anhydride gave a fiber-forming, liquid-crystalline melt at 280 °C.[137]

Wholly aromatic, thermotropic polyesters were prepared by the polycondensation of, e.g., 4.5 m *p*-hydroxybenzoic acid, 2.7 m terephthalic acid, 2.85 m isophthalic acid, 1.2 m 4,4′-dihydroxydiphenyl, and 0.3 m resorcinol in acetic anhydride. In one example in this publication, *m*-aminophenol was substituted for resorcinol, thus introducing a small amount of amide linkages into the copolymer. These materials had excellent dimensional stability, good resistance to chemicals and hydrolysis, had low flammability, and gave products with a smooth, shiny and abrasion-resistant surface, and low color.[138] In a related publication, 5(?)-hexylresorcinol was reportedly used as a comonomer.[139]

Aromatic copolyesters capable of forming an anisopropic melt were prepared, maximizing relatively inexpensive comonomers, including resorcinol or *m*-aminophenol, by a dispersion polymerization process; the products had good strength, stiffness, and processabiity, and could be used for molding powders and high tenacity fibers.[140]

Thermotropic liquid-crystalline polyesters having high tensile strength and modulus, and optical anisotropy at 290 °C, were prepared by transesterifying oligomeric PET with *p*-hydroxybenzoic acid at 260 °C, then adding hydroquinone, resorcinol, terephthalic acid, and acetic anhydride, and heating to 290 °C.[141]

Heat-resistant, liquid crystalline polyesters with good processability and solder heat resistance were prepared, e.g., by the polycondensation of 88 mol% *p*-hydroxybenzoic acid, 2 mol% 6-acetoxy-2-naphthoic acid, 5 mol% resorcinol diacetate, and 5 mol% isophthalic acid, using 0.05% potassium acetate as the catalyst, the temperature of polymerization being programmed from 260–350 °C. The polymer obtained had optical anisotropy at 310 °C and a heat distortion temperature 220 °C.[142]

Melt-processable, heat-resistant, liquid-crystalline polyesters with good mechanical properties were prepared, e.g., by the polycondensation at 270–300 °C of 1152 parts *p*-acetoxybenzoic acid, 353 parts resorcinol diacetate, and 299 parts terephthalic acid in alkylbiphenyl solvent using cobalt(II) acetate tetrahydrate as the catalyst. The polymer thus produced showed optical anisotropy of the melt at 325 °C and had no weight loss at 300 °C.[143]

A chemically and heat- and high temperature-shrinkage resistant, all-aromatic polyester (mp 310 °C, T_g 165 °C) was prepared from 0.300 m *p*-acetoxybenzoic acid, 0.608 m hydroquinone diacetate, 0.030 m resorcinol diacetate, and 0.609 m isophthalic acid; a film was prepared from this polymer by melt processing and biaxial orientation.[144]

9.6. POLYAMIDES

All of the aromatic diamines described in Section 9.2 (monomers) have been used for the preparation of polyamides.

The most important of these products is Teijin's HM-50 high-performance aramid, in which 3,4′-diaminodiphenyl ether, **9-5**, is used as a comonomer with *p*-phenylenediamine and terephthalic acid. Teijin announced a plant for this fiber with 500 mt/yr capacity in mid-1987 and expected to triple production within one year. At the time Teijin said that HM-50 had the highest tensile strength (310 kg/mm^2) of all commercially available fibers (this allowed for the reductions in the weight of composites), excellent impact strength, cut resistance, chemical and thermal resistance, and high dimensional stability.

Here is a glimpse of some other applications of the diamines from *m*-aminophenol described in Section 9.2.

4,4′-Bis-(3-aminophenoxy)diphenylsulfone, **9-14**, was used in preparing aramids useful, e.g., for protective coatings and to make films controlling the orientation of liquid-crystal display elements.[145]

The preparation of amorphous aramids with low (*ca* 128–169 °C) glass transition temperature, offering good processability with high thermoxidative stability, from 4,4′-bis-(*m*-carboxyphenoxy)-diphenylsulfone or -benzophenone and 1,3-bis-(3-aminophenoxy)benzene, **9-9** (X = H), was studied.[146]

Copolyamides were prepared using 3,4′-diaminodiphenyl ether, **9-5**, as one of the two diamine components and an arylenedicarboxylic acid, with some of the amide nitrogen substituted with propargyl groups. These materials were curable, photosensitive, heat and chemical resistant resins, useful, e.g., for electron beam resists; these resins did not evolve gases or water on curing.[147]

An injection molding and casting material was prepared by blending an aramid (containing 3,4′-diaminodiphenyl ether, **9-5**, as a comonomer) and caprolactam with a polymerization initiator. This material could be converted to fine fibers; moldings could be made by the reaction injection method (RIM) to give parts with excellent rigidity, wet dimensional stability, and gloss.[148]

Melt-processible, film-forming, partially aromatic polyamides with $T_g > 150$ °C and $T_m < 350$ °C were made from 3,4′-diaminodiphenyl ether, **9-5**, diphenyl terephthalate, or isophthalate and $C_{6\text{-}12}$-alkylenediamines or -dicarboxylic acids.[149]

Fibers made from terephthaloyl chloride/*p*-phenylenediamine/3,4′-diaminodiphenyl ether (**9-5**) copolymer were used to reinforce epoxy resins; copper foil laminates made with this material showed superb dimensional stability.[150]

A modified polymer based on an acrylic (e.g., methyl methacrylate) matrix polymer containing >50% of an aramid made from 3,4′-diaminodiphenyl ether, **9-5**, and terephthalic acid, where the amide nitrogen was partially substituted by *n*-octyl- and *p*-vinylbenzyl groups, was polymerized using dibenzoyl peroxide catalyst at 60–100 °C. The resulting material had much better physical properties (e.g., bending, tensile, and compressive strength) than poly(methyl methacrylate).[151]

Aromatic polyamide oligomers with crotonylamido end groups were made, reacting isophthaloyl chloride in *N,N*-dimethylformamide with 3,4′-diaminodiphenyl ether, **9-5**, and then with crotonyl chloride in the presence of triethylamine at *ca* 10 °C. The polyamide had mp 165–180 °C and was heat-curable to produce a heat-resistant polymer.[152]

High-performance, amorphous polyamides were prepared, e.g., by the direct solution polymerization of cyclohexane-1,4-dicarboxylic acid and 4,4′-bis-(3-aminophenoxy) diphenylsulfone, **9-14**.[153]

9.7. POLYIMIDES, INCLUDING MALEIMIDE-TYPE POLYMERS

High-performance materials of this type were prepared using *m*-aminophenol and derived diamines, described in Section 9.2, as starting materials or comonomers. An endless variety of products resulted from this work. Even resorcinol has been used in this area, e.g., in the preparation of curing agents for maleimides. A selection of pertinent references follows.

Polyimides with improved elasticity and thermal plasticity were prepared by the condensation of aromatic diamines, such as 1,3-bis-(4-aminophenoxy)benzene, **9-7**, with dianhydrides of aromatic tetracarboxylic acids, such as the dianhydride of resorcinol di-(3,4-dicarboxyphenyl) ether.[154]

The preparation of thermoplastic polyimide adhesives, sufficiently fluid at high temperature for processing, with high glass transition temperatures, good mechanical properties at temperatures up to 230 °C, and with good adhesive strength, was studied at NASA Langley. One of the new polyimides, LARC-STPI, was made from an aromatic tetracarboxylic acid dianhydride with a carbonyl bridge between aromatic rings and both *meta*-substituted and oxygen-bridged aromatic diamines.[155]

Heat-resistant adhesives (good physical properties at 240 °C) were polyimides made from 1,4-bis-[4-(3-aminophenoxy)benzoyl]benzene, **9-11**, and pyromellitic dianhydride.[156] Another, related type of heat-resistant adhesive was made by reacting 1,3-bis-(3-aminophenoxy)-5-chlorobenzene, **9-8** (X = Cl), with pyromellitic dianhydride in *N,N*-dimethylformamide; the resultant polyamic acid solution was cast on glass and heated for 3 hr at 100–300 °C to give a polyimide which was pressed between cold-rolled steel plates at 300 °C to give a composite with high adhesive strength at 200 °C.[157]

A polyimide resin composition, 100 parts, e.g., prepared from 4,4′-bis-(3-aminophenoxy) benzophenone, **9-10**, and the dianhydride of pyromellitic acid, and 5–100 parts fluoropolymer, such as polyvinylidene fluoride or polytetrafluoroethylene, was reported to have very low coefficients of friction, outstanding wear resistance, and excellent high-temperature resistance. These materials could be injection molded to make precision instrument parts such as bearing materials, cams, rollers, piston rings, gears, and others.[158]

Melt-fusible co-polyimides were prepared from pyromellitic dianhydride and two aromatic diamines, e.g., 4,4′-diaminodiphenyl ether and 1,3-bis-(3-aminophenoxy)benzene, **9-9**. Reinforced composites could be prepared from these co-polyimides to make prepregs and tapes, which were then shaped into printed circuit boards or aircraft panels.[159]

Melt-moldable, crystalline polyimides, having a melting point lower than their thermal decomposition temperature and with improved processability, were made by the polycondensation/imidization of aromatic dianhydrides, e.g., benzophenone- tetracarboxylic acid dianhydride and aromatic diamines (optionally, mixtures of diamines), such as 1,3-bis-(3-aminophenoxy)benzene, **9-9**.[160] Similar polyimides, made from dianhydrides of aromatic tetracarboxylic acids and several of the aromatic diamines listed in Section 9.2 above, were colorless materials with good transmission of visible light.[161]

Fiber-forming co-polyimides were synthesized from 3,4′-diaminodiphenyl ether, **9-5**, optionally with added *p*-phenylenediamine, and the dianhydride of biphenyltetracarboxylic acid (optionally with added pyromellitic dianhydride). The polyimidc was dissolved in cresol and the dope was extruded through a nozzle into air; the filament was led into a coagulating bath, wound up, dried, and twice drawn at 250+ °C to give a polyimide fiber with high tensile strength, high elastic modulus, and excellent resistance to moisture and light.[162]

Polyimide fibers of high strength and high modulus of elasticity were obtained by polymerizing/processing of 4,4′-bis-(3-aminophenoxy)biphenyl, **9-31**, and pyromellitic dianhydride.[163]

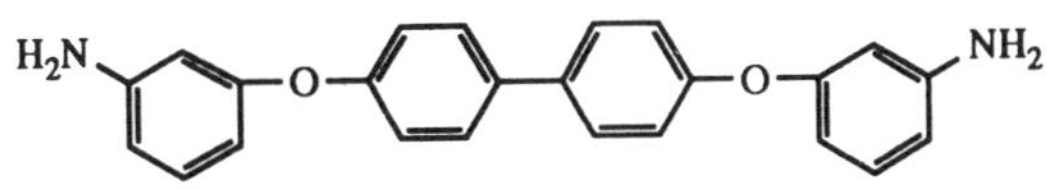

9-31

A polyimide end-capped with ethynyl groups was obtained by the co-polycondensation of the diethyl ester of benzophenone-tetracarboxylic acid, 1,3-bis-(3-aminophenoxy)benzene, **9-9**, and 3-aminophenylacetylene at 100 °C, and conversion of the obtained polyamic acid to the oligomeric polyimide at 100–200 °C; this product, called Thermid A1-600, was developed by Gulf Oil Co. The ethynyl-end-capped polyimide was converted to a crosslinked polyimide at 200–370 °C.[164]

A polyimide prepared from 4,4′-bis-(3-aminophenoxy)biphenyl, **9-31**, and pyromellitic dianhydride was used for insulating copper wire by extrusion coating; the coating had a dielectric strength in water of 20 kV.[165]

Liquid-crystal display devices were made up of a liquid crystal orienting film on a support with attached electrode. The orienting film consisted of a heat-treated polyimide made from diamines such as 1,3-bis-(4-aminophenoxy)benzene, **9-7**, or 4,4′-bis-(3-aminophenoxy)diphenylsulfone, **9-14**, and the dianhydride of an aromatic tetracarboxylic acid.[166]

Polyimides made, e.g., from 4,4′-bis-(3-aminophenoxy)biphenyl, **9-31**, and pyromellitic dianhydride were powdered, pelletized, and compounded with 0.05 m phthalic anhydride (a subliming substance)/100 g pellets and 0.7 g talc/100 g pellets, then extruded at a die temperature of 355 °C to give a polyimide foam with excellent abrasion, heat and fire resistance, and good fabrication qualities.[167]

The polycondensation of 1,3-bis-(3-aminophenoxy)benzene, **9-9**, and 4,4′-diaminodiphenyl ether in *N,N*-dimethylacetamide/pyridine with pyromellitic dianhydride at 0–25 °C for 90 min, then under reflux for 3 hr, gave a polyimide powder which could be compression-molded. The molding showed a 5 wt%-loss temperature 580 °C, bending strength 1460 kg/cm^2, and bending modulus 35,600 kg/cm^2.[168]

Adhesive polyimide films were made by polycondensation of 1.00 m 4,4′-bis-(3-aminophenoxy)biphenyl, **9-31**, in *N,N*-dimethylacetamide with 0.98 m pyromellitic dianhydride, followed by 0.04 m phthalic anhydride and imidization of the intermediate polyamic acid with acetic anhydride. The polyimide was isolated, pelletized, melted at 410 °C, and extruded through a slit die. This film was hot-pressed at 245 °C and 10 kg/cm^2, and showed good adhesion to various substrates.[169]

A hydroxyl-terminated polysulfone was prepared from 4,4′-dichlorodiphenylsulfone and resorcinol. This polysulfone was etherified in dimethylsulfoxide in the presence of NaOH with allyl bromide for 6 hr/40–50 °C. This polysulfone diallyl ether was used to cure the bis-maleimide of 4,4′-diaminobiphenylmethane by B-staging at 190 °C, pressing at 200 °C, and post-curing for 5 hr/230 °C to give moldings with glass transition temperature 239 °C, thermal decomposition temperature 429 °C, flexural strength 12 kg/mm^2, and flexural modulus 431 kg/mm^2.[170]

Heat-stable polyimides useful for impregnants, coatings, or molding resins were prepared by polymerization of one *N,N′*-bismaleimide (e.g., derived from 4,4′-diaminodiphenylmethane) with one alkenyloxyaniline (e.g., the allyl ether of *m*-aminophenol), and, optionally, one *N*-[(meth)allyloxyphenyl]maleimide [e.g., *N*-(*m*-allyoxyphenyl)maleimide], **9-32** (R = H).[171]

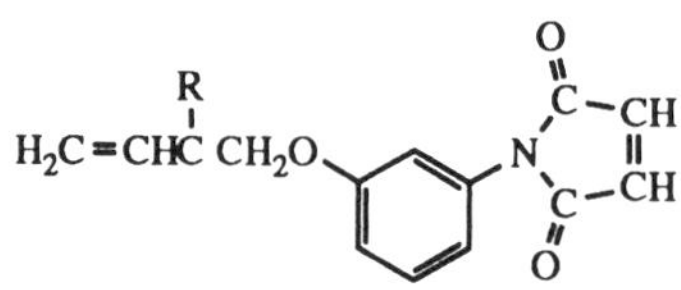

9-32
(R=H or Me)

Copolymers of, e.g., maleic anhydride and an alkyl (meth)acrylate were imidized by reaction with, e.g., *m*-aminophenol; these resins had improved resistance to corrosion, heat, and chemicals, and could be used for paints and coatings.[172]

N-[*m*-(Meth)allyloxyphenyl]maleimides, **9-32**, or similar compounds made from other aminophenols, were prepolymerized at 150 °C with at least one bis-maleimide (e.g., made from 4,4′-diaminodiphenylmethane), then crosslinked with triallyl isocyanurate, using imidazole as the catalyst, at 130 °C to give a resin which was easily moldable and had excellent flexural strength and modulus.[173]

A bis-maleimide was prepared by reacting maleic anhydride with 4,4′-bis-(3-aminophenoxy)diphenylsulfone, **9-14**, followed by imidation with acetic anhydride. This product was a useful polymer intermediate.[174]

Silicone-modified polymaleimide resins were prepared by reaction of a *m*-aminophenol/formaldehyde resin with maleic anhydride, followed by treatment with a chloro-methoxy-terminated siloxane, and curing this modified resin with 4,4′-diaminodiphenylmethane. The cured resin had good heat and crack resistance, bending strength 3.7 kg/mm^2, flexural modulus 340 kg/mm^2, and glass transition temperature 270 °C.[175]

Thermosetting blends of, e.g., 4,4′-bis-(3-maleimidophenoxy)biphenyl (made with *m*-aminophenol) and *o,o*′-diallylbisphenol A were cured by heating at 180 °C. This resin, when press-cured in a mold at 200 °C and 50 kg/cm^2, then post-cured at 250 °C, gave a molding with heat distortion temperature 290 °C, flexural modulus 385 kg/mm^2, Izod impact strength 24 kg-cm/cm, water absorption 0.27%, and flexural strength 18.7 kg/mm^2.[176]

A bis-maleimide, e.g., *N,N*′-(methylenedi-*p*-phenylene)bismaleimide, and *m*-aminophenol were reacted at 160 °C/4 min to give a prepolymer which was storage-stable in the solid state and useful for making fiber-reinforced moldings.[177]

A copolymer of *N*-(*m*-hydroxyphenyl)maleimide (or similar maleimides), maleic anhydride, and isobutylene was blended with a photosensitive aromatic azide. This composition formed flexible films with good adhesion to substrates, high sensitivity, high resolution, good resistance to dry etching, and good thermal stability; it was suitable for positive or negative photoresist products.[178]

Bismaleimides prepared by reaction of maleic anhydride in acetone with 4,4′-bis-(3-aminophenoxy)diphenyl sulfone, **9-14**, followed by the conversion of the maleamic acid to the maleimide with acetic anhydride in the presence of magnesium chloride hexahydrate and triethylamine catalyst, was coreacted with *N*-vinyl-2-pyrrolidinone, optionally with added trimethylolpropane triacrylate or 1-vinyl-3,4-epoxycyclohexane (or other epoxy compounds). A prepreg resin was obtained which was thermally curable to give products with high glass transition temperatures and good physical properties.[179]

The bis-maleimide prepared from 2,6-bis-(*m*-aminophenoxy)pyridine, **9-13**, was used for making heat-resistant polymers with improved flexibity and processability.[180]

Thermosetting resin compositions useful as molding materials comprised a bis-maleimide (from maleic anhydride and a diamine from Section 9.2 above), an epoxy resin, a phenolic novolak resin, and an inorganic filler (e.g., powdered silica or glass fiber).[181]

Thermosetting resins containing polymaleimides, made by reaction of the acid-catalyzed condensation product of aniline with alpha,alpha′-dichloro-*p*-xylene and maleic anhydride, were blended with diamines containing two *m*-aminophenoxy groups [e.g., 4,4′-bis-(3-aminophenoxy) biphenyl, **9-31**], hot-pressed and post-cured to give moldings with good Izod impact strength, bending strength, flexural modulus, and decomposition temperature 391 °C.[182]

Phenolic resin compositions with good heat resistance and mechanical strength were prepared by heating, e.g., *N*-(*m*-hydroxyphenyl)maleimide and azo-bis-isobutyronitrile catalyst in *N,N*-dimethylformamide at 70 °C/6 hr. The polymer obtained was melted and blended with a phenolic novolak, hexamethylenetetramine, zinc stearate, and glass fibers, and the mixture was crushed and transfer-molded to give test pieces having flexural modulus 1695 kg/mm^2 and load-bending temperature 160 °C.[183]

An interesting use of *m*-aminophenol was reported in the production of hydroxyaryl group graft-modified polyolefins. For example, 100 parts polyethylene, 1.5 parts maleic anhydride, and 0.1 part dicumyl peroxide were mixed in an extruder at *ca* 175–260 °C to make a maleic anhydride graft on polyethylene. This product was then mixed with 1.7 parts *m*-aminophenol to give a *m*-hydroxyarylimide-graft material, of which 5 parts were blended with 95 parts polycarbonate. Molded test pieces of this blend showed Izod impact strength 17.2 ft.-lb./in., which was 15% higher than the value of a similar blend made with unmodified polyethylene.[184]

A novel material was a calixarene-based cyclic octaester, **9-33**, which contained norbornenedicarboximide groups. This material was used to make carbon-fiber-reinforced prepregs which were cured under vacuum at 170–330 °C to give a composite with T_g 400 °C.[185]

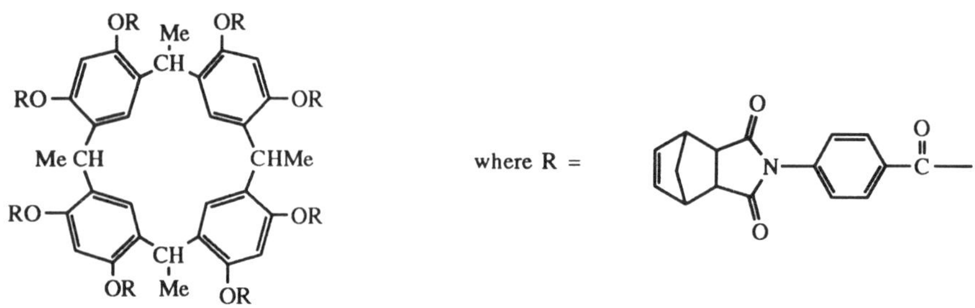

9-33

9.8. POLYAMIDES/IMIDES AND OTHER MULTIFUNCTIONAL POLYMERS

A large variety of ideas in this category have been reported. A few examples follow.

Aromatic imide alkenyl ethers were prepared, e.g., by the condensation on *m*-aminophenol with trimellitic chloride/anhydride in *N,N*-dimethylacetamide, with triethylamine as the hydrogen chloride acceptor, at 50 °C, followed by the etherification of the phenolic hydroxyl group with allyl bromide in the presence of sodium methoxide in dimethylsulfoxide to give compound **9-34**. This product was a monomer useful for making heat-resistant thermosetting resins.[186]

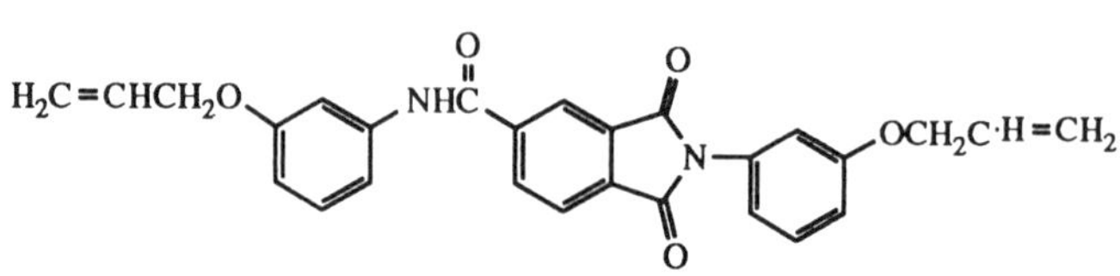

9-34

Cyclic heterocarbonates with, e.g., ester/ether/imide linkages were prepared with *m*-aminophenol as one of the comonomers. The representative product, **9-35**, could be polymerized by ring-opening polymerization using, e.g., tetramethylammonium tetraphenylborate as the catalyst.[187]

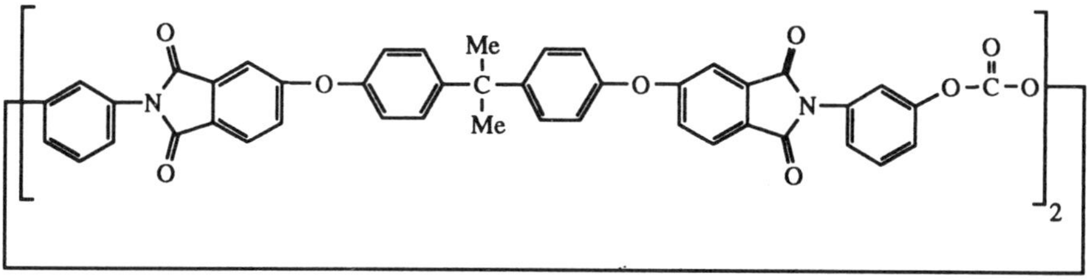

9-35

A phenolic cyanate, **9-36**, of the maleimide obtained by the reaction of *m*-aminophenol with maleic anhydride, followed by imidization and reaction with cyanogen bromide in the presence of triethylamine, could be homopolymerized by heating or copolymerized with 4,4′-diaminodiphenylmethane to give products useful for coatings, castings, and laminates. The imide/cyanate, **9-36**, could also be copolymerized with aromatic polycyanates, optionally with added monomers such as 2-ethylhexyl acrylate or styrene or epoxy resins.[188]

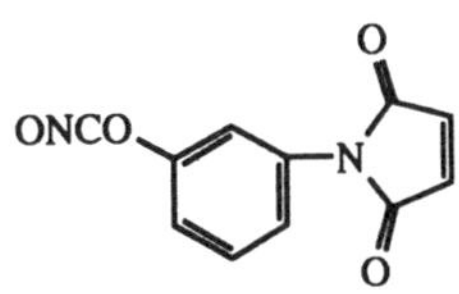

9-36

3,4′-Diaminodiphenyl ether, **9-5**, was reacted with isophthaloyl chloride in a ratio such that an amino-endcapped aryl ether/amide was obtained, which was further reacted with acryloyl chloride to give an acryloylamino-endcapped aryl ether/amide. This product was copolymerized with *N*-phenylmaleimide in the presence of dicumyl peroxide at 120 °C and post-cured at 200 °C to give a heat-resistant, easily processed copolymeric amide/ether.[189]

Another type of polyamide/imide was prepared by the polycondensation of 1,3-bis-(3-aminophenoxy)benzene, **9-9**, with trimellitic anhydride monochloride in the presence of triethylamine in *N*-methyl-2-pyrrolidinone at 190 °C/9 hr. This polymer was used as a hot-melt adhesive for bonding copper foil to polyimide film in the preparation of printed circuit boards.[190]

A polysulfone membrane was coated with an aqueous solution of 2 parts *m*-aminophenol, 0.3 parts sodium laurylsulfate, 2 parts triethylamine, and 4 parts camphorsulfonic acid, then treated with 0.25% trimesoyl chloride in hexane solution, and finally dried at 120 °C/5 min to give a composite reverse osmosis membrane useful for desalting water, showing 98% salt rejection and 0.3 t/m^2-day water flux.[191]

A copolymer was prepared by condensing a mixture of bis-(4-aminophenyl)ether and bis[4-(3-aminophenoxy)phenyl]sulfone, **9-14**, in *N*-methyl-2-pyrrolidinone with trimellitic anhydride chloride in the presence of Et_3N and mixed with pyridine/Ac_2O to give the polyamide/imide. This product was dissolved in *N*-methyl-2-pyrrolidinone, blended with Victrex 4400G, and cured at 230 °C to prepare a coating which had good hardness, adhesion to metals, bending strength, and heat resistance.[192]

9.9. POLYARYL ETHERS

Resorcinol is a starting material for the polyphenyl ethers which are used commercially as small volume/high price/high performance lubricants, and for the poly(cyanoaryl) ethers pioneered by Idemitsu Kosan K.K. (many patents, but information on commercial production, if any, is not available). Polyaryl ether/sulfones and ether/ketones have been shown to be useful as engineering thermoplastics.

9.9.1. Polyphenyl Ethers

A massive development effort for these materials, with 3–7 aromatic rings in their structure, originated in the 1950s when there was an urgent need for high-temperature resistant lubricants for turbojet aircraft engines, used in planes operating at high speed at high altitudes, such as in certain reconnaissance airplanes. With the advent of the globe-circling satellite and improvements in aerial photography, the use of such aircraft declined. The polyphenyl ethers have several important properties, such as high-temperature resistance for use up to *ca* 300 °C, good lubricity and high viscosity index, good corrosion/chemical/oxidation resistance, superior radiation resistance, very low vapor pressure, high flash and autoignition points. The properties of several of these materials are described in detail in the literature.[193,194] These same properties are also valuable for the use of the polyphenyl ethers as base oils for high-temperature greases and hydraulic fluids. The low vapor pressure in particular was a factor in the increasing use of polyphenyl ethers in super-high vacuum pumps in the semiconductor industry and as protective coverings for precious-metal coated connectors used in the electronics industry. A tempering factor in this market was the introduction of oilless vacuum pumps for ultra-impurity-sensitive vacuum operations.

The importance of resorcinol for making polyphenyl ethers is that the *meta*-substituted products have the lowest melting points or freeze points. For commercial products, the best balance of properties may include oligomers with some *p*-substitution, such as in compound **9-37**, designated as pmm-5P4E (where *p* = *para*, *m* = *meta*, P = phenyl ring, and E = ether).

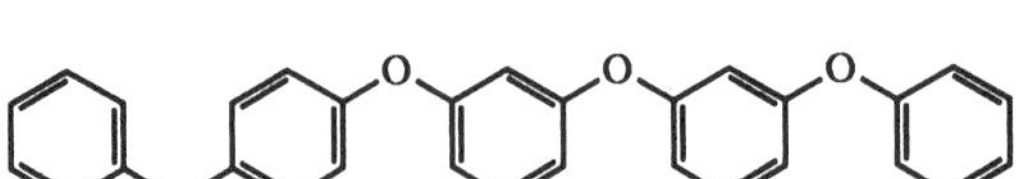

9-37

The synthesis of polyphenyl ethers is generally based on the Ullmann method, involving the condensation of a phenoxide ion with an aryl halide in the presence of a copper catalyst. A glimpse of the early work and the problems with side reactions in the case of resorcinol is given in Ref. 195.

Monsanto Chemical Co. has been the only U.S. producer of polyphenyl ether base stocks for many years; doubtless, the production process has been improved over the initial agonies of synthesis. The price of these polyphenyl ethers is believed to be in the $200–400/lb. range.

Selected references on the production and/or uses of polyphenyl ethers, neat or modified, are given below. Besides lubricants, some of these applications of such material are as fire retardants and plasticizers for plastics and adhesives.

An improved process for the preparation of polyphenyl ether intermediates, such as *m*-phenoxyphenol, *m*-(*m*-bromophenoxy)phenol, or 3-(3-phenoxy)phenoxyphenol, was reported to be the reaction of potassium resorcinate with the appropriate aryl bromide in the presence of copper oxide as the catalyst in *N,N*-dimethylacetamide at 145–185 °C. The desired aryl ether intermediates were obtained in up to 90% yields.[196]

Chlorinated polyaryl ethers, useful as fire retardants for plastics, were obtained by the condensation of hexachlorobenzene in dimethylsulfoxide with, e.g., resorcinol in the presence of alkali hydroxides.[197] Likely, such materials would not be used today because of environmental/health risk concerns.

Amino-substituted polyphenyl ethers were prepared by sequential nitration/reduction of poly-*m*-phenyl ether. The amino-modified ethers were dissolved in a suitable solvent and mixed with aluminum powder and antimony pentasulfide to make pastes used as adhesives for metals such as Mo-stainless steels and titanium-alloy face sheets.[198]

m-Bis-(2,4-diphenoxyphenoxy)benzene was prepared from *m*-phenoxyphenol (resorcinol monophenyl ether) and 2,4-diphenoxy-bromobenzene. This aryl ether oligomer had the good properties required of a synthetic high-temperature lubricant, including low oxygen absorption at 315 °C.[199]

The preparation of *m*-aryloxyphenols from resorcinol and aryl halides was reported to be improved by the use of cuprous oxide, cuprous chloride, or cuprous acetate as the catalyst in the Ullmann synthesis.[200]

Fluorinated polyphenyl ethers were claimed to be useful, improved functional fluids.[201] A process was patented for making poly-*m*-phenyl ethers from an appropriately substituted, resorcinol-derived, monohydric phenol (such as *m*-phenoxyphenol) and *m*-dibromobenzene by the Ullmann method in the presence of KOH and copper powder.[202]

The synthesis of polyphenyl ethers from resorcinol (and other dihydric phenols) and aryl halides was reported to be improved by the use of a copper salt catalyst and a nitrogen-containing solvent, such as 2-ethylhexylamine or pyridine, which dissolves the catalyst by forming coordinate, covalent bonds with copper ions.[203]

Mixtures of polyphenyl thioethers with polyphenyl ethers were claimed as high-performance lubricants, remaining liquid over a wide temperature range, useful for jet engine transmissions.[204]

Bis-(*m*-phenoxy-*m*-phenoxy) long-chain alkylbenzene was claimed to be an improved lubricating oil with, e.g., lower pour point than prior art materials.[205]

An oligomer prepared by the Ullman method from decabromodiphenyl ether and resorcinol was found to be a flame retardant for nylon 66 without showing degradation of the resin at molding temperatures.[206]

9.9.2. Poly(cyanoaryl) Ethers

These thermoplastic polymers are of relatively recent invention and have been under intensive development by Idemitsu Kosan K.K. in Japan. They are made by the condensation of resorcinol with 2,6-dihalobenzonitrile.These polymers were found to have wide potential as high-performance engineering plastics. Selected references follow.

The condensation of 2,6- or 2,4-dihalobenzonitrile with the alkali metal disalt of aromatic diols, including resorcinol, was reported to give di(hydroxyaryl)benzonitriles, intermediates useful for preparing polymers with enhanced flexibility and mechanical strength.[207]

A process for making poly(cyanoaryl) ethers involved the condensation of 2,6- or 2,4-dihalobenzonitrile in an aprotic solvent, optionally with an addition of an aromatic cosolvent such as toluene, in the presence of a lithium salt with sodium or potassium resorcinate. The polymer had good heat resistance and a high initial decomposition temperature, and was useful for making electronic and mechanical parts.[208]

Poly(cyanoaryl) ethers were prepared by the condensation of 2,6-difluorobenzonitriles with silylated aromatic diols, including the di(trimethylsilyl)ether of resorcinol, at 140–300 °C. This method had the advantage that the only byproduct was the volatile trimethylsilyl fluoride. The polymer, **9-38**, was obtained in 97% yield, was crystalline, insoluble in all common solvents, and had a glass transition temperature T_g 147 °C and a melt temperature T_m 335 °C.[209]

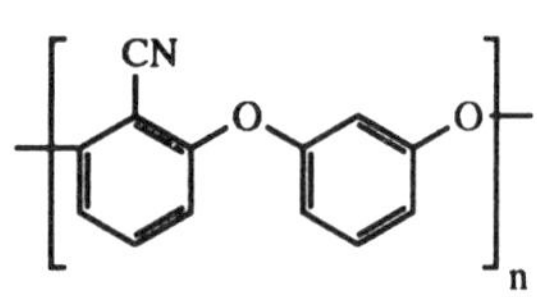

9-38

A cyanoaryl ether copolymer was prepared by the condensation of 2,6-dichlorobenzonitrile with dipotassium salts of hydroquinone and resorcinol in sulfolane/toluene solvent at 160–190 °C. The copolymer had a reduced viscosity, n_{sp}/c 1.51, good heat and flame resistance, and excellent mechanical properties, with a decomposition temperature in air, T_d 490 °C.[210]

Modified poly(cyanoaryl)ethers were prepared by the reaction of dihalobenzonitriles with the dialkali metal salt of resorcinol in *N*-methyl-2-pyrrolidinone at 195 °C to give an oligomer, which was then reacted with pentafluorobenzonitrile; the polymer obtained had high tensile and impact strength.[211]

Another process for the preparation of polymers of structure **9-38** involved the reaction of dialkali metal resorcinate in aprotic solvents first with 2,6-dichlorobenzonitrile in the presence of monohydric phenols as molecular weight controllers, then in a second stage with 2,6-difluorobenzonitrile in the presence of a monohalogenated hydrocarbon as a molecular weight controller; this process gave high molecular weight polymer rapidly.[212]

Poly(cyanoaryl)ethers of type **9-38** were partially hydrolyzed with 48% aqueous sodium hydroxide in tetrahydrofuran at room temperature to give heat-resistant poly(cyanoaryl/carboxyaryl)ethers useful as modifiers for polymers, for ion-exchange resins, and as reactive polymers.[213]

2,6-Difluorobenzonitrile was condensed first with *p*-phenylphenol in *N*-methyl-2-pyrrolidinone/toluene in the presence of potassium carbonate at 170 °C to produce 2-fluoro-6-(phenylphenoxy)benzonitrile, then 2,6-dichlorobenzonitrile and resorcinol were added and copolymerized at 200 °C to give a 96% yield of polymer. This product, in the form of a film, had a 1.8% weight loss after heating at 300 °C for 100 hr and a tensile modulus 33,000 kg/cm^2, yield strength 1,400 kg/cm^2, and breaking strength 1,220 kg/cm^2.[214]

The poly(cyanoaryl)ether, 100 parts, obtained from 2,6-dichlorobenzonitrile/2,6-difluorobenzonitrile and resorcinol was mixed with 20 parts potassium titanate whisker (hardness >6 Mho) and 10 parts Lublon L-2 (detackifying agent) at 355 °C, pelletized and injection molded at 360 °C to give a test piece with 0.7% shrinkage, flexural modulus at 23 °C of 58,000 kg/cm^2, and friction coefficient 7×10^{-5} cm^2-s/kg-m-h (i.e., good sliding properties).[215]

A polymer, **9-39**, was prepared in 94% yield by condensing fluorescein with 2,6-difluorobenzonitrile in the presence of potassium carbonate in *N*-methyl 2-pyrrolidinone at 195 °C. This polymer had T_g 318.9 °C and 5 wt% loss in air at 471 °C.[216]

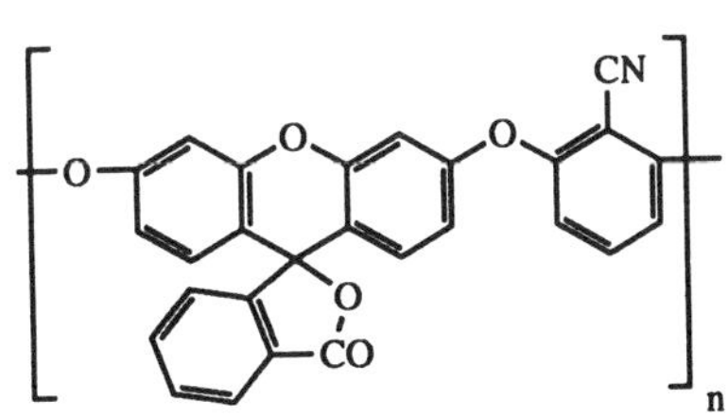

9-39

A poly(cyanoaryl)ether, 70 parts, made from 2,6-dichlorobenzonitrile and resorcinol, was blended with 30 parts Diablack E (carbon black) and 1 part titanium dioxide, then press-molded to a 1-mm-thick sheet which was laminated with Ni foils on both sides, and annealed for 0.5 hr at 230 °C. This composite showed an abrupt resistance increase at 320 °C; it was useful for overcurrent protection devices, high-temperature thermostats, and high-caloric heaters.[217]

A poly(cyanoaryl)ether, 95 parts, prepared from 2,6-dichlorobenzonitrile and resorcinol, was blended with 5 parts fluoroelastomer (Daielthermoplastic T630), pelletized and injection molded to give an impact-resistant test piece with notched Izod impact strength 9 kg-cm/cm and a heat distortion temperature, T_d 161 °C.[218]

Cyclic poly(cyanoaryl)ethers, **9-40**, were prepared by reacting resorcinol 0.3 with 2,6-dichlorobenzonitrile 0.3 and sodium carbonate 0.33 in *N*-methyl-2-pyrrolidinone at 190 °C to give a composition containing 10 parts **9-40** (n = 2), 15 parts **9-40** (n = 3), 15 parts **9-40** (n = 4), 9 parts **9-40** (n = 5), and 1 part higher oligomers. These compositions were useful as fire-retardant agents, e.g., for polycarbonate.[219]

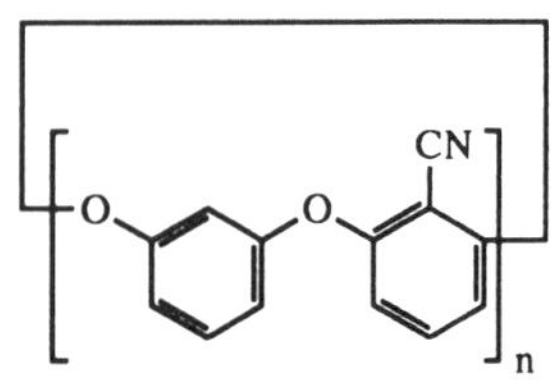

9-40

A poly(cyanoaryl)ether (reduced viscosity 0.7 dL/g) 80 parts was blended with 20 parts silicon nitride powder at 350 °C, pelletized, and easily molded to give test pieces having flexural strength 1590 kg/cm^2, flexural modulus 68,000 kg/cm^2, and heat distortion temperature T_d 212 °C.[220]

Low molecular weight, low viscosity, cyclic poly(cyanoaryl)ethers were made at Dow Chemical Co. from 2,6-difluorobenzonitrile and resorcinol in the presence of alkali hydroxide in dimethylsulfoxide at high temperature by the slow addition of starting materials and maintenance of the correct stoichiometry. The cyclic oligomers could be polymerized by catalysts which reversibly cleaved the carbon–oxygen bonds joining the rings.[221]

The poly(cyanoaryl)ether, 10 vol%, made from 2,6-dichlorobenzonitrile and resorcinol, with 1% of a silane coupling agent added, was dissolved in *N*-methyl-2-pyrrolidinone and blended with 75 vol% iron powder and 15 vol% glass fibers; the slurry was concentrated in vacuo to give a resin-coated powder. This powder was press-molded and heated at *ca* 375 °C to give parts containing 5.95% glass fiber which had a low coefficient of linear expansion, good chemical resistance, and heat distortion temperature T_d 315 °C.[222]

An improved process for the preparation of poly(cyanoaryl)ether, **9-38**, of high molecular weight consisted of the addition of sodium oxalate to the usual charge. For example, 22.02 g resorcinol, 34.35 g 2,6-dichlorobenzonitrile, 23.32 g sodium carbonate, and 1.0 g sodium oxalate in *N*-methyl-2-pyrrolidinone were held at 200 °C for 1 hr, then 0.35 g of 2,6-difluorobenzonitrile was added, and the charge was held at 200 °C for another 5 hr to give a poly(cyanoaryl)ether with reduced viscosity 1.55 dL/g vs 1.29 dL/g for polymer prepared without the addition of sodium oxalate.[223]

A poly(cyanoaryl)ether, **9-38**, of melt viscosity 6000 P at 360 °C was extruded onto a 100-μm-thick copper twisted wire conductor. The abrasion resistance of this insulated wire was 22 times better than that of a PVC-coated wire in a repeated scrape test.[224]

A heat-resistant poly(cyanoaryl)ether composition was prepared by mixing 50.0 g of polymer, **9-38**, as a 20% aqueous cake with 5 g oxalic acid dihydrate and 0.05 g ferric oxide, then drying for 12 hr at 130 °C in vacuo. This product had a thermal decomposition time of 220 min at 350 °C.[225]

A carbon-fiber fabric was impregnated with a 10% solution of the poly(cyanoaryl)ether, **9-38**, in *N*-methyl-2-pyrrolidinone at 200 °C, cooled to 25 °C/1 mm, and press-molded together to give a laminate with interlayer shear adhesion 1130 kg/cm^2, useful for electrical and machinery parts.[226]

Blends of the poly(cyanoaryl)ether, **9-38**, and 1.08/5.40/0.249/0.083 *p*-acetoxybenzoic acid/4,4′-diacetoxybiphenyl/terephthalic acid/isophthalic acid copolymer were prepared in 90/10 ratio at 350 °C/3 min. This blend was injection-molded to give a material which contained dispersed fibrils and had heat distortion temperature, T_d 200 °C and bending strength 1200–1450 kg/cm^2.[227]

9.9.3. Polyaryl Ether/Sulfones and Polyaryl Ether/Ketones

There are many variations on this theme, dependent on the choice of starting materials. A few examples follow.

The polycondensation of 4,4′-difluorobenzophenone with resorcinol in the presence of sodium carbonate in *N*-methyl-2-pyrrolidinone gave a polyaryl ether/ketone with reduced viscosity 0.95 dL/g, glass transition temperature T_g 128 °C, and 5% weight loss at 560 °C. Test specimens, made by injection molding at 310 °C, showed tensile strength 800 kg/cm^2, tensile modulus 25,000 kg/cm^2, bending strength 1200 kg/cm^2, and flexural modulus 48,000 kg/cm^2.[228]

Similarly, the copolymer made in like manner from 4,4′-difluorobenzophenone, resorcinol, and hydroquinone gave a polyaryl ether/ketone with a reduced viscosity 1.12 dL/g, flexural modulus 33,500 kg/cm^2, tensile strength 92 kg/cm^2, tensile modulus 27,800 kg/cm^2, and T_g 145 °C.[229]

The polycondensation of 4,4′-difluorobenzophenone, bis-(4-fluorophenyl)sulfone, and resorcinol in the presence of sodium carbonate in *N*-methyl-2-pyrrolidinone at 195–200 °C gave a polyaryl ether/ketone/sulfone with reduced viscosity 1.20 dL/g, T_g 138 °C, and 5% weight loss at 537 °C.[230]

A very different diol, 3,8-dihydroxy-5*a*,10*b*-diphenylcumarano-2,3,2′,3′-cumaran, **9-17**, was condensed with 4,4′-dichlorodiphenyl sulfone to give a polyaryl ether /sulfone, or condensed with 4,4′-difluorobenzophenone to give a polyaryl ether/ ketone. These thermoplastic polymers had high glass transition temperatures T_g up to

278 °C, they had good mechanical properties, and they were flame resistant, UL 94 classification V-0.[231]

9.10. DICYANATES

These cyanate esters were pioneered at Bayer A.-G. in the 1960s/1970s and later pursued by others (e.g, 1985, Hi-Tek Polymers). These materials homopolymerize by the catalytic cyclotrimerization of (di-)cyanate to (poly)cyanurate. They are targeted for use in structural materials, composites, and electronics applications. The resorcinol dicyanate, **9-41**, can be prepared in 90+% yield, e.g., by adding an aqueous solution of sodium resorcinate, also containing triethylamine, to a solution of excess cyanogen chloride (ClCN) in isopropanol at *ca* 0 °C.[232]

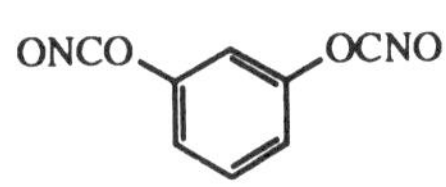

9-41

A comparison of resorcinol dicyanate with three other dicyanates, e.g., bisphenol A dicyanate, showed that the resorcinol diester, **9-41**, was the most reactive and gave the highest crosslink density. The homopolymerization of the resorcinol dicyanate in the presence of a solution of copper naphthenate as the catalyst in nonylphenol showed that this polymer had the best tensile strength (17,300 psi) and tensile modulus (0.70×10^6 psi) of the four dicyanates tested, that it was flame retardant due to intumescence in the UL 94 flammability test, but also that it had the highest water absorption (2.3%) of the four compounds in the comparison.[233]

To date, resorcinol dicyanate itself has not been commercialized, probably in part because its reactivity also includes sensitivity to moisture in storage/use.

9.11. POLYMERS CONTAINING OTHER HETEROCYCLIC UNITS

A phosphonitrilic resin was prepared by the condensation of a dihydric phenol, such as resorcinol, with 2,4,6-trichloro-2,4,6-triphenyl triphosphonitrile in the presence of pyridine in an aromatic hydrocarbon solvent. The resin so obtained was sulfonated with 30% oleum; the product was said to be useful as an ion-exchange resin which can stand up to temperatures above 150 °C.[234]

An oligomeric product was made by condensing 4.4 m monophenyldichloro-*s*-triazine, **9-42**, in dioxane/acetone with 2.2 m resorcinol in the presence of aqueous NaOH at 20 °C, followed by reaction with 4.7 m phenol; the oligomer obtained had a decomposition temperature $T_d > 350$ °C.[235]

Sodium resorcinate in aqueous solution was reacted with cyanuric chloride (2,4,6-trichloro-*s*-triazine) in benzene solution by the interfacial polymerization tech-

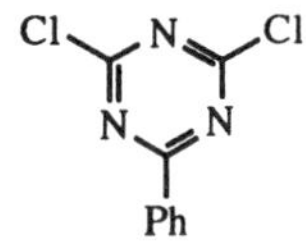

9-42

nique; the film produced could be isolated as a powder which charred at 400–500 °C without melting.[236]

High molecular weight polybenz-1,3-oxazine-2,4-diones, **9-43**, were synthesized from *o,o′*-dihydroxyaromatic dicarboxylic acids such as, e.g., resorcinol-4,6-dicarboxylic acid (4,6-dihydroxyisophthalic acid) with 4,4′-diisocyanato diphenyl ether. These polymers had high heat stability and good mechanical properties.[237]

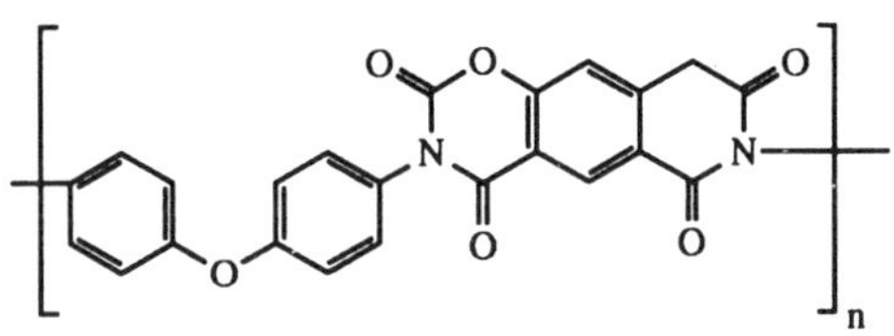

9-43

Poly(*p*-phenylenebenzobisoxazole) (PBO), **9-44**, was made by the condensation of terephthaloyl chloride with 4,6-diaminoresorcinol, **9-6**, hydrochloride in deoxygenated 115% polyphosphoric acid in several stages at 90–200 °C. This polymer formed a liquid-crystalline melt from which oriented fibers could be made which had high cohesive strength and good heat resistance.[238]

This same PBO polymer, **9-44**, was prepared by the condensation of *N,N′,O,O′*-tetramethylsilyl-4,6-diaminoresorcinol with terephthaloyl chloride to give a polyamic acid which was dissolved in methanesulfonic acid containing phosphorous pentoxide and heated at 140 °C for cyclization.[239]

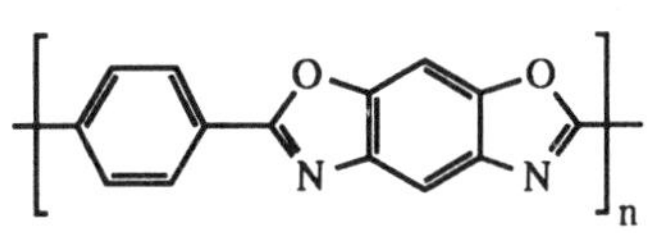

9-44

Another route to these polybenzobisoxazoles, **9-44**, was by the reaction of 4,6-diaminoresorcinol, **9-6**, hydrochloride with 1,4-di-(trichloromethyl)benzene in 65% phosphoric acid with heating and fortification with phosphorous pentoxide in several stages from 90–190 °C; the polymer had an intrinsic viscosity 11.3 dL/g.[240] A similar polymer was prepared starting with 2-methyl-4,6-diaminoresorcinol.[241]

9.12. MISCELLANEOUS POLYMERS

The following selections from the literature are further indication of the variety of compositions which have been studied.

A condensation product of 0.8–8.0 m resorcinol with 1.0 m dicyclopentadiene in the presence of a Lewis acid catalyst, such as aluminum chloride or boron trifluoride, at 60–120 °C gave a resin, **9-45**, which was used as hardener/flexibilizer for other resins used in sealing electronic parts.[242]

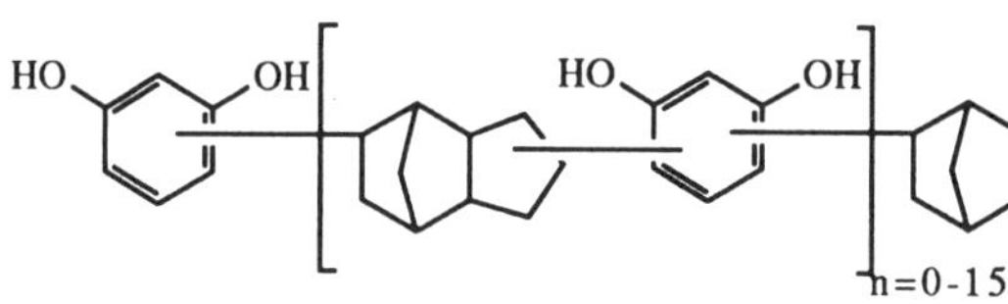

9-45

A polyazomethine resin was produced by condensing 3,4′-diaminodiphenyl ether, **9-5**, or other arylenediamines, in *N*-methyl-2-pyrrolidinone and hexamethylphosphortriamide with terephthaldehyde, optionally in the presence of lithium chloride or calcium chloride as a solubilizing agent and sulfuric or polyphosphoric acid as a catalyst. The polymer obtained was thermosetting and could be molded to give products with good flame and thermal resistance, and good dimensional stability.[243]

A mixture of 4.3 g 1,3-phenylene-bisoxazoline, 3.0 g of a phenol-free novolak, and 1.1 g resorcinol was heated and stirred at 175 °C; 1 wt% triphenylphosphite was then added and the material was post-cured at 225 °C for 1 hr to give a resin with T_g 138.3 °C. The addition of resorcinol improved the toughness of the resin.[244] Similar bisoxazoline-phenolic resin systems, not necessarily containing resorcinol, are being promoted by Ashland Chemical under the tradename Arotech 6000[R] for use in carbon fiber or glass fiber-reinforced composites, which have excellent mechanical properties and low flammability/low smoke emission, suitable for aircraft interiors.[245]

Thermally or UV-light-curable aromatic di- or poly-propenyl ethers were prepared and could be used for coatings, adhesives, printing inks, moldings, wire insulation, etc. For example, the diallyl ether of 2,2′-(*m*-phenylenedioxy)diethanol (**9.2**) was prepared and isomerized to the liquid dipropenyl ether, **9-46**, in the presence of a $RuCl_2[P(C_6H_5)_3]$ catalyst at 120 °C/48 hr. This dipropenyl ether was mixed with 1 wt% diphenyliodonium hexafluorophosphate catalyst and was UV-curable as a 1 mil film to a tack-free coating after 5 sec under a GE H3T7 lamp.[246]

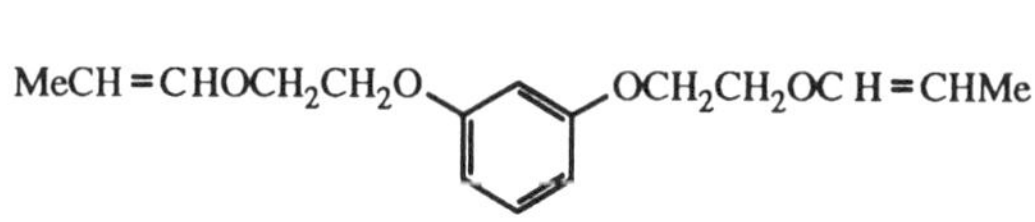

9-46

Stable, bis-aldehyde-functional silyl ethers of resorcinol were prepared by the hydrosilation reaction of the bis-dimethylsilyl ether of resorcinol with 4-allyloxybenzaldehade or with 6-heptenal. These bis-silylaldehydes were acid-curable, e.g., with methanesulfonic acid as the catalyst, or UV-light-curable using a latent-acid sulfonium catalyst (GE 1014). These resin compositions were intended for use as coatings, sealants, adhesives, and potting compositions.[247]

A block polyurethane resin prepared from 0.015 m polyoxyethylene (mol wt 14,000), 0.04 m MDI, 0.015 m hydroquinone di-(2-hydroxyethyl) ether, and 0.01 m resorcinol was useful in melt-forming fibers and nonwoven webs and sheets. The polymer had a melt flow rate (at 190 °C) 480, synthetic urine absorbent capacity 27 g/g, and extractables 5%.[248]

9.13. REFERENCES

1. PCT Int. Appl. WO 92,006,064 (Apr. 16, 1992; I. Nitta and K. Asai to Sumitomo Chemical Co., Ltd.); *C.A., 117*, 7663 (1992).
2. U.S. Pat. 3,644,534 (Feb. 22, 1972; K. G. Reabe and H. Dressler to Koppers Co., Inc.).
3. Jpn. Kokai Tokkyo Koho JP 04,001,151 [92,001,151] (Jan. 6, 1992; M. Kaya and T. Ogino to Mitsu Petrochemical Industries, Ltd.); *C.A., 116*, 173767 (1992).
4. U.S. Pat. 5,059,723 (Oct. 22, 1991; H. Dressler to INDSPEC Chemical Corp.).
5. PCT Int. Appl. WO 91,016,292 (Oct. 31, 1991; B. J. Hitch and B. L. Bernard to Eastman Kodak Co.); *C.A., 116*, 83361 (1992).
6. Jpn. Kokai Tokkyo Koho JP 02,096,545 [90,096,545] (Apr. 10, 1990; O. Kimura and M. Yamashita to Taoka Chemical Co., Ltd.); *C.A., 113*, 77906 (1990).
7. U.S. Pat. 4,546,152 (Oct. 8, 1985; D. F. Koelmel, D. Jamiolkowski, S. W. Shalaby, and R. S. Bezwada to Ethicon, Inc.); the preparation of dimethyl (*m*-phenylenedioxy)diacetate is described but is not part of the claims of this patent.
8. U.S. Pat. 4,922,000 (May 1, 1990; T. R. Walker, Jr. and W. C. Wooten, Jr. to Eastman Kodak Co.).
9. Jpn. Kokai Tokkyo Koho JP 03,204,833 [91,204,833] (Sept. 6, 1991; T. Maki, M. Moriyasu, K. Tanaka, and T. Hirahara to Mitsubishi Kasei Corp.); *C.A., 116*, 83375 (1992).
10. Jpn. Kokai Tokyo Koho JP 04,173,764 [92,173,764] and JP 04,173,765 [92,173,765] (both dated June 22, 1992; Y. Suguro and M. Matsumoto to Mitsubishi Kasei Corp.); *C.A., 117*, 191481 and 171482 (1992).
11. U.S. Pat. 4,238,625 (Dec. 9, 1980; H. Fiege and K. Wedemeyer to Bayer A.-G.).
12. Eur. Pat. Appl. EP 388,732 (Feb. 19, 1991; T. Maki, Y. Asahi, and M. Moryasu to Mitsubishi Kasei Corp.); *C.A., 114*, 246964 (1991).
13. U.S. Pat. 4,791,224 [Dec. 13, 1988; C. E. Sumner, Jr., and E. J. Fugate to Eastman Kodak Co.; a continuous process, starting with resorcinol and chloroacetic acid, is described in U.S. Pat. 4,532,346 (Sept. 30, 1985; K. Rehn and H. J. Westler to Hoechst A.-G.)].
14. U.S. Pat. 4,935,540 (June 19, 1990; to C. E. Sumner, Jr.).
15. Jpn. Pat. 85,000,706 (March 12, 1985; to Mitsui Toatsu Chemical, Inc.).
16. Jpn. Kokai Tokkyo Koho JP 61,140,550 [86,140,550] (June 27, 1986; K. Yamataka, Y. Matsuoka, and A. Shimizu to Asahi Chemical Industry Co., Ltd.); *C.A., 106*, 49770 (1987).
17. Jpn. Kokai Tokkyo Koho JP 61,225,155 (Oct. 6, 1986; K. Yamaguchi, Y. Yoshikawa, K. Sugimoto, Y. Tanabe, and T. Yamaguchi to Mitsui Toatsu Chemicals, Inc.); *C.A., 106*, 66904 (1987).
18. Jpn. Kokai Tokkyo Koho JP 62,081,358 (Apr. 14, 1987; K. Tada and S. Masaki to Sumitomo Chemical Co., Ltd.); *C.A., 107*, 197785 (1987).
19. Jpn. Kokai Tokkyo Koho JP 63,017,852 [88,017,852] (Jan. 25, 1988; N. Kajimoto, T. Nagata, M. Wada, and M. Okasaki to Mitsui Toatsu Chemicls, Inc.); *C.A., 109*, 109999 (1988).

20. Jpn. Kokai Tokkyo Koho JP 01,272,558 [89,272,558] (Oct. 31, 1989; M. Kawamura and K. Mizuno to Mitsui Petrochemical Industries, Ltd.); *C.A., 112*, 216425 (1990).
21. U.S. Pat. 4,745,232 (May 17, 1988; R. J. Schmitt, D. S. Ross, and J. F. Wolfe to SRI International); R. J. Schmitt, D. S. Ross, J. R. Haro, and J. F. Wolfe, *J. Org. Chem., 53*, 5568 (1988).
22. Eur. Pat. Appl. EP 393,559 (Oct. 24, 1990; W. J. Harris, L. K. Kleies, M. B. Liu, E. Lysenko, and S. Rosenberg to Dow Chemical Co.); *C.A., 114*, 123258 (1991).
23. Jpn. Kokai Tokkyo Koho JP 62,116,547 [87,116,547] (May 28, 1987; Y. Yoshikawa *et al.* to Mitsui Toatsu Chemicals, Inc.); equiv. is U.S. Pat. 4,754,068 (Dec. 15, 1988; K. Yamaguchi *et al.* to Mitsui Toatsu Chemicals, Inc.).
24. Jpn. Kokai Tokkyo Koho JP 03,255,058 [91,255,058] (Nov. 13, 1991; K. Sugimoto, K. Yamaguchi, M. Yamazaki, and T. Yamaguchi to Mitsui Toatsu Chemicals, Inc.); *C.A., 116*, 151310 (1992).
25. Jpn. Kokai Tokkyo Koho JP 61,221,158 [86,221,158] (Oct. 1, 1986; Y. Tanabe, K. Yamaguchi, Y. Yoshikawa, K. Sugimoto, and T. Yamaguchi to Mitsui Toatsu Chemicals, Inc.); *C.A., 106*, 84167 (1987).
26. Jpn. Kokai Tokkyo Koho JP 62,042,958 [87,042,958] (Feb. 24, 1987; Y. Tanabe, K. Yamaguchi, Y. Yoshikawa, and K. Sugimoto to Mitsui Toatsu Chemicals, Inc.); *C.A., 107*, 199070 (1988).
27. U.S. 5,026,872 (June 25, 1991; D. K. Kohli to American Cyanamid Co.).
28. Jpn. Kokai Tokkyo Koho JP 62,116,563 [87,116,563] (May 28, 1987; K. Yamaguchi, K. Sugimoto, Y. Yoshikawa, Y. Tanabe, and T. Yamaguchi to Mitsui Toatsu Chemicals, Inc.); *C.A., 108*, 22780 (1988).
29. Jpn. Kokai Tokkyo Koho JP 62,126,172 [87,126,172] (June 8, 1987; K. Yamaguchi, K. Sugimoto, Y. Yoshikawa, Y. Tamabe, and T. Yamaguchi to Mitsui Toatsu Chemicals, Inc.); *C.A., 107*, 198105 (1987).
30. Jpn. Kokai Tokkyo Koho JP 62,142,155 [87,142,155] (June 25, 1987; K. Yamaguchi, K. Sugimoto, Y. Yoshikawa, Y. Tanabe, and T. Yamaguchi to Mitsui Toatsu Chemicals, Inc.); *C.A., 108*, 131586 (1988). The preparation of diamines by the reaction of 2-(3-aminophenoxy)-6-chloropyridine with dihydric phenols is covered by Jpn. Kokai Tokkyo Koho JP 63,154,666 and JP 63,154,664 (both June 27, 1988; both to Mitsui Toatsu Chemicals Inc.).
31. U.S. Pat. 4,661,559 (Apr. 28, 1987; H. C. Gardner, M. J. Michner, G. L. Broda, and R. J. Cotter to Union Carbide Corp.).
32. U.S. Pat. 4,608,404 (Aug. 26, 1986; H. C. Gardner, M. J. Michner, G. L. Broda, and R. J. Cotter to Union Carbide Corp.); also, U.S. Pat. 4,789,722 (Dec. 6, 1988; H. Jabloner and T. T. Nguyen to Union Carbide Corp.).
33. Eur. Pat. Appl. EP 252,726 (Jan. 13, 1988; N. T. Thanh to Hercules, Inc.); *C.A.*, 109,7492 (1988).
34. A. Bunn, M. E. A. Cuddy, and J. C. McGowan, *Recl. Trav. Chim. Pays-Bas, 87*, 599 (1968).
35. U.S. Pat. 4,738,999 (Apr. 19, 1988; D. R. Blenner, K. M. Bond, and A. J. Hannibal to Lord Corp.).
36. U.S. Pat. 5,114,982 (May 19, 1992; M. A. Mendelsohn, F. W. Navish, Jr., and R. Bolton to Westinghouse Electric Corp.).
37. U.S. Pat. 4,910,333 (March 20, 1990; W. E. Slack to Mobay Corp.).
38. Jpn. Kokai Tokkyo Koho JP 01,156,051 [89,156,051] (June 19, 1987; S. Horiki and K. Ito to Nagoya Oilchemical Co., Ltd.); *C.A., 112*, 159990 (1990).
39. Jap. Pat. 21,330 ('65) (Sept 22, 1965; T. Takehide, U. Yuzo, I. Ichiro, M. Takashi, and S. Yoshitaka to Bridgestone Tire Co., Ltd); *C.A., 64*, 3482 (1966).
40. U.S. Pat. 3,980,679 (Sept. 14, 1976; W. Becker to Hoechst A.-G.).
41. Ger. Offen. DE 2,402,358 (Aug. 29, 1974; W. Becker to Reichold-Albert Chemie); *C.A., 82*, 72633 (1975).
42. Brit. Pat. GB 1,485,345 (Sept. 8, 1977; W. Becker to Hoechst A.-G.); equiv. Ger. Offen. DE 2,533,505 (Jan. 8, 1981; same assignment); *C.A., 84*, 179863 (1981).
43. U.S. Pat. 4,284,573 (Aug. 18, 1981; J. F. Arnett and G. A. Doorakian to Dow Chemical Co.).
44. Ger. Offen. DE 3,315,365 (Nov. 3, 1983; I. Ito, Y. Toyoshima, H. Takagishi, and T. Takahashi to Sumitomo Chemical Co., Ltd.); *C.A., 100*, 104396 (1984).
45. Ger. Offen. DE 1,961,888 (Aug. 3, 1970; W. Becker to Hoechst A.-G.); *C.A., 73*, 67204 (1970).

46. Ger. Offen. DE 2,241,393 (Aug. 23, 1972; K. Prater to Bayer A.-G.); *C.A., 81*, 106555 (1974).
47. Fr. Pat. FR 1,398,440 (May 7, 1965; to Peter Spence & Sons); *C.A., 63*, 9914 (1965).
48. U.S. Pat. 4,668,807 (May 25, 1987; J. Darbellay and G. Dessauges to Ciba-Geigy Corp.).
49. Jpn. Kokai Tokkyo Koho JP 01,263,111 (Oct. 19, 1989; K. Yasuda, T. Takada, and T. Suzuki to Mitsui Petrochemical Industries, Ltd.); *C.A., 112*, 199863 (1990).
50. U.S. Pat. 4,313,886 (Feb. 2, 1982; J. Massingill to Dow Chemical Co.).
51. *Chemical Regulation Reporter, March 4, 1983*, reporting studies of the National Toxicology Program.
52. J. P. Seiler, *Chem.-Biol. Interact., 51(3)*, 347 (1984).
53. Belg. Pat. 699,399 (Dec. 4, 1967; to Farbwerke Hoechst A.-G.).
54. Belg. Pat. 699,475 (Dec. 5, 1967; to Farbwerke Hoechst A.-G.).
55. G. J. Fleming, *J. Macromol. Sci.-Chem., A3(3)*, 531 (1969).
56. USSR Pat. SU 479,794 (Apr. 2, 1976; E. N. Teverovskaya, G. A. Blokh, A. N. Sorokina, V. P. Portnyanko, E. S. Derkacheva, and L. V. Ginzburg to Dnepropetrovsk Engineering College); *C.A., 83*, 180472 (1975).
57. U.S. Pat. 3,845,009 (Oct. 29, 1974; R. J. Gander to Johnson & Johnson Co.).
58. U.S. Pat. 4,656,207 (Apr. 7, 1987; H. Jabloner *et al.* to Hercules, Inc.).
59. N. Odagiri and H. Kishi, *Am. Chem. Soc., Polym. Prepr., 33(1)*, 384 (1992).
60. Eur. Pat. Appl. EP 362,853 (Apr. 11, 1990; E. W. Bennett to James River Graphics, Inc.); *C.A., 113*, 116461 (1990).
61. Jpn. Kokai Tokkyo Koho JP 02,120,382 [90,120,382] (May 8, 1990; F. Asakawa and T. Yoshida to Aica Kogyo Co., Ltd.); *C.A., 113*, 154026 (1990).
62. PCT Int. Appl. WO 90,008,168 (July 26, 1990; to D. J. Keehan); *C.A., 113*, 232724 (1990).
63. T. Chukenova, E. E. Ergozhin, R. B. Atshabarova, and K. Kh. Tastanov, *Izv. Akad. Nauk Kaz. SSR, Ser. Khim., 1990(4)*, 57; *C.A., 114*, 7927 (1991).
64. Can. Pat. Appl. CA 2,025,039 (March 28,1991; to J. D. Boyd and L. N. Repecka); *C.A., 115*, 257546 (1991).
65. Jpn. Kokai Tokkyo Koho JP 04,122,715 [92,122,715] (Apr. 13, 1992; S. Ishii, N. Takahashi, M. Hirano, S. Ito, and N. Kami to Nippon Kayaku Co., Ltd.); *C.A., 118*, 82101 (1993).
66. Jpn. Kokai Tokkyo Koho JP 04,337,317 [92,337,317] (Nov. 25, 1992; S. Ishii, N. Takahashi, M. Hirano, S. Ito, and T. Yamazaki to Nippon Kayaku Co., Ltd.); *C.A., 118*, 149166 (1993).
67. E. E. Ergozhin, T. Chukenova, R. B. Atshabarova, and K. Kh. Tastanov, *Izv. Akad. Nauk Resp. Kaz., Ser. Khim., 1992(3)*, 72; *C.A., 117*, 193199 (1992).
68. USSR Pat. SU 1,705,320 (Jan. 15, 1992; V. M. Kuznetsova, M. I. Tokar, O. A. Danilyuk, L. A. Skripko, V. L. Trostyanetskaya, G. V. Shiryaeva, and T. A. Urvelis/Kharkhov Polytechnic Institute); *C.A., 118*, 7871 (1993).
69. USSR Pat. SU 215,366 (Apr. 3, 1968; H. Nurske and E. Raidma/Scientific Research Institute of Shales); *C.A., 69*, 20518 (1968).
70. Jpn. Kokai Tokkyo Koho 51,143,633 (Dec. 10, 1976; to Cemedine Co.); Jpn. Kokai Tokkyo Koho JP 52,083,615 (July 12, 1977; H. Mori, A. Matsukuma, and K. Yoshida to Cemedine Co./Mitsubishi Chemical Industries, Ltd.); *C.A., 87*, 184354 (1977); and Jap. Pat. 59,046,231 (Nov. 10, 1984; Y. Tagaki, H. Mori, A. Matsukuma, and K. Yoshida to Mitsubishi Chemical Co., Ltd.).
71. Jpn. Kokai Tokkyo Koho JP 62,127,315 [87,127,315] (June 9, 1987; M. Yamamoto and Y. Kumitake to Dainippon Ink and Chemicals, Inc.); *C.A., 108*, 38970 (1987).
72. U.S. Pat. 3,773,799 (Nov. 20, 1973; J. J. Schmid to United States of America).
73. Jpn. Kokai Tokkyo Koho JP 59,157,113 (Sept. 6, 1984; to Mitsui Petrochemical Industries, Ltd.); *C.A., 102*, 46742 (1985).
74. U.S. Pat. 3,133,033 (May 12, 1964; W. E. St. Clair and M. J. Waldron to Koppers Co., Inc.).
75. C.-T. Huang and H.-M. Teng, *Ko Fen Tzu T'ung Hsun, 6(1)*, 36 (1964); *C.A., 63*, 11788 (1965).
76. U.S. Pat. 3,406,150 (Oct. 15, 1968; T. S. Boozalis to Dow Chemical Co.); also B. M. Tkacuk *et al.*, *Plast. Massy, 1983(10)*, 56.

77. Jpn. Kokai Tokkyo Koho JP 55,016,047 (Feb. 4, 1980; S. Kanokawa, K. Kawakami, K. Kamio, and Y. Saito to Sumitomo Chemical Co., Ltd.); *C.A., 93*, 27245 (1980).
78. U.S. Pat. 4,614,826 (Sept. 30, 1986; S. Katayama, T. Nakano, and N. Takeda to Mitsui Petrochemical Industries, Ltd.).
79. USSR Pat. SU 1,564,163 (May 15, 1990; V. I. Mikhilov, V. E. Zemlitskii *et al.* to Izhvesk Mechanical Institute; "Mashpriborplastik" Special Construction Technological Bureau, Kharkov); *C.A., 113*, 153753 (1990).
80. USSR Pat. SU 1,636,438 (March 23, 1991; V. M. Satsura, N. N. Tsubulko, A. L. Manarikova, P. K. Lipyanin, and A. A. Shumski to Belorussian Technological Institute); *C.A., 116*, 22479 (1992).
81. U.S. Pat. 3,001,972 (Sept. 26, 1961; R. M. Christenson, M. Wismer, and W. R. Hydro to Pittsburgh Plate Glass Co.).
82. Jpn. Kokai Tokkyo Koho JP 01,198,659 [89,198,659] (Aug. 10, 1989; Y. Okitsu, M. Kitahara, and M. Yorikai to Mitsui Toatsu Chemicals, Inc.); *C.A., 112*, 79022 (1990).
83. Jpn. Kokai Tokkyo Koho JP 01,275,618 [89,275,618] (Nov. 6, 1989; K. Yamaguchi *et al.* to Mitsui Toatsu Chemicals, Inc.); *C.A., 112*, 180274 (1990).
84. Jpn. Kokai Tokkyo Koho JP 63,072,770 (Apr. 4, 1988; S. Kawada, H. Kanbara, T. Yamada, and T. Suzuki to Shikoku Chemical Industries, Ltd.); *C.A., 109*, 201494 (1988).
85. Jpn. Kokai Tokkyo Koho JP 63,048,323 (March 1, 1988; T. Ebara, M. Yamamoto, and M. Yoshizawa to Dainippon Ink and Chemicals, Ltd.); *C.A., 108*, 222632 (1988).
86. Jpn. Kokai Tokkyo Koho JP 02,283,715 [90,283,715] (Nov. 21, 1990; T. Takada to Matsushita Electric Works, Inc.); *C.A., 114*, 248623 (1991).
87. Jpn. Kokai Tokkyo Koho JP 58,208,282 [83,208,282] (Dec. 3, 1983; to Kanegafuchi Chemical Industry Co., Ltd); *C.A., 100*, 103157 (1984).
88. Jpn. Kokai Tokkyo Koho JP 61,186,376 [86,186,376] (Aug. 20, 1986; K. Kubodera and K. Tobukuro to Agency of Industrial Sciences and Technology); *C.A., 106*, 177432 (1986).
89. Eur. Pat. Appl. EP 503,764 (Sept. 16, 1992; W. M. Rolfe and M. R. Thoseby to Ciba-Geigy A.-G.); *C.A., 118*, 103771 (1993).
90. PCT Int. Appl. WO 92,006,128 (Apr. 16, 1992; T. C. Morton, J. H. Hodgkin, and B. N. Dao to Commonwealth Scientific and Industrial Research Organization); *C.A., 117*, 213749 (1992).
91. Fr. Pat. 1,371,648 (Sept. 4, 1964; B. J. Brenner to Dow Chemical Co.); *C.A., 62*, 16466 (1965).
92. Fr. Pat. 1,547,332 (Nov. 22, 1968; D. F. Stackhouse and M. Markovitz to General Electric Co.); *C.A., 71*, 51332 (1969).
93. U.S. Pat. 3,371,008 (Feb. 27, 1968; to E. F. Lopez).
94. U.S. Pat. 4,216,304 (Aug. 5, 1980; F. Gutekunst, F. Lohse, and R. Schmid to Ciba-Geigy Corp.).
95. Ger. Offen. DE 2,251,169 (June 24, 1982; M. Markovitz to General Electric Co.); *C.A., 79*, 19773 (1973).
96. Czech. Pat. 240,069 (July 1, 1987; M. Lidarik, I. Dobas, S. Stary, M. Sima, J. Kaska, V. Ambroz, J. Rajdl, and E. Krejcar); *C.A., 108*, 57230 (1987).
97. Eur. Pat. Appl. EP 230,328 (July 29, 1987; M. F. Conway, G. C. Young, and L. S. Corley to Shell Internationale Research Maatschappij B.V.); *C.A., 108*, 57165 (1987).
98. U.S. Pat. 4,684,700 (Aug. 4, 1987; C. S. Wang, W. N. Chen, R. L. Bowden, and J. R. Berman to Dow Chemical Co.).
99. U.S. Pat. 4,739,028 (Apr. 19, 1988; P. D. McLean, A. Garton, and R. F. Scott to Canadian Patents and Development, Ltd.).
100. U.S. Pat. 4,767,832 (Aug. 30, 1988; E. J. Marx to Shell Oil Co.).
101. Ger. Pat. DE 3,824,689 and DE 3,824,705 (both Feb. 2, 1989; R. Schmid and W. Stauffer to Ciba-Geigy A.-G.); *C.A., 111*, 116286-7 (1989).
102. Jpn. Kokai Tokkyo Koho JP 01,319,519 [89. 319,519] (Dec. 25, 1989; Y. Watanabe and N. Honda to Toshiba Chemical Corp.); *C.A., 112*, 180608 (1990).
103. USSR Pat. SU 1,643,563 (Apr. 23, 1991; A. V. Ragimov, R. I. Mastafaev, A. G. Mamedova, M. Mel'nikova, and B. A. Mamedov to Institute of Chloroorganic Synthesis, Academy of Sciences, Azerbaidzhan SSR); *C.A., 116*, 85016 (1992).

104. U.S. Pat. 4,611,046 (Sept. 9, 1986; M. S. Chattha to Ford Motor Co.).
105. Eur. Pat. Appl. EP 354,510 (Feb. 14, 1990; E. P. Plueddemann to Dow Corning Corp.); *C.A., 113*, 42558 (1990).
106. Eur. Pat. Appl. EP 475,649 (March 18,1992; N. Fujimoto, S. Tanaka, T. Fukui, and Y. Numashima to Shinto Paint Co., Ltd.); *C.A., 116*, 257521 (1992).
107. Jpn. Kokai Tokkyo Koho JP 04,110,369 [92,110,369] (Apr. 10, 1992; K. Matsuo and S. Shibamoto to Shinto Toryo K.K.); *C.A., 117*, 92395 (1992).
108. Jpn. Kokai Tokkyo Koho JP 04,364,147 [92,364,147] (Dec. 16, 1992; M. Ito and S. Iimuro to Mitsui Toatsu Chemicals, Inc.); *C.A., 118*, 233643 (1993).
109. U.S. Pat. 3,047,536 (July 31, 1962; D. A. Gordon to Dow Chemical Co.).
110. U.S. Pat. 4,031,063 (June 21, 1977; A. B. Conciatori and R. W. Stademan to Celanese Corp.).
111. U.S. Pat. 4,435,590 (March 6, 1984; S. W. Shalaby and D. D. Jamiolkowski to Ethicon, Inc.).
112. U.S. Pat. 4,546,152 (Oct. 8, 1985; D. F. Koelmel, D. D. Jamiolkowski, S. W. Shalaby, and R. S. Bezwada to Ethicon, Inc.).
113. Eur. Pat. Appl. EP 335,594 (Oct. 4, 1989; T. Hirahara, M. Nukui, and K. Tanaka to Mitsubishi Kasei Corp.); *C.A., 112*, 140068 (1990).
114. U.S. Pat. 4,956,443 (Sept. 11, 1990; J. R. Overton, S. L. Stafford, and C. E. Sumner, Jr. to Eastman Kodak Co.).
115. Y. Delaviz and H. W. Gibson, *Am. Chem. Soc., Polymer Prepr., 32(3)*, 230 (1991).
116. Jpn. Kokai Tokkyo Koho JP 03,087,233 [91,087,233] (Apr. 12, 1991; K. Yamamoto and K. Niimi to Mitsui Petrochemical Industries, Ltd.); *C.A., 115*, 160847 (1991).
117. Jpn. Kokai Tokkyo Koho JP 03,258,541 [91,258,541] (Nov. 18, 1991; T. Hirahara, M. Nukii, K. Tanaka, and K. Hayashi to Mitsubishi Kasei Corp.); *C.A., 116*, 130794 (1992).
118. *Plastics Technology, June 1990*, 81.
119. Ger. Offen. DE 2,834,536 (Feb. 22, 1979; W. J. Jackson and J. C. Morris to Eastman Kodak Co.).
120. Jpn. Kokai Tokkyo Koho JP 63,189,429 [88,189,429] (May 8, 1988; R. Ueno, K. Shoda, and Y. Hamazaki to Ueno Seiyaku Oyo Kenkyusho); *C.A., 110*, 96033 (1989).
121. U.S. Pat. 4,188,476 (Feb. 12, 1980; R. S. Irwin to DuPont DeNemours Co.).
122. Jpn. Kokai Tokkyo Koho JP 01,201,326 [89,201,326] (Aug. 14, 1989; M. Hirose and T. Kuratsuji to Teijin, Ltd.); *C.A., 112*, 57052 (1990).
123. Jpn. Kokai Tokkyo Koho JP 02,080,688 [90,080,688] (March 20, 1990; N. Matsunaga, K. Kuwabara, K. Kameyama, and B. Hosoi to Japan Exlan Co. Ltd.); *C.A., 113*, 99297 (1990).
124. Jpn. Kokai Tokkyo Koho JP 02,155,924 [90. 155,924] (June 15, 1990; S. Matsuo and N. Yamukai to Idemitsu Kosan Co., Ltd.); *C.A., 113*, 192193 (1990).
125. Jpn. Kokai Tokkyo Koho Jp 02,203,442 [90,203,442] (Aug. 13, 1990; M. Ohashi and T. Fukuda to Toray Industries, Inc.); *C.A., 114*, 218155 (1991).
126. Jpn. Kokai Tokkyo Koho JP 03,014,620 [91,014,620] (Jan. 23, 1991; A. Hara, K. Kimura, T. Koseki, and Y. Kato to Toyobo Co., Ltd.); *C.A., 115*, 10750 (1991).
127. Jpn. Kokai Tokkyo Koho JP 04,011,651 [92,011,651] (Jan. 16, 1992; Y. Hosokawa, Y. Tsuda, T. Amiya, and Y. Yasuda to Kao Corp.); *C.A., 116*, 215712 (1992).
128. Eur. Pat. Appl. EP 448,814 (Dec. 19, 1990; V. Serini, H.-Y. Buysch, and U. Grigo to Bayer A.-G.); *C.A., 116*, 21712 (1992).
129. Jpn. Kokai Tokkyo Koho JP 04,154,835 [92,154,835] (May 27, 1992; S. Koide and K. Nakamura to Showa Shell Sekiyu K.K.); *C.A., 118*, 22825 (1993).
130. PCT Int. Appl. WO 92,006,129 (Apr. 16, 1992; I. Nitta, K. Asai, and H. Harada to Sumitomo Chemical Co., Ltd.); *C.A., 117*, 172752 (1992).
131. Eur. Pat. Appl. EP 492,999 (July 1, 1992; T. Hirahara, K. Tanaka, and K. Mino to Mitsubishi Kasei Corp.); *C.A., 118*, 22822 (1993).
132. Jpn. Kokai Tokkyo Koho JP 04,154,834 [92,154,834] (May 27, 1992; S. Koide and T. Yamauchi to Showa Shell Sekiyu K.K.); *C.A., 118*, 22826 (1993).
133. H. N. Yoon, L. F. Charbonneau, and G. W. Calundann, *Adv. Mater., 4(3)*, 206 (1992).

134. PCT Int. Appl. WO 79,001,040 (Nov. 29, 1979; W. J. Jackson, Jr. and J. C. Morris to Eastman Kodak Co.).
135. U.S. Pat. 4,612,362 (Sept. 16, 1986; Y.-C. Lai, B. T. DeBona, and D. C. Prevorsek to Allied Corp.).
136. A. K. Rao, J.-I. Jin, and R. W. Lenz, *Am. Chem. Soc., Polymer Prepr., 30(2)*, 234 (1989); *Brit. Polym. J., 21(6)*, 473 (1989).
137. Ger. Offen. DE 3,621,519 (Jan. 7, 1988; H. J. Kock, M. Portugall, B. Hisgen, and J. Mortes to BASF A.-G.); *C.A., 108*, 187460 (1988).
138. Ger. Offen. DE 3,629,210 (March 3, 1988; B. Hisgen, H-J. Kock, M. Portugall, E. Seiler, and G. Blinne to BASF A.-G.); *C.A., 109*, 74153 (1988).
139. Eur. Pat. Appl. EP 251,116 (Jan. 7, 1988; H. J. Kock, M. Portugall, B. Hisgen, and J. Merten to BASF A.-G.); *C.A., 108*, 205303 (1988).
140. Eur. Pat. Appl. EP 275,164 (July 20, 1988; W. A. McDonald and T. G. Ryan to Imperial Chemical Industries); *C.A., 110*, 76318 (1989).
141. Jpn. Kokai Tokkyo Koho JP 01,266,132 [89,266,132] (Oct. 24, 1989; O. Kidai *et al.* to Mitsubishi Kasei Corp.); *C.A., 112*, 159227 (1990).
142. Eur. Pat. Appl. EP 337,728 (Oct. 18, 1989; N. Hayashi and K. Hijakata to Polyplastic Co., Ltd.);*C.A., 112*, 159206 (1990).
143. Jpn. Kokai Tokkyo Koho JP 02,051, 523 [90,051,523] (Feb. 21, 1990; H. Sugitomo, Y. Obe, and K. Hayatsu to Sumitomo Chemical Co., Ltd.); *C.A., 113*, 153308 (1990).
144. Jpn. Kokai Tokkyo Koho JP 03,012,429 [91,012,429] (Jan. 21, 1991; A. Hara, K. Kimura, T. Koseki, and Y. Kato to Toyobo Co., Ltd.); *C.A., 115*, 94066 (1991).
145. Jpn. Kokai Tokkyo Koho JP 63,068,672 [88,068,672] (March 28, 1988; H. Minamizawa, T. Morinaga, T. Fukushima, Y. Nomura, K. Hanabusa, and T. Oshima to Hitachi Chemical K.K.); *C.A., 109*, 112218 (1988).
146. T. Abraham, E. J. Soloski, and R. C. Evers, *J. Polym. Sci., Pt. A, Polym. Chem., 26*, 959 (1988).
147. Jpn. Kokai Tokkyo Koho JP 63,275,633 [88,275,633] (Nov. 14, 1988; T. Katayose to Asahi Chemical Industries K.K.); *C.A., 110*, 193632 (1989).
148. Jpn. Kokai Tokkyo Koho JP 63,057,639 [88,057,639] (March 12, 1988; H. Yamaoka to Mitsubishi Monsanto K.K.); *C.A., 109*, 74891 (1988).
149. Eur. Pat. Appl. EP 388,203 (Sept. 19, 1990; W. Sweeney to E. I. DuPont De Nemours and Co.); *C.A., 115*, 160060 (1991).
150. Jpn. Kokai Tokkyo Koho JP 02,253,945 [90,253,945] (Oct. 12, 1990; T. Hirakawa, H. Watanabe, and K. Nishimura to Teijin, Ltd.); *C.A., 115*, 160563 (1991).
151. Jpn. Kokai Tokkyo Koho JP 03,070,735 [91,070,735] (March 26, 1991; S. Yamazaki, A. Yamanaka, T. Shimozato, and N. Mukai to Mitsubishi Rayon Co., Ltd.); *C.A., 115*, 137029 (1991).
152. Jpn. Kokai Tokkyo Koho JP 03,050,219 [91,050,219] (March 4, 1991; E. Takiyama and T. Hosogane to Showa Highpolymer Co., Ltd.); *C.A., 115*, 184122 (1991).
153. R. Srinivasan and J. E. McGrath, *Am. Chem. Soc., Polym. Prepr., 33(1)*, 503 (1992).
154. USSR Pat. SU 257,010 (Apr. 7, 1970; M. M. Koton, F. S. Florinski, M. I. Bessonov, and A. P. Rudakov to Macromolecular Institute, Academy of Sciences, USSR); *C.A., 73*, 4399 (1970).
155. Anon., *Adhesives Age, 29*, 46 (1986).
156. Jpn. Kokai Tokkyo Koho JP 62,050,372 [87,050,372] (March 5, 1987; M. Ota, S. Kawashima, Y. Sonobe, M. Tamai, H. Oikura, and T. Yamaguchi to Mitsui Toatsu Chemicals, Inc.); *C.A., 107*, 155801 (1987).
157. Jpn. Kokai Tokkyo Koho JP 62,199,674 [87,199,674] (M. Tamai, S. Kawashima, M. Ota, H. Oikawa, K. Ogishi, and T. Yamaguchi to Mitsui Toatsu Chemicals, Inc.); *C.A., 108*, 90803 (1987).
158. Eur. Pat. Appl. EP 251,741 (Jan. 7, 1988; N. Yamaya, K. Baba, and N. Koga to Mitsui Toatsu Chemicals, Inc.); *C.A., 108*, 222286 (1988).
159. U.S. Pat. 4,725,642 (Feb. 16, 1988; T. P. Gannett, R. J. Kassal, and R. S. Ro to E. I. DuPont De Nemours and Co.).
160. Eur. Pat. Appl. EP 276,922 (Aug. 3, 1988; K. Kunimune, K. Kikuta, T. Kawamoto, and S. Konotsume to Chisso Corp.); *C.A., 110*, 58341 (1989).

161. Brit. Pat. GB 2,174,399 (Jan. 18, 1986; K. Higashi and Y. Noda to Nitto Electric Industries K.K.); *C.A., 106*, 157086 (1987).
162. Jpn. Kokai Tokkyko Koho JP 60,215,812 (Oct. 29, 1985; Y. Kaneda, T. Katsura, and Y. Nakagawa to Ube Industries K.K.); *C.A., 104*, 131388 (1986).
163. Jpn. Kokai Tokkyo Koho JP 63,211,319 [88,211,319] (Sept. 2, 1988; M. Tamai, S. Kawashima, M. Ota, H. Oikawa, and T. Yamaguchi to Mitsui Toatsu Chemicals, Inc.); *C.A., 111*, 8830 (1989).
164. *Chemtech, 1989*, 117.
165. Eur. Pat. Appl. EP 365,311 (Apr. 25, 1990; M. Saruwatari, M. Nakano *et al.* to Mitsui Toatsu Chemicals, Inc.); *C.A., 113*, 154406 (1990).
166. Jpn. Kokai Tokkyo Koho JP 02,309,322[90,309,322] and JP 02,309,323 [90. 309,323] (both dated Dec. 25, 1990; M. Yamada, M. Fukahori, and S. Kiriyama to Toray Industries, Inc.); *C.A., 114*, 257121 and 257122 (1991).
167. Eur. Pat. Appl. EP 376,592 (July 4, 1990; H. Ezawa, T. Nakakura, T. Watanabe, and H. Tsushima to Mitsui Toatsu Chemicals, Inc.); *C.A., 113*, 232970 (1990).
168. Jpn. Kokai Tokkyo Koho JP 03,045,624 [91,045,624] (Feb. 27, 1991; R. Akahori, H. Iwakiri, K. Iwamoto, and K. Yonezawa to Kanegafuchi Chemical Industry Co., Ltd.); *C.A., 115*, 92709 (1991).
169. Jpn. Kokai Tokkyo Koho JP 03,192,177 [91,192,177] (Y. Honchi, M. Saruwatari, and Y. Ota to Mitsui Toatsu Chemicals, Inc.); *C.A., 116*, 42760 (1992).
170. Jpn. Kokai Tokkyo Koho JP 62,053,319 [87,053,319] (March 9, 1987; S. Kanekawa, Y. Saito, T. Adachi, and K. Okuno to Sumitomo Chemical Co., Ltd); *C.A., 107*, 135264 (1987).
171. U.S. Pat. 4,742,141 (May 3, 1988; R. Dieu and M. Rakoutz to Rhone-Poulenc Specialities Chimiques).
172. Jpn. Kokai Tokkyo Koho JP 62,283,108 [87,283,108] (Dec. 9, 1987; G. Iwamura, Y. Yokoyama, and H. Sakamoto to Dainippon Ink and Chemicals K.K.); *C.A., 108*, 152274 (1988).
173. U.S. Pat. 4,788,295 (Nov. 29, 1988; M. Rakoutz to Rhone-Poulenc Specialities Chimiques).
174. Jpn. Kokai Tokkyo Koho JP 63,141,961 [88,141,961] (June 14, 1988; S. Takeda and H. Kakiuchi to Hitachi Chemical Co., Ltd.); *C.A., 110*, 75067 (1988).
175. Jpn. Kokai Tokkyo Koho JP 02,022,318 [90,022,318] (Jan. 25, 1990; Y. Nakada, K. Nishii, A. Matsura, Y. Takigawa, and N. Surawatari to Japan Atomic Energy Research Institute); *C.A., 113*, 74942 (1990).
176. Eur. Pat. Appl. EP 372,775 (June 13, 1990; N. Yamaya, M. Ohta, and A. Yamaguchi to Mitsui Toatsu Chemical Co., Ltd); *C.A., 113*, 192727 (1990).
177. Eur. Pat. Appl. EP 410,298 (Jan. 30, 1991; R. Peter and L. Schlemmer to BASF A.-G.); *C.A., 114*, 230058 (1991).
178. Ger. Offen. DE 3,701,638 (Jan. 21, 1987; T. Yamaoko, N. Tsujimoto, M. Suzuki, and H. Michinori to Ube Industries K.K.); *C.A., 108*, 177206 (1987).
179. Eur. Pat. Appl. EP 148,534 (July 17, 1985; L. A. Domeier and H. C. Gardner to Union Carbide Corp.); *C.A., 103*, 216354 (1985).
180. Jpn. Kokai Tokkyo Koho JP 62,167,777 [87,167,777] (July 24, 1987; Y. Yoshikawa, M. Yamazaki, K. Yamaguchi, K. Sugimoto, Y. Tanabe, and T. Yamaguchi to Mitsui Toatsu Chemical Co., Ltd.); *C.A., 108*, 168120 (1988).
181. Jpn. Kokai Tokkyo Koho JP 63,034,899 [88,034,899] (July 12, 1988; to Toshiba Chemical Products Co., Ltd.); *C.A., 108*, 168120 (1988).
182. Jpn. Kokai Tokkyo Koho JP 03,172,325 [91,172,325] (July 25, 1991; N. Yamatami, M. Ota, and T. Yamaguchi to Mitsui Toatsu Chemicl Co., Ltd.); *C.A., 116*, 22262 (1992).
183. Jpn. Kokai Tokkyo Koho JP 04,007,313 [92,007,313] (Jan. 10, 1992; A. Matsumoto, K. Hasegawa, T. Harada, A. Fukuda, and K. Otsuki to Daihachi Chemical Industry Co., Ltd.); *C.A., 116*, 215740 (1992).
184. Eur. Pat. Appl. EP 186,060 (July 2, 1986; R. R. Galluci to General Electric Co.); *C.A., 106*, 6045 (1987).
185. Eur. Pat. Appl. EP 425,438 (May 2, 1991; M. R. Southcott to Ciba-Geigy A.-G.); *C.A., 115*, 137518 (1991).

186. Jpn. Kokai Tokkyo Koho JP 61,293,966 [86,293,966] (Dec. 24, 1986; S. Kanekawa, Y. Saito, and T. Adachi to Sumitomo Chemical Co., Ltd.); *C.A., 106*, 213762 (1987).
187. U.S. Pat. 4,696,998 (Sept. 29, 1987; D. J. Brunelle, T. L. Guggenheim, E. P. Boden, T. G. Shannon, and J. W. Guiles to General Electric Co.).
188. U.S. Pat. 4,680,378 (July 14, 1987) and U.S. Pat. 4,769,440 (Sept. 6, 1988) (both by R. E. Hefner, Jr. to Dow Chemical Co.).
189. Jpn. Kokai Tokkyo Koho JP 03,050,218 [91,050,218] (March 4, 1991); *C.A., 115*, 281294 (1991), and JP 03,166,209 [91,166,209] (July 18, 1991); *C.A., 115*, 281394 (1991) (both by E. Takiyama and T. Hosogame to Showa Highpolymer Co., Ltd.).
190. Jpn. Kokai Tokkyo Koho JP 01,282,283 [89,282,283] (May 10, 1988; Y. Nomura, T. Morinaga, T. Fukushima, H. Minamizawa, and K. Hanabusa to Hitachi Chemical Co., Ltd.); *C.A., 112*, 180960 (1990).
191. Jpn. Kokai Tokkyo Koho JP 04,271,820 [92,271,820] (Sept. 28, 1992; H. Hachisuga, S. Shimatani, M. Shimizu, and K. Matsumoto to Nitto Denko Corp.); *C.A., 118*, 109312 (1993).
192. Jpn. Kokai Tokkyo Koho JP 05,005,081 [93,005,081] (Jan. 14, 1993; R. Nirishima, N. Kikuchi, and H. Nishizawa to Hitachi Chemical Co., Ltd.); *C.A., 118*, 256745 (1993).
193. C. L. Mahoney, E. R. Barnum, W. W. Kerlin, K. J. Sax, and W. A. Saari, *J. Chem. Eng. Data, 5*, 172 (1960).
194. D. Klamann, *Lubricants and Related Products*, Verlag Chemie, Weinheim, 1984, pp. 116–121.
195. K. J. Sax, W. S. Saari, C. L. Mahoney, and J. M. Gordon, *J. Org. Chem., 25*, 1590 (1960).
196. U.S. Pat. 3,294,846 (Dec. 27, 1966; J. E. Livalt and E. C. Britton to the Dow Chemical Co.).
197. Belg. Pat. BE 695,512 (Sept. 15, 1967; to Farbenfabriken Bayer A.-G.).
198. U.S. Pat. 3,379,672 (Apr. 23, 1968; S. R. Breshears to Northrup Corp.).
199. U.S. Pat. 3,476,815 (Nov. 4 1969; A. L. Williams and R. E. Kinney to Mobil Oil Corp.).
200. U.S. Pat. 3,487,114 (Dec. 30, 1969; G. Frick, Jr. and G. F. Hawkins to Eastman Kodak Co.).
201. Jpn. Kokai Tokkyo Koho JP 70,013,763 (May 18, 1970; to Monsanto Co.).
202. U.S. Pat. 3,567,783 (March 2, 1971; G. P. Brown to General Electric Co.).
203. U.S. Pat. 3,651,151 (March 21, 1972; R. F. Bridger, R. E. Kinney, and A. L. Williams to Mobil Oil Corp.).
204. Ger. Offen. DE 1,594,501 (May 9, 1974; to Monsanto Co.).
205. Jpn. Kokai Tokkyo Koho JP 61,163,997 [86,163,997] (July 24, 1986; Y. Honda to Nippon Mektron K.K.); *C.A., 105*, 211561 (1986).
206. U.S. Pat. 4,272,429 (June 9, 1981; R. H. Jones to E. I. DuPont De Nemours and Co.).
207. Jpn. Kokai Tokkyo Koho JP 62,145,054 [87,145,054] (June 29, 1987; T. Takoya, T. Tomioka, and H. Yamazaki to Idemitsu Kosan K.K.); *C.A., 108*, 22435 (1988).
208. Jpn. Kokai Tokkyo Koho JP 62,275,125 [87,275,125] (Nov. 30, 1987; S. Matsuo and S. Tagami to Idemitsu Kosan K.K.); *C.A., 108*, 187832 (1987).
209. H. R. Kricheldorf, J. Meier, and G. Schwarz, *Makromol. Chem., Rapid Commun., 8*, 529 (1987).
210. U.S. Pat. 4,703,104 (Oct. 27, 1987; S. Matsuo, T. Murakama, and R. Takasawa to Idemitsu Kosan Co., Ltd.).
211. Jpn. Kokai Tokkyo Koho JP 62,275,124 [87,275,124] (Nov. 30, 1987; S. Matsuo to Idemitsu Kosan Co., Ltd.); *C.A., 108*, 187833 (1987).
212. Jpn. Kokai Tokkyo Koho JP 63,189,435 [88,189,435] (Aug. 5, 1988; S. Matsuo to Idemitsu Kosan Co., Ltd.); *C.A., 110*, 58359 (1989).
213. Jpn. Kokai Tokkyo Koho JP 01,129,033 [89,129,023] (May 22, 1989; T. Asahi, T. Tamai, and Y. Hondo to Tosoh Corp.); *C.A., 112*, 8078 (1990).
214. Jpn. Kokai Tokkyo Koho JP 01,146,858 [89,146,858] (June 8, 1989; N. Yamuki, Y. Kitajima, and S. Matsuo to Idemitsu Kosan Co., Ltd.); *C.A., 112*, 21482 (1990).
215. Jpn. Kokai Tokkyo Koho JP 01,236,269 [89,236,269] (Sept. 21, 1989; T. Murakami to Idemitsu Kosan Co., Ltd.); *C.A., 112*, 120138 (1990).
216. Jpn. Kokai Tokkyo Koho JP 01,259,032 [89,259,032] (Oct. 16, 1989; S. Matsuo, N. Yamukai, and C. Kayano to Idemitsu Kosan Co., Ltd.); *C.A., 112*, 159235 (1990).

217. Jpn. Kokai Tokkyo Koho JP 01,203,462 [89,203,462] (Aug. 16, 1989; M. Miyake and M. Suzuki to Idemitsu Kosan C., Ltd.); *C.A., 112*, 78757 (1990).
218. Jpn. Kokai Tokkyo Koho JP 03,263,460 [91,263,460] (Nov. 22, 1990; T. Bando, K. Tanako, and T. Takeya to Idemitsu Kosan Co., Ltd.); *C.A., 114*, 165717 (1991).
219. Jpn. Kokai Tokkyo Koho JP 02,264,771 [90,264,771] (Oct. 29, 1990; R. Takezawa to Idemitsu Kosan Co., Ltd.); *C.A., 114*, 165717 (1991).
220. Eur. Pat. Appl. EP 420,245 (April 3, 1991; H. Kawato and T. Ishii to Idemitsu Kosan Co., Ltd.); *C.A., 115,* 51115 (1991).
221. M. J. Mullins, E. P. Woo, C. C. Chen, D. J. Murray, M. T. Bishop, and K. E. Balon, *Am. Chem. Soc., Polym. Prepr., 32(20)*, 174 (1991).
222. Eur. Pat. Appl. EP 437,851 (July 24, 1991; H. Kawato and T. Tomioka to Idemitsu Kosan Co., Ltd.); *C.A., 115,* 209451 (1991).
223. Jpn. Kokai Tokkyo Koho JP 03,281,530 [91,281,530] (Dec. 12, 1991; R. Takazawa to Idemitsu Kosan Co., Ltd.); *C.A., 116*, 175000 (1992).
224. Jpn. Kokai Tokkyo Koho JP 03,276,513 [91,276,513] (Dec. 6, 1991; K. Watanabe and T. Yamazaki to Hitachi Cable Co., Ltd.); *C.A., 116*, 196380 (1992).
225. Jpn. Kokai Tokkyo Koho JP 04,050,227 [92,050,227] (Feb. 19, 1992; R. Takazawa, H. Yamazaki, and J. Shiyoa to Idemitsu Kosan Co., Ltd.); *C.A., 117*, 91484 (1992).
226. Jpn. Kokai Tokkyo Koho JP 04,161,450 [92,161,450] (June 4, 1992; K. Yamada and H. Kawahigashi to Idemitsu Kosan Co., Ltd.); *C.A., 117*, 252779 (1992).
227. Jpn. Kokai Tokkyo Koho JP 04,100,856 [92,100,856] (Apr. 2, 1992; T. Bando and T. Takeyo to Idemitsu Kosan Co., Ltd.); *C.A., 117*, 214093 (1992).
228. Jpn. Kokai Tokkyo Koho JP 01,182,322 [89,182,322] (July 20, 1989; S. Matsuo, C. Kayano, and N. Yamukai to Idemitsu Kosan Co., Ltd.); *C.A., 112*, 57461 (1990).
229. Jpn. Kokai Tokkyo Koho JP 01,198,624 [89,198,624] (Aug. 10, 1989; S. Matsuo, N. Yamukai, and C. Kayano to Idemitsu Kosan Co., Ltd.); *C.A., 112*, 57030 (1990).
230. Jpn. Kokai Tokkyo Koho JP 01,221,424 [89,221,424] (Sept. 4, 1989; S. Matsuo, N. Yamukai, and C. Kayano to Idemitsu Kosan Co., Ltd.); *C.A., 112*, 99544 (1990).
231. Ger. Offen. DE 3,804,988 (Aug. 31, 1989; J. Knebel, M. Ude, and J. Vetter to Roehm G.m.b.H.); *C.A., 112*, 99522 (1990).
232. Ger. Offen. DE 2,529,487 (Jan. 20, 1977; G. Rottloff and R. Sundermann to Bayer A.-G.); similar is Ger. Offen. DE 2,507,671 (Sept. 2, 1976; same authors and assignment); *C.A., 85*, 192351 (1971).
233. D. A. Shimp, *Am. Chem. Soc., Proc. Polym. Mat. Sci. Eng., 54*, 107 (1986).
234. U.S. Pat. 3,357,931 (Dec. 12, 1967; R. G. Rice and M. V. Ernest to W. R. Grace & Co.).
235. Brit. Pat. GB 985,607 (March 10, 1965; A. D. Forbes, P. Gould, and J. H. A. Pew to The British Petroleum Co., Ltd.); *C.A., 62*, 14705 (1965).
236. U.S. Pat. 3,297,639 (Jan. 10, 1967; to L. G. Picklesimer and T. F. Saunders).
237. U.S. Pat. 3,451,048 (Nov. 17, 1970; R. Binsack, L. Bottenbruch, and H. Schnell to Farbenfabriken Bayer A.-G.).
238. PCT Int. Appl. WO 87,000,844 (Feb. 12, 1987; J. R. Sybert, J. F. Wolfe, P. D. Sybert, and B. Wilson to Commtech International); *C.A., 107*, 78484 (1987).
239. Jpn. Kokai Tokkyo Koho JP 03,181,521 [91,181,521] (Aug. 7, 1991; H. Nishino to Honda Motor Co., Ltd.); *C.A., 116*, 42267 (1992).
240. PCT Int. Appl. WO 90,006,960 (June 28, 1990; Y. H. So to Dow Chemical Co.); *C.A., 113*, 232313 (1990).
241. U.S. Pat. 5,021,580 (June 4, 1991; Z. Lysenko to Dow Chemical Co.).
242. Jpn. Kokai Tokkyo Koho JP 62,004,720 [87,004,720] (Jan. 10, 1987; H. Namba and M. Hishinuma to Sanyo Kokusaku Pulp K.K.); *C.A., 107*, 97304 (1987).
243. Jpn. Kokai Tokkyo Koho JP 63,234,030 [88,234,030] (Sept. 9, 1988; T. Mita to Teijin Ltd.); *C.A., 111*, 8544 (1989).

244. Eur. Pat. Appl. EP 250,715 (Jan. 7, 1988; O. Tibar and B. M. Culbertson to Ashland Oil, Inc.); see also U.S. Pat. 4,748,230 (May 31, 1988; O. Tibar, B. M. Culbertson, M. L. Deviney, and A. B. Goal to Ashland Oil, Inc.).
245. M. L. Deviney, O. Tibar, B. M. Culbertson, and R. J. Schafer, *Am. Chem. Soc., Polym. Prepr., 32(2)*, 282 (1991).
246. U.S. Pat. 4,864,054 (Sept. 5, 1989; J. V. Crivello and D. A. Conlon to General Electric Co.).
247. Eur. Pat. Appl. EP 448,332 [Sept. 25, 1991; J. G. Woods and P. Coakley to Loctite (Ireland), Ltd.]; *C.A., 115*, 258941 (1991).
248. U.S. Pat. 5,047,456 (Sept. 10, 1991; F. C. Onwumere, S. L. Greene, and S. H. Wright to Kimberly-Clark Corp.).

10

Dyes, Fluorescent Chemicals, Optical Bleaches, Laser Dyes, and Imaging/Recording Technologies

10.1. INTRODUCTION

It is in dyes where resorcinol and its derivatives got their commercial start. From there, new applications in related fields developed. Dyestuff chemists were always pioneering new chemistries. Their inventions not only brought color to our sometimes drab world, they also influenced other endeavors. The history of U.S. dyestuff production has also been very colorful in a different sense—it was shaped by wars, trade battles, fashions, toxicity, and effluent problems and, in the last 25 years or so, by reorganizations and second thoughts, including the exit of several of the major manufacturers from parts or all of dye manufacture. That story will not be belabored here. This chapter is restricted to looking at the uses of resorcinol and *m*-aminophenol, but it shows the progression of dyestuffs from coloring materials to nontraditional applications. These new and emerging technologies require interdisciplinary research and development. A recent book was devoted to many of these new trends in the application of dyestuffs and related chemicals beyond the traditional uses.[1] It is notable that Japanese companies have become particularly robust developers of new technology and new products incorporating these technologies in several categories.

10.2. DYES AND RELATED CHEMICALS IN COMMERCIAL USE

10.2.1. Azo Dyes

These dyes are made by reacting arylamines with nitrous acid ($NaNO_2$/mineral acid) at *ca* 0 °C and, in the simplest case of interest here, coupling the obtained diazonium salt with resorcinol, usually at mildly alkaline pH.

In earlier days, the monoazo dye, Chrysoin S, also called Tropaeolin O, **10-1** (C.I. Food Yellow 8; C.I. stands for the bible of the industry, the Color Index), and the disazo dye, C.I. Acid Yellow 105 (C.I. Food Yellow 109), **10-2**, were used as food colors; they are no longer in such use. Tropaeolin O, **10-1**, is still used as an indicator (yellow at pH 11, orange brown at pH 12.7) and as a plasma stain.

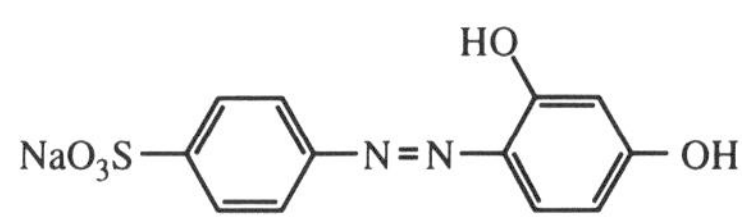

10-1

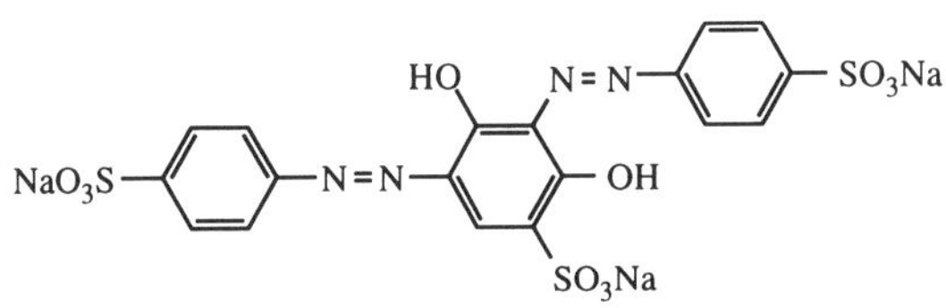

10-2

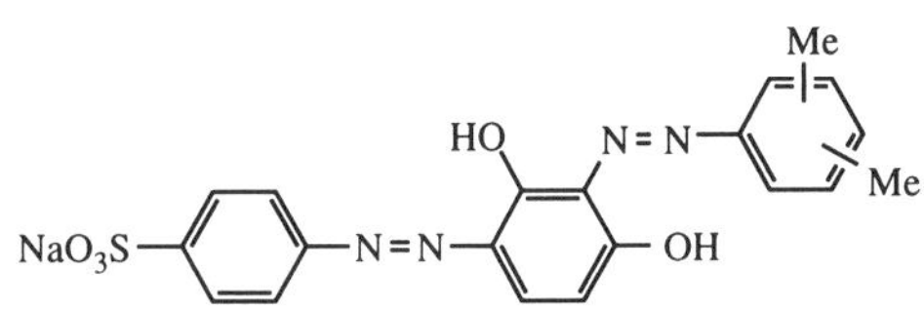

10-3

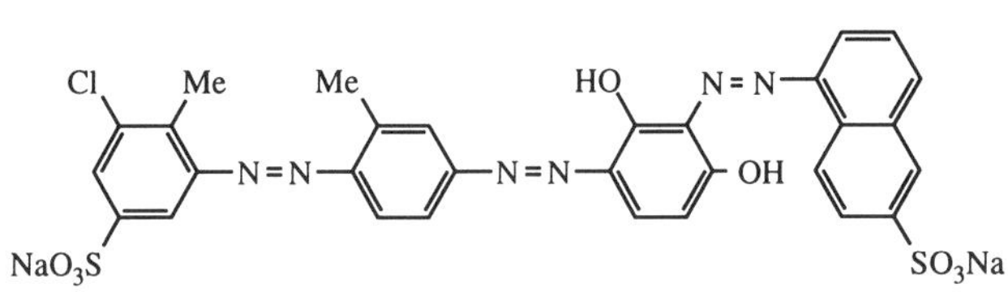

10-4

There are a number of azo dyes for leather which contain resorcinol. Examples are C.I. Acid Orange 24, **10-3**, which dyes leather brown and which had U.S. sales in the early 1970s in the 1 mill lb /yr range (e.g., DuPont Resorcin Brown 5GM), and the trisazo dye, C.I. Acid Brown 105, **10-4**.

A direct dye for cellulosic materials, such as cotton, rayon, or paper, is C.I. Direct Orange 18, **10-5**. C.I. Acid Black 82 (Erganilschwarz C) is the chromium complex of the disazo dye, **10-6**.

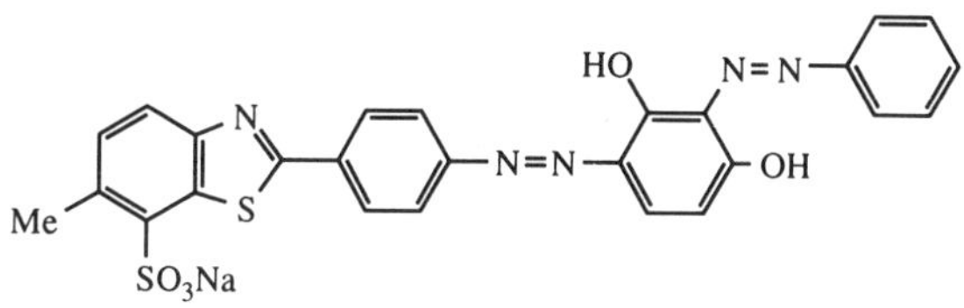

10-5

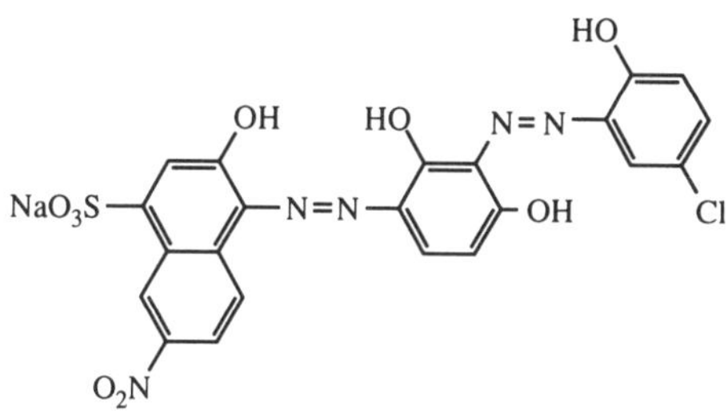

10-6

10.2.2. Coumarins, Xanthenes, and Oxazines

10.2.2.1. Coumarins

The participants here are the 7-hydroxycoumarins, made from resorcinol, and the 7-aminocoumarins, usually made from *m*-aminophenols. Most of the applications of these compounds are based on their fluorescence.

The use of compounds in this group as fluorescent markers in diagnostic products and medical research applications was already mentioned in Chapter 8. Additional examples are 4-bromomethyl-7-methoxycoumarin (Br-MMC), **10-7**, which is used as a fluorescent probe for carboxylic and fatty acids in analysis,[2] particularly of biological fluids, and 7-[chlorocarbonylmethoxy]-4-methylcoumarin (CMMC), **10-8**, which is used as a fluorescent marker by derivatization of hydroxy compounds.[3]

3,3′-Carbonyl-bis-(7-methoxycoumarin), **10-9**, is used as a triplet sensitizer and photoinitiator in chemical synthesis.

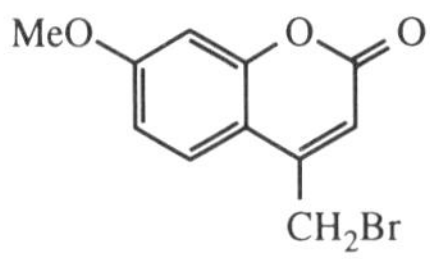

10-7

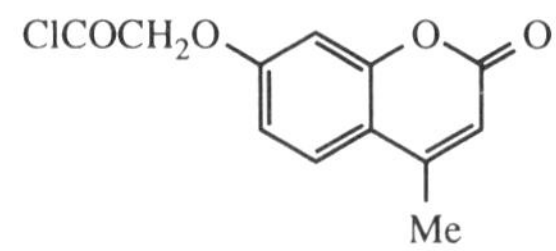

10-8

7-Hydroxy- and 7-amino-coumarins are also useful as laser dyes. These compounds are used in solution, give access to a wide range of spectral properties, and are tunable (using gratings and mirrors) to give coherent light sources of output from the ultraviolet (UV) to the near infrared (IR) range. These laser dyes are used in a variety of instrumental (e.g., spectroscopic) techniques. Two examples are 4-methyl-7-hydroxycoumarin (Coumarin 4), **10-10**, in methanol having an absorption max 322 nm, fluorescence max 386 nm, and laser output (nitrogen pump) max 494 nm; the second example is 4-trifluoromethyl-7-dimethylaminocoumarin (Coumarin 152), **10-11**, in methanol showing an absorption max 394 nm, fluorescence max 496 nm, laser output (Nd:YAG pump) max 524 nm, and laser output (nitrogen pump) max 530 nm. Many additional examples are given in the commercial literature.[4]

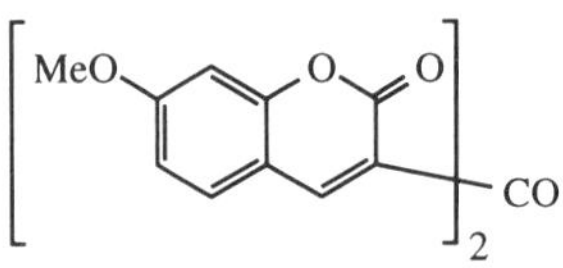

10-9

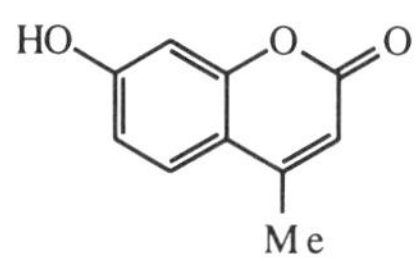

10-10

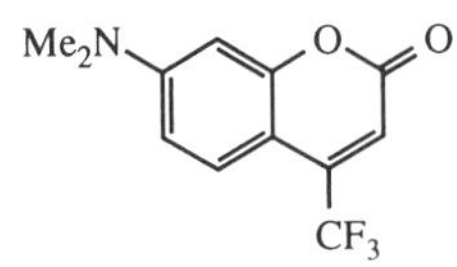

10-11

Derivatives of 7-aminocoumarins are also used as optical bleaches. These products act as whiteners for yellowish/yellowing organic materials via the UV light absorption and blue fluorescence emission of the coumarins. Examples are compound **10-12**, used for plastics, and compound **10-13**, used for polyester fibers. (The starting material for such substituted 7-aminocoumarins may not be 7-aminocoumarin itself but other precursors, such as *p*-nitrosalicylaldehyde.)

There are also fluorescent dyes for organic materials where the dye has the 7-aminocoumarin structure and absorbs visible light. These fluorescers are used for marking, painting, printing, and display applications. An example is compound **10-14** which imparts yellow fluorescence to polyester, polyamide, and polyacrylonitrile articles.

10.2.2.2. Xanthenes

These dyes are essentially derivatives of *o,o'*-dihydroxydiphenylmethane or -triphenylmethane which give, on acid-catalyzed condensation, the basic, condensed-tricyclic xanthene structure, **10-15**. In total, world usage of all xanthene-type dyes in 1980 was in the 11 mill lb./yr range. With new uses of such dyes in imaging

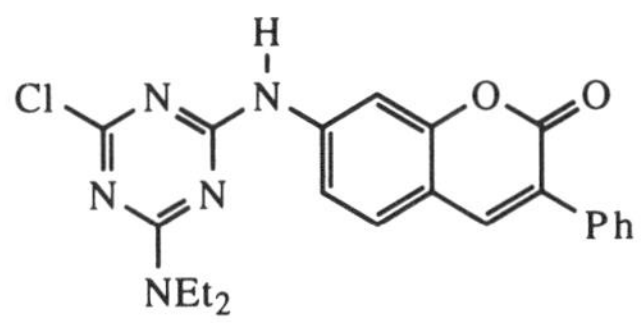

10-12

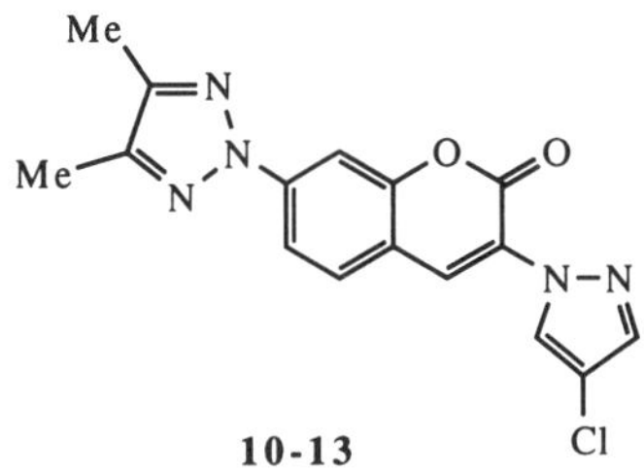

10-13

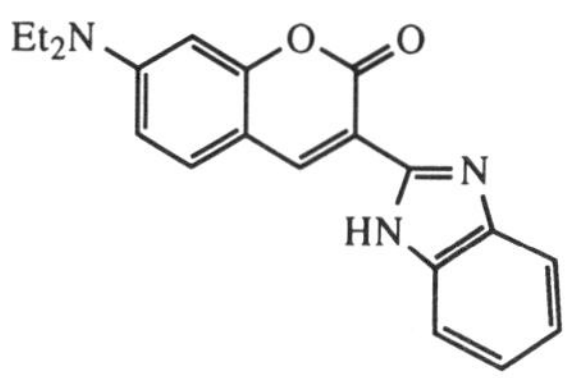

10-14

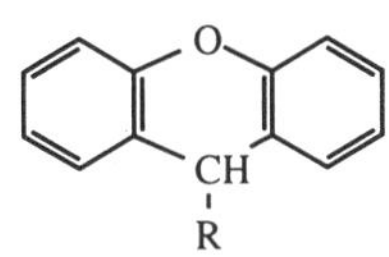

10-15
(where R=H or Ph)

technologies, described later on, the volume of production has increased substantially since then.

The diphenylmethane derivatives are called pyronines. As an example, the condensation of *m*-diethylaminophenol with formaldehyde in the presence of concentrated sulfuric acid, followed by oxidation of the intermediate xanthene with ferric chloride and reaction with HCl, gives Pyronine B, **10-16**, as shown in Eq. (10.1). This dye, **10-16**, is used as a biological wet stain for living cells to aid in their visualization under the microscope.

It is the triphenylmethane-type of xanthene dye that contains many important, commercial members. In the hydroxyxanthene series, the best known example is fluorescein sodium (already mentioned in Chapter 8 for its applications in medicine), **10-17**, which is made by the condensation of resorcinol with phthalic anhydride at 200 °C, followed by cooling of the melt, pulverization, and drowning in dilute sodium hydroxide solution. Fluorescein is used as a fluorescent dye for liquid detergents, degreasing gels, lubricants and coolants, as a marker in air–sea rescue operations, and in leak detection and water tracing. Its bromination product, Eosin G (C.I.Acid Red 87, Pigment Red 90, Solvent Red 43), **10-18**, is used as a dye for stamp pad inks and

(10.1)

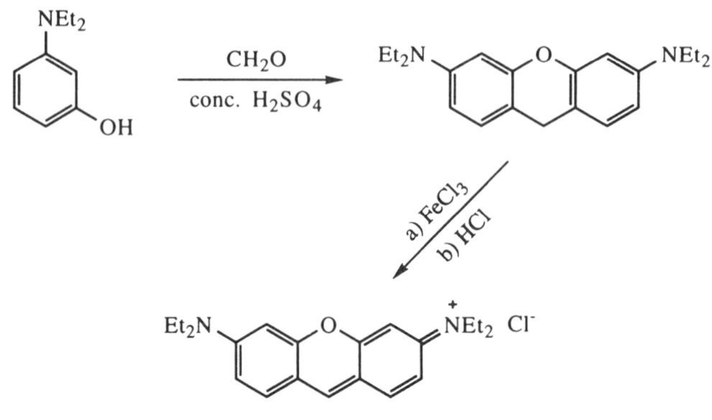

10-16

for cosmetics, e.g., lipstick and nail polish. The analogous tetraiodo-fluorescein is called Erythrosin J (C.I. Acid Red 51, Food Red 14, Pigment Red 172, Solvent Red 140) and is used as a bluish-pink dye for cosmetics and food (C.I. Acid Red No. 3; permitted for food use in the EEC, UK, and USA as E127); in 1978, U.S. usage was about 0.9 mill lb./yr.

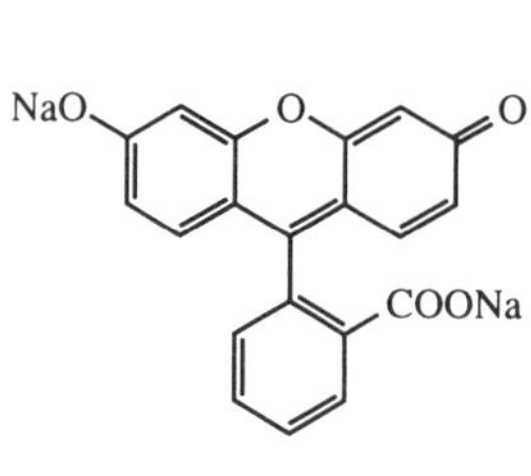

10-17

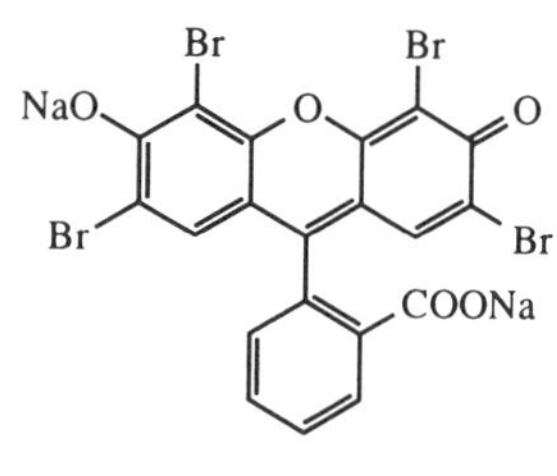

10-18

Even more important are the aminoxanthenes of the triphenylmethane type. For example, the condensation of *m*-diethylaminophenol with phthalic anhydride at 180 °C in the presence of sulfuric acid as the catalyst gives the cationic dye, Rhodamin B (C.I. Basic Violet 10, Solvent Red 49, Pigment Violet 1, Pigment Violet 173), **10-19**, shown as the chloride. This water-soluble compound dyes wool and silk a fluorescent, bluish-red color, polyacrylonitrile fiber is dyed pink, and paper is dyed red. The complexes formed by **10-19** with phosphotungstic acid or phosphomolybdic acid are used in printing inks. Esters of **10-19** can be made and used for dyeing modified polyester and acrylic fibers. A specific example of such esters is Rhodamine 6G (C.I. Basic Red 1, Solvent Red 36), **10-20**, which dyes paper a strong yellowish red; it is also used in printing inks in the form of phosphomolybdic/phosphotungstic acid complexes (C.I. Pigment Red 81) and the copper ferrocyanide complex (C.I. Pigment Red 169).

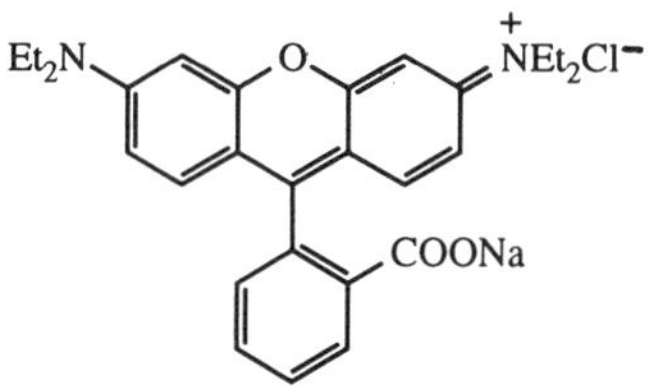

10-19

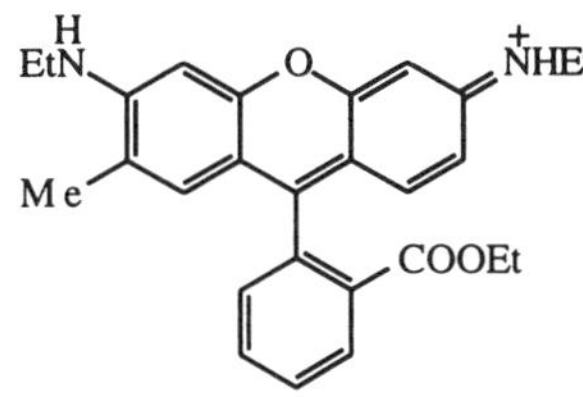

10-20

An alternate preparation of aminoxanthene dyes is exemplified by the condensation of resorcinol with phthalic anhydride in the presence of $POCl_3$ to give the dichlorofluoranlactone, **10-21**. This dichlorocompound is condensed with aniline at 200 °C, followed by sulfonation with 100% sulfuric acid (the "monohydrate" of sulfuric acid) to obtain the anionic dye, Fast Acid Violet B (C.I. Acid Violet 30), **10-22**, which dyes wool a brilliant violet color.

Another example of an anionic, rhodamine-type dye is Sulforhodamine B (C.I. Acid Red 52), **10-23**, obtained by the condensation of *m*-diethylaminophenol with benzaldehyde-2,4-disulfonic acid.

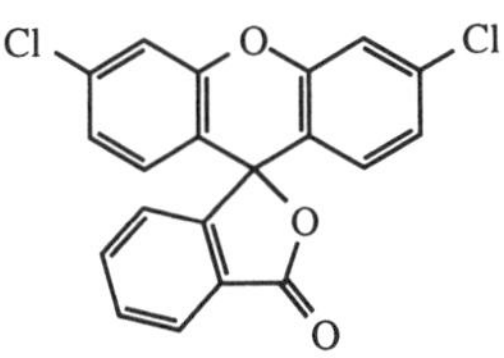

10-21

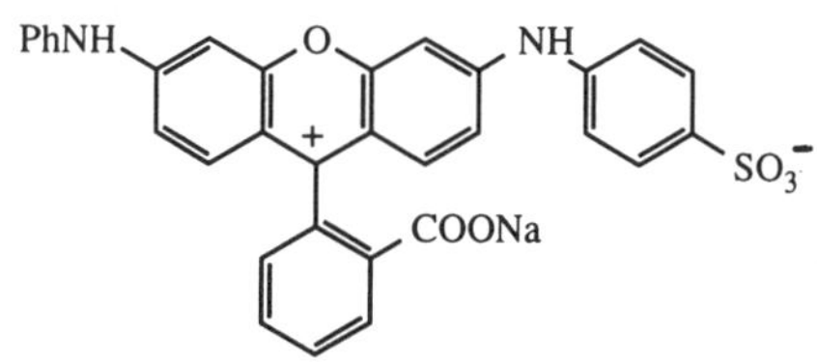

10-22

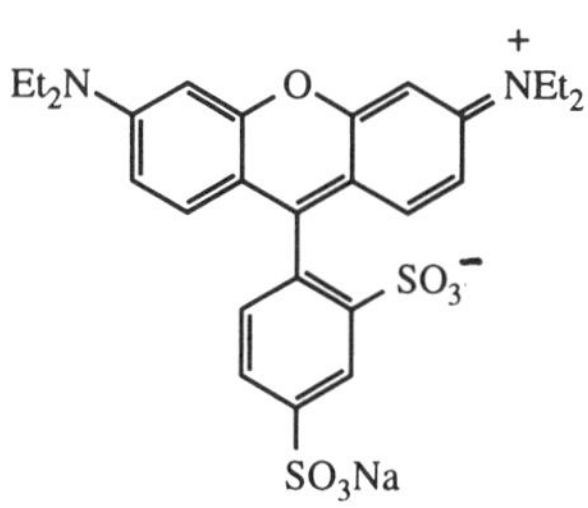

10-23

Obviously, there are many possible variations on the general theme of xanthene dyes. Some additional examples are given later in the sections dealing with chemicals for imaging technologies. For more details, the reference literature on dyestuffs should be consulted.[5]

10.2.2.3. Oxazines

An example of this type of dye is Basic Pure Blue 3G (Rhodulin Pure Blue 3G), **10-24**, produced by the reaction of *m*-diethylaminophenol with 3-ethoxy-4-nitroso-*N,N*-diethylaniline in the presence of hydrochloric acid; this dye is useful for dyeing acrylic fibers.

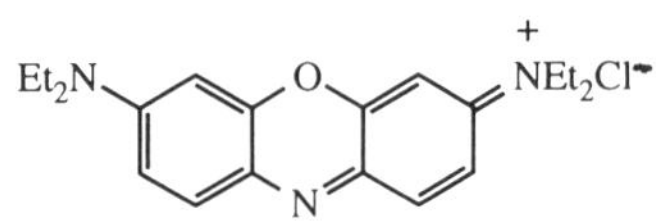

10-24

10.2.3. Plant Extracts As Dyes

One example in this category is brasilein, **10-25**, which contains a resorcinol moiety. It is obtained by extraction/air oxidation from the heartwood of a redwood species of the *Caesalpinia* variety, e.g., Brasilwood. Brasilein was used for mordant dyeing of wool and silk with alum or tin or iron salts. The dye is used as a redox indicator and as a dye in histology. (Additional examples of colorants from plants are given in Chapter 14.)

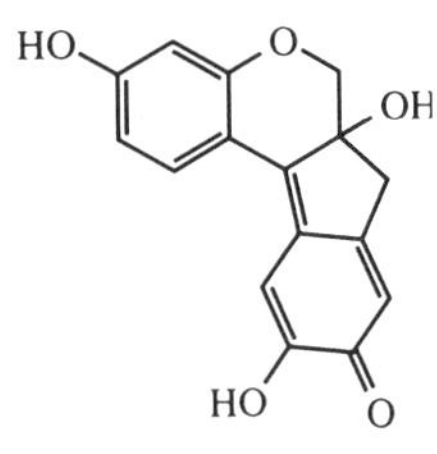

10-25

10.2.4. Chemicals for Diazotype Reprography

This copying process was introduced in the U.S. in the 1920s by Kalle A.-G. In this system, a transparent or translucent sheet of the master picture is superimposed on a paper sheet or film coated with a stabilized diazonium salt, then exposed to high-intensity ultraviolet light from a suitable lamp. Where this light strikes the coated paper or film, the diazonium compound decomposes as shown in Eq. (10.2). In the development step, the diazonium salt remaining in the unexposed areas (corresponding

$$\left.\begin{array}{ll} \text{a)} & Ar\overset{+}{N_2}X^- \text{ (where X = e.g., halide)} \xrightarrow{h\nu} ArX + N_2\uparrow \\ \text{b)} & Ar\overset{+}{N_2}X^- + H_2O \xrightarrow{h\nu} ArOH + N_2\uparrow + HCl \end{array}\right\} \qquad (10.2)$$

to the dark areas of the master) is reacted with a coupler, generally a phenolic compound, to form an azo dye. Thus, diazotype copying is a positive photocopying process.

In the dry development variation, paper coated with the stabilized diazonium salt and the coupler is developed by exposure to ammonia vapor. In the wet development process, paper or film coated with an aryldiazonium salt only is developed with a solution of a coupler at slightly alkaline pH.

The diazotype copying process has lasted for so long because it is relatively low cost, with simple equipment, rapid processing, high image resolution, and the capability to make copies of large documents such as engineering drawings. Diazotype film is mostly used to make copies of microfilm.

The diazonium salts used are *p*-dialkylaminobenzenediazonium salts, 2,5-dialkoxy-4-morpholinobenzenediazonium, and similar salts supplied in stabilized form as zinc chloride complex salts (water soluble), borofluorides or hexafluorophophate complex salts (these are soluble in organic solvents and used to coat [polyester] substrate films), and others.

The chemicals of interest here are the couplers, mostly naphthols or resorcinols. Mixtures of couplers are also used. Suitable resorcinol derivatives are, e.g., beta-resorcylic acid (2,4-dihydroxybenzoic acid), 2,4-dihydroxy-6-bromobenzoic acid, 4,4′-thio-bis-resorcinol, the mono-(2-hydroxyethyl ether) of resorcinol, arylamides and alkylamides of beta-resorcylic acid such as compound **10-26**, and resorcinol still residue; these derivatives give yellow, red-brown, or sepia lines.

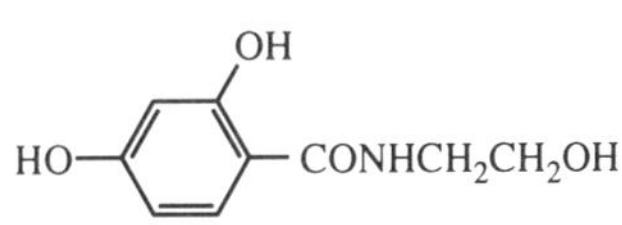

10-26

It has been estimated that the market for resorcinol-type couplers in diazotypy is about 0.3 mill lb./yr in the U.S.[6]

10.2.5. Chemicals for Carbonless Paper Copying

This process, suitable for making single or multiple copies, was pioneered by NCR (National Cash Register Co.) in the 1950s and has achieved large-scale use worldwide. The single-copy carbonless paper consists of two sheets: The top sheet is coated on the bottom with impermeable-walled microcapsules containing a color-forming (chromogenic) leuco (colorless or nearly colorless latent) dye in a solvent or carrier; this top sheet is also called a CB (coated back) sheet. The bottom sheet is coated on its top with the developer, which is an acidic material such as, e.g., an acid clay, kaolin, or a styrene maleic acid copolymer; this bottom sheet is also called the CF (coated front) sheet. Multiple carbonless copy materials contain in addition an interleaved sheet or sheets, coated on top with the developer and on the bottom with the microcapsules containing the solution of the leuco dye; these interleaves are also called CFB sheets. Color is developed by applying pressure (via a typewriter key, pencil, etc.) to the top sheet so that the microcapsules are ruptured, transferring the solution of the leuco dye to contact the bottom sheet/developer, with resulting image formation.

There are several methods for making microcapsules. One of these involves the polycondensation of urea with formaldehyde to form walls of amino resin; resorcinol is often added to the system to enhance the oleophilicity and, probably, to adjust the strength of the capsules. The aqueous solution of the monomers and suitable additives (e.g., emulsifier and modifiers) are intensively mixed and blended with the internal phase components (leuco dye in an organic solvent) at suitable pH/temperature/mixing schedules to form the filled microcapsules of the required particle size.[7]

m-Aminophenol derivatives are used to make the leuco dyes suitable for this process. Some of these dyes are of the xanthene/rhodamine/fluoran-type discussed in general terms above. A specific example is given in Eq. (10.3), where a fluoran-type leuco dye, **10-27**, is made starting with the condensation of *m*-diethylaminophenol with phthalic anhydride.

(10.3)

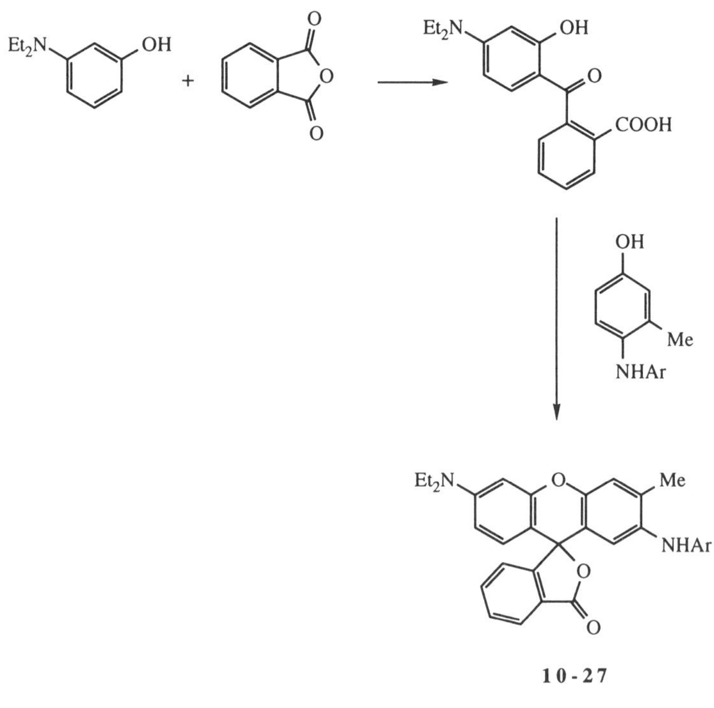

10-27

On contact of the leuco dye with the acidic developer, represented as HX, the lactone ring of **10-27** is opened and the quinonoid dyestuff, **10-28**, is formed. This type of dye will give a black image without the need to blend several colors to obtain black.

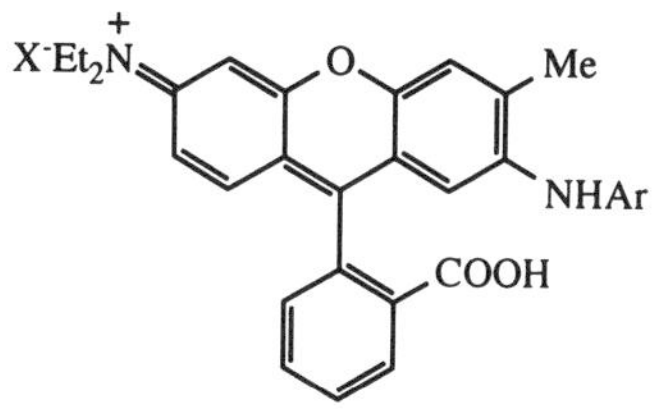

10-28

Numerous leuco dyes are suitable here, and their preparation may or may not start from *m*-aminophenol derivatives (certain of the dyes in use here are of the triphenylmethane type where the O-bridge between aryl rings is absent or replaced by an S-bridge to give phenothiazine-type dyes).

With a use in 1988 of about 800,000 t of carbonless paper in the U.S. alone and with demand still growing, this is still a growth area for the *m*-dialkylaminophenols as starting materials for leuco dyes.[6]

10.2.6. Chemicals for Thermographic Direct and Transfer Recording

Thermographic printing process applications have grown dramatically since their beginnings in the 1950s. Today, the thermal direct process is used in computer printers and word processors, in *X–Y* plotters, printing calculators and adding machines, medical and scientific recorders, fax (facsimile) machines, and in tag, label, and barcode printers. These are fast-growing markets for existing systems, but competition and development of new technology are intense.

The discussion here, too, is limited to systems pertinent to the use of resorcinol and *m*-aminophenol derivatives, but that is a good part of current and developing technologies. Thus, the thermal direct printing process is based on the same principles as the carbonless copy-paper technology. The thermal printing paper is coated with both a leuco (chromogenic) dye and a developer, in this case a di- or polyhydric phenol, as well as binders, carriers, sensitizers, and colorless inorganic pigments. The color development occurs by the application of heat from a printhead which can range from a simple stylus to a sophisticated, solid state, shaped silicon element. The desired reaction preferably takes place in the 65–120 °C range. Papers which record at low temperatures in this range are called high-sensitivity papers. One of the tricks is to prevent prereaction during paper coating, yet to obtain the desired reaction at low temperatures. Examples of other difficulties to overcome are the high melting points of the older leuco dyes and of the lower-cost developers such as 4,4′-isopropylidenediphenol (bisphenol A).

The pertinent thermographic transfer process is the reactive thermal transfer printing technology pioneered by Ricoh. Here, the leuco dye and a melting-point- reducing (eutectic) agent are coated on a ribbon and the paper is coated with a developer and a different melting-point-reducing agent. Heat applied to the printhead transfers leuco dye from the ribbon to the paper where the color-forming reaction occurs.

The high-quality leuco dyes used in these systems employ *m*-(dialkylamino)-phenol as one of the starting materials in their manufacture, the same or similar to the dye, **10-27**, shown above in the section on chemicals for carbonless paper copying. Another example of a fluoranlactone-type dye in commercial use is compound **10-29** (PSD-150), which is also used to produce a black image on contact with the polyhydric phenol/developer. Millions of pounds of these leuco dyes are needed even at the current levels of thermal paper consumption worldwide.

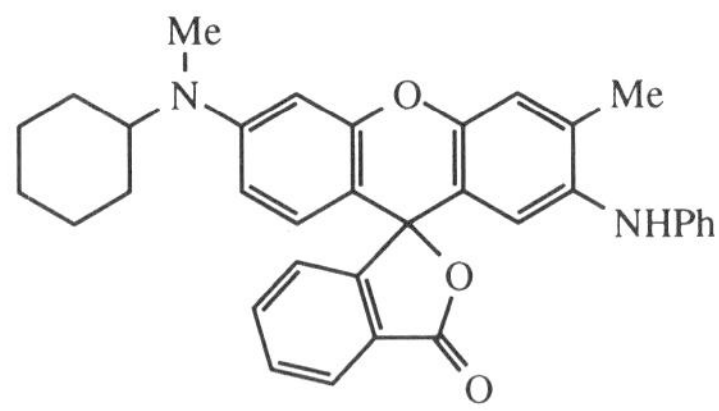

10-29

The (polyhydric) phenols and derivatives commercially used in these systems as developers and sensitizers/melting-point reducers/eutectic agents seem to include resorcinol/derivatives; there is a good bit of development going on, as sketched in Section 10.3 later on.

10.2.7. Applications in Other Imaging Techologies

The patent literature indicates a wide utility of resorcinol and *m*-aminophenol derivatives in imaging techniques, from photography to lithography. However, it is guesswork to judge which of these applications have been commercialized. Therefore, the selection of other applications has been grouped in Section 10.3 below.

For sure, one such technology, Mead's Cycolor color copying process, was commercially launched in 1989 in cooperation with the Japanese equipment manufacturer, Brothers International Corp. A little earlier, in late 1988, Noritsu American Corp. began selling a color copier for 35-mm slides that used Cycolor paper. Mead also has an agreement with Seiko on a color copier that uses the Cycolor process.

The novel Cycolor process appears to work as follows: A paper or film is coated with microcapsules (called cycliths by Mead) which contain a liquid phase of a monomer (e.g., trimethylolpropane triacrylate) or oligomer, an image forming agent (dye), and an efficient photoinitiator(s), sensitive to visible light, for polymerization. Exposure of this paper or film via a mirror image or transparency causes polymerization of the monomer/oligomer where the light strikes the paper, and there the capsules are hardened and become resistant to rupture. The exposed sheet is then pressure-contacted with a recording sheet, likely coated with a developer. The unexposed microcapsules in the light-sensitive sheet rupture and release the (chromogenic) dye to the receiver sheet. For multiple color development, the light-sensitive paper or film is coated with microcapsules containing monomer and either a cyan, magenta or yellow dye, and a photoinitiator sensitive to red, green, or blue light. For an example of color development, exposure of the sheet to red light will harden only the microcapsules containing the cyan dye; on pressure-contact with the recording sheet, the nonpolymerized microcapsules containing the magenta and yellow dye rupture to produce a red print. Similar selective techniques are carried out to produce other colors, so that positive photocopying is obtained. All color gradations can be achieved; where all the microcapsules are polymerized the print is white, and where none of the microcapsules are polymerized the print is black. Obviously, considerable sophistication was required to develop a practical system; e.g., the microcapsules must have transparency for the visible light beam.

As before, the use of resorcinol in the Cycolor system may be as an ingredient in the urea/formaldehyde resin formulation for making microcapsules. As far as the use of *m*-aminophenol derivatives is concerned, this would be in the leuco dyes/acid developer systems, similar to those already described above, which the Mead color printing process was said to employ,[7,8] but this is not mandatory.

There is also a possible use here for resorcinol and *m*-aminophenol derivatives which are photobleachable sensitizers and also improve imaging speed. Many of the ordinary sensitizers that can be used here are slightly yellow and leave a yellowish background on the color print. The photobleachable sensitizers are substituted coumarins such as 3,3′-carbonyl-bis-(5,7-dimethoxycoumarin) or 3-cinnamoyl-7-diethylaminocoumarin, **10-30**. The photobleaching of the sheet is done with UV light (fluorescent lamp) after development to obtain a print with whiter background and improved image quality.[9]

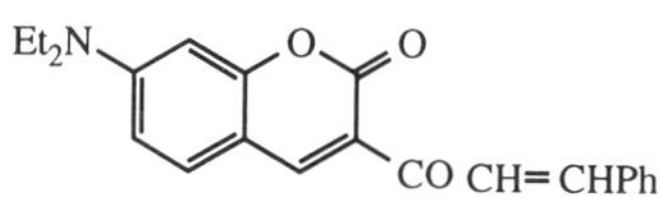

10-30

The market acceptance of the Cycolor technology seems to have been much slower than expected by the manufacturer/inventing company.

10.3. A SAMPLING OF DEVELOPMENTS FROM 1980 ON

There has been a tremendous volume of literature on dyes and imaging technology in the last 20+ years. Even the modest selection of references from this period given here makes the diversity of applications quite clear. Also apparent is the tremendous efforts of Japanese companies in this field, in part the basis of their excellent commercial products/systems.This finding does not belittle the excellent offerings in this field by U.S. manufacturers.

10.3.1. Dyestuff Intermediates and Processes

2-Hydroxycarbazole, **10-32**, was prepared according to Eq. (10.4) in *ca* 70% yield by the reaction of 2-chloro-3′-hydroxydiphenylamine, **10-31**, with aqueous caustic at 270 °C.[10] The intermediate, **10-31**, was prepared in high yield by the acid-catalyzed condensation of *m*-aminophenol with excess 2-chloroaniline.[11]

(10.4)

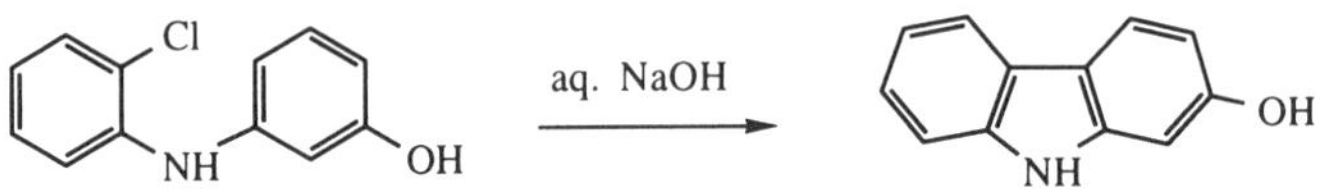

10-31 **10-32**

Certain ethers of *m*-diethylaminophenol, such as **10-33**, were prepared as intermediates for disperse azo- and styryl-type dyes for dyeing acrylic fibers.[12]

A process for the etherification of *m*-dialkylaminophenol with an alkyl halide in methyl isobutyl ketone in the presence of potassium carbonate was patented; e.g., using ethyl bromide, compound **10-34**, a useful dye intermediate, was obtained in 97.5% yield.[13]

A preparation of 7-amino-4-methylcoumarin, **10-35**, was patented, starting with blocking the amino group of *m*-aminophenol with methoxycarbonyl chloride to give *N*-(methoxycarbonyl)-*m*-aminophenol which was condensed with methyl acetoacetate in the presence of sulfuric acid as the catalyst to form the blocked coumarin; finally, the amino group was unblocked by hydrolysis with alkali at 80–100 °C.[14]

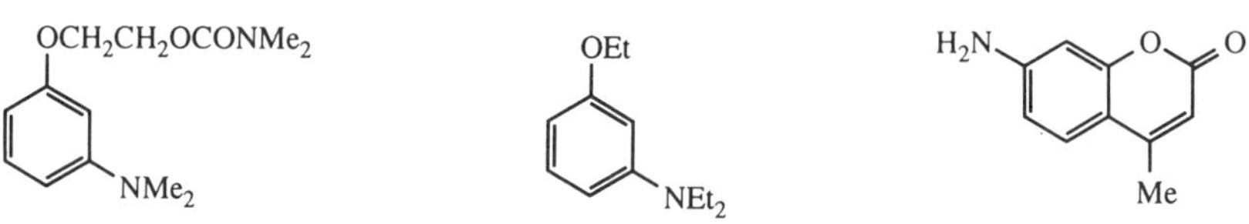

10-33 **10-34** **10-35**

A novel preparation of *m*-hydroxytriarylamines, useful for the preparation of fluoran leuco dyes, by the condensation of diarylamines with resorcinol in the presence of phosphorous compounds was reported. For instance, the condensation of 4-methoxydiphenylamine with resorcinol in the presence of $(PhO)_2POH$ at 200 °C for 10 hr under nitrogen gave the dye intermediate, **10-36**.[15]

A method for the bromination of fluorescein was patented which avoided HBr as a byproduct and gave high utilization of the bromine used. Thus, fluorescein in aqueous ethanol was brominated with bromine at 40–80 °C in the presence of (*ca* 30%) hydrogen peroxide which reoxidized the HBr, as formed, to Br_2; up to 87% yields of eosin, **10-18** (acid form), were obtained.[16]

Esters of 1,3-cyclohexanedione (enol form) were prepared as coupling components for recording materials (or intermediates for herbicides); e.g., the reaction of

5-phenyl-4-methoxycarbonyl-1,3-cyclohexandione with pivaloyl chloride in acetonitrile in the presence of triethylamine gave the ester, **10-37**.[17] Similarly, the substituted 1,3-cyclohexanedione, **10-38**, was prepared and claimed to be useful as a coupling agent for diazonium compounds.[18]

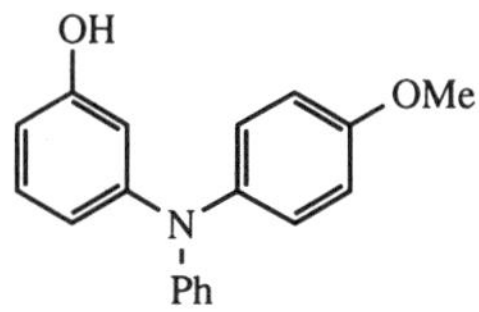

10-36

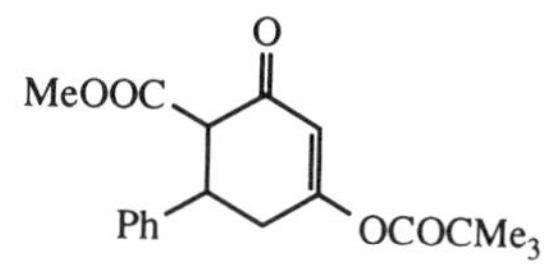

10-37

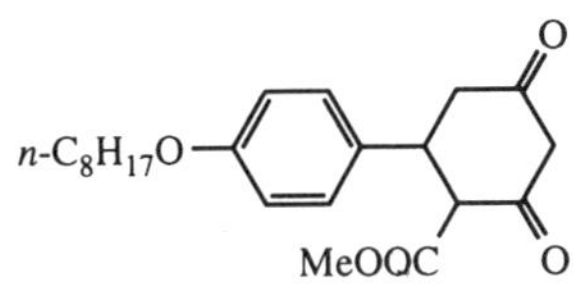

10-38

10.3.2. Azo Dyes

Novel disazo resorcinol dyestuffs were prepared which were useful for dyeing and printing wool, polyamides, and leather; compound **10-39** is an example of the claims of this patent.[19]

Trisazo dyes were prepared by reacting a sulfophenylazonaphthalenediazonium salt with a phenylazoresorcinol. These dyes, such as **10-40**, optionally could be metallized and had high color strength and leveling properties; light- and heat-fast dyeings on leather were obtained and the dyes were also useful for polyamides and paper.[20]

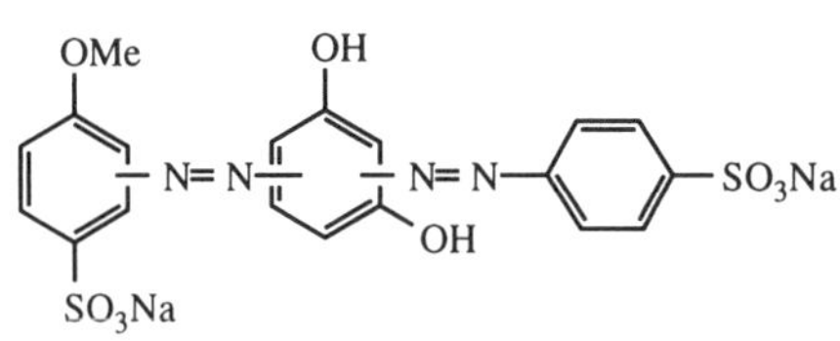

10-39

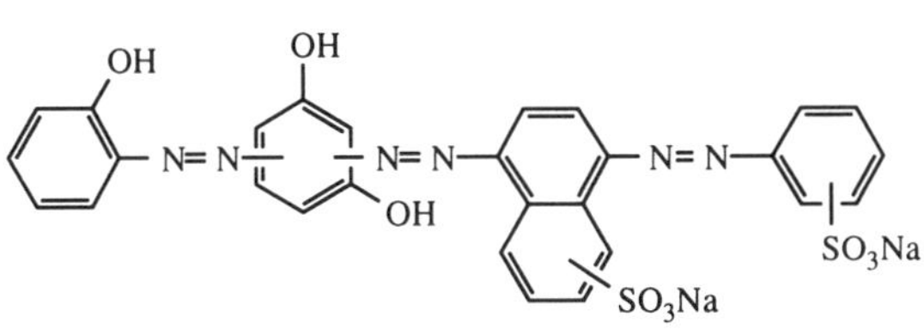

10-40

New (metal-complexed) disazo dyes were prepared by coupling a diazotized *o*-aminophenol and a diazotized aminonaphthalenesulfonic acid with resorcinol to give, e.g., compound **10-41**. These dyes had very good fastness properties on silk, wool, synthetic polyamides, polyurethanes, and leather.[21]

Chromium complexes of resorcinol disazo dyes, e.g., **10-42**, were prepared; these dyes were useful anionic dyes for textile and nontextile substrates, giving fast brown shades on leather.[22]

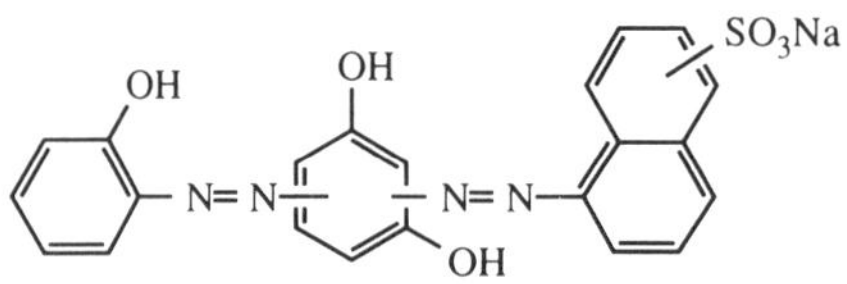

10-41

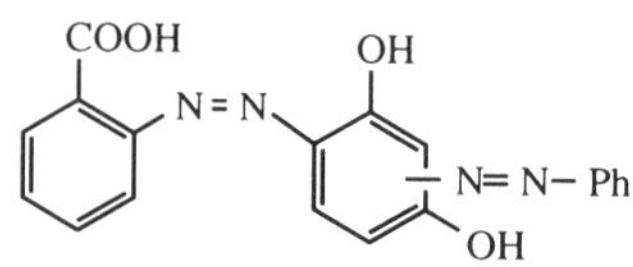

10-42

Water-soluble mono- or poly-azo dyes were made using resorcinol still residue (with 2,4,4′-trihydroxydiphenyl as a major component, although this component was not identified as such in the patent), an oligomeric phenolic material. These dyes were anionic, sulfonated dyes useful for imparting valuable brown shades to leather, fur, cellulosics, silk, wool, or synthetic polyamide textile materials.[23]

Stable, liquid leather dye compositions were developed comprising (form)aldehyde condensation products of mono- or dis-azo compounds of resorcinol; optionally these dyes were metallized.[24]

A novel trisazo dye for the deep black dyeing of leather, natural fibers, and synthetic polyamides was prepared from tetrazotized 4,4′-diaminobenzanilide by sequentially coupling with 1-amino-8-naphthol-3,6-disulfonic acid (H-acid), sulfanilic acid, and *m*-hydroxyphenylglycine to give the dye of structure **10-43**.[25]

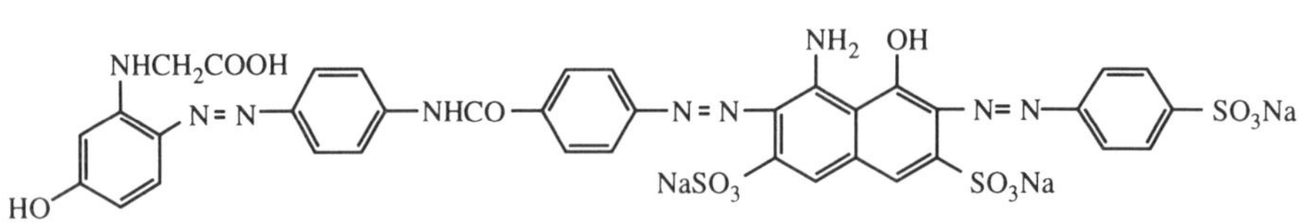

10-43

Copper complexes of stilbene azo and azoxy dyes were prepared which dyed cotton and leather in brown shades. For instance, intermediate **10-44** was condensed with 4,4′-dinitro-2,2′-stilbene-disulfonic acid in the presence of alkali, followed by reaction with copper sulfate.[26]

A novel disazo dye containing a *m*-aminophenol portion, **10-45**, which formed metal (Fe, Co, Cu, Ni) complexes, was patented; e.g., the nickel complex of this dye was applied to leather and gave a light-fast, brown shade.[27]

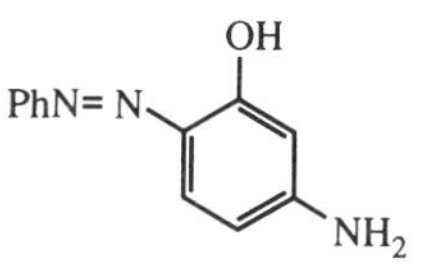

10-44

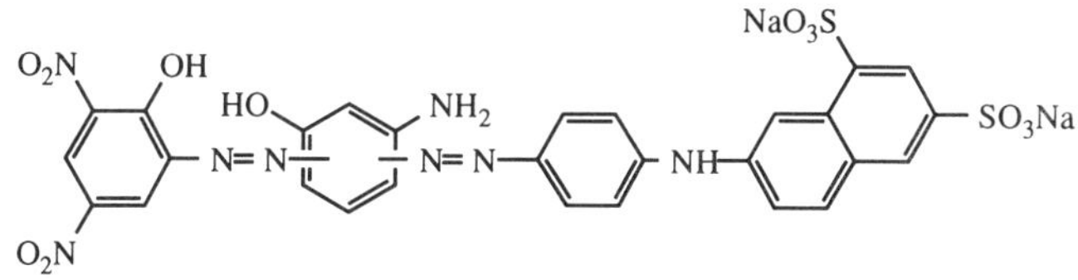

10-45

Fiber-reactive copper complexes of resorcinol disazo dyes, such as **10-46**, were prepared; the fiber-reactive group was a vinylsulfonyl substituent.[28]

A yellow-orange pigment, **10-47**, was prepared by diazotizing 5-amino-6-nitro-2-benzimidazolone and coupling with resorcinol; this pigment was fast to migration and was useful, e.g., for camouflage purposes.[29]

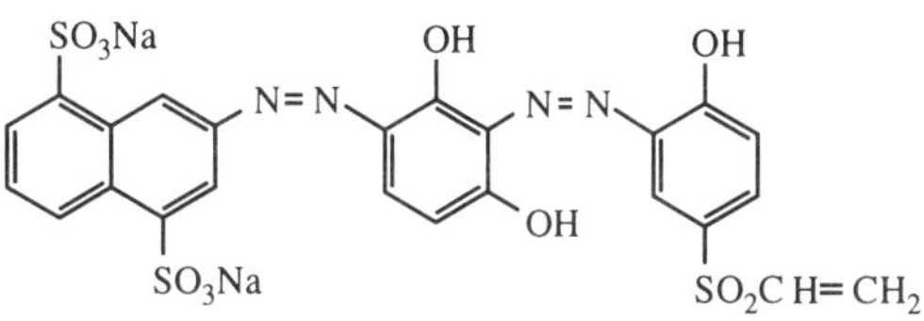

10-46

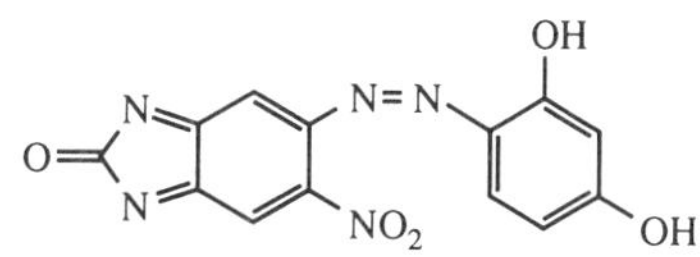

10-47

10.3.3. Fluorescent 7-Hydroxy- and 7-Aminocoumarins, Xanthenes, and Fluorans

3-Phenyl-7-(2*H*-naphthotriazo-2-yl)coumarins, such as **10-48**, were prepared as optical whitening agents for aromatic polyesters (polyethylene or polybutylene terephthalate, PET or PBT).[30]

7-Hydroxycoumarins with a heterocyclic substituent in the 3-position, e.g., compound **10-49**, were used to impregnate paper to prevent fraud; on treatment with an alkaline ink eradicator a strong, very stable, yellow to red color developed.[31]

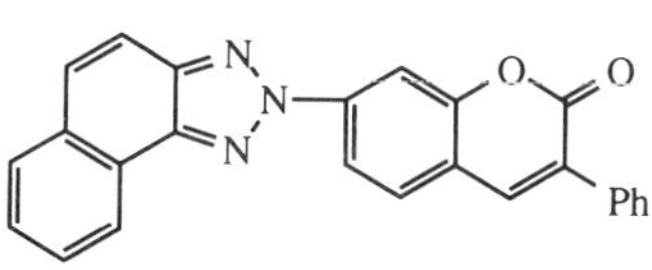

10-48

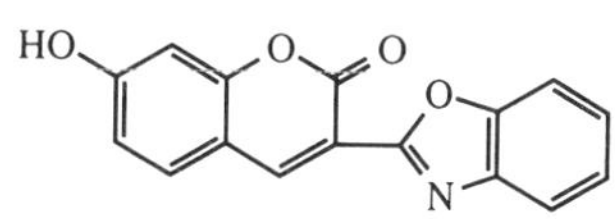

10-49

A color-corrected flashlamp with increased light emission was patented, consisting of a hermetically sealed bulb containing a combustible material, an atmosphere supporting combustion, and an ignition device; the outer surface of the bulb was coated with a blue-emitting fluorescent compound, e.g., the 7-amino-4-methylcoumarin, **10-35**.[32]

3-Carboxyalkyl-4-methyl-7-aminocoumarins, such as **10-50**, were prepared as fluorescent labeling agents for biological materials; casein dissolved in $Na_2B_4O_7$ buffer and reacted with compound **10-50** gave a fluorescent conjugate which could be viewed under UV light of $\lambda = 350$ nm.[33]

A Kokak bulletin described certain 3-ketocoumarins, such as 3-benzoyl-7-methoxycoumarin, **10-51**, and 3-thenoyl-7-diethylaminocoumarin, **10-52**, as efficient triplet sensitizers having high extinction coefficients ($>10^4$) so that a considerable amount of light absorption occurs at low concentration. These compounds were described as useful in organic photochemistry, in flash photolysis experimentation, in the photocrosslinking of polymers for lithographic applications, in photoresists used in the electronics industry, and in combination with activators as photoinitiators for free radical polymerizations; references were given.[34]

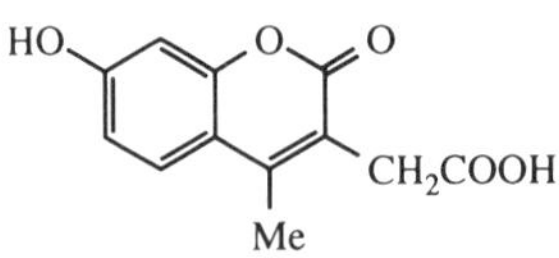

10-50

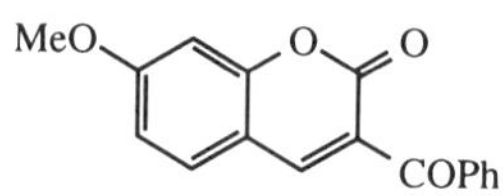

10-51

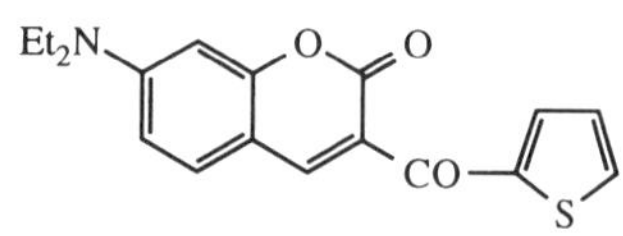

10-52

3-(7-Methoxycoumarin-4-ylmethyl)-5-fluoro-2′-deoxyuridine, **10-53** (and its 3′,5′-di-*O*-acetate), were prepared and patented as useful standards for liquid chromatographic analysis.[35]

Rose Bengal is one of the most efficient and most used singlet-oxygen sensitizers; its bis-triethylammonium salt, **10-54**, was prepared and allowed the use of this photosensitizer in nonpolar solvents (e.g., methylene chloride).[36]

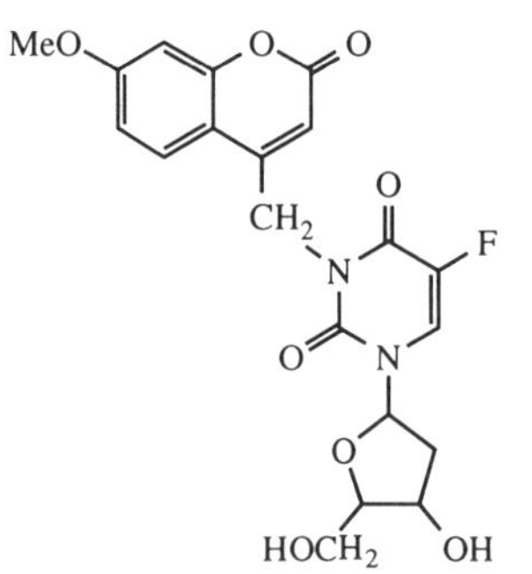

10-53

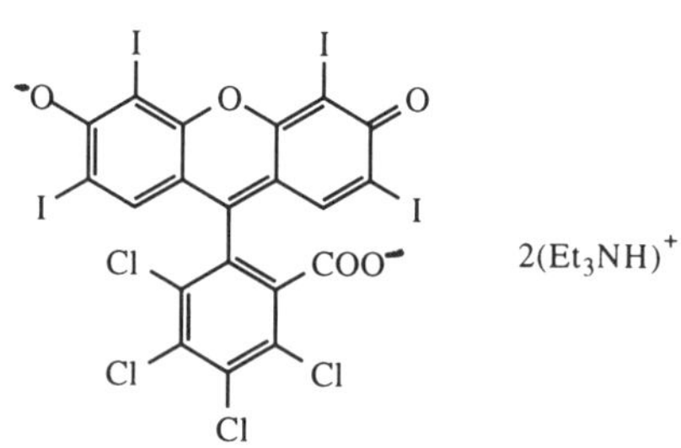

10-54

N-alkylaminomethylated fluoresceins such as compound **10-55** were prepared and patented as fluorescent tracers in fluorescence polarization immunoassays.[37]

Chromogenic and fluorogenic silylation reagents were prepared which were reactive silanes containing coumarin groups. These compounds were useful for the derivatization of protic materials such as amines and alcohols; the derivatives could be recognized/visualized in HPLC (high pressure liquid chromatography), in fluorimetric biochemical assays. These reagents could also be used as permanently bonding optical brighteners for fabrics and plastics. One example of these compounds is **10-56**.[38]

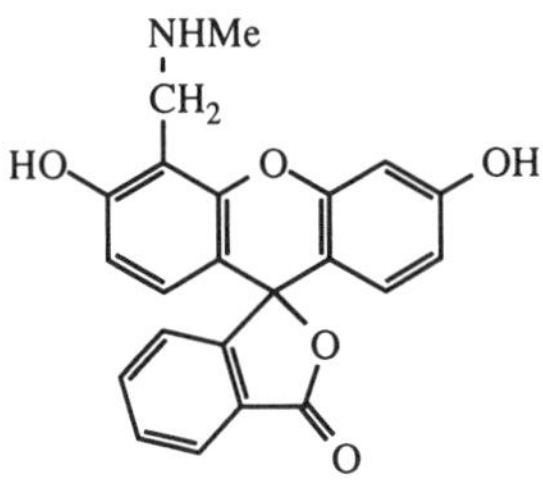

10-55

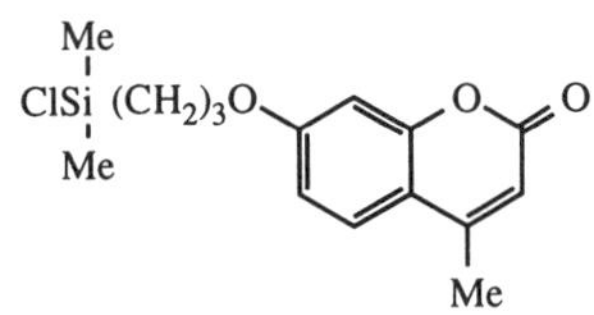

10-56

New fluorescent ethers were prepared. For example, fluorescein was treated with 2-fluorobenzonitrile in the presence of potassium carbonate in *N*-methyl-2-pyrrolidinone at 200 °C to give a 97% yield of compound **10-57**; this compound showed yellow-green fluorescence (524 nm) when irradiated with visible light (504 nm).[39]

Hologram materials were prepared. Thus, a photosensitive film was made from butyl methacrylate, ethyleneglycol dimethacrylate, 1-phenylnaphthalene, benzil dimethyl ketal as the polymerization initiator, and a ketocoumarin, **10-58**, as a visible light sensitizer. This composition gave hologram images having high sensitivity by using an Ar^+ laser.[40]

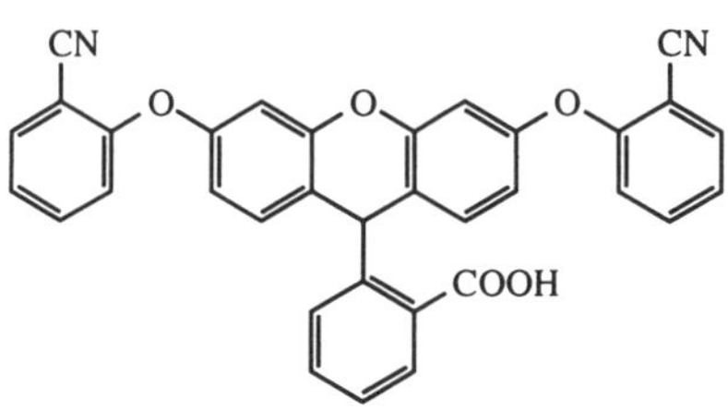

10-57

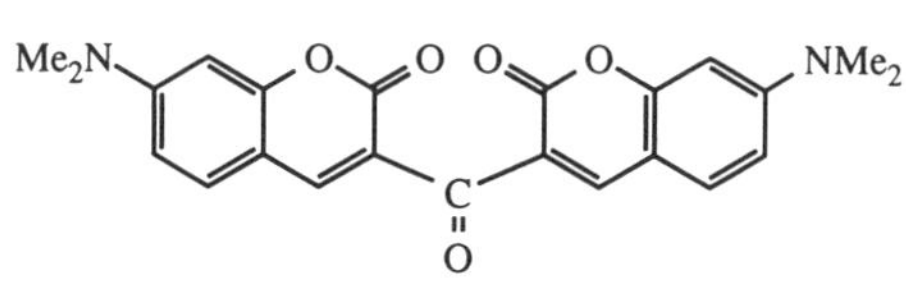

10-58

10.3.4. Chemicals for Diazotypy

A diazo copying paper was obtained by coating the paper with a mixture, e.g., of stabilized 2,5-dibutoxy-4-morpholinobenzenediazonium chloride and, as the couplers, 2-hydroxy-3-[*N*-(2-hydroxyethyl)]naphthamide and 4-(2-hydroxyethyl)-2-methylresorcinol. This mixture formed black images on development in alkaline media. 4-(Morpholinomethyl)-2-methylresorcinol, **10-59**, was another novel coupling component in this patent.[41]

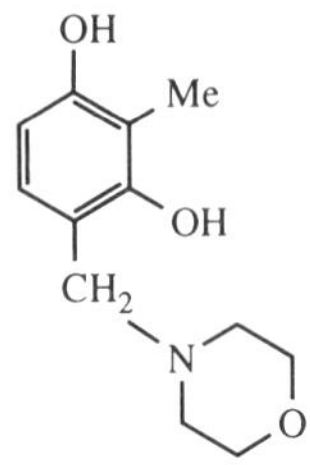

10-59

A related patent on diazotype materials claimed mixtures of blue (amides of 2-hydroxy-3-naphthoic acid), yellow (2-alkyl- or 2-arylresorcinol or a cyanoacetamide), and red-brown (e.g., 4,4′-thio-bis-resorcinol) couplers to provide a black image in the light-sensitive layer.[42]

A similar patent on diazotype copy materials giving high-density, black images covered the use of 4-methyl-, 4-ethyl- or 4-*tert*-butylresorcinol as the coupling component.[43]

A further variation on the theme was the use of two diazo compounds (e.g., 2-methoxy-4-dialkylaminobenzenediazonium salt and a 4-morpholinobenzenediazonium salt) and two couplers (e.g., an alkylresorcinol and 4,4′-thio-bis-resorcinol); these components were said to give sepia-colored images of high density.[44]

A diazo sheet suitable for making master prints was developed which was coated with a binder containing a diazo compound, a stabilizer (as known in the art), and a mixture of two couplers, e.g., *N*-acetyl-*m*-aminophenol and 4,4′-thio-bis-resorcinol. The sheet was said to retain its color-forming properties on aging and was developable by a dry or semidry process.[45]

A diazo-type, heat-sensitive recording sheet was developed. The photo- and heat-sensitive layer contained a fine particle dispersion of a diazonium salt, at least one coupler of the resorcinol type (e.g., resorcinol, methylene-bis-resorcinol or an alkylene-bis-resorcinol ether such as **10-60**), and a hydrophobic organic compound, e.g., **10-61** (a guanidine derivative), which becomes basic when fused with the application of heat.[46]

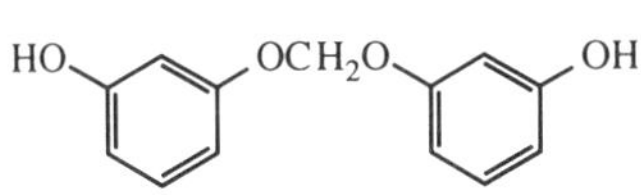

10-60

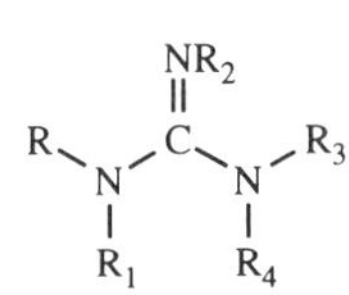

10-61
(where R, R_1, R_2, R_3, R_4 = H or < C_{18} alkyl)

A Russian patent claimed a diazo photocopying material which contained a diazonium salt (e.g., 4-diethylaminobenzenediazonium salt with citric acid, thiourea, and zinc chloride) and a coupler which was the reaction product of resorcinol with alkaline, water-soluble starch. The composition had good storage stability.[47]

A diazo copying process was evolved which used a standard, stabilized diazonium salt and a hydrazone derivative, e.g., the phenylhydrazone of 2,4-dihydroxybenzaldehyde (beta-resorcaldehyde) as the coupling component; on development, black images were formed.[48]

A heat-sensitive recording material was developed consisting of microcapsules (e.g., urea/formaldehyde wall material) formed in anaqueous medium containing sodium naphthalenedisulfonate and core materials of stabilized aryldiazonium salts, coupling agents (for instance resorcinol), and a material which becomes basic on heating. This process was said to give improved, sharp images on thermal development.[49]

A transparent polyethylene terephthalate (PET) film was coated with a composition containing solvents, silica, cellulose acetate propionate, vinyl chloride/vinyl acetate copolymer, methyl acrylate, zinc chloride, tin chloride, sulfosalicylic acid, thiourea, 3-chloro-4-dimethylaminobenzenediazonium PF_6-salt, phenylphenol, the ethanolamide of beta-resorcylic acid, **10-26**, Methyl Violet, and a fluorinated surfactant. This film, on exposure and ammonia development, gave an image which had an UV-shielding effect, useful for photomasks used in printed circuit manufacture.[50]

10.3.5. Leuco Dyes for Carbonless and Thermal Copying Technologies

This was an area of great activity in the 1980–1993 period, with accelerating contributions from Japanese companies. The relatively small selection offered here shows the ingenuity involved in molecular manipulation for specific goals. Factors such as sharper image formation, color variety, higher purity, image stability, better thermal sensitivity, storage stability, improved solubility, lower melting points, and better chemical resistance were achieved. While in the older literature the leuco dye was considered a base, in the newer literature the leuco dye is described as an electron donor and the phenolic developer as an electron acceptor.

New leuco dyes of the fluoran type were developed (such as compound **10-62**), for use in pressure- and heat-sensitive carbonless copying systems.[51]

A dichromatic, heat-sensitive recording paper was developed in which the first coloring layer contained a leuco dye, such as **10-63**, capable of forming color at a relatively high temperature, but no developing agent, and a second coloring layer capable of forming color at relatively low temperature, containing a leuco dye such as

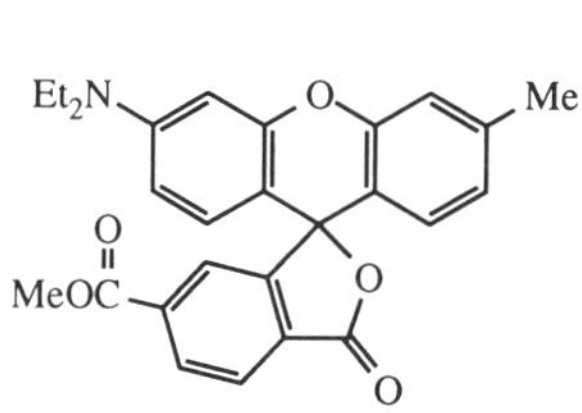

10-62

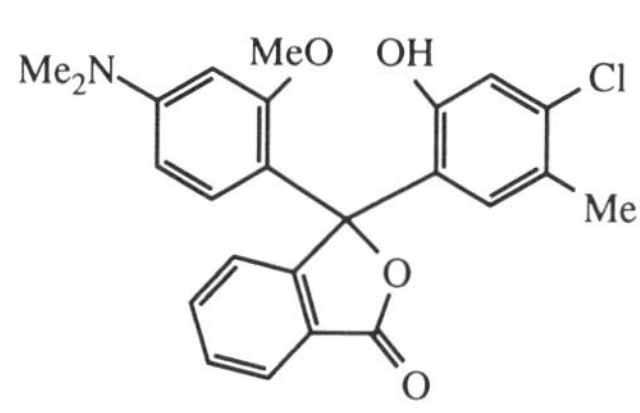

10-63

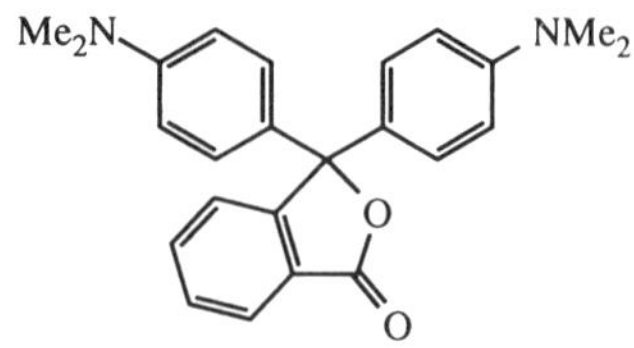

10-64

compound **10-64**, a phenolic substance, and/or an acid such as boric acid, oxalic acid, or maleic acid.[52]

Novel acyl-anilino-fluorans, e.g., **10-65**, were prepared as leuco dyes for carbonless pressure- or thermal-sensitive copy papers. These dyes gave an intense reddish or greenish black color, with low initiating temperature on contact with an acid developer.[53]

A heat-sensitive recording material was developed which contained the phthalide, **10-66**, and a phenolic developer of melting point >90 °C (e.g., 4,4′-isopropylidene-bis-2,6-dibromophenol), either or both dispersed in a binder, on a paper substrate. On thermal development a vermilion recording was obtained.[54]

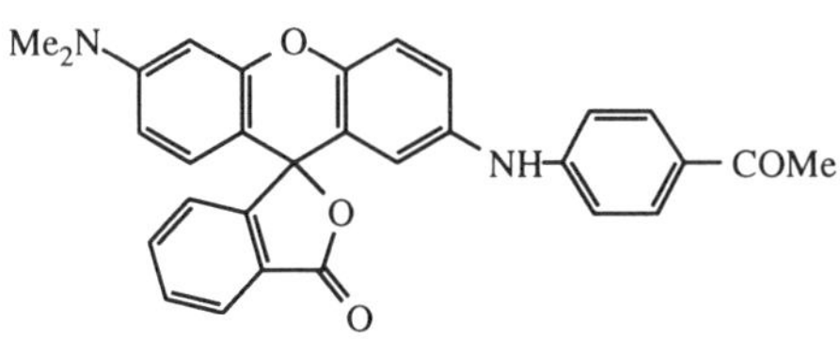

10-65

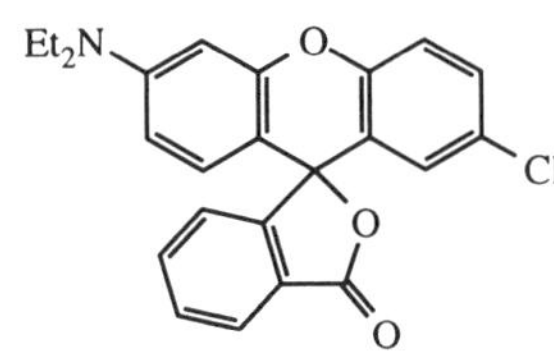

10-66

More variations in leuce dyes for heat-sensitive recording materials were compound **10-67**, which turned reddish black to black on contact with a (phenolic) electron acceptor,[55] and compounds **10-68** and **10-69**, which gave reddish-orange colors on thermal development.[56]

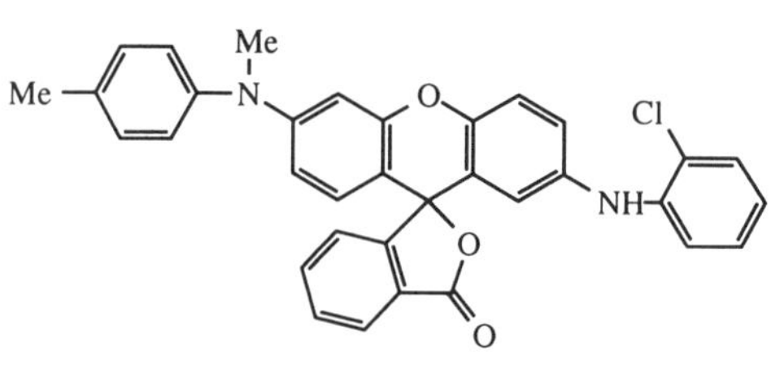

10-67

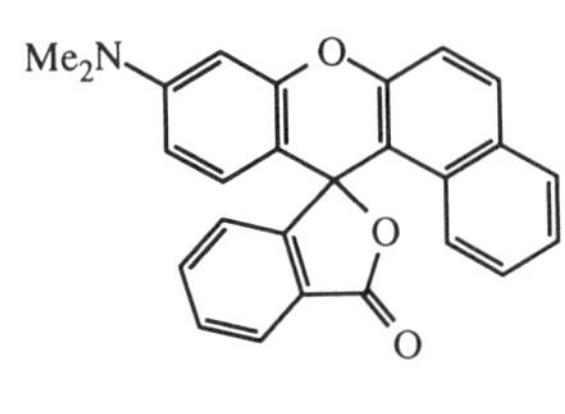

10-68

2-Anilino (**10-70**) or 2-acetamido- or 2-(1-pyrrolyl)- and similar 2-substituted amino-3-methyl-6-diethylamino-5′-carboxy-fluorans were patented as color formers in pressure- and heat-sensitive carbonless marking systems which gave, on contact with acid developers, red to greenish to black images. These dyes had high tinctorial strength, light stability, and resistance to sublimation.[57]

3-Anilino-7-(*N*-isobutyl)ethylaminofluoran, **10-71**, was patented as a leuco dye which developed a green color on contact with an acidic substrate.[58]

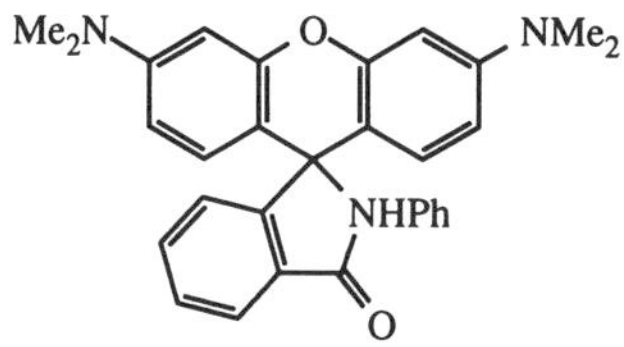

10-69

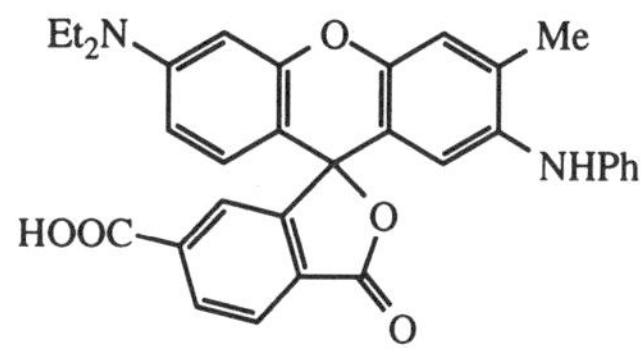

10-70

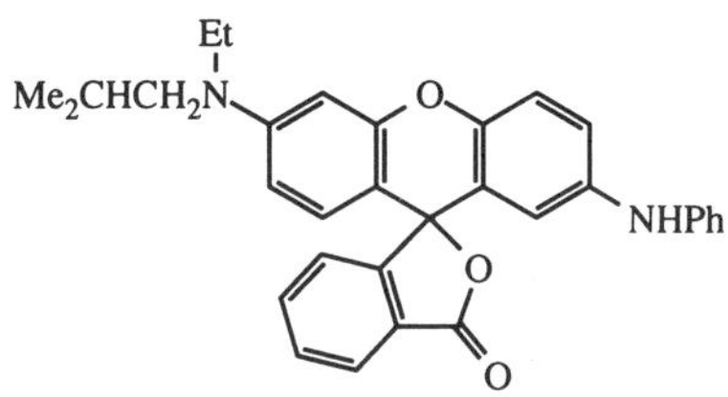

10-71

A process for making 3-(di- or cyclo-)alkylamino-6-(2-carboxybenzoyl)phenols, useful as intermediates for leuco dyes, was patented. For example, *N*-ethyl-*N*-cyclopentyl-*m*-aminophenol was reacted with a slight excess of phthalic anhydride in toluene at 105–110 °C for 20 hr, then stirred with 50% aq. NaOH at 90 °C for 3 hr to decompose byproduct rhodamine dye, then diluted with water to precipitate the sodium salt of the desired product, **10-72**, which was obtained on filtration/acidification in 75% yield, free of the rhodamine impurity.[59]

A recording material was developed based on novel 5′- or 6′-phenoxyfluorans (e.g., compound **10-73**), as the leuco dye and a phenolic developer. These materials gave pure black hues and excellent stability of the developed image.[60]

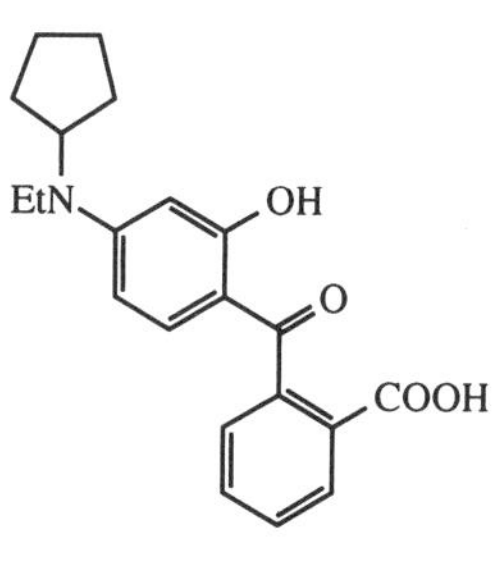

10-72

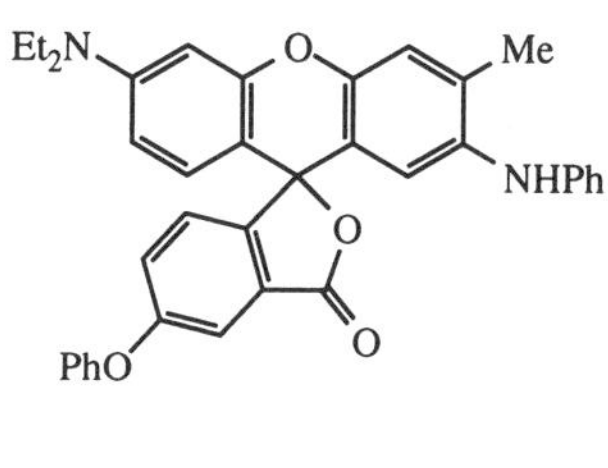

10-73

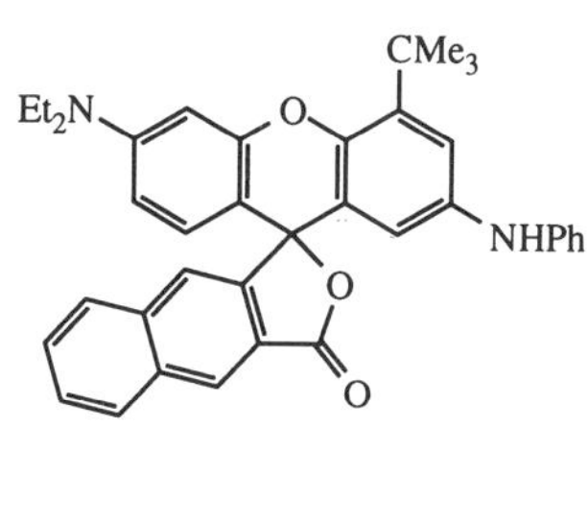

10-74

A large number of novel fluoran derivatives, useful as electron-donating colorants which developed deep black colors, had high light resistance and solubility in organic solvents, were patented. An example of this series of compounds is **10-74**.[61]

A heat-sensitive recording material was developed which contained, in the usual binder media, a fluoran derivative represented by structure **10-75** and a sulfone, such as 4-phenylsulfonylresorcinol, as the developer (a sensitizer, e.g., a higher alkanoic amide, was added also). These materials had good fastness to plasticizers and oils, and had good storage stability without the use of a protective layer.[62]

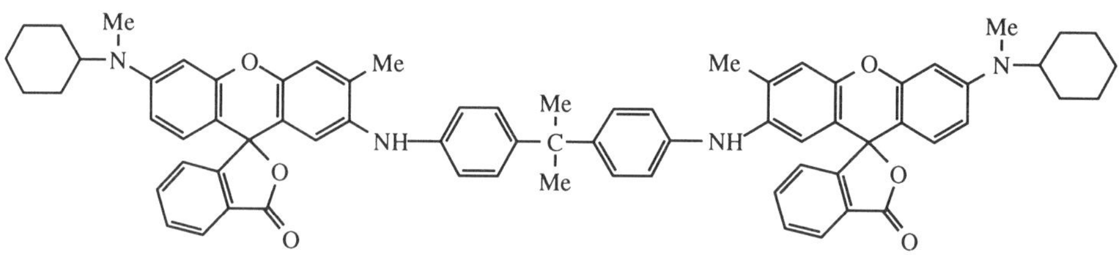

10-75

A recording material useful in pressure-, heat-, or photo-sensitive systems was developed where a novel leuco dye, of type **10-76**, and a monomer or silver halide or developer are contained in microcapsules on a support. Sharp, dense images with good light resistance were obtained.[63]

Three patents to Sumitomo claimed pressure- or heat-sensitive recording materials containing fluoran compounds of type **10-77** as the electron-donating, color-forming component. These leuco dyes were colorless and reacted with developers to give blacks with excellent light and water resistance.[64]

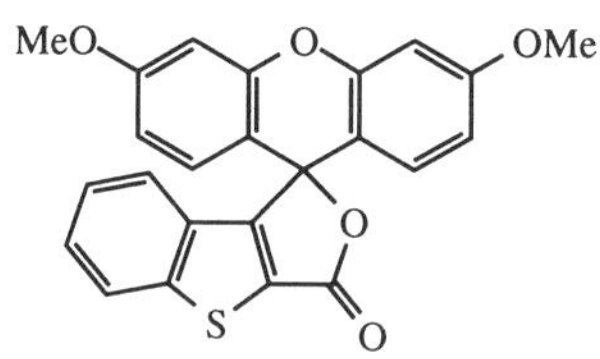

10-76

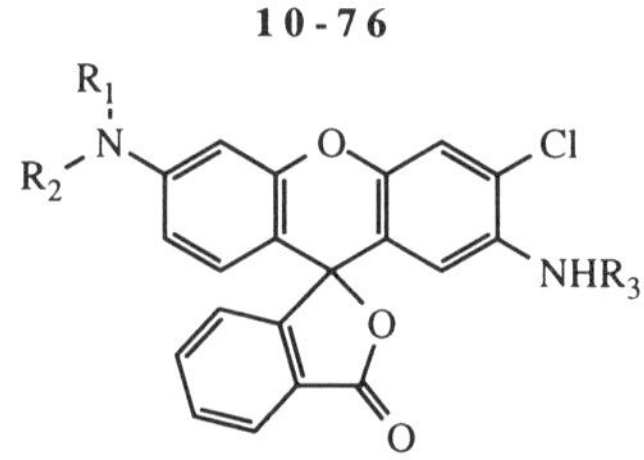

10-77

(where R_1, R_2 = $C_{3\text{-}10}$ alkoxyalkyl or alkoxyalkoxyalkyl R_3 = $C_{1\text{-}8}$ alkyl, $C_{2\text{-}8}$ alkenyl, $C_{3\text{-}10}$ alkoxyalkyl, phenylalkyl)

Thermographic materials containing new fluoran leuco dyes, exemplified by structure **10-78**, were developed. These new dyes had good near-IR absorptions, thus increasing the optical legibility in the near-IR, which is useful for printed bar code labels.[65]

Another variety of leuco dye for pressure- or heat-sensitive copying paper was developed, consisting of the 2-pyridylamino- (e.g., compound **10-79**) or 2-quinolylamino-substituted fluorans which gave black shades on contact with an acidic developer.[66]

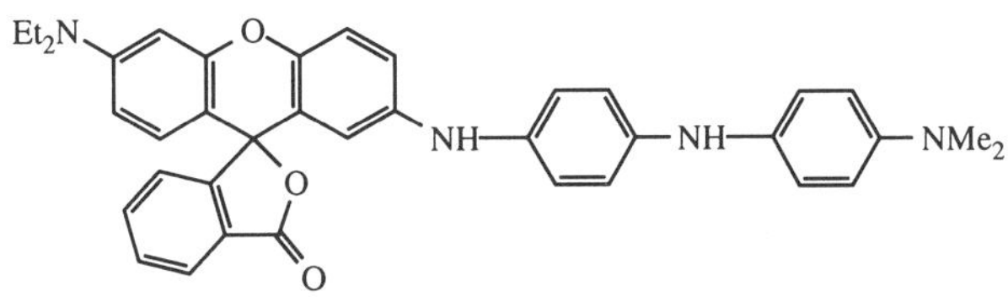

10-78

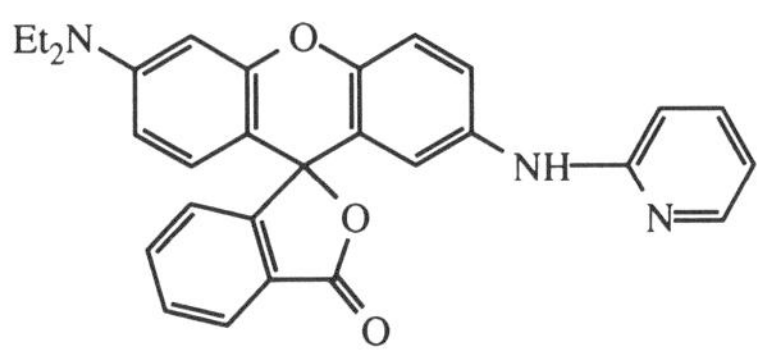

10-79

Violet color-developing fluoran derivatives for pressure or thermally sensitive recording papers had structures typified by compound **10-80**; these dyes had good light resistance and storage stability.[67]

Novel, yellow color-developing leuco dyes of structures **10-81**, **10-82**, and **10-83** were prepared for use in suitable heat- or pressure-sensitive recording materials. The yellow images had high color density and good light resistance.[68]

Substituted benzopyrano-2*H*-pyrazoles, e.g. compound **10-84**, were found to be useful in this field; with an acid clay as the developer, an intense orange color formed.[69]

N-aralkylated aminofluorans, such as compound **10-85**, were prepared and patented as leuco dyes which gave a red color when developed on an acid-modified silica gel.[70]

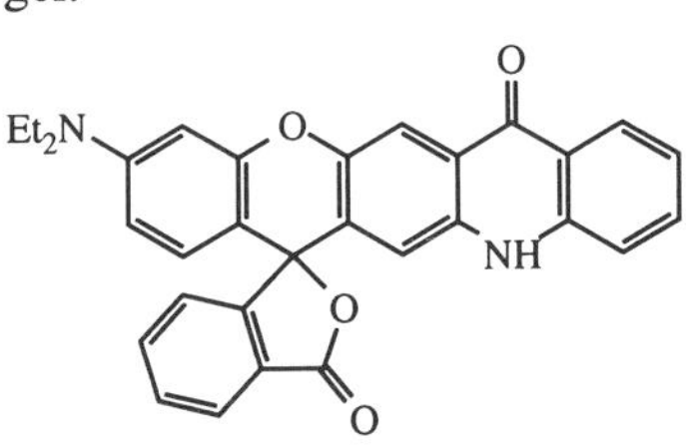

10-80

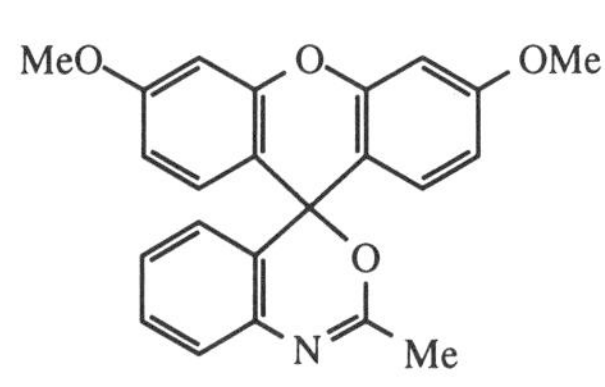

10-81

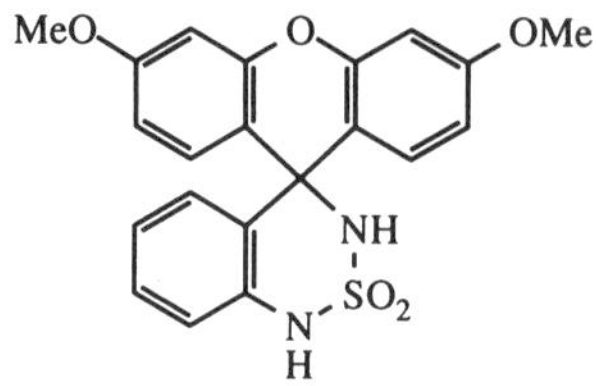

10-82

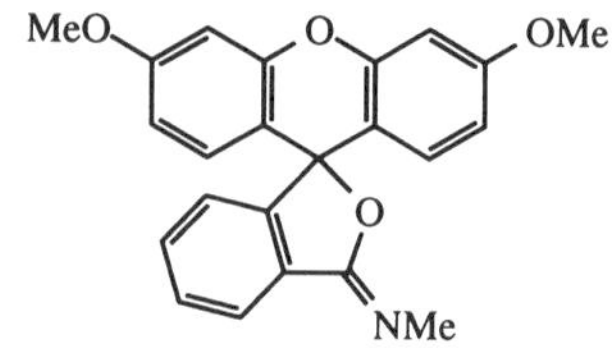

10-83

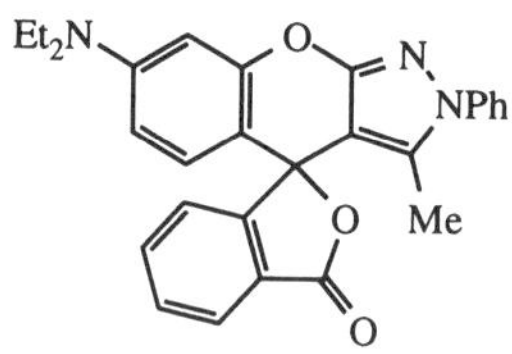

10-84

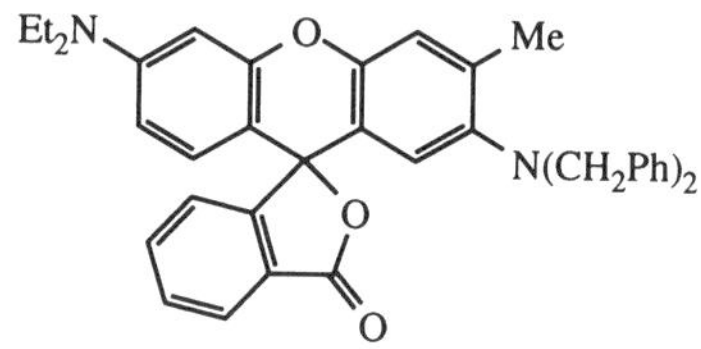

10-85

The preparation of the intermediate, **10-86**, for leuco dyes of the fluoran type, in 55% yield by reacting *N*-cyclohexyl-*N*-propyl-*m*-aminophenol with phthalic anhydride in toluene at reflux, was patented.[71]

Nonfluoran-type, electron-donating leuco dyes, exemplified by compound **10-87**, formed color on reaction with an electron-accepting compound. These materials were used to prepare recording materials of excellent shelf life, high color developability, and stability of the color image. (In one example, a blue color image was mentioned).[72]

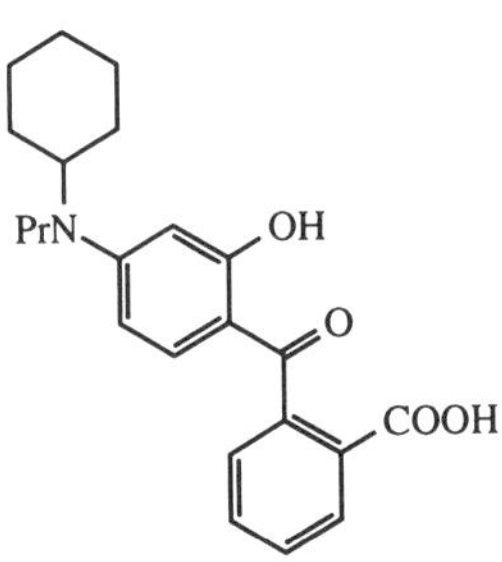

10-86

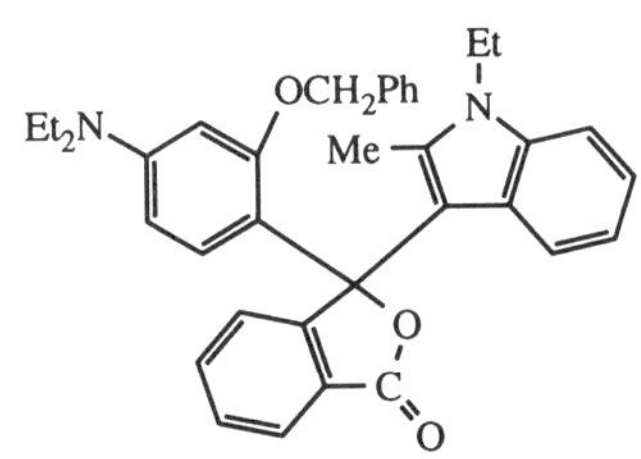

10-87

The 2-bromo-3-methyl-6-dibutylaminofluoran-type leuco dye, **10-88**, was recommended for use in pressure-sensitive recording materials; on contact with acid developers (e.g., silica gel), a high-density vermilion image formed.[73]

2- or 3-Phthalimidylfluorans, such as **10-89**, were prepared and used as leuco dyes for recording materials; on contact with an acid clay a red color image formed.[74]

A novel leuco dye, **10-90**, for thermal printing systems developed a (black) image on contact with bisphenol A. The images had good near-IR absorption and were useful for optical scanning applications.[75]

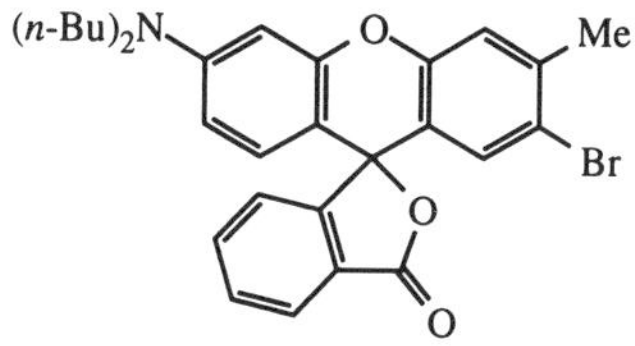

10-88

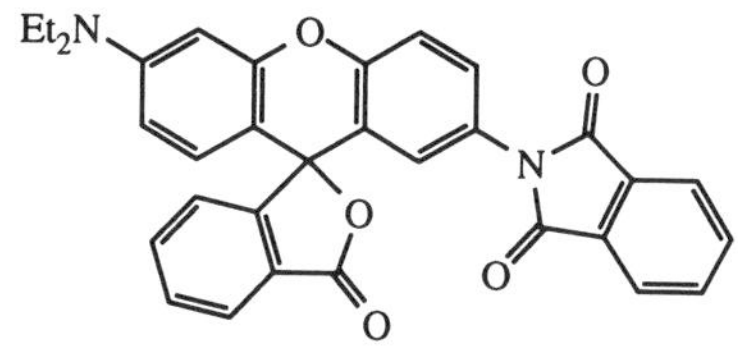

10-89

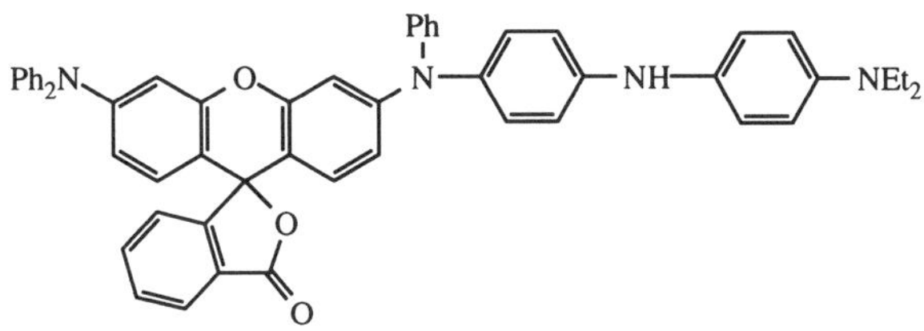

10-90

Fluoran-type leuco dyes, such as compound **10-91**, were developed for use in copying materials; on contact with an acidic developer (e.g., phenolic resins), a green image formed which had high light fastness.[76]

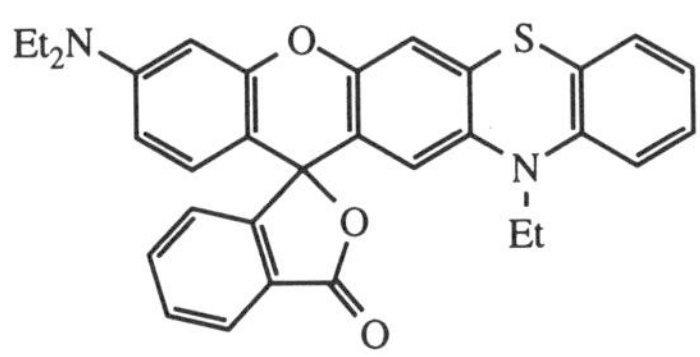

10-91

New fluoran derivatives, exemplified by compound **10-92**, were developed for use in copying materials. On contact of this dye with an acidic developer, a high-density color image formed faster and at lower temperature than with leuco dyes of the prior art.[77]

2-Anilino-3-methyl-6-heterocyclic-substituted fluorans, e.g., **10-93**, were synthesized and used as black color formers by reaction with acidic developers in thermal copy papers.[78]

An optical image-recording material was described which contained a developer, a photooxidizing agent, and an oxidizable leuco dye such as **10-94**, where one of the components was microencapsulated and the other was not. Black-hued images were produced which had excellent stability.[79]

2-Anilino-3-methyl-6-phenoxyalkylaminofluoras, e.g., **10-95**, were used as the leuco dye in carbonless copy paper. On reaction with a color developer sheet containing

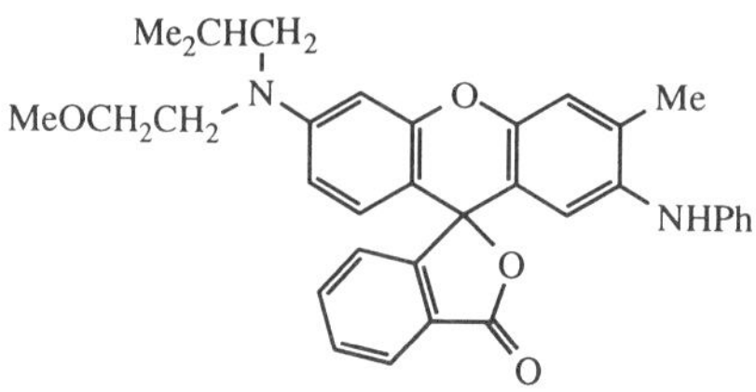

10-92

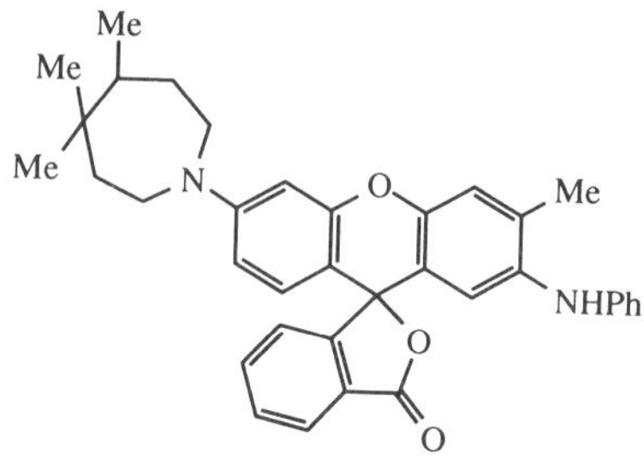

10-93

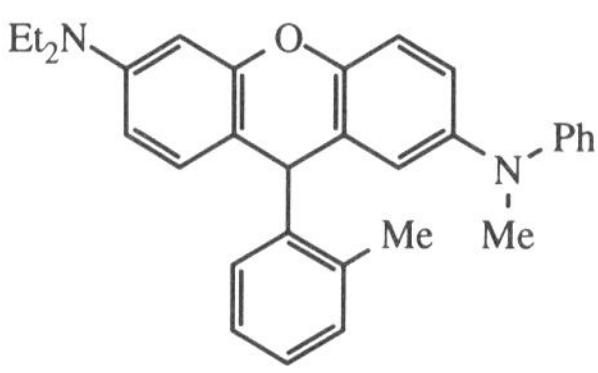

10-94

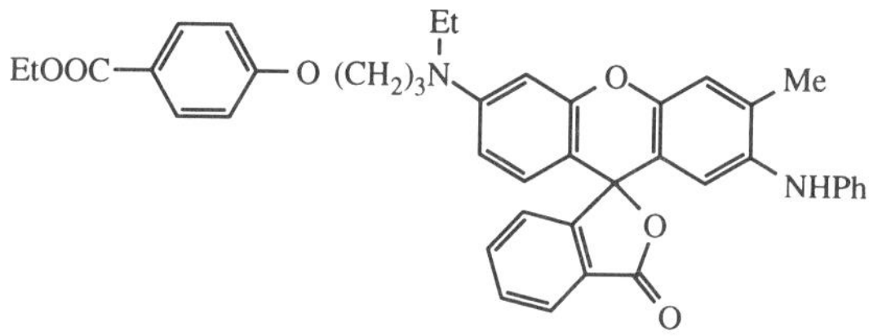

10-95

zinc 3,5-bis-(alpha-methylbenzyl)salicylate, high-density black images were obtained.[80]

Thermal recording materials containing two kinds of fluoran leuco dyes were prepared, one leuco dye having structure **10-96** while the other was 2-anilino-3-methyl-6-(*N*-ethyl-*N*-isopentylamino)fluoran. On contact with bisphenol A a high-density black image was formed which absorbed both visible and near-IR radiation and which had good heat and moisture resistance.[81]

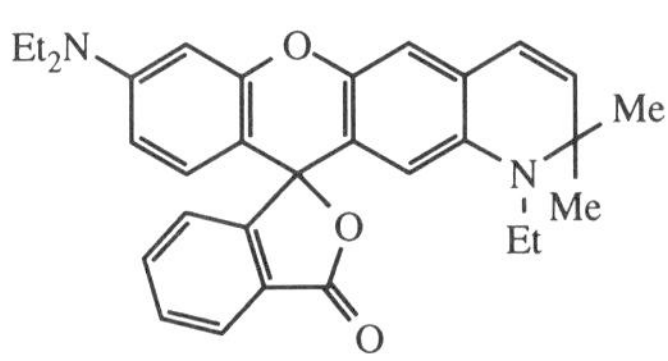

10-96

Higher molecular weight fluoran-type, white leuco dyes, exemplified by compound **10-97**, were used in thermal copy paper. On development of the image with zinc 3,5-bis-(alpha-methylbenzyl)salicylate, high-density black images with low fog were formed which had good light-fastness.[82]

A thermal printing system for forming multicolor images consisted of a sheet coated with leuco dyes **10-98** for black, **10-99** for blue, and **10-100** for red in separate layers, and a thermal transfer sheet containing bisphenol A as the developer/sensitizer.[83]

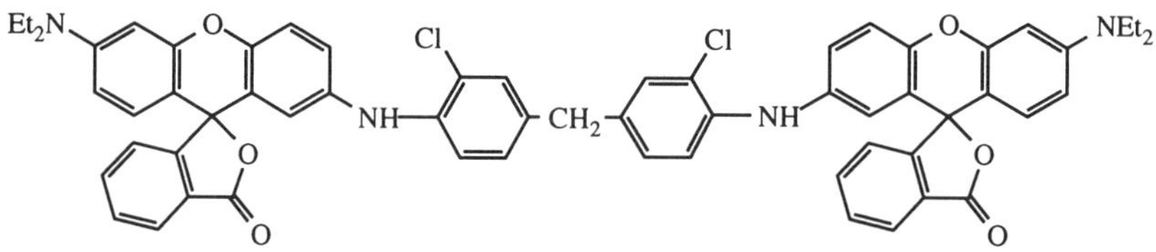

10-97

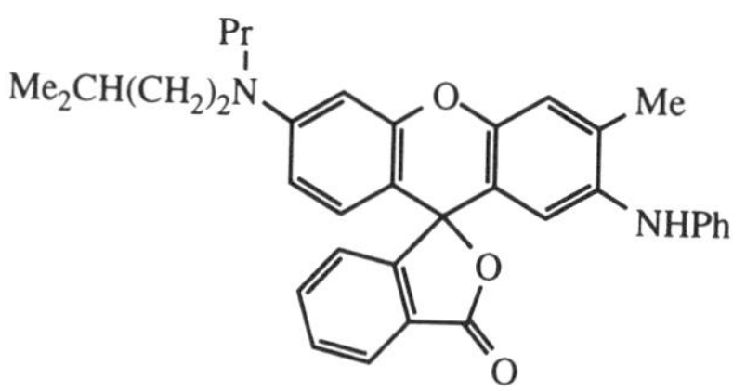

10-98

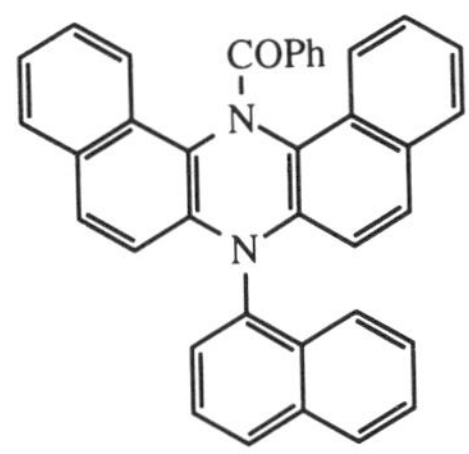

10-99

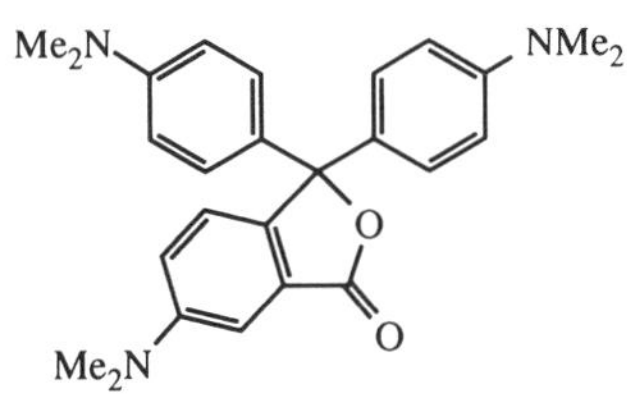

10-100

A new fluoran-type leuco dye, **10-101**, was prepared and used in thermal printing paper. This dye had high solubility in the oils used for microencapsulation and gave high background brightness; on reaction with a phenolic developer, a black image was obtained. The printing paper had good storage stability at 40 °C/60% humidity.[84]

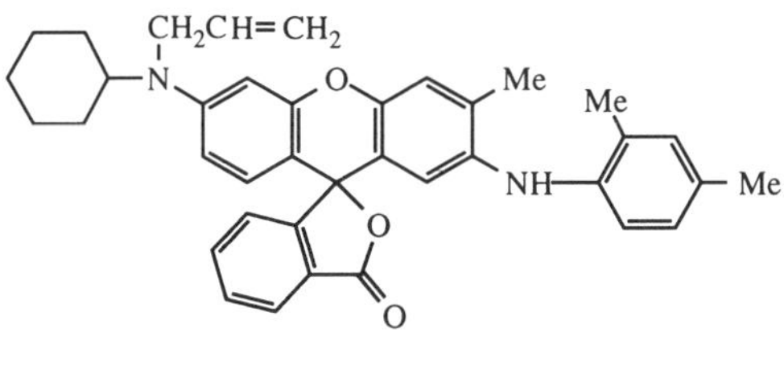

10-101

In further variations on the theme, leuco dyes of type **10-102** were made and used in thermal recording materials, developing color on contact with a phenolic developer.

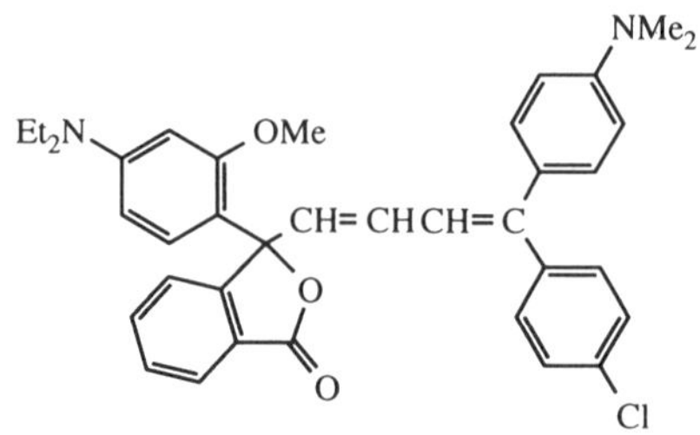

10-102

These dyes gave high-density color images and had absorption in the near-IR spectral region.[85]

The fluoran-type leuco dyes such as compound **10-103** showed high solubility in organic solvents, had high absorption in the visible and near-IR spectral regions, and had good storage stability. They were useful as color formers with developers in thermal- and pressure-sensitive printing materials.[86]

A crystalline, fluoran-type leuco dye, **10-104**, was synthesized and had mp 162–164 °C. A thermographic paper coated with this dye and bisphenol A as the developer showed good storage stability at 60 °C/90% relative humidity, and rapidly developed a black color on heating.[87]

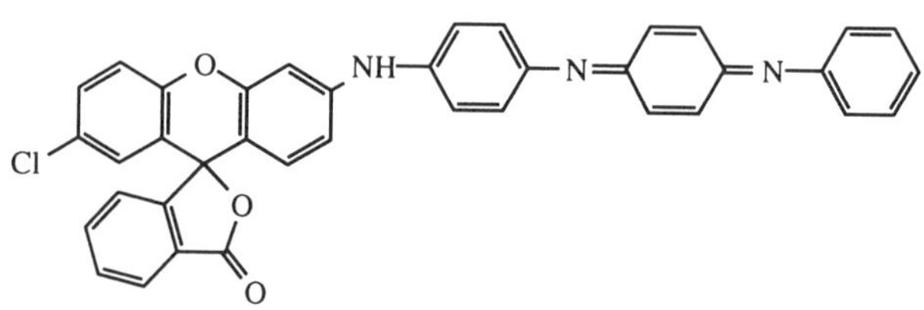

10-103

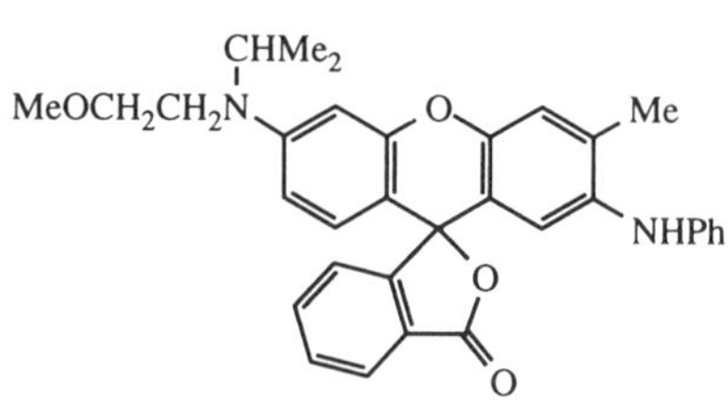

10-104

An improved process for making rhodamines was developed. For instance, *m*-diethylaminophenol was condensed with phthalic anhydride in *o*-dichlorobenzene at 175 °C under nitrogen, with the separation of water, then drowning in 3% aq. NaOH and workup to give the dye, **10-105**, as the hydrochloride in good purity and in 94% yield (in prior practice, using conc sulfuric acid as the condensing agent, an 89% yield of lower-purity product was obtained).[88]

Novel fluoran compounds, such as **10-106**, were prepared. They were useful in thermal- and pressure-sensitive recording papers as color formers which gave images resistant to fading due to NO_x.[89]

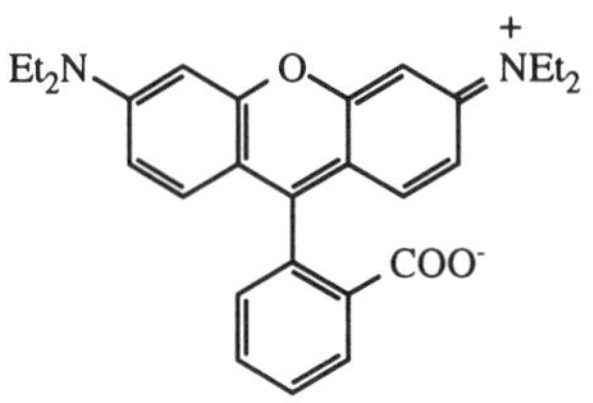

10-105

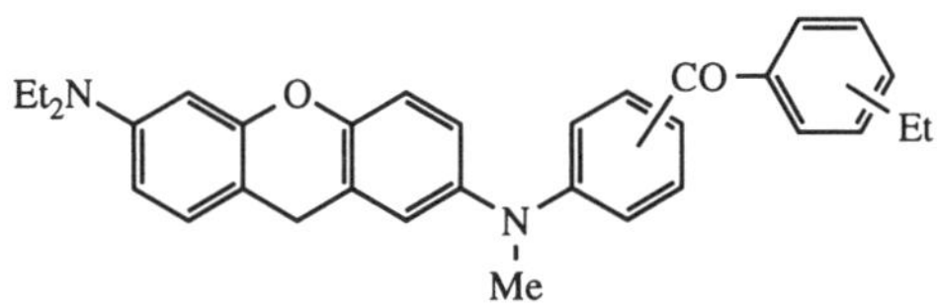

10-106

Fluoran leuco dyes of novel structure, such as **10-107**, were prepared and used in pressure-sensitive copying sheets; the images formed with these dyes could be recognized by an optical character reader at λ~660 nm.[90]

Another new, improved process for making fluoran leuco dyes was patented. For example, the condensation of 4-diethylamino-2-hydroxybenzophenone-2′-carboxylic acid, **10-108**, with *N,N*-dibenzyl-4-anisidine in H_2SO_4/oleum at 30 °C/2 hr, then heating to 60 °C/1 hr, and holding at 60 °C/3 hr, followed by addition of the charge to aq. NaOH/toluene at 80–85 °C gave an 82% yield of compound **10-109**. Compared to prior art processes, this new procedure reduced organic wastes more than fivefold.[91]

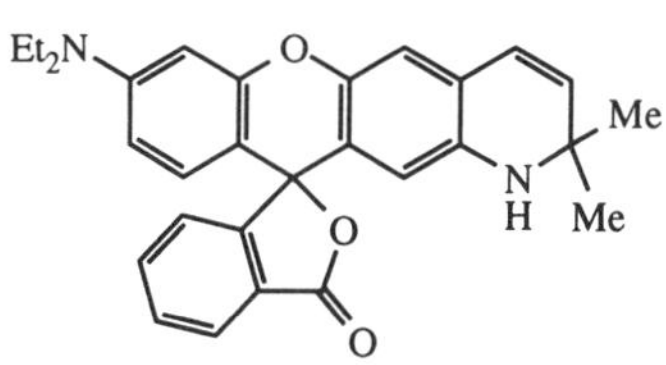

10-107

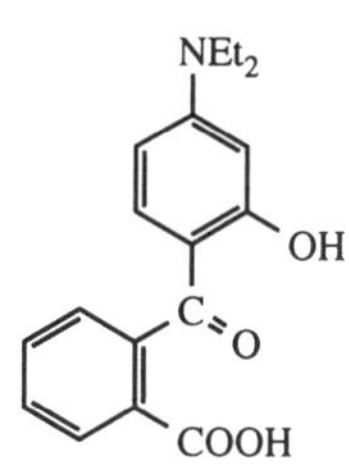

10-108

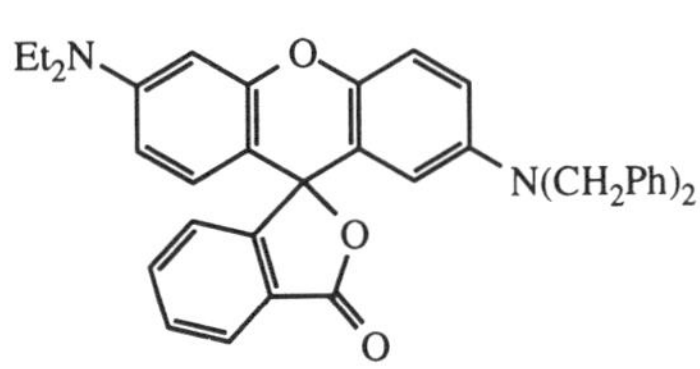

10-109

10.3.6. Developers and Sensitizers for Carbonless and Thermal Printing Paper

Resorcinol and its derivatives likely have been used in this area, as discussed earlier. The developments of new resorcinol and *m*-aminophenol derivatives/applications continued at a good pace, nonetheless, particularly as thermal imaging processes evolved which required developers with a melting point in the 70–120 °C range. Again, the term sensitizer used here appears to mean a compound capable of lowering the response temperature of a relatively high melting phenolic developer (e.g., like bisphenol A, mp 158 °C). Selections from the patent literature from 1980 follow.

Aralkyl esters of 2,4-dihydroxybenzoic acid (beta-resorcylic acid), such as the benzyl ester, **10-110**, were claimed as electron acceptors (developers) for leuco dyes, for use in pressure- and heat-sensitive recording materials.[92]

It was reported that difficulties were experienced in the use of phenolic developers in carbonless paper copying products. For instance, upon exposure to air the phenols gradually oxidized and produced brownish, nonreactive products; where phenolic compounds and a clay were used in coatings, ferric compounds present in the clays reacted with the phenols to develop an undesirable brownish or yellowish color during the preparation and storage of the recording sheets. These problems were said to be avoided by incorporating a phenolic ester in an oily solvent containing the chromogenic leuco dye. When the microcapsules were broken, the esters were hydrolyzed by moisture and the acidic material in the coating; the liberated phenol then stabilized the color image before it faded. Among the phenolic esters said to be useful for this purpose, the benzyl ester of 2,4-dihydroxybenzoic acid, **10-110**, and resorcinol diacetate, **10-111**, were listed.[93]

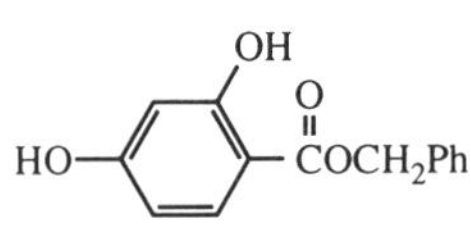

10-110

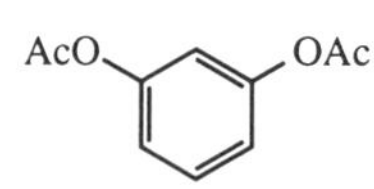

10-111

Alkylene-bis-(dihydric phenol) ethers, such as compound **10-112**, were said to be stable to heat and useful as developers for leuco dyes in thermographic printing materials. These ethers of dihydric phenols were said to cause neither tailing nor sticking to the thermal printing head, and the image had long-term stability.[94]

Resorcinol mono-benzoate, **10-113**, was patented for use as a developer for leuco dyes in thermal printing materials. This developer prevented adhesion to the thermal printing head and improved the whiteness and stability of the background of the

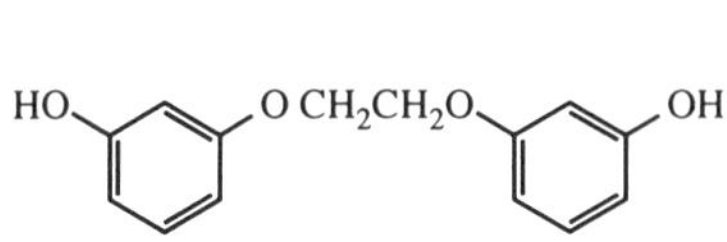

10-112

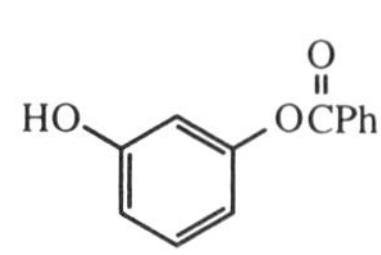

10-113

images. Resorcinol mono-benzoate could also be combined with other phenolic developers to obtain various sensitivities.[95]

Resorcinol derivatives with a substituent capable of forming intramolecular H bonds with a neighboring nuclear OH group, and their Zn and Al salts, were disclosed as developers for leuco dyes in carbonless or thermal copying materials. One example of such a developer was methyl 2,4-dihydroxybenzoate.[96]

Heat-sensitive recording materials were described which contained leuco dyes and bisphenol A in conjunction with, e.g., 4,4′-thio-bis-resorcinol as the developer. The resorcinol derivative was said to prevent fogging on the surface of the thermographic paper.[97]

Novel developers for leuco dyes in thermographic materials were diamides of *m*-aminophenol, made,e.g., by reaction of *m*-aminophenol with succinoyl chloride to give **10-114**. The images produced were color stable and chemical (fats, oils, alcohols) resistant. These materials were especially useful for label/ticket printing.[98]

Esters of 3,5-dihydroxybenzoic acid (alpha-resorcylic acid), such as the butyl ester **10-115**, were found to be good developers for leuco dyes in pressure- and heat-sensitive recording materials. These esters as electron acceptors combined with electron-donating leuco dyes to give images of excellent color density and durability.[99]

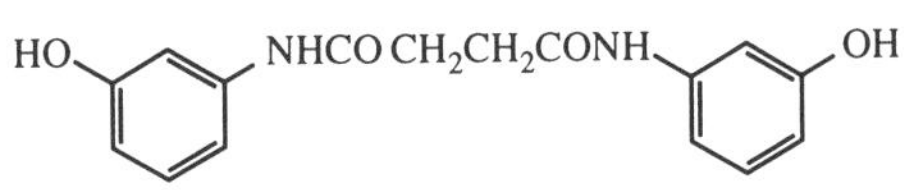

10-114

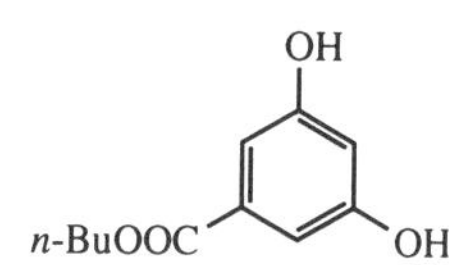

10-115

An improved thermal recording material was reported which contained a leuco dye in the transfer sheet and a hydroxyphenyl aralkyl ether plus a porous filler (e.g., finely divided silica) in the receiver sheet. One example of these developers is compound **10-116**. These materials gave good reproducibility of the middle part of the density gradation (half tone) and excellent image stability.[100]

Hydroxyphenoxy aryloxyalkyl ethers were found to be good developers for leuco dyes in thermographic recording materials. An example of these developers is compound **10-117**. Using these compounds permitted high speed printing and the images did not fade or fog, even when stored in contact with diazotype papers (chemicals).[101]

Sulfonates of dihydric phenols, e.g., **10-118**, were suitable as electron acceptors/developers for electron-donating leuco dyes in pressure- or heat-sensitive, photosensitive, electron beam, electrostatic, pen ink, or crayon recording materials. The images produced had high color density and durability. These developers were said to be easy to synthesize in high purity.[102]

p-Xylylene-bis-2,4-dihydroxybenzoates, **10-119**, were similarly useful developers for thermographic materials containing leuco dyes. The color images formed were said to have excellent fade resistance.[103]

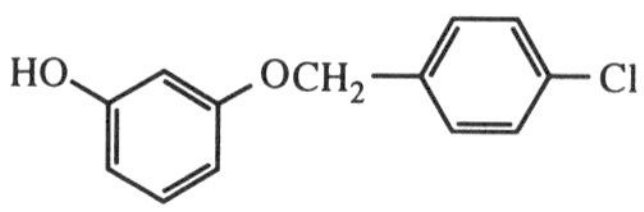

10-116

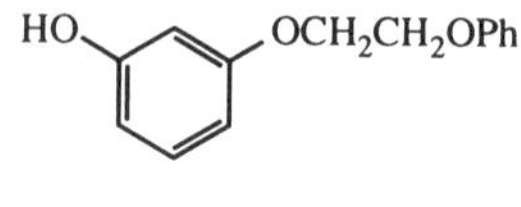

10-117

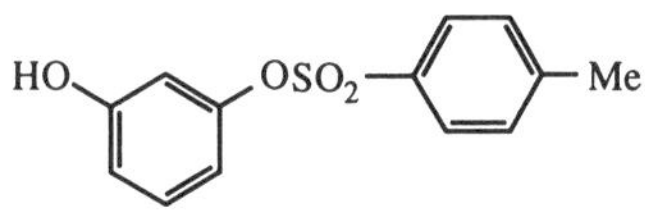

10-118

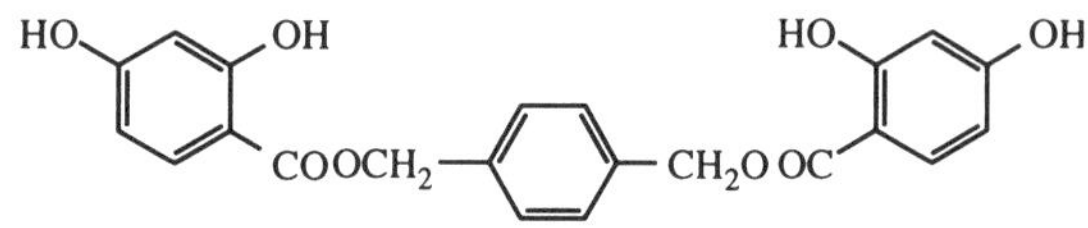

10-119

Ethers of beta-resorcylic acids, such as **10-120** (mp 98–100 °C), and their metal salts were patented as developers for leuco dyes in recording materials. These ethers had suitable heat sensitivity and gave high-density images with no fog development and good chemical resistance.[104]

Another variation was 1-phenoxy-2-propyl beta-resorcylate, **10-121**, which was found to be an excellent developer for leuco dyes in recording materials.[105]

Resorcinol di-(4-acylphenyl) and di-(4-alkoxyphenyl) ethers were used as sensitizers with bisphenol A as the developer for leuco dyes in thermographic materials. An example of these sensitizers is compound **10-122**. Stable, high-density images were obtained with a background of high whiteness.[106]

Phenylthioalkyl ethers of beta-resorcylic acid, such as compound **10-123** (as the Zn salt), were prepared and found to be valuable as color developers for leuco dyes in pressure or thermal printing materials.[107]

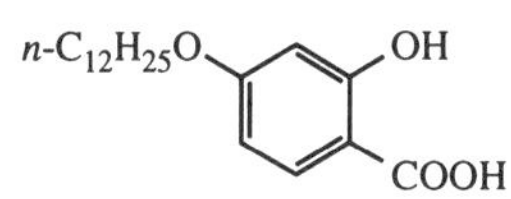

10-120

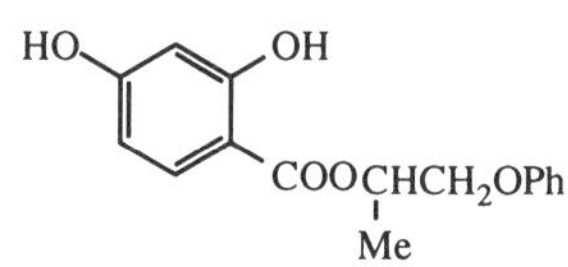

10-121

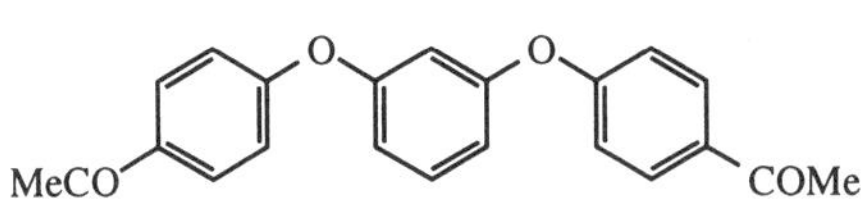

10-122

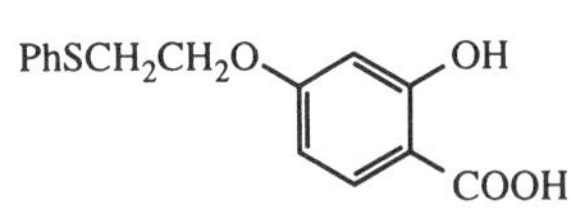

10-123

1,3-Bis-(4′-hydroxyphenoxy)benzene, **10-124**, was patented as a color developer for leuco dyes in thermal printing materials. This developer gave good thermal sensitivity and stable, high-density images.[108]

Also, certain phosphates and phosphites of resorcinol, e.g., compound **10-125**, were electron-accepting compounds for developing color on contact with electron-donating colorless dyes in thermal printing products. Stable color images were obtained.[109]

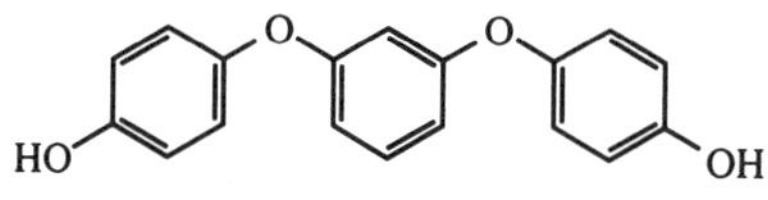

10-124

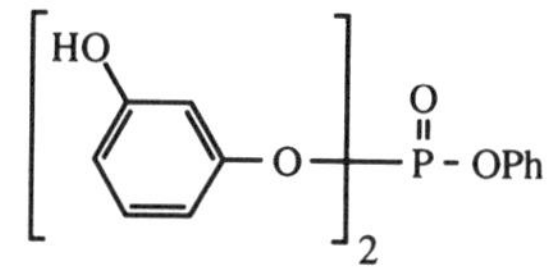

10-125

Alkylenedioxydiphenols such as compound **10-126** were prepared. Thus, charging resorcinol, NaOH, water, and chlorobenzene to an autoclave, stirring and heating to 100 °C, and adding 1,5-dichloropentane dropwise over 30 min at 10 kg/cm^2, then holding the charge at 160 °C for 4 hr gave a 77% yield of compound **10-126** at 28% conversion of resorcinol. This compound was a developer for leuco dyes used in recording materials.[110]

A thermographic copying paper was prepared which contained a leuco dye and a developer containing resorcinol diglycidyl ether, **10-127**, presumably as the sensitizer. This thermal printing paper showed high sensitivity.[111]

The ether, **10-128**, of 2,4-dihydroxybenzoic acid (Zn salt) was prepared and used as a color developer for fluoran-type leuco dyes. Applied to paper coated with spherical, hollow plastic bubbles, a thermal recording material was obtained which had good thermal sensitivity and gave high-density images which had good storage stability.[112]

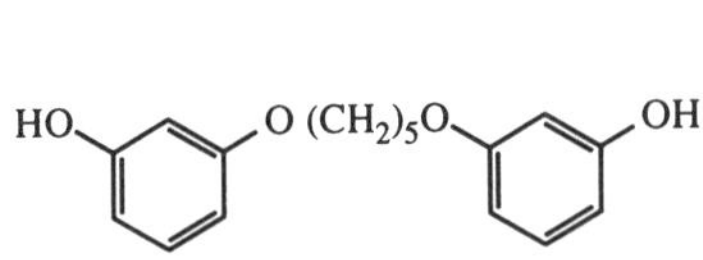

10-126

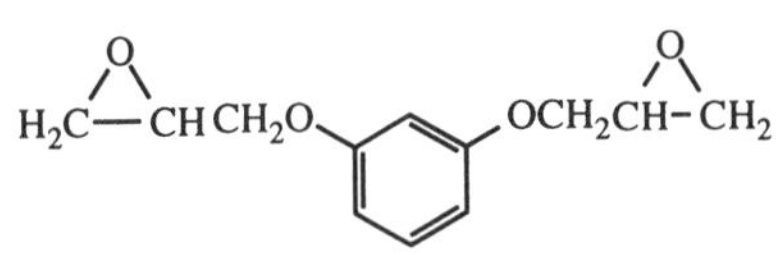

10-127

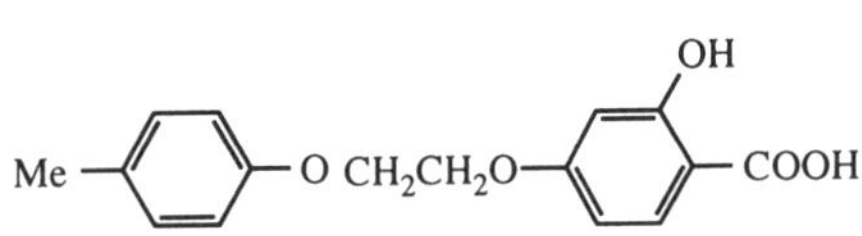

10-128

10.3.7. Silver Halide Photographic Materials

There is a sizeable patent literature on the use of resorcinol and resorcinol derivatives in photographic products, e.g., as agents to strengthen the gelatin emulsion during development, as aids in sharpening images, as secondary reducing agents, and as couplers. Film formulations are kept secret, so that it is not possible to be sure about which additives have been commercialized. The following few selections from the literature published since about 1980 will give an idea of the range of potential applications of resorcinol/derivatives.

A 1977 publication described a black-and-white photographic material which contained a nondiffusible resorcinol coupler, **10-129** (*n*-decyl gamma-resorcylate), and a developer (*N,N*-diethyl-3-methyl-*p*-phenylenediamine·HCl). The coupler provided a stable, neutral dye image which enhanced or replaced the silver image, i.e., it could save some of the cost of the silver halide. (At the time, the Hunt brothers of Texas were cornering the market and had driven silver prices sky-high; a shortage of silver was feared.) In the given example, 57% of the image density was from Ag and 43% from the dye.[113]

Similarly, a photographic silver halide material was described which used, e.g., the substituted *m*-diethylaminophenol, **10-130**, as a coupler and *N,N*-diethyl-*m*-methyl-*p*-phenylenediamine as the developer to produce a neutral black image.[114]

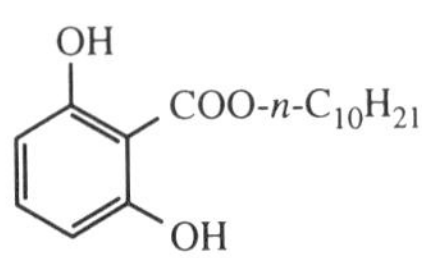

10-129

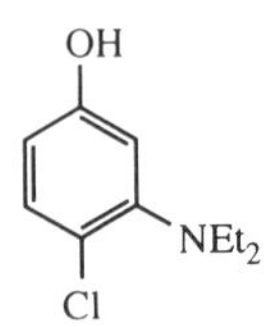

10-130

The stability of a photographic silver halide material, containing color couplers (e.g., a pyrazolone type which is deleteriously affected by formaldehyde), was improved by adding resorcinol and an aminoacid (such as lysine or arginine) to one or more of the gelatine layers.[115]

Esters of carboxyalkylresorcinols, e.g., **10-131**, were prepared and were useful as black color couplers in photosensitive silver halide materials.[116] Similarly, 4-dodecylresorcinol, **10-132**, was patented as a black color coupler for photographic materials; this coupler was incorporated into the silver halide emulsion as an oil dispersion.[117]

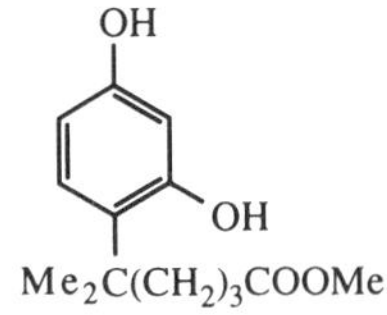

10-131

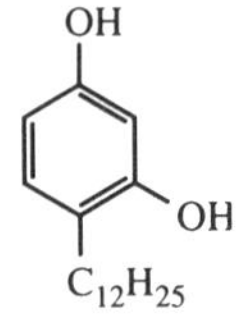

10-132

Black-and-white photographic materials which contained acylresorcinols such as **10-133** or alkylsulfonyl- or arylsulfonylresorcinols such as **10-134** as the black image forming coupler were patented.[118]

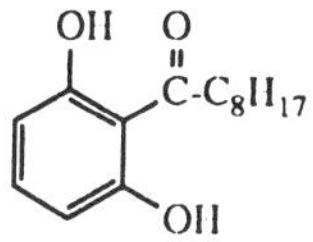

10-133

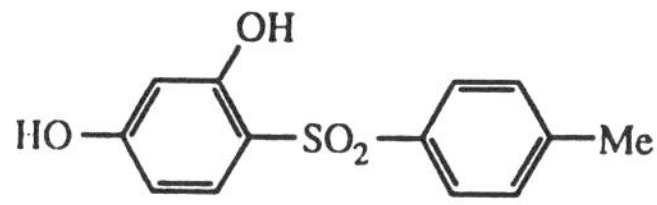

10-134

A new silver halide photographic composition for black-and-white images contained resorcinol or *m*-octadecylaminophenol, **10-135**, as the black coupler which worked with a number of developers (e.g., 4-hydroxy-6-methyl-1,3,3α,7-tetrazaindene).[119]

High-contrast photographic elements exhibiting reduced pepper fog (spots) were developed which contained resorcinol or 4-bromoresorcinol (among a number of suitable polyhydroxybenzenes); an arylhydrazide was added to increase contrast.[120]

A series of new *tert*-alkyl-substituted resorcinols, such as **10-136** or **10-137**, was prepared for use as black color couplers in silver halide photographic compositions.[121]

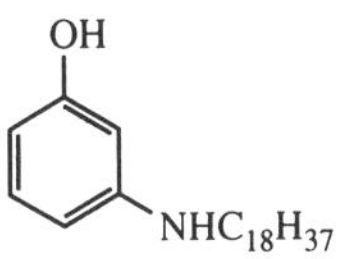

10-135

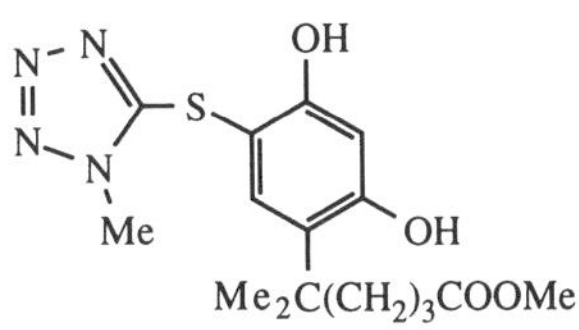

10-136

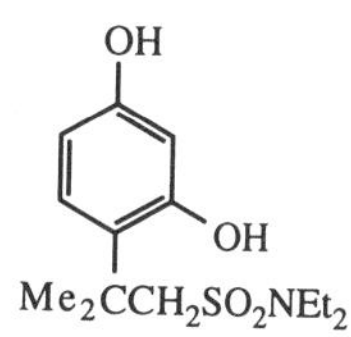

10-137

A novel silver halide photographic composition was patented which had an undercoat of a copolymer containing phenolic hydroxyl groups, e.g., **10-138**; this subcoat provided superior adhesion between the photosensitive layer and the polyester film base.[122]

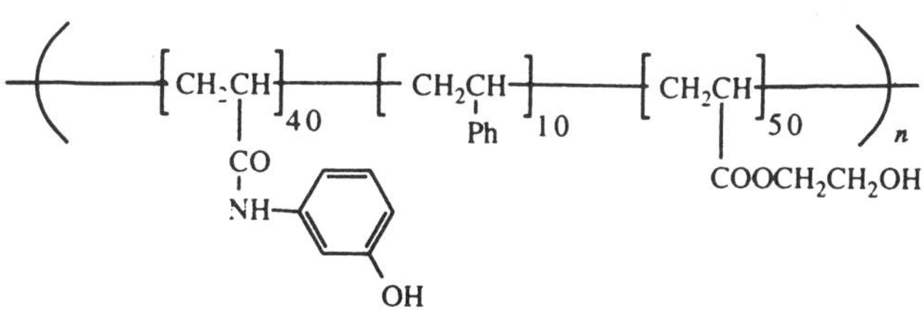

10-138

Color photographic materials were developed using a developer bath free of (the commonly used) benzyl alcohol, but containing aromatic primary amine developer and a phenolic compound, e.g., 2,6-dihydroxybenzoic acid (gamma-resorcylic acid). The composition had improved stability and produced images with less fog.[123]

A silver halide photosensitive material was patented which contained in the sensitive layer silver halide, a reducing agent, a polymerizable compound, and a fluoran leuco dye, such as **10-139**, the last two ingredients being microencapsulated; this product gave sharp images and good light fastness.[124]

A silver halide photographic composition was developed which contained a solid microparticle dispersion of (a substituted) 4-benzoylresorcinol (or hydroxyphenylbenzotriazole) and styrene. This material gave less static marks and no sticking in the manufacturing process.[125]

Esters of 4-benzoylresorcinol (2,4-dihydroxybenzophenone)-type compounds, e.g., **10-140**, were patented for stabilizing magenta couplers (e.g., pyrazoloazoles) in color-photographic silver halide materials.[126]

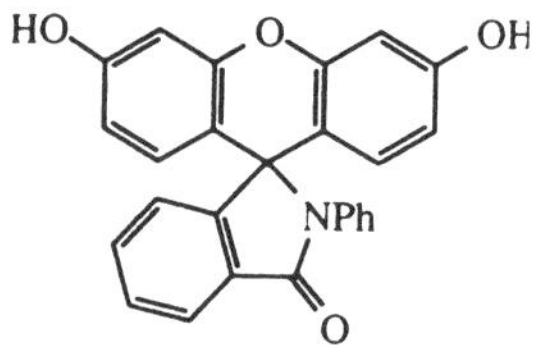

10-139

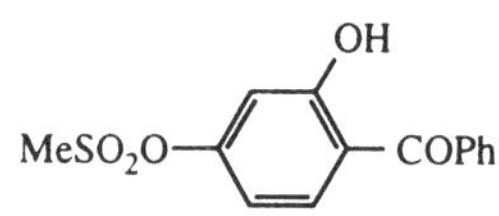

10-140

10.3.8. Lithography and Photoresists

In a simplified way, the commonality of such processes is the preparation of a surface (substrate) by coating with a (UV) light-sensitive material, exposure through a mask (e.g., a transparency), with the image area becoming ink receptive and water repellant, or with development by dissolving the exposed or unexposed material. The term lithography is most often used when a reproduction (printing, copying) process is involved and the oleophilic areas are inked before printing. The photoresist technology/terminology is usually used for microelectronics where the preparation of integrated circuits involves the coating of a silicon (oxidized surface) wafer with a light-sensitive material, (UV-) light exposure through a photomask, etching away of exposed or unexposed areas, depending on the process/material used, and coating of the raised areas with a conductive material. Obviously, a great deal of sophistication is involved in such technologies.

The potential, if not actual, use of resorcinol and resorcinol derivatives here is in coating resins, in the preparation of light-sensitive materials, or in the photosensitizers for photopolymerization of light-sensitive monomers, oligomers, or resins, where the polymerized material is insolubilized and the unreacted material is dissolved away. Indeed, the use of R/F resins in this area was already referred to in Section 5.12.

In lithography, the light-sensitive compound is a diazo compound. In the case applicable here, a positive-working (the coating is solubilized after exposure) printing plate coating is involved. The UV-light-sensitive compound is *o*-naphthoquinone diazide, **10-141**, which is said to be photolyzed, as shown in Eq. (10.5), to produce the alkali-soluble indenecarboxylic acid, **10-142**.

(10.5) 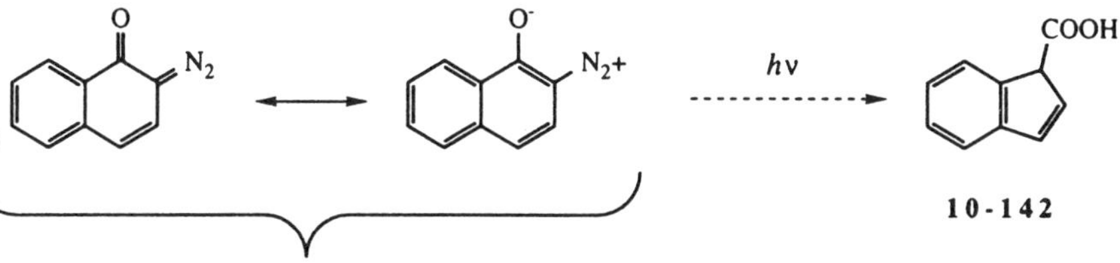

10-142

10-141

The *o*-naphthoquinone diazide in a positive-working coating is attached, via a sulfonyl chloride derivative, to a phenolic (could be resorcinol)/aldehyde novolak resin, as shown in structure **10-143**, or as a disulfonate of a dihydric phenol (bisphenol A) admixed to a phenol/formaldehyde resin.

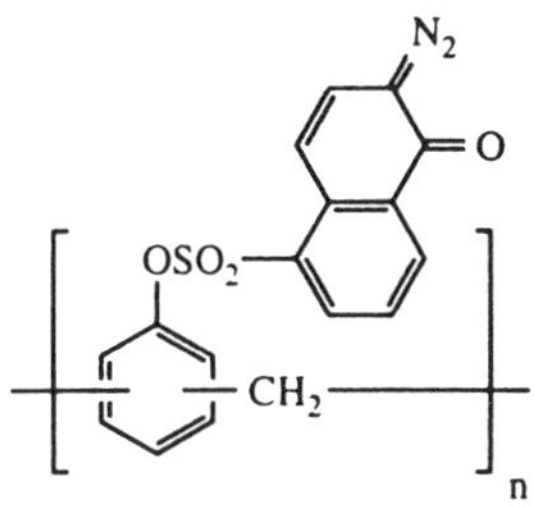

10-143
(n = low number to obtain a soluble resin)

The following, few selections from the literature will give a further glimpse of the possible uses of resorcinol in this technology.

An anodized and sealed Al plate was coated with a mixture of an *o*-naphthoquinone diazide sulfonate of a 2-methylresorcinol/acetaldehyde resin, an *o*-naphthoquinone diazide sulfonate of a *m*-cresol/formaldehyde resin, and an *o*-naphthoquinone diazide sulfonate of a *p-tert*-butylphenol/formaldehyde resin containing Victoria Pure Blue dye (for visualizing the pattern), exposed via a positive original and developed (presumably with dilute, aqueous alkali) to give an offset printing plate with excellent ink receptivity and chemical resistance.[127]

A light-sensitive mixture for use in producing printing plates and photoresists was described, consisting of a water-insoluble binder which was soluble in aqueous alkaline solutions (e.g., a cresol/formaldehyde novolak resin) and *o*-naphthoquinone diazide sulfonate esters of low molecular weight di- or poly-hydric phenolic compounds. Suitable examples of the latter were resorcinol derivatives such as 2,4-dihydroxybenzophenone(4-benzoylresorcinol), 4,4′-thio-bis-resorcinol, 4,6-bis-(2,4-dihydroxyphenylthio)resorcinol, 2,4-dihydroxy-3,5-dibromobenzophenone and similar materials. The new, light-sensitive *o*-naphthoquinone diazide sulfonates were said to be nonexplosive, to have high sensitivity to light, to have good developer resistance, and to be safely storable and transportable. Sulfonates of mixtures of the aforementioned compounds and phenolic novolak resins were also suitable.[128]

A new, silicon-containing positive photoresist (SIPR) was developed which consisted of a partly trimethylsilylmethylated resorcinol/formaldehyde resin mixed with an *o*-naphthoquinone diazide. This resin was used as the thin topcoat being applied onto a thick layer of a polyimide resin on a silicon wafer, in a bilayer resist system. After exposure and development, positive patterns (as small as 0.7-μm lines) were obtained in this thin top layer. These patterns were transferred into the bottom, polyimide layer either by reactive oxygen ion etching or by oxygen ion beam etching. There were no residues on the wafers after the etching transfer.[129]

A deep ultraviolet (<300 nm) lithographic resist composition which was alkali developable was prepared, consisting of a base-soluble polymer (phenolic resin or methyl methacrylate/methacrylic acid copolymer) and, as the photosensitive solubilizing agent, a 2-diazo-1,3-cyclohexanedione, such as compound **10-144**. After exposure and development, improved image resolution was obtained.[130]

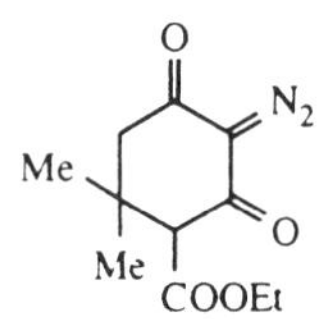

10-144

A positive-type, radiation (UV, far-UV, X-ray, electron beam, etc.) sensitive resin material was patented which was especially suitable as a resist for producing integrated circuits. Prior *o*-naphthoquinone diazide/novolak resin systems were said to give small resin solubility changes on irradiation, so that on alkaline development only low resolution of the resist patterns was obtained. In the new system, resorcinol/acetaldehyde novolaks were used and the tetra-(*o*-naphthoquinone diazide)sulfonate of methylene-bis-resorcinol was employed; after exposure and development, pattern line width in the 0.7–0.9 μm range was obtained and the resist patterns were quite faithful to the pattern mask.[131]

A plasma-developable photoresist composition was disclosed which was photosensitive to light of 436 nm wavelength. These compositions contained *N*-vinyl monomers (e.g., *N*-vinylcarbazole), a polymeric binder (e.g., a polyvinylformal resin), a haloalkene initiator (e.g., tetraiodoethylene), and a photosensitizer (e.g., coumarin **10-145**). After exposure, unpolymerized monomer was dissolved. Good sensitivity and pattern resolution were obtained.[132]

The use of 7-aminocoumarins, such as **10-146**, as light sensitizers in photoresist compositions was said to prevent the stationary wave affect which is caused by the multiple reflection of the incident light; precise resists were obtained.[133]

A light-sensitive polyimide composition for photoresists was patented which contained a combination of a bis-coumarin, e.g. **10-57** (shown earlier) and *N*-phenyl-

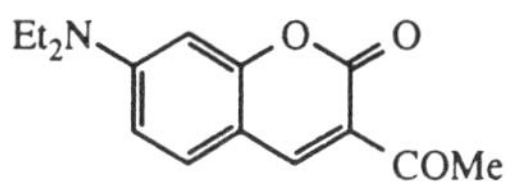

10-145

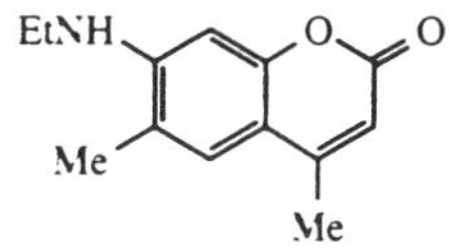

10-146

glycine. These compositions had high sensitivity and heat resistance and were useful for interlayer insulation films of multilayer circuits, for cover coatings for flexible Cu-plated sheets, for solder resist films and liquid-crystal orientation films.[134]

Novel photosensitive lithographic plates were prepared, where sandblasted and anodized Al plates were coated with positive-working photosensitive compositions containing an alkali-soluble resin and photosensitive compounds such as the di-(*o*-naphthoquinone diazide)sulfonate ester of fluorescein. These plates, after exposure and development, gave images from which the film edge marks could be readily removed.[135]

Radiation-sensitive resin compositions for positive-working resists were prepared consisting of a cresol/xylenol/formaldehyde novolak, a cyclic compound prepared by the condensation of resorcinol with an aldehyde (presumably, this is a calixarene compound), and an *o*-naphthoquinone diazide compound such as the tetra-(*o*-naphthoquinone diazide)sulfonate of a tetrahydroxybenzophenone. This composition showed high sensitivity and heat resistance.[136]

The penta-(*o*-naphthoquinone diazide)sulfonate of compound **10-147** was prepared for use in photoactive compounds and in radiation-sensitive mixtures. The pentasulfonate, together with a cresol/formaldehyde novolak resin, was applied to a silicon wafer. This composition showed superior performance in making photoresists.[137]

Similarly, tetra-(*o*-naphthoquinone diazide)sulfonates of alkylidene-bis-resorcinols, such as compound **10-148**, were prepared and applied in conjunction with an alkali-soluble cresol/formaldehyde resin to a suitable substrate to manufacture photosensitive films and photoresists.[138]

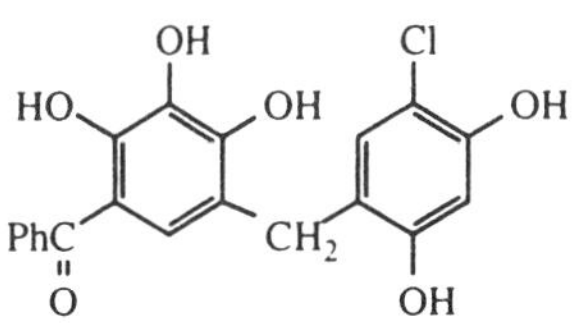

10-147

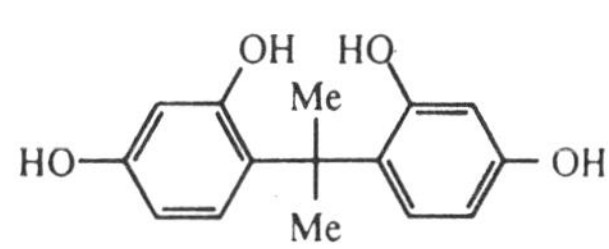

10-148

Positive-working photoresists, useful for making semiconductor devices, were patented in which the photosensitive compound was the tri-(*o*-naphthoquinonediazide)sulfonate of compound **10-149** and the alkali-soluble resin was a *m,p*-cresol/formaldehyde novolak resin.[139] A similar composition contained the tetra-

(*o*-naphthoquinone diazide)sulfonate of compound **10-150** or similar polyhydroxybenzophenones.[140]

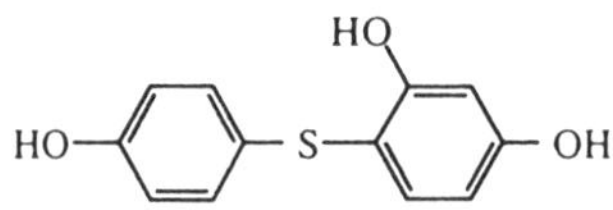

10-149

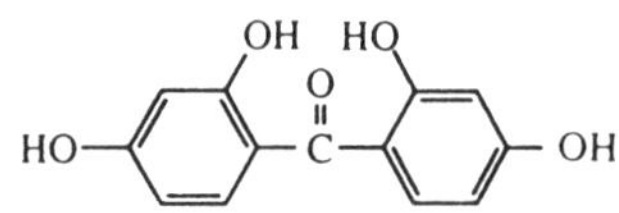

10-150

10.3.9. Liquid Crystals

These compounds are anisotropic (mesomorphic) fluids existing between the boundaries of the solid and an isotropic liquid phase. A variety of structural modifications exist. The ordering in such a fluid is adequate to give it some solid-like properties, but not strong enough to prevent flow. The compounds undergo well-defined phase changes which can be regulated by temperature, pressure, and electric or magnetic fields; thus, they may be used in electronic display devices such as watches, calculators, games, laptop computers, and, perhaps, in flat screens for the TV sets of the future.

Because resorcinol derivatives have not played a big role here, no further discussion is given. However, a few examples from the literature follow.

4′-Alkoxy-2′-hydroxy-4-butylazobenzenes, such as compound **10-151**, were shown to be nematic liquid crystals at room temperature which maintain nematic properties over a wide temperature range. These compounds were said to be useful in display devices or as solvents for ESR and NMR spectroscopy.[141]

4-(2-Hydroxy-4-alkoxybenzylideneamino)cinnamate esters, such as compound **10-152**, were shown to be liquid crystalline materials with good physical and chemical stability, high liquid-crystal temperature, and high dielectric constants. These compounds were blended into nematic liquid-crystal compositions; they were usable for display elements with a memory, i.e., heat or laser writing materials.[142]

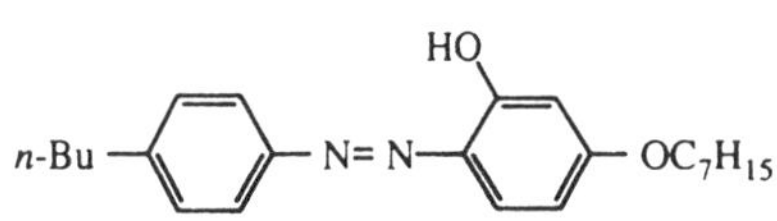

10-151

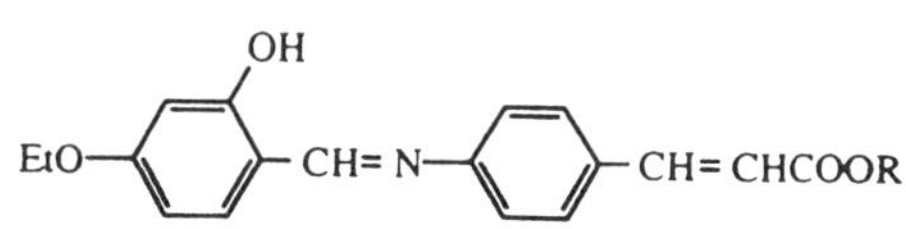

10-152
(R= optically active alkyl or aralkyl)

A ferroelectric liquid-crystal composition was patented which contained compounds such as **10-153** and **10-154**. These compositions had a smectic phase over a broad (even low) temperature range and had excellent display characteristics (e.g., good contrast) and high response speed.[143]

Optically active diesters of aromatic dicarboxylic acids, such as compound **10-155**, were prepared. Incorporated into chiral, smectic liquid-crystal compositions, such products were useful in displays with a very fast response time.[144]

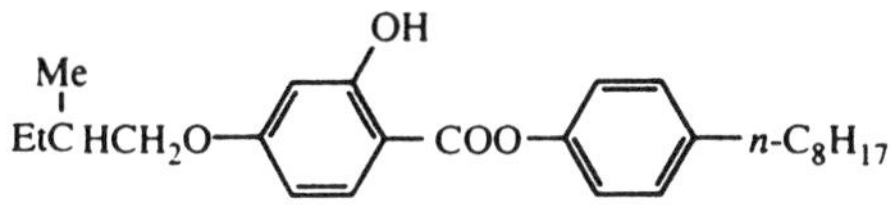

10-153

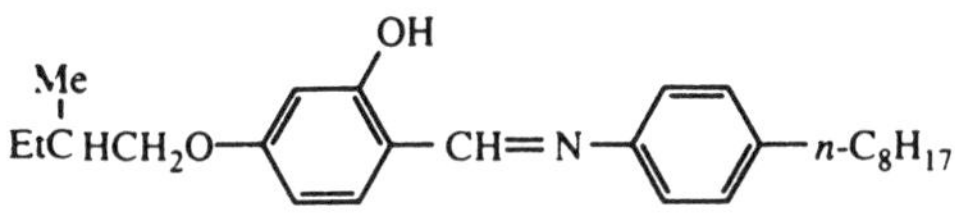

10-154

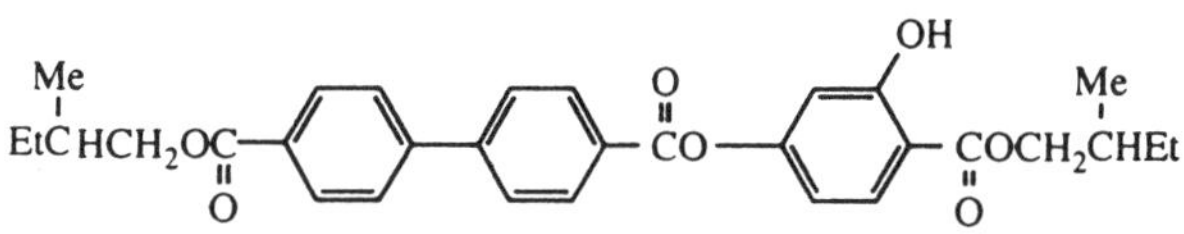

10-155

10.3.10. Miscellaneous Applications in High Technology

A magnetic recording material was prepared by coating a polyester (PET) film base with 2,6-bis-(2′,4′-dihydroxybenzyl)-4-chlorophenol, **10-156**, followed by a topcoat of magnetic paint (e.g., an iron oxide suspension). Excellent adhesion between layers was obtained.[145]

An electrolyte for a photogalvanic solar conversion cell contained a photoreducible thiazine dye, a redox couple, and a photosensitizing compound of structure **10-157** or **10-158**.[146]

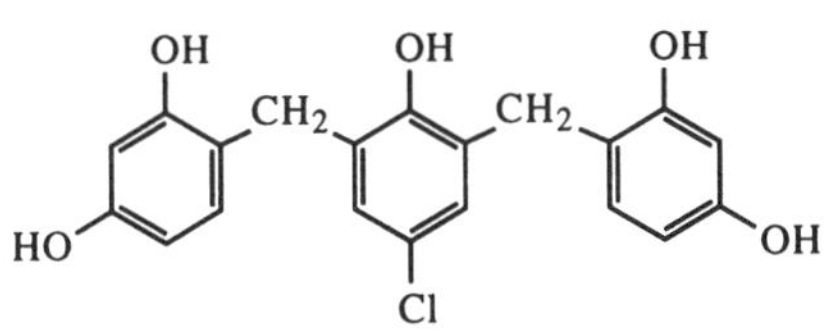

10-156

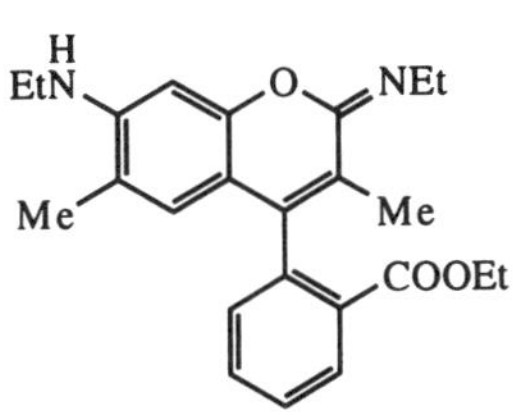

10-157

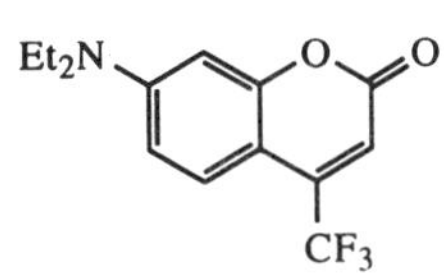

10-158

A light collector system, useful as an energy converter in solar cells, scintillators, and display devices, contained 7-phenylsulfonylaminocoumarins, such as **10-159**. These compounds absorbed diffuse, incident electromagnetic radiation and emitted it into an optically dense medium practically without loss, since most of the emitted light was totally reflected in the medium.[147]

An electrophotographic photoreceptor was prepared, consisting of a charge generating layer (e.g., amorphous silicon in a binder) with a charge-transporting layer containing a 3-phenyl-7-aminocoumarin, e.g., compound **10-160**. This photoreceptor material had improved photoreceptivity and durability on repeated use.[148]

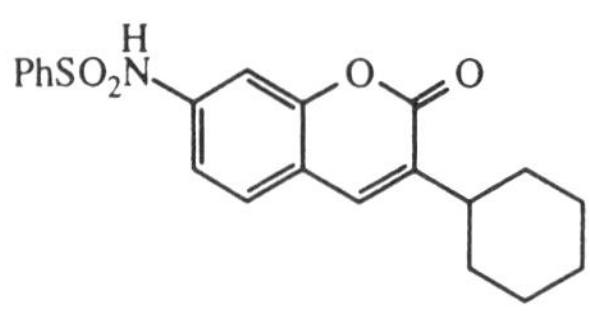

10-159

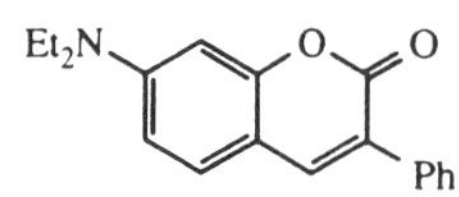

10-160

A photoresponsive device was constructed consisting of a supporting substrate, a photoconductive layer of a squaraine compound, e.g. **10-161**, and a diamine-containing hole transport layer. The preparation of the squaraine compounds by reaction of squaric acid (1,2-dihydroxy-1-cyclobutene-3,4-dione) with *m*-aminophenols in the presence of 5-methylresorcinol and an aliphatic alcohol was described also in this patent.[149]

An optical recording material used for laser discs comprised a leuco dye in monolayers or accumulated layers (e.g., crystal violet lactone), a phenolic compound (e.g., resorcinol) monolayer or accumulated layers, and a light-absorbing material (e.g., Cu-phthalocyanine) on a support.[150]

The preparation of photoconductive, asymmetric squaryliums, similar to **10-161**, from 3,4-dichloro-3-cyclobutene-1,2-dione was patented. These materials were said to be useful in electrophotography.[151]

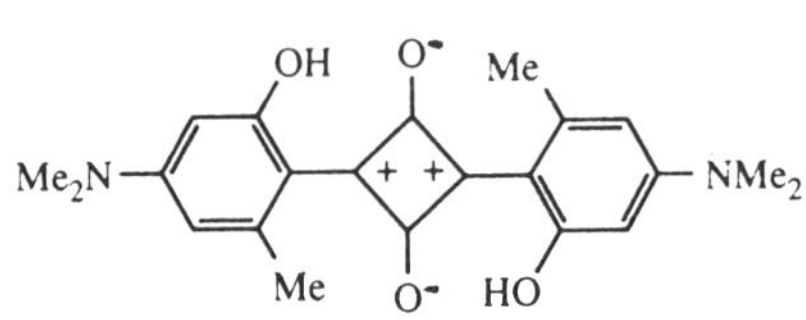

10-161

A toner for developing an electrostatic image in electrophotography contained a binder resin (e.g., polystyrene), coloring agent (e.g., carbon black or aniline blue), and a tri-(2,6-dialkoxyphenyl)phosphine, e.g., compound **10-162**, as a charge-controlling agent. The toner charged stably and positively and provided sharp images of good quality even after repeated copying.[152]

Oligo(benzyl)resorcinols or resorcinol ethers of structures **10-163** and **10-164** were developed as fluids, useful as control films for light valves used in high definition television (HDTV). These fluids were used as thin, dielectric films with electrostatically deformable surface and had improved hydromechanical, hydrodynamic, and electrical properties over prior art materials. The fluids also had increased heat and electrical conductivity, had surfactant properties, and acted as radical and hydrogen getters. The hydrogen-capture (via keto-forms) capability diminished the danger of a pressure increase in the high-vacuum environment of the light valve, an ion getter pump did not have to work as hard or could be eliminated, and the lifetime of the electron gun was increased.[153]

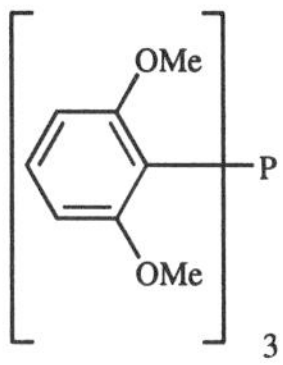

10 - 162

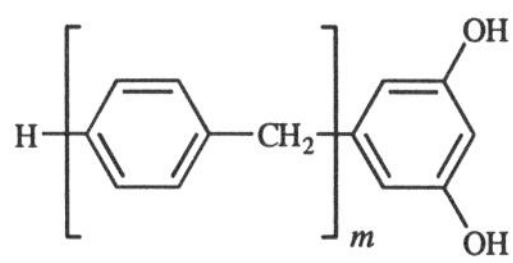

10 - 163
($m \geq 3$)

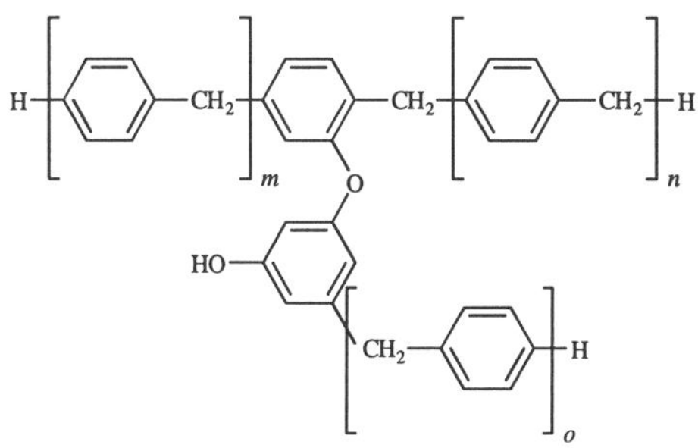

10 - 164
(*m, n, o* = not specified)

Xanthene dyes, e.g., structure **10-165**, were developed for use with the toners for electrophotographic recording.[154] Sterically hindered 1,3-diphenyl-2-pyrazolines were synthesized and used as scintillators. For instance, compound **10-166** had a fluorescence quantum yield of 0.75 at 438 nm.[155]

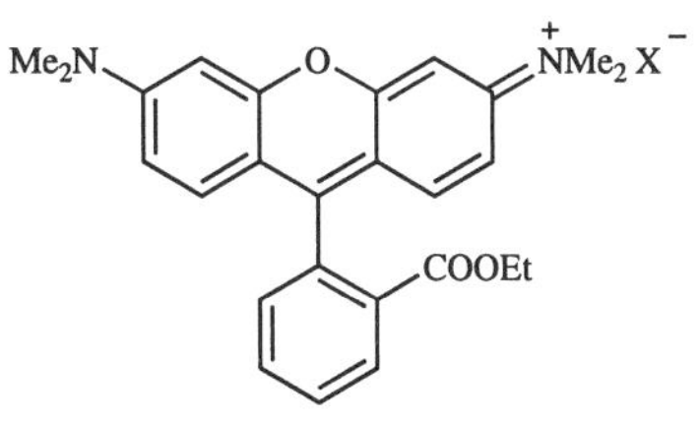

10 - 165

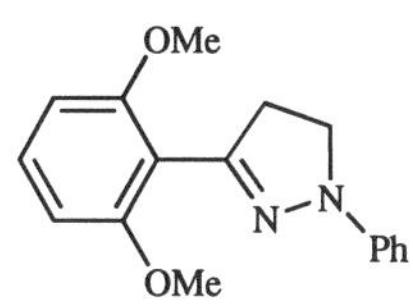

10 - 166

Electrophotographic photoreceptors were prepared, consisting of the squarylium compound, **10-167**, as the charge-generating agent, an interlayer and a charge-transporting layer containing a hydrazone, e.g., $C_6H_4CH{=}CHCH{=}NNPh_2$. These photoreceptors showed good photosensitivity in positive or negative charging processes.[156]

7-Diphenylaminocoumarins, the simplest representative being compound **10-168**, were prepared and were useful as charge-transporting agents for electrophotography, incorporated into the photosensitive layer on an electroconductive support. Among the useful properties of these compositions were excellent sensitivity and durability in multiple use.[157]

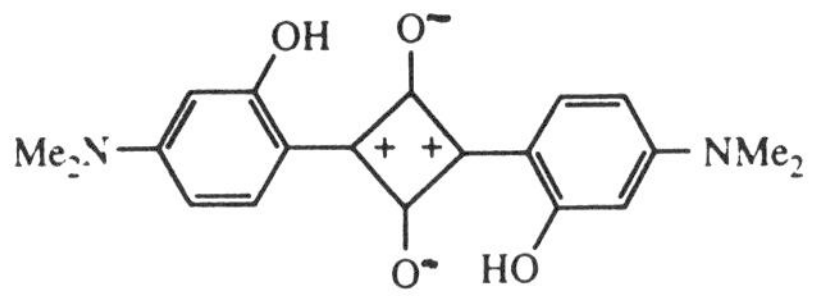

10-167

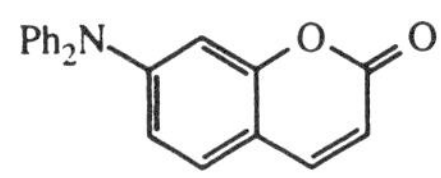

10-168

10. 4. REFERENCES

1. P. Gregory, *High-Technology Applications of Organic Colorants*, Plenum Publishing Corp., New York, 1991.
2. E. Grushka, S. Lam, and J. Chassin, *Anal. Chem., 50*, 1398 (1978).
3. K. E. Karlson, D. Wiesler, M. Alesandro, and M. Novotny, *Anal. Chem., 57*, 229 (1985).
4. An excellent source is *Kodak Laser Dyes*, a pamphlet published in Dec. 1987 by Eastman Kodak Co.
5. As a starting point, extensive reviews of xanthene (and other) dyes are given in Kirk-Othmer's *Encyclopedia of Chemical Technology*, 3rd ed., Vol. 24, p. 662, John Wiley & Sons, New York, 1984, and in *Ullmanns Encyklopaedie Der Technischen Chemie*, 4th ed., Vol. 23, p. 408, Verlag Chemie, Weinheim, Germany, 1983.
6. SRI (Stanford Research Institute) International-Specialty Chemicals Reports, *Reprographic Chemicals—Worldwide*, Oct. 1989.
7. U.S. Pat. 4,399,209 (Aug. 16, 1983; F. W. Sanders, G. F. Hillenbrand, J. S. Arney, and R. F. Wright to the Mead Corp.), and references therein describe the making of urea/resorcinol/formaldehyde microcapsules.
8. J. E. Kassner, *J. Imaging Technol., 11(5)*, 224 (1985). Also U.S. Pat. 4,460,846 (Apr. 3, 1984; F. W. Sanders, G. F. Hillanbrand, J. S. Arney, and R. F. Wright to The Mead Corp.); and U.S. Pat. 4,578,339 (March 25, 1986; D. G. Adkins to The Mead Corp.).
9. U.S. Pat. 4,713,312 (Dec. 15, 1987; P. C. Adair, G. F. Hillenbrand, and P. D. Davis to the Mead Corp.).
10. U.S. Pat. 4,137,238 (Jan. 30, 1979; F. Werner, R. Puetter, and P. Wenzl to Bayer A.-G.).
11. Ger. Offen. DE 2,850,391 (June 4, 1980; F. Werner and H. U. Blank to Bayer A.-G.); *C.A., 93*, 204255 (1980).
12. U.S. Pat. 4,335,246 (June 15, 1982; F. DeFeo, G. Burei, and R. Cipollo to Aziende Colori Nazionali Affini ACNA).
13. Eur. Pat. Appl. EP 258,190 (March 2, 1988; C. Frey and B. Dill to Ciba-Geigy A.-G.); *C.A., 109*, 110010 (1988).

14. USSR Pat. SU 1,325,050 (July 23, 1987; V. F. Pozdnev to Institute of Biological and Medical Chemistry, Academy of Medical Sciences, USSR); *C.A., 108*, 204496 (1988).
15. Jpn. Kokai Tokkyo Koho JP 61,221,155 [86,221,155] (Oct. 1, 1986; K. Iwakura to Fuji Photo Film Co., Ltd.); *C.A., 106*, 121402 (1987).
16. Czech Pat. CS 268,474 (Aug. 31, 1990; to J. Prachar and M. Bednarova); *C.A., 114*, 207030 (1991).
17. Jpn. Kokai Tokkyo Koho JP 04,193,846 [92,193,846] (July 13, 1992; K. Tateishi and N. Yanagihara to Fuji Photo Film Co., Ltd.); *C.A., 117*, 212059 (1992).
18. Jpn. Kokai Tokkyo Koho JP 4,270,239 [92,270,239] (Sept. 25, 1992; K. Tateichi, N. Yanagihara, T. Sugiyama, and S. Ishige to Fuji Photo Film Co., Ltd.); *C.A., 118*, 147223 (1993); and Jpn. Kokai Tokkyo Koho JP 4,270,246 [92,270,246] (Sept. 25, 1992; K. Tateishi and N. Yanagihara to Fuji Photo Film Co., Ltd.); *C.A., 118*, 124108 (1993).
19. Ger. Offen. DE 2,905,974 (Aug. 28, 1980; J. Westphal to Bayer A.-G.); *C.A., 93*, 206151 (1980).
20. Ger. Offen. DE 3,005,940 (Sept. 10, 1981; G. Hansen, H. J. Kolbinger, and G. Zeidler to Bayer A.-G.); *C.A., 95*, 188634 (1981).
21. Ger. Offen. DE 2,926,761 (Jan. 22, 1981; W. Mennicke and J. Westphal to Bayer A.-G.); *C.A., 94*, 158310 (1981).
22. Ger. Offen. DE 3,034,686 (Apr. 2, 1981; H. Hollinger to Sandoz A.-G.); *C.A., 95*, 44723 (1981).
23. Ger. Offen. DE 3,037,157 (May 6, 1982; K. Rosenbusch to Hoechst A.-G.); *C.A., 97*, 57118 (1982).
24. Eur. Pat. Appl. EP 65,282 (Nov. 24, 1982; E. Schaffner and D. Lach to BASF A.-G.); *C.A., 98*, 91031 (1983).
25. Ger. Offen. DE 2,504,068 (Aug. 7, 1975; S. S. Papa and R. Ferrario to Aciende Colori Nazionali Affini ACNA S.p.A.); *C.A., 83*, 165808 (1975).
26. Ger. (East) DD 236, 328 (June 4, 1986; W. Hepp, S. Oschatz, K. H. Steinert, S. Biermann, G. Borchhardt, and E. Richter to VEB Chemiekombinat Bitterfeld); *C.A., 106*, 86218 (1986); and Ger. (East) DD 236,330 (June 4, 1986; G. Knoechel, W. Specht, S. Oschatz, and K. H. Steinert to VEB Chemiekombinat Bitterfeld); *C.A., 106*, 68732 (1986).
27. Ger. Offen. DE 3,503,588 (Aug. 7, 1986; G. Hansen, U. Bergmann, and G. Zeidler to BASF A.-G.); *C.A., 106*, 34617 (1987).
28. U.S. Pat. 4,760,134 (July 26, 1988; F. Meininger and H. H. Steuernagel to Hoechst Celanese Co.).
29. Ger. (East) DD 300,238 (May 27, 1992; F. Kleine, K. Groth, R. Klaus, W. Dassler, and I. Mathes to Chemie A. -G. Bitterfeld-Wolfen); *C.A., 118*, 40815 (1993).
30. Jpn. Kokai Tokkyo Koho JP 55,151,055 [80,151,055] (Nov. 25, 1980; to Toray Industries, Inc.); *C.A., 95, 44670* (1981).
31. Eur. Pat. Appl. EP56,157 (July 21, 1982; H. Harnisch to Bayer A.-G.); *C.A., 97*, 146460 (1982).
32. Ger. Offen. DE 2,924, 413 (Dec. 1, 1983; to General Electric Co.).
33. PCT Int. Appl. WO 87,003,589 (June 18, 1987; D. Robinson to BioCarb AB); *C.A., 108*, 150308 (1988).
34. *Kodak Laboratory Chemicals Bull., 55(1)*, 1 (1984).
35. Jpn. Kokai Tokkyo Koho JP 60,036,495 [85,036,495] (Feb. 25, 1985; to Funai Pharmaceutical Industries, Ltd.); *C.A., 102*, 221154 (1985).
36. J. J. M. Lambert, D. R. Schumacher, and D. C. Neckers, *J. Am. Chem. Soc., 106*, 5879 (1984); also *Aldrichimica Acta, 19(3)*, 77 (1986).
37. Eur. Pat. Appl. EP 201,751 (Nov. 20, 1986; J. R. Fino and M. T. Shipchandler to Abbott Laboratories, Inc.); *C.A., 106*, 192343 (1987).
38. U.S. Pat. 4,918,200 (Apr. 17, 19990; B. C. Ackles to Huels America Inc.).
39. Jpn. Kokai Tokkyo Koho JP 03,120,269 [91,120,269] (May 22, 1991; H. Higuchi, N. Yamukai, and S. Matsuo to Idemitsu Kosan Co., Ltd.); *C.A., 115*, 183100 (1991).
40. Jpn. Kokai Tokkyo Koho JP 04,051,185 [92,051,185] (Feb. 19, 1992; S. Sugawara to Nissan Jishoda K. K.); *C.A., 117*, 181976 (1992).
41. Jpn. Kokai Tokkyo Koho JP 80,164,251 (Dec. 20, 1980; to Ricoh Co., Ltd.); *C.A., 95*, 8789 (1981).
42. Ger. Offen. DE 2,746,551 (Apr. 26, 1979; S. Scheler and H. Schaefer to Hoechst A.-G.); *C.A., 91*, 149457 (1979).

43. Jpn. Kokai Tokkyo Koho JP 56,110,930 [81,110,930] (Sept. 2, 1981; to Taoka Chemical K.K.); *C.A., 96*, 94998 (1982).
44. Jpn. Kokai Tokkyo Koho JP 52,097,725 (Aug. 16, 1977; I. Kawamura and K. Futaki to Mitsubishi Paper Mills, Ltd.); *C.A., 88*, 56963 (1978).
45. Jpn. Kokai Tokkyo Koho JP 60,198,533 [85,198,533] (Oct. 8, 1985; Y. Kasuka to Oji Paper K.K.); *C.A., 104*, 159675 (1986).
46. Jpn. Kokai Tokkyo Koho JP 60,172,584 [85,172,584] (Sept. 6, 1985; K. Yabuta and S. Moriahita to Mitsubishi Paper Mills, Ltd.); *C.A., 104*, 139371 (1986).
47. USSR Pat. SU 1,136,963, SU 1,136,964 and SU 1,136,965 (All Jan. 30, 1985; S. A. Romashkov, V. N. Smirnov, A. M. Idiatullin, *et al.* to Paper Research Institute); *C.A., 102*, 229519-20-21 (1985).
48. Jpn. Kokai Tokkyo Koho JP 86,040,977 (Sept. 12, 1986; H. Tamura and K. Sakai to Ricoh K.K.).
49. Jpn. Kokai Tokkyo Koho JP 62,253,486 [87,253,486] (Apr. 26, 1987; R. Mando, S. Okimoto, K. Ishida, and T. Okamoto to Kanzaki Paper Manufacturing K.K.); *C.A., 108*, 229731 (1988).
50. Jpn. Kokai Tokkyo Koho JP 02,015,253 [90,015,253] (Jan. 18, 1990; H. Shomura, H. Shimogawara, and M. Tomiki to Somar Corp.); *C.A., 113*, 32012 (1990).
51. Ger. Offen. DE 3,018,546 (Nov. 27, 1980; R. J. Schmidt and N. M. Hung to Sterling Drug, Inc.); *C.A., 94*, 123117 (1981).
52. Jpn. Kokai Tokkyo Koho JP 55,133,992 [80,133,992] (Oct. 18, 1980; to Ricoh K. K.); *C.A., 94*, 165636 (1981).
53. Jpn. Kokai Tokkyo Koho JP 57,005,757 [82,005,757] (Jan. 12, 1982; to Hodogaya Chemical Industries K. K.); *C.A., 96*, 182767 (1982).
54. Jpn. Kokai Tokkyo Koho JP 58,065,696 [83,065, 696] (Apr. 19, 1983; to Mitsubishi Paper Mills, Ltd.); *C.A., 99*, 46056 (1983).
55. Jpn. Kokai Tokkyo Koho JP 58,120,670 [83,120,670] (July 18, 1983; to Ricoh K.K.); *C.A., 100*, 23561 (1984).
56. Jpn. Kokai Tokkyo Koho JP 59,127,794 [84,127,794] (July 23, 1984; to Hodogaya Chemical Industries K.K.); *C.A., 101*, 219974 (1984).
57. Ger. Offen. DE 3,018,546 (Nov. 27, 1980; P. J. Schmidt and W. M. Hung to Sterling Drug Co.); *C.A., 94*, 123117 (1981).
58. Belg. Pat. BE 904,878 (June 5, 1986; to Shin Nisso Kako Co., Ltd.); *C.A., 107*, 49668 (1987).
59. Jpn. Kokai Tokkyo Koho JP 62,070,350 [87,070,350] (March 31, 1987; N. Yonese, H. Omura, K. Hirakara, and M. Kondo to Kanzaki Paper Manufacturing K.K.); *C.A., 108*, 23350 (1988).
60. U.S. Pat. 4,728,633 (March 1, 1988; M. Satomura, K. Iwakura, and A. Igarashi to Fuji Photo Film Co., Ltd.).
61. Jpn. Kokai Tokkyo Koho JP 87,061,616 (Dec. 22, 1987; to Sumitomo Chemical Industries K.K.).
62. Jpn. Kokai Tokkyo Koho JP 63,009,573 [88,009,573] (Jan. 16, 1988; T. Nishijima, T. Takakura, T. Yamaguchi, and K. Imamura to Showa Denko K.K.); *C.A., 109*, 139306 (1988).
63. Jpn. Kokai Tokkyo Koho JP 63,112,190 [88,112,190] (May 17, 1988; T. Harada and T. Watanabe to Fuji Photo Film K.K.); *C.A., 109*, 201308 (1988).
64. Jpn. Kokai Tokkyo Koho JP 88,052,590-91-92 (all Oct. 19, 1988; to Sumitomo Chemical Industries).
65. Eur. Pat. Appl. EP 286,116 (Oct. 12, 1988; T. Satake, T. Minami, T. Nagai, and F. Fujimura to Jujo Paper, Co.,Ltd.); *C.A., 110*, 145077 (1989). Similar is Jpn. Kokai Tokkyo Koho JP 63,159,384 [88,159,384] (July 2, 1988; S. Yoshinaka, Y. Onishi, and T. Obitsu to Shin Nisso Kako K.K.); *C.A., 110*, 31502 (1984).
66. Jpn. Kokai Tokkyo Koho JP 82,012,692 (Jan. 22, 1982; to Sumitomo Chemical Industries, Ltd.); *C.A., 97*, 101739 (1982).
67. Jpn. Kokai Tokkyo Koho JP 88,052,589 (Oct. 19, 1988; to Sumitomo Chemical Industries, Ltd.).
68. Jpn. Kokai Tokkyo Koho JP 63,251,278, JP 63,251,279 and JP 63,251,280 (all dated Oct. 18, 1988; to Fuji Photo Film K.K.).
69. Eur. Pat. Appl. EP 340,169 (Nov. 2, 1989; I. -J. Fletcher to Ciba-Geigy A.-G.); *C.A., 113*, 213867 (1990).
70. Ger. Offen. DE 3,832,756 (Apr. 13, 1989; R. Zink to Ciba-Geigy A.-G.); *C.A., 112*, 55832 (1990).

71. Jpn. Kokai Tokkyo Koho JP 03,063,254 [91,063,254] (March 19, 1991; M. Nakatsuka, A. Otsuji, K. Hasegawa, M. Tagaki, and T. Yamaguchi to Mitsui Toatsu Chemicals, Inc.); *C.A., 115*, 161212 (1991).
72. U.S. Pat. 4,800,193 (Jan. 24, 1989; K. Saeki, M. Takashima, K. Iwakura, T. Sugiyama, and M. Satomura to Fuji Photo Film Co., Ltd.).
73. Eur. Pat. Appl. EP 356,199 (Feb. 28, 1990; M. Yamaguchi, K. Kikkawa, and M. Gonda to Hodogaya Chemical Co., Ltd.); *C.A., 113*, 193452 (1990).
74. Eur. Pat. Appl. EP 384,895 (Aug. 29, 1990; R. Zink to Ciba-Geigy A.-G.); *C.A., 114*, 81806 (1991).
75. Jpn. Kokai Tokkyo Koho JP 02,251,484 [90, 251,484] (Oct. 9, 1990; S. Azuma, M. Takashima, and M. Satomura to Fuji Photo Film Co., Ltd.); *C.A., 114*, 124594 (1991).
76. Jpn. Kokai Tokkyo Koho JP 03,114,878 [91,114,878] (May 16, 1991; B. Sawano, M. Matsumoto, and Y. Iwasaki to Yamamoto Kasei K. K.); *C.A., 115*, 185374 (1991).
77. Eur. Pat. Appl. EP 420,024 (Apr. 3, 1991; A. Otsuji, M. Nakatsuka, K. Hasegawa, M. Takagi, and A. Yamaguchi to Mitsui Toatsu Chemicals, Inc.); *C.A., 115*, 171000 (1991).
78. Ger. Offen. DE 4,002,929 (Aug. 8, 1991; T. Gessner and U. Meyer to BASF A.-G.); *C.A., 115*, 183136 (1991).
79. U.S. Pat. 5,051,333 (Sept. 24, 1991; N. Yanagihira, K. Iwakura, T. Endo, and K. Saeki to Fuji Photo Film Co., Ltd.)
80. Jpn. Kokai Tokkyo Koho JP 03,227,289 [91,227,289] (Oct. 8, 1991; K. Araki, N. Yamagihara, M. Takashima, S. Azuma, and M. Satomura to Fuji Photo Film Co., Ltd.); *C.A., 116*, 184659 (1992).
81. Jpn. Kokai Tokkyo Koho JP 03,244,587 [91,244,587] (Oct. 31, 1991; H. Kawai, K. Gendai, M. Taniguchi, and K. Tsunemitsu to Yamada Kagaku Kogyo K. K.); *C.A., 116*, 204615 (1992).
82. Jpn. Kokai Tokkyo Koho JP 03,231,892 [91,231,892] (Oct. 15, 1991; N. Arai, K. Watanabe, and N. Kanda to Kanzaki Paper Manufacturing Co.); *C.A., 116*, 204607 (1992).
83. Jpn. Kokai Tokkyo Koho JP 03,126,579 [91,126,579] (May 29, 1991; K. Tanaka to Tomeogawa Paper Co., Ltd.); *C.A., 115*, 244182 (1991).
84. Jpn. Kokai Tokkyo Koho JP 03,063,172 [91,063,172] (March 19, 1991; M. Nakatsuka, A. Otsuji, K. Hasegawa, M. Takagi, and T. Yamaguchi to Mitsui Toatsu Chemicals, Inc.); *C.A., 115*, 171016 (1991).
85. Jpn. Kokai Tokkyo Koho JP 04,169,287 [92,169,287] (June 17, 1992; T. Ueda and K. Maruta to Ricoh Co., Ltd.); *C.A., 117*, 261731 (1992).
86. Jpn. Kokai Tokkyo Koho JP 04,115,987 [92,115,987] (Apr. 16, 1992; K. Katsura, H. Kubo, and M. Kinoshita to Nippon Soda Co., Ltd.); *C.A., 117*, 181850 (1992).
87. Jpn. Kokai Tokkyo Koho Jp 04,117,384 [92,117,384] (Apr. 17, 1992; M. Nakatsuka, A. Otsuji, K. Hasegawa, K. Yoshikawa, and T. Yamaguchi to Mitsui Toatsu Chemicals, Inc.); *C.A., 117*, 171224 (1992).
88. Eur. Pat. Appl. EP 468,821 (Jan. 29, 1992; K. Abureda and M. Akagi to Taoka Chemical Co., Ltd.); *C.A., 116*, 237270 (1992).
89. Jpn. Kokai Tokkyo Koho JP 04,330,077 [92,330,077] (Nov. 18, 1992; H. Terao, S. Wada, and M. Matsumoto to Yamamoto Kasei K.K.); *C.A., 118*, 263949 (1993).
90. Jpn. Kokai Tokkyo Koho JP 04,224,991 [92,224,991] (Aug. 14, 1992; K. Totani, K. Wakata, Y. Abe, and S. Shioi to Kanzaki Paper Manufacturing Co., Ltd.); *C.A., 118*, 30082 (1993).
91. U.S. Pat. 5,166,350 (Nov. 24, 1992; D. Bedekovic, J. L. Pool. B. J. Williams, R. Garner, and J. B. Henshall to Ciba-Geigy Corp.).
92. Brit. Pat. Appl. GB 2,154,014 (Aug. 29, 1985; K. Iwakura and T. Sugiyama to Fuji Photo Film K.K.); *C.A., 104*, 120093 (1986).
93. U.S. Pat. 4,342,473 (Aug. 3, 1982; C. H. Chang to Champion International Corp.).
94. Jpn. Kokai Tokkyo Koho JP 57,008,194 [82,008,194] (Jan. 16, 1982; to Ricoh K.K.); *C.A., 97*, 47203 (1982).
95. Jpn. Kokai Tokkyo Koho JP 58,162,380 [83,162,380] (Sept. 27, 1983; to Kohjin K.K.); *C.A., 100*, 219110 (1984).

96. Ger. Offen. DE 3,317,559 (Nov. 17, 1983; J. C. Petitpierre, H. Nachbur, and A. Wuermli to Ciba-Geigy A. -G.); *C.A., 101*, 31191 (1984).
97. Jpn. Kokai Tokkyo Koho JP 58,160,189 [83,160,189] (Sept. 22, 1983; to Ricoh K.K.); *C.A., 101*, 15118 (1984).
98. Jpn. Kokai Tokkyo Koho JP 60,222,289 [85,222,289] (Nov. 6, 1985; K. Taniguchi to Ricoh K.K.); *C.A., 104*, 139409 (1986).
99. U.S. Pat. 4,538,164 (Aug. 27, 1985; N. Takigawa, H. Watanabe, K. Kawasaki, K. Kubo, M. Hakiri, and E. Kawamura to Ricoh Co., Ltd.).
100. U.S. Pat. 4,633,277 (Dec. 30, 1986; T. Suzuki, F. Hama, and M. Kondo to Kanzaki Paper Manufacturing Co., Ltd.).
101. Brit. UK Pat. Appl. GB 2,154,014 (Nov. 26, 1986; K. Iwakura and T. Sugiyama to Fuji Photo Film Co., Ltd.); *C.A., 104*, 120093 (1986).
102. Jpn. Kokai Tokkyo Koho JP 61,040,185 [86,040,185] (Feb. 26, 1986; K. Iwakuru, T. Sugiyama, and S. Ishige to Fuji Photo Film Co., Ltd.); *C.A., 105*, 70219 (1986).
103. Jpn. Kokai Tokkyo Koho JP 61,109,757 [86,109,757] (May 28, 1986; H. Harada and H. Maki to Sumitomo Chemical Industries K.K.); *C.A., 106*, 205294 (1987).
104. U.S. Pat. 4,771,034 (Sept. 13, 1988; K. Ikeda, K. Iwakura, and M. Satomura to Fuji Photo Film Co., Ltd.).
105. Jpn. Kokai Tokkyo Koho JP 01,165,550 [89,165,550] (June 29, 1989; K. Nigorikawa and M. Satomura to Fuji Photo Film Co., Ltd.); *C.A., 112*, 76617 (1989).
106. Jpn. Kokai Tokkyo Koho JP 02,202,479 [90,202,479] (Aug. 10, 1990; Y. Takahashi, K. Toyofuku, and A. Iwasaki to Oji Paper Co., Ltd.); *C.A., 115*, 82337 (1991).
107. Jpn. Kokai Tokkyo Koho JP 02,281,992 [90,281,992] (Nov. 19, 1990; T. Tsuchida, F. Semaya, T. Meguro, and M. Kondo to Kanzaki Paper Manufacturing Co., Ltd.); *C.A., 114*, 257033 (1991).
108. Jpn. Kokai Tokkyo Koho JP 03,063,179 [91,063,179] (March 19, 1991; A. Shirai, Y. Takahashi, and K. Toyofuku to Oji Paper Co., Ltd.); *C.A., 115*, 194275 (1991).
109. Jpn. Kokai Tokkyo Koho JP 03,138,195 [91,138,195] (June 12, 1991; K. Hayakawa and H. Shimada to Ricoh Co., Ltd.); *C.A., 116*, 117282 (1992).
110. Jpn. Kokai Tokkyo Koho JP 03,011,030 [91,011,030] (Jan. 18, 1991; M. Tamura, M. Usui, and K. Hamada to Sumitomo Chemical Co., Ltd.); *C.A., 115*, 8291 (1991).
111. Jpn. Kokai Tokkyo Koho JP 03,114,776 [91,114,776] (May 15, 1991; S. Higeta and H. Furuya to Ricoh Co., Ltd.); *C.A., 116*, 184629 (1992).
112. Jpn. Kokai Tokkyo Koho JP 04,153,080 [92,153,080] (May 26, 1992; H. Furuya and H. Shimada to Ricoh Co., Ltd.); *C.A., 117*, 223151 (1992).
113. R. M. Peffer, A. H. Rehn, and M. Grose, *Res. Discl., 162*, 31 (1977); *C.A., 88*, 81776 (1978).
114. Ger. Offen. DE 2,644,194 (Apr. 27, 1977; M. Fujihara, S. Matsuo, M. Kawasaki, T. Masukawa, and Y. Kaneko to Konishiroku Photo K.K.); *C.A., 87*, 109394 (1977).
115. Ger. Offen. DE 3,223,699 (Dec. 29, 1983; G. Krauss to Agfa-Gevaert A.-G.); *C.A., 101*, 46209 (1984).
116. Eur. Pat. Appl. EP 69,066 (Jan. 5, 1983; F. H. Howell to Ciba-Geigy A.-G.); *C.A., 98*, 215312 (1983).
117. Eur. Pat. Appl. EP 71,570 (Feb. 9, 1983; S. R. Postle to Ciba-Geigy A.-G.); *C.A., 99*, 184963 (1983).
118. U.S. Pat. 4,429,035 (Jan. 31, 1984; E. Kato, N. Ohki, and J. Okawa to Fuji Photo Film Co., Ltd.).
119. Jpn. Kokai Tokkyo Koho JP 61,129,640 [86,129,640] (June 17, 1986; H. Sakamoto, T. Yoshizawa, and M. Ishikawa to Konishiroku Photo Industry Co., Ltd.); *C.A., 106*, 25698 (1987).
120. U.S. Pat. 4,618,574 (Oct. 21, 1986; A. Cavallaro to Eastman Kodak Co.).
121. U.S. Pat. 4,661,601 (Apr. 27, 1987; F. H. Howell to Ciba-Geigy A.-G.).
122. Jpn. Kokai Tokkyo Koho JP 63,218,953 [88,218,953] (Sept. 12, 1988; N. Tachibana and S. Nakagawa to Konishiroku Photo K. K.); *C.A., 110*, 85344 (1989).
123. Jpn. Kokai Tokkyo Koho JP 63,044,656 [88,044,656] (Feb. 25, 1988; K. Ando, M. Yagihara, and H. Naruse to Fuji Photo Film Co., Ltd.); *C.A., 109*, 101739 (1988).
124. Jpn. Kokai Tokkyo Koho JP 62,288,827 [87,288,827] (Dec. 15, 1987; T. Harada and K. Sato to Fuji Photo Film Co., Ltd.); *C.A., 109*, 29995 (1988).

125. Jpn. Kokai Tokkyo Koho JP 03,252,652 [91,252,652] (Nov. 11, 1991; M. Kuroishi and S. Ono to Fuji Photo Film Co., Ltd.); *C.A.*, *117*, 160728 (1992).
126. Eur. Pat. Appl. EP 457,723 (Nov. 21, 1991; D. G. Leppard, D. H. Steinberg, and H. Dubas to Ciba-Geigy A.-G.); *C.A.*, *116*, 140028 (1992).
127. Eur. Pat. Appl. EP 54,258 (June 23, 1982; Y. Takeshi, G. Kinoshi, and U. Masafumi to Konishiroku Photo Industry Co., Ltd.); *C.A.*, *98*, 63341 (1983).
128. U.S. Pat. 4,424,270 (Jan. 3, 1984; F. Erdman and U. Simon to Hoechst A.-G.).
129. Y. Saotome, H. Gokan, K. Saigo, M. Suzuki, and Y. Ohnishi (all with NEC Corp.), *J. Electrochem. Soc.*, *132*, 909 (1985); see also Ger. Offen. DE 3,515,210 (Nov. 17, 1988; to NEC Corp.).
130. U.S. Pat. 4,624,908 (Nov. 25, 1986; G. Schwartzkopf to J. T. Baker Chemical Co.); similar light-sensitive compounds were disclosed in Ger. (East) DD 247,759 (July 15, 1987; to Humboldt University Berlin).
131. Eur. Pat. Appl. EP 227,487 (July 1, 1987; I. Nozue, M. Takatori, Y. Harita, and K. Honda to Japan Synthetic Rubber Co., Ltd.); *C.A.*, *108*, 140771 (1988).
132. U.S. Pat. 4,698,286 (Oct. 6, 1987; W. R. Messer to Hercules, Inc.).
133. Jpn. Kokai Tokkyo Koho JP 63,080,247 [88,080,247] (Apr. 11, 1988; A. Sugimoto, K. Ozaki, I. Uehara, and K. Suko to Hitachi K.K.); *C.A.*, *109*, 201511 (1988).
134. Jpn. Kokai Tokkyo Koho JP 03,192,360 [91,192,360] (Aug. 22, 1991; A. Toko, N. Sashita, and E. Takeuchi to Sumitomo Bakelite Co., Ltd.); *C.A.*, *116*, 48931 (1992).
135. Jpn. Kokai Tokkyo Koho JP 03,240,062 [91,240,062] (Oct. 25, 1991; A. Nagashima and K. Uenishi to Fuji Photo Film Co., Ltd.); *C.A.*, *116*, 140155 (1992).
136. Jpn. Kokai Tokkyo Koho JP 03,128,959 [91,128,959] (May 31, 1991; T. Kajita, C. Okuda, T. Mura, and T. Shimokawa to Japan Synthetic Rubber Co., Ltd.); *C.A.*, *116*, 48927 (1992).
137. U.S. Pat. 5,019,278 (May 28, 1991; M. A. Toukhy and A. T. Jeffries, III to Olin Hunt Specialty Products, Inc.).
138. Jpn. Kokai Tokkyo Koho JP 04,149,442 [92,149,442] (May 22, 1992; S. Kobuichi, K. Kasuya, M. Nunomura, M. Hashimoto, K. Kato, and A. Isobe to Hitachi Chemical Co., Ltd.); *C.A.*, *117*, 223116 (1992).
139. Jpn. Kokai Tokkyo Koho JP 04,274,431 [92,274,431] (Sept. 30, 1992; K. Uenishi, F. Nishiyama, and T. Kokubo to Fuji Photo Film Co., Ltd.); *C.A.*, *118*, 202137 (1993).
140. U.S. Pat. 5,124,228 (June 23, 1992; Y. Uetani, M. Hanabata, H. Nakanishi, K. Kuwana, and F. Oi to Sumitomo Chemical Co., Ltd.).
141. Jpn. Kokai Tokkyo Koho JP 59,008,270 [84,008,270] (Feb. 23, 1984; to Philips Gloeilampen, N.V.).
142. Jpn. Kokai Tokkyo Koho JP 60,067,586 [85,067,586] (Apr. 17, 1985; to Ajinmoto K.K.); *C.A.*, *103*, 186952 (1985).
143. U.S. Pat. 4,643,842 (Feb. 17, 1987; to Seiko Instrument and Electronics Co., Ltd.).
144. Jpn. Kokai Tokkyo Koho JP 05,017,405 [93,017,405] and JP 05,017,406 [93,017,406] (both dated Jan. 26, 1993; Y. Fukushima, A. Ishikawa, and Y. Horikawa to Kanebo Ltd.); *C.A.*, *118*, 244746 and 244747 (1993).
145. Jpn. Kokai Tokkyo Koho JP 79,037,803 (Nov. 17, 1979; to Hitachi Maxell K.K.).
146. U.S. Pat. 4,052,536 (Oct. 4, 1977; N. N. Lichtin and P. D. Wildes to University of Boston).
147. Ger. Offen. DE 3,001,877 (July 23, 1981; F. Arndt, U. Claussen, H. Harnisch, and C. W. Schellhammer to Bayer A.-G.); *C.A.*, *95*, 141830 (1981).
148. Jpn. Kokai Tokkyo Koho JP 57,068,455 [82,068,455] (June 2, 1982; to Canon Copy K.K.).
149. U.S. Pat. 4,471,041 (May 9, 1984; G. Baranyi, R. A. Burt, C. K. Hsiao, P. M. Kazmeier, K. M. Carmichael, and A. M. Horgan to Xerox Corp.); U.S. Pat. 4,524,218 (June 18, 1985; G. Baranyi, P. M. Kazmeier, C. K. Hsiao, and R. A. Burt to Xerox Corp.).
150. Jpn. Kokai Tokkyo Koho JP 61,063,482 [86,063,482] (Apr. 1, 1986; Y. Nishimura, H. Kawada, M. Haruta, Y. Hirai, N. Mochizuki, and T. Nakagiri to Canon K. K.); *C.A.*, *105*, 200582 (1986) (and seven consecutively numbered patents).
151. Jpn. Kokai Tokkyo Koho JP 62,249,949 [87,249,949] (Oct. 30, 1987; Y. Akasaki, A. Tokida, I. Kin, S. Saeki, and K. Torigoe to Fuji Xerox Co., Ltd.); *C.A.*, *109*, 170008 (1988).

152. Jpn. Kokai Tokkyo Koho JP 63,257,765 [88,257,765] and JP 63,257,767 [88,257,767] (both Oct. 25, 1988; Y. Nishimura *et al.* to Ricoh Co., Ltd.); *C.A., 110*, 222565 and 125414 (1989).
153. Ger. Offen. DE 3,904,264 (Aug. 4, 1990; S. Mayer and S. Schmidt to Heinrich Hertz-Institute Fuer Nachrichtungstechnik Berlin G.m.b.H.); *C.A., 115*, 18306 (1991).
154. Eur. Pat. Appl. EP 392,356 (Oct. 17, 1990; W. Breitschaft and H. Bruder to BASF A.-G.); *C.A., 115*, 18557 (1991).
155. Ger. Offen. DE 4,023,860 (Jan. 30, 1992; H. Guesten to Kernforschungszentrum Karlsruhe G.m.b.H.); *C.A., 116*, 151757 (1992).
156. Jpn. Kokai Tokkyo Koho JP 04,159,559 [92,159,559] (June 2, 1992; Y. Watanuki and N. Kosho to Fuji Electric Co., Ltd.); *C.A., 117*, 261644 (1992).
157. Jpn. Kokai Tokkyo Koho JP 04,225,364 [92,225,364] (Aug. 14, 1992; A. Karasawa, N. Ito, and T. Ogichi to Mitsui Toatsu Chemicals, Inc.); *C.A., 118*, 29980 (1993).

11

Additives of Many Types

11.1. INTRODUCTION

Resorcinol-based functional additives provide enhanced performance not only for all kinds of organic materials, but they are used in aqueous systems as well (i.e., corrosion inhibitors, electrolytes). This chapter highlights the wide variety of applications. Indeed, it is inevitable that there is an overlap of categories. Where that has occurred, where materials that fit here have been mentioned in other categories, appropriate reference will be made; and the index can be used to locate them all.

11.2. COMMERCIAL PRODUCTS

11.2.1. Ultraviolet Light Absorbers

All polymers undergo a deterioration under the influence of ultraviolet (UV) light and air, modulated by temperature and other environmental conditions. This degradation can be a discoloration or the actual loss of physical properties, e.g., the material may become brittle. In most cases, stabilizers can be added not only to extend the useful life of the polymer, but also, nowadays, to reduce the amount of material to be disposed of in dumps or to reduce the amount of plastics for recycle. Recycling can be costly and the recycled material is usually not quite as good as the virgin material.

The need for stabilizers arose first as light-sensitive plastics, from polystyrene to polyethylene, polypropylene, and many others, achieved acceptance for may uses. Even seemingly inert materials contained impurities such as catalyst residues, processing aids, and pigments which could initiate the formation of radicals, hydroperoxides, ketones, and other causes of eventual deterioration. Indeed, as higher plastics processing temperatures began to be used in this industry, the potential for degradation was accelerated.

The 4-benzoylresorcinols (2,4-dihydroxybenzophenones) were among the first commercial UV stabilizers for plastics, emerging in the 1950s. Their success was based on their high absorption coefficient in the 290–360 nm wavelength range, which is the high-energy part of the sunlight reaching the earth. The UV absorption of the benzoylresorcinols is low at and above 400 nm, so that, at least at the low concentrations at which they are used (usually in the 0.25–1.0% range), they add little color to the polymer. These additives are thermally stable in the use range, they are of little health risk, and they are (designed to be) compatible with the particular polymer. These facets, together with cost/performance considerations, account for continued use as well as the development of a wide variety of compounds based on or related to the 2,4-dihydroxybenzophenones.

The effectiveness of the 2-hydroxybenzophenones has been ascribed mostly to preferential UV-light absorption and a keto–enol equilibrium, the ground state being in the hydrogen-bonded phenol form, with formation of a cyclohexadienone form in the (singlet) excited state which rapidly reverts to the phenol form with dissipation of energy in the form of heat, as sketched in Eq. (11.1) for 4-benzoylresorcinol. The rotation of the hydroxyphenyl group and/or the possibility of energy transfer from the excited state (of oxidized) polymer to the UV stabilizer have also been proposed to account for the effectiveness of these additives.

(11.1)

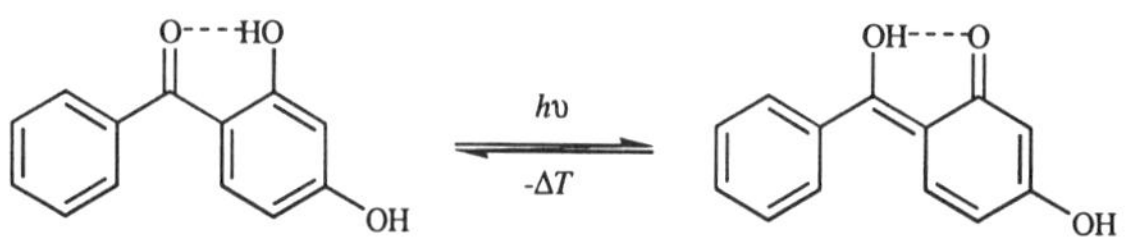

The 4-benzoylresorcinols have been found also to have modest antioxidant activity. All these features have been well discussed in short[1] as well as lengthy[2] reviews. Much academic work has been devoted to this topic. The commercial products of the hydroxybenzophenone type are shown below.

Resorcinol monobenzoate, **11-1**, is a latent stabilizer, active because it undergoes a photo-Fries rearrangement to 4-benzoylresorcinol, **11-2**, on UV-light exposure. The resorcinol monobenzoate is recommended for the stabilization of cellulosics such as nitrocellulose and cellulose acetate, for polystyrene, flexible PVC, and poly(vinylidene chloride).

2,4-Dihydroxybenzophenone (4-benzoylresorcinol), **11-2**, is sold for use in cellulosics, epoxies, polystyrene, and unsaturated polyesters. It minimizes crazing in poly(methyl methacrylate) aircraft windows, it is used in pressure-sensitive adhesives to prevent loss of tack, and it is used to protect alkyd and phenolic varnishes and color photographs.

The sodium sulfonate of benzoylresorcinol, **11-3**, is used to inhibit the yellowing of wool.

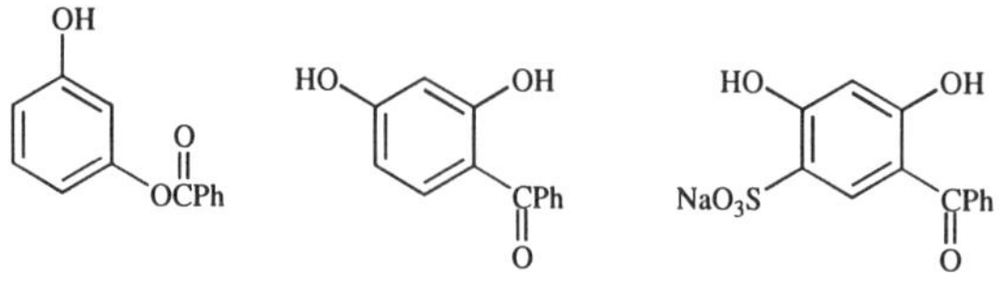

11-1 **11-2** **11-3**

2-Hydroxy-4-methoxybenzophenone, **11-4**, is a UV-light stabilizer used for cellulosics, polyesters, rigid and flexible PVC, polystyrene, poly(vinylidene chloride), acrylics, rubber, paints, and varnishes. The use of the sodium sulfonate of **11-4** in sun screen preparations to protect the skin was already mentioned in Chapter 8.

2-Hydroxy-4-*n*-octoxy- and -4-dodecyloxybenzophenone, **11-5**, are useful in polyolefins such as polyethylene and polypropylene as well as across the whole spectrum of polymers. By weight or dollar volume, these additives are the best sellers in their class.

2-Hydroxy-4-(2-hydroxyethoxy)benzophenone, **11-6**, is useful for the protection of flexible PVC, poly(acrylonitrile/butadiene/styrene) [ABS], polystyrene, polycarbonate, and poly(vinyl butyrate) [PVB]; it provides permanence as well as excellent UV protection.

A polymer of 2-hydroxy-4-(acroylethoxy)benzophenone, **11-7**, is available which is recommended for polyolefins and rigid PVC. This stabilizer shows reduced volatility/migration over the earlier 2-hydroxybenzophenones.

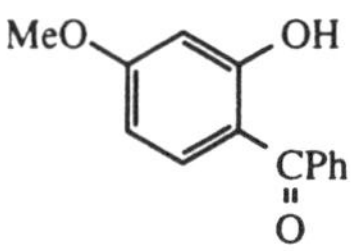

11-4

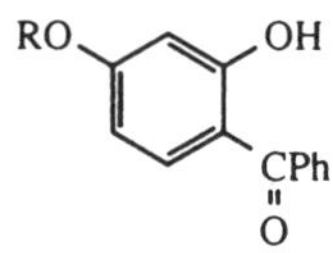

11-5
(where R= n-C_8H_{17} or n-$C_{12}H_{25}$)

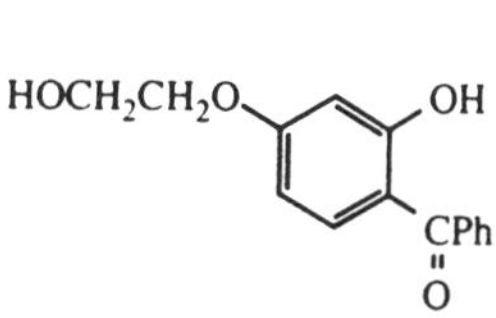

11-6

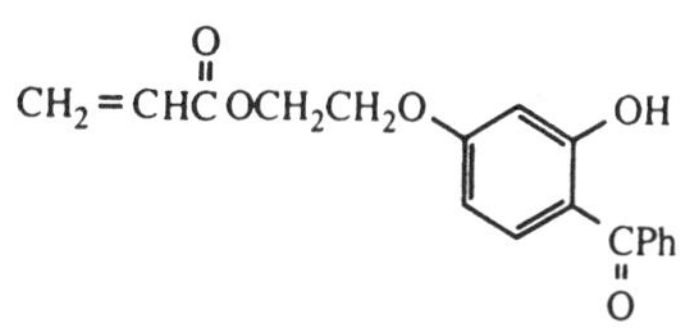

11-7

2,2′,4,4′-tetrahydroxybenzophenone, **11-8**, is a good UV stabilizer for epoxies, polyesters, elastomers, leather, textiles, dyes, pigments, photographic color prints, and cosmetics (e.g., it prevents viscosity increases in surfactant shampoos; in hair sprays it protects dyed/tinted hair from fading).

2,2′-dihydroxy-4-methoxybenzophenone, **11-9**, and 2,2′-dihydroxy-4,4′-dimethoxybenzophenone, **11-10**, are also recommended for epoxies and polyesters.

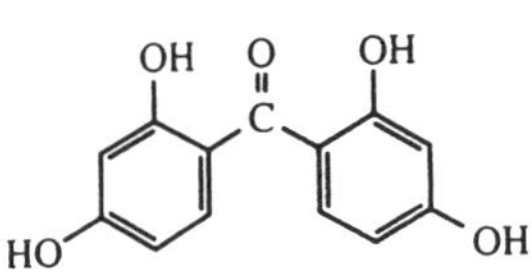

11-8

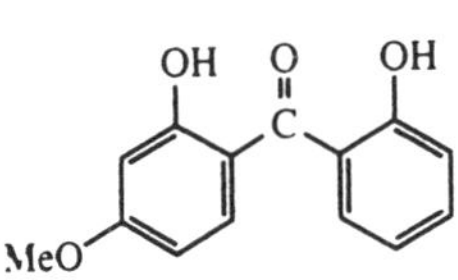

11-9

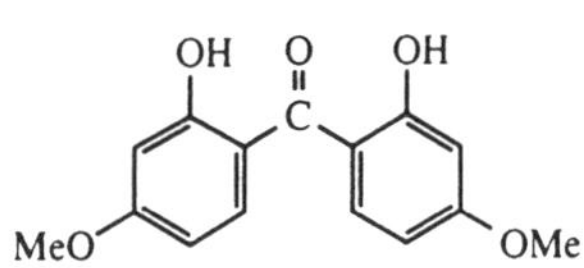

11-10

The dimethoxy compound, **11-10**, imparts a slightly yellow color to plastics; it is useful particularly for chlorine-containing polyesters, polyurethanes, acrylics, PVC, and oil-based paints/varnishes, and for extending the life of fluorescent pigments in coatings. A sodium disulfonate, **11-11**, of the dimethoxy compound is available for use in cosmetics, jute and wool textiles, and in latex paints.

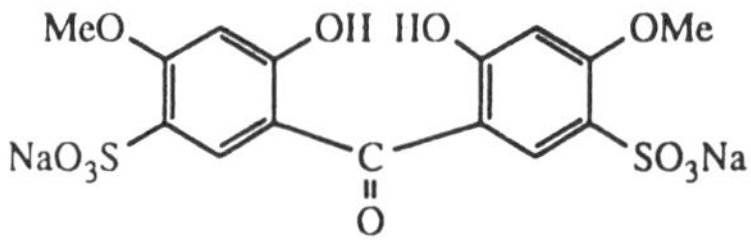

11-11

The choice of a particular stabilizer is based on cost/effectiveness, compatibility, solubility, melting point (ranging from 52–195 °C for the above, unsulfonated compounds), volatility, and other factors, such as the formulation of the total additive package including, e.g., antioxidants. Thus, the UV-stabilizer supplier must develop appropriate information and the user will usually further evaluate recommendations.

In 1991, about 1.5 mill lb. of the hydroxybenzophenone-type UV stabilizers were sold in the U.S. alone, at a list price of *ca* \$7.70–15.90/lb.[3] Resorcinol monobenzoate is less costly and is not included in these figures. Likely, worldwide use of these materials was 3–4 times the U.S. usage.

The largest volume of the 2-hydroxybenzophenone-type UV stabilizers were the 2-hydroxy-4-(long-chain alkoxy)benzophenones, **11-5**, since they were useful in polyolefins, including polypropylene, which have large volume/good growth rates. In general, it is believed that about 1/3 of all polymers produced may be UV stabilized because of their use in outdoor exposure, in filtered daylight, or in exposure to fluorescent lighting.

Of course, the 2-hydroxybenzophenone-type UV stabilizers did not have the market to themselves. Over the years, other UV absorbers were developed (e.g., the benzotriazolyl-alkylphenols and the alkyl 3,3-diphenyl-2-cyanoacrylates); the quenching agents such as nickel 2,2′-thiobis-(4-*tert*-octyl)phenolate (with a *n*-butylamine ligand added), the hydroperoxide decomposers (e.g., metal chelates and dithiophosphates) which overlap in function with antioxidants, and the radical scavenger-type UV-light stabilizers of the hindered amine (HALS) structure [e.g., bis-(2,2′,6,6′-tetramethyl-4-piperidyl)sebacate] which often show excellent performance. The latter, HALS-type, stabilizers were the newest of the products and became available in good variety, including higher molecular weight products, from the 1970s on. They were more expensive but often effective at lower concentration; these materials show the best growth rates of all UV stabilizers.

Nonetheless, the 2-hydroxybenzophenone-type stabilizers, although apparently not showing much growth, continue to be used, alone or in mixture with other stabilizers. The hydroxybenzophenones are useful particularly where the HALS have been problematic, i.e., in halogen-containing polymers (e.g., PVC) and in sulfur-containing polymers, As regards the nickel chelates, concern has been raised, particularly in Europe, as to their potential toxicity in the environment.

11.2.2. Flame Retardants

With increasing, worldwide concerns about the environmental aspects of halogenated flame retardants (e.g., organic bromine compounds synergized by antimony oxide), great effort has been made both by manufacturers of these additives and by plastics producers to develop and test possible alternatives. In the last 10–15 years or so, increasing attention was paid to alkyl and aryl phosphates and phosphonates, including those derived from resorcinol. It appears, for one, that materials containing the tetraphenyl resorcinol diphosphate, **11-12** (the main constituent of commercial products of this type), which was extensively worked on and patent-covered by GE (General Electric Corp.) and others, are excellent fire retardants. They are commercially attractive products with high growth prospects in polymer alloys.

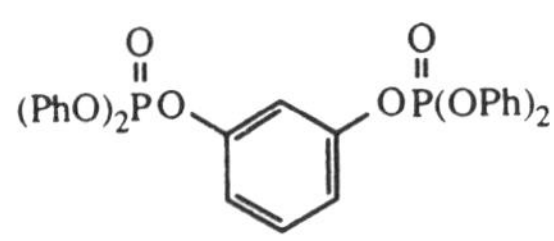

11-12

Such materials have good compatibility with certain polymers, e.g., GE's Noryl resins based on poly(phenylene oxide) [PPO]/polystyrene alloys, at the *ca* 5–30 wt% addition level. Speculatively, the tetraphenyl resorcinol diphosphate compounds may also be compatibilizing agents for different polymers. These diphosphates are liquids of moderate viscosity, they are hydrolysis and heat resistant, they are likely to have low toxicity, they must be EPA approved for this use, they have low color, low volatility (lower than the older triaryl phosphates), and low leachability, they do not give off highly toxic fumes in a fire and release little visible smoke, and they must maintain most of the useful properties of the virgin polymer. The compounds have good melt rheology, thus have favorable processability (possible energy savings). Recyclability of the compounded plastic might also be an important factor, stressed as a development target particularly for GE's polymer departments in recent trade literature.

By early 1993, two commercial versions of tetraphenyl resorcinol diphosphate by U.S. manufacturers were available: Fyroflex RDP, with 11.0% P-content (Akzo), and Kronitex RDP, with 10.8% P-content (FMC Corp.).

Obviously, this is a very important, new product category, where commercial details have not yet been publicized. The voluminous, ongoing developments are sketched below in Section 11.3.6.

11.3. SELECTION OF DEVELOPMENTS IN ADDITIVES

Because of the large volume of literature in this category, a section stressing publications from the 1980–1992 period is presented. It is not possible to be certain, but it is likely that some of the discoveries described here have been commercialized.

11.3.1. Adhesion Promoters

The use of resorcinol in adhesives was discussed already in Chapters 4, 5, and 9. A few references are given here, since the use of resorcinol and derivatives as additives for adhesion promotion falls in the realm of this chapter.

A method was patented for fixing copper salts to synthetic polymers, e.g., textile materials of polyamide, polyester, or polyamide/imide. The fibers were treated with hydrogen sulfide at 3 atm/22 °C in an autoclave for 1 hr, then immersed at 60 °C in an aqueous solution containing 5 wt% copper sulfate and 5 wt% resorcinol, rinsed, and dried. The fibers increased by *ca* 4% in weight and had improved electrical conductivity vs fibers treated without resorcinol. Such materials may be used to produce floor or wall coverings or articles of clothing with improved antistatic properties; they could be made sufficiently conductive to be usable as heating articles.[4]

A high-impact-strength adhesive composition was prepared by adding a small amount of 2,4-dihydroxybenzoic acid (beta-resorcylic acid), 2,6,-dihydroxybenzoic acid (gamma-resorcylic acid), ethyl 2,4-dihydroxybenzoate, or similar materials to alpha-cyanoacrylate.[5]

Storage-stable, self-adhering poly(vinyl chloride) [PVC] compositions were prepared containing 77–142 parts plasticizer (e.g., dibutyl phthalate), 0.5–6.0 parts resorcinol, 2.0–18 parts hexamethylenetetramine, 10–40 parts bisphenol A-based epoxy resin, 70–130 parts PVC, 50–85 parts filler (e.g., calcium carbonate, gypsum, or baryte), and pigments, wetting agents, and viscosity modifiers. These compositions were useful as sealants and coatings; they had good sound-insulating qualities.[6]

Adhesion promotion for electric circuits on thermoplastic poly-ether/sulfone substrates was developed by treating the substrate for 3 min with an isopropanol/resorcinol (or isopropanol/methylene chloride) solution, optionally containing a small amount of poly(cyclohexanone).[7]

A two-component epoxy resin adhesive composition was developed which contained conductive carbon black and iron oxide as an electromagnetic energy-absorbing material. The epoxy resin component contained a phenolic cure accelerator. In one example in this patent, 4 wt% resorcinol was used as the cure accelerator. These compositions were curable with induction heating, using low-frequency induction generators, and were useful for bonding fiber-reinforced thermoset or thermoplastic materials and other plastics.[8]

11.3.2. Antioxidants/Heat and Processing Stabilizers for Organic Materials

In the older (1950s and 1960s) literature, there were mentions of alkylresorcinols and thio-bis-resorcinols as antioxidants for organic materials, e.g., polyolefins. In the Russian literature, in particular, 4,6-di-*tert*-butylresorcinol, **11-13**, was mentioned as

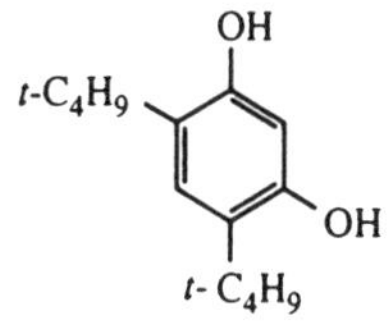

11-13

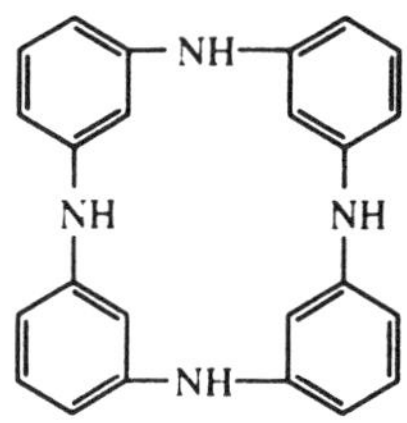

11-14

an antioxidant for resins and rubber (vulcanizates). An interesting example from that period is the cyclic arylene amino compound, **11-14**, prepared by the condensation of resorcinol with *m*-phenylenediamine in aqueous sodium (meta)bisulfite. This cyclic amino compound was said to be useful as an antioxidant for lubricant and fuel oils.[9]

In the 1970s, there was a notable increase in work on resorcinol derivatives useful as antioxidants. A few examples from the patent literature of this period are given in Table 11.1.

In the 1980–1992 period, several additional resorcinol derivatives were shown to be excellent antioxidants/thermal stabilizers for organic materials. In this period, also,

Table 11.1. Examples of Antioxidants Based on Resorcinol, From the 1970s

Compound	Patent	Application
Methylene-, thio-, or C_{2-6}-alkylidene-bis-(3-methoxy)-6-*tert*-butylphenol)	JP 49,102,654 (Sept. 27, 1974; to Sumitomo Chemical Co.)	Antioxidant for rubber and polyolefins
Cyclohexylidine-bis-resorcinol and its di(ar)alkyl ethers	JP 74,109,354 (Oct. 17, 1974; to Sumitomo Chemical Co.)	Thermal stabilizers for polypropylene
2,4,5,6-Tetraethylresorcinol	SU 394,352 (March 5, 1974; to Novosibirsk Org. Chem. Inst.)	Antioxidant
4,4′-Thio-bis-resorcinol	JP 50,158,633 (Dec. 22, 1975; to Dainichi Nippon Cables Co.)	Stabilizer for crosslinked polyethylene
m-Phenylene-bis-(diethylamidothiophosphates)	SU 540,872 (Apr. 15, 1977; to Volzhsk Tire Works)	Anti-aging agents for elastomers; prevent thermooxidative deterioration
Mixture of NaOH-treated rice bran and resorcinol	JP 77,102,477 (Aug. 27 1977; to Daiya Tokkyo Project K.K.)	Oxygen absorbent for packaged food, e.g., potato chips
4-(1-phenethyl)resorcinol and similar compounds	JP 53,072,791 (June 28, 1978; to T. Yamada)	Antioxidant for petroleum, rubber and plastics
2′,6′-Dihydroxy-9-(2,5-dihydroxyphenyl)octylphenone	DE 2,842,221 (May 23, 1979; to Lion Dentrifice K.K.)	Antioxidant for foods, e.g., lard, and for plastics, rubber, paper, cosmetics; has low toxicity
tris-(*m*-Hydroxyphenyl-amide) of thiophosphoric acid (derived from *m*-aminophenol)	SU 642,334 (Jan. 1, 1974; to Minsk BR Tekhnergo)	Thermal stabilizer for polyethylene

it became increasingly costly to develop all the required data on antioxidant packages to show superior efficacy/low toxicity/safety; a potential customer would also have to spend much time and money to reformulate and test his own products. Likely, this was in good part responsible for the few, new antioxidants, resorcinol-derived or not, becoming commercial products in this period. Few of the existing suppliers could afford it, and new entrants would not stand much of a chance.

The following selections of developments from this period mostly addressed needs such as lower volatility, better compatibility, improved efficacy, lower toxicity, etc.

Resorcinol was reacted with 3,5-di-*tert*-butyl-4-hydroxybenzyl alcohol in 1/3 mole ratio in a naphtha solvent, using formic acid as the catalyst, to obtain 2,4,6-tris-(3,5-di-*tert*-butyl-4-hydroxybenzyl)resorcinol, **11-15**. A diacetate of this compound (0.05 wt%) and dioctadecyl thiodipropionate (0.15 wt%) were added to polypropylene and the compound, tested at 150 °C, showed 1528 hr before embrittlement *vs* 50 hr without the diacetate of **11-15**. Several other esters of the new additive were quite effective also for polystyrene, ABS-terpolymer, and polyurethanes.[10]

Russian workers tested several alkylresorcinols vs *tert*-butylhydroquinone, di-*tert*-butylhydroquinone, *tert*-butylcatechol, and di-*tert*-butylcatechol as antioxidants for three grades of Russian, hydrotreated jet fuels. In all three fuels 4,6-di-*tert*-butyl-resorcinol, **11-13**, at 0.005 wt% was the most effective antioxidant and improved the lubricity of the fuels as well.[11]

The ethanolamide of 3,5-dihydroxybenzoic acid, **11-16**, added at 1.0 wt% to crosslinked polyethylene, was seen to be an effective aging/heat stabilizer. When a sheet of this composition was held 4 days/160 °C in a gear-type aging tester, 89% of the tensile strength was retained with the additive vs 33% without it.[12]

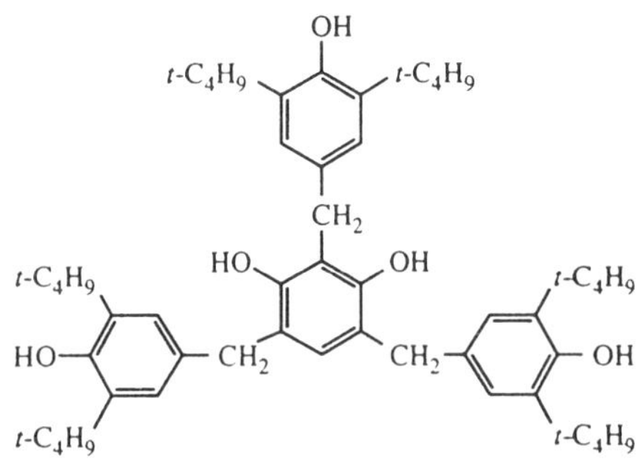

11-15

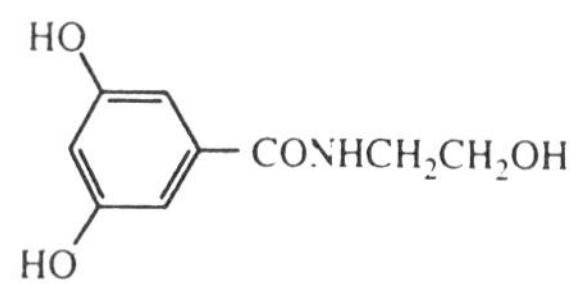

11-16

4-Benzoylresorcinol, **11-2**, was included in a selection of synergistic antioxidant combinations, together with *N*-phenyl-1-naphthylamine or *p,p′*-dioctyldiphenyl-amine, for synthetic lubricants (pentaerythritol esters, neopentyl octyl nonedioate, and other ester types).[13]

Octahydroxanthene-1,8-diones, such as **11-17**, prepared from 5,5-dimethyl-1,3-cyclohexanedione, were claimed to be valuable intermediates for the synthesis of antioxidants.[14]

4-Benzylresorcinol, **11-18**, was reported to have high antioxidant activity in paraffin wax at 160 °C.[15]

N-substituted *m*-aminophenols such as, e.g., compound **11-19** were prepared, tested, and found to be excellent heat stabilizers for PVC and other chlorine-containing thermoplastics. These compounds did not require combination with the commonly used metal-containing stabilizers, and were effective at the 0.1–3.0 wt% level.[16]

4,6-Di-*tert*-butylresorcinol, **11-13** (or 3,5-di-*tert*-butylcatechol), at the 0.4–1.0 wt% addition level was reported to be an outstanding inhibitor of deterioration caused by sunlight/air in hydrocracked lubricating oils.[17] Styrenated *m*-phenylaminophenols such as, e.g., **11-20**, were patented as effective antioxidants for lubricants.[18]

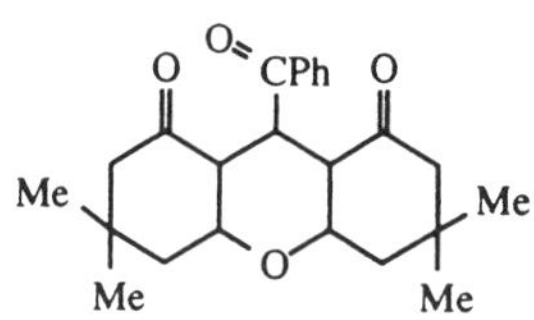

11-17

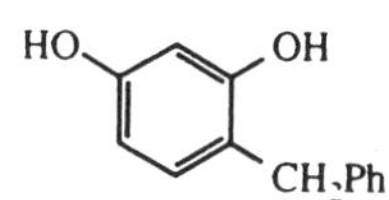

11-18

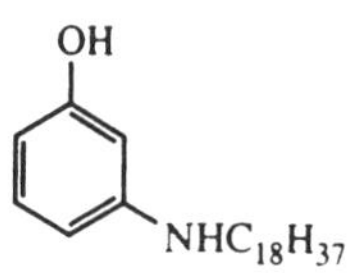

11-19

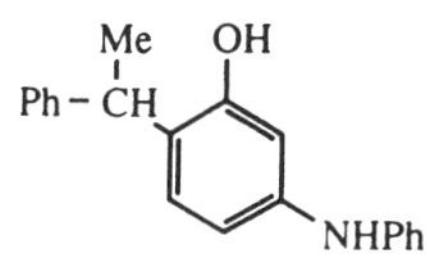

11-20

4,6-Di-(alpha,alpha-dimethylbenzyl)resorcinol, **11-21**, was patented as a new compound useful as an antioxidant and processing stabilizer for polyolefins and paraffinic lubricating oils; the compound also had bactericidal properties. For example, 0.14 wt% compound **11-21** was added to high-density polyethylene (HDPE): after processing for 15 min at 260 °C/80 rpm in a Brabender mill, the melt index (indicator of flowability) of the polymer decreased 41%; with HDPE containing 0.15 wt% of Irganox 1010 commercial stabilizer, the melt index decreased 88% under the same conditions.[19]

Tris-(3-hydroxy-4,6-di-*tert*-butylphenyl)phosphite, **11-22**, was reported as a new compound useful as a polymer processing stabilizer and as an antioxidant for organic materials such as polyolefins and (mineral oil) lubricants. For example, 0.05 wt% of this new additive was blended into polypropylene (Shell 55XX) at 171 °C; the material was submitted to a multiple pass test in a Brabender extruder. After 4 extrusions at 289 °C, the melt index decreased 36% and the color of the polymer was off-white. In a

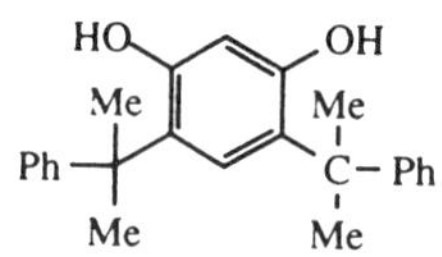

11-21

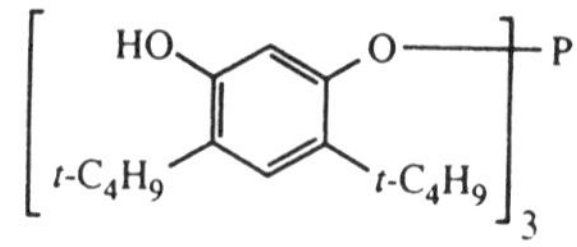

11-22

comparison test using 0.05 wt% Irganox 1010 as the stabilizer, after 4 extrusions at 289 °C the melt index of the polypropylene decreased 69% and the polymer was off-white.[20]

Long-chain alkylresorcinol phosphites such as, e.g., compound **11-23** were prepared and found to be viscous oils with high thermal stability and low flammability, suitable as antioxidants for high-temperature lubricants. For example, compound **11-23** showed a 1% weight loss by TGA (nitrogen flow, heating rate 20 °C/min) at 320 °C, flash point 321 °C, and fire point 334 °C.[21]

The spirobichroman, **11-24**, was prepared by the sulfuric acid-catalyzed condensation of acetone with resorcinol in, e.g., chloroform at >50 °C and was said to be useful, for instance, as a heat stabilizer for hydrocarbons or food.[22]

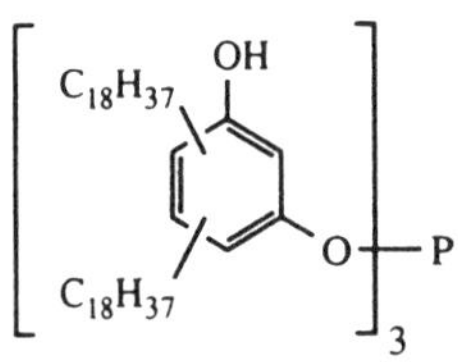

11-23

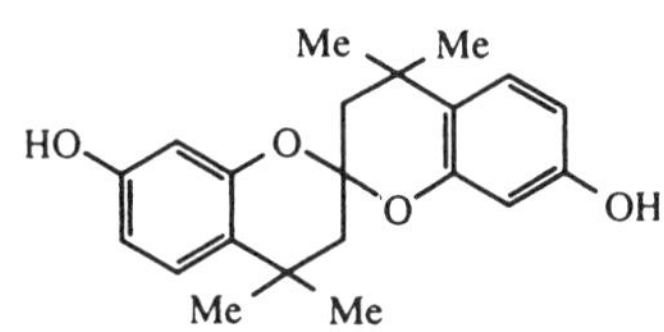

11-24

Calixarenes such as, e.g., **11-25** were found to be (at the *ca* 2.00 phr level) good antioxidants, antiozonants, and aging inhibitors (much better than BHT) for rubber vulcanizates used, for example, as sealants or moldings.[23] Similarly, aminomethylated calixarenes such as, e.g., **11-26** were prepared and found to be useful stabilizers and

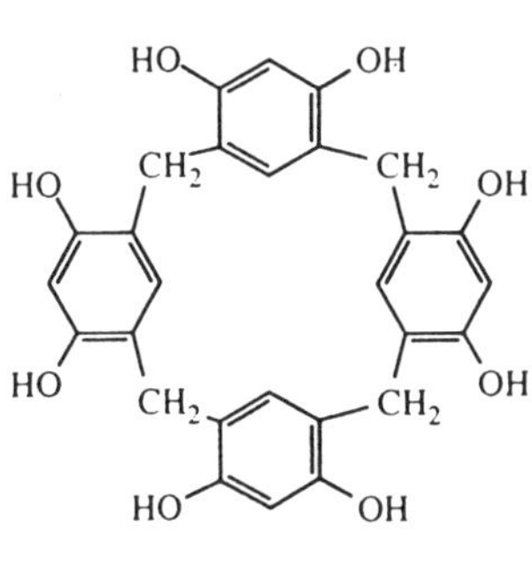

11-25

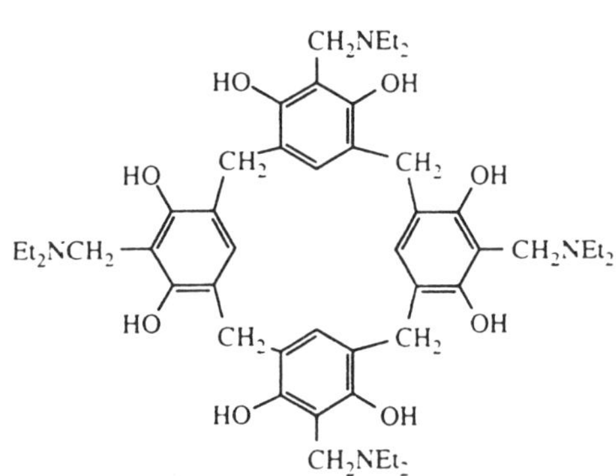

11-26

leach-resistant antiozonants and antioxidants for vulcanized (nitrile) rubber compositions.[24]

The tetramethylcalixarene, **11-27**, at 0.1–2.0 phr was a good antioxidant for styrene/butadiene rubber (SBR; prepared by emulsion polymerization) in storage, transportation, or processing.[25]

Mixed resorcinol-hydroxyester borates, e.g., made by reacting resorcinol, boric acid, Neodol 25-3 (mixed, triethoxylated dodecanol/pentadecanol) in toluene at reflux under a Dean-Stark trap with removal of water, then stripping off the toluene, were claimed to have multifunctional/antioxidant activity in greases, brake and hydraulic fluids, and fuels. These additives enhanced the stability and service life of the fluids, were ashless, and did not contain undesirable metals or chlorine.[26]

Phosphite esters of resorcinol-based calixarenes were synthesized and, e.g., compound **11-28** was proven to be an excellent heat/degradation stabilizer for isotactic polypropylene during extrusion/processing.[27]

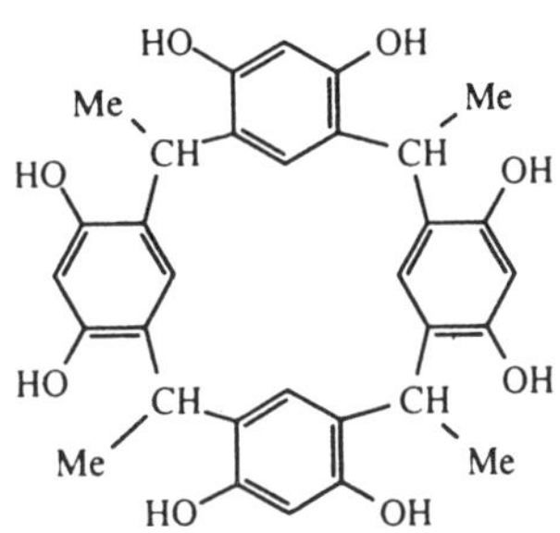

11-27

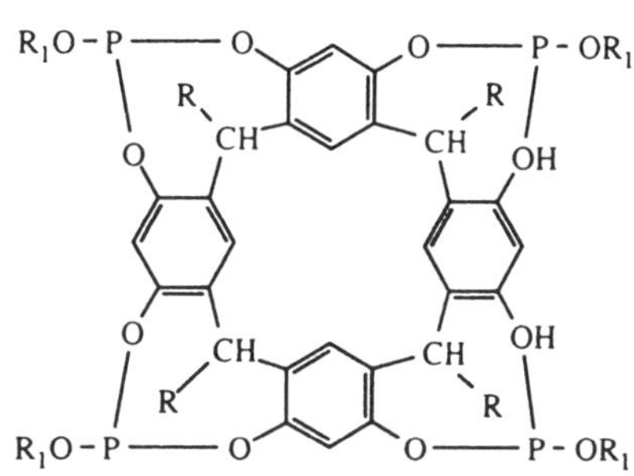

11-28

(where R = n-$C_{11}H_{22}$; R_1 = 2,4,6-tri-*tert*-butylphenyl)

11.3.3. Catalysts

There is a surprising variety of uses of resorcinol in this category. Thus, e.g., in the 1970s it had already been reported that a mixture of quaternary ammonium hydroxide/resorcinol catalyzed the cyclotrimerization of isocyanates. Resorcinol catalyzed the chlorination of C_{2-4} olefins, including butadiene. A reaction product of dialkyl zinc/resorcinol catalyzed the polymerization of epoxides. A mixture of (ketone) peroxide, metal accelerator (e.g., cobalt naphthenate), and resorcinol was used to catalyze polymerizable, or copolymerizable, unsaturated polyester resin compositions. A mixture of Cu–Cr catalyst/resorcinol was used to catalyze the hydration of acrylonitrile to produce acrylamide. The condensation of phenol with acetone to make bisphenol A was catalyzed by HCl/resorcinol (with or without the traditionally used thioglycolic acid)—this system increased the reaction rate and selectivity while reducing the HCl requirement and allowing a lower reaction temperature; substantially colorless bisphenol A, useful for making water-white polycarbonates, was obtained.[28]

In the 1980–1992 period, the range of uses of resorcinol in catalysis was confirmed and extended, as shown in the following selections.

The oxidative (50% O_2/50% N_2), dehydrogenative dimerization of dimethyl phthalate was catalyzed by a small amount of an equimolar mixture of palladium acetate and 1,3-cyclohexanedione (dihydroresorcinol); 5 hr reaction at 140 °C under 50 kg/cm^2 O_2/N_2 gave a 75% yield of tetramethyl biphenyltetracarboxylates at 7% conversion of the dimethyl phthalate.[29]

A monoalkyl aluminum dichloride/benzenediol catalyst was found to be effective for the alkylation of aromatic hydrocarbons with olefins.[30]

Glass fabric-reinforced epoxy resins were cured with diamines (4,4′-methylene-bis-2-chloroaniline), using resorcinol as the catalyst.[31]

A polymer concrete mixture was developed consisting of an unsaturated polyester, cobalt naphthenate catalyst, quartz, triarylimidazoline dimer, and 5,5-dimethyl-1,3-cyclohexanedione (1.5–2.5 parts) as the hardening accelerator. The mix was used for protective, alkali-resistant coatings on metallic, ferroconcrete, and concrete products.[32]

Sodium phenol- and naphthol-sulfonates, including sodium resorcinol-4,6-disulfonate, were found to be "crystallization accelerators" for polyester compositions.[33]

The triethylenediamine salt of 2,6-dihydroxybenzoic acid was developed as a cure accelerator for novolak epoxy resin/novolak resin mixtures. The mixes had 30-day pot life, 31-sec gel time, and T_g 178 °C.[34] Similarly, the 4-dimethylaminopyridine salt of 2,6-dihydroxybenzoic acid was found to be a good cure accelerator for such epoxy resin mixes, giving good pot life and short gel time.[35]

Titanate esters such as [2,2′-bis-{(allyloxy)methyl}butanolato]-tris-(3-aminophenoxy)titanium together with $Bu_2Sn(OAc)_2$, were used as the catalysts for rapid-curing, alkoxysilane-terminated polyurethane sealants.[36]

Novel catalysts for the stereospecific polymerization of olefins were developed. Thus, a mixture of 25 mmol Mg di-(3-methoxyphenoxide), 125 mmol $TiCl_4$, and 4.4 mmol BzOEt in 61.28 ml chlorobenzene was held at 100 °C. The solids were filtered off and washed with isooctane to give a catalyst component which was used, in conjunction with Et_3Al, Et_2AlCl, and ethyl *p*-ethoxybenzoate (electron donor), in the polymerization of propene to give isotactic polypropylene.[37]

A new catalyst for making bisphenol A by the condensation of phenol with acetone consisted of $PHSO_3H$ together with an ion-exchange resin, Duolite LES-9901, C-alkylated with resorcinol in dioxane. By the use of these catalysts, the amount of undesirable *o,p*′-isomer of bisphenol A was reduced (over the prior art).[38]

A new process for the manufacture of elastomeric, primarily isotactic polyolefins involved the use of novel catalysts. For example, propene was polymerized using as the catalyst $AlEt_3$ and a reaction product from $Mg(OMe)_2$/$Si(OMe)_4$/2-methylresorcinol in methanol and $TiCl_4$ in the presence of 2,6-lutidine; the polypropylene obtained had isotacticity 51, syndiotacticity 27, elongation 891%, and tensile strength 1456 psi.[39] Similarly, catalysts made with $AlEt_3$, $TiCl_4$–$Mg(OMe)_2$ $(MeOH)_{10}$ (resorcinolate)$_2$ reaction product, $Si(OEt)_4$, and acridine were used to polymerize propene or butene to give mildly elastomeric products. As an example, the polypropylene obtained showed 64% isotacticity, 31% xylene solubility, residual stretch value 21, tensile set (400%) 204, and elongation at break 785%.[40]

11.3.4. Corrosion Inhibitors

The use of resorcinol as a corrosion inhibitor was mentioned in the older literature. For instance, in the USSR the protection of metals from acid corrosion was described using a condensation product of shale oil-derived alkylresorcinols with *m*-aminophenol and formaldehyde.[41] The post-1980 period brought further examples.

The oxime of 4-benzoylresorcinol alkyl ethers, such as **11-29**, was patented as a rust inhibitor and heavy-metal complexing agent.[42]

Resorcinol (or other polyhydric phenols) was added to lithium bromide solutions, used as (ammonia) absorption liquids in refrigerators, to provide corrosion resistance.[43]

A magnetic recording medium, consisting of Co–Ni alloy (20 wt% Ni) vapor-deposited on PET film, was corrosion-protected with 4-(2-pyridylazo)resorcinol, **11-30**, and related compounds.[44]

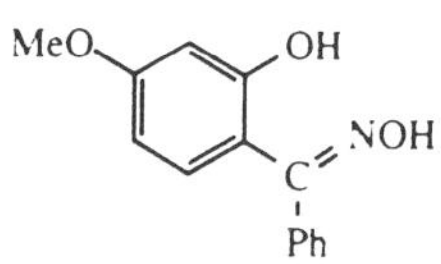

11-29

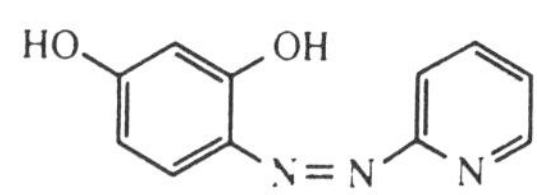

11-30

Several Polish patents described resinous corrosion-protective compositions for steel, with resorcinol mentioned as a resin component.[45]

A salt- and water-resistant coating composition was developed which contained thermally degraded poly(vinyl acetate) (soft point 115–123 °C) 17–31 parts, poly(vinyl butyral) 1.5–5.5 parts, adhesive additive 1.5–3.5 points, low molecular weight polyethylene 1.7–2.1 parts, a reactive additive such as resorcinol (or pyrogallol, urea) 0.01–0.03 parts, and organic solvent (the balance) to make 100 parts.[46]

11.3.5. Electrolytes

A lead battery was developed which held a separator plate (compartment) containing an electrolyte solution with a phenolic additive (resorcinol is mentioned as one) intended to remove antimony ions which can reduce the anode potential and cause overcharging. (The antimony came from the cathode plate.)[47]

A highly conductive electrolyte for condensers with very high specific capacity was developed. For example, the electrolyte contained 75 parts *N,N*-dimethylformamide, 25 parts ethylene glycol, 15 parts 2,6-dihydroxybenzoic acid (gamma-resorcylic acid), 0.7 part phosphate (not specified), and 6 parts diethylamine; this electrolyte had good conductivity at high and low temperatures and was stable. In comparative experiments it was shown that an electrolyte containing 2,6-dihydroxybenzoic acid had higher conductivity than electrolytes containing salicylic acid, 2,4-dihydroxybenzoic acid, or 3,4,5-trihydroxybenzoic acid in similar amounts.[48]

An improved electrolyte for electrolytic capacitors with good high-temperature stability was developed. This electrolyte consisted of 14 wt% ethylene glycol, 58% gamma-butyrolactone, 15% 2,6-dihydroxybenzoic acid, 10% triethylamine, and 3% water. This solution, after 2000 hr charge at 105 °C, showed a –11.2% change in capacitance, a tangent loss angle 0.127, and an impedance (100 kHz) of 118 mΩ; this was much better than the performance of an electrolyte containing maleic acid instead of 2,6-dihydroxybenzoic acid.[49]

An improved electrolyte for capacitors with reduced leak current at high temperature consisted of an ethylene glycol-based solvent with various additives; 2,6-dihydroxybenzoic acid was one of the effective ones.[50]

Another patented electrolyte for capacitors with high spark voltage contained an acid (salt), e.g., boric acid, and a polysaccharide, e.g., dextrin, and an additive such as 2,6-dihydroxybenzoic acid.[51]

11.3.6. Flame Retardants

Aryl phosphates, e.g., made from phenol, cresols, or xylenols, have been known for a long time and have been used as flame retardants. The actual use of resorcinol-derived phosphate esters is only a recent development, discussed in Section 11.2.2. The preparation of tetraphenyl resorcinol diphosphate, **11-12**, was first reported in 1950; as shown in Eq. (11.2), resorcinol was first reacted with excess $POCl_3$ and a small amount of PCl_5 in the presence of a catalytic amount of $AlCl_3$ under reflux until the evolution of HCl ceased, then excess $POCl_3$ was stripped off, and the residue was reacted with phenol at 100–150 °C. The resulting product was well washed with 0.5 N NaOH, followed by stripping of volatiles to give resorcinyl tetraphenyl diphosphate, **11-12**. (At the time, the product was intended for use as a low volatile, fire-retardant plasticizer for nitrocellulose.)[52]

(11.2)

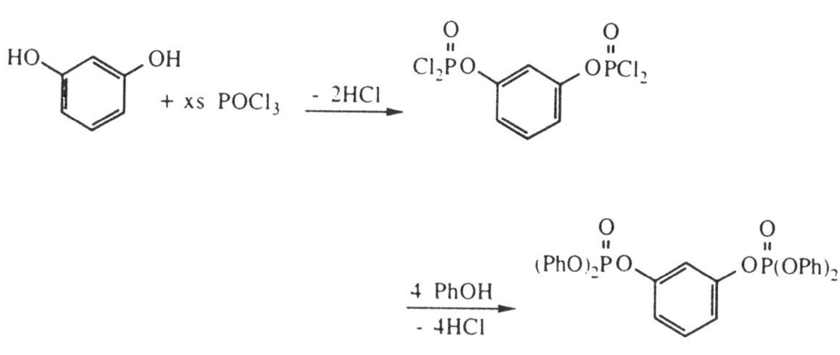

11-12

Work on resorcinol as a starting material for flame retardants was sporadic in the 1960s and early 1970s. Resorcinol phosphates and phosphonates were prepared, as were halogenated aryl ethers and phenoxy-cyclo-triphosphazenes; phosphonates of chlororesorcinols were made. In the 1970s worries were raised about the toxicity of halogenated phenols (or trace impurities therein) and aryl ethers. Indeed, by the late 1980s it appeared that halogenated aryl ethers used as flame retardants would be phased out in Europe for environmental concerns. At any rate, in the 1970s technical interest in resorcinol-derived phosphates and phosphonates increased.

The following selections from the pertinent literature all concern compound **11-12** and similar products; the one exception is a resocinol-derived, halogen-containing calixarene.

Aryl end-capped poly(*m*-phenylene cyclohexylphosphonates), **11-31**, were prepared by reacting resorcinol with the appropriate amount of cyclohexylphosphonic dichloride using $MgSO_4$ as the catalyst, at 150–180 °C, then end-capping with phenol, *p*-phenylphenol, or *p*-bromophenol. The molten products were blended with molten poly(ethylene terephthalate) (PET) and extruded through a spinerette; the filaments were spun into yarns with improved flame-retardant properties.[53]

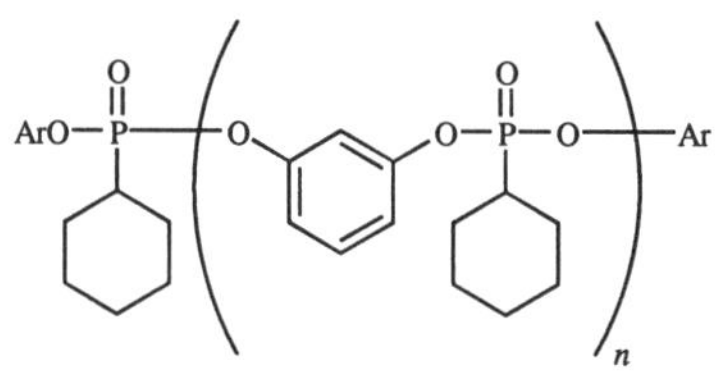

11 - 31

(where Ar= Ph, Ph–Ph, or *p*-BrPh;
$n \sim 4$–8)

The resorcinyl tetraphenyl diphosphate, **11-12**, was patented as a flame retardant for PET polyester oriented films and foils; the preferred phosphate addition level was 8–15 wt%. The compound provided reduced processing temperatures, the color of the film was good, and the product was self-extinguishing (after the removal of the flame).[54]

The resorcinyl tetraphenyl diphosphate, **11-12**, was one of the several related flame retardants claimed to be effective at the 5–15 wt% addition level for glass fiber-reinforced polyphenylene oxide (PPO)/polystyrene blends.[55] Similar fire-resistant PPO compositions, made up of blends of 60 parts PPO with 40 parts rubber-modified polystyrene, containing 8 parts of the tetraphenyl resorcinol diphosphate, 1 part TiO_2 and 0.5 part triphenyl phosphite, were said to resist discoloration, foaming during molding, and blooming.[56]

The tetraphenyl resorcinol diphosphate, **11-12** (and related materials), was also patented as an additive for polystyrene and styrene copolymers with acrylonitrile, (meth)acrylic acid, and /or butadiene to give nonflammable compositions. The resorcinol diphosphate was said to inhibit the dripping of the polystyrene in a flame.[57]

Flame-retardant PVC compositions were prepared containing the tetraphenyl resorcinol diphosphate, **11-12**, and related additives at the 5–70 wt% level. These compositions had good physical and electrical properties, were light-stable, and other, conventional plasticizers could be blended in.[58]

Phenolic resins, including resole types, could be modified with 10–70 parts of the resorcinol diphosphate, **11-12**, or related compounds. Nonflammable compositions were obtained which had excellent solvent resistance and mechanical strength. These products were useful for laminating, molding, and electrical materials which showed soldering heat resistance >60 sec at 260 °C.[59]

Novel, crosslinkable flame-retardant compositions for use as insulation materials for electrical wires and cables were described. These materials consisted of a blend of nonrubbery polyolefin, a poly(phenylene ether) resin, an aryl phosphate such as tetraphenyl resorcinol diphosphate, **11-12**, at the 1–50 parts level, a brominated material such as, e.g., 1,2-bis-(2,4,6-tribromophenoxy)ethane, a crosslinking agent such as alpha,alpha′-bis-(*tert*-butylperoxy)diisopropylbenzene, and other additives (such as antioxidants, fillers, etc.). These mixtures were extruded at 260+ °C, milled at 132–188 °C, then press-cured at 188 °C. The products had good physical properties, did not drip in the vertical burn test, and had excellent self-extinguishing properties.[60]

Blends of aromatic polycarbonates, styrene (graft) copolymers, and 2–20 wt% of the resorcinol diphosphate (**11-12**) type flame retardant were prepared. For example, a blend of bisphenol A polycarbonate 64.5 parts, graft ABS coplymer 25, poly(1,3-phenylene phosphates) (dp 1.3) 10, PTFE 0.5, and sodium trichlorobenzenesulfonate 0.02 part had a UL-94 flammability rating (1.6 mm) V-0, Vicat Temp. 95 °C, and showed no blooming of the flame retardant.[61]

Fireproofed PPO/high-impact polystyrene blends were prepared which did not form cracks during injection molding. For instance, a blend of PPO 35 parts, high-impact polystyrene 65 parts, Cr 7333 (cresyl triphenyl resorcinol diphosphate, **11-32**) 13 parts, and Al methyl methanephosphonate 5 parts had heat-distortion temperature 96 °C, melt viscosity 128 Pas, UL-94 flammability rating V-0, and injection-molded parts showed no cracks.[62]

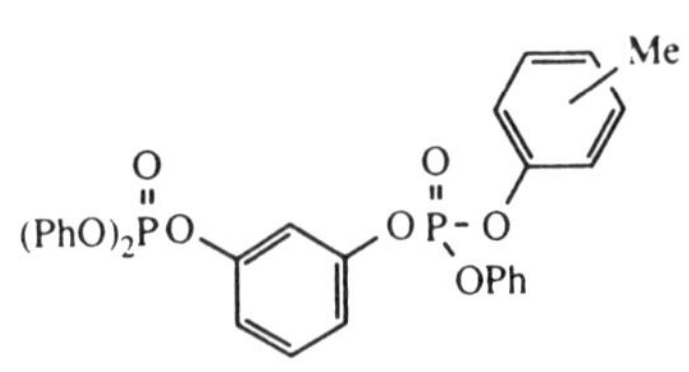

11-32

Polycarbonate resins were also flame-proofed with halogenated, resocinol-derived calixarenes. As an example, bisphenol A 84 g, calixarene, **11-33**, 1.37 g, NaOH 65.2 g and $Na_2S_2O_6$ 0.02 g were dissolved in 650 ml water, 6.3 ml 0.5 N aq. NEt_3 was added, and 1.7 g *p*-(*tert*-butyl)phenol in 1.3 L methylene chloride; 44 g $COCl_2$ was bubbled in during 30 min and the charge was stirred for 2 hr to give a polymer having UL-94 flammability rating V-0. (The calixarene, **11-33**, was prepared by the condensation of resorcinol with *p*-chlorobenzaldehyde in equimolar ratio.)[63]

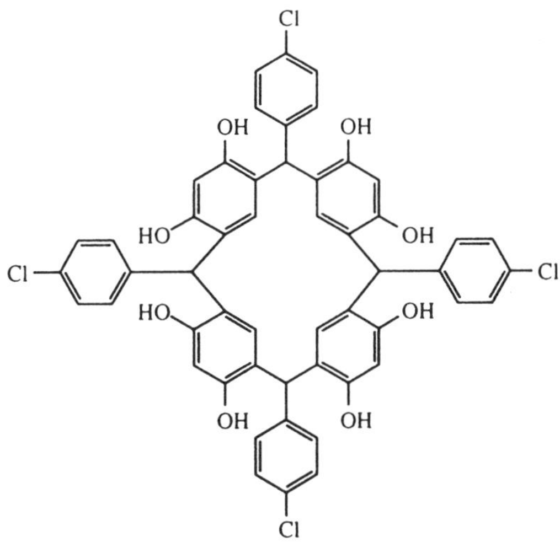

11-33

An improved process for preparing resorcyl tetraphenyl diphosphate, **11-12**, was developed in which resorcinol was reacted with excess $POCl_3$ in xylene using, e.g., $MgCl_2$ as the catalyst, at reflux until the evolution of HCl stopped. The excess $POCl_3$ and the xylene were then distilled off to 142 °C head temperature, phenol was added and reacted until the evolution of HCl stopped. The product was washed with dilute sodium hydroxide and water, finally vacuum-stripped to remove the remaining solvent and water. The final product contained *ca* 98% of the desired compound, **11-12**, plus some oligomers and <2% triphenyl phosphate.[64]

Fire-resistant poly(butylene terephthalate) compositions were patented also. As an example, PBT 44.35 parts, bisphenol A polycarbonate 40 parts, Irganox 1076 (antioxidant) 0.15 part, and tetraphenyl resorcinol diphosphate 15 parts were compounded and extruded. This compound had a UL-94 flammability rating of V-0 and

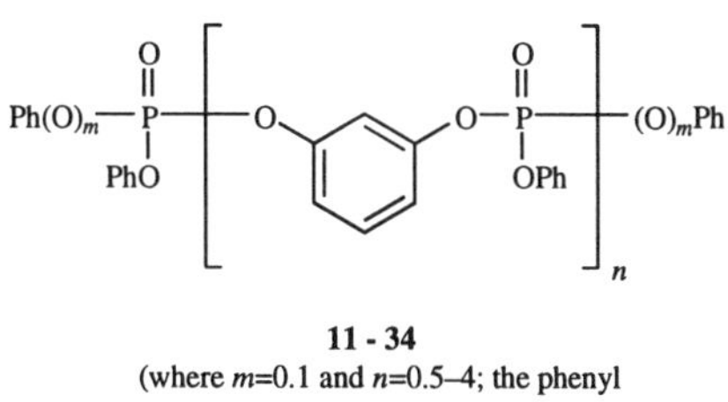

11-34
(where m=0.1 and n=0.5–4; the phenyl rings may contain alkyl or alkoxy substituents)

Izod notched impact strength (room temperature) 29 J/m. In this patent the fire retardant was described by formula **11-34** to include oligomers and phosphonate esters as well as the main component,**11-12**.[65]

A recent review of P-containing flame retardants presented their mechanism of action, the environmental problems and opportunities, and the several types of newer products, including the resorcinol diphosphate,**11-12**.[66]

11.3.7. Additives for Functional Fluids

Antioxidants for lubricants were mentioned already in Section 11.2.2 above.

In the older literature there were references, e.g., to alkylresorcinols as additives for spindle oils and of thio- and dithio-bis-resorcinols as additives for greases to improve their load-carrying and antiwear properties. There were sporadic mentions of resorcinol derivatives as lubricant additives in the 1970s and 1980s. Examples follow.

4,4′-Bis-(4-octadecylcarbamoyloxy-2-hydroxyphenylazo)-3, 3′-dimethyl-biphenyl, **11-35**, was prepared by coupling of resorcinol with tetrazotized *o*-toluidine and treatment of the azo product with octadecyl-isocyanate and related compounds. These products were thickening agents with good high-temperature bearing-life properties in lubricating greases.[67]

Lubricating agents for synthetic fibers were obtained by adding an epoxy alkane to resorcinol (or catechol) and esterifying the di-(hydroxyalkyl)ether with a C_{8-18} saturated or unsaturated fatty acid. An example of such products is compound **11-36**. These lubricants could be modified with emulsifiers, antistatic agents, and others.[68]

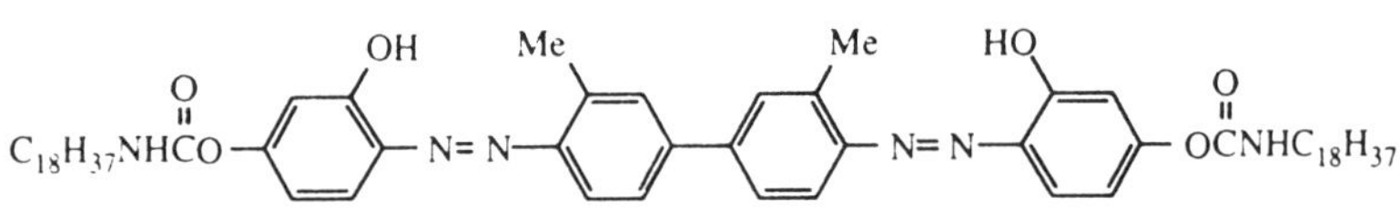

11-35

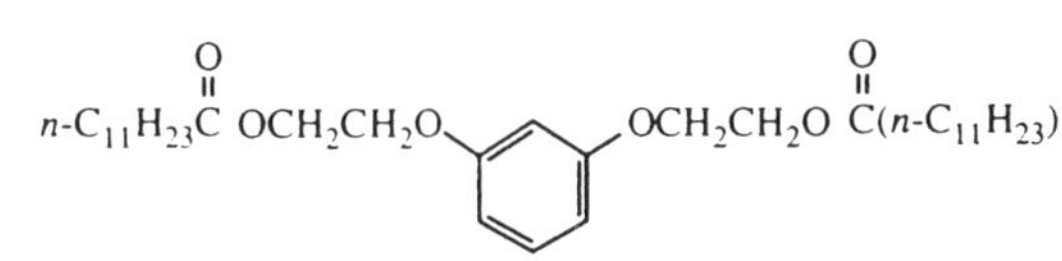

11-36

Long-chain alkyl-(*m*-)aminophenols were suitable stabilizing additives for lubricants and fuels.[69]

Water-based hydraulic fluids containing poly(ethylene oxide/propylene oside)-type thickening agents and diethylene glycol as a freeze and pour point depressant were developed which contained conventional additives (e.g., aliphatic fatty acids as acidic lubricity agents and corrosion inhibitors) and two-component antiwear agents consisting of resorcinol (or other phenols)/*p*-nitrobenzoic acid (or 2,4,6-trinitrophenol).[70]

11.3.8. Ultraviolet-Light Stabilizers

In the 1970s and 1980s, a large amount of work was done on new UV absorbers with a resorcinol moiety in their structure, including the 2,4-dihydroxybenzophenone type, even though the market for these products was not large and showing only modest, if any, growth due to the availability of several other types. The work described below was driven by the search for better-performing light stabilizers, with improved compatibility in specific systems and/or reduced volatility/migration, often via reactive or homo- or co-polymerizable compounds. Several new processes for known UV absorbers emerged and many new applications were developed. The persistence of these efforts is testimony to the practical importance of UV-light stabilizers and, perhaps, to a proliferation of companies interested in getting into specialty chemicals.

2-Hydroxy-4-alkoxybenzophnone was prepared by reacting resorcinol with benzotrichloride, $PhCCl_3$, in the presence of a solvent containing 25–100 wt% *n*-octanol and 0–75 wt% water at reflux. In the second stage the 4-hydroxy group of the benzophenone made in the first step was etherified by adding $NaHCO_3$, KI catalyst, and a C_{1-10} alkyl chloride or bromide.[71]

2-Hydroxy-4-alkoxybenzophenones were reacted with a secondary amine and formaldehyde to give 2-hydroxy-4-alkoxy-5-aminomethylbenzophenones, such as, e.g., compound **11-37**. This product could be condensed with additional 2-hydroxy-4-alkoxybenzophenone, using sodium methoxide as the catalyst at 125–135 °C, to produce 5,5′-methylene-bis-(2-hydroxy-4-alkoxybenzophenones), e.g., **11-38**, as shown in Eq. (11.3). Both products were useful for preventing the photodegradation of synthetic resins such as polyolefins, EVA (ethylene/vinyl acetate copolymer), polystyrene, polyacrylates, ABS, and many others.[72]

(11.3)

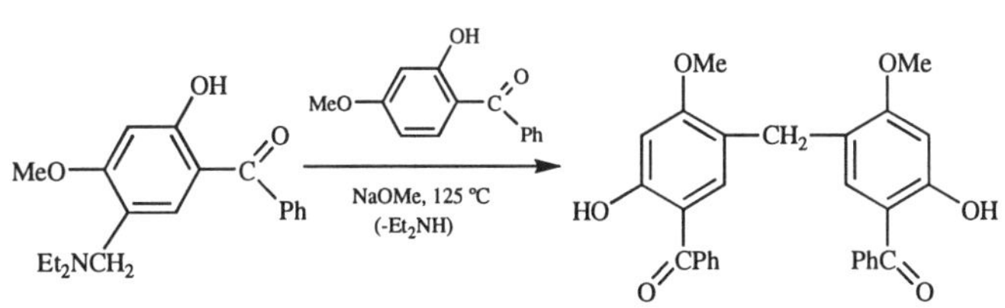

11 - 37 **11 - 38**

Linear (co)polyesters, e.g., PET, incorporating 0.05–10 wt% 3,6-bis-(2-hydroxyethoxy)xanth-9-one, **11-39**, had good stability to UV light. These polymers could be converted to fibers or films. The transparent films could be used, e.g., as windows for greenhouses or as coverings for solar cells. (The xanthone, **11-39**, was made starting from resorcinol.)[73]

Mixtures of the nickel salt of an organic ester of phosphoric acid, such as **11-40**, and a 2-hydroxy-4-alkoxybenzophenone were patented as synergistic heat and light stabilizers for polyolefins.[74]

Ultraviolet light-resisting coatings were prepared from 20–80 wt% melamine wholly or partly N-substituted with hydroxymethyl or methoxymethyl groups, a polyol, and 2,4-dihydroxybenzophenone. These coatings were resistant to thermal aging. They could be used to impart UV-light resistance to polycarbonates.[75]

UV-light-resistant ink jet compositions were prepared containing a water-soluble dye, water, wetting agent, and a water-soluble UV stabilizer such as **11-41**. Optionally, $NaCO_3$ was added; these inks could be used for a long time in ink jet (computer) printers without causing clogging of the printer nozzle and the prints had good UV-light stability.[76]

Alkoxysilylalkyl- or alkanoyloxy-silylalkyl ethers of 2,4-dihydroxybenzophenones, such as, e.g., compound **11-42**, were prepared; they were useful to impart UV-light resistance/protection from discoloration to polysiloxane-coated (polycarbonate) materials. These UV stabilizers had a much reduced tendency to escape or oxidize during thermal processing. The products were useful for transparent glazing.[77]

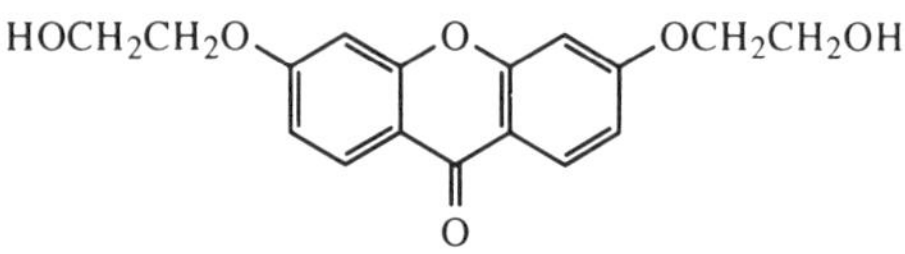

11-39

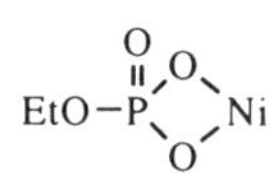

11-40

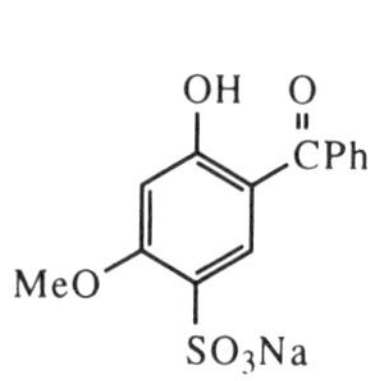

11-41

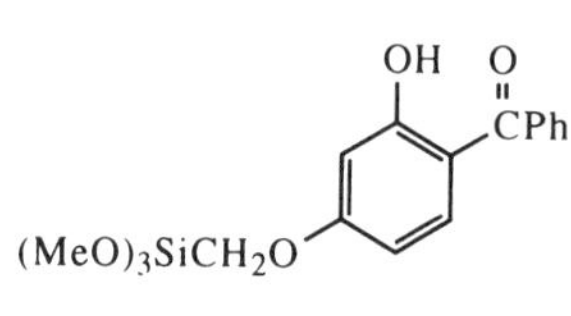

11-42

An improved process for making 2-hydroxy-4-alkoxybenzophenones was developed in which the dihydroxybenzophenones were O-alkylated with an alkyl halide in the presence of both an alkali and a trialkylamine. For example, 840 kg of 2,4-dihydroxybenzophenone, **11-2**, 705 kg *n*-octylchloride, 172 kg Na_2CO_3, 170 kg $NaHCO_3$, 20 kg tributylamine, and 16 kg KI in $MeO(CH_2)_3OH$ as the solvent were reacted for 10 hr/110–125 °C to give a 90% yield of the 2-hydroxy-4-*n*-octoxybenzophenone, **11-5**, vs only an 84.9% yield when the tributylamine was omitted.[78] In another, similar process *N*-methyl-2-pyrrolidinone was used instead of the tributylamine and the yield of **11-5** was 96%.[79]

2,4-Dihydroxy-4′-fluorobenzophenone, **11-43**, and similar materials were prepared and claimed to be useful as UV-light-protecting agents for plastics and cosmetics.[80]

The UV-light stability or weatherability of cellulose esters, such as cellulose acetate butyrate, was substantially improved by an additive combination consisting of 2-(2′-hydroxy-5′-methylphenyl)benzotriazole (0.1–0.2 wt%) and resorcinol monobenzoate (2 wt%), together with an antioxidant of the hindered phenol type.[81]

High molecular weight, hindered spiropiperidine compounds were prepared and found to be useful stabilizers for polymers. Thus, resorcinol was condensed with an equimolar amount of 2,2,6,6-tetramethyl-4-piperidone in methanol saturated with HCl at 56 °C for 12 hr to give compound **11-44**. A test specimen of polypropylene containing 0.25 phr of this compound, **11-44**, and 0.1 phr antioxidant was tested in the Xenon Weatherometer test and took 1160 hr before the tensile strength was reduced by 50%.[82]

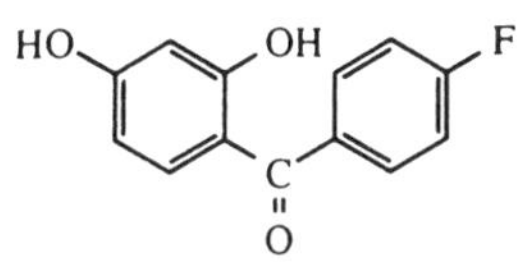

11-43

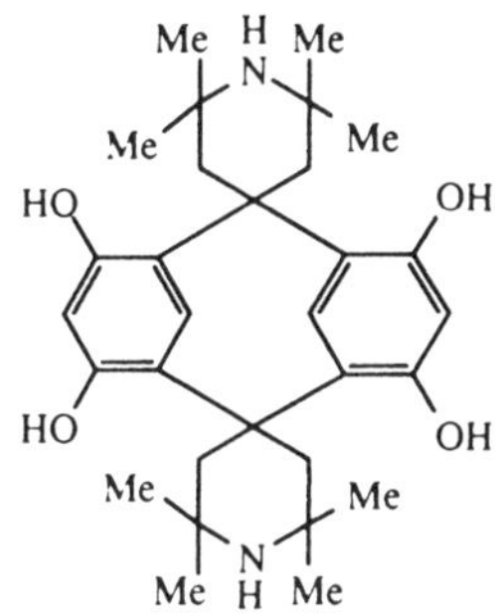

11-44

Biaxially oriented polyester (PET) film was stabilized with 1.0% of the low-volatile ether of 2,4-dihydroxybenzophenone, **11-45**. The film showed 115% elongation at break; on testing in a weatherometer it lasted 13,000 hr to 0% elongation at break and 5000 hr to 50% haze. The film was suitable for greenhouse applications.[83]

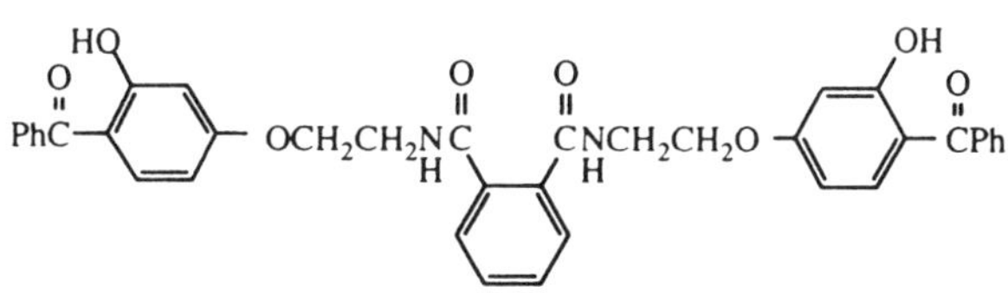

11-45

3-(3-Trimethoxysilylpropoxy)phenyl benzoate, **11-46**, was one of several silyl-functionalized UV stabilizers prepared. Under UV irradiation it rearranged to a 2-hydroxybenzophenone. These products were useful for stabilizing polycarbonates used in glazing materials.[84]

A preparation of 5,5′-methylene-bis-(2,4-dihydroxybenzophenone), **11-38**, was developed which involved heating 2,4-dihydroxybenzophenone with paraformalde-

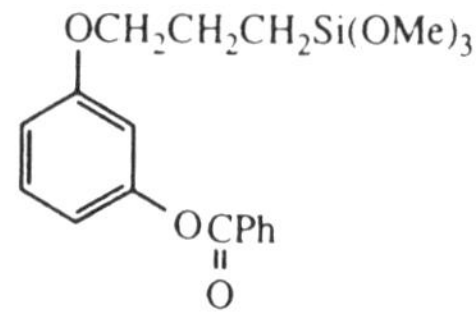

11-46

hyde in cyclohexanol at reflux for 9 hr. This product was said to be a UV stabilizer for PVC, PE, PP, ABS, and nylon.[85]

The adduct, **11-47**, of 2,4-dihydroxybenzophenone with 1,3,5-triglydcidylisocyanurate was made and reported to be a good UV stabilizer for polymers. This additive showed little migration/bleeding and had excellent water resistance and weatherability.[86]

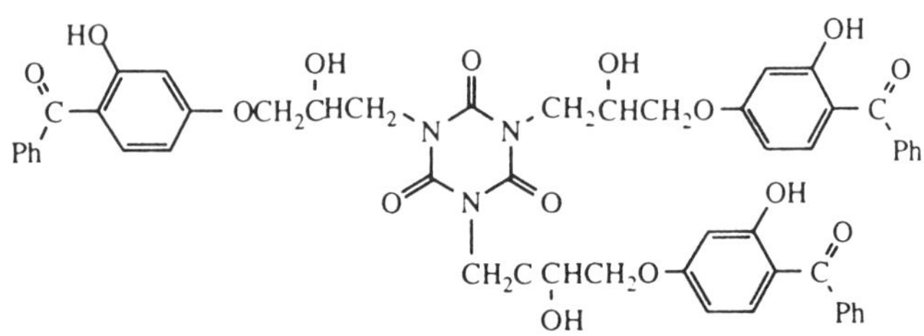

11-47

5,5′-Methylene-bis-(2-hydroxy-4-methoxybenzophenone), **11-48**, was claimed to be a good UV stabilizer for PET polyester; this additive could be added during the polymerization or before melt extrusion of the film. The oriented PET film, treated at least on one side with a water-repelling polymer, such as PVA, was said to last outdoors as long as 13 years. The film was suitable for use in greenhouses.[87]

An improved process for the preparation of 2-hydroxy-4-alkoxybenzophenone was patented, where the O-alkylation of 2,4-dihydroxybenzophenone with an alkyl halide was carried out in the presence of a polyethylene glycol which increased reaction rates. For instance, 2,4-dihydroxybenzophenone was reacted with *n*-octylchloride in the presence of sodium carbonate, polyethylene glycol (9 oxyethylene units), and KI as the catalyst at 155 °C for 2 hr to obtain a 93.3% yield of 2-hydroxy-4-octyloxybenzophenone.[88]

A novel, benzotrizole-type UV stabilizer containing a resorcinol moiety was developed, exemplified by compound **11-49**. This product, alone or in combination with other UV stabilizers, was useful for the protection of thermosetting or thermoplastic coatings from deterioration due to UV light.[89]

Cyanoacrylate derivatives of hydroxybenzophenones, e.g., compound **11-50**, were prepared and used as UV-light stabilizers for polycarbonate resins. The new compounds could be applied as thin surface coatings to the polymers.[90]

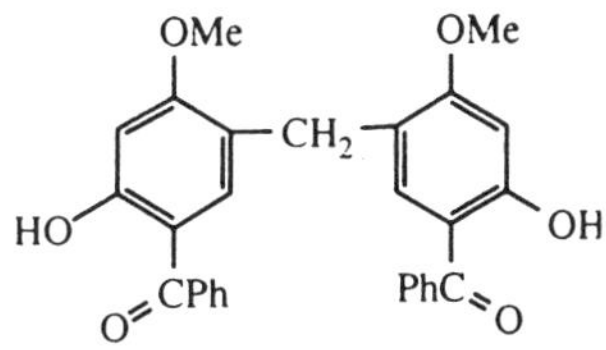

11-48

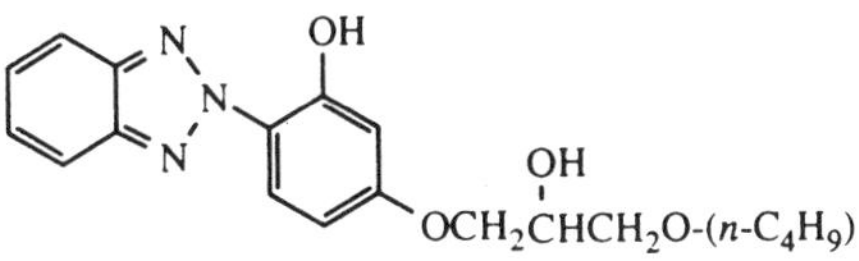

11-49

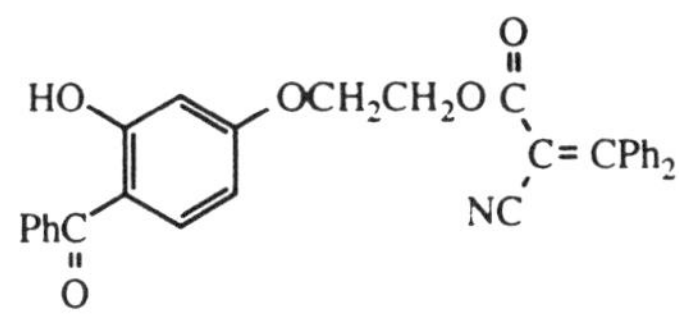

11-50

A guest–host-type liquid-crystal cell useful for displays contained a two-tone coloring agent, a nematic liquid crystalline material, and a 2,4-dihydroxybenzophenone-type UV-light stabilizer.[91]

A photoconductor layer used in laser printers, CRT printers, copiers, and electrophotographic plate-producing systems was improved by adding a 2,4-dihydroxybenzophenone-type UV stabilizer to the electric charge-carrying layer. The deterioration of the electrophotographic characteristics in this layer caused by UV light was prevented.[92]

New UV-light stabilizers containing both a hindered piperidine and a 2-hydroxy-4-alkoxybenzophenone moiety were patented. For example, the esterification of 2,2,6,6-tetramethyl-4-hydroxypiperidine with 2-hydroxy-4-carbomethoxy benzophenone produced compound **11-51**, useful for the stabilization of many polymers, including polyolefins.[93]

Hydroxybenzophenone (poly)glyceryl ethers, such as compound **11-52**, were prepared by the addition of excess glycidol to 2,4-dihydroxybenzophenone or by the reaction of 2-hydroxy-4-(2,3-epoxypropoxy)benzophenone with glycerol. These products were water soluble and nonirritating UV absorbers.[94]

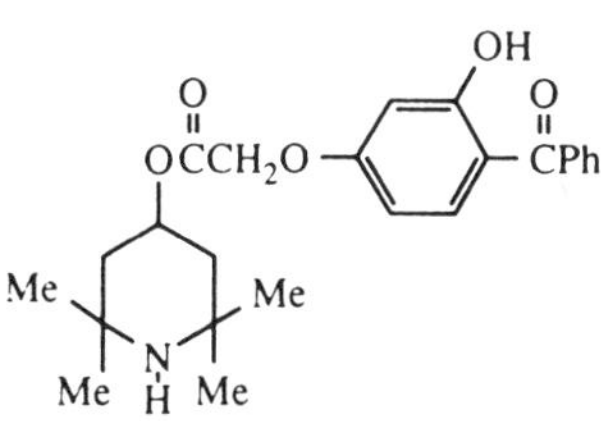

11-51

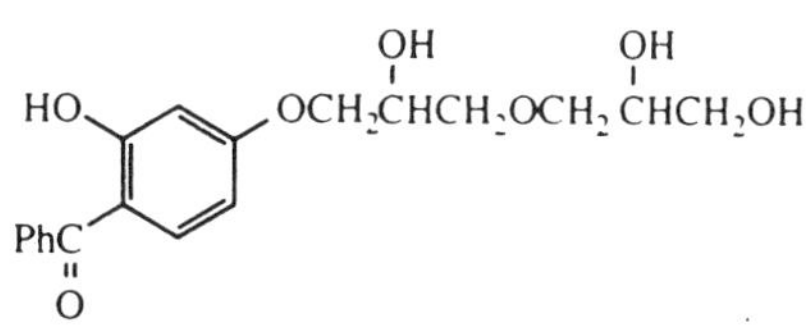

11-52

Resorcinol (and phloroglucinol) with two benzotriazole substituents was prepared. Resorcinol was coupled with 2-nitrobenzenediazonium chloride, followed by reductive cyclization with Zn–NaOH to give compound **11-53**. This product had high extinction coefficients in the UV range and UV absorption cutoff below 400 nm; it was an excellent candidate for UV stabilizers for polymers.[95]

1,3-Dihydroxy-4,6,-bis-4′-(2′,2′,6′,6′-tetramethylpiperidyl)benzene, **11-54**, was prepared by the acid-catalyzed condensation of resorcinol with triacetoneamine, followed by catalytic hydrogenation of the unsaturated intermediate. The new compound, its N-alkyl derivatives, and polymeric ethers prepared from it were useful as UV-light stabilizers for polymers.[96]

Two-coat automobile lacquers, consisting of a metallized base coat and a clear top coat, were improved by adding alkyl-substituted 2,4-dihydroxybenzophenone, **11-55**, to the top coat. This improved the light and weather fastness of organic pigments, reduced the tendency to form hairline cracks, and improved luster retention.[97]

Oximes of 2-hydroxy-4-octoxybenzophenone, **11-56**, and similar compounds were found to be more effective stabilizers for synthetic resins than the parent compound.[98]

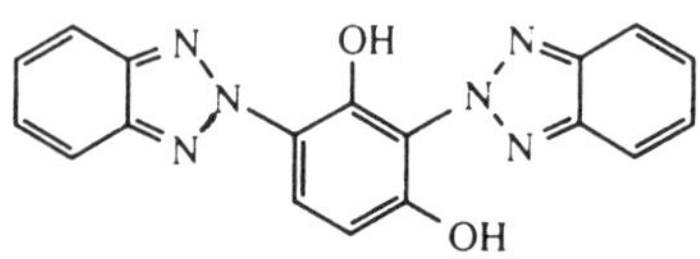

11-53

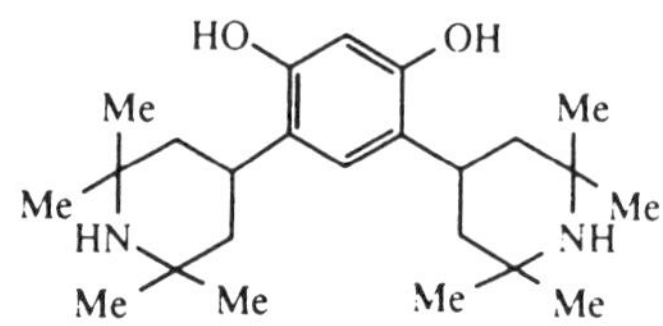

11-54

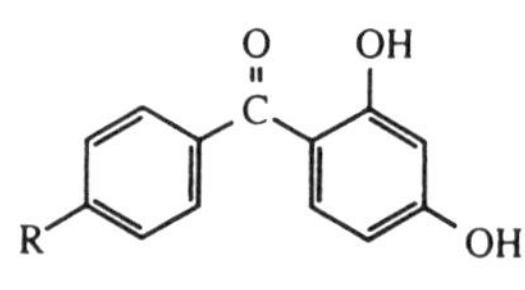

11-55
(where R= C_{3-12} alkyl)

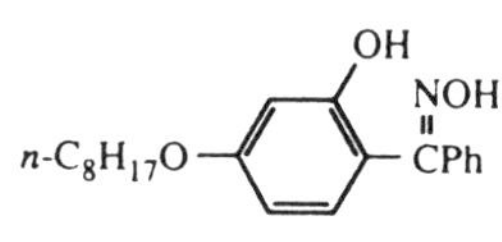

11-56

2,4-Dihydroxy-4′-(2-carboxyvinyl)benzophenone, **11-57**, was prepared. This compound had good compatibility and dispersibility in many polymers, and dimerized or polymerized on UV irradiation so that the compound did not volatilize from the plastic and retained its UV stabilizer action for a long time.[99]

Polyhydroxybenzophenones useful, e.g., as UV-light stabilizers were prepared. For example, the reaction of 38.5 g 2,4-dihydroxybenzoic acid with 27.5 g resorcinol in chlorobenzene in the presence of 25.5 g BF_3 for 4 hr/20 °C and 4 hr/65–75 °C gave an 86% yield of 2,2′4,4′-tetrahydroxybenzophenone, **11-8**.[100]

Polycarbonate resins were stabilized against the adverse effects of UV light by incorporating 2-methylresorcinol dibenzoate, **11-58**. This additive is a latent UV stabilizer more effective at lower concentration than either resorcinol monobenzoate

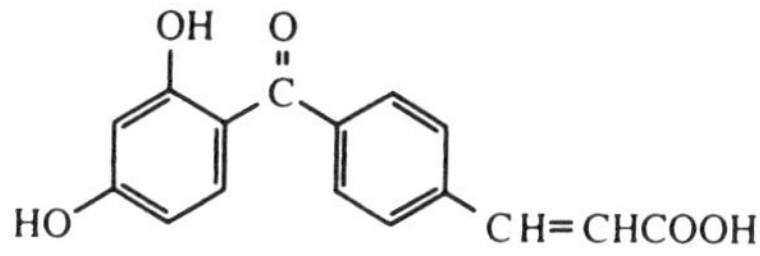

11-57

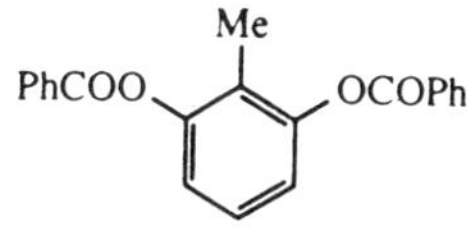

11-58

or dibenzoate, and can be applied to the polycarbonate surface in a UV-curable carrier such as a methyl methacrylate prepolymer.[101]

2,4-Dihydroxybenzophenone was prepared by heating resorcinol monobenzoate with Nafion XR (fluorinated, sulfonic acid ion exchange) resin for 4 hr/160 °C.[102]

Coatings compositions useful for clear top coats on automobile finishes were developed, which contained large amounts of polymer-bound UV-light absorbers and a crosslinker. Resorcinol monobenzoate and 2,4-dihydroxybenzophenone and its derivatives were suitable UV stabilizers which could be bound to a reactive polymer. As an example, a butyl acrylate/hydroxyethyl acrylate/glycidyl methacrylate (49/4/37 weight ratio) copolymer was reacted with 2,4-dihydroxybenzophenone or resorcinol monobenzoate to give a polymer containing 20–50% bound UV stabilizer. A typical enamel contained 50–80 parts of this acrylic UV stabilizer resin, 20–50 parts hexamethoxymethylmelamine (crosslinking agent), 0.1–13 parts *p*-toluenesulfonic acid, 0–50 parts pigment and/or metallic flake, and 20–400 parts solvent. The coating was applied to the substrate and baked for 30 min/120 °C to give a hard, glossy finish. Many variations on this theme were described. These coatings resisted intercoat adhesion failure and flaking off the surface layer, and they retained their gloss.[103]

Sunscreen formulations for cosmetics were prepared which contained 2,4-dihydroxybenzophenone or 2,2′,4,4′-tetrahydroxybenzophenone absorbed in montmorillonite clay as one of the UV absorbers, as well as other, conventional components. The lattice compound of hydroxybenzophenone in clay did not induce skin irritation and could be easily dipersed in almost transparent form.[104]

Novel, polymerizable 3-aroyloxyphenyl carbamates, such as, e.g., compound **11-59**, were prepared which had the feature of latent UV stabilizers. These compounds formed addition polymers useful as surface coatings for resins (e.g., polycarbonate) which were resistant to discoloration due to UV irradiation. The coatings were also suitable as primers for abrasion-resistant coatings such as silicone hardcoats.[105]

Adducts of UV-light stabilizers having phenolic hydroxyl groups with aliphatic glycidyl ethers solubilized the stabilizer in the polymeric matrix, were nonfugitive,

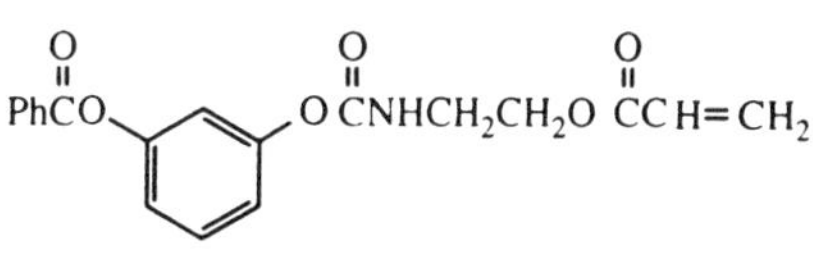

11-59

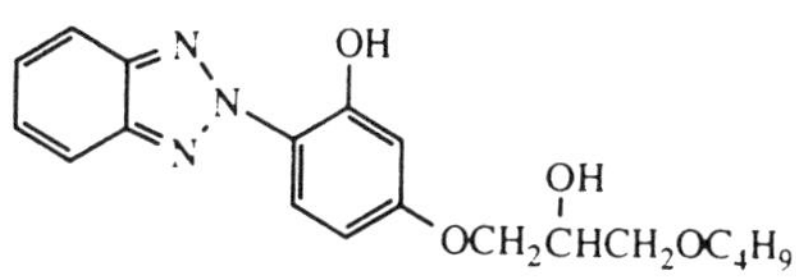

11-60

and showed superior performance as stabilizers. One example of such materials was compound **11-60**.[106]

A transparent polyester film, for use in color photography, was stabilized with 5,5′-methylene-bis-(2-hydroxy-4-methoxybenzophenone), **11-38**. This support film was used for color diffusion transfer and thermal development. The film had high UV-light resistance, good coloring properties, and heat resistance.[107]

Hydroxyphenyltriazines such as, e.g., compound **11-61** were synthesized and found to be useful as UV absorbers/stabilizers in color photographic materials.[108]

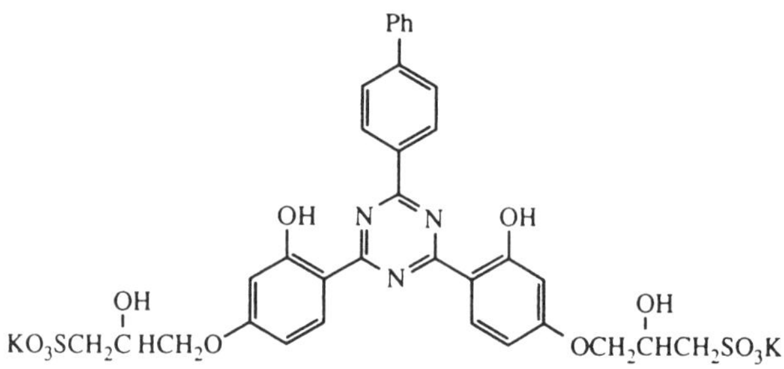

11-61

A new process for making 2,2′,4,4′-tetrahydroxybenzophenone was developed, consisting of the reaction of resorcinol in nitrobenzene with phosgene ($COCl_2$) in the presence of $ZrCl_4$ as the catalyst. The conversion of resorcinol was 85% and a 74% yield of the desired product was obtained.[109]

Another process for the production of 2-hydroxy-4-octyloxybenzophenone was invented in Hungary. Thus, a mixture of 300 mL *n*-butanol, 150 g 2,4-dihydroxybenzophenone, 78 g KI, 5 g triethylamine, 112 g sodium carbonate, and 130 g *n*-octyl chloride was stirred and refluxed for 15 hr, then cooled to 100 °C; 250 mL water was added and 7 g $Na_2S_2O_4$ and the mixture was refluxed for another 15 min to give a 90% yield of 2-hydroxy-4-octyloxybenzophenone.[110]

Yet another process for the production of 2,4-dihydroxybenzophenone itself was patented. An equimolar amount of benzotrichloride was added to a mixture of resorcinol, $ZnCl_2$, conc HCl, and water to give a 92% yield of dihydroxybenzophenone.[111]

A polymer (such as an acid-catalyzed thermoset acrylic or alkyd) film, coating, or molded product was stabilized against degradation by UV light, moisture, and air by adding 0.1–5 wt% each of a synergistic mixture of a hindered amine (HALS) light stabilizer and, as the UV absorber, a triaryltriazine such as, e.g., compound **11-62**.[112]

Esters of 1,3-bis-(3-hydroxy-4-benzoylphenoxy)-2-propanol with unsaturated carboxylic acids, exemplified by compound **11-63**, could be homo- or co-polymerized and were useful, nonfugitive UV-light stabilizers for, e.g., poly(methacrylate esters), polystyrene, or ABS.[113]

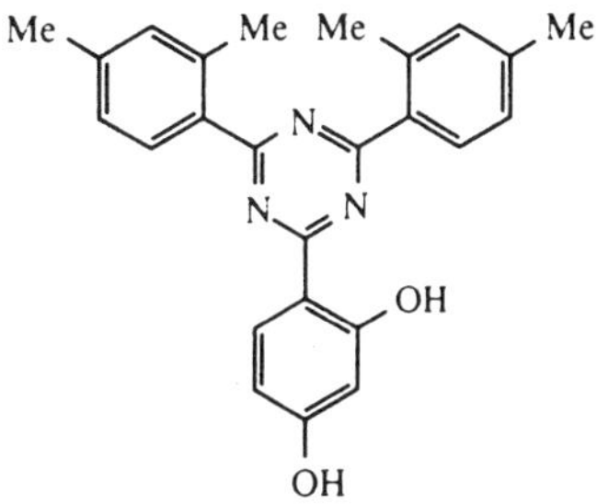

11-62

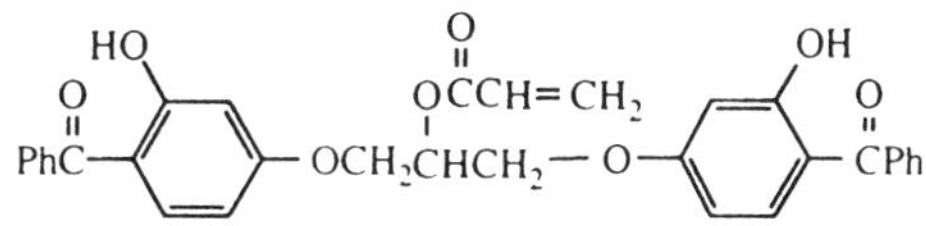

11-63

Polyphenylene oxide (PPO) and PPO/polystyrene blends were photostabilized by a mixture of 2,4-dihydroxybenzophenone-type UV absorbers, a hindered amine light stabilizer (HALS), and a cycloaliphatic epoxide synergist.[114]

The Zn, Ca, Mg, or Ba salts of 2-hydroxy-4-methoxybenzophenone-5-sulfonic acid, and related compounds, were useful as sunscreening agents in cosmetic preparations. These divalent metal sulfonate salts were insoluble in water, organic solvents, oils, fats, or sebum and were not absorbed through the skin.[115]

An abrasion-resistant coating composition with improved weatherability was developed, consisting of a partial condensate derived from an organotrialkoxysilane (e.g., methyl trimethoxysilane) and containing at least about 30 wt% methyltrisilanol, a colloidal, aqueous dispersion of silica, a colloidal aqueous dispersion of a metal (metal alloy or salt or oxide, such as antimony oxide), isopropanol/*n*-butanol, and 2-hydroxy-4-[gamma-(trimethoxysilyl)propoxy]benzophenone, which is a silylated UV absorber capable of coreacting with the organotrialkoxysilane.[116] (This reference/patent gives a lot of background information and prior art.)

A product of the addition of an acrylic prepolymer containing glycidyl methacrylate to the hydroxyl group of resorcinol monobenzoate was prepared and was used in coating compositions and films to impart strong UV absorbent properties. This patented preparation is similar to prior art which is well reviewed there.[117]

A polymer based on acrylic and/or methacrylic acid was modified by reaction with 2-hydroxy-4-glycidoxybenzophenone or similar compounds. This polymer was resistant to degradation by UV light and was useful as a soil conditioner and water retention aid.[118]

Another type of reactive UV absorber was prepared by esterifying 2-hydroxy-4-(2-hydroxyethoxy)benzophenone with 2-ethylhexyl monomaleate in toluene, using *p*-toluenesulfonic acid as the catalyst, to produce compound **11-64**. This additive was homo- or co-polymerizable to make it nonfugitive.[119]

Radiation-sterilizable polyolefins for use in the medical field were modified with a hindered piperidine compound (HALS) and resorcinol monobenzoate to protect them from degradation by high-energy radiation.[120]

The lightfastness of dyes on polyester and cellulose ester fiber materials was improved by adding an *o*-hydroxyphenylpyrimidine such as, e.g., compound **11-65** to the dyebath at 0.5–7.5 wt% (based on the fiber).[121] Suitable for the same application were said to be certain 2,4-dihydroxybenzophenone derivatives, such as compound **11-66**. These stabilizers exhausted well from the dyebath on polyester, had a low tendency to sublime, and did not affect the brightness of the dyeing.[122]

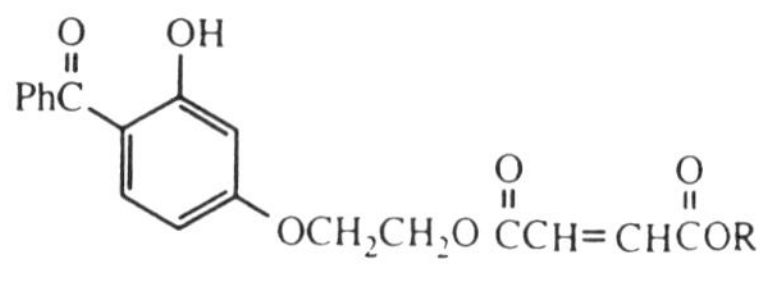

11-64

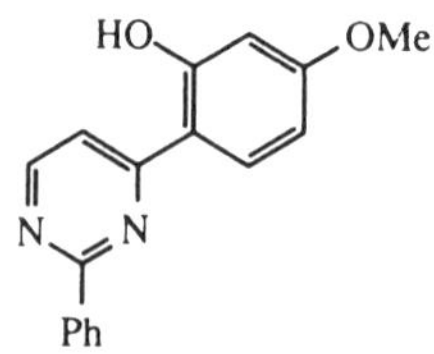

11-65

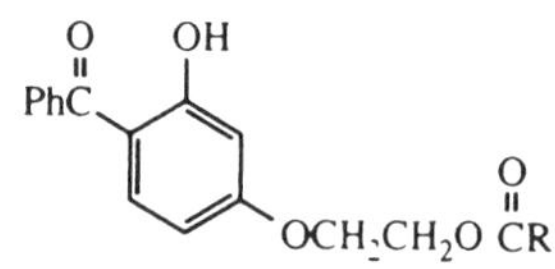

11-66
(where R= C_{1-12} alkyl, hydroxyalkyl, cycloalkyl, aryl, etc.)

Perfume compositions containing monoterpene compounds (e.g., alpha-pinene, limonene, citronellal, etc.) and natural, orange essential oils (e.g., lemon oil, orange oil, or bergamot oil, etc.) were stabilized against deterioration by light or heat by the addition of a 2,4-dihydroxybenzophenone or derivative.[123]

A process for the surface impregnation of thermoplastic polymers with UV absorbers was patented. For instance, a 6% solution of 2-hydroxy-4-octyloxybenzophenone in 1/1/1 propanediol/2-butoxyethanol/4-hydroxy-4-methyl pantanone was sprayed onto a Lexan polycarbonate film and dried to give a film having the same optical properties before and after washing with isopropanol.[124]

Hydrazides of 2,4-dihydroxybenzophenone derivatives, such as compound **11-67**, were found to be excellent heat/light/oxidation stabilizers for polymers. A coatings formulation containing this hydrazide showed excellent gloss retention.[125]

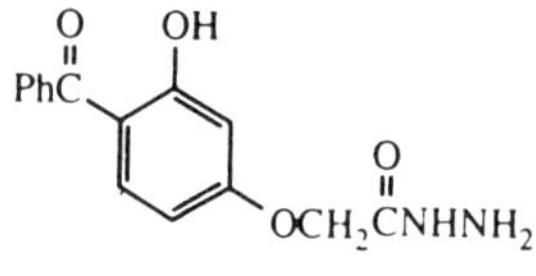

11-67

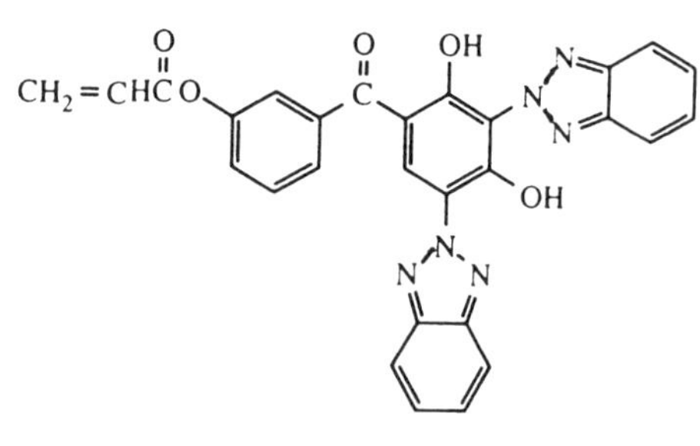

11-68

Bis-(benzotriazolyl)dihydroxybenzophenones or their derivatives, such as, e.g., compound **11-68**, were prepared and used as UV absorbers for contact and eyeglass lenses. In one formulation, compound **11-68** was used at the 0.15-ppm level in poly(methyl methacrylate) used for intraocular lenses.[126]

The UV-light stabilizer, 2-hydroxy-4-(2′-hydroxyethoxy)benzophenone, **11-6**, was prepared by a new process, consisting of the reaction of 2,4-dihydroxybenzophenone with ethylene carbonate at 140–150 °C, using benzyl triethylammonium chloride as the catalyst, to give a 93% yield of the ethoxylated product of 97% purity.[127]

3-Ethoxycarbonyl-7-hydroxycoumarin, **11-69**, and related compounds, and 3-(2-benzoxazolyl)-7-hydroxycoumarin, **11-70**, were prepared and reacted or copolymerized with linear, thermoplastic polyesters. These UV stabilizers were thus nonextractable from the polymer and were stable during processing.[128]

A hexamethoxymethylmelamine/2,4-dihydroxybenzophenone acid-catalyzed reaction product, containing compound **11-71** and related materials, was synthesized and shown to be an excellent UV stabilizer for curable fluoropolymers, such as a 62/12/16 C2F4/allyl alcohol/vinyl propionate copolymer, cured with the additive for 40 min/130 °C.[129]

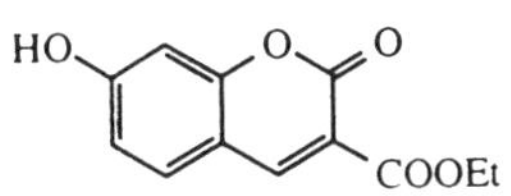

11-69

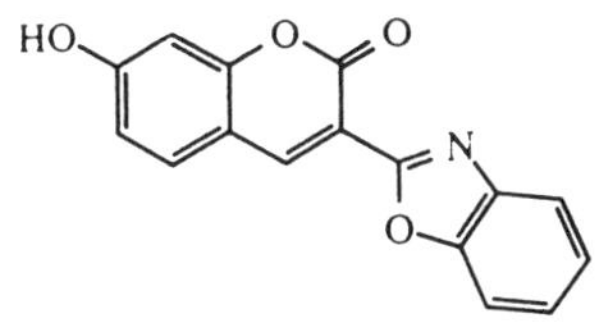

11-70

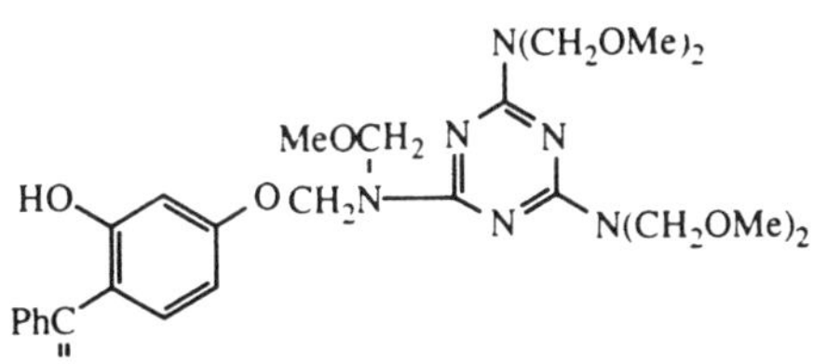

11-71

Japanese Cedar wood panels were coated with a sealer, hardener, thinner, and a UV absorber, 2,2′-dihydroxy-4,4′-dimethoxybenzophenone, **11-33**. These panels had good resistance to discoloration or bleaching on UV-light exposure.[130]

An intraocular lens was prepared from a copolymer made from 95 parts methyl methacrylate, 5 parts ethylene glycol dimethacrylate, 0.2 part 2,2′-azobisisobutyrate (initiator), and 5 parts 2-hydroxy-4-(2-methacryloxybutoxy)benzophenone. This material was colorless, transparent, and the UV stabilizer was practically nonextractable in aqueous solutions.[131]

UV-absorbing polyester composition useful for packaging materials, such as film and containers, was prepared. For example, 254 parts bis-(2-hydroxyethyl)terephthalate, 83 parts terephthalic acid, and 1.01 parts 2-hydroxy-4-(2-hydroxyethoxy)benzophenone were esterified/polycondensed at 255–285 °C to produce a UV-absorbing polymer with good stability.[132]

Sunscreen cosmetic compositions were patented which contained 2-hydroxybenzophenone-type UV absorbers, e.g., 2,2′,4,4′-tetrahydroxybenzophenone, **11-8**, in silica-microencapsulated form. Approximately 30 wt% of these microcapsules could be incorporated into the sunscreen formulations, which were more compatible with the skin than conventional, prior art compositions.[133]

Homo- and co-polymerizable unsaturated derivatives of UV absorbers were prepared. Thus, 2,4-dihydroxybenzophenone was reacted with vinyl chloroacetate in the presence of $NaHCO_3$ and methylhydroquinone (polymerization inhibitor) in acetonitrile at 80 °C to produce compound **11-72**.[134]

UV-absorbing resin compositions were developed which were useful in plastics sheets, films, and laminates. As an example, a UV stabilizer made by reacting 2,4-dihydroxybenzophenone with glycerine polyglycidyl ether was mixed with polypropylene glycol, butyl acetate, a small amount of acetic acid (catalyst), and a polyisocyanate (Coronate L). This mix was applied to unstretched polypropylene film. Laminates were made which had good adhesion and transparency, and the UV absorber was not extractable by solvents.[135]

Weather-resistant poly(vinyl chloride) compositions useful in agricultural applications were prepared. For instance, 100 parts PVC and 0.3 part 2-hydroxy-4-(methacryloxyethoxy)benzophenone were suspension-polymerized and blended with a plasticizer, heat stabilizer, and other, conventional additives. The product was converted to a 0.1-mm film which showed weatherability 4600 hr vs 2800 hr when a blend of PVC and the noncopolymerizable 2-hydroxy-4-octoxybenzophenone were used instead.[136]

Compounds containing both a 2-hydroxybenzophenone and a 1,4-dihydropyridine segment, such as compound **11-73**, were prepared and shown to be useful for stabilizing polymers (e.g., plasticized, impact-resistant PVC film) against heat and light.[137]

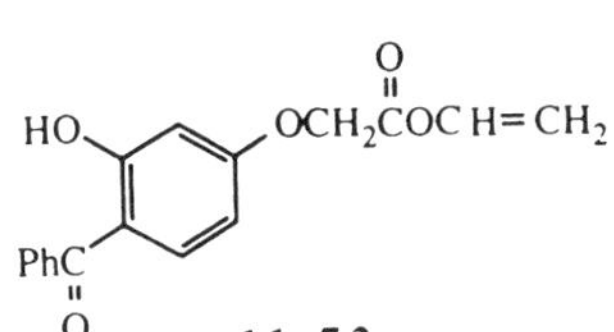

11-72

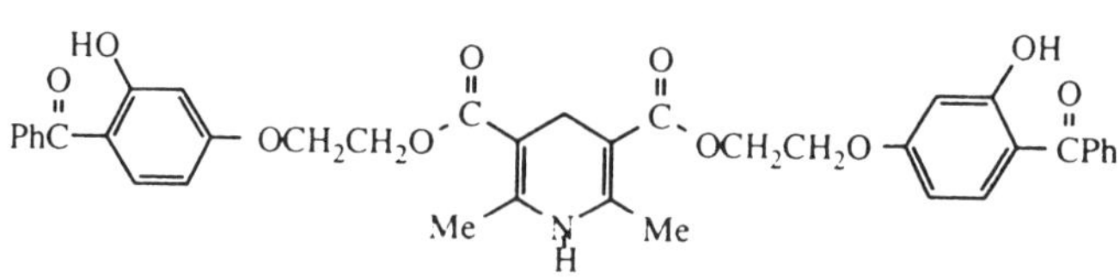

11-73

An electrophotographic color toner, based on polyester resin, C.I. Pigment Red 81, Zn-salicylate, and iron powder was prepared with 2,4-dihydroxybenzophenone added as an image stabilizer.[138]

A jet printing ink containing water, glycerine, and C.I. Acid Yellow pigment was UV-light-stabilized with the 2-hydroxyphenyl-*s*-triazine derivative, **11-74**.[139]

Benzotriazole-type reactive UV absorbers were prepared by reacting the sodium salt of 2-(2,4-dihydroxyphenyl)-2*H*-benzotriazole with gamma-butyrolactone at 180–200 °C to obtain 4-[4-(2*H*-benzotriazol-2-yl)-3-hydroxyphenoxy]butanoic acid. This acid was esterified with *n*-dodecanol to give the dodecyl ester, or reacted with glycidyl methacrylate to give a product containing compound **11-75** as the major component. These products were useful as UV absorbers/stabilizers for cosmetics, coatings, films, contact lenses, etc.[140]

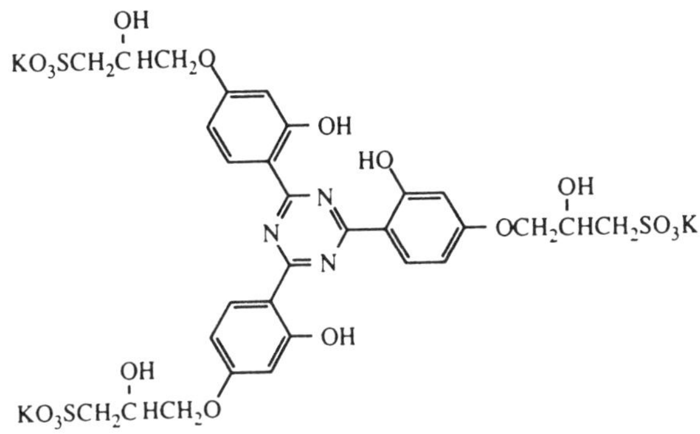

11-74

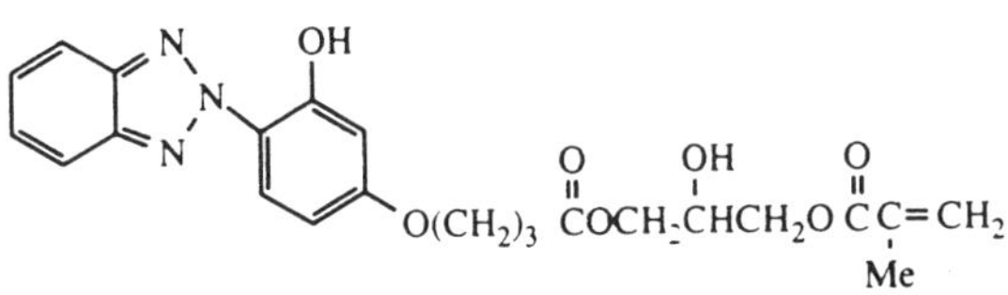

11-75

Light-resistant spandex (polyurethane elastomer) fiber blend fabrics were prepared by treatment of the fiber with UV absorbers. For instance, a knit nylon 6/spandex fabric was dyed and treated with a mixture of a synthetic tannin and 2,2′,4,4′-tetrahydroxybenzophenone/2,2′-dihydroxy-4,4′-dimethoxybenzophenone for UV-light stabilization.[141]

Novel 2-hydroxybenzophenone derivatives of structure **11-76** were developed as UV-light stabilizers for aromatic polycarbonates and polyester/carbonates. The effectiveness of these additives was proven in accelerated weathering tests.[142]

p-Phenylenediacrylic acid esters of 2,4-dihydroxybenzophenone were prepared and shown to be excellent UV absorbers/stabilizers for coatings, fibers, cosmetics, and so on. An example is compound **11-77** which, in methylene chloride solution, showed high UV absorption in the 280–350 nm range, with a maximum at 331 nm at which the molecular extinction coefficient was 7.1×10^4.[143]

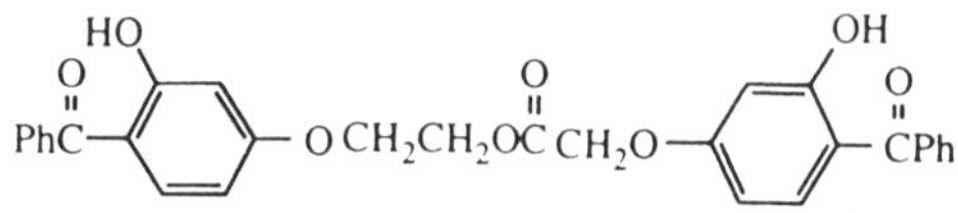

11-76

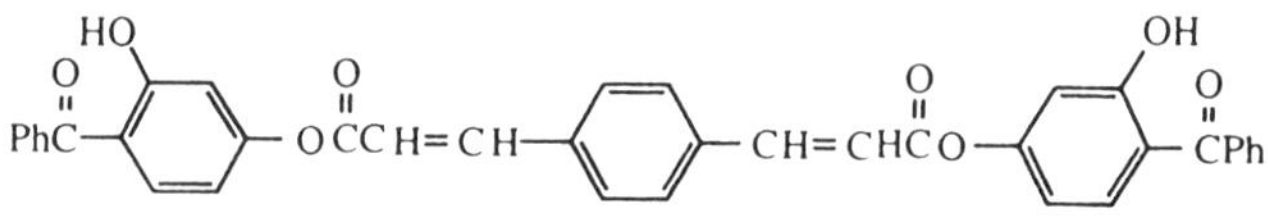

11-77

The acetate, benzoate, and methacrylate, **11-78**, of 1,3-bis-(3-hydroxy-4-benzoyl)-2-propanol were prepared and found to be effective for improving the lightfastness of diperse dyes on polyester yarn.[144]

Benzotrizolylresorcinols such as, e.g., compound **11-79** were excellent additives to improve the light fastness of dye images and whiteness of background in light-sensitive silver halide, color photographic paper.[145]

Silyl derivatives of 2-hydroxybenzophenones were prepared for use as UV absorbers for sunscreens. Thus, $Me_3SiCH_2CH_2COCl$ was reacted with 2,4-dihydroxybenzophenone in diethyl ether to give compound **11-80**. This new product was formulated into sunscreen lotions and creams which effectively blocked UV light from the skin.[146]

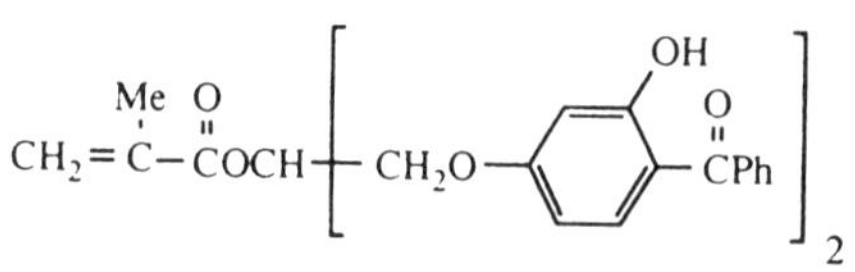

11-78

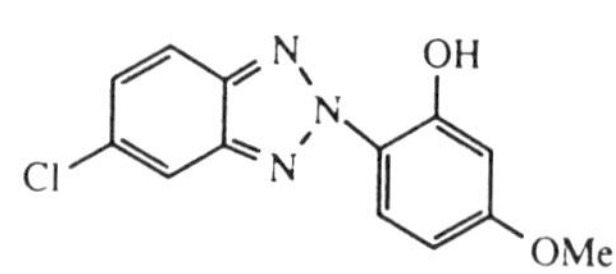

11-79

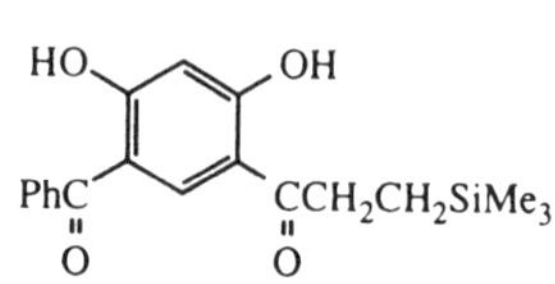

11-80

Another new process for making 2,2′,4,4′-tetrahydroxybenzophenone was developed. Resorcinol, 2,4-dihydroxybenzoic acid, and $ZnCl_2$ in sulfolane were treated with $POCl_3$ over 2 hr/50 °C and stirred for another 1 hour to obtain a 92% yield of the target compound.[147]

Dyes on polyamide fibers were stabilized with a hindered phenol/UV stabilizer combination. For instance, a nylon-6 fiber was dyed in a dyebath containing 0.2% red azo dye, 1% antioxidant (a sulfonated hindered phenol), and 1% of compound **11-81**. The dyed fibers had excellent physical properties and xenon lamp light fastness of the dyeing.[148]

Still another process for making polyhydroxybenzophenones was patented. For example, resorcinol was reacted with 2,4-dihydroxybenzoic acid in the presence of $POCl_3$ and $ZnCl_2$ in methanesulfonic acid at 70 °C/2 hr to get a 70.5% yield of 2,2′,4,4′-tetrahydroxybenzophenone.[149]

Ethoxylated and propoxylated 2-hydroxybenzophenones were prepared. Thus, an autoclave was charged with 2,4-dihydroxybenzophenone and a catalytic amount of NaOH and, at 160 °C, ethylene oxide was added dropwise. The charge was held for another 1 hr/160 °C to give a 99.6% yield of a product containing 97.4% 2-hydroxy-4-(2-hydroxyethoxy)benzophenone, **11-6**.[150]

Novel hydroxyphenyltriazines were developed as UV stabilizers. For instance, compound **11-82** was added at 1.5 wt% to a clear melamine resin/crosslinked acrylic resin coating on a metal base. The coating showed good gloss retention after a 5000-hr Weatherometer test.[151]

Another version of the preparation of tetrahydroxybenzophenones was patented. For example, resorcinol monomethyl ether in xylene was reacted with $COCl_2$ in the presence of $AlCl_3$ at 0–30 °C, then heated to 40–110 °C to remove the methyl groups, diluted with gamma-butyrolactone to eliminate agitation problems and worked up to obtain 2,2′,4,4′-tetrahydroxybenzophenone in high yield.[152]

A copolymer of compound **11-83**/methyl methacrylate/triethylene glycol dimethacrylate/alpha-methylstyrene was prepared. This copolymer was suitable for making intraocular lenses with improved stability.[153]

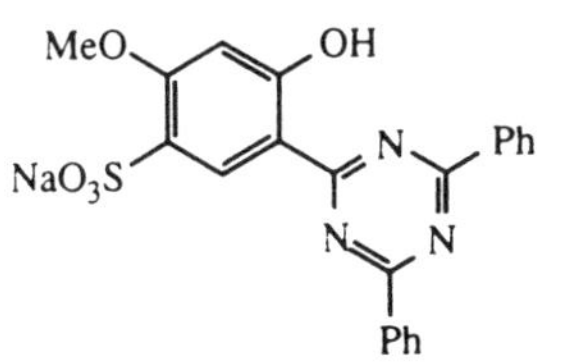

11-81

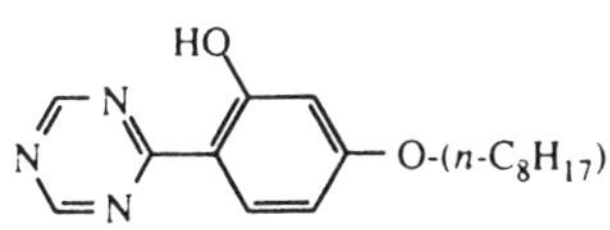

11-82

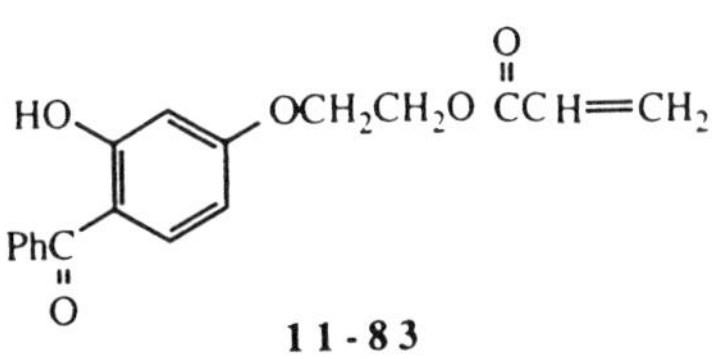

11-83

11.3.9. Food Additives

Examples of products in this category, from the 1980–1992 period, include synthetic sweeteners and browning inhibitors. Sweeteners for edibles of the dihydrochalcone type were prepared starting from 4-acetylresorcinol (or from phloroglucinol derivatives). Examples of these sweeteners are structures **11-84** and **11-85**.[154]

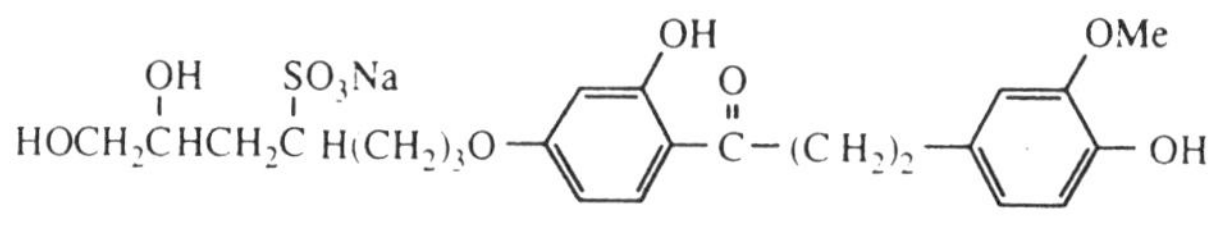

11-84

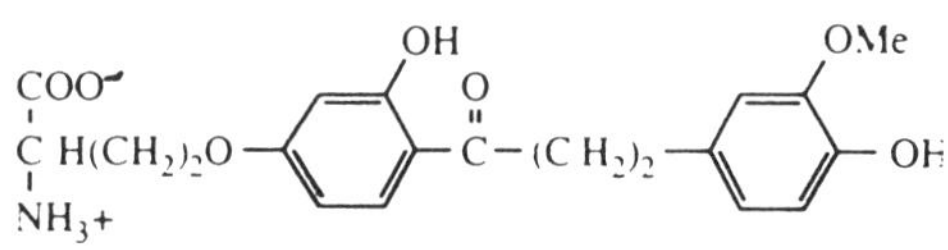

11-85

Flavone-type sweeteners were also made. For instance, compound **11-86** was synthesized, starting from 2-acetylresorcinol. This product had 350 times the sweetening power of cane sugar.[155]

The 3-hydroxyphenoxytetrazole, **11-87**, was prepared. It was a sweetener for foods named Melizame by Lilly.[156]

Compositions for preventing or inhibiting the oxidative darkening (browning) of foods and beverages, such as shrimp, cut potatoes, apples, avocados, fruit juices, and wines, were developed. These compositions contained at least one resorcinol derivative, e.g., 4-*n*-hexylresorcinol, **11-88**, or compound **11-89**, and a second component such as sodium metabisulfite, ascorbic acid, citric acid, and chelating agents such as

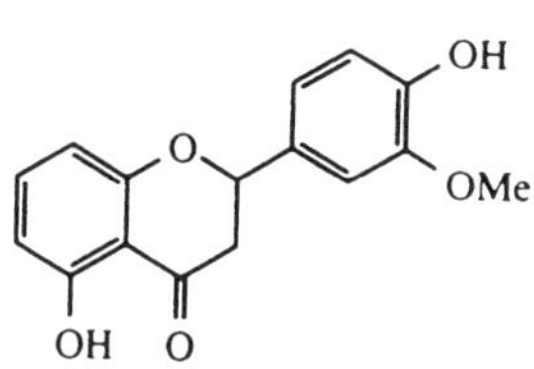

11-86

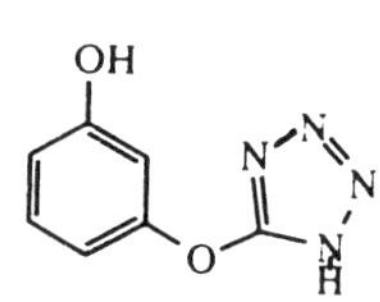

11-87

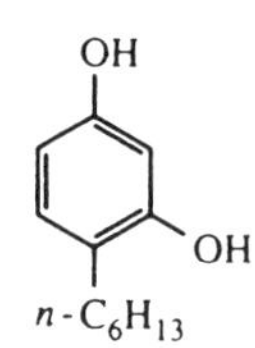

11-88

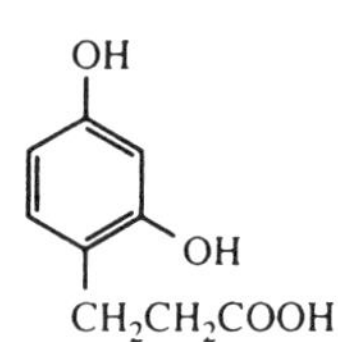

11-89

ethylenediaminetetraacetic acid (EDTA) or a salt such as sodium chloride or bicarbonate.[157]

11.3.10. Miscellaneous Additives

The various examples of the use of resorcinol/derivatives as additives for organic materials presented here are a further demonstration of their unusual range of applications.

Alkoxysilyl ethers of cyclic 1,3-diones were made, such as compound **11-90** (the dione is shown in the enol form). These products were useful as vulcanizing agents for silanol-terminated poly(diorganosiloxanes).[158]

The workability of concrete mixes was improved by adding the sodium salt of a sulfonated naphthalene/formaldehyde condensation product (a known plasticizer/water reducing agent for concrete) and *ca* 0.1 wt%, based on the cement content, of a polyhydric phenol such as resorcinol. This additive prevented a lowering of the slump value of the concrete.[159]

Triglycidyl-*m*-aminophenol (and *o*- or *p*-aminophenol) was reacted with acrylic acid to give compound **11-91**. This adduct was useful as a component of UV-light-curable paints, varnishes, printing inks, adhesives, etc.[160]

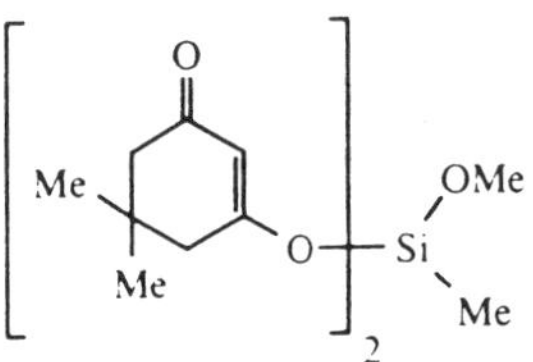

11-90

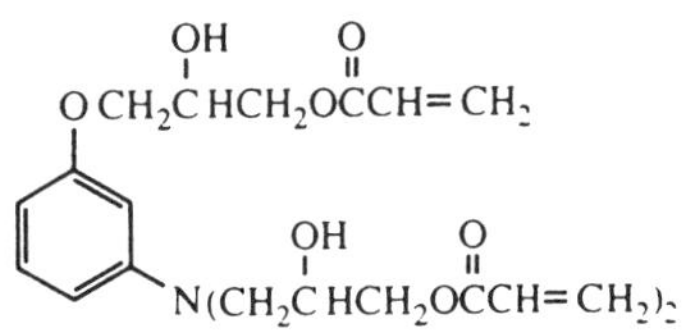

11-91

A hydraulic (brake) fluid was developed which contained borates of arenediols. As an example, resorcinol, triethylene glycol monomethyl ether, and boric acid at 1/5/2 mole ratio were refluxed at 110 °C and the water of reaction was removed by vacuum distillation. The dehydrated product was mixed with 60.6 wt% triethylene glycol monomethyl ether and the resultant hydraulic fluid had a suitably high bp 225 °C, could bind water and remain high-boiling, and had a favorable viscosity (2.2 mm^2/sec at 100 °C, 1015 mm^2/sec at –40 °C).[161]

An adduct of 100 parts 4,4′-diphenylmethane diisocyanate with 2.4–3.0 parts resorcinol was prepared at 55–90 °C, using triethylamine/dibutyltin laureate as the catalysts. The melt was solidified by cooling and crushed to make a free-flowing, dry powder which remained free-flowing on storage. This product contained 27–31 wt% isocyanate groups and was used to prepare thermoplastic polyurethanes and cast polyurethane elastomers, etc.[162]

Self-bonding, dimedone (5,5-dimethyl-1,3-cyclohexanedione) room-temperature-vulcanizable (RTV) silicone rubber compositions were prepared. These consisted of a silanol endcapped with methyl-3,3,3-trifluoropropyl-polysiloxane, fumed silica filler, and iron oxide, to which was added a catalyst solution of hexyltri-dimedonylsilane and bis-(trimethoxysilylpropyl)maleate or 1,3,5-tris-(3-trimethoxysilylpropyl)isocyanurate plus dibutyltin diacetate. This mix cured to a silicone elastomer on exposure to atmospheric moisture, with final cure from 24–27 hr at room temperature. The product showed good adhesion to aluminum and stainless steel.[163]

Wood pulp from the Kraft process was bleached with oxygen and alkali (4% on the fiber weight) and 0.3–1.5% resorcinol to give an improved yield of bleached pulp with good whiteness.[164]

A sulfite pulping process for hardwood or softwood chips was improved by the addition of "sulfite-containing waste from resorcinol production" (15% Na_2SO_3, 0.2% resorcinol/sodium bisulfite adduct, the rest water). Better delignification of the cellulose than without the resorcinol byproduct was obtained.[165]

A 3% hydrogen peroxide solution for medicinal use was stabilized with 0.05% sodium benzoate, 0.05% sodium salicylate, and 0.01% resorcinol. This solution had satisfactory activity after 2 yr storage.[166]

Molding compositions were prepared from cellulose acetate waste. The waste was regenerated from spent filter cloth obtained from acetate fiber production and was plasticized with a condensate prepared from resorcinol with dimethyl phthalate in 1/20 to 10/29 ratio.[167]

Rapid-curing thermoset polymers and prepolymers were obtained by the condensation of 2,4-di- or 2,6-di-, or 2,4,6-tri-methylpyridine with terephthalaldehyde, with resorcinol (*ca* 10 wt%) as a cure accelerator. Other dihydric phenols were less effective than resorcinol. These resins had good mechanical properties at high temperatures (up to 450 °C). They were suitable for the fabrication of composite materials.[168]

A method for improving the color of tall oil rosin esters was described involving sequentially (a) heating and stirring tall oil rosin in the presence of a Lewis acid catalyst (e.g., $ZnCl_2$) to form a mixture, (b) distilling the mixture to provide a tall oil rosin distillate, and (c) esterifying this distillate with a polyol in the presence of a phenol sulfide (e.g., thio-bis-resorcinol). The products were useful for making paper and textile sizes, plasticizers for polyolefin films, paints, varnishes, hot-melt adhesives, and pressure-sensitive adhesives.[169]

Poly(phenylene oxide) (PPO)/ polyamide blends were developed which had improved resistance to water absorption and swell, by incorporating therein a phenolic compound (several phenols worked; resorcinol at 2 parts/100 blend performed well).[170]

Thermoplastic, chopped glass fiber-reinforced polyester (PET) compositions were modified with tetrahydroxybenzophenone metal salts, such as **11-92**, to give products with good hydrolysis and heat resistance and a quick rate of crystallization.[171]

Wood chips were pulped by cooking in aqueous ethanol containing 1–5 wt% nonylphenol, resorcinol, or pyrocatechol. These additives increased the degree of delignification of the pulp.[172]

Oxalic acid diarylamides such as, e.g., compound **11-93** were synthesized and demonstrated to be useful as photochemical and thermal stabilizers for textile dyes.[173]

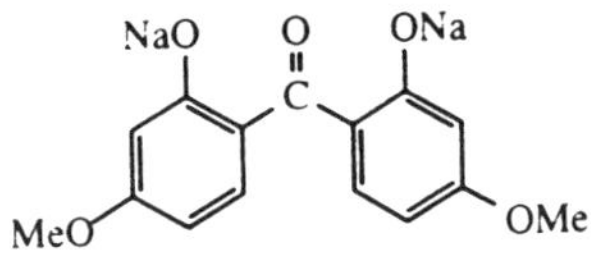

11-92

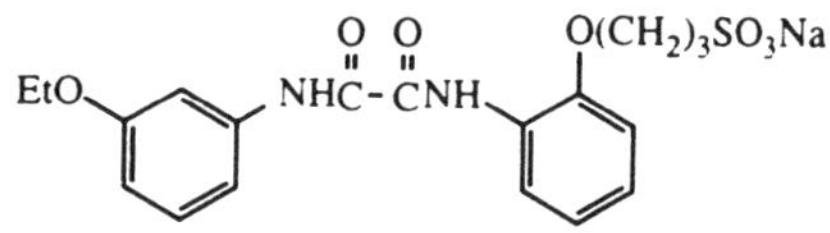

11-93

A blend of poly(ethylene terephthalate) (PET) with 30 phr resorcinol dibenzoate was converted to a stretched film which had very low gas permeability. This blend was suitable for packaging films and containers.[174]

Cellulose triacetate solutions in EtOH/CH_2Cl_2 with added 0.02–0.06 mass% resorcinol and 0.10–0.14% 2-mercaptobenzothiazole were patented for their improved filterability.[175]

C_{4-8}-alkyl diesters of *m*-phenylene diacetic acid, such as compound **11-94**, were prepared and shown to be effective plasticizers for PVC.[176]

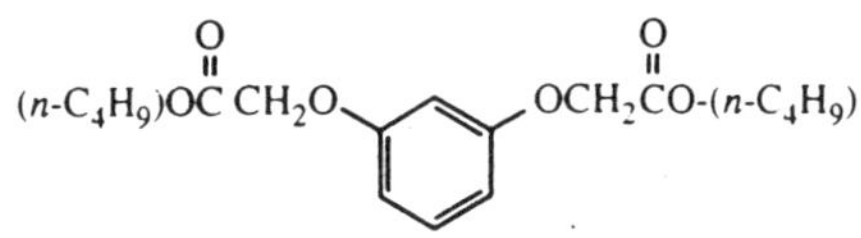

11-94

A mixture of 35 parts acrylic resin (NT-100), 20 parts Cu thiocyanate, 3 parts 4-*n*-hexylresorcinol, and 70 parts xylene was demonstrated to be suitable as an antifouling coating, nontoxic to humans, comparable in performance to the toxic, tin-compound-modified acrylic resin coatings.[177]

Glass was protected with a material having high tensile strength and large rupture elongation, which consisted of 100 parts poly(vinyl butyral) and 0.6–8.3 parts dihydroxybenzene (e.g., resorcinol).[178]

11.4. REFERENCES

1. E. g., H. Lind, *Kunststoffe, 69(7)*, 397 (1979); M. Dexter, in *Kirk-Othmer's Encyclopedia of Chemical Technology*, 3rd ed., Vol. 23, p. 615, John Wiley & Sons, New York, 1983.
2. E. g., G. Scott, ed., *Developments in Polymer Stabilization*, Vol. 6, Elsevier Applied Science Publishing, Amsterdam, 1983; a complete list of commercial UV stabilizers and their properties is given in: E. W. Flick, *Plastics Additives: An Industrial Guide*, Noyes Data Corp., Park Ridge, N.J., 1986.
3. SRI (Stanford Research Institute) International, *Specialty Chemicals Update Program, Plastics Additives—Jan. 1992.*
4. U.S. Pat. 3,983,286 (Sept. 28, 1976; A. Arsac to Rhone-Poulenc Textile).
5. Jpn. Kokai Tokkyo Koho JP 55,104,375 [80,104,375] (Aug. 9, 1980; to Taoka Chemical K.K.); *C.A., 94*, 4737 (1981).
6. Ger. (East) DD 271,908 (Sept. 20, 1989; E. Born, M. Goritzka, G. Bageritz, and G. Mehl to VEB Chemische- und Lackfabrik Doellnitz); *C.A., 112*, 120792 (1990).

7. PCT Int. Appl. WO 90,012,846 (Nov. 1, 1990; G. A. Salensky and T. S. Thoman to Amoco Corp.); *C.A., 114*, 385282 (1991).
8. U.S. Pat. 5,075,034 (Dec. 24, 1991; M. A. Wanthal to the Dexter Corp.).
9. Brit. Pat. GB 989,560 (Apr. 22, 1965; T. V. Cullum and E. V. Whitehead to British Petroleum Co., Ltd.); *C.A., 63*, 415 (1965).
10. Belg. Pat. BE 875,027 (July 16, 1979; K. R. Molt to Cincinnati Milacron Chemicals, Inc.); *C.A., 92*, 23542 (1980); also, Ger. Offen. DE 2,917,761 (July 9, 1981; same author and assignment).
11. O. P. Lykov, T. P. Vishnyakova, N. V. Tumar, E. P. Seregin, V. N. Prokudin, and I. A. Konishcheva, *Khim. Technol. Topliv i Masel*, 1980 (5), 18 [English translation by Plenum Publ. Corp., *Chemmotechnology Fuels Lubr.*, 1980, 316.]
12. Jpn. Kokai Tokkyo Koho JP 82,004,663 (Jan. 27, 1982; H. Kato, A. Murata, and S. Irioka to Dainichi Nippon Cables); *C.A., 84*, 136692 (1976).
13. T. S. Chao and M. Kjonaas, *Am. Chem. Soc., Div. Petr. Chem. Prepr., 27(2)*, 362 (1982).
14. USSR Pat. SU 895,985 (Jan. 7, 1982; E. Bisenieks, J. Uldrikia, and G. Duburs to Institute of Organic Synthesis, Academy of Sciences, Latvian SSR); *C.A., 96*, 219097 (1982).
15. F. Yamada, T. Nishiyama, and Y. Nakai, *J. Am. Oil Chem. Soc., 60(9)*, 1651 (1983).
16. U.S. Pat. 4,385,147 (May 24, 1983; H. O. Wirth, J. Buessing, and H. H. Friedrich to Ciba-Geigy Corp.).
17. U.S. Pat. 4,498,998 (Feb. 12, 1985; D. A. Hutchison to Atlantic Richfield Co.).
18. Ger. Offen. DE 3,422,428 (Dec. 19, 1985; R. Baur, K. Oppenlaender, and H. H. Vogel to BASF A.-G.); *C.A., 104*, 186112 (1986).
19. U.S. Pat. 4,599,464 (July 8, 1986; H. Dressler to Koppers Co., Inc.).
20. U.S. Pat. 4,656,302 (Apr. 7, 1987; H. Dressler to Koppers Co., Inc.).
21. U.S. Pat. 4,705,879 (Nov. 10, 1987; H. Dressler to Koppers Co., Inc.).
22. Jpn. Kokai Tokkyo Koho JP 62,103,085 [87,103,085] (May 13, 1987; H. Harada and M. Usui to Sumitomo Chemical Industries K.K.); *C.A., 109*, 92832 (1988).
23. Ger. (East) DD 291,088 (June 20, 1991; D. Ehrhardt, S. Hauptmann, G. Mann, W. Mertens, B. Noll, F. Weinelt, and H. Weinelt to Karl-Marx-Universitaet Leipzig); *C.A., 115*, 234431 (1991). Also, Ger. (East) DD 290,427 (May 29, 1991; same authors and assignment); *C.A., 115*, 258104 (1991).
24. Ger. (East) DD 291,902 (July 18, 1991; D. Ehrhardt, S. Hauptmann. G. Mann, W. Mertens, B. Noll, F. Weinelt, and H. Weinelt to Karl-Marx-Universitaet Leipzig); *C.A., 115,* 258105 (1991). A related patent is Ger. (East) DD 290,430 (May 29,1991; same authors and assignment); *C.A., 115*, 281981 (1991).
25. Ger. (East) DD 290,429 (May 29, 1991; D. Ehrhardt, S. Hauptmann, G. Mann, F. Weinelt, B. Noll, H. Weinelt, H. G. Fuchs, I. Solf, and L. Weber to Karl-Marx-Universitaet Leipzig); *C.A., 115*, 281840 (1991). See also: D. Ehrhardt, B. Noll, F. Weinelt, H. Weinelt, and G. Mann, *Gummi, Fasern, Kunst., 45*, 234 (1992); *C.A., 118,* 126231 (1993).
26. U.S. Pat. 5,006,270 (Apr. 9, 1991; L. O. Farng and A. G. Horodysky to Mobil Oil Corp.).
27. Eur. Pat. Appl. EP 487,036 (May 27, 1992; M. Fon and S. Strologo to Himont, Inc.); *C.A., 117*, 21376 (1992).
28. U.S. Pat. 4,052,466 (Oct. 4, 1977; K.K. Sun to Upjohn Co.).
29. Jpn. Kokai Tokkyo Koho JP 80,079,324 (June 14, 1980; to Ube Industries, Ltd.); *C.A., 94*, 121052 (1981).
30. French Pat. FR 2,106,975 (June 9, 1972; B. Torck and L. Sajus to Institute Francais Du Petrole); *C.A., 78*, 15747 (1973).
31. USSR Pat. SU 958,435 (Sept. 15, 1982; M. Z. Tsirkin, A. S. Zhukhova, T. A. Lykova, and V. V. Fedorova to "Elektrosila" Leningrad Industrial Electrical Machine-Building Enterprises); *C.A., 98*, 161768 (1983).
32. USSR Pat. SU 1,146,290 (March 23, 1985; to Mordovsk University).
33. Jpn. Kokai Tokkyo Koho JP 60,056,180 [85,056,180] (Dec. 9, 1985; to Asahi Chemical Industries K.K.); equiv. JP 56,057,825 [81,057,825] (May 20, 1981; same assignment); *C.A., 95*,162382 (1981).

34. Jpn. Kokai Tokkyo Koho JP 01,009,962 (Jan. 13, 1989; A. Ora and H. Kondo to Toray Industries, Inc.); *C.A., 111*, 79312 (1989).
35. Jpn. Kokai Tokkyo Koho JP 64,003,171 (Jan. 6, 1989; A. Ora, H. Kondo, and S. Shintani to Toray Industries, Inc.); *C.A., 111*, 79300 (1989).
36. U.S. Pat. 4,889,903 (Dec. 26, 1989; J. Baghdachi to BASF Corp.).
37. Eur. Pat. Appl. EP 429,128 (May 29, 1991; A. Villena, R. P. C. Van Gaalen, J. J. Chadwick, and T. K. Jurriens to Shell Internationale Research Maatschappij B. V.); *C.A., 115*, 184156 (1991).
38. Eur. Pat. Appl. EP 445,050 (Sept. 4, 1991; J. R. Desmurs and F. Pierre to Rhone-Poulenc Chimie); *C.A., 115,* 255798 (1991).
39. U.S. Pat. 5,118,768 (June 2, 1992; R. C. Job and D. F. Haas to Shell Oil Co.).
40. U.S. Pat. 5,118,767 (June 2, 1992; R. C. Job to Shell Oil Co.).
41. USSR Pat. SU 216,411 (Jan. 5, 1978; R. Metsik, L. Kalda, and E. Rohumagi to Scientific Research Institute of Shales); *C.A., 88*, 140626 (1978).
42. Jpn. Kokai Tokkyo Koho JP 58,144,364 [83,144,364] (Aug. 27, 1983; to Sumitomo Chemical Co., Ltd.); *C.A., 100*, 85402 (1984).
43. Jpn. Kokai Tokkyo Koho JP 58,210,986 [83,210,986] (Dec. 8, 1983; to Showa Denka K.K.); *C.A., 100*, 141513 (1984).
44. Jpn. Kokai Tokkyo Koho JP 58,194,133 [83,194,133] (Nov. 12, 1983; to Matsushita Electric Industries K.K.); *C.A., 101*, 162474 (1984).
45. Pol. Pat. PL 126,160 (Jan. 31, 1986; H. Jatymovicz *et al.*); *C.A., 106*, 88529 (1986). PL 128,062 (Oct. 2, 1986; A. Krolikowska *et al.*); *C.A., 108*, 75762 (1987). PL 128,054 (Jan. 2, 1986; T. Zak *et al.*); *C.A., 108*, 7574 (1987). PL 152,076 (Nov. 30, 1990; J. Iwanov *et al.*); *C.A., 115*,11013 (1991). All assigned to Instytut Mechanik Precyzyjnej.
46. USSR Pat. SU 1,645,273 (Apr. 30, 1991; P. Kaikaris, J. Blinas, A. Kaminskas, D. Matkenajte, A. Jurgauskas, E. Jakubonis, and R. Sh. Derevyanko to All-Union Scientific-Research Institute of Insulation Materials and Products); *C.A., 116*, 131295 (1992).
47. Jpn. Kokai Tokkyo Koho JP 55,080,267 [80,080,267] (June 17, 1980; to Nippon Muki Seni Ko).
48. Ger. Offen. DE 3,338,184 (May 2, 1985; A. Guentner to Roederstein and Tuerk KG Fabrik elektrischer Bauelemente); *C.A., 103,* 15595 (1985).
49. Jpn. Kokai Tokkyo Koho JP 62,193,111 [87,193,111] (Aug. 25, 1987; H. Shimizu and H. Yamada to Elna Co., Ltd.); *C.A., 108*, 47736 (1987).
50. Jpn. Kokai Tokkyo Koho JP 01,114,022 [89,140,022] (May 2, 1989; T. Tsunetsugu, H. Shimamato, K. Mori, and S. Yoshida to Matsushita Electric Industrial Co., Ltd.); *C.A., 111*, 185900 (1989).
51. Jpn. Kokai Tokkyo Koho JP 03,177,012 [91,177,012] (Aug. 1, 1991; T. Myashita to Shin-Ei Tsushin Kogyo Co., Ltd.); *C.A., 115,* 221316 (1991).
52. U.S. Pat. 2,520,090 (Aug. 22, 1950; G. R. Barrett to Monsanto Chemical Co.).
53. U.S. Pat. 4,035,442 (July 12, 1977; W. P. Dunworth to E. I. DuPont de Nemours Co.).
54. U.S. Pat. 4,203,888 (May 20, 1980; R. B. Rashbrook to ICI Americas Inc.).
55. Ger. Offen. DE 2,909,442 (Sept. 18, 1980; W. R. Haaf, D. R. Heath, and F. F. Holub to General Electric Co.); *C.A., 94*,16615 (1981).
56. Jpn. Kokai Tokkyo Koho JP 57,207,642 [82,207,642] (Dec. 20, 1982; to Adeka Argus Chemical Co., Ltd.); *C.A., 99,* 106272 (1983).
57. Jpn. Kokai Tokkyo Koho JP 59,024,736 [84,024,736] (Feb. 8, 1984; to Adeka Argus Chemical Co., Ltd.); *C.A., 101*, 152895 (1984).
58. Jpn. Kokai Tokkyo Koho JP 59,030,845 [84,030,845] (Feb. 18, 1984; to Adeka Argus Chemical Co., Ltd.); *C.A., 101*, 111934 (1984).
59. Jpn. Kokai Tokkyo Koho JP 60,063,251 [85,063,251] (Apr. 11, 1985; to Adeka Argus Chemical K.K.); *C.A., 103*, 124470 (1985).
60. U.S. Pat. 4,808,647 (Feb. 28, 1989; V. Abelins, J. E. Betts, F. Holub, and G. F. Lee, Jr. to General Electric Co.).
61. PCT Int. Appl. WO 90,003,417 (Apr. 5, 1990; J. C. Gosens, H. B. Savenije, C. F. Pratt, and C.A. A. Claesen to General Electric Co.); *C.A., 113*, 60708 (1990).

62. Eur. Pat. Appl. EP 356,633 (March 7, 1990; C.A. A. Claesen and J. H. G. Lohmeijer to General Electric Co.); *C.A., 113*, 79811 (1990).
63. Eur. Pat. Appl. EP 350,092 (Jan. 10, 1990; A. Petri to Enichem Tecnoresine S.p.A.); *C.A., 112*, 236060 (1990).
64. Eur. Pat. Appl. EP 485,807 (May 20, 1992; L. T. Gunkel and H. J. Barda to FMC Corp.); *C.A., 117*, 111806 (1992).
65. Eur. Pat. Appl. EP 491,986 (July 1, 1992; F. Tabankia and G. DeWit to General Electric Co.); *C.A., 117*, 235108 (1992).
66. A. M. Aaronson, *Phosphorous Flame Retardents for a Changing World*, presented at the 202nd ACS Natl. Meeting, The Fourth Chemical Congress of North America, Aug. 25–30, 1991 (Inorg. Chem. Sect., paper #346).
67. U.S. Pat. 4,022,700 (May 10, 1977; H. A. Harris to Shell Oil Co.).
68. Jpn. Kokai Tokkyo Koho JP 59,106,574 [84,106,574] (June 20, 1984; to Matsumoto Yushi Seiyaku Co., Ltd.); *C.A., 102*, 26790 (1985).
69. Ger. Offen. DE 3,308,742 (Sept. 22, 1983; M. Holuber to Lubrizol Corp.); *C.A., 99*, 197814 (1983).
70. U.S. Pat. 4,569,778 (Feb. 11, 1986; P. A. Miller, R. M. Mullins, and R. J. Bucko to Olin Corp.).
71. Eur. Pat. Appl. EP 32,275 (July 22, 1981; A. Paucot, P. Malfroid, and J. Mulders to Solvay and Cie.); *C.A., 96*, 6386 (1981).
72. Eur. Pat. Appl. EP 6,091 (Jan. 9, 1980; N. Kubota, T. Shibata, and K. Sugibuchi to S.A. Argus Chemical N.V.).
73. Eur. Pat. Appl. EP 6,686 (Jan. 9, 1980; to Imperial Chemical Industries); *C.A., 92*, 58622 (1980).
74. Ger. Offen. DE 2,713,376 (Nov. 3, 1977; K. R. Moult to Cincinnati Milacron Chemical Co.); *C.A., 88*, 38583 (1978).
75. U.S. Pat. 4,197,392 (Apr. 8, 1980; J. E. Moore to General Electric Co.).
76. U.S. Pat. 4,256,493 (March 17, 1981; S. Yokoyama, T. Toyoda, and N. Okamura to Dainippon Toryo K.K.).
77. U.S. Pat. 4,278,804 (July 14, 1981; B. A. Ashby and S. H. Schroeter to General Electric Co.).
78. Jpn. Kokai Tokkyo Koho JP 81,022,742 (March 3, 1981; to Shipuro Kasei K.K.); *C.A., 95*, 6830 (1981).
79. Jpn. Kokai Tokkyo Koho JP 81,002,536 (Jan. 20, 1981; to Mitsubishi Chemical Industries K.K.); equiv. JP 49,015,545 (Jan. 8, 1974; T. Haga, H. Fukutani *et al.*, same assignment); *C.A., 81*, 3596 (1974).
80. Ger. Offen. DE 3,206,129 (Nov. 4, 1982; K. Eiglmeier and J. Schulz to Riedel de Haen A.-G.); *C.A., 98*, 106978 (1983).
81. U.S. Pat. 4,269,629 (May 26, 1981; J. S. Zannucci to Eastman Kodak Co.).
82. U.S. Pat. 4,251,435 (Feb. 17, 1981; P. -N. Son and J. T. Lai to The B. F. Goodrich Co.).
83. Jpn. Kokai Tokkyo Koho JP 82,067,652 (Apr. 24, 1982; to Teijin, Ltd.); *C.A., 97*, 183610 (1982).
84. FR Demande 2,487,837 (Feb. 5, 1982; R. H. Chung and P. M. Turner to General Electric Co.); *C.A., 97*, 6517 (1982).
85. Jpn. Kokai Tokkyo Koho JP 57,035,220 [82,035,220] (July 28, 1982; to Adeka Argus Chemical K.K.).
86. Jpn. Kokai Tokkyo Koho JP 57,042,742 [82,042. 742] (March 10, 1092; to Sumitomo Chemical K.K.); *C.A., 97*, 73406 (1982).
87. Jpn. Kokai Tokkyo Koho JP 57,006,470 (Feb. 4, 1982; to Toray Industries, Ltd.); *C.A., 97*, 24876 (1982).
88. U.S. Pat. 4,323,710 (Apr. 6, 1982; A. Wexler, P. M. Chakrabarti, and M. J. Brown to GAF Corp.).
89. U.S. Pat. 4,414,393 (Nov. 8, 1983; M. Dexter and R. A. E. Winter to Ciba-Geigy Corp.).
90. U.S. Pat. 4,366,207 (Dec. 28, 1982; B. F. Anthony to General Electric Co.).
91. Jpn. Kokai Tokkyo Koho JP 58,152,087 [83,152,087] (Sept. 9, 1983; to Citizen Watch K.K.); *C.A., 100*, 165527 (1984).
92. Jpn. Kokai Tokkyo Koho JP 58,163,945 [83,163,945] (Sept. 28, 1983; to Canon K.K.); *C.A., 100*, 52599 (1984).

93. Jpn. Kokai Tokkyo Koho JP 58,159,460 [83,159,460] (Sept. 21, 1983; to Sumitomo Chemical Co., Ltd.); *C.A.*, *100*, 52599 (1984).
94. Jpn. Kokai Tokkyo Koho JP 58,110,535 [83,110,535] (July 1, 1983; to Pola Chemical Industries, Ltd.); *C.A.*, *99*, 212282 (1983).
95. S. Li, A. Gupta, and O. Vogl, *Monatsh. Chem.*, *114 (8–9)*, 937 (1983).
96. U.S. Pat. 4,415,687 (Nov. 15, 1983; L. Avar, E. Kalt, and H. Reinshagen to Sandoz Ltd.).
97. Ger. Offen. DE 3,320,615 (Dec. 22, 1983; L. Avar and E. Hess to Sandoz Patent G.m.b.H.); *C.A.*, *100*, 193628 (1984).
98. Jpn. Kokai Tokkyo Koho JP 58,144,364 [83,144,364] (Aug. 27, 1983; to Sumitomo Chemical., Ltd.); *C.A.*, *100*, 85402 (1984).
99. Jpn. Kokai Tokkyo Koho JP 58,179,245 [83,179,245] (Oct. 20, 1983; to Nippon Telegraph and Telephone Public Corp.); *C.A.*, *100*, 140262 (1984).
100. Jpn. Kokai Tokkyo Koho JP 59,190,943 [84,190,943] (Oct. 29, 1984; to Shipuro Kasei K.K.); *C.A.*, *102*, 166465 (1985).
101. Eur. Pat. Appl. EP 97,841 (Jan. 11, 1984; T.-Y. Ching to General Electric Co.); *C.A.*, *101*, 8225 (1984).
102. Jpn. Kokai Tokkyo Koho JP 60,185,743 [85,185,743] (Sept. 21, 1985; T. Sekiguchi and M. Tanaka to Agency of Industrial Sciences and Technology); *C.A.*, *104*, 69598 (1986).
103. U.S. Pat. 4,495,325 (Jan. 22, 1985; M. DeBergalis and R. P. O'Fee to E. I. DuPont de Nemours and Co.). A similar patent is U.S. Pat. 4,522,971 (June 11, 1985; M. DeBergalis to DuPont).
104. Jpn. Kokai Tokkyo Koho JP 60,081,124 [85,081,124] (May 9, 1985; to Pola Kasei Kogyo K.K.); *C.A.*, *103*, 59109 (1985).
105. U.S. Pat. 4,520,074 (May 28, 1985; T. W. O'Donnell and D. R. Olson to General Electric Co.).
106. U.S. Pat. 4,544,691 (Oct. 1, 1985; M. Dexter and R. A. E. Winter to Ciba-Geigy Corp.).
107. Jpn. Kokai Tokkyo Koho JP 60,166,947 [85,166,947] (Aug. 30, 1985; H. Hayashi and K. Kato to Fuji Photo Film K.K.); *C.A.*, *104*, 79139 (1986).
108. Eur. Pat. Appl. EP 165,608 (Dec. 27, 1985; M. Fryberg, G. Jan, R. Mariacs, and E. Kramp to Ciba Geigy A.-G.); *C.A.*, *105*, 124162 (1986).
109. Jpn. Kokai Tokkyo Koho JP 61,293,946 [86,293,946] (Dec. 24, 1986; Y. Inoue, K. Hata, and T. Oishi to Sumitomo Chemical Co., Ltd); *C.A.*, *107*, 39398 (1987).
110. Hung. Teljes HU 39,708 (Oct. 29, 1986; L. Knapp, T. Wein, E. L.-Tarjanyi, F. Varfalvi, L. Ivadi, and L. Ersek to Szerves Vegyipari Fejleszto Kozos Vallalat; Borsodi Vegyi Kombinat); *C.A.*, *107*, 39400 (1987).
111. Hung. Teljes HU 37,588 (Jan. 23, 1986; L. Knapp, F. Varfalvi, T. Wein, J. Hertelendy, L. Ivadi, E. Ludwig-Tarjanyi, and L. Ersek to Szerves Vegyipar Fejleszto Kozos Vallat; Borsodi Vegyi Kombinat); *C.A.*, *105*, 114738 (1986).
112. U.S. Pat. 4,619,956 (Oct. 28, 1986; P. V. Susi to American Cyanamid Co.).
113. Eur. Pat. Appl. EP 182,056 (May 28, 1986; K. Eiglmeier and J. Schulz to Riedel DeHaen A.-G.); equiv. Ger. Offen. DE 3,438,190 (Apr. 24, 1986; same author and assignment); *C.A.*, *105*, 61576 (1986).
114. Eur. Pat. Appl. EP 240,723 (Oct. 14, 1987; R. C. Bopp to General Electric Co.); *C.A.*, *108*, 39108 (1988).
115. Jpn. Kokai Tokkyo Koho JP 62,161,753 [87,161,753] (July 17, 1987; H. Hotta and M. Akasaka to Kao Corp.); *C.A.*, *108*, 43840 (1988).
116. U.S. Pat. 4,680,232 (July 14, 1987; A. Factor and G. A. Patel to General Electric Co.).
117. U.S. Pat. 4,636,431 (Jan. 13, 1987; M. DeBergalis to E. I. DuPont de Nemours and Co.).
118. Jpn. Kokai Tokkyo Koho JP 63,165,437 [88,165,437] (July 8, 1988; K. Ito, T. Shibano, and T. Nakamura to Mitsubishi Petrochemical K.K.); *C.A.*, *110*, 39941 (1989).
119. Jpn. Kokai Tokkyo Koho JP 63,253,050 [88,253,050] (Oct. 20, 1988; M. Minafuji and S. Mori to Mitsubishi Petrochemical Co., Ltd.); *C.A.*, *110*, 172879 (1989) .
120. Jpn. Kokai Tokkyo Koho JP 60,094,438 (May 27, 1985; to Becton Dickinson Co.); *C.A.*, *103*, 143099 (1985).

121. Eur. Pat. Appl. EP 280,654 (Aug. 31, 1988; G. Reinert and K. Burdeska to Ciba-Geigy A.-G.); *C.A., 110*, 77445 (1989).
122. Ger. Offen. DE 3,625,355 (Feb. 4, 1988; P. Neumann *et al.* to BASF A.-G.); *C.A., 109*, 56512 (1988).
123. Jpn. Kokai Tokkyo Koho JP 63,048,249 (Feb. 29, 1988; to Kao Corp.); *C.A., 109*, 75220 (1988).
124. Eur. Pat. Appl. EP 322,557 (July 5, 1989; J. G. Allbright and B. Van Dam to General Electric Co.); *C.A., 112*, 57418 (1990).
125. Eur. Pat. Appl. EP 303,283 (Feb. 15, 1989; T. N. Meyers to Pennwalt Corp.); *C.A., 111*, 57305 (1989).
126. Eur. Pat. Appl. EP 431,868 (June 12, 1991; A. Gupta to Ioptex Research, Inc.); *C.A., 115*, 183312 (1991).
127. U. S. Pat. 4,885,396 (Dec. 5, 1989; E. Hahn and P. Neumann to BASF A.-G.).
128. U.S. Pat. 4,882,412 (Nov. 21, 1989; M. A. Weaver, C. A. Coates, W. P. Pruett, and S. D. Hilbert to Eastman Kodak Co.).
129. Jpn. Kokai Tokkyo Koho JP 01,287,160 [89,287,160] (Nov. 17, 1989; T. Takayanagi, M. Miyazaki, and Y. Saaso to Asahi Glass Co., Ltd.); *C.A., 112*, 200807 (1990).
130. Jpn. Kokai Tokkyo Koho JP 01,284,368 [89,284,368] (Nov. 15, 1989; H. Yamaguchi to Earth Chemical Co., Ltd.); *C.A., 112*, 160700 (1990).
131. Jpn. Kokai Tokkyo Koho JP 01,299,559 [89,299,599] (Dec. 4, 1989; T. Harada, H. Iwashita, Y. Nakahara, and H. Yamanoi to Nippon Contact Lens, Inc.; Akeda Argus Chemical Co., Ltd.); *C.A., 113*, 46343 (1990).
132. Jpn. Kokai Tokkyo Koho JP 02,199,120 [90,199,120] (Aug. 7, 1990; K. Onishi to Kanebo, Ltd.); *C.A., 114*, 83497 (1991).
133. Jpn. Kokai Tokkyo Koho JP 02,002,867 [90,002,867] (Jan. 8, 1990; Y. Shimai and I. Yoshikawa to Pias Chemical, Ltd.); *C.A., 113*, 217799 (1990).
134. Jpn. Kokai Tokkyo Koho JP 02,180,909 [90,180,909] (July 13, 1990; S. Kumagai, K. Kashiwai, and A. Suga toIpposha Oil Industries Co., Ltd.); *C.A., 113*, 232242 (1990).
135. Jpn. Kokai Tokkyo Koho JP 02,058,571 [90,058,571] (Feb. 27, 1990; K. Hayama, Y. Kitani, T. Yazaki and M. Noro to Mitsubishi Petrochemical Co., Ltd.); *C.A., 113*, 98750 (1990).
136. Jpn. Kokai Tokkyo Koho JP 02,070,711 [90,070,711] (March 9, 1990; K. Takatori, R. Kimura, and H. Yamanoi to Adeka Argus Chemical Co., Ltd.); *C.A., 113*, 79886 (1990).
137. Eur. Pat. Appl. EP 401,134 (Dec. 5, 1990; M. Gay and S. Levault to Rhone-Poulenc Chimie); *C.A., 114*, 248540 (1991).
138. Jpn. Kokai Tokkyo Koho JP 02,264,966 [90,264,966] (Oct. 29, 1990; H. Yashina, Y. Sato, Y. Kamoshita, and A. Yanogihori to Mitsubishi Kasei Corp.); *C.A., 114*, 237621 (1991).
139. Can. Pat. Appl. CA 2,023,918 (Feb. 26, 1991; H. S. Laver to Ciba-Geigy A.-G.); *C.A., 115*, 161389 (1991).
140. Jpn. Kokai Tokkyo Koho JP 03,139,589 [91,139,589] and JP 03,139,590 [91,139,590] (both dated June 13, 1991; M. Minafuji and Y. Nomura to Mitsubishi Petrochemical Co., Ltd.); *C.A., 115*, 233839 (1991).
141. Jpn. Kokai Tokkyo Koho JP 03,152,271 [91,152,271] (June 28, 1991; T. Suzuki and N. Sadamasa to Seiren Co., Ltd.); *C.A., 115*, 282026 (1991).
142. Eur. Pat. Appl. EP 446,740 (Sept. 18, 1991; P. Neumann, H. Trauth, and A. Aumueller to BASF A.-G.); *C.A., 115*, 281355 (1991).
143. Jpn. Kokai Tokkyo Koho JP 03,038,547 [91,038,547] (Feb. 19, 1991; A. Onishi, M. Takahashi, and T. Ishihara to Mitsubishi Petrochemical Co., Ltd.); *C.A., 115*, 71144 (1991).
144. Ger. Offen. DE 3,930,516 (March 21, 1991; J. Schulz and G. Bartels to Riedel de Haen A.-G.); *C.A., 115*, 91845 (1991).
145. Eur. Pat. Appl. EP 451,813 (Oct. 16, 1991; T. Nahijima and M. Tanji to Konica Co.); *C.A., 116*, 140025 (1992).
146. Jpn. Kokai Tokkyo Koho JP 03,287,588 [91,287,588] Dec. 18, 1991; M. Yoshida, Y. Yokogawa, K. Umishio, and K. Uehara to Shiseido Co., Ltd.); *C.A., 116*, 194587 (1992).
147. Jpn. Kokai Tokkyo Koho JP 03,167,151 [91,167,151] (July 19, 1991; Y. Abe and K. Okawa to Shipuro Kasei K.K.); *C.A., 116*, 6240 (1992).

148. Eur. Pat. Appl. EP 459,950 (Dec. 4, 1991; G. Reinert and F. Fuso to Ciba-Geigy A.-G.); *C.A., 116*, 216285 (1992).
149. Jpn. Kokai Tokkyo Koho JP 04,021,651 [92,021,651] (Jan. 24, 1992; S. Koyama, O. Kimura, M. Yamashita, and K. Morino to Taoka Chemical Co., Ltd.); *C.A., 116*, 255325 (1992).
150. Jpn. Kokai Tokkyo Koho JP 04,198,148 [92,198,148] (July 17, 1992; K. Kashiwai and M. Endo toIpposha Oil Industries Co., Ltd.); *C.A., 117*, 233587 (1992).
151. Eur. Pat. Appl. EP 502,816 (Sept. 9, 1992; A. Valet, J. L. Birbaum, and M. Slongo to Ciba-Geigy A.-G.); *C.A., 118*, 104874 (1993).
152. Ger. Offen. DE 4,134,773 (Nov. 12, 1992; K. Schmeidl and E. Schaffner to BASF A.-G.); *C.A., 118*, 212674 (1993).
153. Jpn. Kokai Tokkyo Koho JP 05,003,912 [93,003,912] (Jan. 14, 1993; T. Honda and I. Kaetsu to Tokyo Keikaku K.K.); *C.A., 118*, 198260 (1993).
154. U.S. Pat. 4,204,007 (May 20, 1980; P. C. Wang, R. E. Wingard, and G. A. Crosby to Dynapol); U.S. Pat. 4,290,957 (Sept. 22, 1981; G. E. DuBois, R. A. G. Stephensons, and G. A. Crosby to Dynapol).
155. Jpn. Kokai Tokkyo Koho JP 56,045,912 (Oct. 29, 1981; to Daiwa, Ltd.); *C.A., 95*, 133600 (1981).
156. S. African Pat. 68,002,405 (Oct. 16, 1967; W. L. Garbrecht to Eli Lilly and Co.); *C.A., 72*, 120260 (1970).
157. PCT Int. Appl. WO 92,022,213 (Dec. 23, 1992; A. J. McElvily, R. Iyengar, and A. Gross to Opta Food Ingredients, Inc.); *C.A., 118*, 146573 (1993).
158. U.S. Pat. 4,238,401 (Dec. 9, 1980; J. A. Cella and T. D. Mitchell to General Electric Co.).
159. Jpn. Kokai Tokkyo Koho JP 57,175,765 [82,175,765] (Oct. 28, 1982; to Denki Kagaku Kogyo, Ltd.); *C.A., 98*, 131373 (1983).
160. Jpn. Kokai Tokkyo Koho JP 57,040,448 [82,040,448] (March 6, 1982; to Nippon Oils & Fats K.K.); *C.A., 97*, 56375 (1982).
161. Ger. Offen. DE 3,037,236 (May 6, 1982; K. Ulm and W. Knoblauch to Hoechst A.-G.); *C.A., 97*, 95328 (1982).
162. U.S. Pat. 4,344,892 (Aug. 17, 1982; F. Sanns, Jr. to Mobay Chemical Corp.).
163. U.S. Pat. 4,395,507 (July 26, 1983; J. J. Dziark and K. Shinohara to General Electric Silicones).
164. USSR Pat. SU 1,060,732 (Oct. 3, 1984; N. E. Rikhter, V. M. Nikitin, and L. P. Bichevaya to Leningrad Forestry Academy); *C.A., 100*, 158471 (1984).
165. USSR Pat. SU 1,333,732 (March 31, 1986; L. N. Malysheva, B. I. Bobrov *et al.* to Central Scientific Research Institute for Paper); *C.A., 108*, 7750 (1988).
166. L. F. Sokolova, L. I. Arefeva, and M. E. Parkach, *Farmatsia (Moscow), 36(4)*, 74, (1987); *C.A., 107*, 205149 (1987).
167. USSR Pat. SU 1,331,867 (Aug. 23, 1987; to A. Sh. Sharipdzhanov, D. Tillakhodzaev, and I. N. Kim); *C.A., 107*, 238055 (1987).
168. U.S. Pat. 4,642,264 (Feb. 10, 1987; M. Ropars and B. Bloch to Societe Nationale des Poudres et Explosifs; Office National d'Etudes et de Recherches Aerospatiales).
169. U.S. Pat. 4,657,703 (Apr. 14, 1987; S. C. Durkee to Hercules, Inc.).
170. PCT Int. Pat. Appl. WO 88,006,168 (Feb. 12, 1987; R. R. Gallucci to General Electric Co.); *C.A., 110*, 76746 (1989).
171. Jpn. Kokai Tokkyo Koho JP 02,292,356 (Dec. 3, 1990; J. Kumaki, S. Nakamura, and K. Yonetani to Toray Industries, Ltd.); *C.A., 115*, 51130 (1991).
172. USSR Pat. SU 1,694,752 (Nov. 30, 1991; A. V. Burov, Z. P. Elnitskaya, R. G. Aliev, and L. B. Luferova to Leningrad Technological Institute of the Cellulose Paper Industry); *C.A., 117*, 173644 (1992).
173. Eur. Pat. Appl. EP 507,732 (Oct. 7, 1992; J. Kaschig, G. Reinert, and G. Metzger to Ciba-Geigy A.-G.); *C.A., 118*, 38603 (1993).
174. Jpn. Kokai Tokkyo Koho JP 03,252,449 [91,252,449] (Nov. 11, 1991; M. Hirose and T. Karatsuji to Teijin, Ltd.); *C.A., 116*, 107589 (1992).

175. USSR Pat. SU 1,686,048 (Oct. 23, 1991; E. I. Berenstein, G. M. Kozin, B. I. Aikhodzhaev, and V. Grabauskas/Scientific-Research Institute of the Chemistry and Technology of Cotton Cellulose); *C.A., 117*, 235737 (1992).
176. L. E. Mirci, *Mater. Plast. (Bucharest), 27*, 186 (1990); *C.A., 116*, 153062 (1992).
177. Jpn. Kokai Tokkyo Koho JP 04,277,570 [92,277,570] (Oct. 7, 1992; to E. Uchimoto); *C.A., 118*, 126623 (1993).
178. USSR Pat. SU 1,724,614 (Apr. 7, 1992; I. E. Sokolov, V. M. Evzhenko, P. B. Mel'nikov, and A. V. Rogachev to Belorussian Institute of Railroad Transport Engineers); *C.A., 118*, 239370 (1993).

12

Other Uses for Resorcinol

This is the last of the chapters strictly on the technological side of resorcinol, last but not least—the applications cited here are quite important and fascinating in their own right. Several commercial applications are included, e.g., the use of lead beta-resorcylate as an explosive/propellant modifier, the use of 2-methylresorcinol and *m*-aminophenol in hair dyes, and 4-methyl-7-ethoxycoumarin as a perfume ingredient. The selected references are mostly from the 1980–1992 period.

12.1. EXPLOSIVES

The manufacture of explosives and many ingredients for explosives, their handling and transport are hazardous, well regulated by governments, and best left to the experts in the field.

The role of resorcinol in this field is outlined by the following selections of examples. Binders for nitrate-fuel explosive compositions, comprising resorcinol/formaldehyde resins, have been reported in the older literature. DATB, 1,3-diamino-2,4,6-trinitrobenzene, **12-1**, is a heat-resistant, insensitive explosive, manufactured from resorcinol dimethyl ether (1,3-dimethoxybenzene) by sulfonation and trinitration, followed by amination with ammonia or ammonium acetate.[1]

Lead styphnate (lead 2,4,6-trinitroresorcinate, Trizinat), **12-2**, as well as lead dinitroresorcinate, are initiators for explosives.

The direct nitration of resorcinol (and other phenols) is rapid, highly exothermic, and difficult to control. Therefore, the reaction usually is performed by first sulfonating resorcinol and then displacing the sulfo by nitro groups. Styphnic acid (2,4,6-trinitroresorcinol), **12-3**, was also prepared by treating resorcinol with sodium nitrite/nitric acid, i.e., by nitrosation/nitration (in this case the nitroso groups likely were oxidized to nitro groups) at 5–97 °C; the yield of styphnic acid was 94%.[2] The effluent from such processes requires special treatment; trinitroresorcinol can be recovered by extraction or removed by electrolysis at pH 1.0 in the presence of chloride ion.[3] The lead styphnate was prepared, e.g., by first adjusting a styphnic acid solution with magnesium carbonate to pH 6.8–8.0, then adding a soluble lead salt at >50 °C.[4]

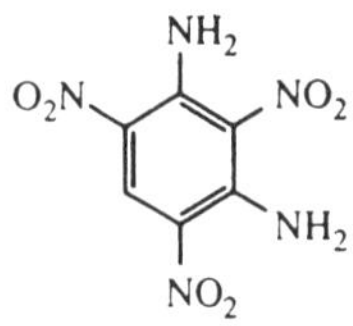

12-1

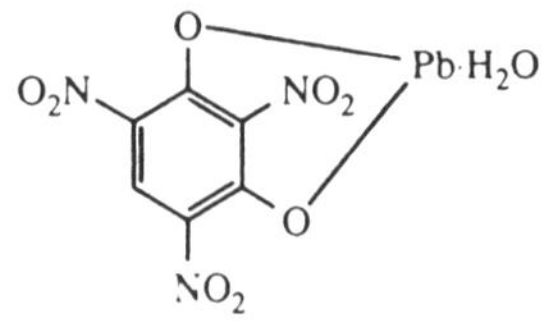

12-2

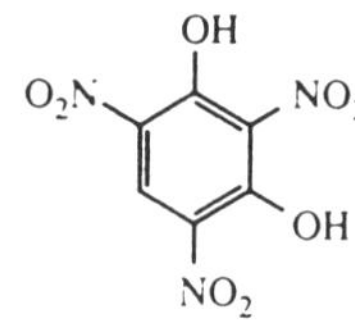

12-3

A continuous process for the manufacture of crystalline lead styphnate was patented and said to be safer and to give a better quality of product.[5]

Lead (2,4- and/or 4,6-)dinitroresorcinate is prepared and used in similar fashion. For example, a suspension of 4,6-dinitroresorcinol was treated at 90 °C with the appropriate amount of lead acetate and ammonium hydroxide solution, then worked up to give a compound, $3[C_6H_5N_2O_6Pb]\cdot 2[Pb(OH)_2]$, which was free-flowing, dust-free and very suitable in primers and fast delay compositions for fuses and detonators.[6]

Moist granules of lead styphnate compositions were prepared which were safer to handle because they were insensitive to shock, flame, and static electricity.[7]

A priming charge for detonators, consisting of lead azide and lead styphnate plus a binding agent, was patented.[8]

Novel polybasic lead salts of nitroresorcinols, such as $C_6H(NO_2)_3(OPbOH)_2$-2PbO, $C_6H(NO_2)_3(OPbOH)_2$4PbO, $C_6H_2(NO_2)_2(OPbOH)_2$PbO, $C_6H_2(NO_2)_2$-$(OPbOH)_2$ 2PbO, and similar products, optionally containing minor proportions of finely divided boron dispersed throughout individual crystals, were prepared. These products were said to be primary explosives used as initiatory or fast delay explosives for fuses, detonators, etc., which explode under mild stimulus to produce a detonation wave capable of exploding larger amounts of less sensitive explosives.[9]

An improved process was developed for producing the known primer consisting of lead styphnate/barium nitrate plus water. Styphnic acid, **12-3**, was reacted with barium hydroxide or barium carbonate and lead nitrate in water; no waste was produced, the cost was improved, and safety was improved.[10]

The production of lead styphnate, useful as a component of ignition mixtures for ammunition, was improved by mixing aqueous solutions of Mg styphnate and lead nitrate; a 94% yield of lead styphnate of uniform, suitable crystal size was obtained.[11]

A dinitroresorcinoldiazonium salt, useful in detonators for explosives, was made by treating styphnic acid with $SnCl_2$ in acetic acid at 30 °C, followed by diazotization; the Pb or Ba salts of this compound were more sensitive detonators than Pb styphnate.[12]

In the older literature lead dinitrosoresorcinate was reported to be used as an initiator which was much less sensitive to impact than lead azide and somewhat less sensitive to friction than mercury fulminate, and to have a low ignition temperature. However, it was less heat-stable than lead dinitroresorcinate.

Lead (and copper) beta-resorcylate, **12-4** (monobasic form) and **12-5** (dibasic form), are useful as ballistic modifiers in rocket propellant formulations. Free beta-resorcylic acid (2,4-dihydroxybenzoic acid), if present as an impurity, was said to have a tendency to bloom to the surface of the compositions and to act as an ignition

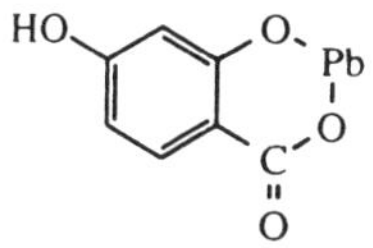

12-4

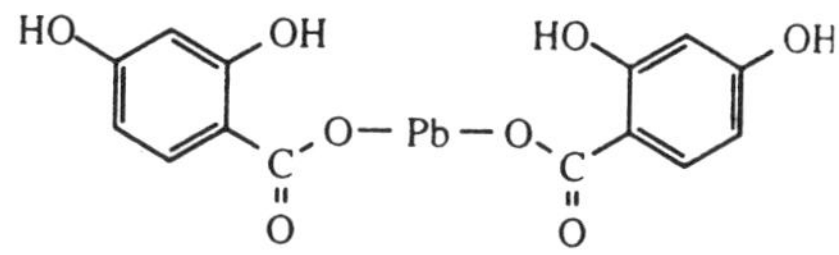

12-5

inhibitor. A process was developed to make these Pb resorcylate products in high purity by reacting beta-resorcylic acid with PbO in the appropriate stoichiometry in absolute or aqueous ethanol at 25–70 °C.[13]

Smokeless, double-base propellants of nitrocellulose and nitroglycerin were prepared, containing lead beta-resorcylate for improved exhaust, high burning rate, and higher operating pressure than previously obtained by conventional propellants for use in propulsion systems.[14]

Although nitroglycerin (which contains a secondary nitroxy group) has excellent plateau burning characteristics, propellant plasticizers containing only primary nitroxy groups, i.e., MTN (metriol trinitrate) and TEGDN (triethylene glycol dinitrate), were recognized as superior plasticizers in regard to health, stability, and storage. Useful double-base propellants were developed based on plasticizers with only primary nitroxy groups, such as MTN, TEGDN, PETRIN (pentaerythritol trinitrate), and mixtures of these, containing 2-nitrodiphenylamine as a stabilizer and monobasic cupric beta-resorcylate as a ballistic modifier. These compositions showed little performance change in the –40 to 82 °C range, had constant burning rates over a pressure range of 1000–3000 psia, contained only *ca* 0.15 wt% solids in the exhaust gas (reduced visibility and toxicity), and had superior stability. These compositions could be used as gun or missile propellants, or in auxiliary gas-producing devices.[15]

A major difficulty in double-base propellants containing lead and cupric salicylates or beta-resorcylates as modifiers was said to be their hydrolytic instability, requiring their addition at a late stage of propellant compounding. New modifiers were developed which were the Cu, Pb, Sn, or Bi chelates of 2,4-dihydroxybenzophenone, 2,2′,4,4′-tetrahydroxybenzophenone, 2-hydroxy-4-alkoxybenzophenone, or 2,2′-dihydroxy-4,4′-dialkoxybenzophenone. An example was the copper chelate of 2,2′,4,4′-tetrahydroxybenzophenone, shown in the patent as structure **12-6**. The new compositions had extremely low susceptibility to hydrolysis, gave improved plateau and mesa burning rate–pressure functions in double-base propellants, and they had better dispersibility throughout the propellant than prior art modifiers.[16]

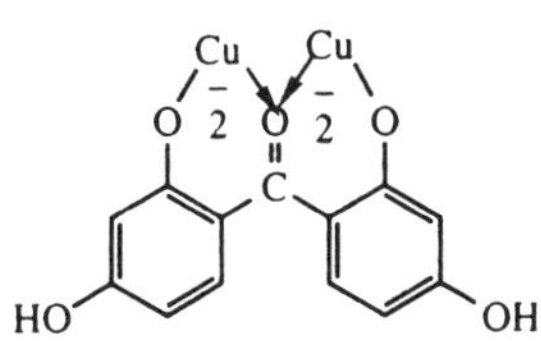

12-6

Ultra-ultrahigh burning rate composite, modified, double-base propellants were developed for use in the propulsion subsystems of ballistic missile interceptors. These compositions contained nitrocellulose, nitroglycerin, ammonium perchlorate, aluminum powder, aluminum staples, triacetin as an inert plasticizer, and small amounts (0.5–1.5 wt%) of resorcinol and 2-nitrodiphenylamine as stabilizers.[17]

12.2. BEAUTY AIDS AND FLAVORANTS

12.2.1. Hair Dyes

Oxidation dyes are used in hair dyeing preparations composed of *o*- or *p*-phenylenediamines or aminophenols as the precursor (developer) and a dihydroxybenzene such as resorcinol as the coupler. On addition of the oxidant, usually hydrogen peroxide solutions, azine- and oxazine-type dyes are formed. For example, in the oxidative dyeing system 2-methylresorcinol gives reddish tints with *p*-phenylenediamine and an orange-brown with *p*-aminophenol.[18] *m*-Aminophenol is also used in such compositions.The color is produced inside the hair fiber.

In 1981, several hundred hair dye formulations containing resorcinol or 2-methylresorcinol at the *ca* 1–5% level were available.[19]

This old art is still being improved upon, as evidenced by the following few selections from the literature of the 1970s to 1980s. Novel N-substituted aminophenols, **12-7**, were prepared for use as coupling components in oxidative hair dyes; these couplers were more weather-resistant and lightfast than *m*-aminophenol hydrochloride.[20]

Another oxidative hair dyeing composition was patented which contained *p*-phenylenediamine and a halogenated *m*-aminophenol, **12-8**, as the coupler; lightfast shades such as purple green and blue could be obtained.[21]

Another type of hair coloring used an amine such as 4-aminoresorcinol, a mineral or organic acid, and metal salts of, e.g., cobalt or iron; air oxidation of the hair developed the color.[22]

New N-acylated 2,6-dimethyl-3-aminophenols, such as **12-9**, were synthesized and used as the coupling agent in oxidative hair dyes to give very fast reddish-blue shades.[23]

N-Mesylaminoalkyl-*m*-aminophenols such as, e.g., **12-10**, were used as color couplers in oxidative hair dyeing compositions; the dyes were light- and weather-stable and had low toxicity.[24]

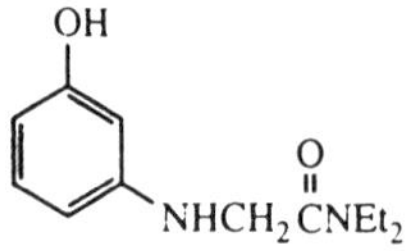

12-7

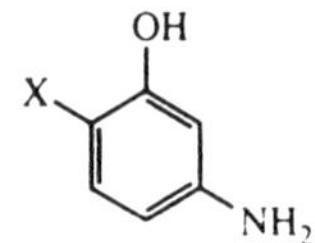

12-8
(where X= F, Cl or Br)

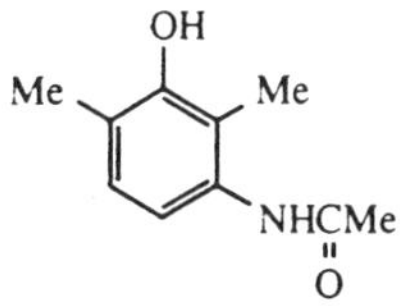

12-9

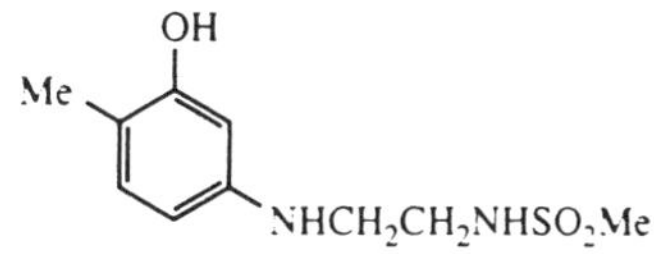

12-10

A new oxidative hair dyeing composition was developed which contained a standard developer such as *p*-phenylenediamine or 2,4,5,6-tetraaminopyrimidine and a 5-substituted resorcinol, **12-11**, as the coupler. Brilliant brown shades were obtained which were fast to heat, light, washing, and rubbing.[25]

Diresorcyl sulfide, sulfoxide, or sulfone, **12-12**, were used as the coupling component in oxidation hair dye compositions, using *p*-phenylenediamine or 2,4,5,6-tetraaminopyrimidine as the developer. Color tones in the brown range with good fastness properties were obtained and there was a negligible tendency to skin staining.[26]

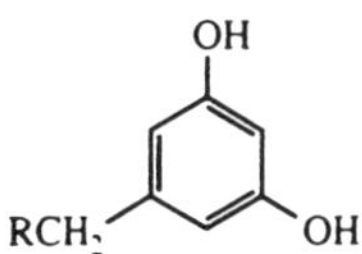

12-11
(where R= NH_2, OH or COOH)

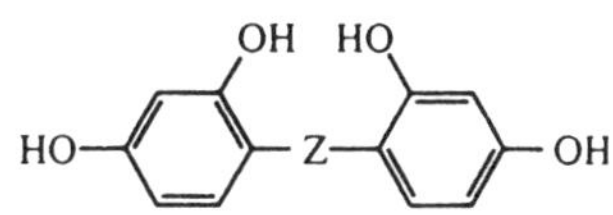

12-12
(where Z= S, SO or SO_2)

Oxidative hair dyeing compositions were developed which contained, e.g., *p*-phenylenediamine 1.00, *p*-aminophenol 2.95, *N*-methylaminophenol 2.14, *o*-aminophenol 0.30, resorcinol 0.79, *m*-aminophenol 0.60, 6-hydroxyethoxy-1,3-diaminobenzene 0.78, and 3-(2-hydroxyethylamino)-6-methylphenol 3.00 parts, as well as water-soluble bio-heteropolysaccharides in cosmetically compatible carriers. Before application, the composition was mixed with 6% hydrogen peroxide; natural white hair was dyed a blond-coppery color.[27]

m-(2,2,2-Trifluoroethylamino)phenol, **12-13**, was patented as a new color coupler in oxidative hair dyeing compositions. With *p*-phenylenediamine as the developer, blue to black shades could be obtained; with *p*-aminophenol as the developer, a mahogany shade was achieved . The dyeings were fast, chemically stable, and resisted abrasion.[28]

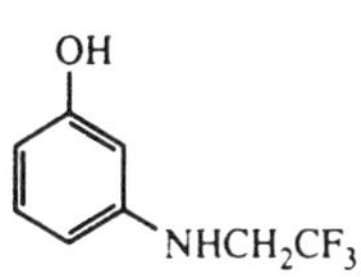

12-13

12.2.2. Perfumes and Aromas

Resorcinol ethers do occur in nature (see also Chapter 14). For example, 3,5-dimethoxytoluene, **12-14**, is one of the volatile components in yellow tea roses.[29]

Resorcinol dimethyl ether (1,3-dimethoxybenzene) has been used as a flavorant in foods in fruit, nut, and vanilla compositions at the 3–8 ppm level for beverages, ice cream, candy, and baked goods.[30]

Marianol (4-methyl-7-ethoxycoumarin), **12-15**, was listed in the Givaudan Index, one of the bibles of the trade, as a perfumery ingredient with an odor of walnut character, useful in fougere and chypre compositions.

The biggest use of resorcinol derivatives in this category is as alkylated resorcylic esters useful as synthetic substitutes for natural oakmoss (a lichen extract), which is an important and basic part of chypre and lavender fragrances. These synthetics make up a good part of the following selections from the literature.

Thus, a process was developed for aromatizing dialkyl-substituted 1,3-cyclohexanedione-4-carboxylic acids, or their esters, by oxidation with chlorine. (The appropriate cyclohexanediones were made by the Mannich condensation of an alkyl alpha-, beta-alkenoate with an alkyl beta-ketoalkanoate; such processes were described in Chapter 4.) The products were 3,5-dialkylresorcylic acids and esters such as, e.g., **12-16** (methyl 2,4-dihydroxy-3-*n*-butyl-6-methylbenzoate) were obtained that were useful as substitutes for natural oakmoss fragrance (mousse olfactory notes).[31]

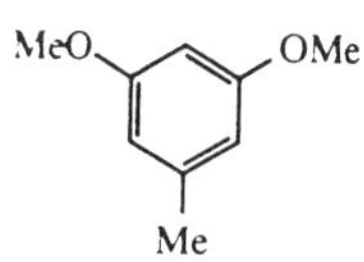

12-14

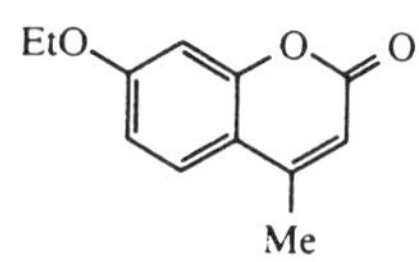

12-15

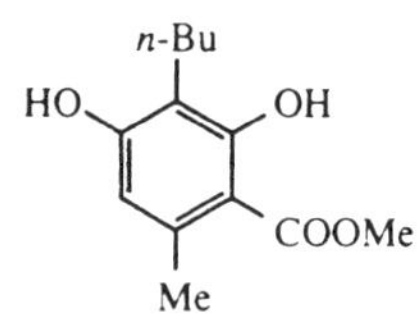

12-16

Methyl and ethyl 2,4-dihydroxy-3,6-dimethylbenzoates were prepared by carboxylating 2,5-dimethylresorcinol with K_2CO_3 under CO_2 pressure at 150 °C, then esterifying the resorcylic acid intermediate; these products were intended for use in perfumery as well.[32]

7-Alkoxycoumarins were prepared. These compounds had sweet to medicinal odors and were used for improving, enhancing, and modifying the organoleptic properties of tobacco products at the 0.0001 to 0.1 wt% addition level. Thus, 7-methoxycoumarin, **12-17**, gave a smoke with a hay-like sweet note, 7-ethoxycoumarin gave a cocoa-sweet note, 7-isobutoxycoumarin produced a waxy, sweet-sour note with decreased tobacco taste, and 7-*n*-hexoxycoumarin lent an initially maple-sweet note which later became spicy.[33]

2,5-Dimethyl-6-cyano-1,3-cyclohexanedione, **12-18** (shown in the enol form), 2,4-dihydroxy-3,5-dimethyl benzonitrile, **12-19**, and 2,4-dihydroxy-3,6,-dimethylacetophenone, **12-20**, were recognized as valuable components for enhancing the organoleptic properties of consumable materials such as soaps, detergents,

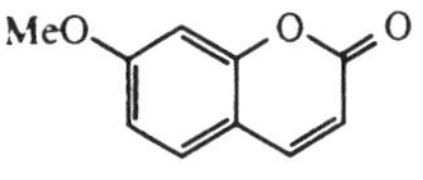

12-17

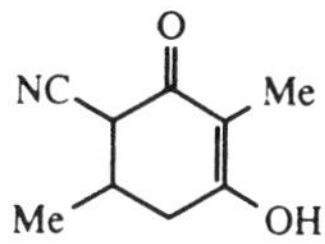

12-18

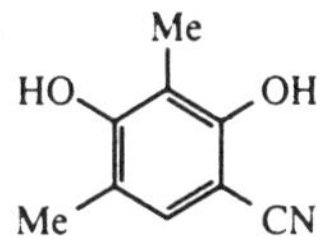

12-19

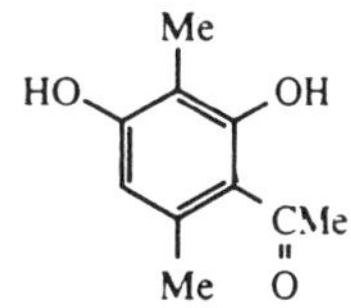

12-20

and tobacco products. These compounds were formulated with other perfume ingredients to achieve the odor quality of natural oak moss.[34]

Substituted spirocyclic derivatives of 1,3-cyclohexanediones such as, e.g., compound **12-21** were prepared and shown to be useful in perfumery, food flavorings, and tobacco products.[35]

A process for the aromatization of assorted 1,3-cyclohexanedione derivatives to the corresponding resorcinol derivatives (useful in perfumery) was patented and involved treating the precursors with Cu_2O catalyst and thionyl chloride or HCl and air at *ca* 50 °C in a solvent (methanol, tetrahydrofuran, or methyl *tert*-butyl ether). For example, 2,5-dimethyl-6-carbomethoxycyclohexane-1,3-dione was converted in this manner to methyl 2-hydroxy-3,6-dimethyl-4-methoxybenzoate, **12-22**, in 84% yield plus a 6% yield of methyl 2,4-dihydroxy-3,6-dimethylbenzoic acid; this mixture had the odor of oak moss.[36]

Another process for the preparation of alkyl 2,4-dihydroxy-3-alkyl-6-methylbenzoates consisted of treating the appropriately substituted alpha-pyrones with a base. For instance, 4-methyl-alpha-pyronyl-6-(alpha-isopropyl)acetic acid methyl ester, **12-23**, was refluxed in methanol in the presence of an equimolar amount of sodium methylate for 3 hr, then cooled and acidified with 2N H_2SO_4 to give methyl 2,4-dihydroxy-3-isopropyl-6-methylbenzoate, **12-24**. These ester products were useful as scents having the character of oak moss.[37]

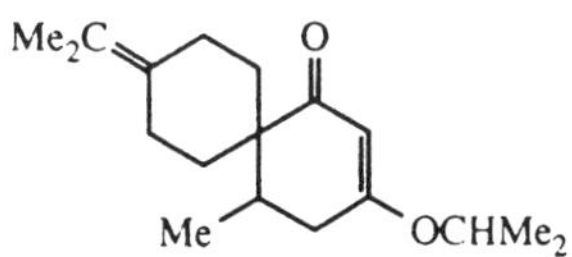

12-21

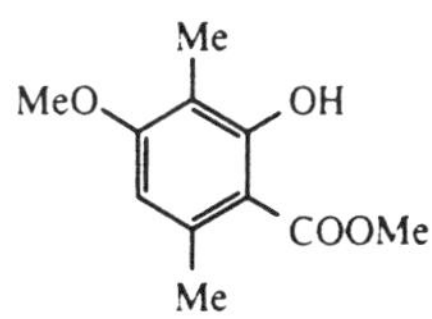

12-22

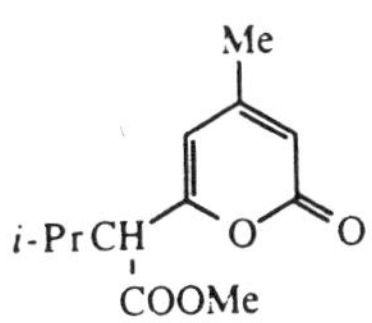

12-23

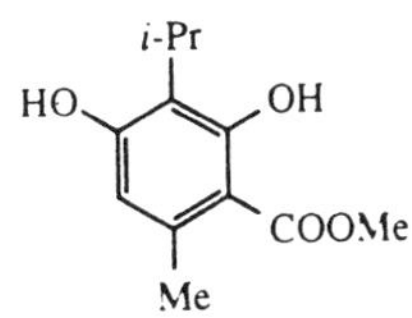

12-24

In a further variation, shown in Eq. (12.1), methyl 2,4-dihydroxy-6-methylbenzoate was reacted with morpholine and formaldehyde to give methyl 2,4-dihydroxy-

3-(*N*-morpholinomethyl)-6-methylbenzoate, **12-25**; this intermediate was converted by catalytic hydrogenation to methyl 2,4-dihydroxy-3,6-dimethylbenzoate, **12-26**, useful in scents.[38]

(12.1) 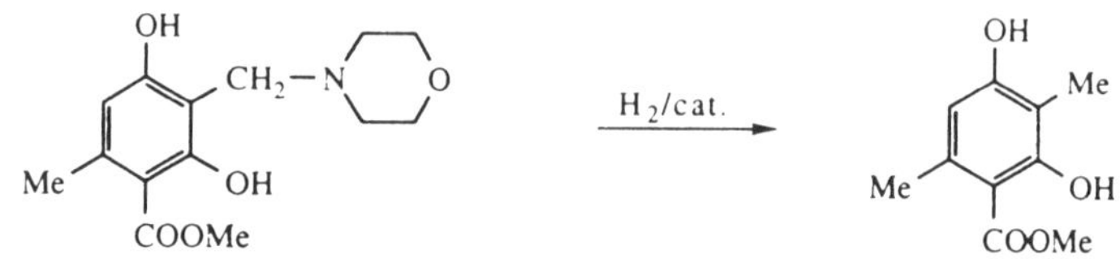

12-25 12-26

The treatment of diketene derivatives with alkyl beta-oxo-alkanoates in the presence of alkaline earth oxides also gave alkyl-substituted 2,4-dihydroxybenzoates. For instance, a 48/52 mixture of 4-methylene-2-oxetanone/3-methyl-4-ethylidene-2-oxetanone was reacted in tetrahydrofuran with methyl acetoacetate in the presence of an equimolar amount of CaO to give a 46/54 mixture of methyl 2,4-dihydroxy-6-methylbenzoate/methyl 2,4-dihydroxy-3,6-dimethylbenzoate in 88% yield. This mixture had a mossy, woody, phenolic odor note similar to oak moss.[39]

12.2.3. Other Applications in Cosmetics

Just a few examples are given in this subcategory.

Eosin (tetrabromofluorescein) was mentioned earlier in Chapter 9 as a dye used in lipsticks. As an example, a lip rouge formulation was patented consisting of a base of wax, oil, and soybean lecithin, a halogenated fluorescein (e.g., eosin) dye at 5–10 wt% of the soybean lecithin, and an amine, e.g., triethanolamine at 26–46 wt% of the halogenated fluorescein.[40]

A chemical skin-tanning composition and method was invented which comprised a suitable phenolic dye precursor (*m*-aminophenol was one of many) and the enzyme, tyrosinase, to form melanin-like dyes which tanned the skin.[41]

A method and composition was developed for modifying the moisture retention of hair, thus improving its settability. The process comprised applying to hair an aqueous solution containing a heximinium salt such as, e.g., **12-27**, and resorcinol at neutral to mildly acid pH, whereby condensation polymers were formed which imparted improved stiffness and settability to the hair. The improved settability was resistant to high humidity but was removed by normal shampooing.[42]

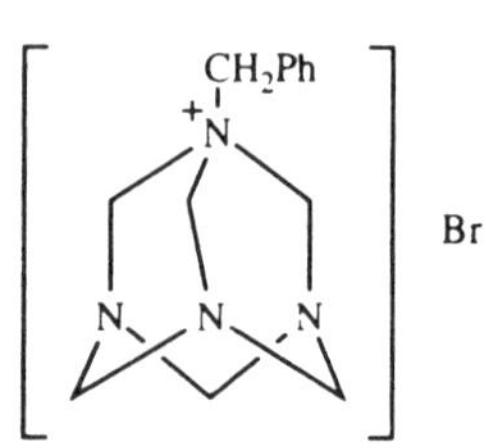

12-27

A hair-growth-promoting scalp dressing for the treatment of baldness was proposed, consisting of boric acid, resorcinol, salicylic acid in glycerine/glycerine thymol compound/surgical spirit, and perfume.[43] (Wish it were true—but it is not known if such preparations were marketed with success.)

4,4′-Thio-bis-resorcinol was found to be an excellent dandruff fighter which had good solubility in water and was readily incorporated in suitable cosmetic preparations such as shampoos, lotions, and sprays. This compound was also suitable as a coupler in oxidation hair dyes using *p*-phenylenediamines as developers.[44]

12.3. MISCELLANEOUS OTHER USES

Most of the selections grouped here have as a common thread the versatility of resorcinol for the preparation of intermediates (sometimes their use was not given in the abstracts) and products for a wide variety of applications. Two references to the stabilization of resorcinol are included.

Hydroxybenzoxazoles were obtained from dihydroxyacetophenone oximes. For example, 2,4-dihydroxyacetophenone oxime, **12-28**, was refluxed in formic acid to give 2-methyl-6-hydroxybenzoxazole, **12-29**, in 44% yield.[45]

m-Mercaptophenol, **12-30**, was obtained in modest yield by passing resorcinol with H_2S at 550 °C over a V_2O_5-on-silica catalyst.[46]

3-Hydroxy-4′-methoxydiphenylamine, **12-31**, was obtained in 91% yield by condensing resorcinol with *p*-anisidine in the presence of sulfanilic acid at 250–260 °C.[47]

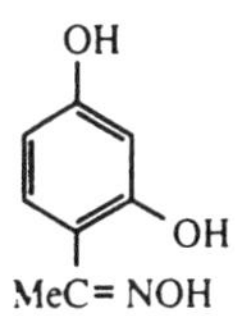

12-28

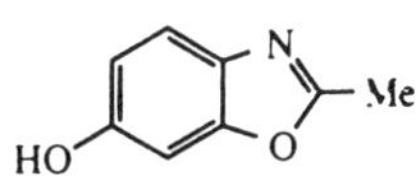

12-29

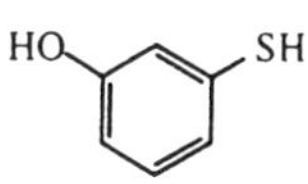

12-30

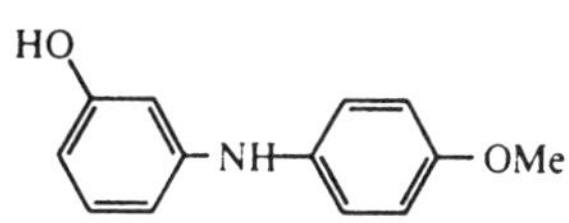

12-31

2-Methylresorcinol (2,6-dihydroxytoluene), **12-32**, was obtained in 91% yield by autoclaving a mixture of 7 parts resorcinol, 2 parts ammonium chloride, and 20 parts methanol for 2 hr/330 °C; the conversion of resorcinol was 100%.[48]

7-Hydroxy-4-methylcoumarin, **12-33**, was prepared in 90% yield by the condensation of resorcinol with ethyl acetoacetate in conc H_2SO_4 containing $ZnCl_2$ at 5–20 °C, then drowning the mix in water.[49]

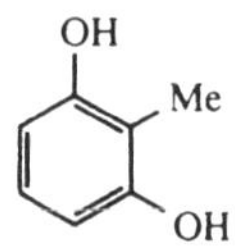

12-32

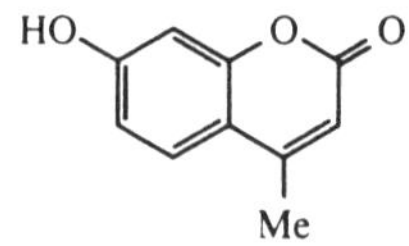

12-33

7-Hydroxycoumarin (umbelliferone), **12-33** without the methyl group, was also obtained by separating undifferentiated cells from the cultivated callus of Hydrangea plants and extraction with an organic solvent (e.g., ethanol or ethyl acetate), followed by column chromatography on Mg or Al silicate or alumina for purification.[50]

Hydroxyflavans were obtained by the reaction of 1.0 m aliphatic ketone with 3.0 m resorcinol in the presence of 0.5–1.5 m mineral acid. For instance, the reaction of acetone with resorcinol in this mole ratio, in 7% HCl first at 44 °C and then 4 hr/50–55 °C, gave a 96% yield of compound **12-34**; the use of an excess of resorcinol prevented side-reactions such as the self-condensation of the ketones.[51]

On the other hand, spirochroman derivatives were said to be obtained when 2.0 m resorcinol was condensed with 1.0 m aliphatic ketone in the presence of conc HCl. For instance, the spirochroman, **12-35**, was obtained in 90% yield when 110 g resorcinol was dissolved in 100 g 36% HCl at 43–44 °C and 29 g acetone was added in 40 min, then holding the charge for 2 hr/50–55 °C.[52]

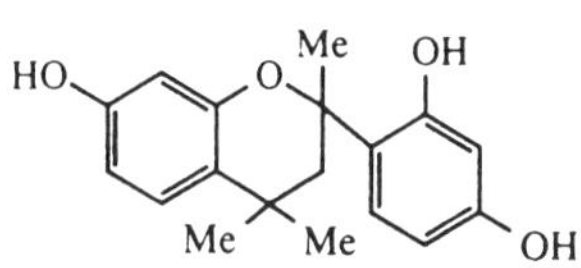

12-34

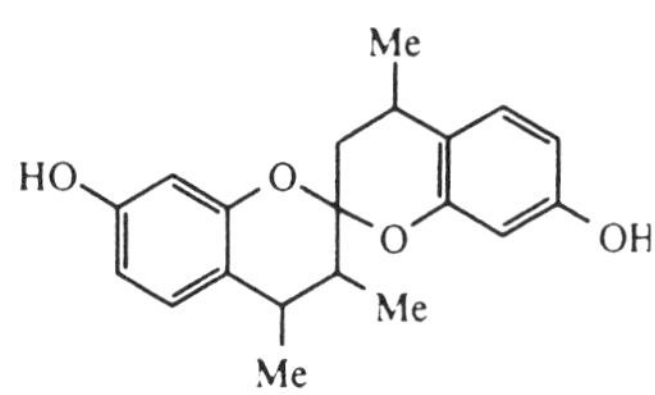

12-35

The discoloration of resorcinol on storage, solid or molten, at elevated temperature was said to be inhibited, even in the presence of iron, by the addition of 0.00001–0.1 wt% of a hydroxycarboxylic acid, such as tartaric acid.[53]

The condensation of resorcinol with a threefold excess of 2-chloroaniline in the presence of *p*-toluenesulfonic acid for 18 hr/180 °C gave a 100% yield of the hydroxydiphenylamine, **12-36**; several other substituted anilines (e.g., *p*-anisidine) were used in this reaction and high yields of the substituted hydroxydiphenylamines were obtained.[54]

The discoloration of resorcinol (or 2-naphthol) on storage was inhibited by keeping it in the presence of an activated iron oxide as an oxygen absorbing agent which was packed in a separate cellophane bag.[55]

2,6-Diaminotoluene, **12-37**, was prepared in 97% yield by autoclaving 2,6-dihydroxytoluene (5 parts) with 28% ammonia (20 parts), $CoCl_3$ (1 part) and ammonium chloride (1 part) for 5 hr/300 °C.[56]

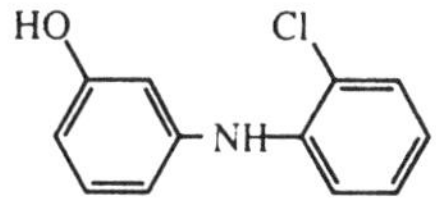

12-36

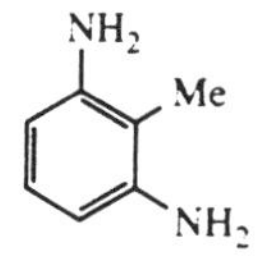

12-37

A heat-generating (thermogenous) composition, suitable for a disposable pocket heater, was developed comprising 100 parts metal (e.g., iron) powder, 20–300 parts water, 1–300 parts carbonaceous material (e.g., char of sawdust, coconut husk or wood bark, or a coke [from coal]), and 0.1–100 meq/100 g of carbon of a reducing agent (e.g., Na_2S, NaHS, Na_2HPO_3, and phenols, including resorcinol) packed in a gas-impermeable container; when brought in contact with oxygen or air, heat was generated effectively.[57]

3-[2-(Diethylamino)ethyl]-4-methyl-7-hydroxycoumarin hydrochloride, **12-38**, was prepared by condensing resorcinol with ethyl 2-(2-diethylaminoethyl)acetoacetate in an organic solvent in an HCl atmosphere.[58]

Novel phenoxyacetic acid derivatives were prepared. For example, 2-hydroxy-4-methallyloxyacetophenone was reacted with methyl chloroacetate in the presence of K_2CO_3 and a catalytic amount of NaI in acetone at 60 °C under N_2 to give an 86% yield of compound **12-39**.[59]

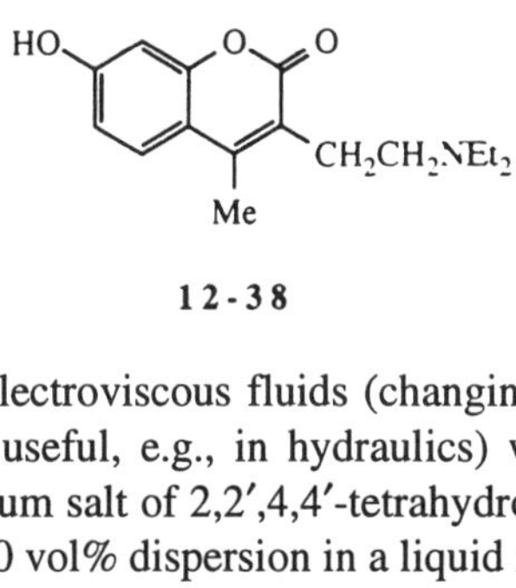

12-38

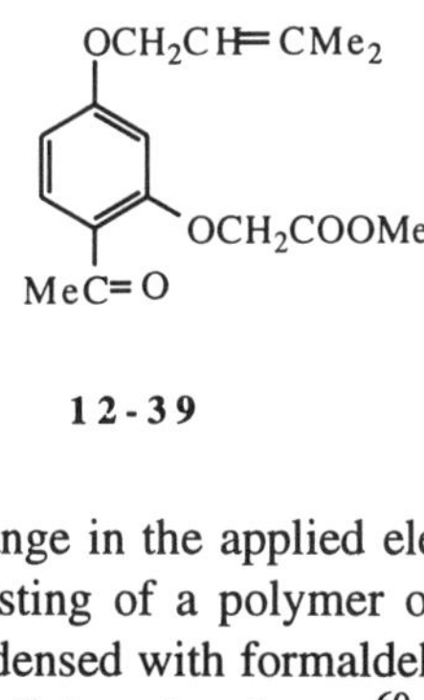

12-39

Electroviscous fluids (changing viscosity with a change in the applied electric field, useful, e.g., in hydraulics) were developed, consisting of a polymer of the dilithium salt of 2,2′,4,4′-tetrahydroxybenzophenone condensed with formaldehyde, as a 30 vol% dispersion in a liquid mixture of brominated diphenylmethanes.[60]

New tetrahydric phenols were prepared. Thus, the condensation of 0.3 m resorcinol with 0.1 m of the dicarbinol, **12-40**, in the presence of conc HCl and benzene as the solvent at 5 °C/15 hr gave a 78% yield of the tetrahydric phenol, **12-41**.[61]

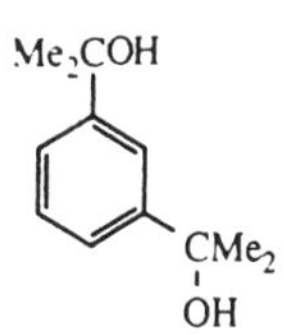

12-40

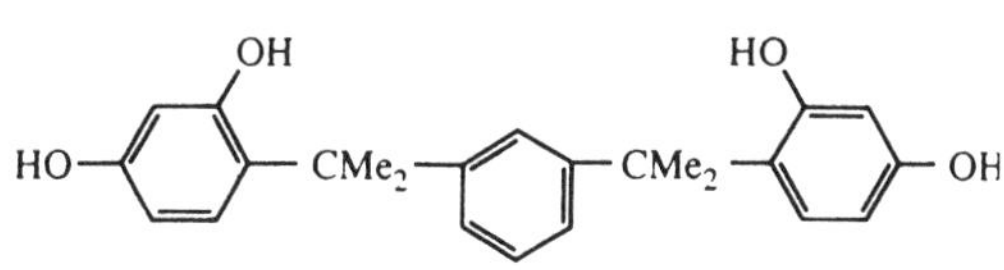

12-41

A process for the catalytic alkylation of resorcinol with alcohols was developed. For instance, a solution of resorcinol in methanol (1/5 ratio) was passed at 15 ml/hr over an iron oxide catalyst at 370 °C to give a product containing 25% 2-methylresorcinol, 54% 4-methylresorcinol, 10% 2,4- and 2,6-dimethylresorcinol, and 8% trimethylresorcinol; the conversion of resorcinol was 55%.[62]

Phenols, including resorcinol, were removed from solutions in organic solvents by adsorption on a crosslinked poly(vinylpyridine) resin. The phenols could be desorbed by treating the resin with aqueous alkali, methanol, hot water, or hot organic solvent.[63]

4,6-Dichlororesorcinol was prepared in 99% purity and 75% yield by reacting 0.05 m resorcinol in methyl acetate with 0.11 m SO_2Cl_2 for 4.5 hr, then removing the solvent, dissolving the crude product in hot toluene, filtration, cooling and collecting the purified product by filtration, and drying. Prior art processes for dichlororesorcinol were said to have been impractical.[64]

A detergent composition containing a photoactivated bleaching system was developed. Thus, a textile was contacted with a solution containing as soluble (di)salt of fluorescein, **12-42**, and a *tert*-amine; the textile was then (UV) irradiated and rinsed with water, preferably under irradiation. The system removed colored stains (e.g., tea or wine) from textiles at room temperature and did not damage the fabrics.[65]

Phenylene mono- and di-ester peracid precursors for textile bleaching were prepared. For instance, the reaction of equimolar quantities of resorcinol and hexanoic anhydride at 100 °C/3 hr in the presence of a catalytic amount of H_2SO_4 gave a 90% yield of resorcinol dihexanoate, **12-43**. This diester in combination with H_2O_2 and Neodol 25-12 was much more effective in removing Crystal Violet from stained cotton than a prior art bleach.[66]

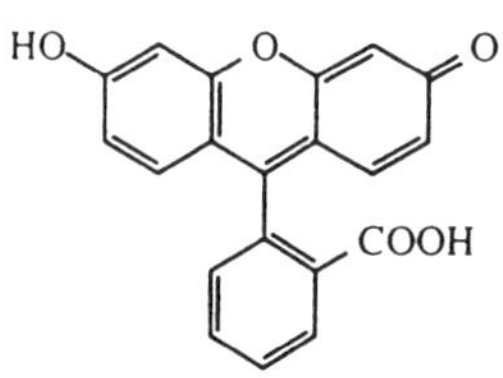

12-42

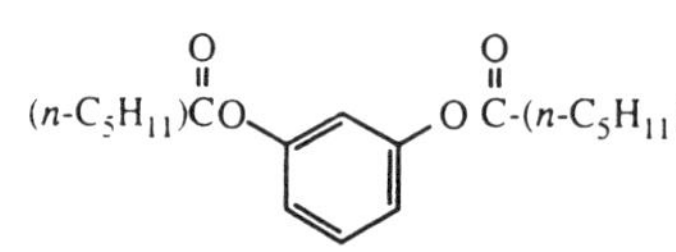

12-43

A process for the preparation of oxyflavans was developed. For example, 58.0 g (1.0 m) acetone was added in 45 min to a mixture of 220 g (2.0 m) resorcinol, 93.3 g 36% HCl and 291.5 g water at 50 °C; the mixture was held at 50 °C for another 4 hr and neutralized to give 113.6 g of compound **12-34**.[67]

A process for the preparation of monoalkylated resorcinol comprised, e.g., the dropwise addition of alpha-methylstyrene to an aqueous solution of resorcinol containing H_2SO_4 at 80 °C to obtain a 90% yield of 4-(alpha, alpha-dimethylbenzyl)resorcinol, **12-44**, of 99.9% purity.[68]

The preparation of tris-(2,6-dimethoxyphenyl)phosphine, **12-45**, and similar compounds was reported. A toluene dispersion of phenyl sodium was mixed with resorcinol dimethyl ether, stirred at 10 °C/1 hr, then PCl_3 was added over 30 min at –20 °C; the mixture was held at 70 °C/1 hr to give a 92% yield (based on the PCl_3) of the phosphine, **12-45**.[69]

In the cellulose pulp bleaching process involving chlorine treatment, alkalization, and treatment with sodium hypochlorite, 0.05–0.5% eosine (tetrabromofluorescein) was added to reduce the degradation of the cellulose during this process.[70]

Trihydroxybenzenes were prepared by the hydroxylation of resorcinol. Thus, 110 g (1.0 m) of resorcinol was heated to 110 °C and treated with 0.05 m 70% H_2O_2; in 10 min, 98% of the H_2O_2 was converted to give 5.5 g 1,2,4-trihydroxybenzene (hydroxyhydroquinone) and 0.85 g 1,2,3-trihydroxybenzene (pyrogallol).[71]

3-Alkoxy-4-nitrosophenol, **12-46**, was obtained by the nitrosation of resorcinol dimethyl ether at –15 to 0 °C with nitrosylsulfuric acid.[72]

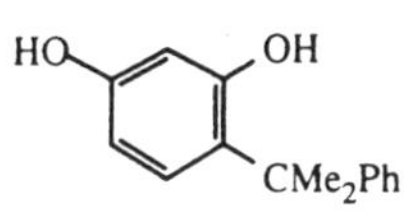

12-44

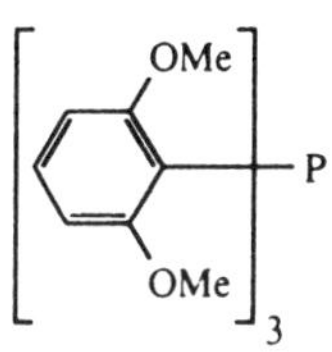

12-45

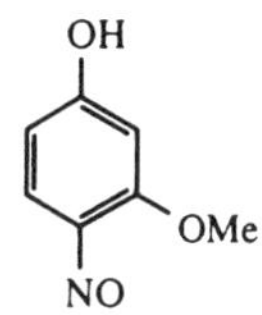

12-46

Cobalt could be extracted from dilute aqueous solutions of its salts using 4-nitroso-2,5-dimethylresorcinol or 4-nitroso-5-methylresorcinol; after 24-hr contact, the cobalt complex was extracted with isooctyl alcohol containing dodecylamine hydrochloride. The extraction efficiency was 93%.[73]

Resorcinol and substituted resorcinols were converted to trihydroxybenzenes by oxidation with a benzene solution of peroxypropionic acid at 80–110 °C. For instance, 55 g (0.5 m) resorcinol was heated to 110 °C and, in 15 min, 20.5 g of 22% peroxypropionic acid (0.051 m) in benzene was added with stirring. The temperature rose to 120 °C and, in 30 min, the conversion of the peracid was 100%; the product contained 4.85 g 1,2,4-trihydroxybenzene and 1.16 g 1,2,3-trihydroxybenzene (95% yield of trihydroxybenzenes, based on the peracid used).[74]

Mixed esters of resorcinol were prepared. Thus, 13.8 g (0.125 m) resorcinol, 0.7 g (0.006 m) 4-dimethylaminopyridine (catalyst), and 2.5 g (0.025 m) triethylamine were dissolved in 50 ml acetone and a solution of 6.7 g (0.025 m) of octanoic anhydride in 15 ml acetone was added during 10 min; the solutions was stirred for 1.75 hr and then the acetone was removed under vacuum. The residue was dissolved in 200 ml diethyl ether and extracted with 200 ml 3% HCl to remove the catalyst and with 4X 100 ml 5% $NaHCO_3$ to remove the octanoic acid. The ether solution was dried with 40 g Na_2SO_4, filtered, and the ether was evaporated; the residual oil was redissolved in 100 ml 1/1 (by vol) of benzene/cyclohexane and cooled to 10 °C to obtain crystals of unreacted resorcinol which were filtered off. The filtrate was washed with 5% $NaHCO_3$ and the organic phase was evaporated to leave 5.9 g of resorcinol monooctanoate. Finally, this monoester was esterified with acetic anhydride, using a small amount of methanesulfonic acid as the catalyst and, similar to the preceding purification procedure, gave the resorcinol acetate octanoate, **12-47**. This, and related mixed esters of resorcinol, were useful in dry bleaching compositions.[75]

A process for the preparation of polyalkyl-2-alkoxy-7-hydroxychromans was developed. As an example, resorcinol was reacted with acetone and methanol in the presence of H_2SO_4 to give a 63% yield of compound **12-48**.[76]

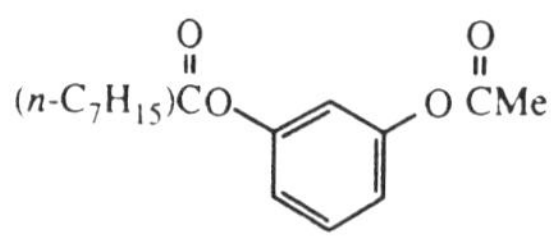

12-47

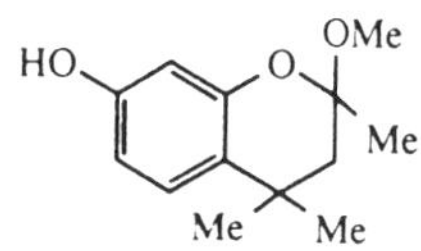

12-48

A chemical sensor for the determination of hydrogen-binding organic compounds was developed, comprising an electrode system containing an electrometer. The electrode was coated with a calixarene containing four resorcinol rings, **12-49**, to form a chemical sensor for determining compounds such as ribose, xylose, and galactose. The method was said to be selective and highly sensitive.[77]

4,6-Bis-(10-hydroxy-9-anthryl)resorcinol tetraacetate derivatives were prepared. For instance, compound **12-50** was made starting with the condensation of 3-(2,4-dihydroxyphenyl)-3-phenylphthalide with 2-(4-methylbenzoyl)benzoic acid in the presence of zinc chloride.[78]

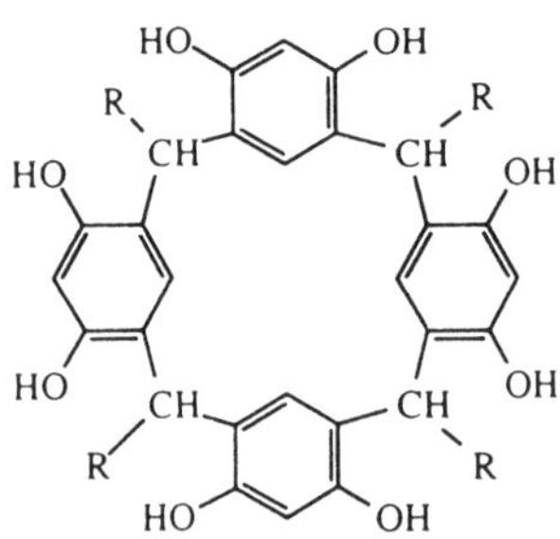

12-49
(where R= unspecified, presumably H, alkyl, or aryl)

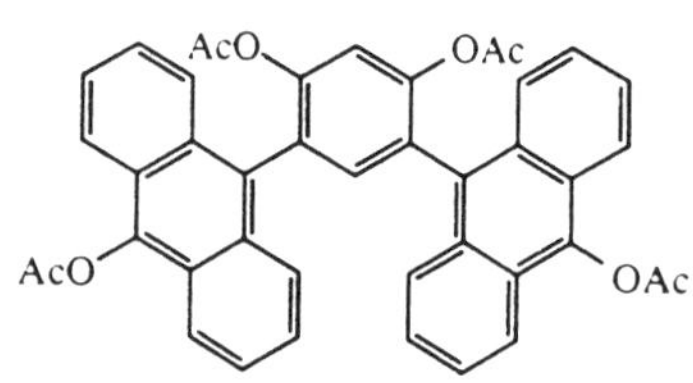

12-50

An improved method of producing 4,6-di-*tert*-alkylresorcinol, such as, e.g., compound **12-51**, was developed and gave the desired products in better selectivity and yield than previous procedures. The process consisted of reacting resorcinol/appropriate *tert*-alcohol/95–98% H_2SO_4 in a mole ratio of 1.0/3.0–3.3/0.45–0.55 at 60–70 °C; 87–97% yields of the dialkylresorcinols were obtained.[79]

Trimethylresorcinols were obtained by passing 5-methylresorcinol dimethyl ether in benzene over gamma-alumina at 410–430 °C. When methanol or toluene was added to the feed, tetramethylresorcinol, **12-52**, was obtained.[80]

Reactors for the bulk or suspension polymerization of vinyl chloride were coated with a methanol solution of a resorcinol/SCl_2 condensation product (molecular weight *ca* 1000). This coating was antifouling, i.e., it reduced the sticking of PVC to the reactor surface.[81]

m-Fluorophenol, **12-53**, was prepared in 63% yield via the hydroxybenzenediazonium fluoride by treating *m*-aminophenol with $NaNO_2$/80% HF in toluene at 0 °C, then holding at 25–30 °C overnight.[82]

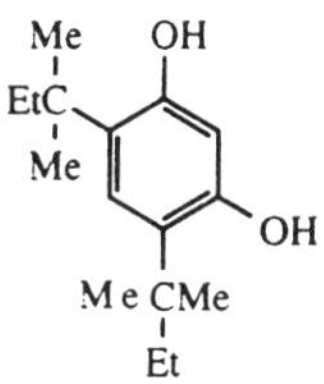

12-51

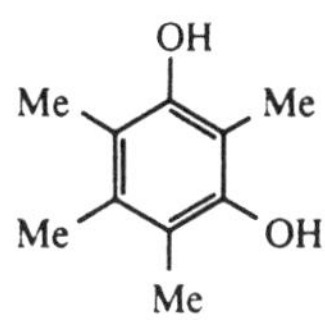

12-52

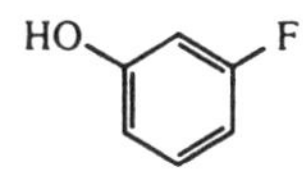

12-53

A carbonaceous binder for refractory bricks was developed. The binder consisted of carbohydrates (polymerized and dehydrated), resorcinol or resorcinol resin, water, an acid-forming phosphate ester, carbon black, boric acid, and maleic and/or phthalic anhydride. This binder and refractory aggregates were blended, shaped, and pyrolyzed to produce the bricks.[83]

12.4. REFERENCES

1. U.S. Pat. 3,394,183 (July 23, 1968; J. C. Dacons, M. J. Kamlet, and J. C. Hoffsommer to the U.S.A., Secretary of the Navy).
2. Ger. Offen. DE 1,959,930 (June 25, 1970; D. A. Salter and R. J. J. Simkins to Minister of Technology in Her Britannic Majesty's Government of the United Kingdom of Great Britain and Northern Ireland); *C.A., 73*, 76860 (1970).
3. USSR Pat. SU 867,885 (Oct. 3, 1981; to A. N. Khaskin); *C.A., 96*, 57281 (1982).
4. Belg. Pat. BE 800,570 (Oct. 1, 1973; to UK Secretary of State for Defence).
5. U.S. Pat. 3,929,405 [Dec. 23, 1975; J. H. Hobbs and W. Melville to Imperial Metal Industries (Kynoch), Ltd.].
6. U.S. Pat. 3,803,190 (Apr. 9, 1974; G. W. C. Taylor, A. T. Thomas, and R. J. E. Williams to Minister of Aviation, UK).
7. Ger. Offen DE 2,531,997 (Feb. 5, 1972; J. F. Kenney to Remington Arms Co.); *C.A., 85*, 80510 (1976).
8. FR Demande 2,014,445 (Apr. 17,1970; J. Prior and A. Florin to Dynamit Nobel A.-G.); *C.A., 74*, 14703 (1971).

9. U.S. Pat. 3,954,822 (May 4, 1976; K. J. Holloway, G. W. C. Taylor, and A. T. Thomas to Ministry of Defence, London).
10. U.S. Pat. 4,336,084 (June 22, 1982; V. S. Urs to Olin Corp.).
11. Pol. Pat. PL 118,885 (Apr. 15, 1983; J. Grzesiek, J. Guga, and J. Piecuch to Zaklady Tworzyw Sztucznych "Nitron-Erg"); *C.A., 99*, 53372 (1983).
12. Belg. Pat. BE 874,187 (May 29, 1979; to Dynamit Nobel A.-G.); *C.A., 91*, 157432 (1979).
13. U.S. Pat. 3,748,349 (July 24, 1973; D. R. Satriana to the U.S.A. as represented by the Secretary of the Army).
14. U.S. Pat. 3,868,282 (Feb. 25, 1975; J. W. Lavitt to the U. S.A. as represented by the Secretary of the Army).
15. U.S. Pat. 4,239,561 (Dec. 16, 1980; A. T. Camp, E. R. Csanady, and P. R. Mosher to the U.S.A. as represented by the Secretary of the Army).
16. U.S. Pat. 4,420,350 (Dec. 13, 1983; A. T. Camp and E. R. Csanady to the U.S.A. as represented by the Secretary of the Army).
17. U.S. Pat. 4,944,816 (July 31, 1990; D. C. Sayles to the U.S.A. as represented by the Secretary of the Army).
18. I. Schwartz, J. Kravitz, and A. D'Angelo, *Cosmet. Toilet., 94*, 47 (1970).
19. Anon., *J. Am. Coll. Toxicology, 5*, 169 (1986).
20. Ger. Pat. DS 1,492,069 (June 19, 1975; to L'Oreal S.A.).
21. U.S. Pat. 3,918,896 (Nov. 11, 1975; G. Kalopissis and A. Bouquet to L'Oreal S.A.).
22. Jpn. Kokai Tokkyo Koho JP 53, 072,836 (June 28, 1978; to T. Yoshida); *C.A., 89*, 168955 (1978).
23. Ger. Offen. DE 1,939,063 (Feb. 19, 1970; G. Kalopissis, H. Gaston-Breton, and A. Bugaut to L'Oreal S.A.); *C.A., 73*, 46678 (1970).
24. Jpn. Kokai Tokkyo Koho JP 57,043,260 (Oct. 12, 1981; to L'Oreal S.A.).
25. Ger. Offen. DE 3,233,541 (March 15, 1984; D. Rose and E. Lieske to Henkel KGaA); *C.A., 101*, 28102 (1984).
26. U.S. Pat. 4,552,565 (Nov. 12, 1985; M. Maak and E. Lieske to Henkel KGaA).
27. Ger. Offen. DE 3,545,371 (Dec. 20, 1985; J. F. Grollier to L'Oreal S.A.); *C.A., 105*, 158608 (1986).
28. Ger. Offen. DE 3,627,398 (Feb. 18, 1988; H. J. Braun, E. Konrad, and H. Mager to Wella A.-G.); *C.A., 110*, 44733 (1989).
29. Anon., *Chem. Week, Aug. 24, 1988*, 64.
30. Natl. Acad. Sci., Natl. Res. Council Publ. 1274, *Chemicals Used in Food Processing*, 109, Washington 1965.
31. U.S. Pat. 3,634,491 (Jan. 11, 1972; J. D. Grossman, R. S. DeSimone, and L. G. Heeringa to International Flavors and Fragrances, Inc.).
32. Jpn. Kokai Tokkyo Koho JP 77,122,336 (Oct. 14, 1977; T. Hagihara, H. Tsuruta, and T. Yoshida to Takasago Perfumery Co., Ltd.); *C.A., 88*, 136324 (1978).
33. U.S. Pat. 4,163,453 (Aug. 7, 1979; A. G. Kallianos and M. I. Simpson to Liggett Group, Inc.).
34. U.S. Pat. 4,155,867 (May 22, 1979; J. B. Hall, M. A. Sprecker, E. J. Shuster, F. L. Schmitt, and J. F. Vinals to International Flavors & Fragrances Inc.).
35. U.S. Pat. 4,281,204 (July 28, 1981; B. J. Willis and R. G. Eileman to Fritzsche Dodge and Olcott, Inc.).
36. Ger. Offen. DE 3,116,913 (Nov. 18, 1982; F. Thoemel and W. Hoffmann to BASF A.-G.); *C.A., 98*, 125646 (1983).
37. U.S. Pat. 4,420,629 (Dec. 13, 1983; H.-G.Schmidt to Dynamit Nobel A.-G.).
38. Jpn. Kokai Tokkyo Koho JP 58,039,819 [83,039,819] (Sept. 1, 1983; to Takasogo Perfumery K. K.).
39. Eur. Pat. Appl. EP 446,925 (Sept. 18, 1991; R. Hirsenkorn to Consortium fuer Elektrochemische Industrie G.m.b.H.); *C.A., 115*, 255818 (1991).
40. U.S. Pat. 3,662,061 (May 9, 1972; F. Lachampt and C. Zviak to L'Oreal S.A.).
41. U.S. Pat. 4,515,773 (May 7, 1985; W. C. Herlihy to Repligen Corp.).
42. U.S. Pat. 4,588,760 (May 13, 1986; J. Jachowicz and L. J. Wolfram to Clairol Inc.).
43. Brit. UK Pat. Appl. GB 2,170,405 (Jan. 31,1986; to J. M. Jackson); *C.A., 106*, 9219 (1987).

44. Ger. Offen, DE 3,330,623 (March 14, 1985; E. Konrad, H. Mager, and M. Rudi to Wella A.-G.); *C.A., 103*, 11449 (1985).
45. Jpn. Kokai Tokkyo Koho JP 76,054,552 (May 13, 1976; H. Yamaguchi and S. Isimoto to Teijin, Ltd.); *C.A., 86*, 5440 (1977).
46. U.S. Pat. 4,088,698 (May 9, 1978; N. A. Fishel and D. E. Gross to Monsanto Co.).
47. Rom. Pat. RO 70,037 (Oct. 20, 1979; T. Panea and M. Moldovan to Intreprinderea de Medicamente, Cluj-Napoca); *C.A., 99*, 139474 (1982).
48. Jpn. Kokai Tokkyo Koho JP 79,088,277 (July 13, 1979; S. Enemoto, T. Kamiyama, and N. Mitsui to Seitetsu Kagaku Co., Ltd.); *C.A., 91*, 157443 (1979).
49. Pol. Pat. PL 104,738 (Dec. 31, 1979; H. Jozwiak and S. Kotlicki to Pabianickie Zakldy Farmaceutyczne "Polfa"); *C.A., 95*, 61993 (1981).
50. Jpn. Kokai Tokkyo Koho JP 52,116,500 (Sept. 29, 1977; M. Noguchi, T. Matsumoto, T. Ikeda, and H. Susuki to Japan Monopoly Corp.); *C.A., 88*, 126344 (1978).
51. Jpn. Kokai Tokkyo Koho JP 55,139,375 [80,139,375] (Oct. 31, 1980; to Sumitomo Chemical Co.); *C.A., 94*, 103168 (1981).
52. Jpn. Kokai Tokkyo Koho JP 55,139,383 [80,139,383] (Oct. 31, 1980; to Sumitomo Chemical Co.); *C.A., 94*, 121358 (1981).
53. Ger. Offen. DE 3,021,453 (Dec. 18, 1980; M. Takenaka, H. Harada, H. Maki, and S. Sasaki to Sumitomo Chemical Co.); *C.A., 94*, 191911 (1981).
54. Ger. Offen. DE 2,850,391 (June 4, 1980; F. Werner and H. U. Blank to Bayer A.-G.); *C.A., 93*, 204255 (1980).
55. Jpn. Kokai Tokkyo Koho JP 56,043,228 (Apr. 21, 1981; to Mitsui Toatsu Chemicals Inc.); *C.A., 95*, 132487 (1981).
56. Jpn. Kokai Tokkyo Koho JP 56,002,939 [81,002,939] (Jan. 13, 1981; to Seitetsu Kagaku Co., Ltd.); *C.A., 94*, 208515 (1981).
57. Jpn. Kokai Tokkyo Koho JP 57,190,558 [82,190,558] (Nov. 24, 1982; to Takeda Chemical Industry K.K.); *C.A., 98*, 129329 (1983).
58. USSR Pat. SU 956,479 (Sept. 7, 1982; A. B. Sidorov, A. R. Birins, J. Dzenitis, and J. Melbergs to Institute of Organic Synthesis, Academy of Sciences, Latvian SSR); *C.A., 98*, 71931 (1983).
59. Jpn. Kokai Tokkyo Koho JP 58,126,833 [83,126,833] (July 28, 1983; to Dainippon Ink and Chemicals, Ltd; Kawamura Physical and Chemical Research Institute); *C.A., 100*, 22418 (1984).
60. Ger. Offen. DE 3,310,959 (March 25, 1982; J. E. Stangroom and I. Harness to National Research and Development Corp., London); *C.A., 99*, 213604 (1983).
61. Jpn. Kokai Tokkyo Koho JP 58,024,531 [83,024,531] (Feb. 14, 1983; to Sumitomo Chemical Co., Ltd.); *C.A., 99*, 22098 (1983).
62. Jpn. Kokai Tokkyo Koho JP 58,072,530 [83,072,530] (Apr. 30, 1983; to Mitsui Petrochemical Industries, Ltd.); *C.A., 99*, 122003 (1983).
63. Jpn. Kokai Tokkyo Koho JP 60,059,887 [85,059,887] (Dec. 27, 1985; to Sumitomo Chemical Industries K.K.).
64. U.S. Pat. 4,518,808 (May 21, 1985; R. H. S. Wang and G. L. Meyers to Eastman Kodak Co.)—the patent claims include only the similar preparation of 2,6-dichlorohydroquinone.
65. Eur. Pat. Appl. EP 165,115 (Dec. 18, 1985; F. Ricchiero to Rhone-Poulenc Chimie de Base); *C.A., 104*, 209206 (1986).
66. Eur. Pat. Appl. EP 185,522 (June 25, 1986; R. A. Fong and S. B. Kong to Clorox Co.); *C.A., 105*, 172058 (1986).
67. Jpn. Kokai Tokkyo Koho JP 61,027,980 [86,027,980] (Feb. 7, 1986; F. Matsunaga and M. Kondo to Mitsubishi Petrochemical Co., Ltd.); *C.A., 105*, 42541 (1986).
68. Jpn. Kokai Tokkyo Koho JP 61,005,037 [86,005,037] (Jan. 10, 1986; H. Harada, H. Maki, and K. Hamada to Sumitomo Chemical Co., Ltd.); *C.A., 105*, 42469 (1986).
69. Jpn. Kokai Tokkyo Koho JP 62,234,089 [87,234,089] (Oct. 14, 1987; H. Tsumaki, H. Suzuki, and N. Matsura to K. I. Kasei K.K.); *C.A., 109*, 54956 (1988).

70. USSR Pat. SU 1,305,227 (Apr. 23, 1987; E. I. Chupka and C. S. Pak to Siberian Cellulose Carton Research); *C.A., 107*, 156774 (1987).
71. Ger. Offen. DE 3,607,924 (Aug. 20, 1987; G. Prescher, G. Ritter, and H. Sauerstein to Degussa A.-G.); *C.A., 107*, 175659 (1987).
72. USSR Pat. SU 1,210,399 (Sept. 23, 1987; E. Yu. Belyaev, I. V. Klimova, A. A. Fedin, Ya. I. Shpinel, and G. A. Gareev to Siberian Technological Institute); *C.A., 109*, 37596 (1988).
73. USSR Pat. SU 1,397,525 (May 23, 1988; I. Iokhannes, L. Melder, and V. R. Akhelik to Institute of Chemistry, Academy of Sciences, Estonian SSR); *C.A., 109*, 132592 (1988).
74. U.S. Pat. 4,751,332 (June 14, 1988; G. Prescher, G. Ritter, and H. Sauerstein to Degussa A.-G.).
75. Eur. Pat. Appl. EP 252,724 (Jan. 13, 1988; R. A. Fond and R. R. Rowland to The Clorox Co.); *C.A., 110*, 75038 (1988).
76. Eur. Pat. Appl. EP 406,758 (Jan. 9, 1991; S. Fukuoka, T. Ishitoku, K. Takahashi, T. Tashiro, J. Imuta, H. Tan, M. Ishiguro, N. Kihara, and T. Mukaiyama to Mitsui Petrochemical Industries, Ltd.); *C.A., 114*, 143147 (1991).
77. Jpn. Kokai Tokkyo Koho JP 02,236,437 [90,236,437] (Sept. 19, 1990; K. Kuriharam, Y. Aoyama, and T. Kunitake to Research Development Corp. of Japan); *C.A., 114*, 243915 (1991).
78. Pol. Pat. PL 151,590 (Sept. 28, 1990; J. Gronowska, K. Aleksandrzak, Jos. Heldt, and Jan. Heldt to Universytet Mikolaja Kopernika); *C.A., 115*, 01870 (1991).
79. USSR Pat. SU 1,583,207 (Aug. 7, 1990; S. M. Korneev, V. A. Nikolaev, and I. K. Korobitsyna to Leningrad State University); *C.A., 114*, 81226 (1991).
80. USSR Pat. SU 1,657,485 (June 23, 1991; V. F. Timofeev and I. P. Soloveichik to Leningrad Technological Institute); *C.A., 116*, 105821 (1992).
81. Eur. Pat. Appl. EP 440,452 (Aug. 14, 1991; S. Masuko, I. Nakamura, K. Fukuda, Y. Hatakeyama, A. Yamaguchi, and K. Yamaguchi to Mitsui Toatsu Chemicals, Inc.); *C.A., 115*, 233155 (1991).
82. Jpn. Kokai Tokkyo Koho JP 03,232,829 [91,232,829] (Oct. 16, 1991; K. Momota, T. Yonezawa, and Y. Inoue to Morita Kagaku Kogyo Co., Ltd.); *C.A., 116*, 105801 (1992).
83. PCT Int. Appl. WO 92,012,198 (July 23, 1992; R. Matalon, K. Bridger, and F. W. Henry, Jr. to Martin Marietta Magnesia Specialties, Inc.); *C.A., 118*, 65500 (1993).

13

Selected Business Aspects

This chapter presents a collection of snapshots intended to show that a challenging climate existed even 50 years ago in the resorcinol business, and that, if anything, the entrepreneurial ingenuity was being challenged even more in the 1980s and 1990s. Even the very limited information given below indicates that correctly estimating investment and operating cost in a manufacturing facility, estimating the capability of the competition, and estimating the size and growth of the market for resorcinol and its derivatives at a given time requires the best information, talent, and savvy. Misestimates are costly and can lead to the death of a venture or to missed opportunities.

13.1. HISTORY OF ACTUAL/POTENTIAL PRODUCERS OF RESORCINOL

This information is split into regions. Table 13.1 summarizes the actual producers of resorcinol in the U.S. since World War I started and also lists several companies which had declared serious intentions, according to trade newspapers, to build resorcinol plants.

The small size of the older (to about 1950) plants in the U.S. was due to the limited industrial use of resorcinol, mainly in dyes, nonprescription medications, and marine applications of wood adhesives. In the 1950s, resorcinol began to be used in dips for textiles for rubber reinforcements, and soon after in bonding agents for rubber, and demand increased. Indeed, from the 1950s onward, Koppers Co. expanded its capacity for resorcinol in several stages, as only implied in Table 13.1.

Two of the early operators (Pennsylvania Coal Products and Hayden Chemical) exited the business because of severe production problems. In the 1960s, U.S. Pipe & Foundry Co. started up the second U.S. resorcinol plant with respectable capacity, using their own benzene disulfonation/caustic fusion process. Production costs are tied to many factors, e.g., the efficiency of the process, capacity and capacity utilization/sales (the economy of scale). When U.S. Pipe & Foundry Co. found it difficult to reach profitable capacity utilization, they filed a complaint against Koppers Co. for monopolistic practices (e.g., long-term supply contracts with major users, lower

Table 13.1. Chronology of Actual/Potential U.S. Producers of Resorcinol

Company	Year	Status	Capacity (mt/yr)	Process[a]	Remarks
Barrett (later Allied Chemical)	1914–1918	Operational	*ca* 1000	Bz.sulfo/fusion	
DuPont	1930–1950	Operational	*ca* 1000	Bz.sulfo/fusion	
Hayden Chemical	1947–1951	Operational	*ca* 150	Bz.sulfo/fusion	Destroyed by explosion
Pennsylvania Coal Products	to 1948	Operational	*ca* 1000	Bz.sulfo/fusion	Damaged by fire in 1948. Bought by Koppers in 1948
Koppers	1948–1988	Operational	1,000–20,000	Bz.sulfo/fusion	Koppers taken over by Beazer in 1988
Beazer PLC	1988–1989	Operational	20,000	Bz.sulfo/fusion	
INDSPEC Chemical	1989–	Operational	22,000	Bz.sulfo/fusion	INDSPEC Chemical bought from Beazer
U.S.Pipe & Foundry (subsid. Jim Walters)	1967–1972	Operational	3000	Bz.sulfo/fusion	Shut down because not profitable
Signal Chemical	*ca* 1972	Near operational	3000	DIPB hydroperox.	Production problems; sold[b]
Hooker Chemical (later Occidental Petroleum)	*ca* 1972	Planned	2500	Bz.sulfo/fusion	Licensed by Koppers; never built
Texas Chemical & Plastics (TCP)	1974	Planned	2500	Bz.sulfo/fusion	Licensed by Koppers; never built
Resorcinol Puerto Rico (RPI)	1977	Planned	2500	Bz.sulfo/fusion (Halliday contin.)	TCP/Halliday (UK) joint venture; never started

[a]Bz.sulfo/fusion=benzene disulfonation/caustic fusion process; DIPB hydroperox.= *m*-diisopropylbenzene hydroperoxidation/cleavage process.

[b]Goodyear Tire & Rubber Co. first leased, then bought this plant from Big Three Industries, and produces hydroquinone there.

prices) with the Federal Trade Commission (FTC). Although Koppers refuted the allegations, the FTC decreed, among several orders, that Koppers had to license its process knowhow to up to five other companies interested in entering the resorcinol market. Indeed, in about 1972, Hooker Chemical and Texas Chemical & Plastics expressed their intention to license knowhow but never seemed to have gotten beyond the engineering stage. In 1977, Texas Chemical & Plastics Corp. and L.B. Holliday (a one-time modest-scale producer of resorcinol in Great Britain) formed a joint venture, Resorcinol Puerto Rico, Inc., to produce resorcinol in Puerto Rico (presumably, tax and other incentives were available for locating there), going as far as announcing the suppliers of their equipment in the press; for unknown reasons, this plant, too, failed to materialize.

Table 13.2. Chronology of Actual/Potential Non-U.S. Producers of Resorcinol

Company	Year[a]	Status	Capacity (mt/yr)	Process[b]	Remarks
Hodagaya Chemicals Japan	1972	Operational	Not available	Bz.sulfo/fusion	Ceased production
Mitsui Toatsu Chemicals, Japan	1972	Operational	1200	Bz.sulfo/fusion	Production phased out in 1980, when Mitsui Petrochem. started new plant
Taoka Chemicals, Japan	1972	Operational	1200	Bz.sulfo/fusion	Shut down in 1981, when Sumitomo started new plant
USSR (and successor)	1976	Operational	230–1000	Bz.sulfo/fusion	
Hoechst A.-G., Germany	1978–1992	Operational	*ca* 9000	Bz.sulfo/fusion	Stopped production in Dec. 1992
Mitsui Petrochemicals, Japan	1980–present	Operational	*ca* 3000	DIPB hydroperox.	
Sumitomo Chemicals, Japan	1981–present	Operational	5000–9000 (18,000 by late 1993)	DIPB hydroperox.	
Hindustani Chemicals, India	mid-1980s	Operational	*ca* 750	Bz.sulfo/fusion	
China	mid-1980s	Operational	*ca* 750	Bz.sulfo/fusion	
Jamil Vergara, Brazil	1986	Planned	2000	Not available	Never constructed
Copesul, Brazil	1987	Planned	1000	Not available	Never constructed

[a]This is the year when information in the trade newspapers was accessed. Note that, e.g., Hoechst A.-G. was one of the oldest and largest producers of resorcinol.
[b]Bz.sulfo/fusion = benzene disulfonation/caustic fusion process; DIPB hydroperox.= *m*-diisopropylbenzene hydroperoxidation/cleavage process.

For about 10 years, the U.S. scene remained stable, until in 1988, in the midst of increasing turbulence in the chemical industry and a sharp rise in the takeover of U.S. companies by foreign companies, Koppers Co. was acquired by Beazer PLC, a British construction company interested in Koppers's large building materials/construction business. In order to reduce the very large debts incurred in the takeover, Beazer sold off most of the businesses which were not construction-related. Thus, in 1989, INDSPEC Chemical Corp. was formed to acquire the resorcinol business of Koppers/Beazer in a leveraged buyout (LBO) by a former Koppers management group—meaning that a major portion of the financing came from borrowing/outside investors. As an aside and sign of the times, not long after, Beazer PLC faced financial difficulties and was itself acquired by Hanson PLC, the large British conglomerate.

The history of resorcinol production outside of the U.S. is also quite varied, but not easily accessible; the available information is summarized in Table 13.2. The production year quoted in the table refers to the date a press release was found.

It will be noted that by about 1972, the two Japanese producers, Mitsui Toatsu and Taoka Chemicals, who used the benzene disulfonation/caustic fusion process, had a combined capacity of about 2400 mt/yr, which was adequate for supplying domestic needs. In 1980–1981, Sumitomo Chemicals and Mitsui Petrochemicals, both using their newly developed *m*-diisopropylbenzene hydroperoxidation/cleavage technology, built resorcinol plants of 5000 and 3000 mt/yr capacity, respectively, together several times larger than the Japanese market. Mitsui was reported to be able to produce either hydroquinone by hydroperoxidation of *p*-diisopropylbenzene (DIPB) and/or resorcinol from *m*-DIPB in its plants, depending on demand. On the other hand, in 1981, a good portion of Sumitomo's 5000 mt/yr resorcinol capacity in the new plant was said to be aimed at export opportunities and the company planned to phase-in production from 2000 mt in 1981, to 3000 mt in 1982 and 4000 mt in 1983.[1] Both companies developed broad research programs on uses for resorcinol—as evidenced in the preceding chapters—involving all areas, from chemicals to polymers to rubber and new products. They also did their market research well and built up exports of resorcinol. As described in Section 3.6 and later below, they also built the first plants for *m*-aminophenol from resorcinol.

In Europe, the old-time producer of resorcinol by benzene disulfonation/caustic fusion, Hoechst A.G. (one of the companies stemming from the splitup of I.G. Farbenindustrie, the giant conglomerate, after World War II), had a resorcinol capacity of about 9000 mt/yr in the late 1970s and a distinguished record of achievement across the board in chemicals, including research and development. It came as a surprise to many when Hoechst was reported in December 1991 to be stopping production of resorcinol and 2-naphthol, made by the sulfonation/caustic fusion of benzene and naphthalene, respectively. In the trade magazines, it was said that this shutdown was due to environmental problems, presumably in the byproduct area. About five years ago, Hoechst was said to have invested DM55 million ($35 million) "in a new salt factory for purifying the waste waters from the plant" (presumably, this included the oxidation of byproduct sodium sulfite, for which there was poor demand in Europe, to sodium sulfate).[2]

Shortly after Hoechst's exit from resorcinol production, Sumitomo Chemicals announced that they were expanding capacity of their resorcinol plant in Japan by 9000 mt/yr, to a total capacity of 18,000 mt/yr; the expansion was expected to require about 18 months, to be complete by mid-1993 [as of this writing, there has been some delay].

As shown in Table 13.2, there were relatively modest-scale plants for resorcinol in the USSR, China, and India in the 1980s. The sudden disintegration of the USSR in 1991 and subsequent political and financial turmoil have paralyzed trade, including the trade in resorcinol, in the region. China and India were in a transition period in the early 1990s, aiming to modernize their economies and trade. Plans to produce resorcinol by Brazilian companies in the mid-1980s have remained at the planning stages so far.

A historical perspective on the price of resorcinol from 1940 to 1991 is given in Table 13.3. It is interesting to note that the list price for the technical grade resorcinol was in the $0.55 to $0.84/lb price range from 1940 to 1971—with lower prices in years of low demand, a low being reached in 1971, presumably because of lowered

Table 13.3. U.S. List Prices of Technical Grade Resorcinol[a]

Year	$/lb.
1940	0.75
1946	0.64
1951	0.84
1956	0.78
1961	0.78
1966	0.66
1971	0.55
1976	1.20
1981	1.56
1986	1.95
1991	2.20

[a]This refers to U.S. prices; INDSPEC Chemical Corp.'s technical grade resorcinol is now 99.3% minimum purity.

demand and excess capacity. However, by the mid-1970s, when U.S. Pipe & Foundry Co. left the business, there was a brief period of short supply, and spot prices of resorcinol in the $5.00/lb range were rumored to have been asked by some traders. However, the shortage of resorcinol did not last long.

The price of resorcinol, or rather, the cost of manufacture, was also impacted by political events and the rise of environmentalism. The mid-1970s brought the Arab oil crisis, with skyrocketing crude oil prices which translated into rising petrochemical prices, including the cost of benzene, the starting material for resorcinol. On top of that were worries about the continued supply/availability of feedstocks. At the same time, environmental issues and concern about the toxicity of chemicals escalated. Analytical techniques became so refined that parts per trillion of particular entities became detectable. Pollution detection and control became not only major issues, but a major business. The costs of remediation and control, including, e.g., litigation, filing of reports to government agencies, and filing requests for permits of all sorts, had to be recovered eventually in higher prices if an operator were to stay in business. Another example of the effect of the rising alarm about health risks attributed to chemicals is that this lead to a decrease in the production of chlorinated hydrocarbons, which lowered demand for chlorine, which escalated the price of sodium hydroxide copro-duced with chlorine. Sodium hydroxide is used for the caustic fusion of sodium *m*-benzenedisulfonate to make resorcinol. Additionally, inflation, e.g., interest rates on borrowed funds, began to escalate fast in the late 1970s into the 1980s, escalating the cost of raw materials, equipment, and other supplies as well as the cost of labor. In this context, it is surprising that the price of resorcinol only quadrupled from its lowest price in 50 years in 1971 to 1991—many other chemicals, including commodities such as phenol and styrene, increased much more in this period—and the price of many other goods, such as a loaf of bread, increased as much to much more.

Table 13.4. Japanese Production of *m*-Aminophenol(s) from Resorcinol

Producer	Year	Estimated capacity (mt/yr)
Sumitomo Chemicals	1987	2000
Mitsui Petrochemicals	late 1980s	*ca* 2000

Finally, a historical perspective on *m*-aminophenol. In the past, this product was made by the sulfonation/reduction/caustic fusion of nitrobenzene, as mentioned in Chapter 6. Of the several producers of *m*-aminophenol by this route, ACNA of Italy was the largest with a capacity of about 4000 mt/yr. In the late 1980s, Sumitomo Chemicals and Mitsui Petrochemicals began manufacture of *m*-aminophenols [including *m*-(*N,N*-dialkalkylamino)phenols] from resorcinol, as shown in Table 13.4. The *m*-aminophenols are believed to be the largest-volume derivative of resorcinol. Much of the increase in demand for the *m*-aminophenols in recent years has come from their use as starting materials for leuco dyes used in imaging technologies (see Chapter 10).

Also, in 1987 it was reported that Sumitomo Chemicals was building an 800 mt/yr. N-alkylated *m*-aminophenol plant in Japan, for completion by the end of 1987.[3]

A hint of the value-added dimension of the (*N,N*-dialkylamino)phenol-derived leuco dyes was an announcement in 1990 that Sofix Corp., a joint venture of Yamada Chemicals (Kyoto) and Nagaso (Tokyo), a trading company, would build a $35 million plant in the U.S. (mid-1993) for dyes for fax and heat- and pressure-sensitive copier paper.[4] This plant was in operation in mid-1993.

13.2. MARKETS/MARKETING

The worldwide production capacity for resorcinol in late 1993 will likely be 40,000–45,000 metric tons/year—Sumitomo Chemicals's capacity addition of 9000 mt/yr will have made up for the shutdown of Hoechst's plant of similar capacity in late 1991, as mentioned above. The largest producer, in the U.S. and, by a modest margin, worldwide, will still be INDSPEC Chemical Corp. with a capacity of about 22,000 mt/yr. Exports are an important factor in U.S. as well as Japanese capacity utilization. The broad categories of use for resorcinol are given in Table 13.5. The data given are the best available to the author.

The predominant use of resorcinol in the U.S. in rubber products, particularly tires, was due to the large number of motor vehicles of all types in the U.S.—tied to their affordability, the vast road system, the low cost of fuel compared to most other regions, etc. It appears, however, that in the U.S. and the other regions, the "other" category of use is increasing due to the newer, specialty uses of resorcinol in agricultural, pharmaceutical, polymer, additives for polymers, and dye intermediate applications (e.g., the manufacture of *m*-aminophenols from resorcinol in Japan).

Table 13.5. Large Categories of Use (% of Total) of Resorcinol[a]

Category	U.S. 1988	U.S. 1992	Europe 1988	Japan 1988
Rubber	69	53	50	55
Wood adhesives	16	25	25	20
Other[b]	15	22	25	25

[a]Estimates of Koppers Co. in 1988; INDSPEC Chemical Corp. for U.S. in 1992.
[b]Includes pharmaceuticals, agricultural chemicals, imaging chemicals, dyes, UV absorbers, and others.

Indeed, the many actual uses of resorcinol, as shown in Chapters 4–12, indicate that accurate market analysis and forecasting is very important and complex. The many potential uses of resorcinol require an open mind, experience, and selective contact with potential customers for accurate assessment. Even then it is not easy to determine which inputs are true and which information is false, faulty, or misleading, intentionally or not.

A good example of the complexity of sales analysis and forecasting is the use of resorcinol in the tire industries. Interpretation was required of the impact of changes since the 1940s in tire construction from inner tubes to steel cord reinforcement, from the simple balloon to the steel-belted radial tire (see also Chapter 5). On top of that, the U.S. tire business was spearheaded and grown for decades by U.S. companies; but, in the turbulent 1980s, several of the large U.S. tire makers were taken over by non-U.S. manufacturers. Firestone was acquired by Bridgestone of Japan, General Tire by Continental of Germany, B.F. Goodrich's and Uniroyal's tire businesses were bought by Michelin of France and Armstrong Tire by Pirelli of Italy. Only Goodyear Tire & Rubber Co. of the big makers and Cooper Tire of the medium-sized companies stayed under American ownership/management. There were other technical threats to the use of resorcinol in rubber products, such as the about 20-year-old development of the injection-molded polyurethane tire in several countries which has not reached large-scale use to date. Other types of threats to the use of resorcinol in tires were a campaign in the 1970s to use 2-naphthol instead of resorcinol by the U.S. manufacturer, American Cyanamid, and to use catechol instead of resorcinol by Crown Zellerbach. Neither of these substitution proposals succeeded, essentially because of the technical merits of resorcinol, even at a higher price. [As an aside on competition in the chemical industry, American Cyanamid had developed a new process for 2-naphthol from 2-isopropylnaphthalene by hydroperoxidation/cleavage and built a new plant, reportedly a $30 million investment, using this process in the 1970s; this plant was shut down a few years later, probably in the main because of low-priced imports from overseas producers using the naphthalene sulfonation/fusion process for 2-naphthol.]

Good market research and intelligence has been and will continue to be a key ingredient in the life of any business, not only for the U.S., but increasingly on a worldwide scale.

13.3. MANAGEMENT TASKS AND CHALLENGES

13.3.1. General

The task of top management of any company is to build a lean but effective organization, so that profitable operations can be maintained. Profits are needed to pay bills of all sorts, to reward shareholders in the enterprise, to pay employees in the form of a paycheck, and to pay incentives to the key employees (who isn't, in that lean and effective organization) so they will stay with it. That sounds fairly simple, but the agonizing reappraisals and restructurings, from cost cutting to sell-offs and lay-offs, of the last 10–15 years in the chemicals industry show that excellent performance by top management is very tough to achieve; it is easier to go astray.

To achieve that profitable operation still requires the virtues of old: Effective sales management and salespersons, with components in program (short-term and long-term) preparation, in-house and hired market research, technical service, customer relations, etc.; production management, with components in labor relations, preventative maintenance, engineering, quality control, power generation, packaging, etc.; purchasing management, not only to assure a reliable supply of high-quality raw materials at the best available price, but also to know and anticipate changes in sources, prices of raw materials, etc.; financial management has to forecast and control budgets and insurance needs, bad debts, inflation, etc.; benefits such as health care require constant watching. Nowadays, the chemical industry's exports are strong; currency fluctuations can be wide and have sizable impacts. Thus, there is even "insurance"—currency hedges—to be considered for such losses.

All the departments report to top management. The information is no longer by plain paper or mouth (telephone)—now there is fax and electronic mail, by computer. Information has become management information systems (MIS), faster if not more accurate than before. Problems can by minimized by astute management. Analysis and control is the name of the game.

Old virtues and common sense are being alternated with actions such as "the customer comes first"; "R&D must be customer-responsive and market-oriented"; "we deliver a high-quality, in-specification product the first time and every time"; "equal opportunity employer"; "total quality commitment"; "environmental responsibility/responsible care"; "noncompliance with rules and regulations will be corrected if not punished immediately"; etc. All are good and laudable objectives, often encoded in laws formulated by lawyers, congressmen, and senators with their large staffs, and state and federal agencies. The analysts and formulators are being watched over and prodded by environmental activist groups, all well-meaning and idealistic. It seems that they find it difficult to believe that most people in industry also want pure water and air, pristine forests and fields, a safe workplace, and management of resources not only for profit but with reasoned concern for the common good and for the next generation.

The following two examples are intended to sketch some of the complex problems that the chemical industries, even small producers, have to manage every day.

13.3.2. Environmental and Health Regulations

The conduct of business is closely regulated, be it in the office, the laboratories, the plants, the processes, the products (old and new), the employees, the shipments (in and out), and the byproducts/effluents. Since regulations change frequently, past practices which were legal and, often, state-of-the-art at the time become punishable if not remediated quickly, usually at high cost. A recent trade paper story gave a good example—Hoechst (a former producer of resorcinol), the German chemical giant, was reported to have spent a giant amount (almost $1 billion) on environmental improvements and safety analysis in 1992 and was expected to spend more than that in 1993.[5]

The rules and regulations effecting the sale of resorcinol (and derivatives) have already been implied in Chapter 2. Material Safety Data Sheets (MSDS) have to accompany every shipment of every product, and MSDS writing has become a specialty as well as a sizable expense. Label writing is another specialty that may require a legal opinion before printing.

The regulatory agencies are usually dual, at the state and the federal levels, usually with different regulations. Local regulations may be imposed on top of that.

Here are some regulations at the federal level that have to be met: The Occupational Safety and Health Act (OSHA); the Toxic Substances Control Act (TSCA) and the related Federal Hazardous Substance Act; the Federal Food, Drug and Cosmetic Act; the Consumer Product Safety Act; the Pollution Prevention Act of 1990, the related Resource Conservation and Recovery Act (RCRA), the related Comprehensive Environmental Response, Compensation and Liability Act (CERCLA) and the related Emergency Planning & Community Right-to-Know Act (part of Superfund Amendments and Reauthorization Act—SARA); the Hazardous Materials Transportation Act; the Ports and Waterways Safety Act, the related Poison Prevention Packaging Act and the related Marine Protection, Research and Sanctuaries Act; and others.

Before a new product can be introduced, it may have to be tested exclusively for health and safety, even if it is made from known starting materials and/or if it is similar to existing products. A premanufacturing notification (PMN) has to be filed with the Environmental Protection Agency (EPA) for amounts of material exceeding 2000 lb., which is not just a notification, but an application for permission to manufacture; the EPA may require further, extensive tests before giving or denying permission.

An added requirement for regulatory approval for (chemical) production in existing plants was recently introduced by OSHA (Occupational Safety and Health Administration): A process hazard analysis (PHA) must be done and filed with OSHA.

From the outside, the environmental activist organizations have staffs of lawyers to sue, whomever it may concern including the regulators, if they feel that their idealistic standards are not met.

In the U.S. there is a sort of double taxation to meet regulations. One is the cost of the specialist personnel and tests needed to be able to file the mountainous paperwork required. The other is the tax monies, business and personal, spent to maintain the growing staffs of state and federal environmental/health agency personnel which test, file, issue, etc.

But that is not all. Exporters have to meet the regulations applicable to their products in different countries. Often, their regulations differ from U.S. regulations. However, to understand these regulations a translator or consultant may be needed, perhaps one who has an up-to-date dictionary of the terms used. And the regulation of chemicals may be even more advanced than in the U.S. in some overseas regions. For example, the European Economic Community (EEC), through the International Organization for Standardization (ISO), has formulated a program of quality standards/assessment to meet ISO 9000 series quality standards. To meet ISO 9000–9004, a comprehensive review and analysis has to be done, aided by independent consultants, of the design, production, installation, inspection, packaging, and marketing of their products. Roughly, ISO 9000 introduces quality and the selection and use of standards, ISO 9004 reviews quality management and quality systems, ISO 9003 concerns inspection and testing controls, ISO 9002 covers manufacture of product from raw materials, and ISO 9001 includes product development and service. The goal is to have well-defined quality assurance in all its aspects. In fact, ISO 9000 is a new procedure, likely to be tuned up (changed) as it is practiced. Indeed, different customers may insist on different facets of the ISO 9000 series.

13.3.3. Research & Development and Patents

The above sections imply how difficult it is to do profitable R&D. The lean and efficient organization has little tolerance for long-range research, and development of a "new" product is always risky because of the difficulty of accurately forecasting markets, because it is difficult to know what the customer really wants (the greatest immediate want is lower prices for, e.g., resorcinol); and the cost of R&D is high, the laboratory instruments are becoming more sophisticated and much more expensive, and trying to introduce new products to new fields requires new expertise, more new customer contact, and establishing relationships, new tests, and more people. And when there is that first customer with modest needs for a new product, can the investment in new equipment be justified/recouped? Is the market really large enough to have an impact on company performance? Can the cost of toxicity testing ever be recovered? What is the risk of litigation if the product does not perform as predicted? How valuable is that new product in terms of return on investment (ROI)? The critics have a field day, because the process is so iffy, because the sure thing is not so sure, because every chemical company now has its cases of old and recent failures, because success stories are harder to come by. If the consensus or edict on a new product is to go ahead, there begins the costly process of doing scaleup, pilot plant work, of refining cost estimates and specifications, of applying for regulatory permits at the state and local level, all requiring stout-hearted, competent staff.

One might think that the initial quantities of a new product might be best handled via a toll producer with lots of flexible equipment. In fact, often these specialists may lack a certain piece of equipment or expertise. They too have to make a living—they may have to charge seemingly high prices per unit of product since the batch volumes contracted for are usually fairly small; they, too, have to apply for regulatory permits before making these batches of new products.

What about new, let's say less polluting or more efficient, processes for old products. To be sure that the new process works well, to familiarize engineers with the process so they can develop and better assess plant design and production costs, the laboratory process has to be scaled up in a pilot plant. This will take more precious time and money under the best of circumstances.

Then there is always the possibility that the new product is copied by someone else very quickly, or that someone else develops a variation of the product which cuts into sales which had been expected to grow very quickly and to remain unchallenged for years. It is possible that someone else obtains a patent on the new process which was held proprietary by the first, innovating company—that may scramble or scuttle plans very quickly.

Well then, there is the patent system to resort to, to protect that new product or new process from use by others, in the U.S. for 17 years. Not only that, there is also the potential to license a patented process to others and to collect fees and royalties, hopefully. There is also advertising value in patents. The patent portfolio of a company is one of the proofs of its virility, its ingenuity, its desirability as a vendor or as a partner for joint developments. Yet, in the financial world that patent portfolio is considered to be an intangible asset!

Perhaps this is so because only a small percent of all patents, maybe 5–10%, actually goes on to market. The patent ostensibly protects that profitable, new product from competition for 17 years in the U.S. Actually, a patent may be obtained early on and product development may take so long now that the patent protection may be much less than 17 years. And there are always the feisty competitors out there developing their own, similar aces.

The cost of that patent portfolio is also a factor that has to be considered. Obtaining a U.S. patent may cost from \$3,000–\$10,000, depending on how costs are allocated, on the use of an in-house or external patent lawyer, and on the length of the application process which may require answers to rejections, etc.; but that U.S. patent only gives protection for that new product in the U.S. The ability to copy, produce, and sell that new product is now worldwide, and not confined to the so-called developed countries any more. Indeed, some of the countries formerly considered to be underdeveloped suddenly have leapfrogged, becoming aggressive, knowledgeable, quick-moving competitors. The answer is to apply for foreign patents as well, say in the 5 or 10 countries most likely to be a consumer or potential producer of one's hard-won, new product. Now costs multiply rapidly, with patent application fees in these foreign lands being close to the fees in the U.S. and the process of obtaining a patent often being slower because a published Patent Application system may be used to allow the domestic competitors to come up with objections to the originality of the application, before a final patent is granted. In Japan that can take up to 10 years. But that is not the only cost; in many countries, annual patent maintenance fees are charged, in the

\$500–\$1000 range per patent per country. A portfolio of 20, 50, or 100 patents then costs millions of dollars to get and maintain.

As with environmental/health regulations, so with the patent system. One can rail against it, one can decide to quit participating in it, one can follow the rules at whatever the cost, go with the flow, or decide to cut down on that patenting activity, restrict it only to the most clear-cut cases. One can excuse the giving up on patents by saying they give away more than one receives. However, there are all these able competitors out there who use the world's patent systems to the hilt, not only to protect and promote their inventions, but also to restrict and confuse the other parties.

In the face of all these challenges, there are out there well-managed companies, run by stout-hearted and knowledgeable people, that prosper by their style, their foresight, their ability to cope with the rules, the regulations, the patent systems, by their ingenuity and risk-taking ability. Let them be properly rewarded.

13.4. REFERENCES

1. *Chem. Age, June 19, 1981*, 10.
2. *Chem. Week, Dec. 4, 1991*, 6.
3. *Japan. Chem. Week, Nov. 26, 1987*, 2.
4. *Chem. Week, Apr. 25, 1990*, 8.
5. *Chem. Eng. News, 71(14)*, 20 (1993).

14

Occurrence in Nature—A Domain of Academic Researchers

14.1. INTRODUCTION

The isolation, structure determination, and proof of structure by synthesis of natural products from plants, microorganisms, and animals have been a focus of organic chemists in universities since the beginnings of chemistry. The plant phenolics alone, of which the resorcinol ring-containing constituents are a part, are ubiquitous in nature and are the subject of an immense literature.

Many plants and crude plant extracts have been known to have medicinal properties and were/are being used as folk medicines in many countries. Many dozens of pharmaceutical companies have had and/or intensified longstanding efforts in the chemistry and utilization of such products, and in structure modification to improve on nature. Biotechnology is beginning to play a part. Also, for decades now, much thought and work has been devoted to elucidating biological pathways by which the natural products arise. All this is intricate and wonderful chemistry, very much alive—the task is far from finished.

These natural products come in all sizes and shapes, their structures range from simple to challengingly complex, and their numbers are endless. New compounds are being identified continuously. The functions of these chemicals, if known, range from foods to attractants to repellants, from vitamins to very toxic materials. In plants, such toxic compounds occur in large numbers and they are believed to function as bactericides, fungicides, and deterrents to predators; these chemicals are the so-called phytoalexins. Whether on contact with the skin or on ingestion by man, the toxic plant constituents can have effects ranging from irritants to carcinogens and mutagens. A key review of this subject by the eminent Bruce Ames, "Dietary Carcinogens and Anticarcinogens,"[1] presented a balanced view not only of the known toxics in plants, such toxics being ingested in the normal diet of people in amounts far exceeding the toxic synthetics that man may be exposed to in the modern environment, but also the antimutagens and anticarcinogens found in plants which may help us to withstand the assault of the toxics. Another example of a balanced discussion of diet and cancer was given by Wong.[2]

Resorcinol was first discovered in 1864 in the alkali fusion product of *Galbanum* resin.[3] This resin is obtained from certain Asian plants of the genus *Ferula*. It was used in incense and had medicinal use as an antispasmodic, as was the related *Asefitada* gum, obtained from the same sources.

Since then, the resorcinol moiety has been found in a wide variety of natural products, described in a massive literature. Of course, this literature was not focussed on the relationship to resorcinol. The compounds which contain a resorcinol segment, often interrelated, encompass the chalcones, coumarins, coumaranones, chromenes, chromanes, chromanones, benzofurans, flavanoids, aurones, psoralens, condensed tannins, 5-alkylresorcinols, mold metabolites, and many others. Many have served as starting points for pharmaceutical development.

An example of the specialist literature in the field is the book on chromenes, chromanones, and chromenes published in 1977 in the series, *The Chemistry of Heterocyclic Compounds*, which has 1196 pages.[4]

In this chapter, the intent is to demonstrate the variety of natural compounds containing a resorcinol segment by the following selections and examples taken from the literature, in good part from 1980–1992, highlighting current research.

14.2. RESORCINOL AND RELATIVELY SIMPLE DERIVATIVES

Resorcinol was among the low molecular weight polyphenols found in the seeds of 51 selected cultivars of the broad bean, *Vicia faba*.[5] Resorcinol was one of the flavor-forming compounds in fresh, honey-colored mushrooms, *Armillaria mellea*.[6] Resorcinol, as well as catechol and hydroquinone, were found in an acidic fraction of cigaret smoke condensate.[7] Resorcinol was among the phenolic substances extracted from the needles of *Pinus thunbergii*.[8] Resorcinol and resorcinol monomethyl ether were among the phenolic constituents obtained from extracts of the Indian Natu tobacco.[9]

It was mentioned earlier that resorcinol dimethyl ether (1,3-dimethoxybenzene) is a component of the fragrance of yellow tea roses (Section 12.2). Also, 1,3-dimethoxybenzene is the fragrant compound, at 0.4 wt% of the dried specimen, of the yellow mushroom (Agaricales), *Rhodophyllus icterinus*.[10] 1,3-Dimethoxybenzene was also found in raw filberts (hazelnut).[11]

beta-Resorcylic acid (2,4-dihydroxybenzoic acid) and gamma-resorcylic acid (2,6-dihydroxybenzoic acid) were among the phenolic acids found in the needles of the Central European pine, *P. mugo*, *P. sylvestris*, and *P. uliginosa*.[12] Earlier, *o*-orsenillic acid (2,4-dihydroxy-6-methylbenzoic acid) had been isolated in the depside from the fungus, *chaetomium cochliodes*[13]; the related sparasol, methyl 6-hydroxy-4-methoxy-2-methylbenzoate, **14-1**, is an antibiotic substance which had been isolated from the fungus, *sparassis ramosa*.[14]

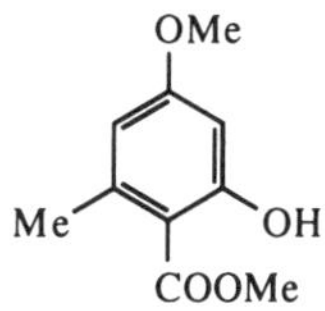

14-1

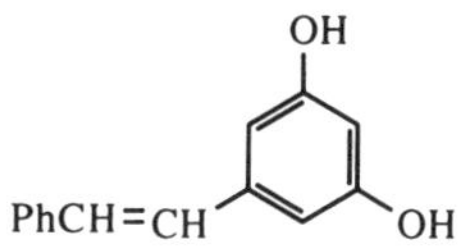

14-2

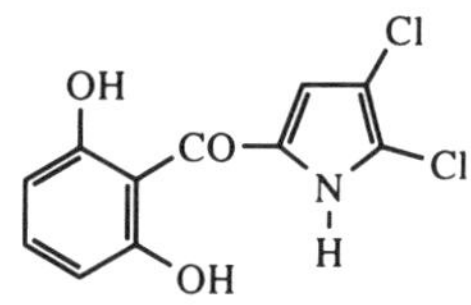

14-3

Pinosylvine, 5-styrylresocinol, **14-2**, and its monomethyl ether were found first by Erdtmann in the heartwood of several pine species[15]; it has antifungal activity.

Pyoluetin, **14-3**, is a metabolite of *Pseudomonas fluorescens* Pf-5, which is effective in controlling damping-off in cotton seedlings by *Pythium ultimatum.*[16]

The long-chain alk(en)ylresorcinols have been found in a variety of plants and bacteria, and have been the subject of analysis not only of their detailed structure but also their possible functions. Bilobol, **14-4**, was found to be a constituent of the fruit of *Ginko biloba.*[17] The monomethyl ethers of 5-*n*-heptadecyl- and 5-*n*-pentadecyl-4,6-dimethylresorcinol were part of a bacterial lipid fraction obtained from *Mycobacterium leprae.*[18] Mixtures of 5-*n*-alk(en)ylresorcinols were isolated from the woods of several members of the family Proteaceae.[19] A component of the lipid-soluble extract from the brown alga, *cystophora torulosa*, was a complex mixture of alkenylresorcinols.[20] 5-Nonadecylresorcinol monomethyl ether, **14-5**, was reported in 1992 to have been isolated from the aerial parts of *Artemisia annua.*[21] The 5-tridecenylresocinols, **14-6a** and **14-6b**, were obtained from *Ardisia japonica* Blume and had antitubercular activity.[22] *Azobacter vinlandii*, when encysted by initiation with beta-hydroxybutyrate, formed a mixture of 5 long-chain alkylresorcinols.[23] Dialkylresorcinols, i.e., 2-*n*-hexyl-5-*n*-pentyl- and 2-hexyl-5-*n*-heptylresorcinols, were isolated from *Pseudomonas aureofaciens.*[24] The cell lipids of the nitrogen-fixing bacterium, *Azobacter chroococcum*, contained 5-*n*-(C_{19},C_{21}) alkylresorcinols.[25] Mango (*Mangifers indica*

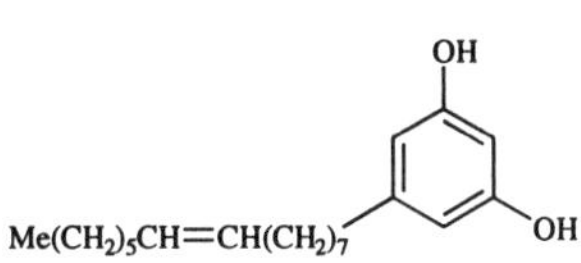

14 - 4

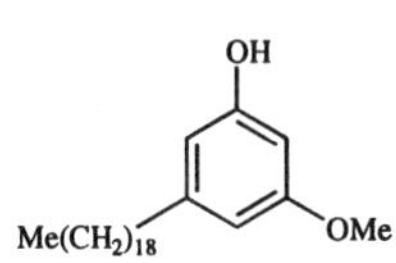

14 - 5

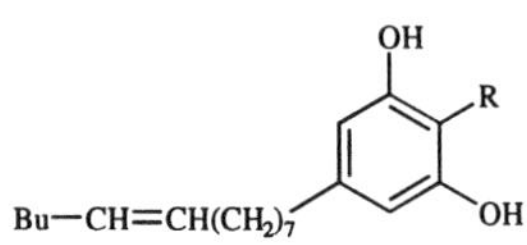

14 - 6

(where **a**: R=Me; **b**: R=H)

L.) latex was found to contain 5-[2(*Z*)-heptadecenyl]resorcinol, which was considered to be a mango dermatitis allergen.[26]

Allergenic long-chain alk(en)ylresorcinols were also isolated from several *Philodendron* species.[27] An antifungal agent consisting of 5-heptadecenyl- and 5-pentadecenylresorcinols was isolated from the peel of unripe mango.[28] A mixture of long-chain 5-alk(en)ylresorcinols extracted from *Hakea trifurcata* was found to mediate $\Phi\chi174$ DNA cleavage. After some fractionation/purification, the newly isolated material exhibited a diminished ability to mediate DNA strand scission; the activity increased on storage of the sample or by aeration of the aqueous solution (oxygenation of the benzene nucleus was suspected).[29]

The cell lipids of the nitrogen-fixing bacterium, *Azotobacter chroococcum*, which are a mixture of 5-*n*-alkyl(C_{19-25})resorcinols, were found to inhibit the hydrolysis of phosphatidylcholine catalyzed by the phospholipase A2 in the venom of the cobra, *Naja naja oxiana*.[30]

Long-chain alkyl-/alkenylresorcinols were also found in the seed oils of three Iris species.[31]

There is a considerable literature on the detection and study of long-chain 5-*n*-alkylresorcinols in cereal grains. For example, a mixture of 5-*n*-(C_{11-36})alkylresorcinols was found in the testa and pigment strand of decorticated barley grains.[32] Long-chain 5-alkylresorcinols were isolated also by extraction of rye (*Secale cereale* L.) grains.[33] The 5-alkylresorcinols were found to occur primarily in the outer layers of rye and wheat kernels, and were associated with the rejection of feed, especially of byproducts from rye milling, by animals. Bran and shorts had much higher alkylresorcinol levels than the whole grain; commercial rye and wheat flowers had lower levels of alkylresorcinols and baking reduced the alkylresorcinol levels of rye breads by 32–78%.[34] Two reviews of alkylresorcinols in cereal grains and their biological activity were published.[35] 5-*n*-Alk(en)ylresorcinols from rye were found to produce permeability changes of erythrocytes and liposomes.[36] A screening study (small number of test animals) of the effect of long-chain 5-alkylresorcinols on albino rats was made; neither the use of rye grain freed of alkylresorcinols by extraction nor fortification of the rye fodder with 5-alkylresorcinols influenced fodder acceptance or weight gain.[37]

Chromatographic analysis of the 5-*n*-alk(en)ylresorcinols in rye grains showed that over 50% were alkenylresorcinols.[38] In a study of samples of triticale from seven countries, grown on the same site, it was found that their content of 5-alkylresorcinols (antifeedants) in the grain differed depending on the weather.[39] Cereal 5-*n*-alk(en)ylresorcinols had hemolytic properties which were related to their affinity to the red blood-cell membrane (much higher for the alkylresorcinols).[40] By a refined extraction and chromatographic procedure, it was found that rye grains contained not only 5-*n*-alkyl- and 5-*n*-alkenyl- but also 5-*n*-alkadienylresorcinols.[41]

In a paper on the effect of nonadec(en)ylresorcinol on the liposome and erythrocyte membranes, showing their modulating properties,[42] it was mentioned that alkenylresorcinols showed remarkable antitumor activity in mice.[43]

U.S. and Canadian wheat samples were analyzed and contained 317.5–655.0 μg/g dry wt of alkylresorcinols; flours from these wheat samples contained only 0.025–14.45 μg/g dry wt of alkylresorcinols. Hard wheat had higher protein and lower alkylresorcinol content, soft wheat had low protein and higher alkylresorcinol assay; as kernel maturity increased, the alkylresorcinol content was lower.[44]

Long-chain alkylresorcinols and alkylphenols isolated from cereal grains showed inhibitory activity on the binding of fibrinogen by human platelets, while 5-methylresorcinol (orcinol) and amylresorcinol showed no effect or very weak inhibitions; these results were attributed to the ability of long-chain alkylresorcinols to decrease membrane fluidity.[45] The amount and composition of the long-chain alk(en)ylresorcinols in rye, triticale, wheat, durum, and barley was determined and compared.[46]

The 5-(C_{13-27})alk(en)ylresorcinols in cereal grains were shown to be effective in protecting fatty acids against peroxidation; the phospholipid bilayers also appeared to be protected by the resorcinolic lipids.[47]

Long-chain acylresorcinols have also been found in nature. For instance, malabaricone A, **14-7**, and several similar compounds were found in the fruit of *Myristica malabarica* Lam, which is a source of Indian folk medicines.[48] The fruits of the Brazilian Myristeaceae (nutmeg tree), *Virola sebifera* and *V. elongata*, contain 2-hexadecanoylresorcinol and several similar compounds.[49] The rhododendron lace bug, *Stephanitis rhododendron* Horvath (Hemiptera; Tingidae), secreted two resorcinol diketones, **14-8a** and **14-8b**, accompanied by the corresponding chromones, **14-9a** and **14-9b**.[50]

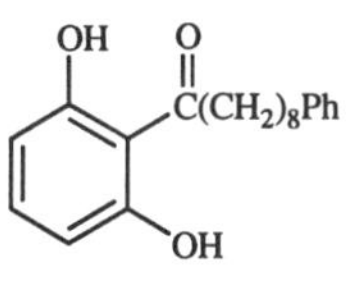

14 - 7

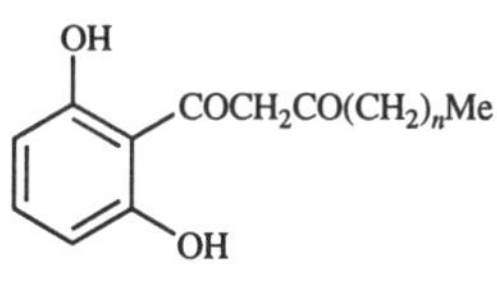

14 - 8

(where **a**: n=10; **b**: n=12)

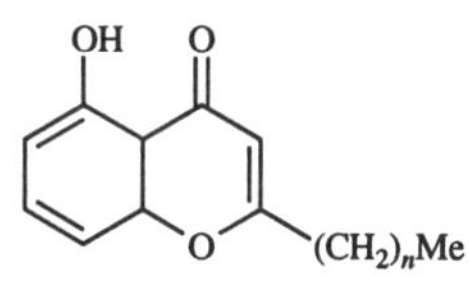

14 - 9

(where **a**: n=10; **b**: n=12)

2,4-Dihydroxy-6-(10-phenyldecyl)acetophenone, **14-10**, was found in extracts of the stem bark of *Knema laurina* and *K. tenuinervia ssp. setosa*, together with several related phenols.[51]

Three acetoxy-substituted long-chain alkylresorcinol derivatives, such as compound **14-11**, were found in extracts of the aerial parts of *Ononis viscosa*.[52]

Finally, it should be mentioned that a good review was published of the literature to the mid-1970s on natural, nonisoprenoid, long-chain alkylphenols, covering much of the chemistry of long-chain alkylresorcinols and their occurrence.[53]

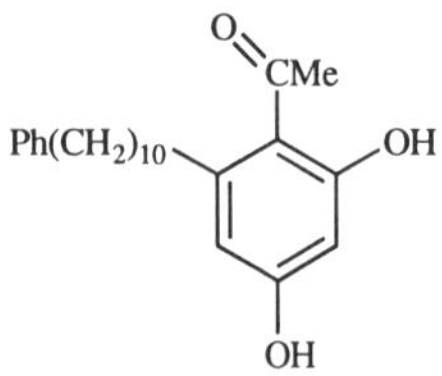

14 - 10

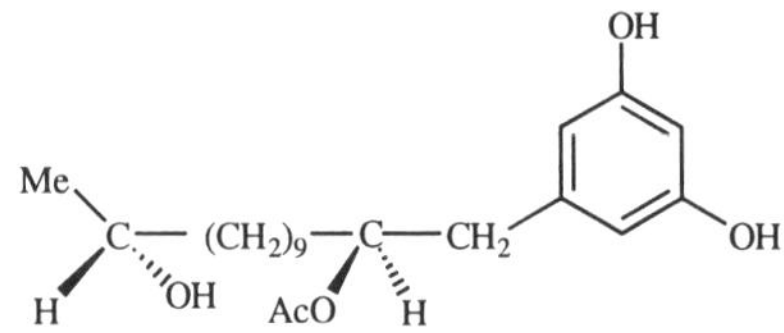

14 - 11

14.3. CASHEW NUT SHELL LIQUID (CNSL)

A review of this natural product by Tyman appeared in 1980.[54] Another, extensive (605 references) review of CNSL by Gedam and Sampathkumaran appeared in 1986.[55]

CNSL contains long-chain 5-alk(en)ylresorcinols as a lesser constituent, and is discussed separately here because it is a commercial product.

CNSL is obtained by heating/extraction from the outer shell of the cashew nut produced by the tree, *Anacardium occidentale*. In the 1970s the main producers of cashews were India, Mozambique, and Tanzania, with small amounts coming from Kenya and Brazil, all together producing about 600,000 t/yr of the nuts; the potential for byproduct CNSL was >125,000 t/yr, but only about 8000 t of CNSL were sold in 1978.[54]

The raw CNSL contains about 80% anacardic acid (a long-chain alkylsalicylic acid), **14-12**, 15% of cardol [a mixture of alk(en)ylresorcinols], **14-13**, and small amounts of minor components such as 2-methylcardol, **14-14**. During thermal processing at *ca* 180–200 °C and extraction of the cashew nuts, the anacardic acid is largely decarboxylated and some polymerization occurs to give technical CNSL containing about 63% cardanol, **14-15**, 11% cardol, **14-13**, 2% 2-methylcardol, **14-14**, 1% anacardic acid, **14-12**, and 23% "polymer." This technical oil is dark colored; even after distillation it darkens again, supposedly due to its cardol content. Therefore, the technical CNSL cannot be used in the production of light-colored coatings. Among the methods proposed to overcome that was the treatment of technical CNSL with amines (or stronger bases), e.g., with triethylenediamines or triethylamine, before distillation. The amines were said to form amine salts preferentially with the cardol components; distillates as high as 98% cardanol, **14-15**, and only 2% cardol, **14-13**, were obtained. The distillation residue contained *ca* 44% cardanol, **14-15**, 45% cardol, **14-13**, and 11% 2-methylcardol, **14-14**.[56]

For a low-cost material, however, the technical CNSL is used. The main applications were in friction dusts for making automotive brake linings and clutch facings; at the time, these materials contained asbestos as the filler. The phaseout of asbestos, at least in the U.S., may have had an effect on the usage of CNSL. There are smaller uses in resins for waterproofing and laminating; derivatives such as epoxy compounds are available. The handling of CNSL products requires safety precautions to avoid skin irritation to allergic individuals.

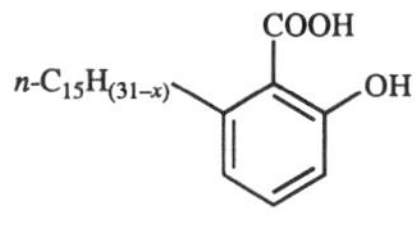

14 - 12

(where x = 0, 2, 4, or 6)

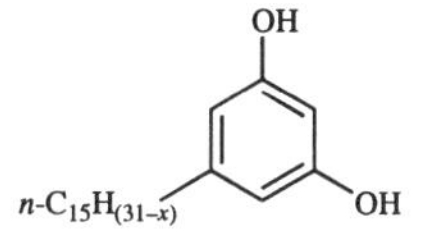

14 - 13

(where x = 0, 2, 4, or 6)

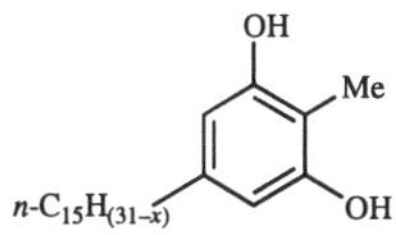

14 - 14

(where x = 0, 2, 4, or 6)

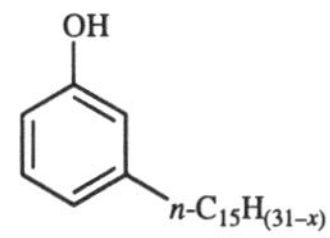

14 - 15

(where x = 0, 2, 4, or 6)

Work on the characterization of and possible biogenetic routes to the 5-alk(en)ylresorcinols in Anacardiaceae plants is continuing. For instance, 2,4-dihydroxy-6-*n*-pentadecylbenzoic acid, **14-16**, was recently synthesized from noncyclic precursors via the Michael addition/Claisen cyclization route to the appropriate 1,3-cyclohexanedione precursor, which was aromatized to **14-16**.[57]

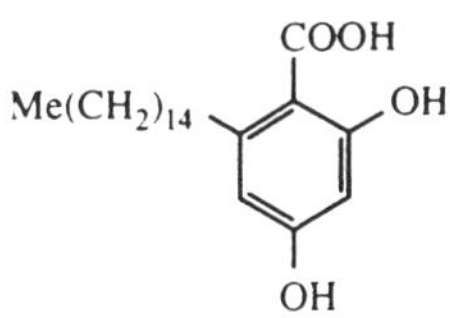

14 - 16

14.4. COUMARINS

These compounds occur widely in nature, particularly in plants of the order *Rutaceae* and *Umbelliferae*; 7-hydroxycoumarin (umbelliferone), **14-17**, which is a resorcinol derivative found in carrots and parsnips, has been regarded as a parent compound for many structurally more complicated products. Many of these have biological activities ranging from antifungal and bactericidal to antitumor activity. Because of these facets, the coumarins, including the 7-hydroxycoumarin type, have been well reviewed. A book by Murray *et al.* at Glasgow covered the coumarins from their discovery in 1812 to the mid-1980s and listed 786 compounds known to occur in nature at the time[58]; this review was updated with 426 references (158 resorcinol-related) from the mid-1980s to mid-1988.[59] Other reviews in English of naturally occurring coumarins have come from India, long a center of research on these compounds.[60]

Since extensive and recent reviews of the coumarins are available, the following selections and examples from the literature into 1992 are given just to impart the flavor and structural variety of this area of research.

Geipaverine, **14-18**, extracted from the leaves of the Australian shrub, *Geijera parviflora* Lindl. (Rutaceae), has antitumor activity (in vitro) against human tumors of the nasopharynx; its first synthesis provided the proof of structure.[61] The structure of wairol, **14-19**, a coumestan found in *Medicago sativa*, was proven by synthesis as well.[62]

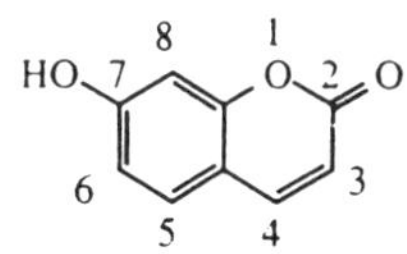

14-17

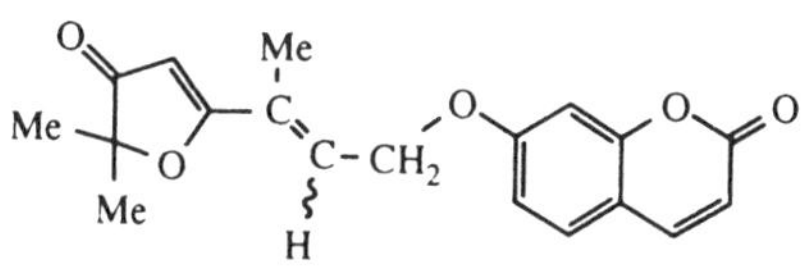

14-18

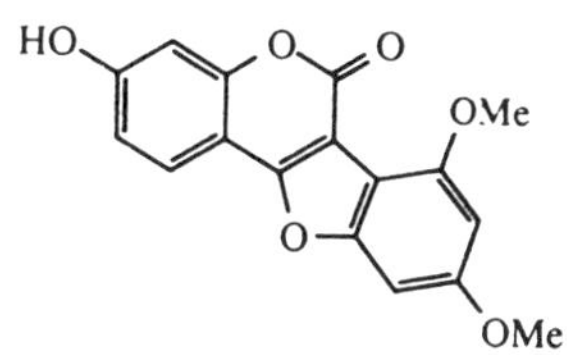

14-19

The structure of seselin, a steam-volatile component of *Carum roxburghianum*, was shown to be **14-20**.[63] Lasioerin, **14-21**, a coumarin isolated from *Lasiosiphon eriocephalus* Decne (Thymelaeaceae), was assigned its structure on the basis of spectral data and for biogenetic reasons.[64]

New syntheses of xanthyletin, **14-22**, a known benzodipyran from *Atalantia racemosa*, were developed.[65] The structure and stereochemistry of angelol B, **14-23**, isolated from the root of *Angelica pubenscens* (Umbelliferae), was determined by chemical, spectral, and X-ray analysis.[66]

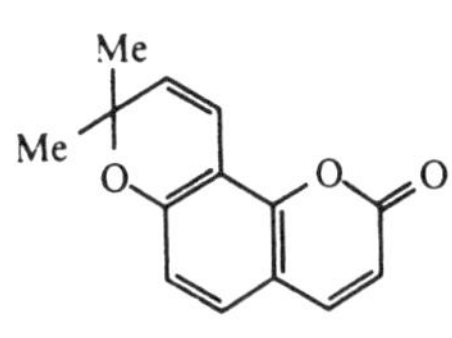

14-20

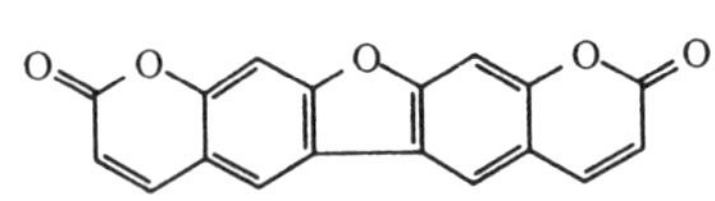

14-21

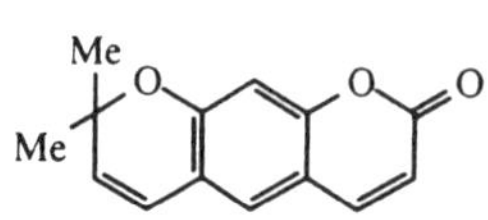

14-22

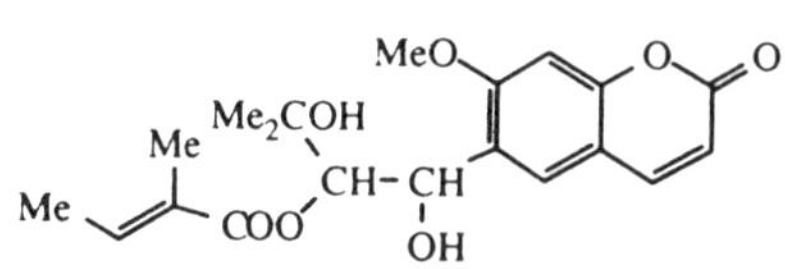

14-23

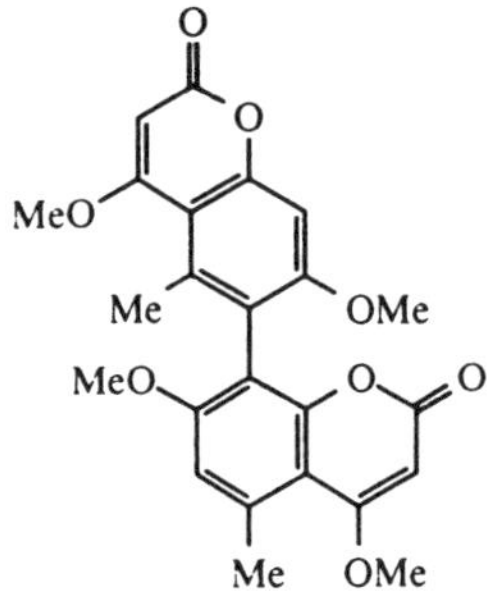

14-24

The structure of desertorin C, **14-24**, a bicoumarin obtained from the fungus *emericella desertorum*, was proven by synthesis.[67]

3-Aryl-4-hydroxycoumarins such as, e.g., **14-25**, and related isoflavanoids, were found in the subfamily Papilionoideae (Leguminosae). These compounds were said to have considerable potential as therapeutics; they were synthesized in good to high yields by the direct arylation of 4-hydroxycoumarin precursors with aryllead-IV triacetate.[68]

The structure of cyclobisuberodiene, **14-26**, a constituent of Australian *Zanthoxylum* species, was confirmed by synthesis.[68]

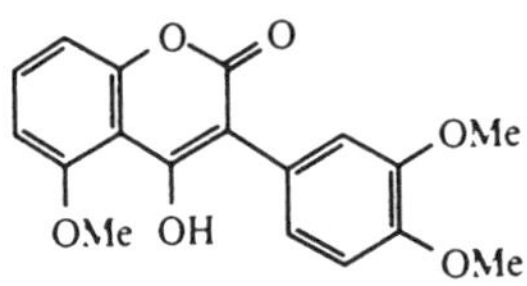

14-25

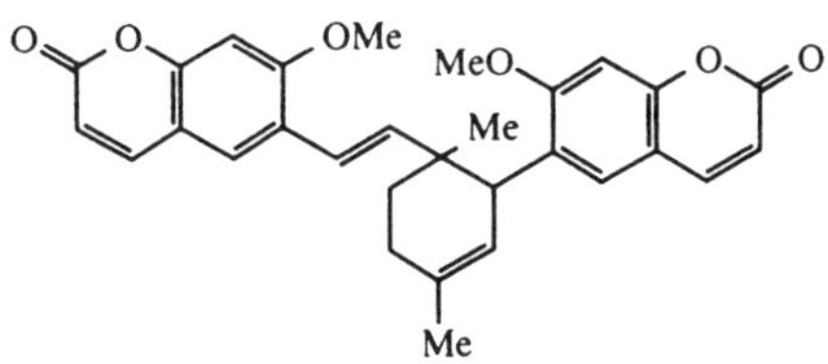

14-26

A total synthesis of microminutin, **14-27**, a cytotoxic compound from *Micromelum minutum*, was reported.[69] A new 3-acetylcoumarin, **14-28**, was isolated from the bark of the ash and its structure was deduced by NMR and differential NOE spectroscopy.[70]

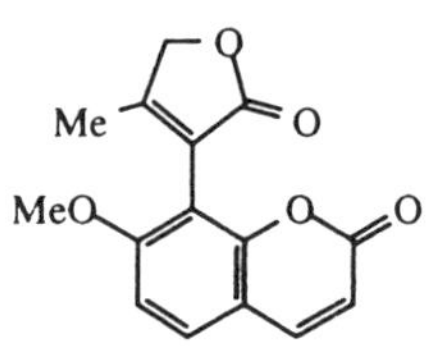

14-27

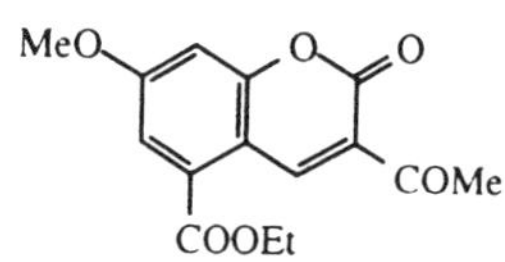

14-28

14.5. PRECOCENE-1, AN INSECT ANTIJUVENILE HORMONE

The discovery of this compound, **14-29a** (resorcinol-derived), and the related precocene-2 (1,2,4-benzenetriol-derived), **14-29b**, by extraction/chromatographic fractionation from the common bedding plant, *Ageratum houstonianum*, and their screening as antijuvenile hormones (anti-JH) to control bugs was described in 1976 in a review by Bowers *et al.*[71] Precocene-1 at 3.9 μg/cm^2 (Petri dish test) gave 100% precocious metamorphosis in the milkweed bug and resulted in sterilization by suppressing the function of the corpora allata gland. Although precocene-1 was not as active as precocene-2, the potential of insect hormones as fourth-generation pesticides prompted a sustained interest in the sources, chemistry, modification, and use of such compounds.

Selections from the literature of 1980–1992 relating to precocene-1, also called 7-methoxy-2,2-dimethylchromene or 7-methoxy-2,2-dimethyl-2*H*-1-benzopyran, are presented below.

Encecalin, **14-30**, and encecalol ethyl ether, **14-31**, were extracted from the brittle bush which grows profusely in Baja California; these compounds were found to have true insecticide activity.[72]

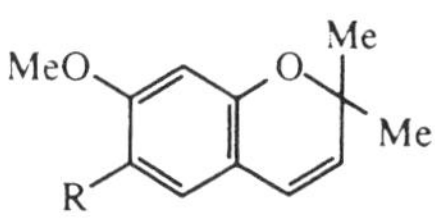

14-29

(where **a**: R = H; **b**: R = OMe)

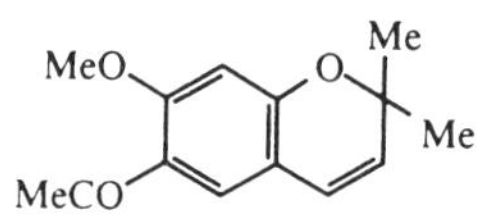

14-30

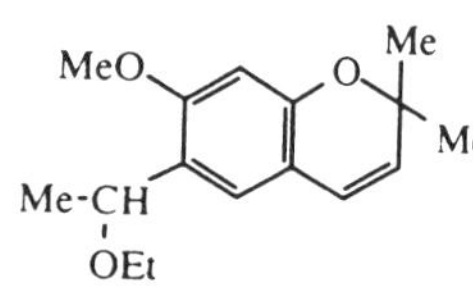

14-31

The synthesis of precocene-1, **14-29a**, by the cyclocondensation of resorcinol monomethyl ether with isoprene and purifications/dehydrogenation of the intermediate benzopyrans was described.[73]

Precocene-1 dimer, **14-32**, a potential antijuvenile hormone, was synthesized in 36% yield by passing precocene-1 over a column of silica gel impregnated with trichloroacetic acid.[74]

The ^{13}C NMR spectra of several (3-, 4-, and/or 5-) methyl-substituted derivatives of precocene-1 were obtained and related to their antijuvenile hormone activity. A ^{13}C chemical shift at C-3 of <130 ppm correlated to precocene activity.[75]

The anti-JH activity of the precocenes was thought to be the formation of highly reactive precocene epoxides in the insect's corpora allata glands (where the juvenile hormones are synthesized) by oxidation of the precocenes by the cytochrome-P-450-linked JH epoxidase; *cis*- and *trans*-3,4-diols of precocene-1 were the sole, low molecular weight metabolites produced. The diol metabolites were synthesized, resolved, and their absolute configuration was determined.[76]

Several analogs of the precocenes were prepared and tested. 6-Amino- (**14-33a**) and 6-nitroprecocene-1 (**14-33b**) did not have anti-JH activity, but had significant insecticidal activity, and showed both antagonism and synergism with permethrin.[77]

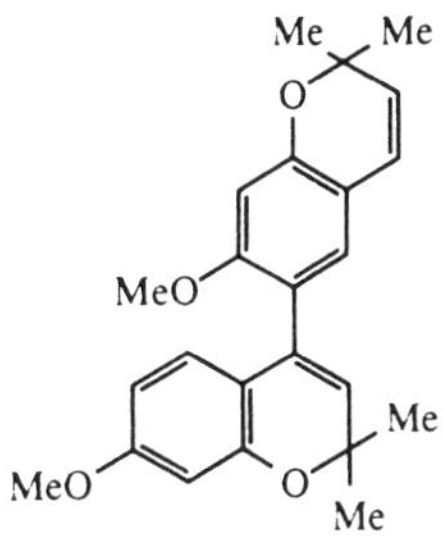

14-32

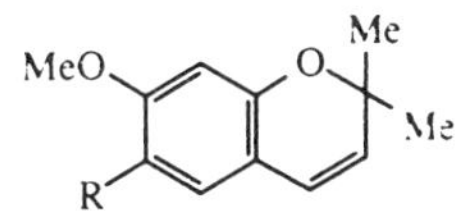

14-33
(where **a**: R = NH_2; **b**: R = NO_2

Treatment of precocene-1 with *N*-bromosuccinimide in aq. THF at 0 °C and then warming to 25 °C gave the bromohydrin, which was treated with NaH in THF to obtain precocene-1 epoxide, **14-34**. This is the highly reactive metabolite responsible for the biological effects of the precocenes.[78]

The addition of Cl_2C: to precocene-1 gave the cyclopropyl analog, **14-35**, which had insecticidal activity against *Aphis fabae*.

2,2-Dimethyl-5-hydroxychromene, **14-36**, was prepared, starting from 2,6-dihydroxyacetophenone and acetone, as an analog of precocene-1.[80]

Further, alkyl, cycloalkyl, benzyl, allyl, prenyl, propargyl, and 1-naphthylmethyl ethers of 2,2-dimethyl-7-hydroxychromene were prepared as analogs of precocenes. These compounds had allatotoxic and nematocidal activity, with the methyl ether (precocene-1) and the isobutyl ether, **14-37**, being the most active.[81] The 5- and 8-methyl derivatives of these compounds were also prepared.[82]

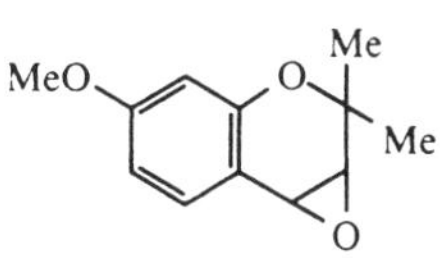

14-34

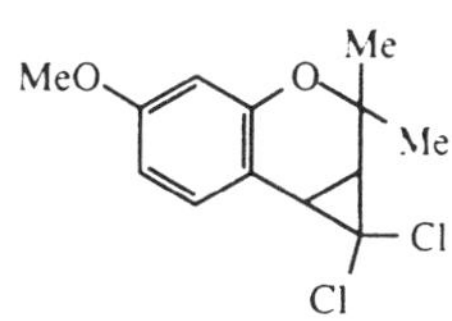

14-35

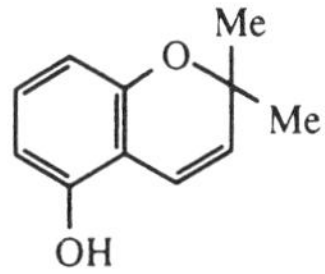

14-36

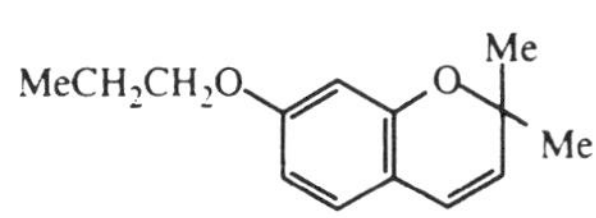

14-37

Similar chromenes, with 6-chloro or 6-*tert*-butyl substituents were made also.[83]

The first direct epoxidation of precocene was attained by their reaction with dimethyldioxirane; nearly quantitative yields of the 3,4-epoxide were obtained.[84]

Precocene-1 was prepared in 20% yield from the prenylated resorcinol monomethyl ether, **14-38**, by a $PdCl_2/CuCl$-catalyzed cyclization in aq. EtOH at 60 °C.[85]

2,2-Dimethyl-7-hydroxychroman, **14-39**, a precursor of precocene-1, was prepared in 50% yield by the reaction of resorcinol with isoprene in benzene in the presence of $MgI_2–Et_2O$.[86]

Cyclic analogs of precocene-1 such as, e.g., **14-40** were prepared, starting with the Friedel-Crafts condensation of resorcinol with cycloalkylideneacetic acid.[87]

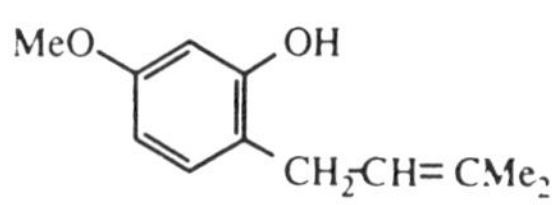

14-38

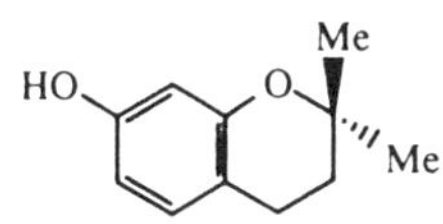

14-39

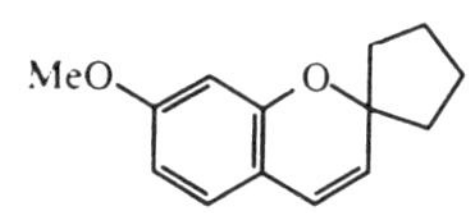

14-40

Precocene-1 was prepare in 40% yield in a single step by refluxing resorcinol monomethyl ether (3-methoxyphenol) with 3-methyl-2-butenal in pyridine for 2 hr/140 °C.[88]

A report, published in 1992, on a route to precocene-2 (or precursors) by introduction of a formyl group in the 6-position of precocene-1, followed by oxidation with H_2O_2, SeO_2/CH_2Cl_2, also summarized the syntheses of precocenes in the prior art.[89]

14.6. LICHEN CONSTITUENTS

The lichens consist of an alga in symbiotic union with a fungus, growing on rocks (see Fig. 14.1), trees, etc., as gray, green, brown, or black, undifferentiated plants. From various species of lichens has been derived litmus (lacmus), an acid–base indicator used for preparing litmus papers, red at pH 4.5 and blue at pH 8.3, and lacmoid (Resorcin blue), used in 0.2% solution as a an acid–base indicator, red at pH 4.4 and blue at pH 6.4. A red-violet dye had been known since ancient times to be obtained when the light-gray lichens were contacted with urine or ammonia and air. This red-violet dye was named Orseille and used for dying silk and wool till the late 19th century; a purified alcohol extract of Orseille was called orcein, which still has some uses today as a dye in histology. Many famous chemists, starting with Liebig in the early 19th century and continuing in the 20th century, tried to determine the structure of these dyes and failed, until Musso in the 1955–1968 period solved these puzzles; much resorcinol chemistry was involved in his work.

When this book was in the planning stage, Musso's work had been selected for description as an intricate piece of academic work. During the writing stage, however, in 1992 Hopf published, in German, a 24-page, great memorial to Prof. H. Musso (he had died in 1988), including an appreciation of his scientific work with a complete list

Figure 14.1. Lichen growing on a rock (Zion National Park, Utah).

of references and a glimpse of the makeup of the man was well.[90] Musso's development through education by good teachers, hard work and dedication, powers of concentration and insight, excellence in theory and practice, a love and knowledge of nature, perseverance against all odds, his drive to arrive at a conclusion, his high skills in the separation techniques of paper and column chromatography, his use of the best available instrumental techniques (e.g., IR and UV spectroscopy), led to valuable research results and to a steady progression of appointments in the German university system at Goettingen, Marburg, then at the new Ruhr-University in Bochum. Once his own research program was well established, his hard work led to the publication of 3–10 scientific papers every year of his working life. Recognition followed in the form of medals and honors. Many of these patterns and qualities are followed by the best in academia, and in applied fields of chemistry as well.

Just a few examples of Musso's work may be appropriate here. He separated orcein by chromatography into 14 dyes and determined their constitution and the existence of stereoisomers. The main structures of these dyes were shown to be 7-amino- and 7-hydroxy-2-phenoxazones, **14-41**. The orcinol (5-methylresorcinol) segment in this structure is readily recognizable. The main component of litmus was shown to be a compound of molecular weight about 3000, containing four phenoxazone units. In all, Musso and his co-workers published 26 papers on the orcein-type dyes. These investigations then led to work on hydrogen bonding in phenols and

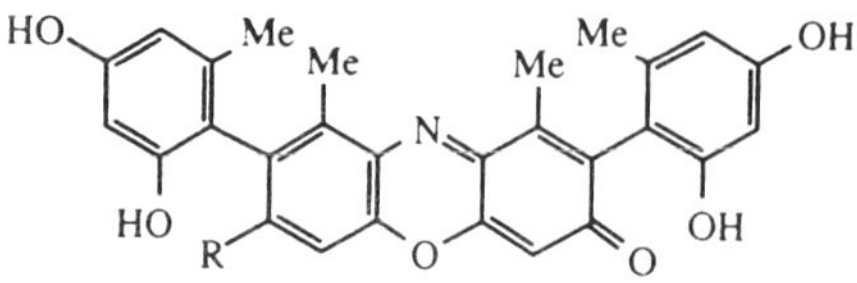

14-41
(where R=NH_2 or OH)

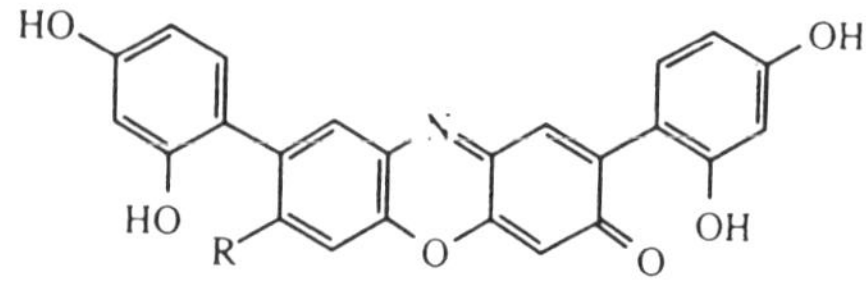

14-42
(where R=NH_2 or OH)

amines, and to studies of phenol (resorcinol and resorcinol derivatives) oxidation leading to bi- and poly-aryls. (A summary of his work on phenol oxidation reactions appeared in English.[91]) The autoxidation of resorcinol in ammoniacal solution was shown to give 7-hydroxy- and 7-amino-3,6-bis-(2,4-dihydroxyphenyl)-phenoxazone-2, **14-42**, identical with the dye, lacmoid/resocinol blue, used as an indicator and quicker made by using H_2O_2 as the oxidant, or by the fusion of resorcinol with nitrite.[92]

Examples of Musso's work which added to the understanding of phenol oxidation to form hydroxyquinones and biaryls, and phenoxazines in the presence of ammonia, are the following: (1) As shown in Eq. (14-1), the oxidation of 4,6-di-*tert*-butylresorcinol in

(14-1) 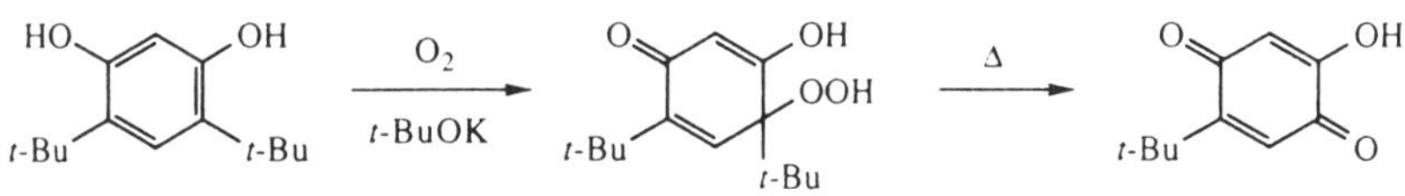

14-43 **14-44**

the presence of K-*tert*-butoxide at –20 °C gave the hydroperoxide, **14-43**, in 96% yield.[93] This hydroperoxide decomposed thermally to give the quinone, **14-44**. (2) As shown in Eq. (14-2), 5-*tert*-butylresorcinol on oxidation gave the hydroxyquinone,

(14-2) 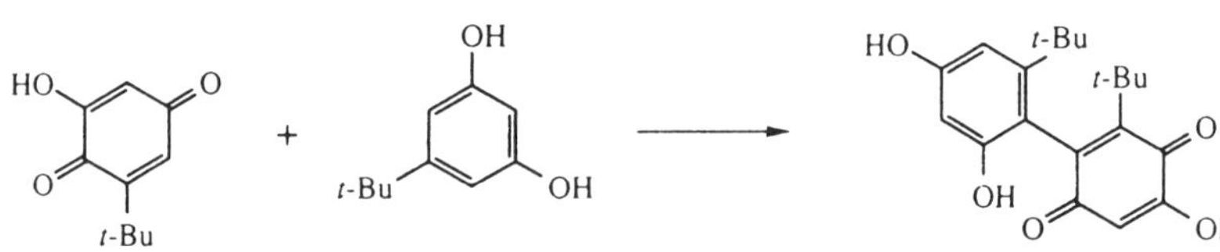

14-45 **14-46**

14-45, which coupled with 5-*tert*-butylresorcinol to give the biphenyl-quinone, **14-46**.[91] (3) On the other hand, an oxidation product of 5-*tert*-butylresorcinol in the presence of KOH was shown to be 2,6-di-*tert*-butyl-3-acetyl-4-hydroxybenzofuran, **11-47**.[93]

Some examples of the work of others in this field follows:

Dyes (precursors) of related structures were not only obtained from lichens. For example, a quinonoid pigment, phoenicine, **14-48**, and related quinones were isolated from *Penicillium phoeniceum* and chaetomium aureum; a biosynthetic pathway, involving orsellinic acid as an intermediate, was described.

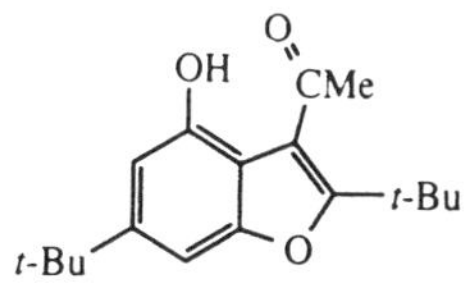

14-47

Other compounds with a resorcinol moiety were isolated from lichens and include depsides (aryl esters derived from hydroxybenzoic acids), dibenzofurans, and diphenyl ethers. Thus, a new depside, 2-*O*-methylobtusatic acid, **14-49**, was obtained from the lichen, *Xanthoparmelia tusconensis*.[95]

The lichen depside, atranorin, **14-50**, which contains a formyl group, was characterized; the two benzene rings were demonstrated to be nearly perpendicular.[96]

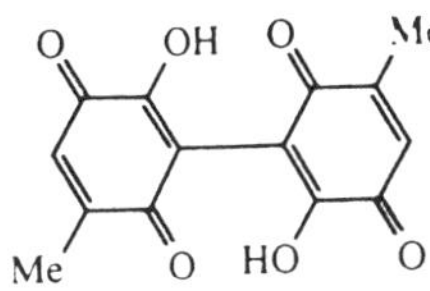

14-48

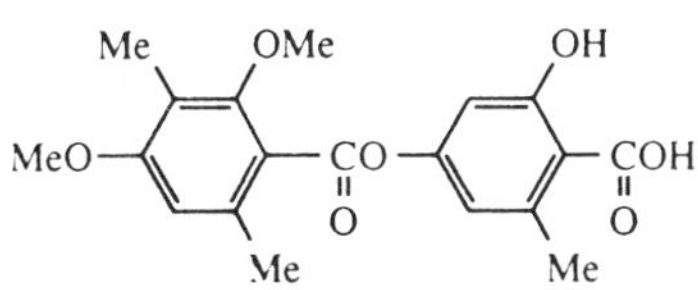

14-49

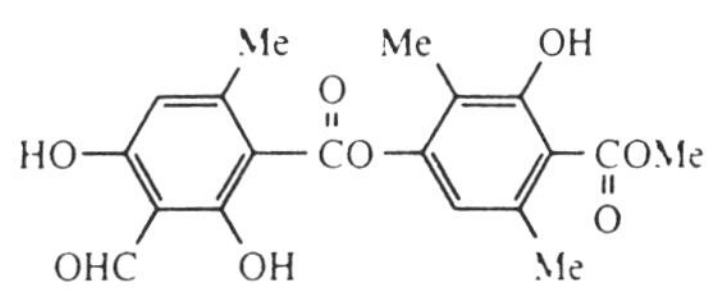

14-50

The structure of schizopeltic acid, **14-51**, a dibenzofuran occurring in lichens, was confirmed by synthesis.[97] Isodidymic acid, **14-52**, another dibenzofuran from lichens, was synthesized also to prove the structure.[98]

The depsides from the lichen *Lecidea leucophea*, miriquidic acid, **14-53a**, and normiriquidic acid, **14-53b**, were prepared.[99] Another lichen depside, sphaerophorin,

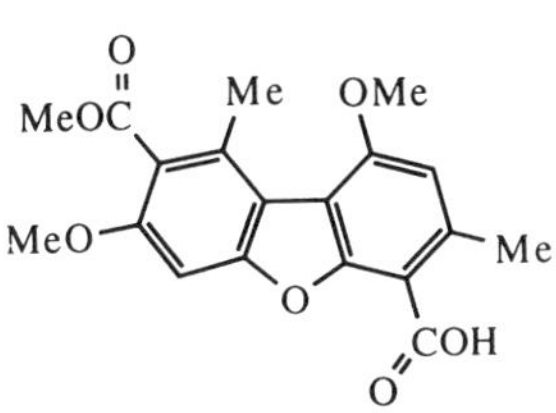

14-51

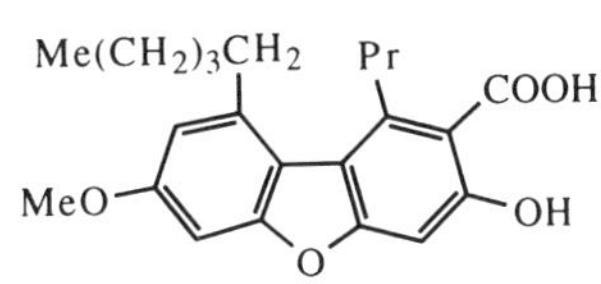

14-52

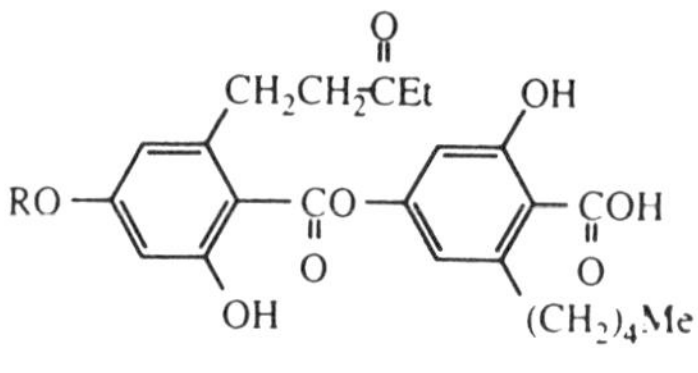

14-53
(where **a**: R = Me; **b**: R = H)

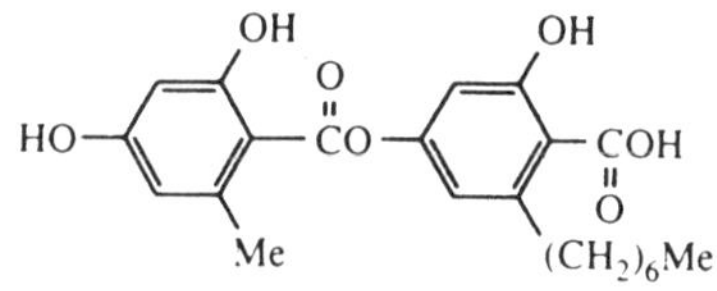

14-54

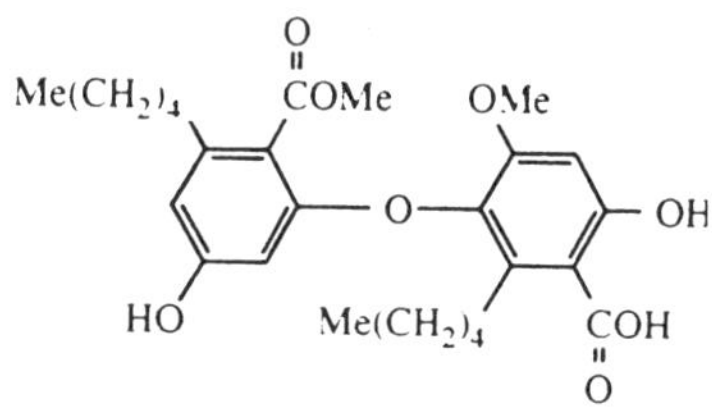

14-55

14-54, and related compounds were described and synthesized.[100] The structure of epiphorellic acid, **14-55**, a diphenyl ether isolated from lichen, was proven by an unambiguous route.[101]

14.7. CANNABIS/MARIJUANA

This drug, also called Indian hemp, bhang, ganja, kif, hasach, and pot, is the dried flowering tops of the female plant, *Cannabis sativa* L., *Moraceae*. Its active constituents are the isomeric tetrahydrocannabinols (THCs). The most active components are **14-56**, cannabinol, **14-57**, and cannabidiol, **14-58**. Hashish is a purified alcoholic extract of cannabis.

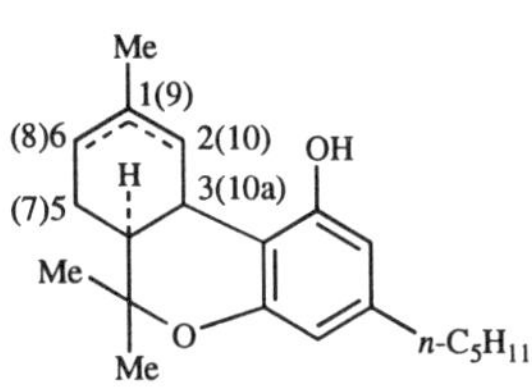

14 - 56

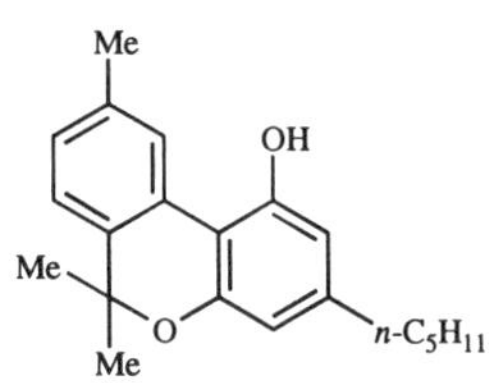

14 - 57

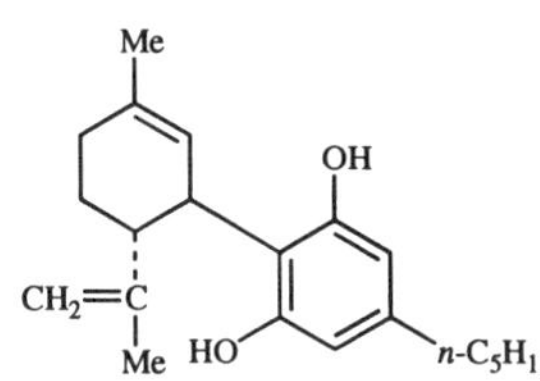

14 - 58

[alternate ring numbering in parentheses; mainly Δ^1-THC (Δ^9-THC); ~1% Δ^6-THC(Δ^8-THC)]

The drug has a long history of use and abuse. It is a controlled substance (hallucinogen).[102] There is a large body of work on the isolation and chemistry of the cannabis components. A book published in 1973 summarized earlier work.[103] The toxic effects of cannabis, ingested or inhaled in smoke, can include not only hallucinations but also drowsiness, delirium, euphoria, weakness, and hyporeflexia.[104] Academic researchers and the pharmaceutical companies have made major efforts to develop modifications of the cannabinoids, and a few products have been marketed (see Chapter 8). The following few examples of academic and industrial work on this chemisty in recent years show that these efforts are far from over.

From the Lilly Research Laboratory came a paper on synthetic approaches to 9-ketocannabinoids and on the total synthesis of the drug, nabilone (Chapter 8).[105]

N-heterocyclic derivatives of olivetol (5-*n*-pentylresorcinol), such as the pyrrolobenzoxazine, **14-59**, were prepared as analogs of the cannabinoids.[106] Another type of synthetic analog of cannabinoids contained a 7-membered ring, i.e., compound **14-60**.[107]

As a precursor to labeled cannabinoid standards, the deuterated olivetol, **14-61**, was made.[108]

Δ^6-THC was synthesized by a novel route, using cyclobutenone as a precursor.[109]

Mercapto- and thioacetyl derivatives of Δ^8-THC, e.g., **14-62a** and **14-62b**, were prepared for pharmacological evaluation.[110]

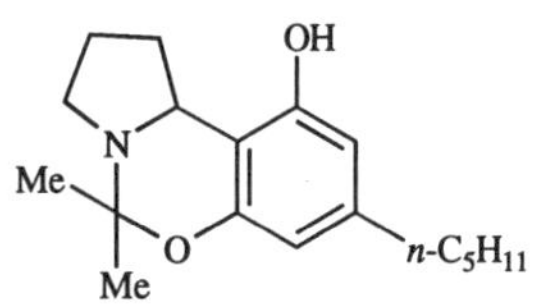

14 - 59

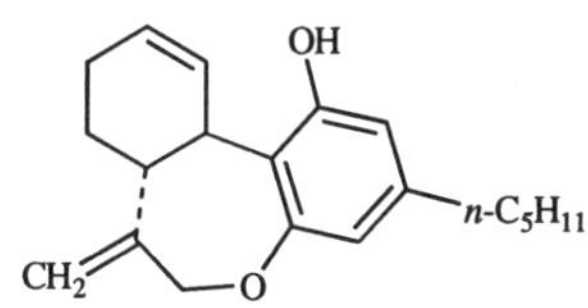

14 - 60

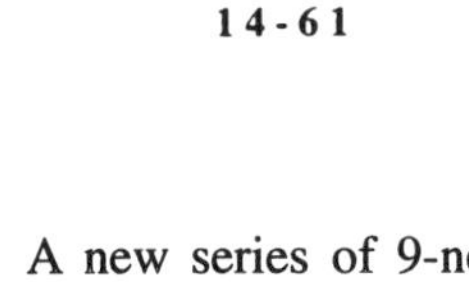

14 - 61

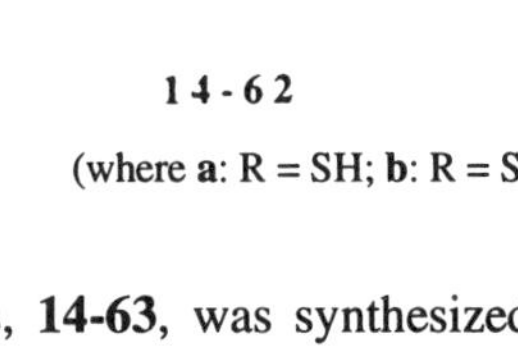

14 - 62

(where **a**: R = SH; **b**: R = SAc)

A new series of 9-nor-9β-hexahydrocannabinoids, **14-63**, was synthesized to weigh the importance of the phenolic hydroxyl function in cannabinoids for analgetic properties.[111]

The in vitro metabolism of Δ^9-THC and several other cannabinoids was studied in liver microsomal preparations obtained from rats, mice, rabbits, guinea pigs, hamsters, gerbils, and a cat. Several species differences in metabolism (hydroxylation) were observed.[112]

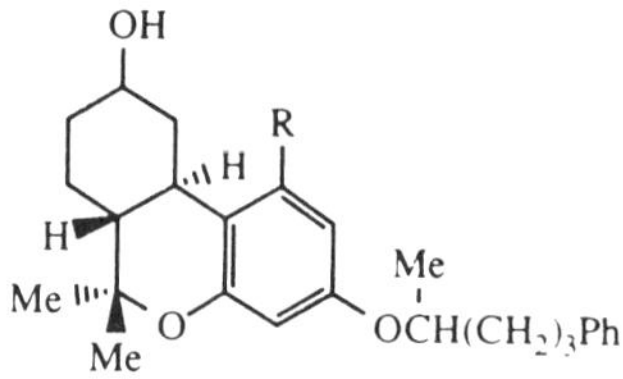

14-63
(where R=H, OH, NH_2 or CH_2OH)

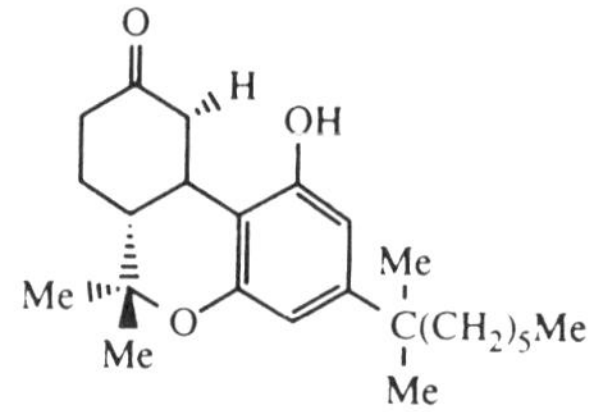

14-64

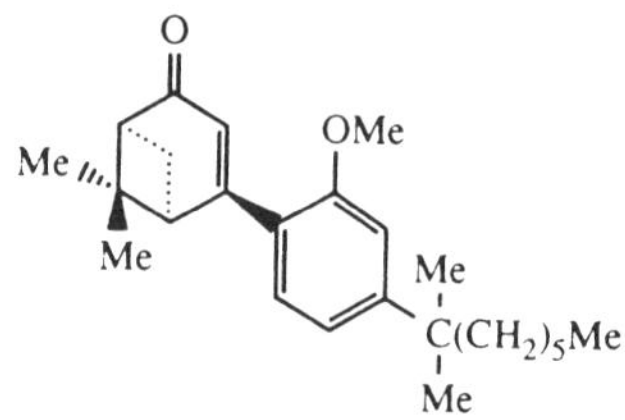

14-65

Two enantiomers of the drug, nabilone, **14-64**, were synthesized from the enone intermediate, **14-65**. This represented an enantiodivergent synthesis of cannabinoids.[113] The deuterated cannabinoid, **14-66**, was synthesized as well.[114]

Cannabinoids such as compound **14-67** were made which had potent antiinflammatory, analgesic, and leukocyte antiadhesion activities.[115]

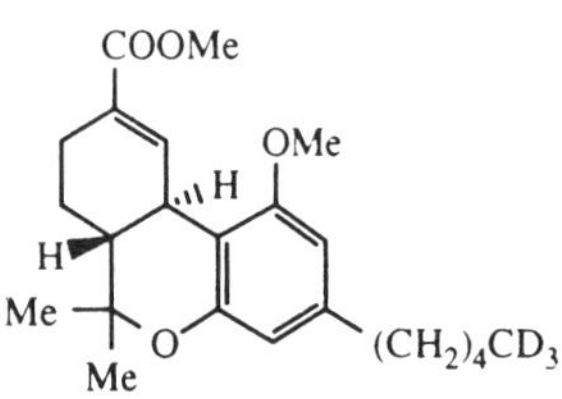

14-66

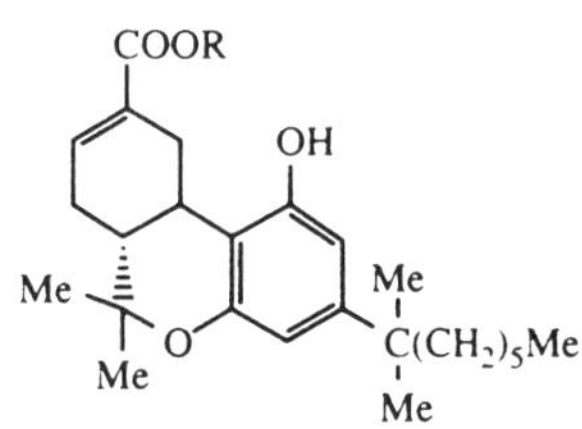

14-67
(where R = H or Me)

5′-Azido-Δ^8-THC, **14-68**, was prepared as a novel photoaffinity label for the cannabinoid receptor in the rat forebrain. This compound was 6.5–19-fold more potent than (–)-Δ^8-THC itself in producing sedation, catalepsy, and antinociception (against pain reception), and in lowering body temperature.[116]

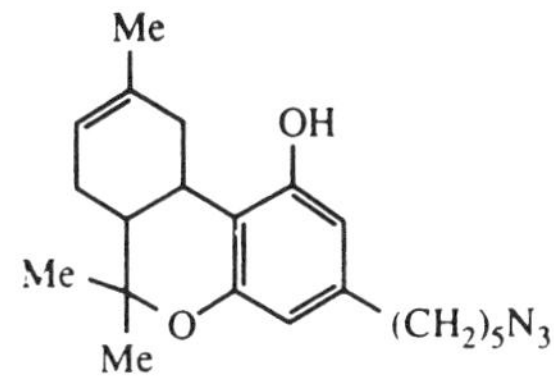

14-68

14.8. PSORALENS AND RELATED COMPOUNDS

The psoralens are linear furocoumarins. They, too, are phytoalexins which help plants to fight fungi and insects. Psoralen itself, **14-69**, is one of the compounds containing a resorcinol moiety while others are derivatives of benzenetriols. They occur in Umbeliferae (e.g., parsnips, celery, carrots), Rutaceae (e.g., limes, cloves, bergamot), Leguminosae (e.g., the pea and bean family), and Moraceae (e.g., figs). A relative, coriandrin, **14-70** (a furo[2,3-*g*]benzopyran), was recently isolated from coriander,[117] which is used as a spice in curry, as an aroma component in perfumery, and as a carminative (to relieve colic or flatulence). Angelicin, **14-71**, has been found in angelica, a herb of the carrot family.

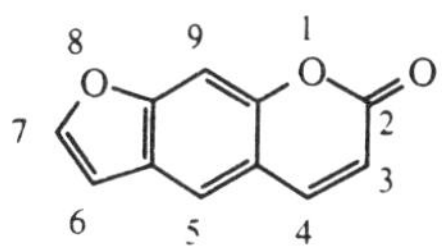

14-69

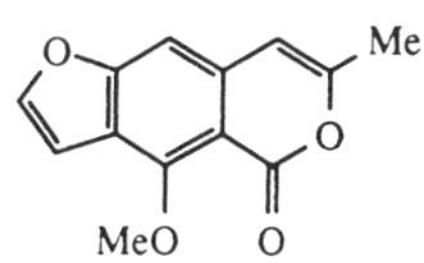

14-70

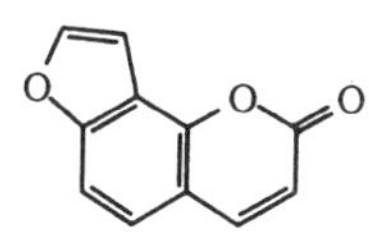

14-71

The psoralens are photoactive compounds which had been used in suntanning agents because they produced melanization of the skin in sunlight. They were also used in the PUVA (psoralen/UV-A light) treatment of psoriasis. However, the compounds were found to be acutely toxic in animals and there were indications of photocarcinogenicity to the skin. Thus, exposure to parsnip roots might present some toxicological risk, the carrot much less so.[118]

Psoralen and analogs were said to damage cells by reacting with nucleic acids after UV-light exposure (300–360 nm), forming adducts with pyrimidine bases in DNA, possibly creating crosslinks within DNA strands. This has been related to the health risks of vegetables.[119] Cells or viruses with psoralen-bound nucleic acid are incapacitated and may die; however, psoralen-killed cells may be useful in diagnostics or as therapeutics (vaccines).[120] (A few references to the interest in psoralens as therapeutics, including commercial items, are given in Chapter 8).

A few additional, selected references on work with psoralens in the 1980–1992 period follow:

A synthesis of linear furanochromones, e.g., **14-72**, from the appropriate allylchromones by cyclization/dehydrogenation was reported.[121]

Using laser flash photolysis or pulse radiolysis, triplet decay, the extinction coefficient, and quantum efficiency of the formation of the triplet excited state of 3-carboethoxypsoralen, **14-73**, was studied. The dark binding of this compound to DNA was also examined.[122]

Other variations on this theme, at Hoffmann-LaRoche's laboratories, led to a synthesis of the trioxsalen analog, **14-74**.[123]

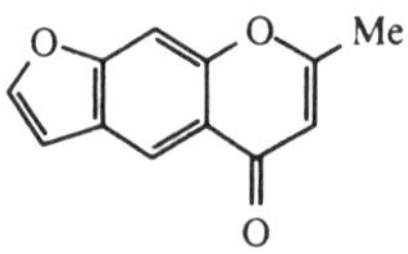

14-72

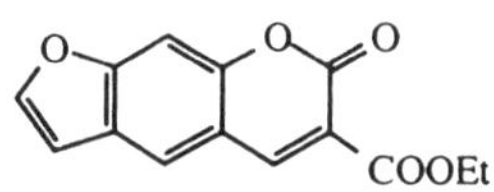

14-73

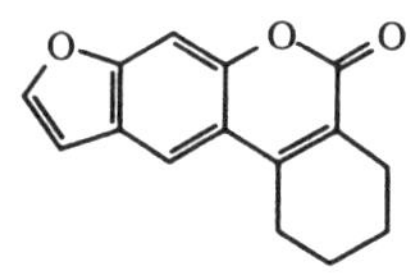

14-74

A new synthesis of psoralen from 7-hydroxycoumarin was developed, involving the blocking of the 8-position with iodine, O-allylation, Claisen migration, and oxidative cyclization.[124]

New derivatives of psoralen, such as **14-75**, were prepared and their potential photosensitizing activity was studied. For example, compound **14-75** showed no photodynamic action on the bacterium, *E. coli*, had a moderate photolethal effect on *Bacillus subtilis*, and photosensitized the skin of albino guinea pigs.[125]

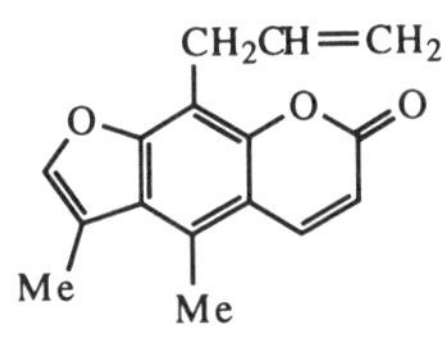

14-75

14.9. MACROLIDES

Mainly, as related to resorcinol, these are large-ring lactones of beta-resorcylic acid (2,4-dihydroxybenzoic acid), also known as RALs or resorcylic acid lactones. Zearalenone, obtained from the fungus *Gibberella zeae* (*Fusorium graminearum*), and zearenol (Ralgro) attained considerable commercial success as anabolics (weight gain promoters) for cattle, and were already mentioned in Chapter 7).

Monorden, **14-76**, is a chlorinated macrolide antibiotic isolated in the 1950s from *Monosporium bonorden*. The interest in macrolides in recent years is demonstrated by the following, selected references. Studies of the fine structure of the macrolides were of special interest.

A simplified synthesis was developed of the plant growth inhibitor lasiodiplodin, **14-77**, a macrolide which is a metabolite of the fungus *Botrysdiplodia theobromae*.[126]

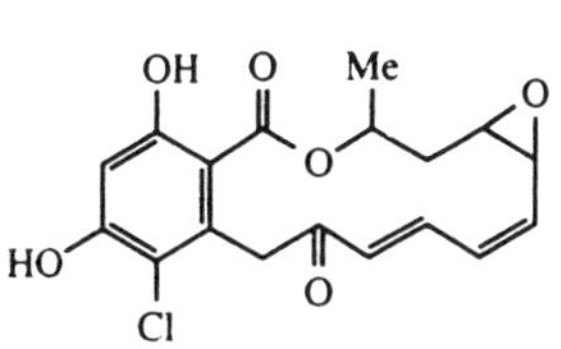

14-76

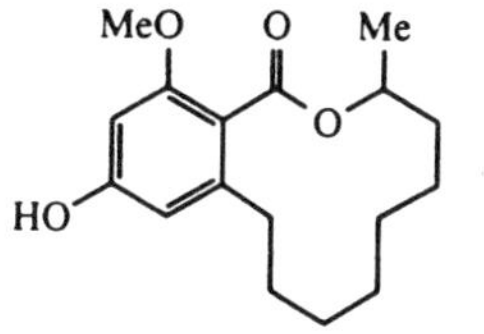

14-77

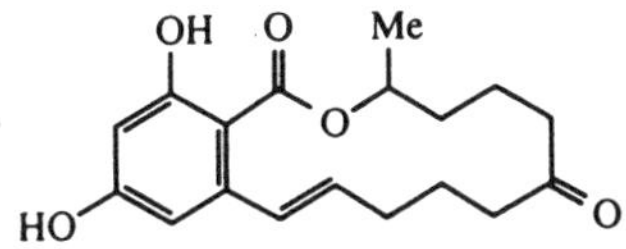

14-78

A new synthesis of (+)-zearalenone, **14-78**, starting from 1-iodo-2-phenylthiomethyl-4,6-dimethoxybenzene, in 8 steps, was devised.[127]

Cobalt and molybdenum carbonyl clusters, **14-79**, of derivatives of zearalenone were prepared. These compounds had a high binding affinity for antibodies and might be used for a new type of nonisotopic immunoassay.[128]

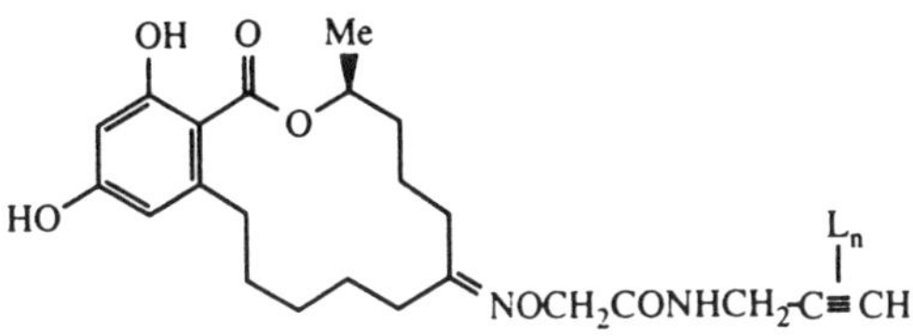

14-79

[where L_n = $Co_2(CO)_6$, $Mo_2Cp_2(CO)_4$]

NMR and molecular modeling work was done to study the geometry/conformation of zearalenone in solution.[129] A concise synthesis of (–)-zearalenone via allylic radical intermediates was reported.[130]

A 9-step synthesis of the macrolide, (*R*)-lasiodiplodin, from poly(hydroxybutyric acid), a biopolymer, via (*R*)-1,3-butanediol was developed.[131] An asymmetric synthesis of (*S*)-zearalenone dimethyl ether was developed, using chiral sulfoxides.[132]

A CD (circular dichroism) study of the RALs, zearalenone, and zearalenol was done to determine the conformation of the *E* and *Z* isomers.[133]

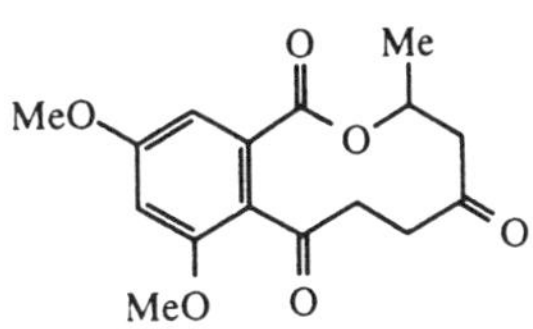

14-80

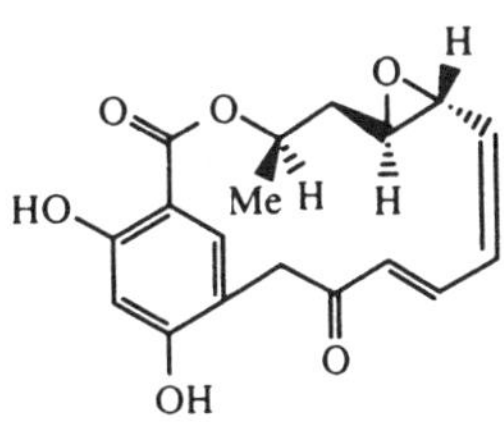

14-81

A synthesis of (+)-di-*O*-methyl-12-oxocurvularin, **14-80**, was reported. This compound is related to metabolites obtained from hybrid strains of *Penicillium citreo-veride* B.[134]

The first stereospecific, total synthesis of monorden, **14-76**, and monocillin I, **14-81**, has been published.[136]

14.10. OTHERS

Several special topics on the occurrence of resorcinol in nature were presented in the above sections 14.2–14.7. A few more diverse occurrences of resorcinol in natural products are given here in chronological order, as random selections from the literature of 1980–1992, to enhance the picture.

The use of condensed tannins in wood adhesives was already mentioned in Chapter 5. These compounds are polyflavanoids. Those obtained from the wattle tree (e.g., in South Africa and Australia) or the quebracho tree (e.g., South America) contain a resorcinol moiety and have received considerable attention as a renewable resource material. In addition to the references given in Chapter 5, there are two sections in other books containing reviews of this subject.[137,138]

Tannin extract from the black wattle tree, *Acaci mearnsii*, was sulfited to generate resorcinol and to prepare weather- and boil-resistant laminating adhesives for wood. Optimization of the adhesive showed that only half of the previously practiced resorcinol addition to the tannin extract was required to get good results.[139]

Ruscodibenzofuran, **14-82**, isolated from *Ruscus aculeatus*, a European evergreen shrub, was synthesized starting from resorcinol monomethyl ether.[140]

The structure of cristacarpin, **14-83**, a phytoalexin of the Erythrina (a tropical shrub/tree), *E. crista-galli*, was determined from chemical and spectral data.[141]

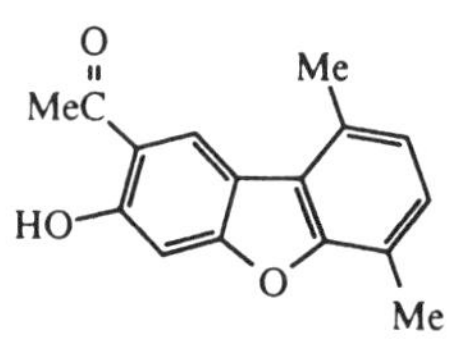

14-82

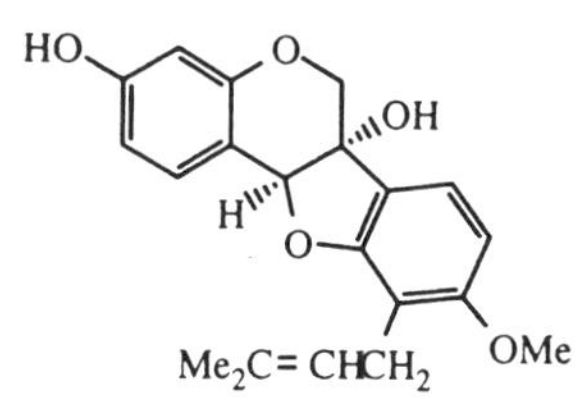

14-83

6-Demethylvignafuran, **14-84a**, and isopterofuran, **14-84b**, were phytoalexins isolated from *Coronilla emerus* (a European herb of the pea family) leaflets inoculated with *Helminthosporium carbonum*. The structures were deduced from chemical and spectral data, and confirmed by synthesis.[142]

The structures of the kuwanons G and H (albanins F and G), **14-85a** and **14-85b**, isolated from *Morus alba* (the white mulberry tree), were confirmed by partial synthesis.[143]

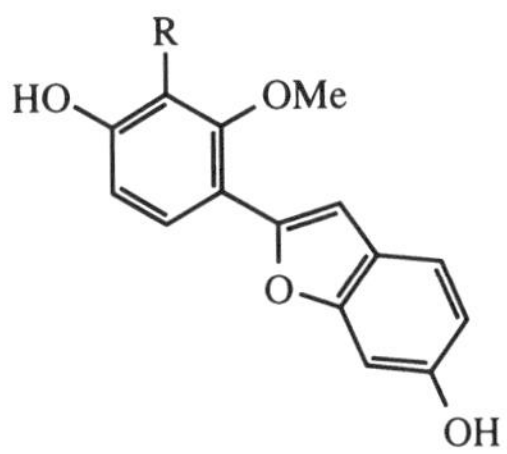

14-84
(where a: R=H; b: R=OMe)

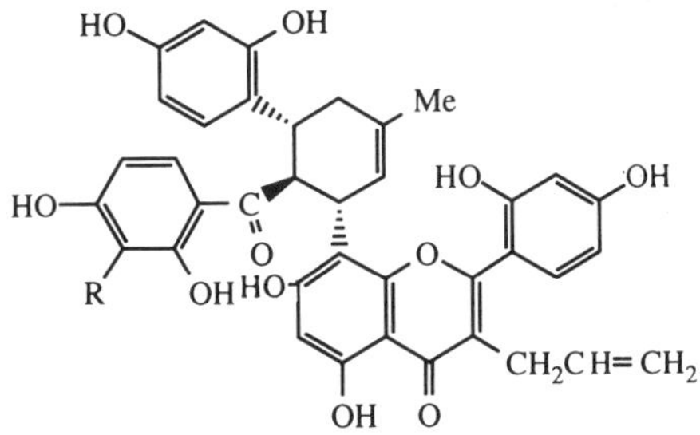

14-85
(where a: R=H; b: R=prenyl)

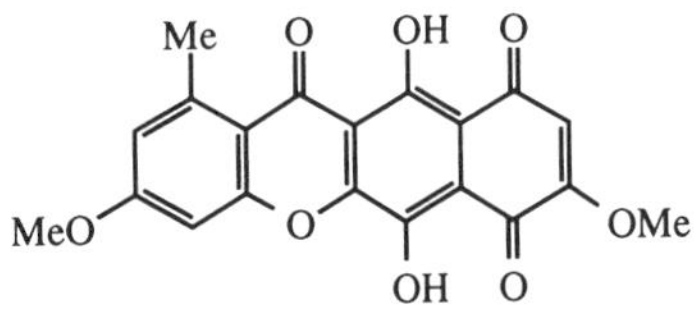

14-86

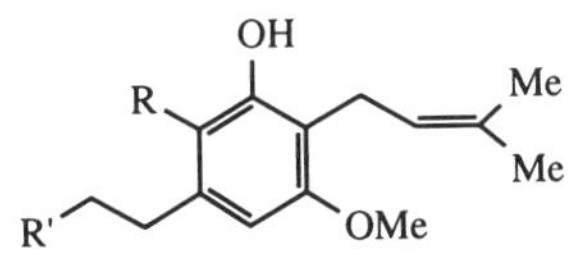

14-87
(where R=H or COOH; R'=Pr or Ph)

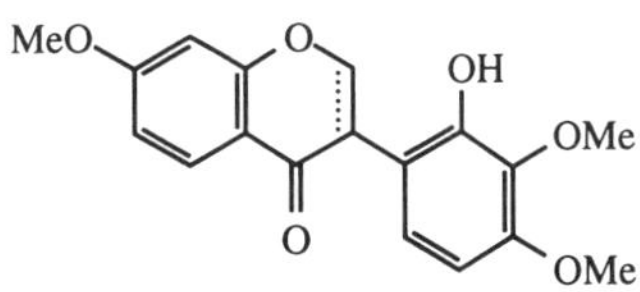

14-88
(where the dotted line signifies a single or a double bond)

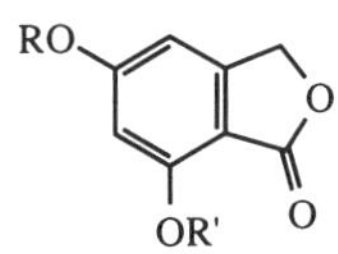

14-89
(where R=H or Me; R'=H, Me or $CH_2CH=CMe_2$)

A total synthesis of bikaverin, **14-86**, a red fungal pigment, was reported; one of the starting materials was 3,5-dihydroxybenzoic acid (alpha-resorcylic acid).[144]

Several isopentenylresorcinol derivatives, **14-87**, were isolated from *Glycyrrhiza acanthocarpa*, a herb of the licorice family. The structures were derived from chemical and spectral data.[145]

The syntheses of two isoflavanoids, **14-88**, isolated from the tree, *Myroxilon peruiferum*, was published.[146]

The structure of three natural phthalides, **14-89**, found in *Anaphalis contorta* (a herb of the thistle family) and in *Helichrysum arenarium* (a marigold), was confirmed by synthesis.[147]

The thielavins A, **14-90a**, B, **14-90b**, and C, **14-90c**, were assigned these formulas on the basis of ^{1}H NMR and ^{13}C NMR spectral data and the synthesis of some derivatives.These compounds are prostaglandin synthetase inhibitors from fungi.[148]

Peniolactol, **14-91**, produced by the fungus *Peniophora sanguinea* in rotting wood, was assigned this constitution on the basis of spectral data and the synthesis of a precursor.[149]

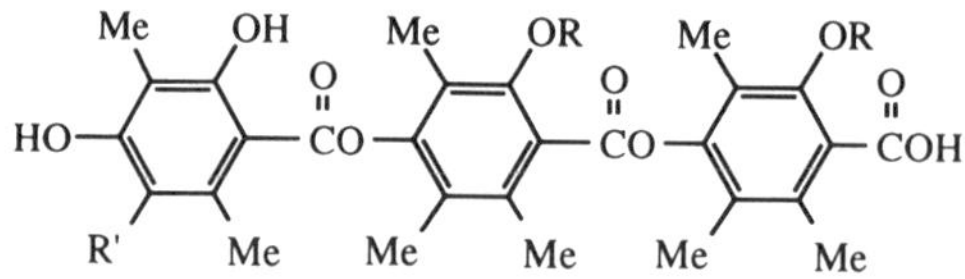

14-90
(where a: R=R'=H; b: R=Me, R'=H;
c: R=R'=Me)

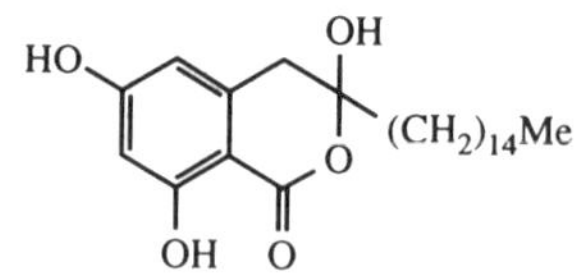

14-91

Sanggenon O, **14-92**, was identified in the acetone extract of the Chinese crude drug, Sang Baipi (named Sohakuhi in Japan), from the root bark of the mulberry (*Morus* sp.).[150]

A condensed tannin extract from Southern pine bark was blended with resocinol in 2/1 weight ratio and used to formulate a room-temperature-setting wood laminating adhesive which met the standards of the American Institute of Timber Construction.[151]

A new biflavanone, semecarpuflavanone, **14-93**, was isolated from the nut shells of the Indo-Malayan tree, *Semecarpus anacardium* Linn.[152]

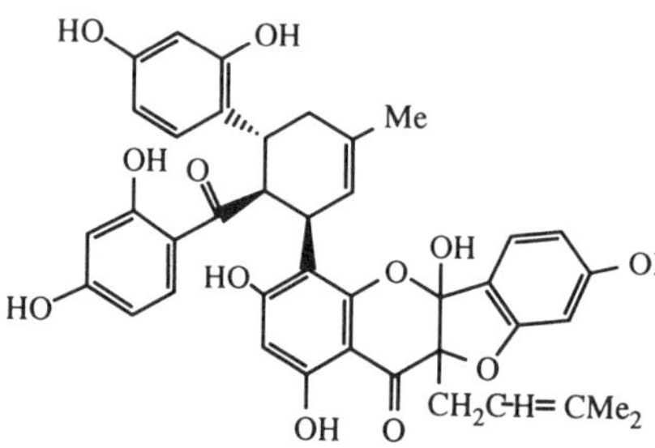

14-92

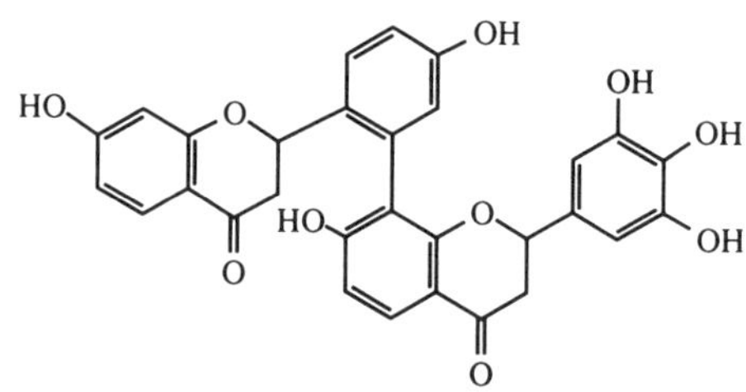

14-93

An orange pigment was isolated from *Brackenridges zanguebarica*, a small tree found in Northern Transvaal and Zambia, and was shown to have the structure of a complex benzofuran, **14-94**.[153]

An overview was published of microbial nitrogen fixation in the soil, showing a model of the molecular basis of nodule induction on legumes by the *rhizobia* bacteria. The isoflavone daidzein, **14-95**, a compound in the root exudate of the soy bean, was shown to induce the *Bradyrhizobium japonicum* early nodulation gene (nod ABC) with the help of Nod D protein; root hair curling was produced and, ultimately, bacterial nitrogen fixation.[154]

The structure of the neurotoxin NSTX-3, **14-96**, obtained from the venom of *Nephila maculata*, a Papua New Guinean spider, was unequivocally determined by total synthesis.[155] Another, similar toxin, JSTX-3, was isolated from *Nephila clevata*, the Joro spider. The proposed structure of the toxin was confirmed by synthesis. Structure modification of these toxins to produce insecticides and drugs was under consideration.[156]

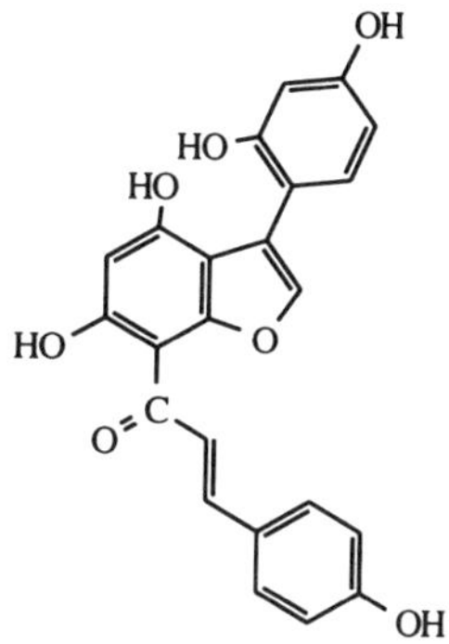

14-94

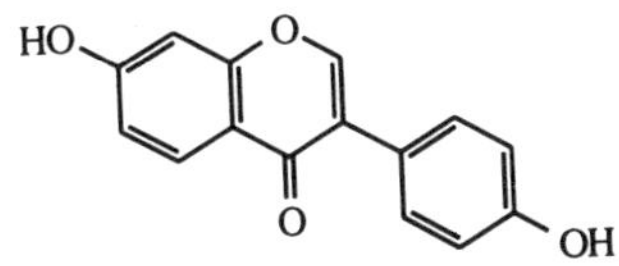

14-95

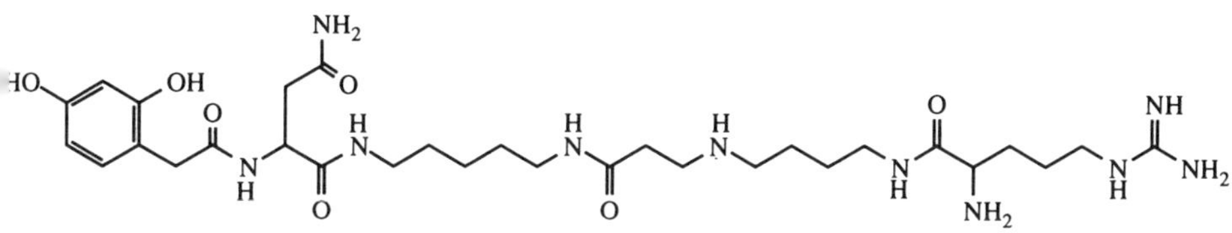

14-96

Bonducellin, a 3-benzylidene-benzopyranone-4, **14-97**, isolated from *Caesalpinia bonducella* (a tropical tree), was synthesized.[157]

The fungal depsidone, emeguisin A (and two related compounds), was isolated from the mycelium of *Emericalla unguis*, strain IFM 42017; the structure of this compound was shown to be **14-98** from chemical and spectral data.[158]

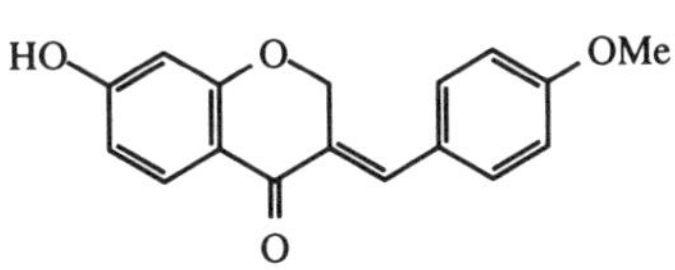

14-97

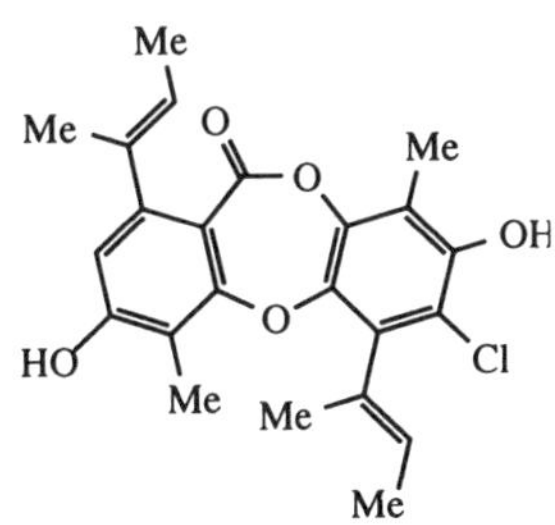

14-98

A total synthesis proved the structure of argiotoxin 636, **14-99**, a polyamine spider toxin found in the venom of *Arigiope aurantia.*[159]

6-Methoxybenzoxazolin-2(3*H*)-one (MBOA), **14-100**, found in extracts from graminaceous plants like *Coix lachryma-jobi*, wheat, and maize, was synthesized. This compound is implicated as a resistance factor against fungi and insects.[160]

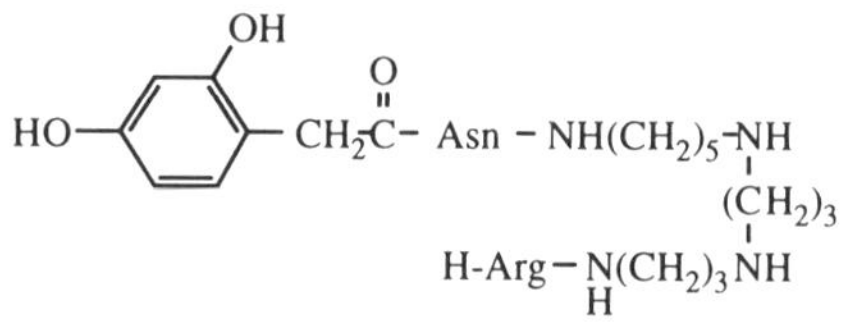

14-99

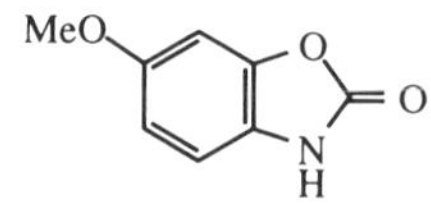

14-100

Hormothamnione, **14-101**, first isolated in minute amounts from the marine plant *Chrysopheum Halyori*,[161] was found to be a potent cytotoxin to several human leukemia cell lines in vitro. A total synthesis of this compound was developed which made more material and analogs accessible.[162]

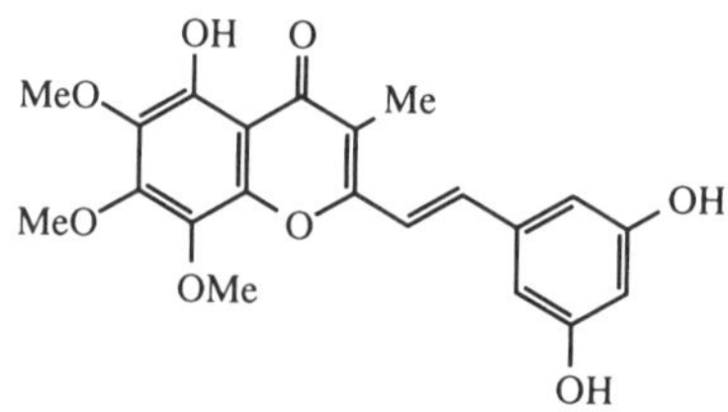

14-101

The curing characteristics of a phenol/resorcinol/formaldehyde resin (Borden's LT-75) with Borden's hardener FM-260, fortified with a condensed tannin extracted from finely ground bark in various proportions, was studied by dynamic mechanical analysis (Rheometrics RSA-II instrument). These adhesive compositions showed promise for bonding wood and had potential for bonding tire cords.[163]

Prenylated bibenzyl derivatives were synthesized, such as **14-102**, which were known components of bryophytes (mosses and liverworts) and known to have antibacterial activity; broader biological screening of such compounds was said to be made possible.[164]

The hydrogenation of vermistatin, a metabolite of *Penicillium vermiculatum*, gave compound **14-103**, and some other products. This compound, **14-103**, was a strong inhibitor of RNA synthesis of P388 leukemia cells.[165]

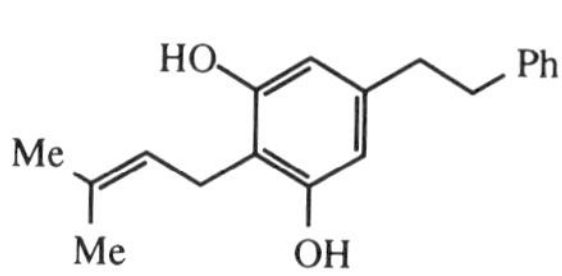

14-102

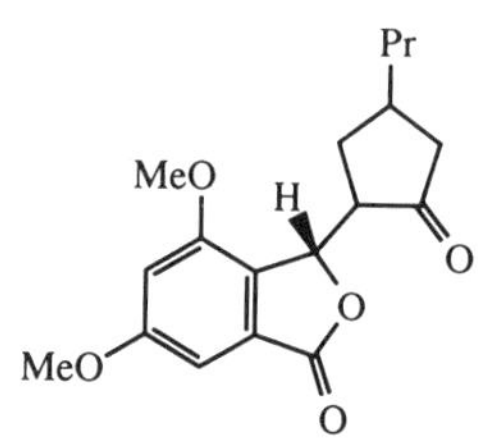

14-103

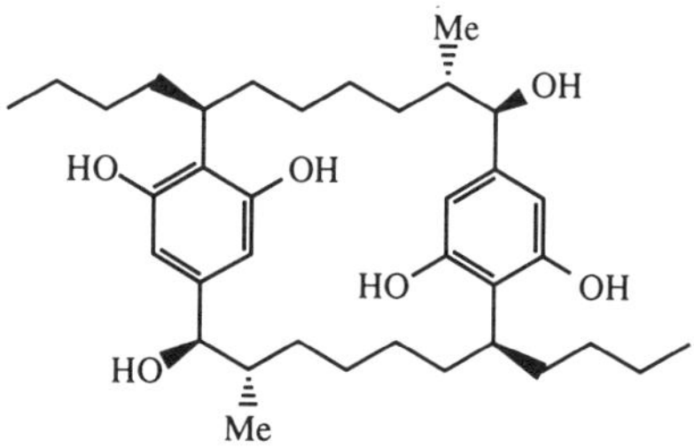

14-104

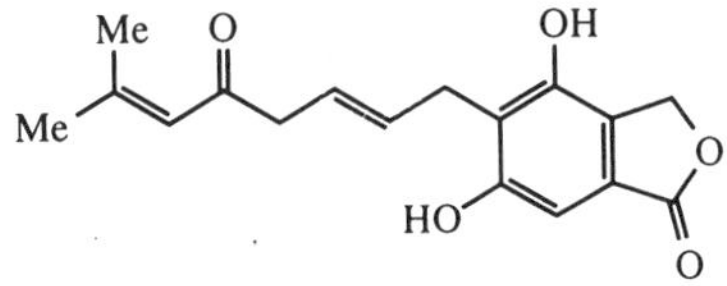

14-105

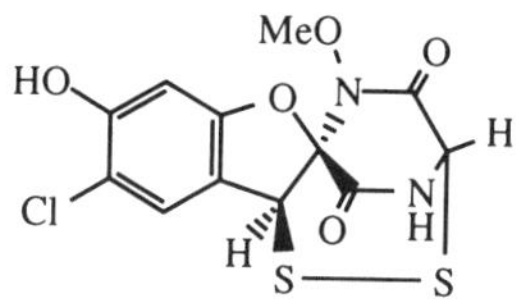

14-106

The cylindrocyclophane B, **14-104**, and related compounds, were isolated from three strains of the terrestrial blue-green algae, *Cylindrospermum licheniforme*.[166]

Hericenone, **14-105**, first isolated from the mushroom *Hericium erinaceum* in Japan,[167] was synthesized to prove this structure with a revision of the original proposal.[168]

(+)-Aspirochlorine, **14-106**, is produced by *Aspergillus tamari*, *A. flavus*, and *A. oryzae*; it has modest antifungal properties. This material was synthesized by a diastereoselective route in 13 steps, starting from 4-chlororesorcinol.[169]

14.11. REFERENCES

1. B. N. Ames, *Science (Washington, D.C.)*, *221*, 1256 (1983).
2. J. L. Wong, *Chemtech*, *1986*, 100; *ibid.*, 436.
3. H. Hlasiwetz and L. Barth, *Liebigs Ann. Chem.*, *130*, 354 (1864).
4. G. P. Ellis, ed., *Chromenes, Chromanones and Chromones*, part of the series, *The Chemistry of Heterocyclic Compounds* (A. Weissberger and E. C. Taylor, eds.), John Wiley & Sons, New York, 1977.
5. J. Lachman, V. Rehakova, V. Pivec, and J. Hubacek, *Rostl. Vyroba*, *28(2)*, 211 (1982); *C.A.*, *96*, 177976 (1982).
6. E. N. Stepanova and I. E. Tsapalova, *Izv. Vyssh. Uchebn. Zaved. Pishch. Tekhnol.*, *1982 (5)*, 154; *C.A.*, *98*, 33403 (1983).
7. W. S. Schlotzhauer, D. B. Walters, M. C. Snook, and H. C. Higman, *J. Agric. Food Chem.*, *26(6)*, 1277 (1978).
8. D. K. Lee, S. H. Hong, and K. S. Kim, *Yongu Pogo-Immok Yukchong Yonguso*, *17*, 24 (1982); *C.A.*, *97*, 36233 (1982).
9. G. Nagaraj and M. K. Chakraborty, *Tob. Res.*, *5(2)*, 147 (1979); *C.A.*, *92*, 143522 (1980).
10. J. A. Schmitt and W. Klose, *Liebigs Ann. Chem.*, *1973*, 544.

11. W. Schroll, S. Nitz, and F. Drawer, *Chem., Mikrobiol., Technol. Lebensm., 10 (5–6)*, 188 (1987); *C.A., 106*, 155011 (1987). Also, W. Schroll, R. Gottfried, and F. Drawer, *ibid., 11(3)*, 95 (1987); *C.A., 108*, 54664 (1988).
12. T. Krzaczek and L. Urbaniak, *Acta Soc. Bot. Pol., 54(4)*, 429 (1985); *C.A., 105*, 187654 (1986).
13. Z. Mosback, *Z. Naturforsch., 14B*, 69 (1959).
14. R. Falck, *Chem. Ber., 56*, 2555 (1923); E. Wedekind and K. Fleischer, *ibid., 56*, 2556 (1923).
15. H. Erdtmann, *Liebigs Ann. Chem., 539*, 116 (1939).
16. Anon., *Chem. Brit., 28*, 42 (1992).
17. S. Furakawa, *Sci. Pap. Inst. Phys. Chem. Res. Tokyo, 26*, 178 (1935); *C.A., 29*, 2948 (1935).
18. J. D. Bu'Lock and A. T. Hudson, *J. Chem. Soc. (C), 1969*, 61.
19. K. A. Cirigotlis, L. Cleaver, J. E. T. Corrie, R. G. Grasby, G. H. Green, J. Mock, S. Nimgirawath, R. W. Reed, E. Ritchie, W. C. Taylor, A. Vadasz, and W. R. G. Webb, *Aust. J. Chem., 27*, 345 (1974).
20. R. P. Gregson, P. Kazlauskas, P. T. Murphy, and R. J. Wells, *Aust. J. Chem., 30(11)*, 2527 (1977).
21. G. D. Brown, *J. Nat. Prod., 55*, 1756 (1992); *C.A., 118*, 143398 (1993).
22. Y. Hu, W.-S. Chen, B.-H. Huang, L.-Z. Lin, and R.-S. Xu, *K'o Hsueh Tung Pao, 24(19)*, 907 (1979); *C.A., 92*, 124878 (1979).
23. R. Reusch and H. L. Sadoff, *J. Bacteriol., 139(2)*, 448 (1979).
24. H. Budzikiewicz, H. Scholl, W. Neuenhaus, G. Pulverer, and H. Korth, *Z. Naturforsch., B: Anorg. Chem., Org. Chem., 35B(7)*, 909 (1980); *C.A., 93*, 182464 (1980).
25. S. G. Batrakov, N. N. Pridachina, E. B. Kruglyak, E. D. Novogrudskaya, and E. V. Chesakina, *Izv. Akad. Nauk SSSR, Ser. Biol., 1984 (2)*, 177; *C.A. 206221 (1984)*.
26. V. R. Mamdapur, *J. Agric. Food Chem., 33(3)*, 377 (1985).
27. T. Reffstrup and P. M. Boll, *Phytochemistry, 24(11)*, 2563 (1985).
28. S. Droby, D. Prusky, B. Jacoby, and A. Goldman, *Physiol. Mol. Plant Pathol., 29(21)*, 173 (1986); *C.A., 106*, 30163 (1987).
29. R. T. Scannel, J. R. Barr, V. S. Murty, K. S. Reddy, and S. M. Hecht, *J. Am. Chem. Soc., 110*, 3650 (1988).
30. K. Yu. Gordeev, V. V. Bitkov, N. N. Pridachina, V. A. Nenashev, and S. G. Batrakov, *Bioorg. Khim., 17(10)*, 1357 (1991); *C.A., 117*, 22281 (1992).
31. F. J. Marner and W. Horper, *Helv. Chim. Acta, 75(5)*, 1557 (1992).
32. D. E. Briggs, *Phytochemistry, 13(6)*, 987 (1974).
33. W. Mejbaum-Katzenellenbogen, F. Tluscik, A. Kozubek, A. Sikorski, and Z. Maresz, *Acta Soc. Bot. Pol., 44(4)*, 479 (1975); *C.A., 85*, 90202 (1976).
34. D. Weipert and A. W. ElBaya, *Getreide, Mehl Brot, 31(9)*, 225 (1977); *C.A.,88*, 61233 (1978).
35. M. Salek, *Postepy Nauk Roln., 26(2)*, 81 (1979); J. Musehold, *Getreide, Mehl Brot, 34(11)*, 304 (1980).
36. A. Kozubek and R. A. Demel, *Biochim. Biophys. Acta, 603(2)*, 220 (1980).
37. H.-D. Bock, W. Flamme, and S. Kesting, *Nahrung, 25(9)*, 805 (1981).
38. A. Kozubek, F. Tluscik, and W. Mejbaum-Katzenellenbogen, *Acta Soc. Bot. Pol, 50(4)*, 637 (1981); *C.A., 98*, 122826 (1983).
39. V. E. Shevchenko, V. V. Karpachev, and T. I. Siridova, *S-kh. Biol., 17(3)*, 365 (1982); *C.A., 97*, 69397 (1982).
40. A. Kozubek, *Z. Naturforsch., C: Biosci., 39C (11–12)*, 1132 (1984); *C.A., 102*, 107643 (1985).
41. A. Kozubek, *Acta Aliment. Pol., 11(2)*, 185 (1985); *C.A., 105*, 189542 (1986).
42. A. Kozubel, A. Jezierski, and A. F. Sikorski, *Biochim. Biophys. Acta, 944*, 465 (1988).
43. H. Itokoawa, N. Totsuka, K. Nakahara, K. Takeya, J. P. Lepoitteoin, and Y. Asakawa, *Chem. Pharm. Bull, 35*, 3015 (1987).
44. P. Hengtrakul, K. Lorenz, and M. Mathias, *Cereal Chem., 67(5)*, 413 (1990).
45. A. Kozubek and Z. Wroblewski, *Stud. Biophys., 139(3)*, 177 (1990) [Publ. 1991]; *C.A., 116*, 39902 (1992).
46. P. Hengtrakul, K. Lorenz, and M. Mathias, *J. Food Compos. Anal., 4(1)*, 52 (1991); *C.A., 116*, 5433 (1992).

47. D. G. J. Struski and A. Kozubek, *Z. Naturforsch., C: Biosci., 47(1–2)*, 47 (1992); *C.A., 116*, 123408 (1992).
48. K. K. Purushothaman, A. Sarada, and D. J. Connelly, *J. Chem. Soc., Perkin I, 1977*, 587.
49. M. J. Kato, L. M. X. Lopes, H. F. P. Fo, M. Yoshida, and O. R. Gottlieb, *Phytochemistry, 24 (3)*, 533 (1985).
50. J. E. Oliver, J. W. Neal, Jr., and W. R. Lusby, *J. Chem. Ecol., 13(4)*, 763 (1987); *C.A., 107*, 36896 (1987).
51. A. Kijjoa, M. J. T. G. Gonzalez, M. M. M. Pinto, and I. O. Monanondra, *Planta Med., 57(6)*, 575 (1991); *C.A., 116*, 170200 (1992).
52. A. F. Barrero, J. F. Sanchez, F. Reyes, and I. Rodriguez, *Phytochemistry, 30(2)*, 641 (1991).
53. J. H. P. Tyman, *Chem. Soc. Rev. 8(4)*, 499 (1979).
54. J. H. P. Tyman, *Chem. Ind. (London), 1980*, 59.
55. P. H. Gedam and P. S. Sampathkumaran, *Prog. Org. Coat., 14*, 115 (1986).
56. Brit. Pat. Appl. GB 2,066,820 (Jan. 6, 1981; J. H. P. Tyman, M. S. Patel, and A. P. Manzara to Minnesota Mining and Manufacturing Co.).
57. I. Bruce, R. Spencer, and J. Tyman, *J. Chem. Res. (S), 1992*, 224.
58. R. D. H. Murray, J. Mendez, and S. A. Brown, *The Natural Coumarins, Occurrence, Chemistry and Biochemistry*, John Wiley & Sons Ltd., Chichester, New York, 1982.
59. R. D. H. Murray, *Nat. Prod. Rep., 6*, 591 (1989).
60. P. Bhandari and R. P. Rastogi, *J. Sci. Ind. Res., 42*, 383 (1983) and references cited therein.
61. P. J. Jerris and A. B. Smith, III, *J. Org. Chem., 46(3)*, 577 (1981).
62. G. J. Shaw, M. K. Yates, and D. R. Biggs, *Phytochemistry, 21(1)*, 249 (1982).
63. A. Sattar, M. Ashraf, M. K. Bhatty, and N. H. Chisti, *Phytochemistry, 17(3)*, 559 (1978).
64. S. Sengupta and S. C. Das, *Chem. Ind. (London), 1978*, 954.
65. P. Waykole, S. Shaikh, and R. N. Usgaonkar, *Ind. J. Chem., Sect. B, 19B (3)*, 238 (1980); *C.A., 94*, 192182 (1981).
66. K. Baba, Y. Matsuyama, T. Ishida, M. Inoue, and M. Kozawa, *Chem. Pharm. Bull., 30(6)*, 2036 (1982); *C.A., 97*, 162673 (1982).
67. M. A. Rizzacasa and M. V. Sargent, *J. Chem. Soc., Perkin Trans. 1, 1988 (8)*, 2425.
68. J. Reisch, H. M. T. B. Herath, and N. S. Kumar, *Liebigs Ann. Chem., 1990 (7)*, 931.
69. S. Allison, S. J. Burks, and R. T. Taylor, *Tetrahedron, 47 (47)*, 9737 (1991).
70. D. Ge, M. Kong, X. Liang, and J. Xie, *Bopuxue Zazhi, 8(3)*, 253 (1991).
71. W. S. Bowers, T. Ohta, J. S. Cleere, and P. A. Marsella, *Science (Washington, D.C.), 193*, 542 (1978).
72. Anon., *Chem. Eng. News, 58*, 42 (1980).
73. V. K. Ahluwalaia, R. S. Jolly, and S. Bala, *Int. Conf. Chem. Biotechnol, Biol. Act. Nat. Prod. [Proc.], 1st, 3(2)*, 591 (1981); *C.A., 97*, 127367 (1982).
74. A. V. Vyas and N. B. Mulchandani, *Synth. Commun., 11(8)*, 663 (1981).
75. A. P. Ottridge, R. C. Jennings, and G. T. Brooks, *Dev. Endocrinol. (Amsterdam), 15* (*Juv. Horm. Biochem.)*, 381 (1981); *C.A., 96*, 181050 (1982).
76. R. C. Jennings, *Tetrahedron Lett., 23(26)*, 2693 (1982) and references cited therein; R. A. Halpin, S. F. El-Naggar, K. M. McCombe, K. P. Vyas, D. R. Boyd, and D. M. Jerina, *ibid., 23(16)*, 1655 (1982).
77. P. Anastasis and P. E. Brown, *J. Chem. Soc., Perkin Trans. 1, 1982, (9)*, 2013.
78. R. C. Jennings and A. P. Ottridge, *J. Chem. Soc., Perkins Trans. 1, 1984 (8)*, 1733.
79. J. Mann and P. D. Kane, *Tetrahedron Lett., 26(38)*, 4677 (1985).
80. J. W. ApSimon, L. W. Herman, and C. Huber, *Can. J. Chem., 63(10)*, 2589 (1985).
81. T. Timar, S. Hosztafi, J. C. Jaszberenyi, K. E. Kover, and G. Batta, *Acta Chim. Hung., 125(2)*, 303 (1988); *C.A., 110*, 192507 (1989).
82. T. Timar, S. Hosztafi, and J. C. Jaszberenyi, *Acta Chim. Hung., 125(4)*, 457 (1988); *ibid., 125(4)*, 617 (1988); *C.A., 111*, 39046 and 39047 (1989).
83. P. Sebok, T. Timar, J. C. Jaszberenyi, and J. Jeko, *Acta Chim. Hung., 126 (4)*, 471 (1989); *C.A., 114*, 114873 (1991).

84. J. Bujons, F. Camps, and A. Messeguer, *Tetrahedron Lett., 31(36)*, 5235 (1990).
85. M. Iyer and G. K. Trivedi, *Synth. Commun., 20(9)*, 1347 (1990).
86. P. K. Chowdhury, *J. Chem. Res., Synop., 1990 (12)*, 390.
87. S. Y. Dike, J. R. Merchant, and N. Y. Sapre, *Tetrahedron, 47*, 4775 (1991).
88. M. Tiabi and H. Zamarlik, *Tetrahedron Lett., 32(49)*, 7251 (1991).
89. S. A. Kulkarni and M. V. Paradkar, *Synth. Commun., 22(11)*, 1555 (1992).
90. H. Hopf, *Chem. Ber., 125*, I-XXIV (1992).
91. H. Musso, *Angew. Chem., Int. Ed. Engl., 2(12)*, 723 (1963).
92. H. Musso, V. I,Zahorszky, D. Maasen, and I. Seeger, *Chem. Ber., 96*, 1579 (1963).
93. H. Alles, D. Bormann, and H. Musso, *Chem. Ber., 103*, 2526 (1970).
94. E. Steiner and E. Charollais, paper presented to the Chem. Soc. Basel, abstracted in *Chimia, 28 (1)*, 25 (1974).
95. D. O. Chester and J. A. Elix, *Aust. J. Chem., 32(6)*, 1399 (1979).
96. C. Brassy, B. Bachet, B. Bodo, and D. Molho, *Acta Crystallogr., Sect. B, B38(12)*, 3126 (1982).
97. M. V. Sargent and P. O. Stransky, *J. Chem. Soc., Perkins Trans. 1*, 1982 (10), 2373.
98. C. F. Carvalho and M. V. Sargent, *Aust. J. Chem., 39(11)*, 1765 (1986).
99. J. A. Elix and V. K. Jayanthi, *Aust. J. Chem., 39(5)*, 791 (1986).
100. J. A. Elix and J. H. Wardlaw, *Aust. J. Chem., 40(2)*, 425 (1987).
101. J. A. Elix and U. A. Jenie, *Aust. J. Chem., 42(6)*, 987 (1989).
102. *U.S. Code of Federal Regulations, Title 21*, Part 1308.11 (1985).
103. R. Mechoulam, ed., *Marijuana: Chemistry, Pharmacology, Metabolism and Clinical Effects*, Academic Press, New York, 1973.
104. R. E. Gosselin, *Clinical Toxicology of Commercial Products; Acute Poisoning*, 5th ed., William Wilkins, Baltimore, 1984.
105. R. A. Archer, W. B. Blanchard, W. A. Day, D. W. Johnson, E. J. Lavignino, and C. W. Ryan, *J. Org. Chem., 42(13)*, 2277 (1977).
106. A. Arnoldi, N. Bizzozero, L. Merlini, E. Ragg, and A. C. Rossi, *J. Chem. Res., Synop., 1983 (5)*, 107.
107. V. S. Jorapur, Z. H. Khalil, R. P. Duffley, R. K. Razdan, B. R. Martin, L. S. Harris, and W. L. Dewey, *J. Med. Chem, 28(6)*, 783 (1985).
108. M. Girard, D. B. Moir, and J. W. ApSimon, *Can. J. Chem., 62(1)*, 189 (1987).
109. C. J. Kowalski and G. S. Lal, *J. Am. Chem. Soc., 110*, 3693 (1988).
110. D. R. Compton, P. J. Little, B. R. Martin, J. K. Saha, J. Gilman, H. Sard, and R. K. Razdan, *Eur. J. Med. Chem., 24(3)*, 293 (1989).
111. L. S. Melvin, J. Bordner, W. A. Hada, and M. R. Johnson, *J. Heterocycl. Chem., 27(3)*, 535 (1990).
112. D. J. Harvey and N. K. Brown, *Pharmacol., Biochem. Behav., 40(3)*, 533 (1991); *C.A., 118*, 75078 (1993).
113. J. W. Huffman, H. H. Joyner, M. D. Lee, R. D. Jordan, and W. T. Pennington, *J. Org. Chem., 56(6)*, 2081 (1991).
114. M. A. Tius and G. S. K. Kannangara, *Tetrahedron, 48(42)*, 9173 (1992).
115. S. H. Burstein, C. A. Audette, A. Breuer, W. A. Devane, S. Colodner, S. A. Doyle, and R. Mechoulam, *J. Med. Chem., 35(17)*, 3135 (1992).
116. A. Charalambous, G. Yan, D. B. Houston, A. C. Howlett, D. R. Compton, B. R. Martin, and A. Makriyannis, *J. Med. Chem., 35(16)*, 3076 (1992).
117. G. A. Poulton, M. J. Ashwood-Smith, and G. C. Cross, *poster at the 70th Can. Chem. Conf.*, Quebec, 1987.
118. G. W. Ivie, D. L. Holt, and M. C. Ivey, *Science (Washington, D.C.), 213*, 909 (1981).
119. Anon., "Vegetable Health Risk," *Chem. Brit., 24*, 203 (1988) and references cited therein.
120. P. Wesley, *Genetic Eng. News, July/Aug. 1982*, 12.
121. V. K. Ahluwalia, K. K. Arora, and C. Prakash, *Gazz. Chim. Ital., 11(3–4)*, 103 (1981); *C.A., 95*, 80776 (1981).

122. J. C. Ronfard-Haret, D. Averbeck, R. V. Bensasson, E. Bisagni, and E.J. Land, *Photochem. Photobiol., 35(4),* 479 (1982).
123. P. N. Confalone and D. L. Confalone, *Tetrahedron, 39(8),* 1265 (1983).
124. O. M. Abdel-Hafez, E. A. M. El-Khrisy, S. A. Meshaal, and E. A. Abu-Mustafa, *Egypt. J. Chem., 30(6),* 453 (1987) [publ. 1989]; *C.A., 114,* 42308 (1991).
125. H. Abu-Shady, S. T. Hassib, K. M. Amin, and A. E. Farag, *Egypt. J. Pharm. Sci., 31(3–4),* 453 (1990).
126. S. Danishefsky and S. J. Etheredge, *J. Org. Chem., 44(25),* 4716 (1979).
127. T. Takahashi, H. Ikeda, and J. Tsuji, *Tetrahedron Lett., 22(14),* 1363 (1981).
128. M. Gruselle, P. Deprez, A. Vessieres, S. Greenfield, G. Jaouen, J. P. Larue, and D. Thouvenot, *J. Organomet. Chem., 359(2),* C53 (1989).
129. L. Muniz and R. Ricardo, *Magn. Reson. Chem., 28(8),* 665 (1990); C. Cordier, M. Gruselle, G. Jaouen, D. W. Hughes, and M. J. McGlinchy, *ibid., 28(10),* 835 (1990).
130. S. A. Hitchcock and G. Pattenden, *Tetrahedron Lett., 31(25),* 3641 (1990).
131. M. Braun, U. Mahler, and S. Houben, *Liebigs Ann. Chem., 1990 (6),* 513.
132. G. Solladie, M. C. Carreno, and J. L. Garcia Ruano, *J. Org. Chem., 56(7),* 2317 (1991).
133. M. Gelo, Z. Raza, V. Sunjic, J. Guo, and G. Schnatzke, *Tetrahedron: Asymmetry, 2(10),* 1005 (1991).
134. S. Lai, Y. Shizuri, S. Yamamura, K. Kawai, Y. Terada, and H. Furukowa, *Tetrahedron Lett., 30,* 2241 (1989).
135. R. A. Kasar, R. A. Khan, V. H. Deshpande, and N. R. Ayyangar, *Tetrahedron Lett., 32(12),* 1599 (1991).
136. M. Lampilas and R. Lett, *Tetrahedron Lett., 33(6),* 777 (1992).
137. A. Pizzi, in *Renewable-Resource Materials* (C. E. Carraher and L. H. Sperling, eds.), p. 323, Plenum Press, New York, 1986.
138. R. Kreibich and R. Hemingway, in *Adhesives from Renewable Resources* (R. Hemingway and A. Conner, eds.), p. 203, American Chemical Society, Washington, D. C., 1989.
139. A. Pizzi and G. M. E. Darling, *Holzforsch. Holzverwert., 32(3),* 64 (1980).
140. R. T. Scannel and R. Stevenson, *J. Chem. Soc., Chem. Commun., 1980 (22),* 1103.
141. J. L. Ingham and K. R. Markham, *Phytochemistry, 19(6),* 1203 (1980).
142. P. M. Dewick and J. L. Ingham, *Phytochemistry, 19(2),* 289 (1980).
143. T. Nomura, T. Fukai, T. Narito, S. Terada, J. Uzawa, Y. Iitaka, M. Takasugi, S. Ishikawa, S. Nagao, and T. Masamune, *Tetrahedron Lett., 22(23),* 2195 (1981).
144. N. Katagiri, J. Nakano, and T. Kato, *J. Chem. Soc., Perkin Trans. 1, 1981 (10),* 2710.
145. E. L. Ghisalberti, P. R. Jefferies, and D. McAdam, *Phytochemistry, 20(8),* 1959 (1981).
146. D. K. Bhardwai, M. S. Bisht, R. K. Jain, and A. Munjal, *Proc. Indian Natl. Sci. Acad., Part A, 48 (1),* 103 (1982); *C.A., 98,* 34407 (1983).
147. B. Talapatra, M. K. Roy, and S. K. Talapatra, *J. Indian. Chem. Soc., 60(12),* 1169 (1983); *C.A., 101,* 191502 (1984).
148. N. Kitahara, H. Haruyama, T. Hata, and S. Takahashi, *J. Antibiot., 36(5),* 599 (1983).
149. M. Ala-Peijari and T. Hase, *Finn. Chem. Lett., 1984(4–5),* 118; *C.A. 103,* 6122 (1985).
150. Y. Hano and T. Nomura, *Heterocycles, 23(10),* 2499 (1985).
151. R. E. Kreibich, *Forest Prod. J., 35(3),* 23 (1985).
152. S. S. N. Murthy, *Proc. Indian Acad. Sci., Chem. Sci., 97(1),* 63 (1986); *C.A., 106,* 196083 (1987).
153. S. E. Drewes, N. A. Hudson, R. B. Bates, and G. S. Linz, *J. Chem. Soc., Perkin Trans. 1,* 1987, 2809.
154. H. Hennecke, *Chimia, 41(11),* 370 (1987).
155. T. Teshima, T. Wakamiya, Y. Aramaki, T. Nakajima, N. Kawai, and T. Shiba, *Tetrahedron Lett., 28(30),* 3509 (1987); a more detailed description was given in *Tetrahedron, 47,* 3305 (1991).
156. Y. Hashimoto, Y. Endo, K. Shudo, Y. Aramaki, N. Kawai, and T. Nakajima, *Tetrahedron Lett., 28(30),* 3511 (1987).
157. S. Malhotra, V. K. Sharma, and V. S. Parmar, *J. Chem. Res., Synop., 1988 (6),* 179.
158. N. Kawahara, K. Nozawa, S. Nakajima, K. -i. Kawai, and M. Yamazaki, *J. Chem. Soc., Perkin Trans. 1, 1988,* 2611.

159. V. J. Jasys, P. R. Kelbaugh, D. M. Nason, D. Phillips, N. A. Saccomano, and R. A. Volkmann, *Tetrahedron Lett., 29(48)*, 6233 (1988).
160. D. Sicker, *Synthesis, 1989*, 875.
161. W. H. Gerwick, A. Lopez, G. D. Van Duyne, J. Clardy, W. Ortiz, and A. Baez, *Tetrahedron Lett., 27*, 1979 (1986); W. H. Gerwick, *J. Nat. Prod., 52*, 252 (1989).
162. L. W. McGarry and M. R. Detty, *J. Org. Chem., 55(14)*, 4349 (1990).
163. T. G. Rials, *Am. Chem. Soc., Div. Polymer Chem., Polymer Prepr., 32(3)*, 356 (1991).
164. T. Eicher, K. Tiefensee, R. Doenig, and R. Pick, *Synthesis, 1991*, 98.
165. B. Proska, S. Uhrinova, J. Adamcova, and J. Fuska, *Monatsh. Chem., 123(3)*, 251 (1992).
166. B. S. Moore, J.-Lu Chen, G. M. L. Patterson, and R. E. Moore, *Tetrahedron, 48(15)*, 3001 (1992).
167. H. Kawagishi, M. Ando, and T. Mizuno, *Tetrahedron Lett., 31(3)*, 373 (1990).
168. A. V. R. Rao and R. G. Reddy, *Tetrahedron Lett., 33(28)*, 4061 (1992).
169. G. F. Miknis and R. M. Williams, *J. Am. Chem. Soc., 115(2)*, 536 (1993).

15

Other Examples of Mostly Academic Work with Resorcinol

15.1. INTRODUCTION

This is the final chapter on the chemistry/uses of resorcinol and its derivatives. It outlines selections of works which are snapshots of the great range of topics/experimentation in universities, other than those on natural products dealt with in the preceding Chapter 14. There are a few references from industrial laboratories in this collection. Again, here are illustrations of the makeup of the cream of the crop: Starting with apparently simple/complex concepts and theories, they and their students/staff bring the concept to reality by experiment, utilizing the best instrumentation, enthusiasm, hard work, and so on. Here is more proof that chemistry is alive, building on the past, growing, and taking new flights of imagination.

15.2. HOST–GUEST CHEMISTRY AND MOLECULAR RECOGNITION

15.2.1. Calixarenes, Baskets, Carcerands, Cavitands, Hemicarcerands, Velcraplexes, Enzyme Mimics, and Related Topics in Supramolecular Chemistry

The calixarenes are $[1_n]$ metacyclophanes with a half-bowl shape made by the condensation of *p*-substituted phenols or resorcinol with aldehydes. The long title of this subsection is meant to convey how this chemistry has evolved from its beginning into an exciting field of its own. Many prominent chemists have pioneered in this area and have written cogent views/reviews of it: for example, D.J. Cram (the Nobel laureate, together with Pederson and Lehn, for his work on crown ethers and spherands) in 1983[1]; C.D. Gutsche in 1989 (monograph) and 1991 (giving interesting historical and philosophical aspects as well as straight chemistry)[2]; V. Boehmer and

M.A. McKervey (in German) in 1991[3]; a monograph edited by J. Vicens and V. Boehmer in 1991[4]; L.C. Groenen and D.N. Reinhoudt in 1991[5]; A. McKervey and V. Boehmer (in English) in 1992[6]; a review of molecular container compounds (with a biological perspective) by D.J. Cram in 1992[7]; and also a new journal, started in 1992, with a scope including the chemistry of calixarenes.[8] These extensive reviews cover the literature into 1991.

The purpose here is not another review of calixarenes, but to show that the high reactivity of resorcinol makes it a useful starting material for this chemistry. The older literature is discussed briefly and the later, outstanding developments are sketched, with a few examples from the literature. For details, the original literature and/or reviews should be consulted.

The nomenclature terms and depictures are used interchangeably to accustom the reader to the nonunified descriptions and structures given in the literature. A little more detail is given for papers from the 1991–1993 period since they have, for the most part, not been covered by the prior reviews.

Baeyer[9] and Michael[10] had already found in the 19th century that resorcinol and benzaldehyde reacted under acid catalysis in equimolar proportions. In 1940 Niederl and Vogel suggested a macrocyclic structure for the crystalline condensation products of resorcinol and acetaldehyde, propionaldehyde, or isobutyraldehyde, based on the 4/4 mole ratio of reagents deduced from molecular weight determinations.[11] Resorcinol and formaldehyde gave only resins.

In 1968, Erdtman and Hoegberg studied the condensation of resorcinol with benzaldehyde, obtained a crude crystalline product in 90–95% yield, and suggested a macrocyclic tetramer, **15-1**, which in theory could exist in four configurations; by fractional crystallization of the tetramer butyrate they obtained two compounds which had different NMR spectra, indicating steric differences. A different butyrate was obtained from the resorcinol/*p*-bromobenzaldehyde condensation product and X-ray analysis showed a macrocycle with the four aldehyde-derived carbons being planar, the four *p*-bromophenyl groups being all on the same side of that plane, and the three resorcinol rings connecting two aldehyde-derived carbons being essentially perpendicular to each other.[12] Twelve years later, A.G.S. Hoegberg reported that the acid-catalyzed reaction of resorcinol with benzaldehyde produced two stereoisomers of the macrocyclic tetramer, **15-1**, now called 1,8,15,22-tetraphenyl [1_4] metacyclophane-

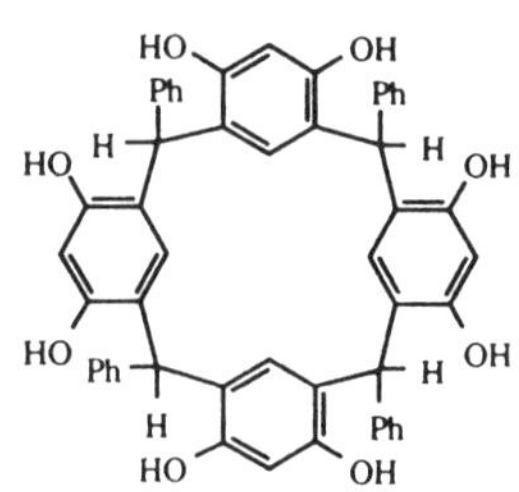

15-1

3,5,10,12,17,19,24,26-octol. On the basis of molecular models, symmetry considerations and dynamic NMR spectra of the benzaldehyde and *p*-bromobenzaldehyde reaction products with resorcinol, and their octabutyrates, Hoegberg established that the kinetically favored isomer had a chair-like conformation with the phenyl groups in pairs, in axial positions on each side of the macrocyclic ring. The thermodynamically more stable isomer (obtained in high yield after longer reaction times) had a boat-like conformation with all four phenyl groups in the axial positions below the plane of the macrocyclic ring.[13] Hoegberg also found two similar stereoisomeric macrocyclic resorcinol/acetaldehyde condensation products.[14]

Abis and co-workers studied the acid-catalyzed condensation of resorcinol with heptanal and dodecanal, made the acetates of the macrocycles, and showed by NMR spectral analysis, NOE (nuclear Overhauser effect) enhancements, and T_1 experiments the existence of one isomer with a boat conformation, one with a chair conformation, and, in addition, a new isomer with a diamond-like conformation. In these three isomers the alkyl substituents were in the less hindered axial position and their relative configuration was all-*cis*, *cis-trans-trans*, and *cis-trans-cis*.[15]

Gutsche renamed these macrocycles calix[4]arenes, calix (from the Greek) in reference to a bowl shape.[2]

In the late 1980s, D. J. Cram and co-workers made an extensive study of the cyclocondensation of resorcinol and 2-substituted resorcinols with various aldehydes and the crystalline products were found to have, at most, only two stereoisomers, **15-2** and **15-3**, both having bowl-shaped conformation. With aliphatic aldehydes, only the **15-2** isomer was found (45–92% yield); varying results were obtained with *p*-substituted benzaldehydes—e.g., *p*-nitrobenzaldehyde gave no cyclic tetramer, and *p-tert*-butylbenzaldehyde gave mainly **15-3** isomer. No octol was obtained from 2-nitro or 2-bromoresorcinol and acetaldehyde or benzaldehyde.[16] The octols (calixarenes) were starting materials for making cavitands and carcerands.

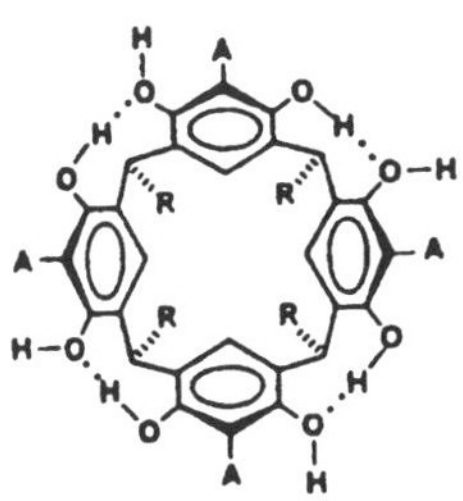

15-2
(C_{4v} symmetry)

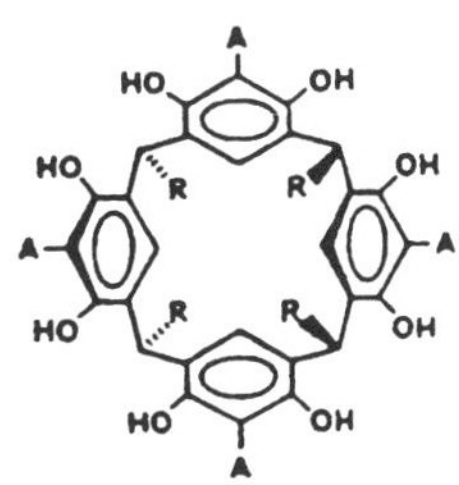

15-3
(C_{2v} symmetry)

[where A=H or Me; R=alkyl or aryl]

Thus, in Cram's terms the cavitands are host compounds with concave surfaces and are able to form caviplexes with organic guests. For example, when the resorcinol hydroxyl groups in octol **15-2** (R=Me) were bridged with alkylene groups, a bowl-shaped cavitand, **15-4**, was formed in which the methyl groups formed the feet of the bowl. These compounds formed solvates with simple guest molecules such as CH_2Cl_2, SO_2, MeCN, benzene, and others.[17]

Other cavitands, with cylindrical wells, were made from the octol, **15-2** (R=Me), by bridging the resorcinol hydroxyl groups with, e.g., dimethyl dichlorosilanes to give **15-5** which, in $CDCl_3$, complexed with the linear CS_2, $CH_3C{\equiv}CH$, or O_2.[18]

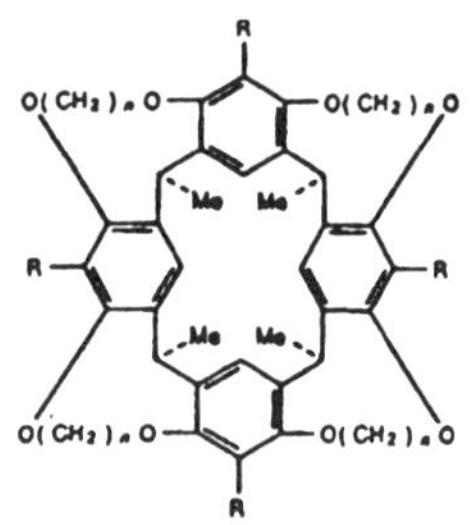

15-4
(n = 1,2)

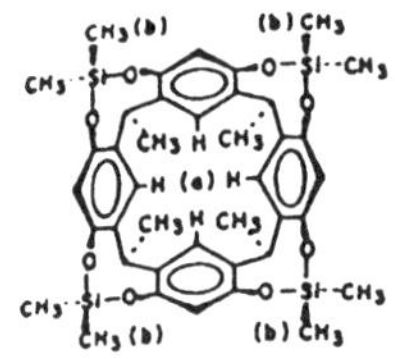

15-5

Carcerands such as **15-6**—the first closed molecular container compounds—were prepared by Cram's group by combining/shellclosing two bowl-shaped cavitands. The carcerands formed carciplexes by trapping simple guests (organic compounds, gases, or inorganic ions).[19] Astoundingly, a chemicarcerand complex of cyclobutadiene, itself quite unstable, was prepared in situ and this hemicarciplex, **15-7a** (G = cyclobutadi-

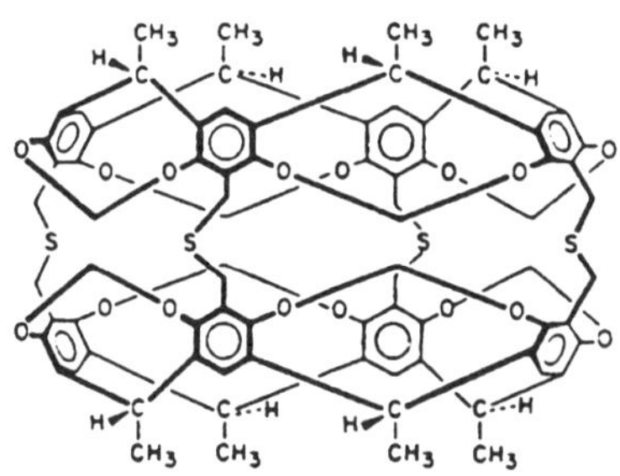

15-6

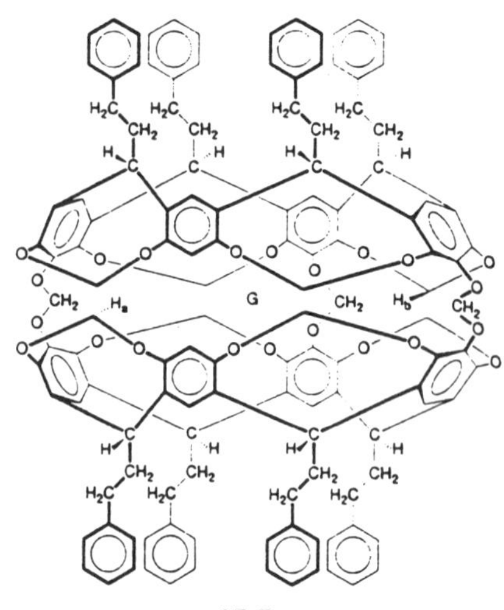

15-7
(where **a**: G = cyclobutadiene;
b: G = (*Z*)-OCH=CHCHO)

ene), was stable and produced acetylene upon photolysis. With oxygen, hemicarciplex **15-7a** was converted to another complex, **15-7b** [where G=(*Z*)-OCHCH=CHCHO).[20]

Cram and his group also prepared velcraplexes of the calixarene, **15-8**, which were fourfold, lock–key dimers in organic media.[21] (This was paper #62 from Cram's laboratories on host–guest complexation.)

Another example of the attractive chemistry coming from Cram's laboratories was the reaction of the macrocyclic octols with 2,3-dichloroquinoxaline to make the substituted cavitands, **15-9**. When R^2=H, the cavitand existed only in the vase form at room temperature or higher, but when R^2=Me, Et, or Br, the cavitands existed only in the kite conformation.[22]

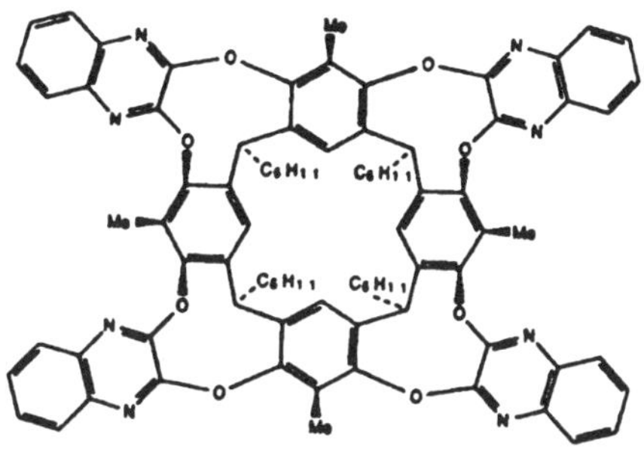

15-8

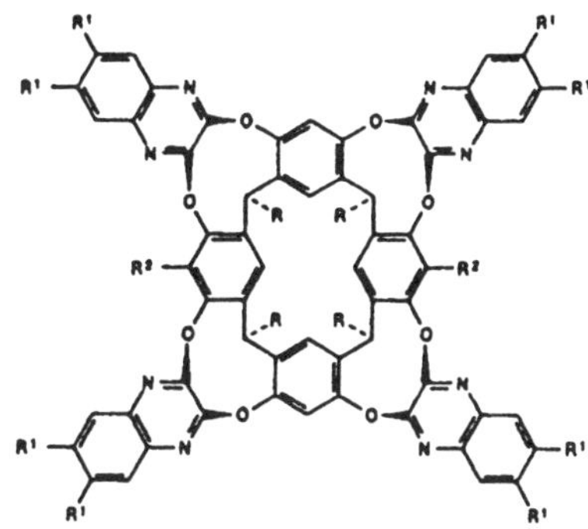

15-9

(where R = Me, $(CH_2)_4Me$, $(CH_2)_5Cl$, $(CH_2)_5I$, CH_2CHMe_2; R^1, R^2 = H, Me)

Similar cavitands and their carciplexes were also studied (X-ray crystal structure, ^1H NMR complexation studies) by an Italian group. A $-CH_2OH$ group was attached on one quinoxaline ring which was said to be easily modifiable with functional groups, aimed at possible high-technology applications, e.g., in molecular electronics.[23]

Other research groups were quite active in this field. For instance, Weinelt and Schneider studied the macrocycle formation from resorcinol and aldehyde from a mechanistic viewpoint, using analysis/integration of 50 individual rate constants. Three epimeric cyclophanes (calixarenes) were found and their isomerization rates were studied.[24] (This was paper #27 in the Saarbruecken series in host–guest chemistry.)

P.D. Beer (a Meldola medalist) designed redox-responsive calixarenes containing metal redox centers. For example, using the acid-catalyzed condensation of ferrocenecarboxaldehyde with resorcinol, followed by bridging of the hydroxyl groups with $BrCH_2Cl$ in the presence of a base, the first redox-active cavitand, **15-10**, was obtained. Another example was the formation of macrocycles by reaction of resorcinol with ferrocenecarboxaldehyde, followed by esterification of the octol produced with fero-

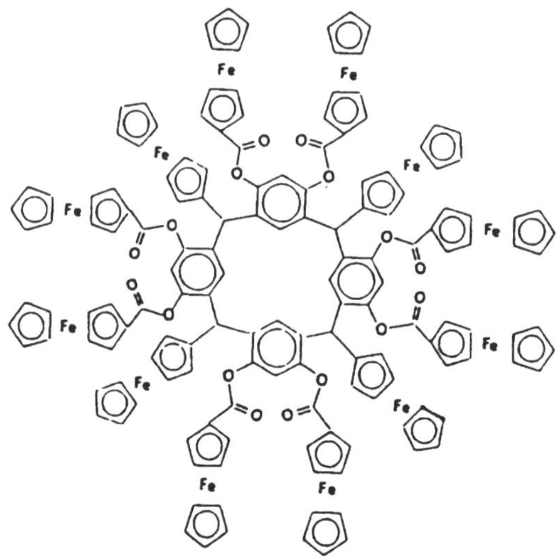

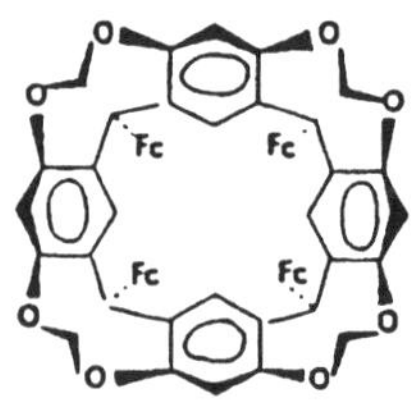

15-10
(where Fc = ferrocenyl

15-11

cenyl chloride to obtain compound **15-11**; such compounds were said to be prototypes of molecular sensors able to electrochemically detect the capture of an organic guest in the cavitand.[25]

Benzo-crown ether/calixarene combinations, such as **15-12**, were prepared, which were capable of binding bipyridinium and sodium cation guests.[26]

A thiophenecarboxaldehyde/resorcinol cyclocondensate was prepared (octaacetate) by a Dutch group and was determined by X-ray analysis to have a box-like cavity.[27]

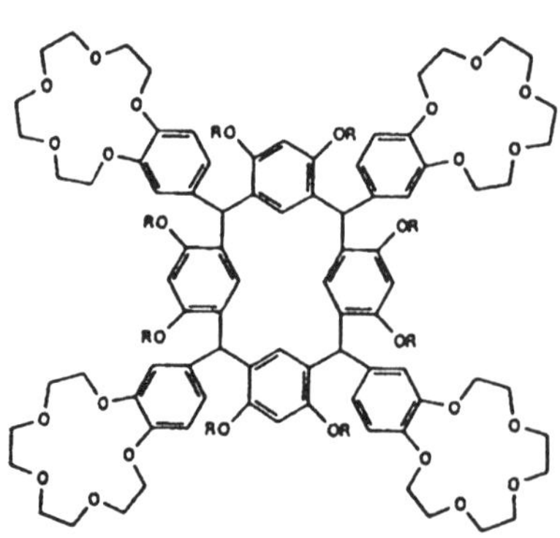

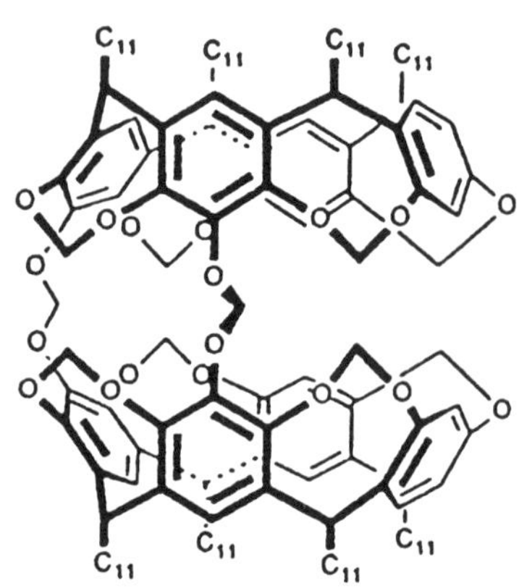

15-12

15-13

Phosphorylated calix[4]resorcinarenes were prepared.[28] The four possible *cis–trans* isomers of tetramethylcalix[4]arenoctol and derivatives were separated by a reversed-phase HPLC method.[29]

A new hemicarcerand, **15-13**, was synthesized; this was said to be a planned approach to frameworks in which functional groups can be introduced for intermolecular interactions and, later on, to build up larger structures.[30]

An improved procedure for preparing a calix[4]arene from 2-methylresorcinol/-formaldehyde was developed. Using paraformaldehyde and HCl catalyst, a 90% yield of the cyclophane was obtained. Resorcinol itself did not give calixarenes under these conditions with either paraformaldehyde or acetaldehyde.[31]

Coupling of the calixarene prepared from resorcinol/acetaldehyde with *p*-sulfonatobenzenediazonium salt gave a water-soluble host with a deep cleft, **15-14**. This host accepted large guest molecules such as adamantane, pyrene, and coronene in the cleft in aqueous media.[32]

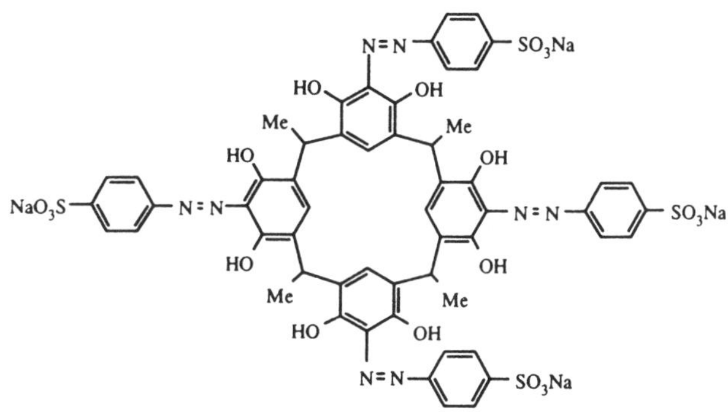

15-14

The BF_3 etherate-catalyzed tetramerization of alkyl 2,4-dimethoxycinnamates was developed as a new route to the calix[4]resorcinarenes, e.g., **15-15** (two isomers).[33]

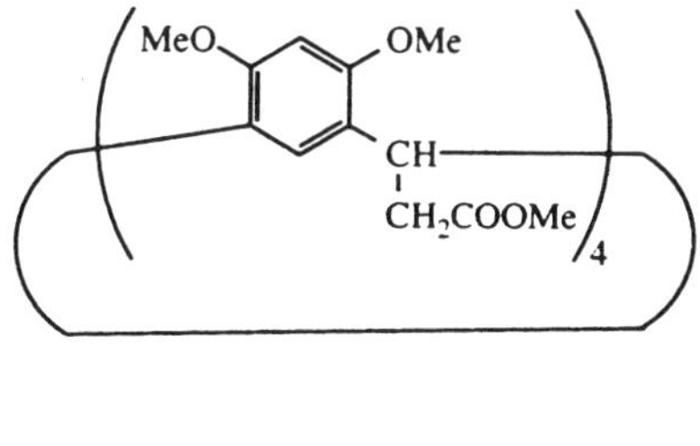

15-15

Other fascinating aspects of the chemistry of calix[4]resorcinarenes were found. Aoyama and co-workers established that the lipophilic calixarene prepared form resorcinol and dodecanal was able, in apolar media, to bind polar guests of biological origin, such as glycerol, water, D-ribose, riboflavin, and vitamin B at varying host/guest ratios. Insoluble guests were solubilized. The polar host–guest interaction was aided by H-bonding and could compete well with host–host, host–water, and guest–water interactions. This enabled molecular recognition of polar, biological compounds.[34] These studies were expanded and extended to the extraction/solubilization of alpha-ribopyranose and 1,4-cyclohexanediol (the *cis*-isomer was preferentially extracted). The binding of aldopentoses, aldohexoses, and their deoxy derivatives was also studied.[35]

From Aoyama's laboratories also came the finding that the cyclophane tetramers of resorcinol/dodecanal or resorcinol/hexanal were able to bind glutaric acid, with the calixarene acting as a multidentate host. Evidence of a two-point H-bonding interaction with the carboxyl groups was developed. The longer-chain-length pimelic acid (C_7 dicarboxylic acid) and valeric acid [$Me(CH_2)_3COOH$] showed much less binding affinity to the calix[4]resorcinarene.[36]

It was found that the cyclotetramer of resorcinol/dodecanal formed a stable monolayer at the air–water interface and this monolayer interacted with sugars. Various electrochemical and spectroscopic techniques were used to study this system.[37]

Furthermore, it was found that the lipophilic cyclotetramer from resorcinol/dodecanal enabled the direct glycosidation of ribose in an apolar solvent. The calixarene host acted as a sugar solubilizer and as an acid, protonating the bound sugar; host/sugar selectivities were also studied.[38]

Aoyama wrote two reviews (in Japanese) of molecular recognition of bivalent molecules via hydrogen bonding which included the calix[4]arenes.[39]

The highly cooperative binding of alkyl glucopyranosides to the cyclotetramer of resorcinol/dodecanal was studied in detail and ascribed to intracomplex guest–host hydrogen bonding. Methyl and *n*-octyl glucoside behaved differently—the solvophobicity/solvophilicity control by the alkyl group of the geometry, stoichiometry, stereoselectivity, and cooperativity were detailed.[40]

The cyclotetramer obtained from the condensation of resorcinol/dodecanal, as an achiral host in $CHCl_3$, was found to form H-bonded complexes with different, chiral diols or polyols, such as steroids and sugars. These complexes were studied by induced CD (circular dichroism) spectroscopy. The calixarene host was said to be usable as a supramolecular probe for the assignment of stereochemistry in the chiral guests.[41] (This was paper #18 in a series on molecular recognition from Ayama's laboratories.)

Tetrasulfonated calix[4]resorcinarenes, such as **15-16**, were water soluble but not surface active or micelle forming, had a symmetrical (C_{4v}) bowl-shaped conformation, and bound simple, aliphatic alcohols as well as hydrophobic monosaccharide derivatives such as aldopentoses, deoxy, and methylated sugars, and the nucleoside cytidine and its 5′-phosphate. The guest–host CH-π interaction of the host's electron-rich benzene rings as π-bases was suggested to account, at least in part, for the complexation of the hydrophobic sugars and nucleosides in water.[42]

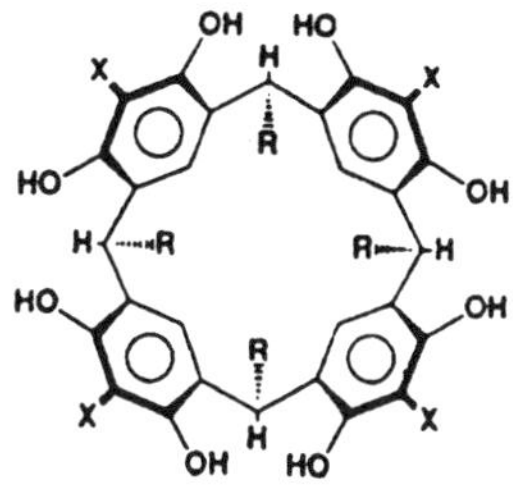

15-16
(wnere X = H; R = $(CH_2)_2SO_3Na$)

A paper from Aoyama and his group, published in 1993, reported their study of the binding of simple alcohols (monools) and actylated alcohols to the resorcinol/dodecanal cyclotetramer by proton NMR and CD spectroscopy. It was concluded that a combination of CH-π and hydrogen-bonding interactions was involved in the binding. Binding constants were determined ($C_2 < C_3 < C_4$ alcohol). The largest binding constant, $K = 54\ M^{-1}$, was obtained for borneol which has three Me groups participating in the CH-π bond interaction. Chiral secondary alcohols such as terpenes and steroids showed induced circular dichroism from guest-to-host chirality transfer. Acetyl groups in a guest alcohol, such as bornyl acetate, promoted complexation.[43]

Other researches published in early 1993 included a report of a study by conductometric titration and UV spectrometry of the interaction of amines with the resorcinol/dodecanal cyclotetramer, indicating a 3/1 cyclotetramer/triethylamine stoichiometry.[44] From Cram's group came a paper on the further characterization of cavities in carcerands using space-filling models and X-ray analysis of crystals, giving further definition to hemicarcerands (they have holes in their shells which allow entry and departure of guests, but permit the isolation of complexes—now called chemicarciplexes), and velcrands which form fourfold, lock–key dimers (velcraplexes) with large surfaces of contact between monomers; a new host molecule was described which incarcerated ferrocene.[45]

There are hints of practical applications for resorcinol-based calixarenes in the above paragraphs. Indeed, this fascinating chemistry has not been missed by the applied chemists and a number of interesting application patents have appeared. Such references were dispersed in various chapters of this book and are best accessed through the index (entry: calixarenes). Here is one additional example: Italian scientists patented a process for recovering lactic acid from solutions, such as fermentation liquors and cornsteep liquors, by vigorously stirring these solutions with a solution of a resorcinol/C_{6-19} aliphatic aldehyde cyclotetramer to form an adduct.[46]

15.2.2. Other Macrocyclic Hosts

Only a few examples from the literature of the 1980–1992 period are given here.

Metacyclophanes, **15-17**, were prepared by the reaction of resorcinol in DMSO–water (99/1 volume ratio) with alpha,omega-dibromoalkanes in the presence of KOH

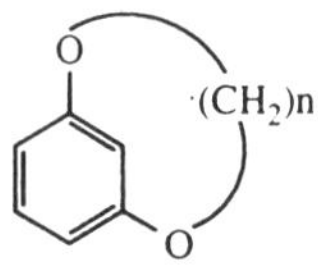

15-17
(where n = 8, 10, 12)

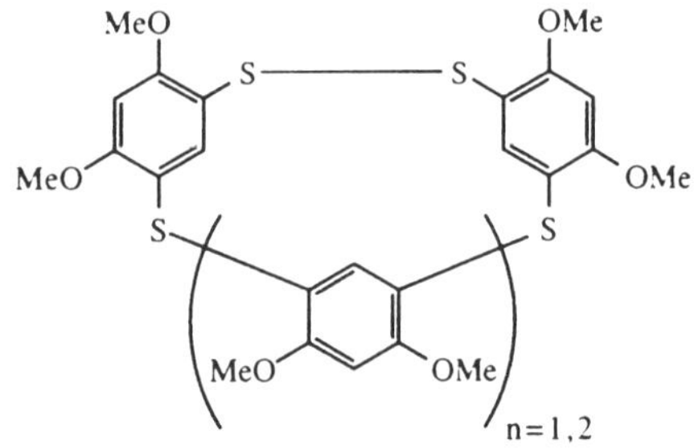

15-18

[via m-$KOC_6H_4O(CH_2)_nBr$]. The kinetics of the reaction was studied and the reactivity decreased rapidly with decreasing n values/increasing ring strain.[47]

Some odd-membered disulfide-bridged metacyclophanes derived from resorcinol dimethyl ether (m-dimethoxybenzene) were prepared. From ^{1}H-NMR data and suitable open-chain and cyclic models it was deduced that **15-18** ($n = 1$) and **15-19** assumed preferentially a saddle-shaped configuration.[48]

The 30-membered macrocyclic compound, **15-20**, was prepared and shown to have a basket-like cavity. This host, in perdeutero-acetic acid, effected a complete exchange of the H-atoms in the 4- and 6-position of resorcinol with deuterium in a few minutes, while phenol, catechol, and 2,6- and 2,7-naphthalenediols did not show H/D exchange even after several days. It was concluded that the host, **15-20**, was a guest regiospecific deuteration catalyst.[49] (Note that in this example resorcinol was the guest.)

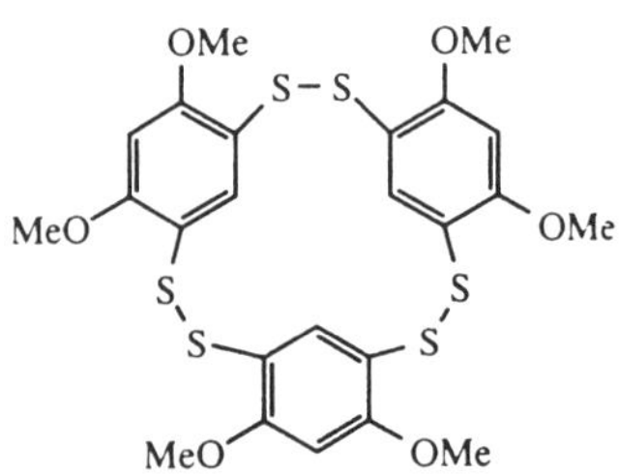

15-19

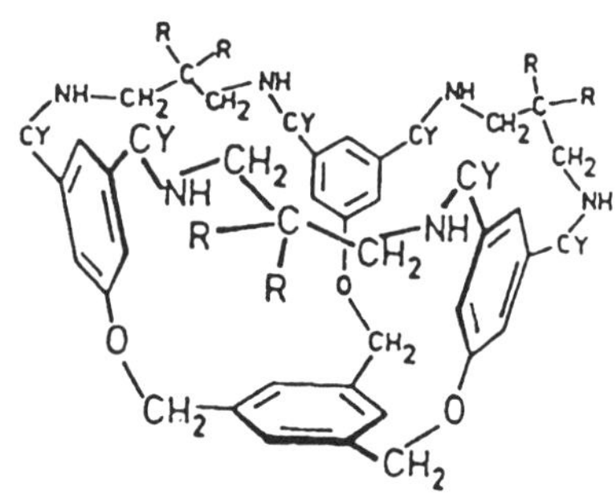

15-20
(where R = Me, Y = O)

The reaction of the terphenyl tetrol, **15-21**, with m-xylylene dichloride in the presence of a base gave the cupped oxacyclophane, **15-22**.[50]

A novel host, **15-23**, was prepared which was capable of binding resorcinol (or catechol) in solution.[51] (Resorcinol was the guest in the cleft of **15-23**.)

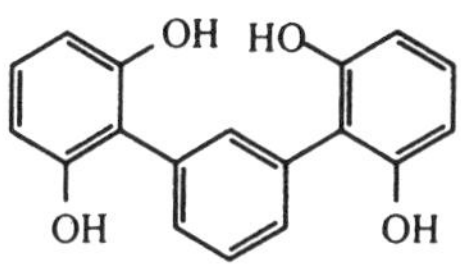

15-21

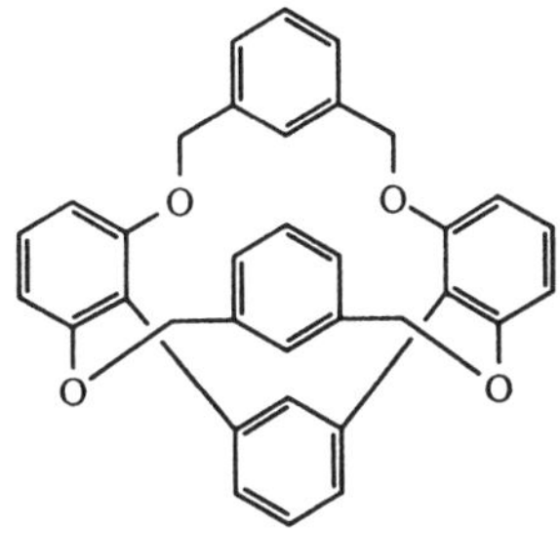

15-22

The dissociation constants of the inclusion complex of beta-cyclodextrin (the host) with resorcinol (and other phenolics) were determined by fluorescence intensity measurements.[52]

A macrocyclic hemiacetal cleavage catalyst, **15-24**, was synthesized and acted through H-bonding, and was effective in accelerating glycolaldehyde dimer dissociation and tetramethylglucose mutarotation.[53]

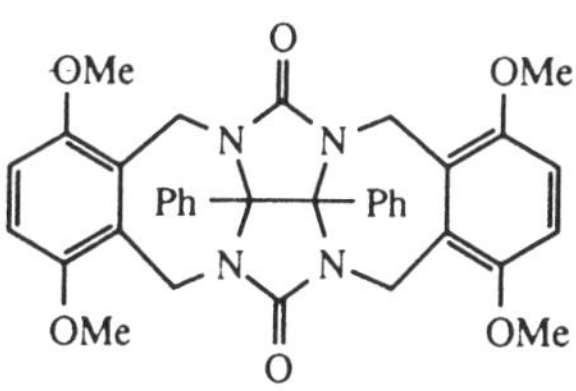

15-23

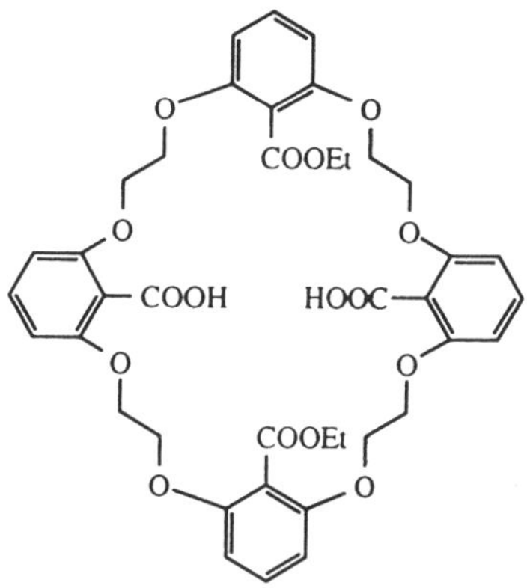

15-24

New bis-(*m*-phenylene)-32-crown-10 macrocycles, such as compound **15-25**, were synthesized.[54]

A new keystone for the construction of rigid molecular cavities, named tricornan, was devised. For instance, starting from resorcinol dimethyl ether the trioxatricornan, **15-26**, was made.[55]

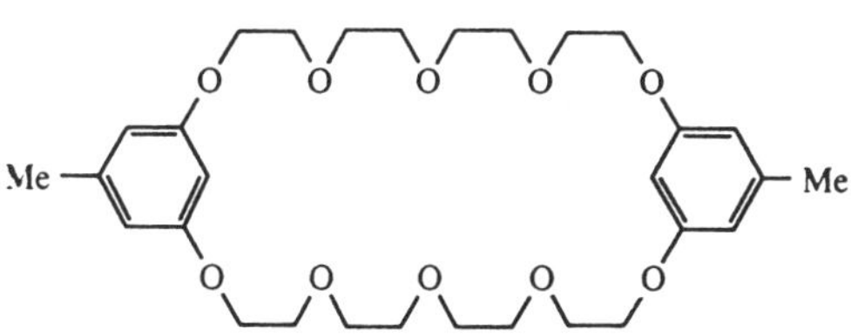

15-25

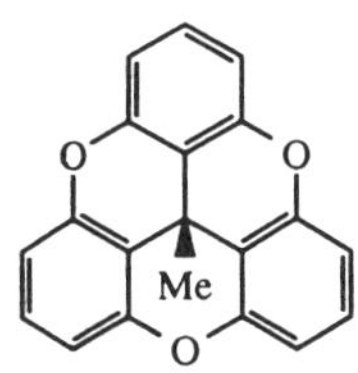

15-26

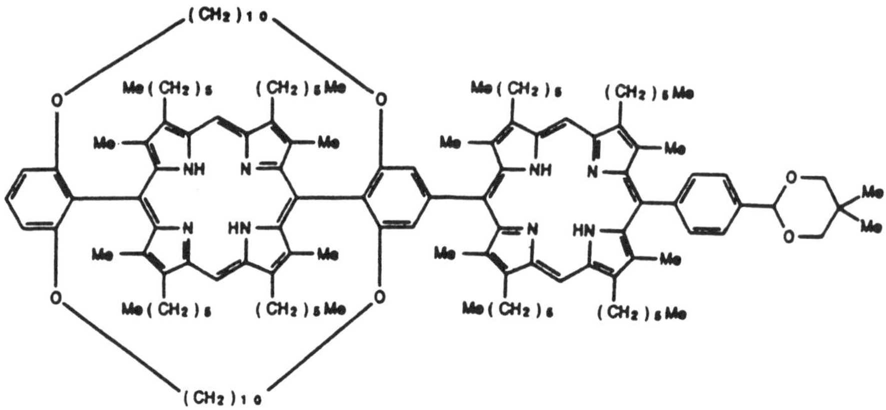

15-27

A doubly-strapped porphyrin, **15-27**, was synthesized which reacted with zinc acetate by insertion of zinc into the nonstrapped rings, only. This work was aimed at photosynthetic model compounds.[56]

15.2.3. Noncyclic Hosts

A few examples, involving resorcinol chemistry, are given below on the preparation of such materials.

The monomeric Schiff base, **15-28** Co-complex, was prepared and treated with D-phenylalanine under oxidative (air) conditions to give a complex, Δ-β_2-[Co[**15-28**](D-Phe)]. This complex was copolymerized with styrene and divinylbenzene. D-Phenylalanine was cleaved from the copolymer complex with 3 M HCl; the remaining copolymer complex was capable of complexing D-phenylalanine preferentially when exposed to an excess of D,L-phenylalanine.[57]

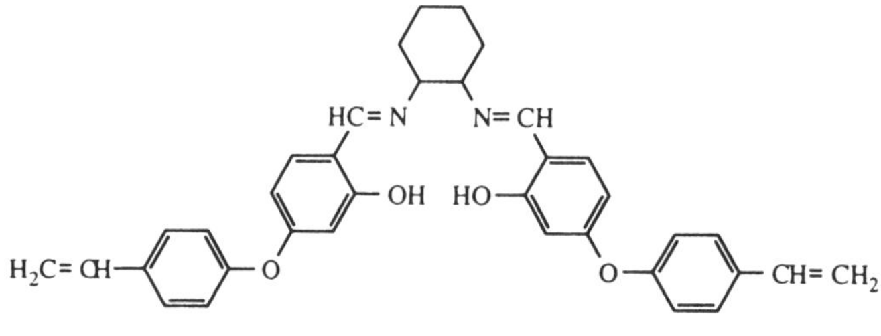

15-28

1,3-Cyclohexanedione (dihydroresorcinol), in benzene, was found to form two stereoisomeric, H-bonded supermolecules in the solid state. The nonsolvated, crystalline form consisted of chains of 1,3-cyclohexanedione. The second crystalline form was a cyclamer which consisted of six 1,3-cyclohexanedione moieties entrapping one benzene molecule. This was a novel host–guest species.[58]

The pyridinium compound, **15-29**, which has a bis-resorcinol moiety, was prepared. This compound formed a two-point adduct with the mono-anion of flavin mononucleotide in chloroform containing a small amount of methanol. This is an example of molecular recognition of nucleotides.[59]

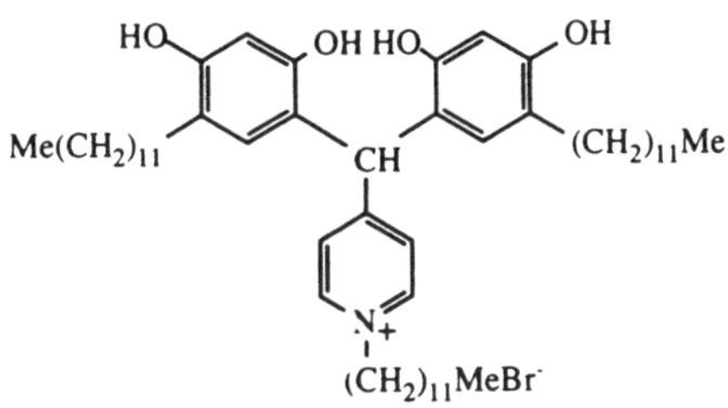

15-29

In Aoyama's laboratories the imidazole derivative, **15-30**, which has a bis-resorcinol moiety, was prepared. It catalyzed the hydrolysis of *p*-nitrophenyl acetate in 9/1 water/acetonitrile. This was said to involve transition-state stabilization via dynamic molecular recognition—a mechanism involving concerted acid–base bifunctional catalysis. The nucleophilic approach of the imidazolyl moiety on the ester substrate was assisted by the acidic hydroxyl groups of the bis-resorcinol component.[60]

The 3,5-bis-[(3,5-benzyloxy)benzyloxy]benzyl bromide, **15-31**, was prepared and reacted first with 3,5-bis-(benzyloxy)benzyl bromide and with 4-$(OHC_6H_4)_3CMe$ to give a dendritic starburst polyether of low polydispersity.[61] (Could such polymers "bind" other materials?)

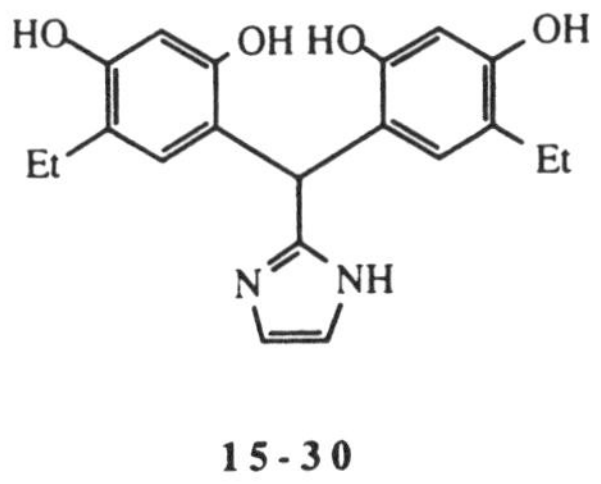

15-30

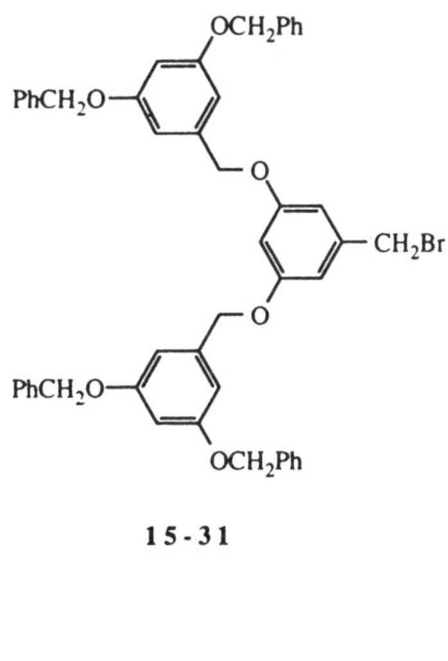

15-31

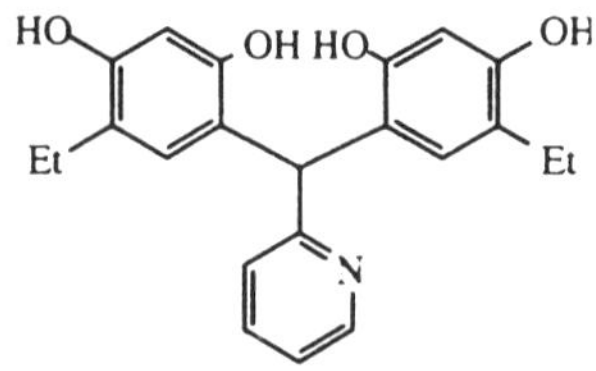

15-32

The pyridyl-bis-resorcinol compound, **15-32**, was prepared and formed a pyridinium phosphate salt with dodecyl phosphate. This salt was stabilized in hydroxylic media (H_2O–MeOH, MeOH, EtOH, *i*-PrOH) via hydrogen-bonding interaction of the bis-resorcinol segment with the bound phosphate anion. This work was said to have implications for the fixation of biological phosphates, such as nucleotides and nucleic acids.[62]

15.3. PHYSICAL AND ANALYTICAL CHEMISTRY

These topics have been grouped together here because they are interrelated and relate to much material in the earlier chapters as well. In addition, there is an increasingly important link to environmental chemistry.

15.3.1. Studies in Physical Chemistry

A detailed study of the distribution of phenolic solutes, including resorcinol, between water and butyl acetate or methyl isobytyl ketone (MIBK) was made. The distribution coefficients at room temperature were determined. The reason for the study was said to be that solvent extraction is an effective method for removing smelly and/or toxic pollutants from waste water.[63]

A conformational analysis of the benzenediols, including resorcinol, was undertaken using molecular calculations to determine structural parameters, stable conformations, potential barrier heights, dipole moments, and charges. Resorcinol had three stable isomers (pyrocetechol had only one, hydroquinone two).[64] Based on these assignments, literature data, and heats of formation (–274.89 kJ mol^{-1} for resorcinol), the ideal gas thermodynamic properties of the benzenediols were determined.[65]

The H-bonding of resorcinol to the proton acceptors, Et_2O, THF, Bu_2S, and tetrahydrothiophene in CCl_4 was investigated by IR spectroscopy (OH stretching frequency at *ca* 3μ) at various temperatures. The results indicated that the H-bonding of the first OH did not significantly alter the enthalpy of reaction of the second OH of resorcinol with the proton acceptor.[66]

The formation of molecular complexes of styphnic acid (2,4,6-trinitroresorcinol) with 1- and 2-substituted naphthalenes was studied and the free energies and enthalpies of complexation were determined. Charge transfer was a minor factor in the stability of these complexes.[67]

Hydroxyphenyltellurium(IV) halides were prepared from the benzenediols, including resorcinol, and were studied. The chlorides were prepared directly by reaction of the diols with $TeCl_4$ in chloroform; the bromides and iodides were prepared by halogen exchange with the chloride. Conductivity, cryoscopic, and ^{1}H-NMR spectroscopic data were produced and discussed.[68]

The heats of fusion of orcinol (5-methylresorcinol) and resorcinol were determined, and extrapolated to infinite dilution, in dibutyl and diisopropyl ether. The heat of solution of resorcinol in water was determined, also.[69]

The UV spectra of resorcinol in 10–96% sulfuric acid were taken; in 70–95% H_2SO_4 after 24 hr, an absorption band appeared at 233–241 nm (dependent on the sulfuric acid concentration) which was ascribed to sulfonated resorcinol. Protonated resorcinol had an absorption peak at 312 nm (independent of the sulfuric acid concentration). Protonated resorcinolsulfonic acid had an UV absorption peak at 322–324 nm. Resorcinol was disulfonated in 96% H_2SO_4 and the disulfonate was insoluble at that acid concentration.[70]

Some lanthanide metal ternary complexes, with EDTA as the primary and resorcinol as the secondary ligand, in 1/1/1 mole ratio (also with catechol or phloroglucinol in place of resorcinol) were prepared.[71]

The extraction of resorcinol from the quenched (and presumably neutralized) caustic fusion product of sodium *m*-benzenedisulfonate with dihexyl, dioctyl, diphenyl, and petroleum sulfoxide was studied and distribution coefficients between the aqueous and organic phases determined. These sulfoxides were efficient extractants for resorcinol.[72]

The kinetics of the base-catalyzed isotope exchange of ring-H in resorcinol was measured in water and methanol using NaOD. By ^{1}H NMR it was found that the H-2 is exchanged faster than H-4 or H-6, while H-5 was not exchanged; ketonic intermediates likely were involved.[73]

Resorcinol-2-^{13}C was prepared in 5 steps starting from 2-methoxybenzoic acid. In water, resorcinol-2-^{13}C reacted with Cl_2 and Br_2 to give $^{13}CHCl_3$ and $^{13}CHBr_3$, respectively, with enrichment proportional to the concentration of resorcinol-2-^{13}C. The proposed reaction mechanism involved 1,3-cyclohexanedione intermediates.[74]

Several 4-alkoxy-4′-(alkyl or alkoxy)-2-hydroxy-alpha,alpha′-dimethylbenzalazines, such as **15-33**, were prepared and their thermodynamic properties and mesophase ranges determined. The dialkoxy-substituted compound showed polymorphism in the solid state.[75]

The chelate-forming, polymerizable enamino-ketone, **15-34**, was prepared and copolymerized with styrene.[76]

A STO-3G molecular orbital calculation was made for resorcinol (also for hydroquinone), using fully optimized molecular structures for the various planar conformers. The results were compared with prior reports. The theoretical model also yielded information related to crystal structure determination.[77]

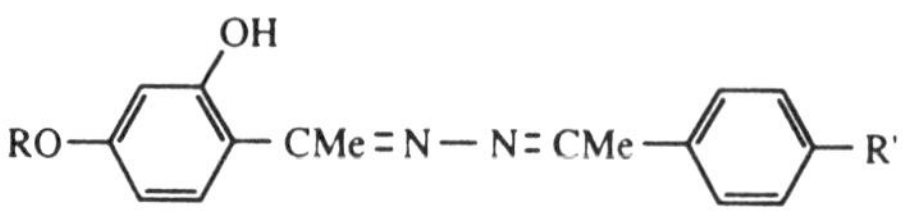

15-33
($R=C_1$-C_9 *n*-alkyl;
R'= Bu, RO or *n*-C_mH_{2m+1}),
where m=1,2,5,8)

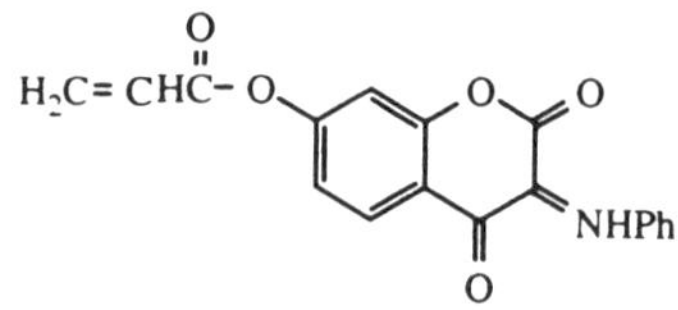

15-34

Binary single crystals of resorcinol and progesterone were oxidized by X-irradiation at 4.2 K to generate the semiquinone of resorcinol. The progesterone trapped electrons and was reduced.[78]

The heats of dilution at 298.15 K of aqueous catechol, resorcinol, or pyrogallol solutions were determined and compared. It was indicated that very different mechanisms were involved in the association of the different phenolics.[79]

A review of the photochemistry and photophysics of the furocoumarins (psoralens and angelicins) was made in the context of their photosensitization of biological substrates.[80]

The solubility of the isomeric benzenediols, including resorcinol, and also of the isomers of hydroxybenzoic acid, in supercritical CO_2 at 373 K and 200–400 bar was determined. The lower the melting point, the higher was the solubility of the isomer. The large differences in solubility between the *ortho*- and *para*-isomers suggested that supercritical extraction might be used to separate their mixtures.[81]

In relation to the mechanism of inhibition of the (metal-catalyzed) autoxidation of sulfite by the benzenediols and their sulfonates, the kinetics of the oxidation of the benzenediols by $\cdot SO_3^-$ radicals (the initial radical produced on one-electron oxidation of sulfite) and the peroxysulfate radical, $\cdot SO_5^-$ (from $\cdot SO_3^- + O_2$) were studied. Rate constants for the oxidation of the benzenediols by $\cdot SO_3^-$ were small in neutral solution and much larger in alkaline solutions. Resorcinol and phenol radicals were able to oxidize sulfite (SO_3^{2-}) in neutral solution.[82]

The steam distillation of 42 phenols, including resorcinol and *m*-aminophenol, was studied using UV spectroscopy. The distillates gave an 0.4% recovery of resorcinol and an 0.2% recovery of *m*-aminophenol (some decomposition occurred during the steam distillation).[83]

The alpha–beta crystal phase transitions in resorcinol were studied, using Raman spectroscopy and calorimetry. Optically clear crystals, stable at room temperature, of the alpha-modification of resorcinol were prepared by sublimation of resorcinol which had been repeatedly crystallized from acetone plus vacuum-sublimed six to eight times. The beta-form crystals were prepared in the same manner but maintained at 373 K (100 °C). The enthalpy and entropy change for the solid alpha–solid beta phase transition were 1.370 + 0.007 kJ mol^{-1} and 3.71 + 0.05 J K^{-1} mol^{-1}, respectively. The heat of fusion of beta-resorcinol at 382.8 K was found to be 20.89 + 0.07 kJ mol^{-1}, with an entropy change of 54.59 + 0.18 J K^{-1} mol^{-1}. The pressure-induced alpha–beta phase transition at room temperature was observed in a diamond anvil cell at 5 kbar.

m-$C_6H_4(OD)_2$ was prepared by repeated deuteration of resorcinol in D_2O and was the subject of similar studies.[84]

The ^{13}C-NMR spectra of several azo-derivatives of resorcinol (and also of phenol) in organic solvents and in acid, neutral, and alkaline aqueous solutions were studied. The resorcinol derivatives existed in the azo form in all solutions except in aqueous alkali, where they had a quinonoid structure with intramolecular H-bonding.[85]

The mono- and di-protonation of resorcinol and several derivatives in superacid was shown to give benzeneonium ions, **15-35**, and cyclohexyl dications, **15-36**.[86]

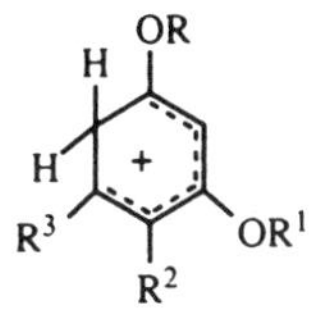

15-35

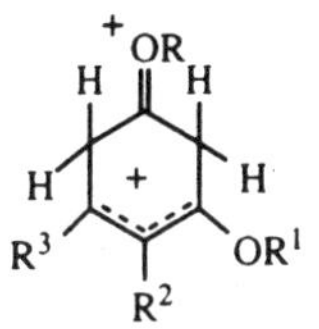

15-36

(where R, R^1=H, Me; R^2=H, Cl; R^3=H, Me, Ph, Cl)

It was established that the high-silica, synthetic zeolite H-ZSM-5 catalyzed the isomerization/equilibration of the three benzenediols. The reaction was so slow that equilibrium could not be attained in 2 hr/350 °C. The three-component equilibrium could not be determined because the resorcinol isomerized too slowly (at equilibrium the resorcinol content was estimated as "less than 70%"). A severe side reaction was the dehydroxylation to give phenol. Acid-catalyzed phenolic condensation products, such as diphenyl ether, dibenzofuran, and several hydroxydiphenyl ethers, were observed also.[87]

A study of the percents of the enol-form of 1,3-cyclohexanedione and 5,5-dimethyl-1,3-cyclohexanedione (dimedone) in 9 solvents was made. The tautomers were separated by low temperature (*ca* –60 °C) HPLC. The amount of enol form was low (5–30%) in CH_2Cl_2, $CHCl_3$, and CCl_4, and highest in ethanol and 1-propanol, which was explained in terms of solute–solvent and solute–solute hydrogen bonding.[88]

Several cyclic bis-(cyclopentadienyl) zirconium aromatic oxide derivatives, such as **15-37** from resorcinol, were prepared and characterized.[89]

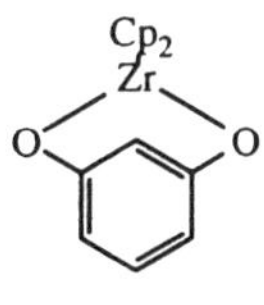

15-37
(where Cp=cyclopentadienyl)

The relative electronic energies of the diaminobenzenes, the difluorobenzenes, and the benzenediols, including resorcinol, were calculated using *ab initio* molecular orbital theory. The relative energies of catechol, resorcinol, and hydroquinone were 0.0, 0.9 and 2.4 kcal mol^{-1}. Thus, the *o*-isomer is the most stable.[90]

A theoretical study of the conformation and rotational barriers in the benzenediols, including resorcinol, was made using *ab initio* SCF (molecular orbital) calculations. The relative stabilities of the structural isomers, including the three resorcinol isomers, were determined and correlated with prior data from free-jet molecular beam spectra.[91]

The kinetics of the nucleophilic attack of sodium *m*-methoxyphenolate at *p*-nitrophenyl acetate was determined in the absence and presence of three different crown ethers (15-crown-5, 18-crown-6, and dicyclohexano-18-crown-6) in MeCN. The crown ethers complexed the Na^+ and retarded the acetyl transfer to the *m*-methoxyphenolate. The mechanistic interpretation was that the inhibitory effect of the crown ethers involved a less dispersed charge in the transition state than in the ground state.[92]

The complexing/rheological behavior of poly(vinylpyrrolidone) [PVP] with resorcinol in chloroform was studied. The contraction of the coils of PVP upon complexation was proportional to the product of the degree of polymerization of PVP and the concentration of resorcinol, and this was quantitatively related to the maximum possible number of intramolecular bridges.[93]

A precursor of the strong first-order structural phase transition in a resorcinol crystal was detected by fluorescence-lifetime measurement.[94]

The structure and pressure-induced phase transition of 1,3-cyclohexanedione (enol form) crystals as a function of temperature was studied by X-ray analysis.[95]

The protonation of several hydroxybenzenes, including resorcinol, in perchloric, sulfuric, and phosphoric acid were correlated to UV absorption peaks and H_0 (highest, 4.5).[96]

The solid–liquid equilibria for the catechol–resorcinol binary system were studied by differential scanning calorimetry (DSC) at 1 atm.[97]

The crystal structures of methanol-solvated 1,3-cyclohexanedione (enol form) and 2,2,2-trifluoroethanol-solvated lithium 1,3-cyclohexanedione enolate were studied. Conversion of the 1,3-cyclohexanedione to the alcohol-solvated Li-enolate caused the π-system to delocalize completely.[98]

An X-ray study of the structure of crystalline 4,6-diacetylresorcinol was made. The molecule was almost planar and intramolecular H-bonds with O were detected. The molecules were packed in layers, held together by van der Waals interactions.[99]

The rate constants and reaction orders for the electrochemical thiocyanation of several phenolics, including resorcinol, were determined with a rotating ring–disk electrode. The rate constants increased in the order catechol<pyrogallol<resorcinol<phloroglucinol.[100]

An NMR analysis of the intramolecular hydrogen bonding in alkylidene-bis-dimedone (5,5-dimethyl-1,3-cyclohexanedione) compounds, **15-38**, was reported and gave evidence for a fast tautomeric equilibrium and a slow rotation about the C–C bond linking the dimedone rings to the bridging carbon atom. Reference was made to

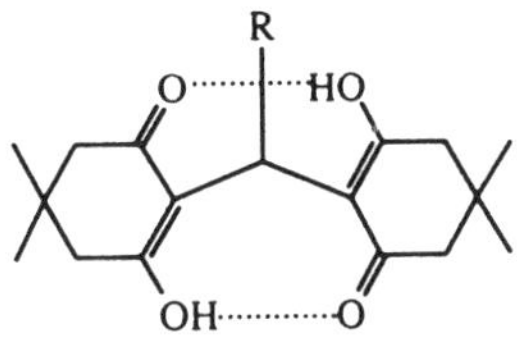

15-38
(wnere R= lower alkyl)

previous NMR, IR, and UV studies showing that the keto–enol equilibrium of the dimedone was solvent, concentration, and temperature dependent.[101]

The polymorphism and phase transitions of crystals of 1,3-cyclohexanedione were studied by X-ray crystallography; a new polymorph was found.[102]

The molar enthalpy of formation in the gas state at 298.15 K of 1,3-cyclohexanedione was –335.6+1.6 kJ mol^{-1}, calculated from the enthalpies of combustion and sublimation of the crystals. Theoretical studies were also made of the isomerization of 1,3-cyclohexanedione to its enol form and of other phenomena. Similar data were obtained for dimedone (and for 1,4-cyclohexanediones).[103]

15.3.2. Topics in Analytical Chemistry

These studies cover both the analysis of resorcinol (see also Chapter 2) and the use of resorcinol/derivatives in analysis. They are played out in a time of rapid progress and change in analytical methods/instrumentation. It is also a time of new, big impetus from the rise in environmental chemistry and increasing awareness of the nonchemist public of the environment and pollution. It is a time of explosive growth in biochemistry/biotechnology. One example of the use of resorcinol derivatives in this new environment is the use of fluorescent compounds, e.g., fluorescein, to tag or observe and analyze drugs and compounds of biochemical significance. The previous Chapter 8 (drugs) and Chapter 10 (dyes) have already made reference to the patent literature on fluorescent tags for biosignificant compounds; many additional applications are described here.

An early example is the use of 4-(2-pyridylazo)resorcinol (PAR), **15-39**, and 4-(2-thiazolylazo)resorcinol (TAR), **15-40**, in the complexometric/calorimetric analysis of metal ions which was of intense interest in the 1960s and 1970s. With the advent of highly sensitive AA (atomic absorption) and ICC (inductively coupled plasma) spectroscopy/instrumentation for many of the metals, the older techniques paled somewhat but are still important where the newer instrumentation/techniques are not available or applicable. Therefore, work on complexometric methods of analysis for metal ions continues to this day.

In this subsection then, snapshots of analytical topics are given to show the sweep of interests, with a few references to the pre-1980 literature and emphasis on the later literature.

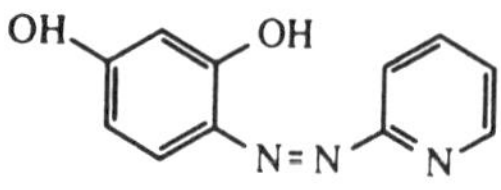

15-39

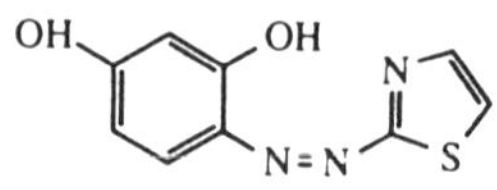

15-40

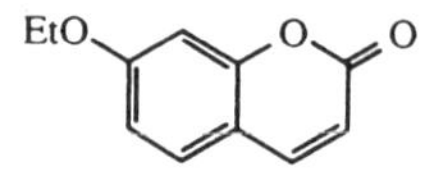

15-41

A review of the literature up to December 1967 on the use of PAR, **15-39**, as a versatile analytical reagent for the determination of metal ions was published.[104] The use of TAR, **15-40**, in the analysis of metal ions was reviewed about the same time.[105]

7-Ethoxycoumarin, **15-41**, was used as a substrate for a rapid, sensitive, fluorometric method to determine microsomal monooxygenase activity. In this assay, O-dealkylation occurs and the highly fluorescent 7-hydroxycoumarin is formed.[106]

An NMR spectroscopic method was developed for the determination of the approximate reagent ratios in phenol/resorcinol/formaldehyde resins.[107]

Analytical methods were published for resorcinol/formaldehyde resins: monomers, dimers, and trimers were determined by GLC and HPLC; average chain length by GPC (gel permeation chromatography) and by ^{1}H-NMR spectroscopy; the R/F ratio was verifiable by ^{1}H-NMR analysis as well; structural information, relative amounts of free resorcinol, and the relative degree of condensation were assayed by ^{13}C-NMR spectra.[108]

Phenol and resorcinol in P/R/F resins could be determined by catalytic thermometric titrimetry.[109] 4-(2-Thiazolyl)resorcinol (TAR), **15-40**, was used in a rapid and very sensitive assay of free fatty acids in biological fluids.[110] A sensitive fluorometric method for the determination of nitrate with resorcinol was developed; it was capable of assaying as low as 5–70 ng of nitrate nitrogen.[111]

The Zn-catalyzed autoxidation or H_2O_2 oxidation of resorcinol (forming colored products) as an assay for the spectrophotometric determination of Zn concentration was optimized. The method was useful to the 0.9–8 ppm Zn level.[112]

A bromination method for the determination of phenol in the presence of resorcinol with excess bromine water was examined. Phenol gave the cycloheadienone, **15-42**, which precipitated and was filtered off and analyzed for active Br. Resorcinol gave compound **15-43** under these conditions and this product did not precipitate. The method was suitable for determining 8 mg or more of phenol in the presence of not more than 25 mg resorcinol.[113]

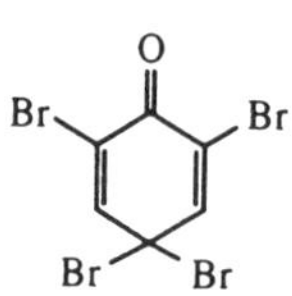

15-42

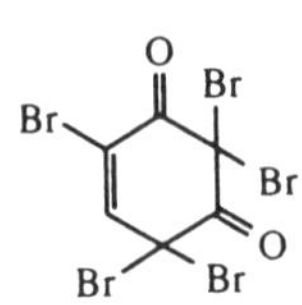

15-43

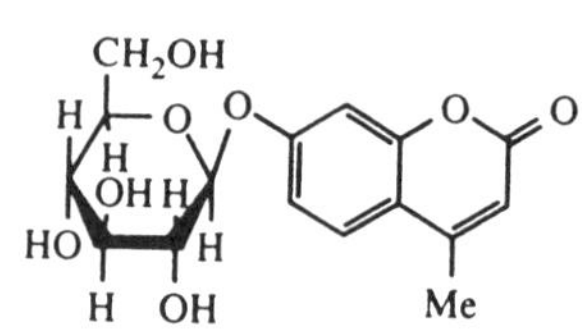

15-44

4-(2-Pyridylazo)resorcinol (PAR) was used as a sensitive reagent for the amperometric, ultramicro determination of Pr^{3+} and Nd^{3+}.[114]

An improved, automated periodate-resorcinol method for the determination of bound sialic acids (amino sugars widely occurring in animals) was developed. As little as a <0.2 mL sample was needed and 70 samples/hr could be processed.[115]

An improved method for the detection of *Escheria coli* bacteria in food and water was developed, using 4-methylumbelliferone-beta-D-glucuronide, **15-44**, as a fluorogenic substrate for the bacterial beta-glucuronide.[116]

A colorimetric method for the identification of nitrate and nitrite (e.g., in fertilizer solutions, soils and wheat seedlings) with phenolic reagents, e.g, with resorcinol, was developed. The sample solutions were mixed with phenolic-H_2SO_4 solution and a yellow color formed if nitrite was present. To determine nitrate, the sample solutions were first treated with urea and H_2SO_4 to destroy nitrite, then with Zn powder and phenolic-H_2SO_4 solution—if nitrate was present, a yellow color formed. The detection limit for nitrate and nitrite was 0.75 μg with resorcinol as the phenolic reagent.[117]

A large number of esters of 7-hydroxycoumarin derivatives, such as **15-45**, were synthesized. These compounds were fluorescent substrates for the direct and continuous kinetic assay of carboxylesterases, phosphatases, and sulfatases.[118]

A rapid, UV-spectrophotometric method for the determination of nitrate in mineral waters with resorcinol was developed. The assay had good reproducibility and low interference from common inorganic ions.[119]

The 7-methoxycoumarincarbonyl azides, **15-46** and **15-47**, were synthesized and used as highly fluorescent labeling agents for alcohols in fluorescence HPLC.[120]

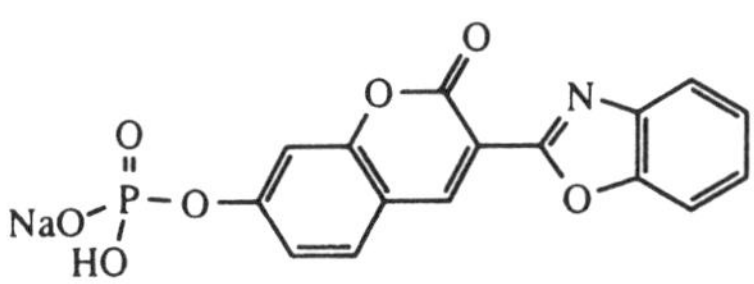

15-45

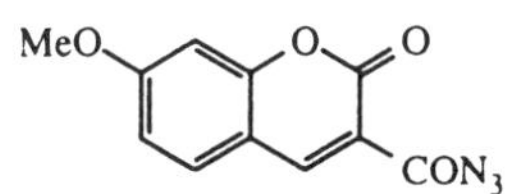

15-46

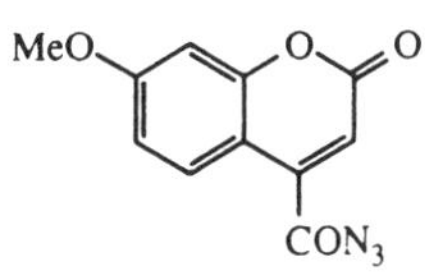

15-47

An extractive, spectrophotometric method was developed for analyzing iron in titanium base alloys, using 4-(2-pyridylazo)resorcinol (PAR) and a long-chain quaternary ammonium salt to extract the iron complex anion into $CHCl_3$.[121]

A photometric method for assaying resorcinol, based on the treatment of the sample with excess *p*-sulfophenyldiazonium salt, was developed and patented.[122]

The kinetics of the ozonation of resorcinol in water was studied (related to effluent treatment).[123] An extractive, spectrophotometric procedure was worked out for determining molybdenum in steels and nickel-base high-temperature alloys as a ternary complex with hydroxylamine and 4-(2-pyridylazo)resorcinol (PAR).[124]

A rapid, spectrophotometric method for assaying nitrate in chrome plating or zinc plating solutions involved first treating the solutions with hydrazine sulfate to reduce the chromic acid, then adding an H_2SO_4 solution of resorcinol for reaction with the nitrate.[125]

A selective, spectrophotometric method for analyzing tantalum(V) in ores was developed, involving extraction of Ta(V) from a hydrochloric acid solution with *N*-phenylhydroxamic acid into toluene, then adding a solution of 4-(2-pyridylazo)resorcinol (PAR) in DMF/pyridine to develop a colored metal complex.[126]

Titanium in weakly acid solution was analyzed by adsorption voltammetry with 4-(2-pyridylazo)resorcinol (PAR), using a stationary Hg electrode.[127]

The spectrophotometric analysis of Pb(II) in aqueous solution was studied using preconcentration of lead by sorption on silica and later complexing with 4-(2-pyridylazo)resorcinol (PAR). The lower detection limit for Pb(II) was 0.01 μg/mL.[128]

A sensitive and selective method for Pd(II) involved the formation of a red complex with 4-(2-thiazolylazo)resorcinol (TAR) and extraction of this complex from water into chloroform assisted by tetraphenylphosphonium chloride.[129]

The complexes of Fe(II and III), Co(II and III), and nickel(II) with dinitrosoresorcinol were synthesized and characterized (potential use in colorimetric analysis).[130]

A chiral dihydrofluorescein-type dye, **15-48**, was used to measure the optical purity of chiral compounds (drugs) by HPLC.[131] (There was little detail in this reference.)

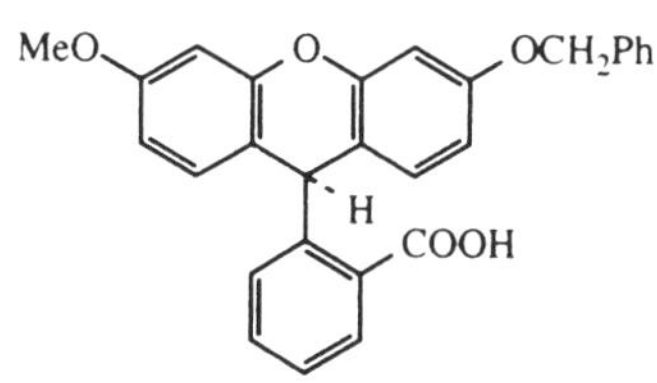

15-48

Oxine (8-hydroxyquinoline)/resorcinol/ formaldehyde resins (also similar hydroquinone resins) were prepared and used as column packings applicable to the preconcentration of copper from solution, followed by stripping with 0.1 M HCl and atomic absorption spectrometry of the eluate; copper could be determined at the ng level in this fashion.[132]

A spectrophotometric method was developed for the separation and determination of trace quantities of V(V), Co(II), and Cr(III) by reverse-phase HPLC, using precolumn chelate formation with 4-(2-thiazolylazo)resorcinol (TAR).[133] A similar method was used for analyzing Fe, Co, and Ni in ores, anode mud, and recovered catalyst solutions.[134]

A highly sensitive, stripping voltammetry analysis of uranium was reported, based on an adsorptive accumulation of its 4-(2-pyridylazo)resorcinol (PAR) complex at the static mercury drop electrode. The method was applicable to U analysis in fresh and sea water.[135]

A spectrophotometric method for the analysis of resorcinol in the presence of phenol was developed, based on the difference of absorption coefficient ratios at multiwavelengths.[136]

A kinetic spectrophotometric procedure was developed for the determination of Cr(III) with 4-(2-pyridylazo)resorcinol (PAR) which was suitable for the determination of Cr(III) in the presence of Cr(IV) and other impurities, and which could be used for the analysis of waste water.[137]

Recently (1991), the first use was reported of chelates of Os, Ru, and Rh with 4-(2-thiazolyazo)resorcinol (TAR) in the reverse-phase HPLC separation and determination of these metals.[138]

Fe, Co, and Ni were determined by reverse-phase HPLC following ion-pair (counter-cation) extraction of the metal-4-(2-pyridylazo)resorcinol (PAR) complexes into chloroform. The detection limits were 0.50 ppb for Co and 1.50 ppb for Fe and Ni.[139]

Additional kinetic studies of the ozonation of resorcinol (and phloroglucinol) in water were reported.[140]

A new, rapid and sensitive spectrophotometric method for the determination of traces of Ce(IV) as its yellow complex with resorcylaldehyde guanylhydrazone was reported.[141]

An extraction-spectrophotometric analysis of trace amounts of cadmium was developed in which Cd(II) was first complexed with *N*-hydroxy-*N,N′*-diphenylbenzamidine, extracted with chloroform, then contacted with 4-(2-pyridylazo)resorcinol to give a reddish-orange, ternary complex. The lower detection limit in this method was 0.05 $\mu g\ cm^{-3}$ Cd, useful for the analysis of environmental samples.[142]

A very good separation of resorcinol and hydroquinone in thin-layer chromatography (TLC) was achieved by using silica gel layers impregnated with $CuSO_4 \cdot 5H_2O$ (2 wt%). The isomeric aminophenols also separated well on this medium.[143]

A new complex-forming azo dye, 3-(2′,4′-dihydroxyphenylazo-1′)-1,2,4-triazole (TRIAR), was prepared; it formed water-soluble (organic solvent-insoluble), colored chelates with Co(II), Cu(II), Pb(II), Ni(II), Fe(II), Fe(III), and Bi(III), suitable for spectrophotometric analysis. A similar dye, pyrocatechinoazotriazole (TRIAP), complexed in addition Hg(I), Hg(II), and Cd(II).[144]

The thin-layer chromatography (TLC) of catechol, resorcinol, and hydroquinone, with aqueous beta-cyclodextrin solution as the mobile phase and a polyamide as the stationary phase, was developed.[145]

The phase transfer catalyzed reaction/extraction/removal of resorcinol, phenol, and *m*-cresol from alkaline, aqueous streams by contact with benzoyl chloride in toluene in the presence of hexadecyltributyl phosphonium bromide was developed. In a batch agitated reactor, >99.5% conversion of the phenolics was attained.[146]

15.4. 1,3-CYCLOHEXANEDIONE IN ORGANIC SYNTHESIS

The chemistry of 1,3-cyclohexanedione (1,3-CHD; dihydroresorcinol) involves many challenges, e.g., O vs C alkylation. Many ways to overcome such problems become apparent in the following review.

The versatility of 1,3-CHD as a starting material for aliphatic, five- and six-membered cycloaliphatic, aromatic, and heterocyclic compounds is so great that a sizable book could and should be written on the subject, but that has not happened so far. Reviews have appeared in Russian on the synthesis of heterocyclic compounds from 1,3-CHD, covering this literature to the mid-1980s.[147]

Some actual and potential commercial uses of 1,3-CHD have been described in previous chapters, e.g., under processes for making resorcinol (Chapter 3), agricultural chemicals (Chapter 7), pharmaceuticals (Chapter 8), and aroma chemicals (Chapter 12). These can be accessed through the index.

In this section, the topic is addressed by presenting a few selections from the literature and a few patent references from 1980 onward, in chronological order. Just one example of the wealth of chemistry in the older literature is this: Buechi developed routes to the perfume components, jasmone, **15-49**, and methyl jasmonate, **15-50**, from 1,3-CHD in several steps in 1971.[148]

Here, then, are the selections from the newer literature. 1,3-CHD mesylates, **15-51**, were prepared in high yield at room temperature by the reaction of 1,3-CHD in CH_2Cl_2 with mesyl chloride in the presence of K_2CO_3. The 1,3-CHD mesylate reacted with a variety of nucleophiles at room temperature to give 77–93% yields of the mesylate displacement products, **15-52**. To produce the halo-substitution products, benzyltrialkylammonium halides were the reagents and BF_3-etherate was used to accelerate the reaction. The products were useful intermediates for synthesis.[149]

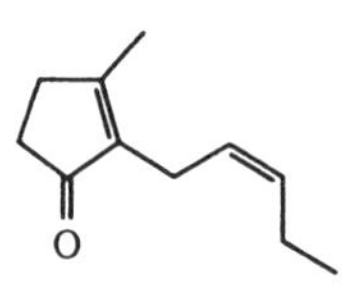

15-49

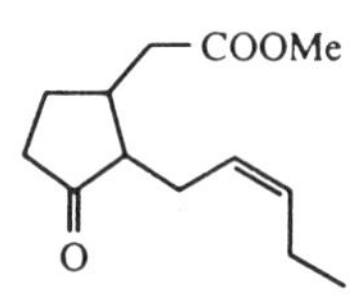

15-50

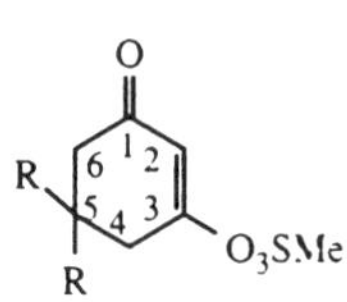

15-51
(where R=H or Me)

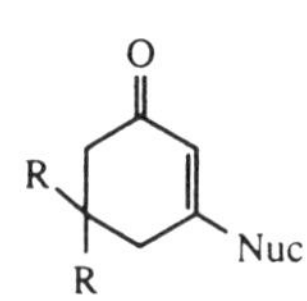

15-52
(where R=H or Me;
Nuc=Cl, Br, I, OEt, NMe_2, SCH_2Ph)

The Michael addition of 2,3,5-trimethylbenzoquinone to 1,3-CHD in the presence of a base (piperidine or sodium methoxide) gave a 52% yield of the tetrahydrobenzofuran, **15-53**. This product was used further to make, e.g., the carbazole derivative, **15-54**, by reacting the methyl ether of **15-53** with phenylhydrazine (the Fischer indole synthesis).[150]

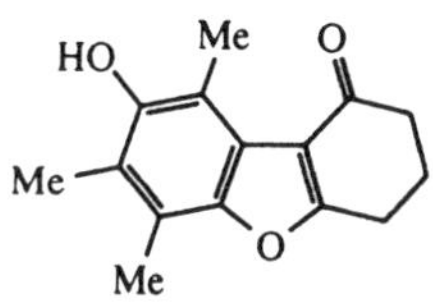

15-53

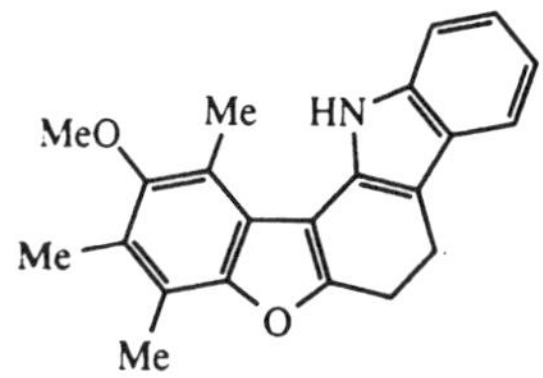

15-54

The reaction of the condensation product, **15-55**, of 5,5-dimethyl-1,3-CHD with glyoxylic acid, OHCCOOH, with ammonia in ethanol provided access to the acridine derivative, **15-56**.[151]

The condensation of 1,3-CHD with (*R*)-citronellal, catalyzed by NaOMe, followed by acidification and isolation by column chromatography, gave compound **15-57**, which has the tetrahydrocannabinol skeleton, in 33%yield.[152]

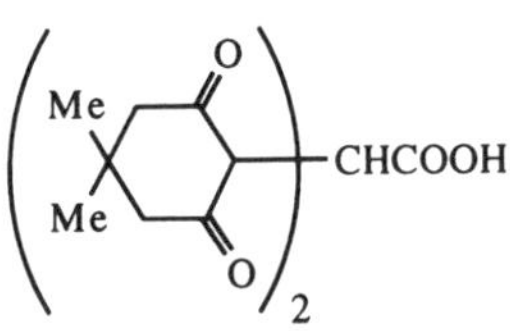

15-55

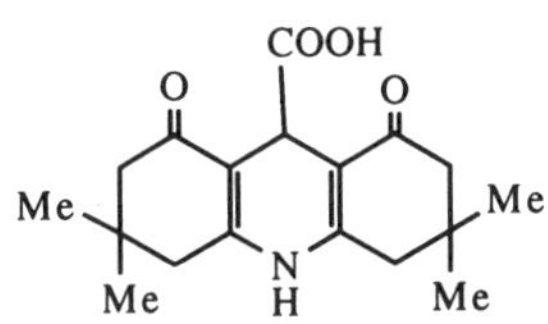

15-56

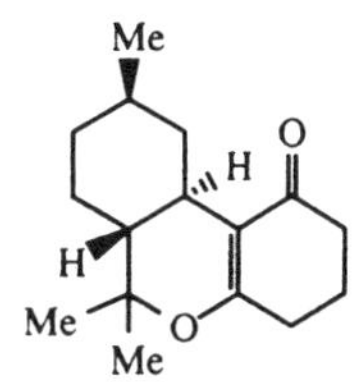

15-57

The reaction of D-glycero-L-manno-2-amino-2-deoxyheptose with 5,5-dimethyl-1,3-CHD gave the indolone, **15-58**. Several related compounds were prepared in a similar manner.[153]

The condensation of 2-nitro-5-chlorobenzaldehyde with 1,3-CHD in HCl/AcOH produced the *N*-hydroxy-dihydroacridinedione, **15-59**. A series of similar compounds were prepared and many of these had strong coccidiostatic and malaricidal activity in vivo.[154]

The condensation of 1,3-CHD with phenylhydrazone, followed by cyclization, gave 1,2,3,4-tetrahydro-4-oxo-carbazole which was aromatized using a Raney-Ni catalyst to produce 4-hydroxycarbazole, **15-60**.[155]

The spiro-ketone, **15-61**, was made from 3-ethoxy-2-cyclohexen-1-one, as shown in Eq. (15.1).[156]

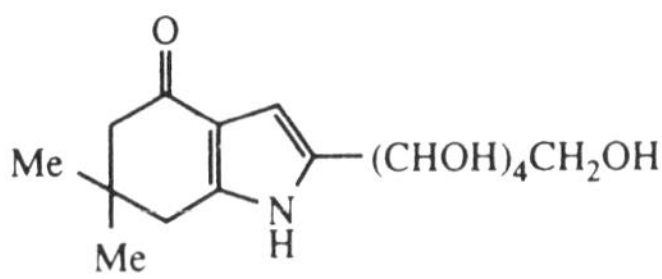

15-58

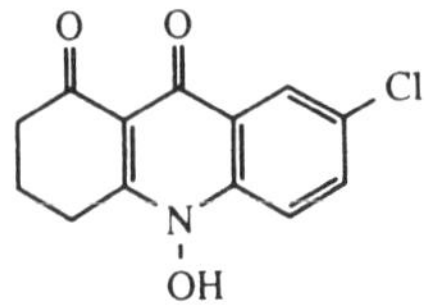

15-59

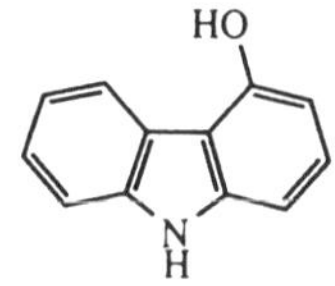

15-60

(15.1) 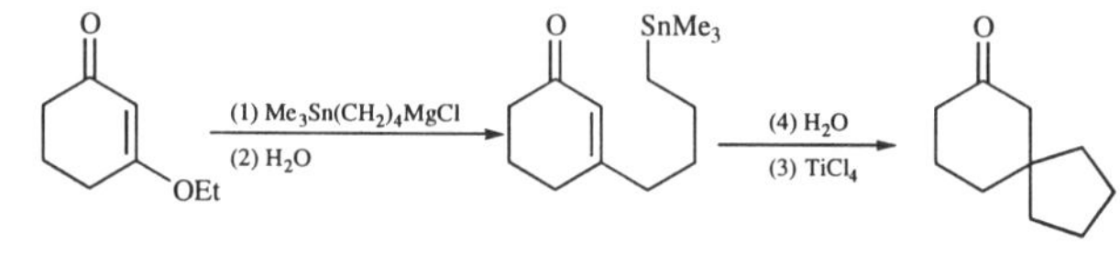

15-61

The UV irradiation of the 3-alkenyloxy-2-cyclohexen-1-one, **15-62**, produced the oxatricycloundecanone, **15-63**, by intramolecular [2+2] photocycloaddition.[157]

The condensation of 5,5-dimethyl-1,3-CHD with primary or secondary (di)alkylamines, catalyzed with BF_3-etherate, gave high yields of 3-(di)alkylamino-2-cyclohexen-1-ones. For instance, compound **15-64** was obtained by the reaction of 5,5-dimethyl-1,3-CHD with *n*-butylamine in 96% yield. Such compounds were said to have potential uses in synthetic/medicinal chemistry.[158]

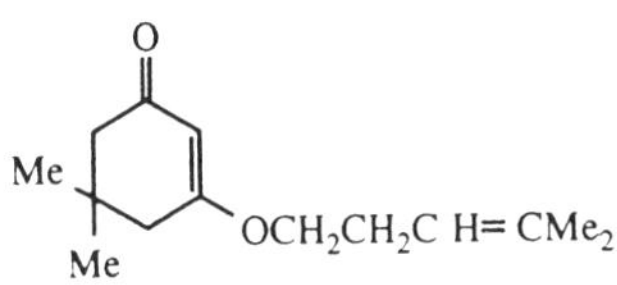

15-62

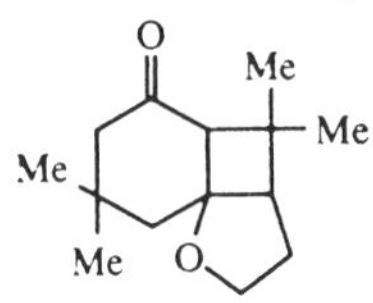

15-63

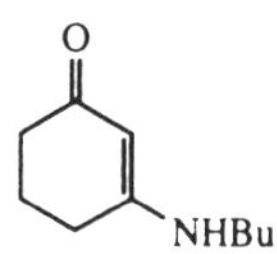

15-64

The disulfide, **15-65**, was obtained by treating 5,5-dimethyl-1,3-CHD with sulfur in amines. On heating above its melting point, the disulfide was converted to the corresponding sulfide; this sulfide, on refluxing in Ac_2O, cyclized to compound **15-66** and this product reacted with aniline to give compound **15-67**.[159]

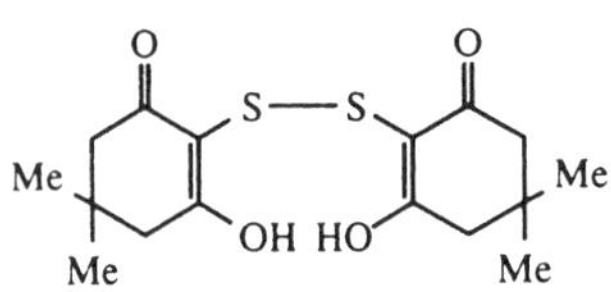

15-65

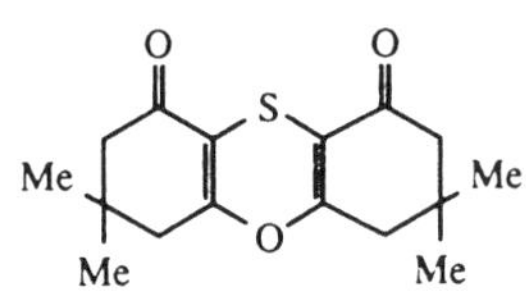

15-66

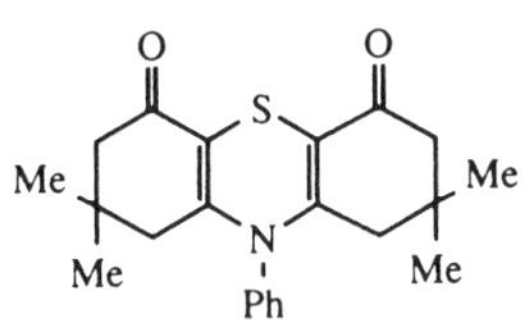

15-67

Treatment of 2-allyl-2-hydroxyalkyl-1,3-CHD, **15-68**, with NaH in THF gave an intramolecular rearrangement to produce the macrocyclic 5-ketolactones, **15-69**, in 11–56% yield (byproducts were dimeric, cyclic 5-ketolactones).[160]

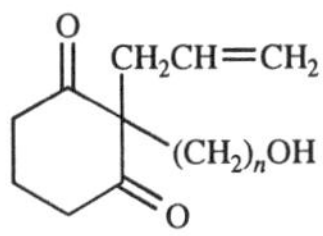

15 - 68

(where n=4, 6, 9, 20)

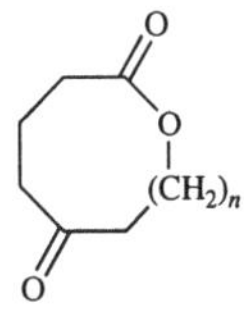

15 - 69

(where n=4, 6, 9, 20)

The reaction of 1,3-CHD with triethyl orthoformate and urea in DMF/AcOH at 90 °C gave 2-ureidomethylene-1,3-CHD, **15-70**, which was condensed with methyl cyanoacetate/*tert*-BuOK in DMF to obtain the 5-oxo-5,6,7,8-tetrahydrocoumarin, **15-71**, in good overall yield. Several similar compounds were prepared by this route.[161]

Ring-opening was observed when 2-chloro-2-methyl-1,3-CHD was treated with NaOAc in MeOH and $ClCH(Me)CO(CH_2)_3COOMe$ was obtained. 2-Allyl- and 2-benzyl-2-chloro-1,3-CHD gave the corresponding chloro-substituted esters.[162]

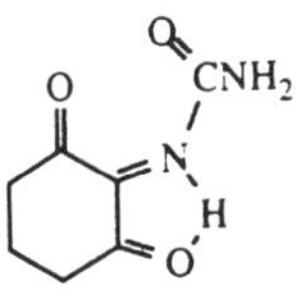

15 - 70

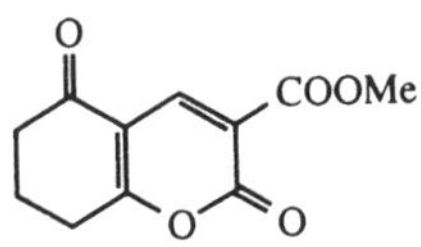

15 - 71

5,5-Dimethyl-1,3-CHD was first silylated with hexamethyldisilazane/imidazole at 126 °C, then treated with MeLi at 25 °C to give 3,5,5-trimethyl-2-cyclohexen-1-one, **15-72**. Several other 3-alkyl-2-cyclohexenones were prepared in a similar manner.[163]

3-Bromoquinoline-N-oxide reacted with 1,3-CHD in DMF/Ac_2O to give the 2-substituted quinoline, **15-73**. Pyridine-N-oxide reacted with 1,3-CHD to give 3-(2,6-dioxocyclohexylidene)-2-(4-pyridyl)cyclohexanone. Several similar sequences were reported.[164]

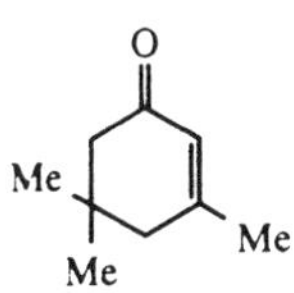

15 - 72

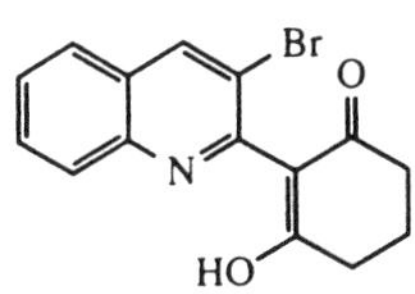

15 - 73

The 1,3-CHD derivative, **15-74**, underwent acid-catalyzed cyclization to produce the benzofuran derivative, **15-75**; the latter was reacted with aniline to produce the 4-oxo-4,5,6,7-tetrahydroindole derivative, **15-76**.[165]

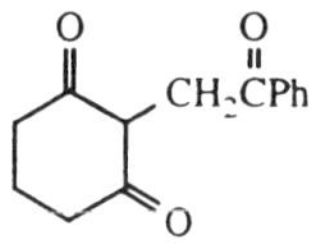

15-74

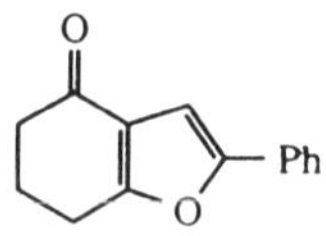

15-75

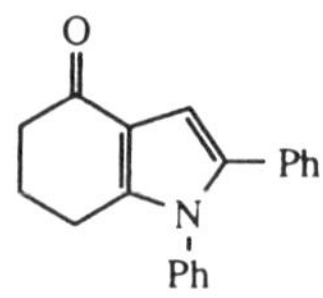

15-76

The dye-sensitized, fluoride-promoted photooxidation of 5,5-dimethyl-1,3-CHD gave the 5,5-dimethyl-1,2,3-cyclohexanetrione, **15-77** (shown in the enol form), in 72% yield.[166]

1,3-CHD was reacted with tosyl azide to give 2-diazo-1,3-CHD. This compound was reduced with $NABH_4$ to give the diazo compound, **15-78**. Similar reactions with 5,5-dimethyl-1,3-CHD and the photochemical, catalytic, and thermal decomposition of these materials were studied also.[167]

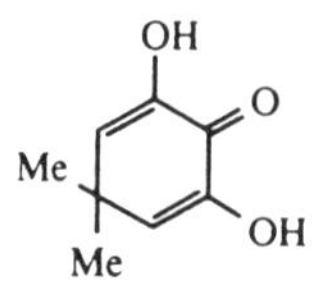

15-77

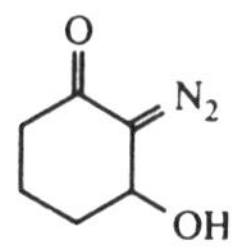

15-78

Treatment (deprotonation) of the dimethyl ether of methyl orsenillate, **15-79**, with lithium diisopropylamide (LDA) in THF and condensation with 3-ethoxy-5,5-dimethyl-2-cyclohexen-1-one at –78 °C produced the strongly fluorescent, tricyclic ketone, **15-80**, in 64% yield. (This was part of a study on approaches to a synthesis of the antitumor aureolic acid antibiotics.)[168]

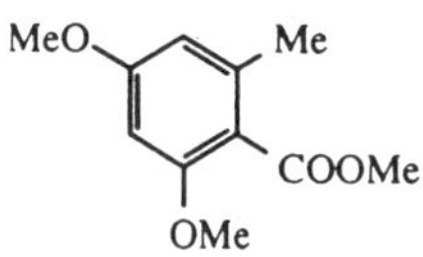

15-79

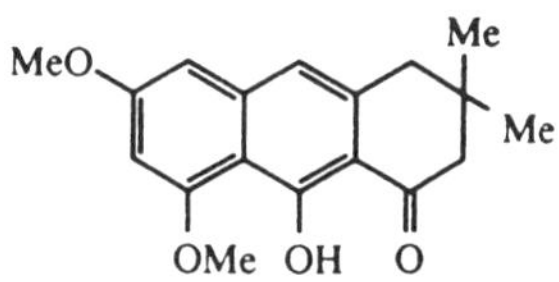

15-80

An examination of the Michael condensation of 5,5-dimethyl-1,3-CHD with aldehydes in the presence of amines was made. For example, in CH_2Cl_2 the reaction of the dione with 2-methylpropanal and piperidine produced compound **15-81** in 95% yield; its reactions were studied. (This was part of an extensive study of the synthesis of plant growth regulators found in *Eucalyptus grandis*.)[169]

The reaction of 1,3-CHD with the ethyl ester of glycine gave an enamine which added phenylisothiocyanate to give compound **15-82**.[170]

o-Mercaptoanilines were condensed with 1,3-CHD(s) to give good yields of phenothiazine derivatives, such as **15-83**.[171]

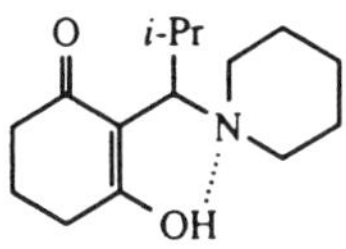

15-81

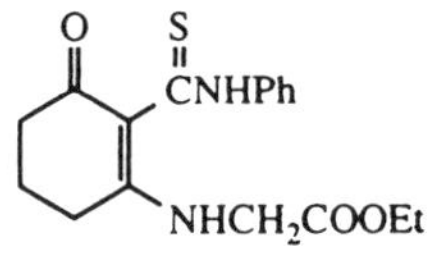

15-82

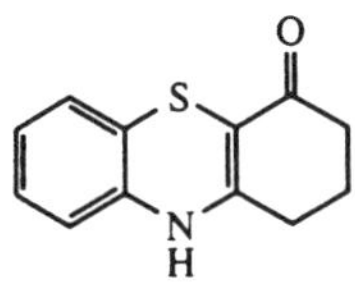

15-83

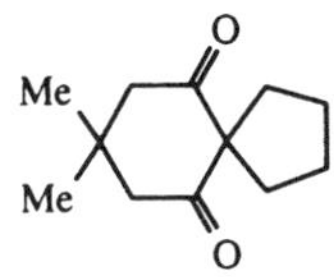

15-84

The alkylation of 5,5-dimethyl-1,3-CHD with 1,4-dibromobutane in Me_2SO in the presence of excess K_2CO_3 gave the spiro compound, **15-84**, in 60% yield. Similar reactions with 1,2-dibromoethane and 1,3-dibromopropane gave O-alkylation products.[172]

The *p*-toluenesulfonic acid catalyzed condensation of 1,3-CHD with 2-aminobenzamide gave the enamine, **15-85**; fusion of this enamine at 250–260 °C/0.5 hr gave the 3,4-dihydroquinazoline derivative, **15-86**, in 94% yield.[173]

1,3-CHD was reacted with ethanolamine in diglyme for 30 min/20 °C and the resultant enamine was cyclized by refluxing for 3 hr in the presence of dichlorotris(triphenylphosphine)ruthenium(II) to give a 75% yield of 4-oxo-4,5,6,7-tetrahydroindole, **15-87**.[174]

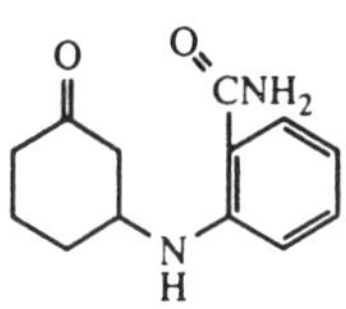

15-85

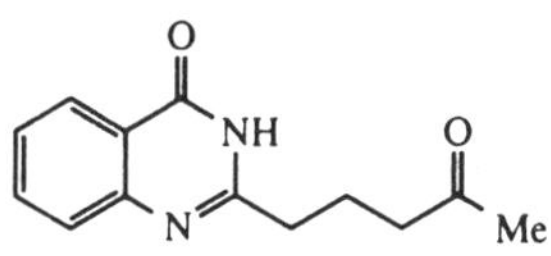

15-86

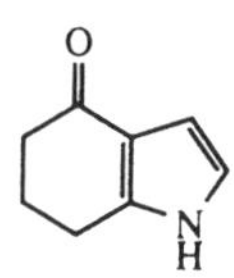

15-87

The base-catalyzed reaction of 1,3-CHD with *p*-benzoquinone mono-methyl acetal gave the adduct, **15-88**. The $ZnCl_2$-catalyzed reaction of 1,3-CHD with *p*-benzoquinone bis-(dimethyl acetal) gave compound **15-89**.[175]

The reaction of 2,2-dibromo-1,3-CHD with styrene in the presence of copper powder activated by iodine gave the dihydrofuran derivative, **15-90**, in a regioselective synthesis.[176]

The *p*-toluenesulfonic acid catalyzed reaction of 5,5-dimethyl-1,3-CHD with 4-phenylcyclohexanone gave an octahydrodibenzofuranone, which was dehydrogenated in *p*-cymene at reflux using a Pd/C catalyst to obtain the dibenzofuran derivative, **15-91**.[177]

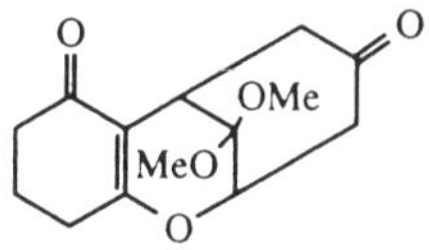

15-88

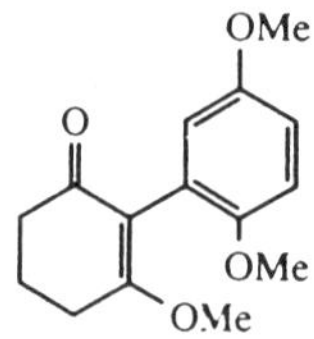

15-89

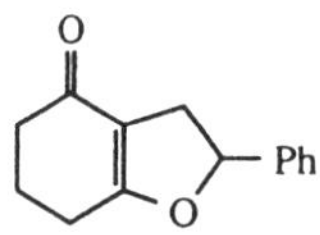

15-90

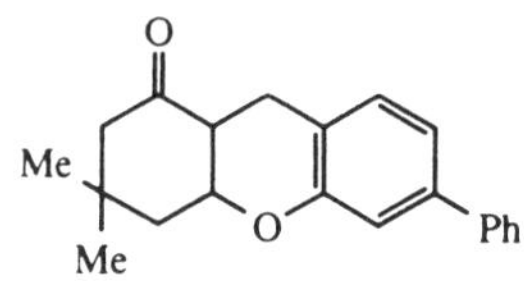

15-91

The condensation of 5,5-dimethyl-1,3-CHD with benzaldehyde gave the octahydroxanthenedione derivative, **15-92**, in good yield.[178]

A convenient synthesis of indazoles was reported. For instance, the condensation of 2-acetyl-1,3-CHD with hydrazine produced the tetrahydroindazole, **15-93**. Its keto group was reduced to the alcohol with sodium borohydride and the product aromatized with *p*-toluenesulfonic acid/10% Pd/C to obtain the indazole, **15-94**.[179]

The reaction of 5,5-dimethyl-3-amino-2-cyclohexen-1-one with ethyl acrylate in THF in the presence of NaH at reflux gave the N-alkylation product, but the same reaction conducted in diglyme gave a 49% yield of the C-alkylation product, **15-95**.[180]

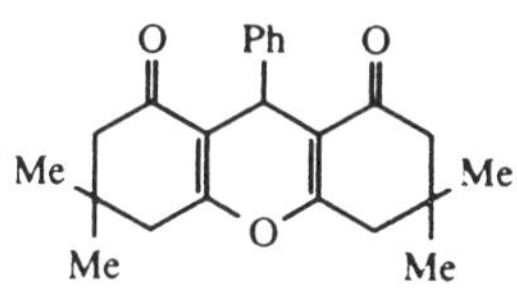

15-92

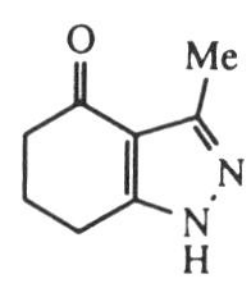

15-93

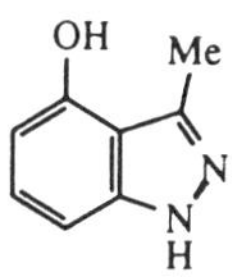

15-94

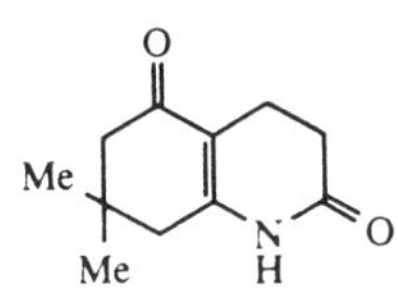

15-95

Treating various 2,2-disubstituted 1,3-CHDs, such as 2-propyl-2-methyl-1,3-CHD, with actively fermenting baker's yeast produced good yields of monoreduced, chiral ketols—this was an example of asymmetric microbial reduction.[181]

A regioselective synthesis of 5-hydroxychroman-4-ones was developed. As an example, 1,3-CHD in CH_2Cl_2 was first reacted with Ti(IV) chloride and then with

$Me_2C{=}CHCOCl$ to produce the dioxo-tetrahydrochroman, **15-96**; this was dehydrogenated with 10% Pd/C catalyst in refluxing diethylene glycol dimethyl ether to obtain the 5-hydroxy-2,2-dimethyl-4-chromanone **15-97**.[182]

A convenient synthesis of 4-hydroxybenzo(*b*)furan (karanjol), **15-98**, was developed, starting with the cyclocondensation of 1,3-CHD with $BrCH_2COCOOH$ to obtain a tetrahydrobenzofuran derivative which was aromatized/decarboxylated.[183]

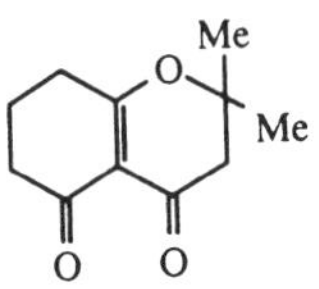

15-96

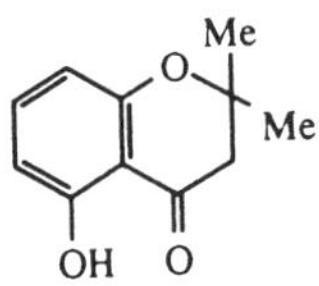

15-97

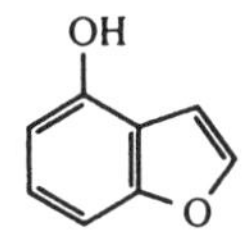

15-98

Treatment of the 2,2-disubstituted 5,5-dimethyl-1,3-CHD, **15-99**, with ammonium acetate in refluxing AcOH gave a 96% yield of the 1-acyl-1*H*-pyrrole, **15-100**, via a rearrangement involving transannular interaction.[184]

The benzothiazolone derivative, **15-101**, was obtained in 87% yield by treating thiourea first with chlorine in water, then with 5,5-dimethyl-1,3-CHD, and heating to 90–100 °C.[185]

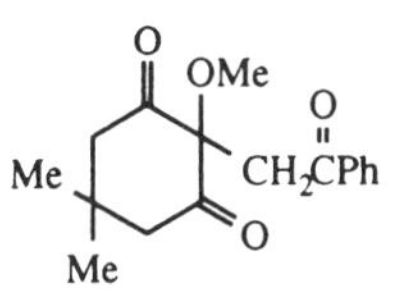

15-99

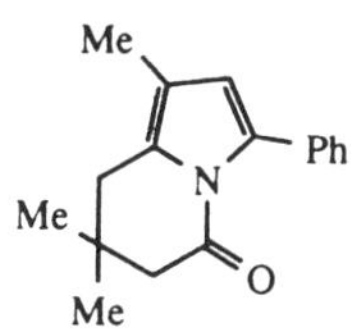

15-100

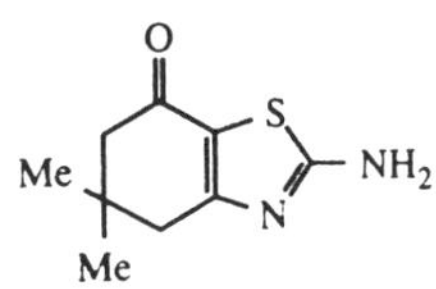

15-101

The 1,3-CHD derivative, **15-102**, was aromatized by treatment with $CuCl_2{\cdot}2H_2O$ and $MgCl_2{\cdot}6H_2O$ in acetonitrile at reflux to give a 76% yield of methyl 2,4-dihydroxy-3,6-dimethylbenzoate, **15-103**.[186]

The enamine obtained by reacting 5,5-dimethyl-1,3-CHD with *o*-(benzoylamino)aniline was cyclized in polyphosphoric acid to give the diazepinone derivative, **15-104**. This compound had moderate analgesic properties in mice at the 50 mg/kg dose level.[187]

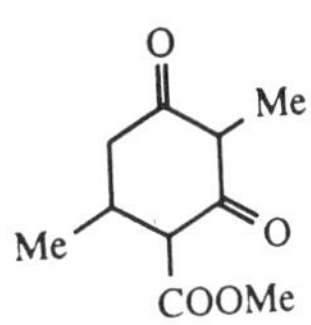

15-102

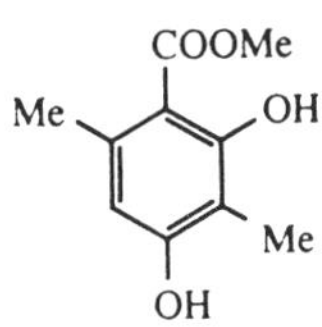

15-103

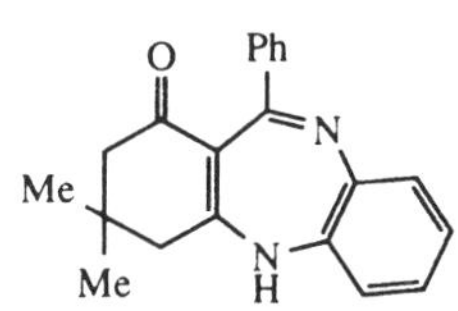

15-104

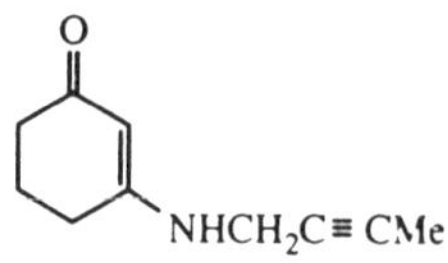

15-105

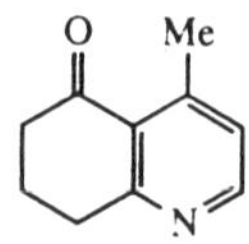

15-106

The enamine, **15-105**, was thermally cyclized to produce the tetrahydroquinolone, **15-106**.[188]

The condensation of the sodium salt of 1,3-CHD in DMF with isatoic anhydride led to the hexahydroacridine-1,9-dione, **15-107**. Similar compounds were prepared from substituted CHDs and isatoic anhydrides.[189]

A simple access to 1,2,3,4-tetrahydrophenanthrenones was found. For instance, 2-(1-phenacetyl)-1,3-CHD was chlorinated with oxalyl chloride to give the intermediate, **15-108**, which was cyclized in the presence of $AlCl_3$ to obtain compound **15-109**. Starting from 2-(1-naphthylacetyl)-1,3-CHD in a similar reaction sequence, 1,2,3,4-tetrahydrochrysenone derivatives were obtained.[190]

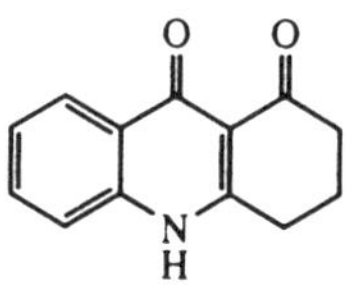

15-107

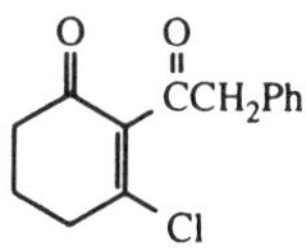

15-108

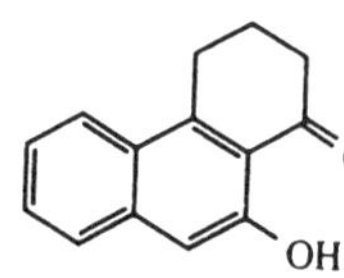

15-109

When 1,3-CHD was reacted in DMF with excess Vilsmeyer reagent ($DMF/POCl_3$) for 24 hr/20 °C, then worked up by neutralization with Na_2CO_3 in CH_2Cl_2/H_2O, evaporation of the organic layer, followed by chromatography on alumina of the residue redissolved in petroleum ether, a 24% yield of the 1,4-cyclohexadiene-1,5-dicarboxaldehyde derivative, **15-110**, was obtained. This product is a nonaromatic tautomer of a simple benzene derivative.[191]

5,5-Dimethyl-1,3-CHD, as its enol acetate, was reacted with $TeCl_4$ in AcOH at 100 °C to give a 60% yield of the phenoxatellurin, **15-111**. This product was dechlorinated by Na_2SO_4 to give compound **15-112** in 92% yield; the reactions of this end product were studied also.[192]

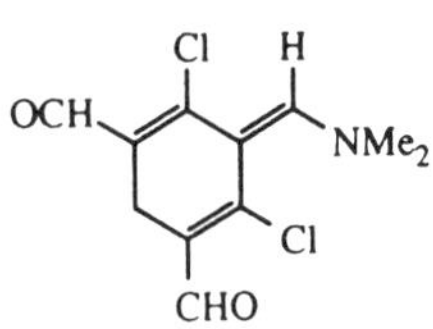

15-110

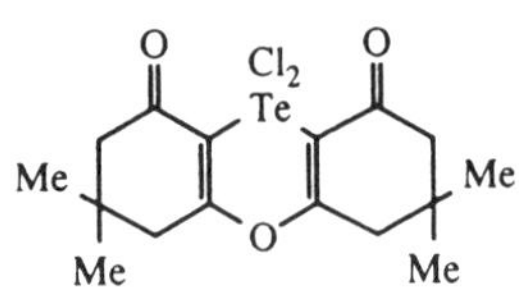

15-111

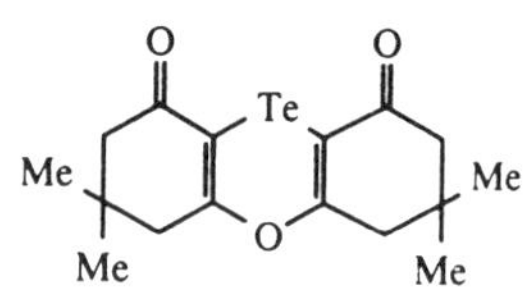

15-112

The D-β-phenylalanine-catalyzed asymmetric aldol condensation of the 2,2-disubstituted 1,3-CHD, **15-113**, with perchloric acid in acetonitrile gave the optically active octahydronaphthalenedione, **15-114** (a Wieland–Miescher ketone analog), in 86% yield.[193]

The intramolecular cyclocondensation of 2-chloroacetyl-1,3-CHD with sodium acetate in acetone gave a 99% yield of the hexahydrobenzo{*b*}furan-dione, **15-115**.[194]

The thermolysis of 2-diazo-1,3-CHD formed a carbene which was captured by PhIO or by PhI in the presence of $Rh(OAc)_2$ catalyst to obtain the iodonium ylide, **15-116**. Several similar compounds were prepared, e.g., from 5,5-dimethyl-2-diazo-1,3-CHD and $2\text{-}MeC_6H_4IO$.[195]

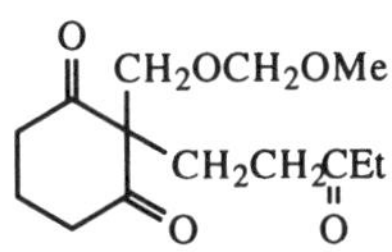

15-113

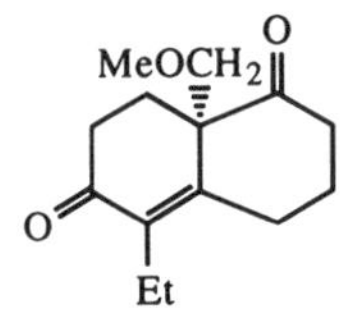

15-114

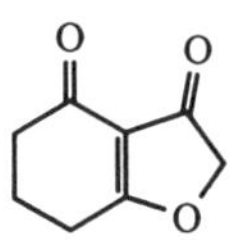

15-115

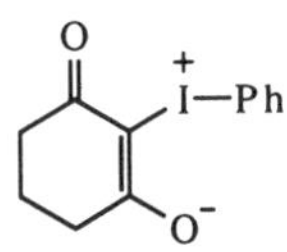

15-116

The condensation of 1,3-CHD with methyl propargyl carbonate, $HC{=}CH_2OCOOMe$, in THF at 80 °C/4hr in the presence of $Pd_2(dba)_3CHCl_3$ and 1,2-bis-(diphenylphosphino)ethane in catalytic amounts gave the tetrahydrofuranone, **15-117**, in 39% yield.[196]

The *tert*-BuOK-catalyzed addition of methyl vinyl ketone to 5,5-dimethyl-1,3-CHD in THF gave an 80% yield of 5,5-dimethyl-2,2-di-(2-acetylethyl)-1,3-CHD, which cyclized when treated with tolenesulfonic acid to the acetyl-spiroundecenedione, **15-118**; further treatment with the sulfonic acid produced the tricyclo[7.3.1.0]tridecane derivative, **15-119**.[197]

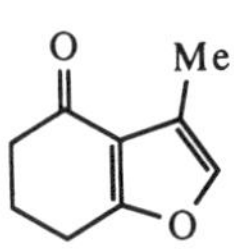

15-117

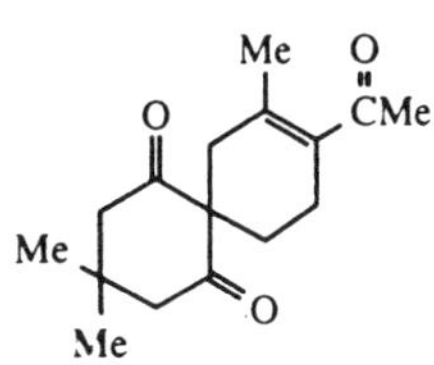

15-118

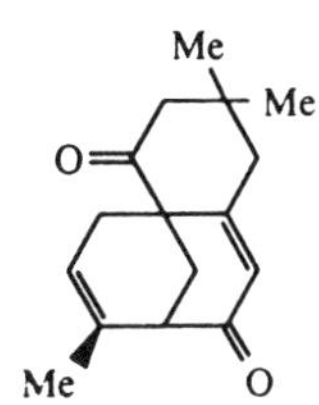

15-119

The direct 4-alkylation of 1,3-CHDs was achieved via a dianionic species. For instance, when 1,3-CHD was treated with 2.2 equiv lithium diisopropylamide (LDA) and MeBr in hexamethylphosphoramide/THF, 4-methyl-1,3-CHD, **15-120**, was obtained in 41% yield. Several variations of this reaction were demonstrated.[198]

A novel synthesis of hexahydroquinolines and hexahydroacridines was reported. For instance, the condensation of *N*-methylimidazole-2-carboxaldehyde with 1,3-CHD and ethyl acetoacetate in EtOH with NH_4OAc gave the hexahydroquinolone carboxylate, **15-121**, in 78% yield. The reaction of *N*-methylimidazole-2-carboxaldehyde with two equivalents of 1,3-CHD in the presence of NH_4OAc gave the acridine-dione derivative, **15-122**, in good yield. Variations of the starting materials and combinations were explored.[199]

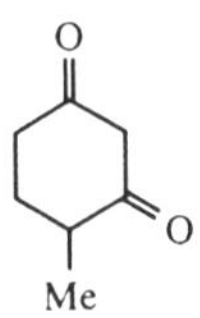

15-120

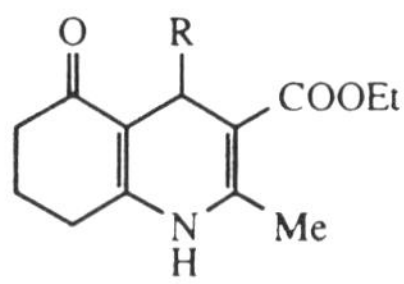

15-121
(where R= *N*-methyl-2-imidazolyl)

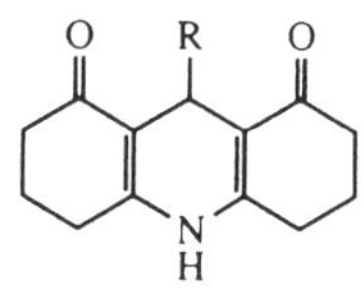

15-122
(where R= *N*-methyl-2-imidazolyl)

The photochemical (Hg lamp irradiation) alkylation of 1,3-CHD with alkyl vinyl ethers in benzene or benzene/ethanol gave 40–73% yields of the 2-(2-alkoxyethyl or $CH_2CHMeOMe$ or 3-tetrahydropyranyl)-1,3-CHD.[200]

The direct monofluorination of 5,5-dimethyl-1,3-CHD was achieved by reacting its enoltrimethylsilylether with 5% F_2 in N_2 in $CFCl_3$ as the solvent; a 42% yield of 2-fluoro-5,5-dimethyl-1,3-CHD, **15-123**, was obtained.[201]

The addition of 1,3-CHD to 5,6,7-trimethoxy-3,4-dihydroisoquinoline at 90 °C/0.5 hr gave a 77% yield of compound **15-124**.[202]

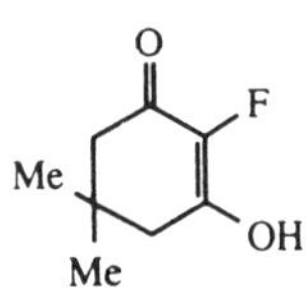

15-123

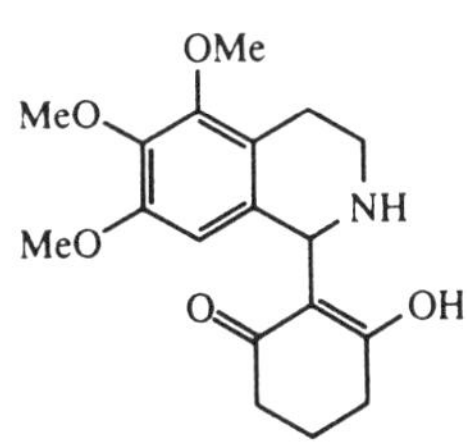

15-124

2-[1-(1-Alkylamino)alkylidene]-1,3-CHDs, e.g., compound **15-125**, were synthesized as photosynthetic electron transport (PET) inhibitors. These compounds resembled the potent new cyanoacrylate PET inhibitors.[203] (See also Chapter 7 for many similar compounds.)

A one-pot synthesis of pyran derivatives was reported. For instance, the reaction of 1,3-CHD with 2-methylene-propane-1,3-diol in THF in the presence of $Pd(OAc)_2/PPh_3$ and DBU (1,8-diazabicyclo[5.4.0]undec-7ene) in catalytic amounts gave the interesting pyran derivative, **15-126**, in 52% in-hand yield.[204]

A study of the condensation of alpha,beta-unsaturated aldehydes with 5,5-dimethyl-1,3-CHD was made. For example, with *E*-PhCH=CHCHO the methylene-bis-1,3-CHD product, **15-127**, was obtained; CH_2=CHCHO (acrolein) as the aldehyde component gave the pyran, **15-128**, by intramolecular cycloaddition.[205]

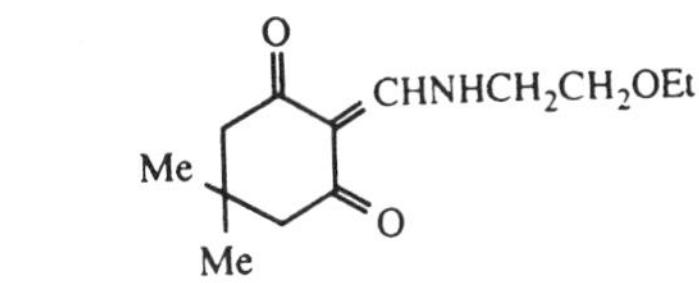

15-125

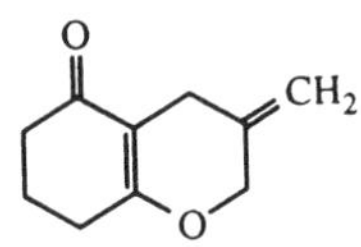

15-126

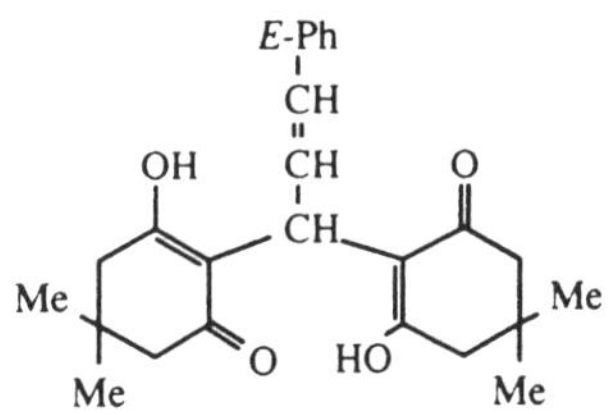

15-127

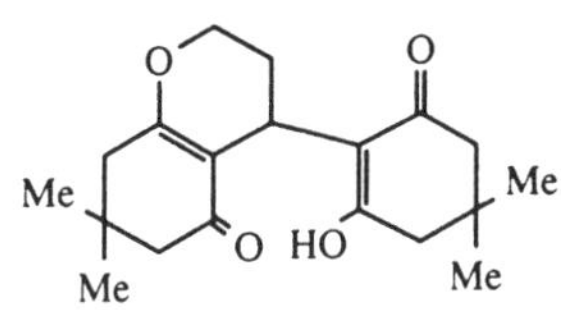

15-128

1,3-CHD reacted with the sulfur transfer reagent, dithio-bis-phthalimide, in the presence of pyridine in DMF at 90 °C/3 hr to give a 64% yield of 2,2′-thio-bis-1,3-CHD.[206]

In a study of a safe and efficient procedure for the diazo-transfer reaction in neutral medium, 5,5-dimethyl-1,3-CHD was treated with NaN_3 and 1-ethyl-2-chloropyridinium tetrafluoroborate to give the 2-diazo-5,5-dimethyl-1,3-CHD, **15-129**, in 79% yield.[207]

Heating the esters of 1,3-CHD, **15-130**, at 150–190 °C gave 66–91% yields of the 2-acyl-1,3-CHDs, **15-131**; these were kairomonal compounds of Lepidoptera.[208]

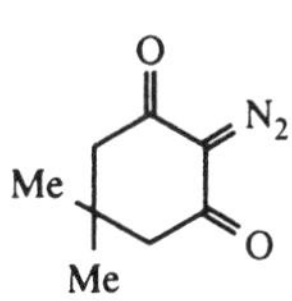

15-129

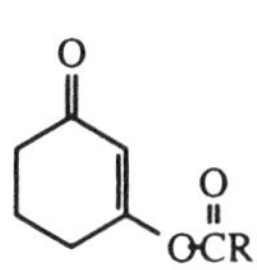

15-130

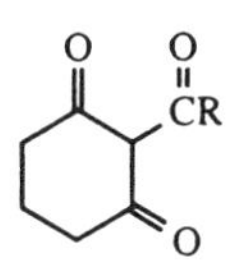

15-131

(where RCO= palmitoyl, oleyl, linolyl)

The alkylation of 1,3-CHD with 2-(chloromethyl)-3,5-dioxa-1-hexene in the presence of $PdCl_2/PPh_3$ catalyst and DBU gave compound **15-132** in 45% yield. This is an example of the catalytic acetonylation of cyclic 1,3-diketones.[209]

A gentle method for the preparation of ylides from diazo compounds was developed, involving the use of bis-(hexafluoroacetylacetonato)copper(II)[$Cu(hfa)_2$] as the catalyst. For example, the reaction of 2-diazo-5,5-dimethyl-1,3-CHD with Ph_2Te in toluene in the presence of $Cu(hfa)_2$ catalyst at reflux for 1 hr gave a 23% yield of the telluronium ylide, **15-133**.[210]

The electrochemical [3+2] cycloaddition of 1,3-diketones and olefins was studied. For instance, 1,3-CHD was reacted with styrene in such a manner, using MeCN as the solvent and quaternary ammonium salts as the supporting electrolyte, to give the cycloadduct, **15-134**, in 85% yield.[211]

A series of pyridinyl-substituted 1,3-CHDs were prepared and their herbicidal activity was studied. Significant activity against weed seedlings was exhibited by 2-(2-pyridinyl)-1,3-CHDs. As an example, 2-(2-pyridinyl)-5,5-dimethyl-1,3-CHD, **15-135**, had high pre- and post-emergence activity, especially against broadleaf weeds.[212]

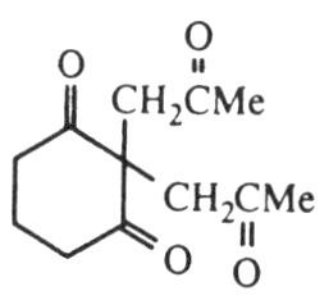

15-132

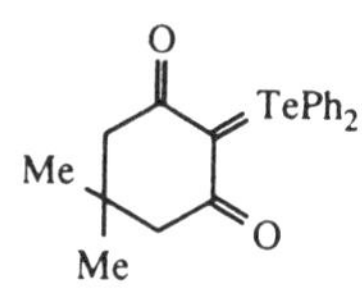

15-133

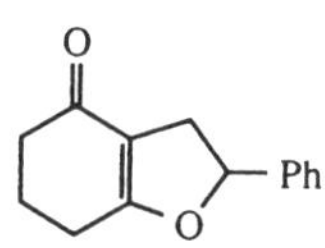

15-134

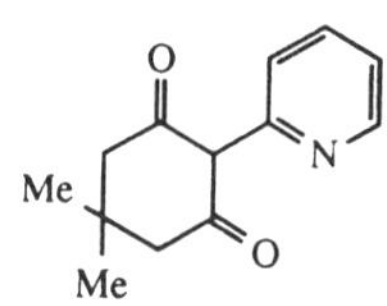

15-135

A one-pot synthesis of substituted 3-(benzoylamino)-5-oxo-5,6,7,8-tetrahydrocoumarins was reported. Thus, 1,3-CHD was reacted with hippuric acid and triethyl orthoformate to obtain compound **15-136**.[213]

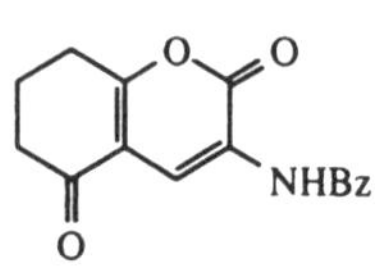

15-136

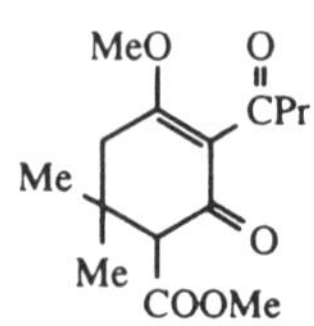

15-137

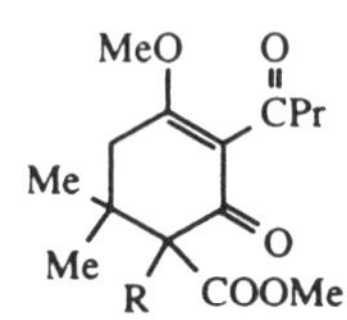

15-138
(where R= Cl, OAc)

A method for the hydroxylation of 1,3-CHDs in the 4-position was developed. For instance, treating the 2-acyl-CHD enol ether, **15-137**, with $Pb(OAc)_4$ and Me_3C-OCl or *m*-chloroperoxybenzoic acid gave a mixture of the 4-chloro- and 4-acetoxy-2-acyl-enol ethers, **15-138**, which were both convertible by hydrolysis/decarboxylation to the 2-butanoyl-4-hydroxy-5,5-dimethyl-1,3-CHD.[214]

A new C–C bond-forming reaction of 3-hydroxyenamides and arenes was invented. For example, the enamine prepared from 1,3-CHD and 2-pyrrolidinone was reduced with $LiBH_4$ in THF to give the 3-hydroxyenamide, **15-139**; this product was reacted with anisole in CH_2Cl_2 and $CF_3COOH/(CF_3CO)_2O$ at 0–25 °C and treated with water to obtain the 4-substituted anisole, **15-140**, in 40% yield (based on the 3-hydroxyenamide).[215]

The reaction of 5,5-dimethyl-1,3-CHD with 2-naphthylamine and benzaldehyde in EtOH at reflux yielded the benzacridinone derivative, **15-141**.[216]

The long-chain 2-alkanoylresorcinols, **15-142**, which are natural products found in insects and medicinal plants, were prepared from the corresponding 2-acyl-1,3-CHDs by aromatization with mercuric acetate.[217]

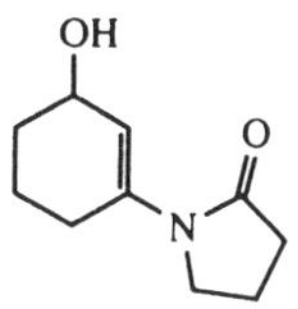

15-139

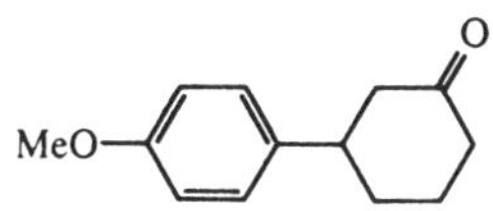

15-140

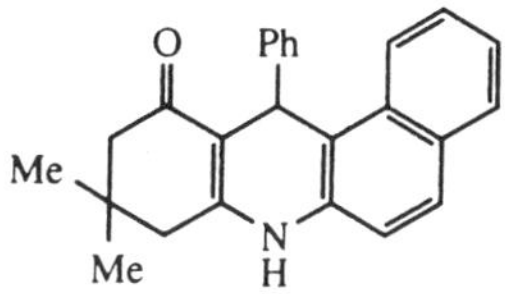

15-141

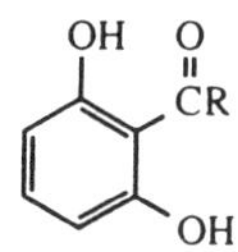

15-142
[where R= *n*-$C_{11}H_{23}$, *n*-$C_{13}H_{27}$, $(CH_2)_7CH_2Ph$ or $(CH_2)_9CH_2Ph$]

The condensation of 1,3-CHD with aldehydes and methyl beta-aminocrotonate gave the hexahydroquinolones, **15-143**, which showed calcium-modulatory effects at the electrically stimulated left atrium, and also caused relaxation of the KCl-stimulated aortic muscle in guinea pigs.[218]

The reaction of 1,3-CHD in 20% aq KOH with allyl bromide gave a 65% yield of 2-allyl-1,3-CHD; this intermediate was dehydrogenated in cymene at reflux with 10% Pd/C catalyst to obtain 2-propylresorcinol in 66% yield. (This product is a starting material for drugs—see Chapter 8).[219]

The reaction of 3-amino-2-cyclohexen-1-one with methyl isothiocyanate in the presence of NaH gave the new spiro-dihydrotriazine, **15-144**. Other reactions of

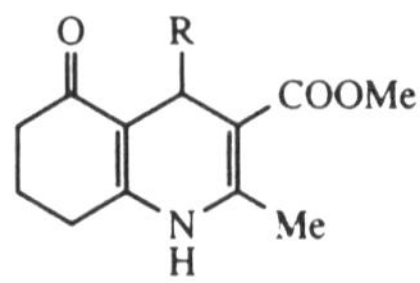

15-143
(where R= substit. Ph, 2-furyl, 1-naphthyl, 2- or 3-pyridyl)

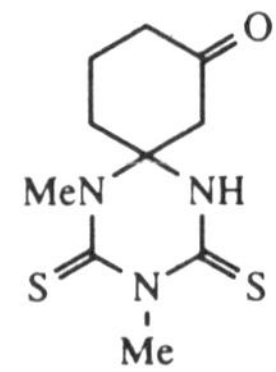

15-144

enamines derived from 1,3-CHDs with methyl or phenyl isothiocyanate were explored.[220]

The condensation of 5,5-dimethyl-1,3-CHD with $Me_2C{=}NO(CH_2)_3COOH$ and hydrolysis of the intermediate gave the isoxazoline derivative, **15-145**.[221]

When 5,5-dimethyl-1,3-CHD was treated first with NaH and then with beta-bromostyrene/CuI in HMPA (hexamethylphosphoramide) at 120 °C, a 95+% yield of the bis-CHD compound, **15-146**, was obtained; the latter was treated with H_2SO_4/benzene to give the corresponding pyran derivative, **15-147**, in high yield.[222]

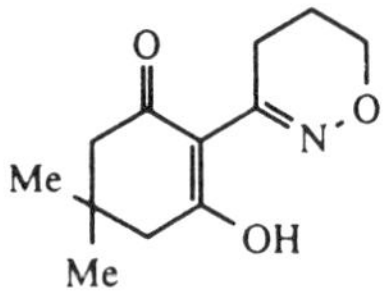

15-145

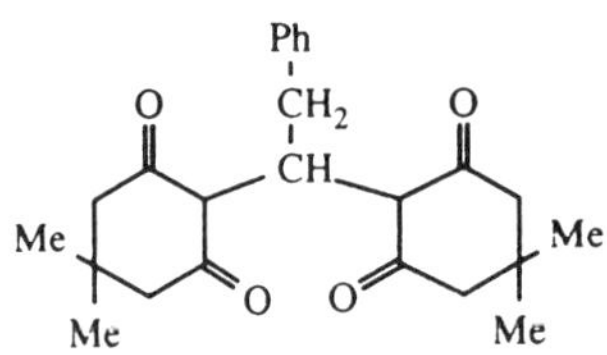

15-146

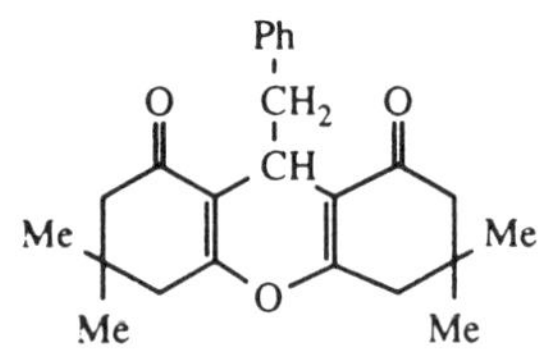

14-147

Decahydrophenanthridinediones, such as **15-148**, were obtained from the reaction of 3-amino-2-cyclohexen-1-one with 2-acetyl-2-cyclohexen-1-one at –10 °C. In refluxing toluene containing pyrrolidine the kinetic product **15-148** isomerized. In refluxing toluene containing *p*-toluenesulfonic acid, the kinetic product was dehydrated and an 81% yield of compound **15-149** was obtained. Byproducts and similar reaction were studied also.[223]

The condensation of 1,3-CHD with $PhCH{=}C(CN)_2$ lead to the 2-amino-3-cyanopyran derivative, **15-150**.[224]

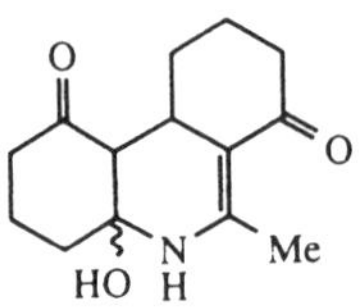

15-148

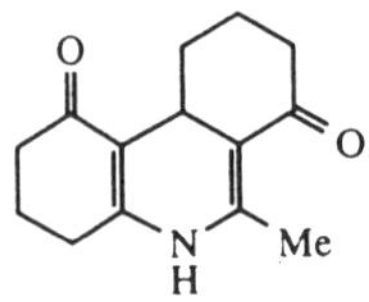

15-149

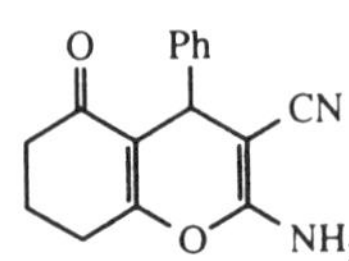

15-150

The acylation of 5,5-dimethyl-1,3-CHD with propionic anhydride in pyridine in the presence of POLYDMAP (4-dimethylaminopyridine supported on an insoluble polymer) gave the trione, **15-151**, in 99.5% yield. This triketone is an important intermediate for herbicides.[225]

Trans-4,6-dimethyl-1,3-cyclohexanedione was prepared in 69% yield by the methylation of 1,3-CHD in the presence of lithium diisopropylamide (LDA) and PhH_3 in a mixed solvent of THF/HMPA/hexane.[226]

1,3-CHD and 5,5-dimethyl-1,3-CHD were acylated in polyphosphoric acid with acetic and propionic acid to give high yields of the 2-acetyl- and 2-propionyl-1,3-CHD, respectively.[227]

A new route to building blocks for the synthesis of indole alkaloids was developed. For example, 2-(*o*-nitrophenyl)-1,3-CHD was allylated in the 2-position to give compound **15-152**; this intermediate was first ozonized at –78 °C, then reductively aminocyclized by reaction with $MeNH_2 \cdot HCl/NaBH_3CN$ in methanol at 25 °C to give the octahydroindol-4-one derivative, **15-153**.[228]

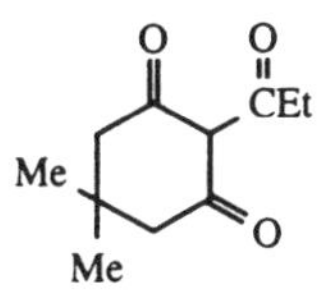

15-151

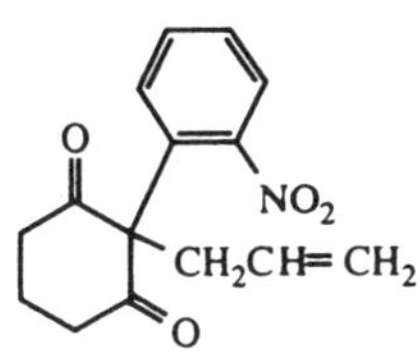

15-152

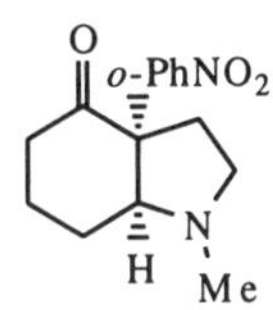

15-153

The rearrangement of pyrrole-2-carboxylic acid and indole-2-carboxylic acid esters of 1,3-CHD (enol form) was investigated. For instance, the pyrrolecarboxate, **15-154**, gave the enamino acid, **15-155**, in the presence of triethylamine. In the presence of cyanide, the indolecarboxylate gave the expected 2-(2-indoloyl)-1,3-CHD, **15-156**.[229]

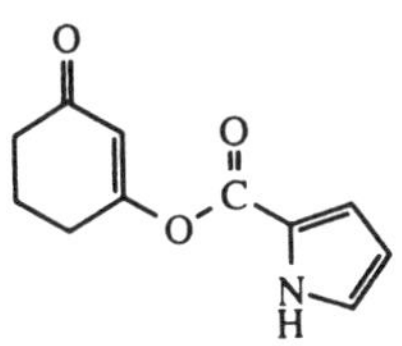

15-154

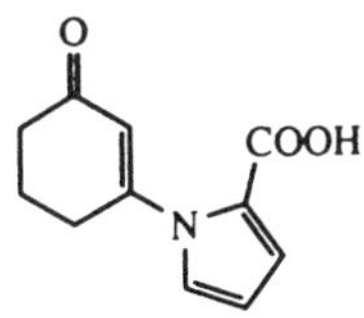

15-155

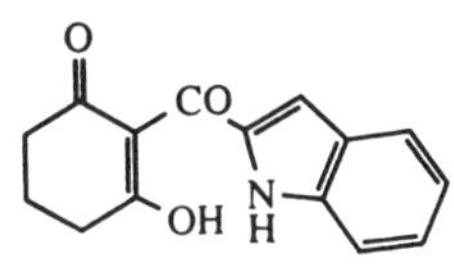

15-156

Synthetic approaches to the condensed, pentacyclic triterpenes of the arborae family were investigated, starting from 2-methyl-1,3-CHD.[230]

A direct synthesis of racemic demethoxyaflatoxin B2, **15-157** (an aflatoxin analog), was devised, starting with the addition/cyclization of 1,3-CHD with dihydrofuran.[231]

A variety of alkyl- and dialkyl-1,3-CHDs were converted to (di)alkylresorcinols with iodine and methanol. [A good list of references regarding the importance of such intermediates for the synthesis of natural products and leukotriene antagonists (drugs) was given by Kotnis in this paper.][232]

Approaches to angularly fused tri- and tetracyclic aromatic compounds (some of which are antibiotics) were studied. Initial model reactions involved the condensation of 2-acetyl-1,4-benzoquinone with the trimethylsilyl ether of 1,3-CHD to yield compound **15-158**. The enolate-OH of this product was esterified with MeI/K_2CO_3 in boiling acetone, then treated with *tert*-BuOK in DMF at 100 °C to obtain the angular, tricyclic product, **15-159**. Similar reactions of 2-acetyl-1,4-naphthoquinones with the trimethylsilyl enol ethers of 1,3-CHDs led to angular, tetracyclic compounds.[233]

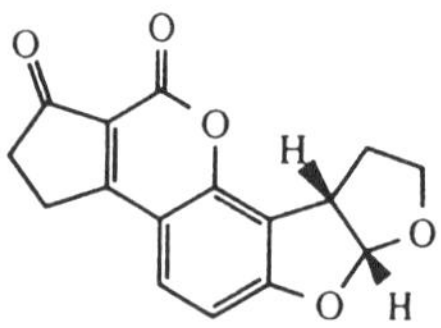

15-157

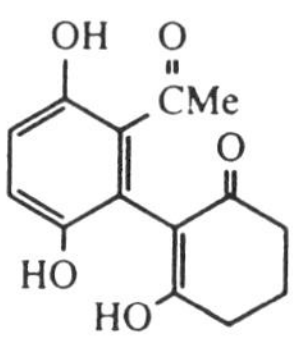

15-158

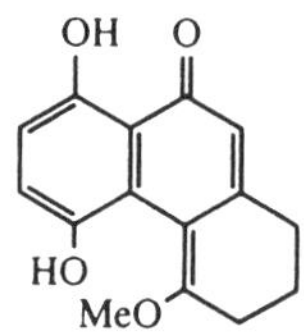

15-159

Syntheses were developed for (*d,l*)-*cis*-chrysanthemic acid, **15-160a**, and 2,2-di-desmethylchrysanthemic acids, **15-160b**, starting from 2,2,5,5-tetramethyl-1,3-CHD and 2,2-dimethyl-1,3-CHD, respectively. These compounds were intermediates for insecticides.[234]

2-Diazo-4,6-di-*tert*-butyl-1,3-CHD (*cis*- and *trans*-isomers), **15-161**, were prepared and photolyzed to give relatively stable oxo-ketenes, **15-162**. The reactions of these ketenes with nucleophiles (water, MeOH, Me_2NH) and the photolysis of these ketenes were studied also and gave di-*tert*-butylcyclopentanones.[235]

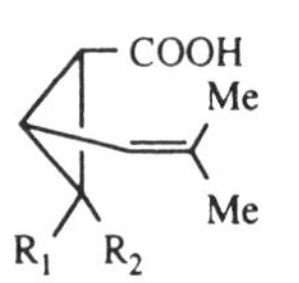

15-160
(a: R_1,R_2=Me
b: R_1,R_2=H)

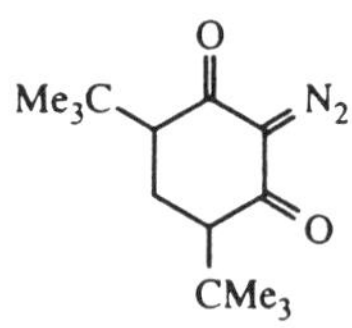

15-161

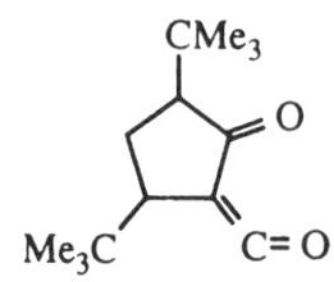

15-162

The oxidation of 1,3-CHD in aq H_2SO_4 at a Pt anode was reported to give a 76% yield (up to 54% current efficiency) of 1,2,3-cyclohexanetrione, **15-163**.[236]

An investigation was made of the preparation of resorcinol derivatives from alkyl-1,3-CHDs by treatment with one equivalent of H_2SO_4 in $(CF_3CO)_2O$ at reflux. Related reactions, such as the aromatization of 3-chloro-5,5-dimethyl-2-cyclohexen-1-one, were studied also. For instance, treating 5,5-dimethyl-1,3-CHD in this way gave, interestingly, a 91% yield of 4,5-dimethylresorcinol, **15-164**.[237]

The regioselectivity of the alkylation with MeI of 2-acetyl-5,5-dimethyl-1,3-CHD after deprotonation with 3.2 equiv lithium diisopropylamide (LDA) was studied. At

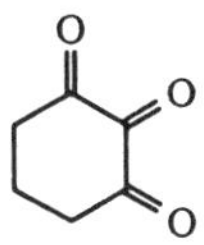

15-163

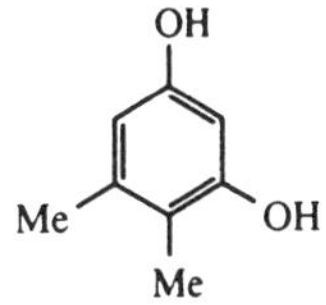

15-164

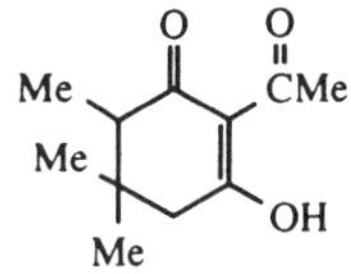

15-165

–78 °C, both 2-acetyl-5,5,6-trimethyl-1,3-CHD, **15-165**, and 2-propionyl-5,5-dimethyl-1,3-CHD, **15-166**, were obtained. Alkylation at 0–4 °C gave mostly 2-propionyl-5,5,6-trimethyl-1,3-CHD, **15-167**. Other reagent ratios/conditions were also investigated.[238]

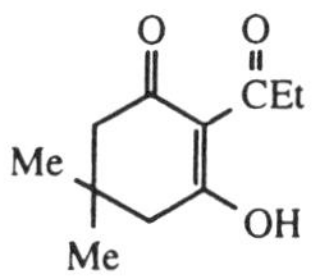

15-166

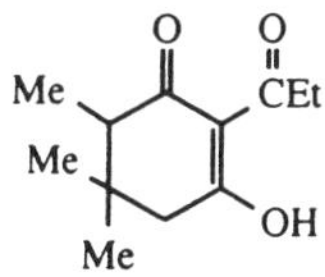

15-167

3,4-Dihydro-1-methylisoquinoline was annelated with 2-acetyl-1,3-CHD to obtain the condensed tetracyclic diketone, **15-168**.[239]

The double Michael addition of 3-methoxy-6-methyl-2-cyclohexen-1-one, **15-169**, with methyl crotonate produced diastereoselectively the bicyclo[2.2.2]octanone, **15-170**; similar reactions were studied as well.[240]

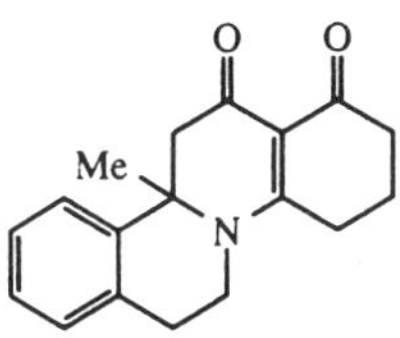

15-168

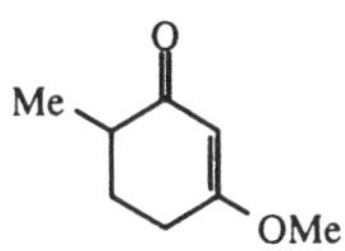

15-169

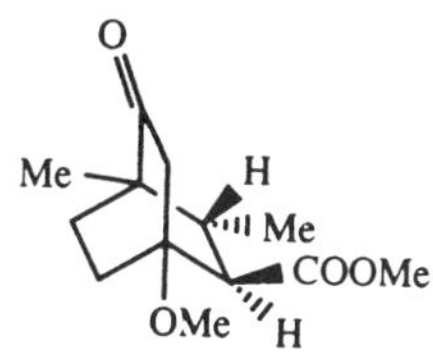

15-170

5,5-Dimethyl-3-phosphoryloxy-2-cyclohexen-1-one, **15-171**, was obtained in 72% yield from the reaction of 5,5-dimethyl-1,3-CHD in chloroform (containing a catalytic amount of THF) with phosphorous pentoxide.[241]

Finally, a tie-in to previous references (e.g., Chapter 3) to the commercial potential of 1,3-CHD as a precursor to alkylresorcinols: A 1993 patent to Sumitomo/Sumika Fine Chemicals claimed a process for the preparation of 2-methylresorcinol, **15-173**, by the Mannich reaction of 1,3-CHD with CH_2O/Me_2NH to give the adduct, **15-172**; this intermediate was hydrogenated to obtain 2-methyl-1,3-CHD, which was dehydrogenated in mesitylene over a Pd/C catalyst to obtain **15-173** in good yield.[242]

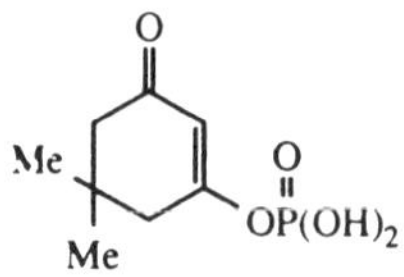

15-171

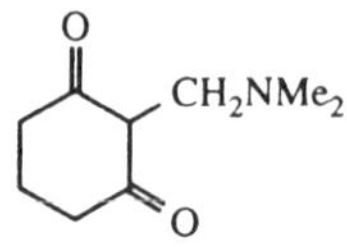

15-172

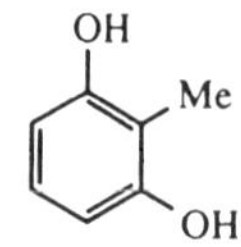

15-173

15.5. REFERENCES

1. D. J. Cram, *Science (Washington, D.C.), 219*, 1177 (1983).
2. C. D. Gutsche, *Calixarenes*, Monograph in Supramolecular Chemistry, Royal Society of Chemistry, London, 1989; C. D. Gutsche in *Inclusion Compounds*, Vol. 4, (J. L. Atwood, J. E. D. Davies, and D. D. McNicol, eds.), pp. 27–63, Oxford University Press, 1991.
3. V. Boehmer and M. A. McKervey, *Chemie Unserer Zeit, 25(4)*, 195 (1991).
4. J. Vicens and V. Boehmer (eds.), *Topics in Inclusion Science*, Vol. 3, *Calixarenes: A Versatile Class of Macrocyclic Compounds*, Kluwer Academic Publishers, Dordrecht, 1991.
5. L. C. Groenen and D. N. Reinhardt, in *Supramolecular Chemistry* (V. Balzani and L. DeCola, eds.), p. 51ff, Kluwer Academic Publishers, Dordrecht, 1991.
6. A. McKervey and V. Boehmer, *Chem. Brit., 28(8)*, 724 (1992).
7. D. J. Cram, *Nature, 356*, 29 (March 5, 1992).
8. J. L. Atwood, G. N. Gokel and Y. Murakami (eds.), *Supramolecular Chemistry* (first issue Apr. 1992), Gordon and Breach Science Publishers, London, 1992.
9. A. v. Baeyer, *Chem. Ber., 5*, 25, and 280 (1872).
10. A. Michael, *J. Am. Chem. Soc., 5*, 338 (1883).
11. J. B. Niederl and H. J. Vogel, *J. Am. Chem. Soc., 62*, 2512 (1940).
12. H. Erdtman and S. Hoegberg, *Tetrahedron Lett., 14*, 1679 (1968).
13. A. G. S. Hoegberg, *J. Am. Chem. Soc., 102*, 6046 (1980).
14. A. G. S. Hoegberg, *J. Org. Chem., 45(22)*, 4498 (1980).
15. L. Abis, E. Dalcanale, A. DuVosel, and S. Spera, *J. Org. Chem., 53*, 5475 (1988).
16. L. M. Tunstead, J. A. Tucker, E. Dalcanale, J. Weiser, J. A. Bryant, J. C. Sherman, R. C. Helgeson, C. B. Knobler, and D. J. Cram, *J. Org. Chem., 54*, 1305 (1989).
17. D. J. Cram, S. Karbach, H. E. Kim, C. B. Knobler, E. F. Maverick, J. L. Ericson, and R. C. Helgeson, *J. Am. Chem. Soc., 110(7)*, 2229 (1988).
18. D. J. Cram, K. D. Stewart, I. Goldberg, and K. N. Trueblood, *J. Am. Chem. Soc., 107*, 2574 (1985).
19. D. J. Cram, S. Karbach, Y. H. Kim, L. Baczynski. K. Marti, R. M. Sampson, and G. W. Kalleymeyer, *J. Am. Chem. Soc., 110*, 2554 (1988).
20. D. J. Cram, M. E. Tanner, and R. Thomas, *Angew. Chem., Int. Ed. Engl., 30(8)*, 1024 (1991).
21. D. J. Cram, H. J. Choi, J. A. Bryant, and C. B. Knobler, *J. Am. Chem. Soc., 114(20)*, 7748 (1992).
22. J. R. Moran, J. L. Ericson, E. Dalcanale, J. A. Bryant, C. B. Knobler, and D. J. Cram, *J. Am. Chem. Soc., 113(15)*, 5707 (1991).
23. P. Soncini, S. Bonsignore, E. Dalcanale, and F. Ugozzoli, *J. Org. Chem., 57(17)*, 4608 (1992).
24. F. Weinelt and H.-J. Schneider, *J. Org. Chem., 56(19)*, 5527 (1991).
25. P. D. Beer, *Chem. Soc. Rev., 18*, 409 (1989).
26. P. D. Beer, E. L. Tite, and A. Ibbotson, *J. Chem. Soc., Chem. Commun., 1989(24)*, 1874.
27. R. J. M. Egberink, P. L. H. Cobben, W. Verboom, S. Harkema, and D. N. Reinhoudt, *J. Inclusion Phenom. Mol. Recognit. Chem., 12(1–4)*, 151 (1992).
28. L. N. Markovskii, V. I. Kalchenko, D. M. Rudkevich, and A. N. Shivanyuk, *Mendeleev Commun., 1992(3)*, 106; *C.A., 117*, 234117 (1992).
29. G. Mann, V. Tittelbach-Helmrich, T. Lippmann, and K. Mueller, *Chromatographia, 34(9–10)*, 453 (1992).

30. P. Timmermann, M. G. A. van Mook, W. Verboom, G. J. van Hummel, S. Harkema, and D. N. Reinhoudt, *Tetrahedron Lett., 33(23)*, 3377 (1992).
31. H. Konishi, Y. Iwasaki, O. Morikawa, T. Okano, and J. Kiji, *Chem. Express, 5(11)*, 869 (1990); *C.A., 114*, 121691 (1991).
32. O. Manabe, K. Asakura, T. Nishi, and S. Shinkai, *Chem. Lett., 1990(7)*, 1219; *C.A., 114*, 6001 (1991).
33. B. Botta, P. Iacomacci, C. DiGiovanni, G. Delle Monache, E. Gacs-Baitz, M. Botta, A. Tafi, F. Corelli, and D. Masili, *J. Org. Chem., 57(12)*, 3259 (1992).
34. Y. Aoyama, Y. Tanaka, H. Toi, and H. Ogoshi, *J. Am. Chem. Soc., 110(2)*, 634 (1988).
35. Y. Aoyama, Y. Tanaka, and S. Sugahara, *J. Am. Chem. Soc., 111(14)*, 5397 (1989).
36. Y. Tanaka, Y. Kato, and Y. Aoyama, *J. Am. Chem. Soc., 112*, 2808 (1990).
37. K. Kurihara, T. Kunitake, K. Ohto, Y. Tanaka, Y. Aoyama, and T. Kunitake, *J. Am. Chem. Soc., 113(2)*, 444 (1991).
38. C. Khare and S. Sugahara, *J. Indian Chem. Soc., 68(1)*, 42 (1991); *C.A., 115*, 232624 (1991).
39. Y. Aoyama, *Nippon Kagaku Kaishi, 1991(8)*, 1041; *C.A., 115*, 278987 (1991). Also: Y. Aoyama, *Kagaku Kogyo, 42(4)*, 327 (1991); *C.A., 114*, 207600 (1991).
40. Y. Kikuchi, Y. Tanaka, S. Sutarto, K. Kobayashi, H. Toi, and Y. Aoyama, *J. Am. Chem. Soc., 114(26)*, 10302 (1992).
41. Y. Kikuchi, K. Kobayashi, and Y. Aoyama, *J. Am. Chem. Soc., 114(4)*, 1351 (1992).
42. K. Kobayashi, Y. Asakawa, Y. Kato, and Y. Aoyama, *J. Am. Chem. Soc., 114(4)*, 1351 (1992).
43. K. Kobiyashi, T. Asakawa, Y. Kikuchi, H. Toi, and Y. Aoyama, *J. Am. Chem. Soc., 115(7)*, 10307 (1993).
44. A. F. Danil de Namor, P. M. Blackett, M. T. Garrido Pardo, D. A. Pacheco Tanaka, F. J. Sueros Velarde, and M. C. Cabaleiro, *Pure Appl. Chem., 66(3)*, 415 (1993).
45. H.-J. Choi, D. J. Cram, C. B. Knobler, and E. F. Maverick, *Pure Appl. Chem., 66(3)*, 539 (1993).
46. Eur. Pat. Appl. EP 375,463 (June 27, 1990; E. Dalcanale, S. Bosignore, and A. DuVosel to Instituto Guido Donegani S.p.A.); *C.A., 114*, 61544 (1991).
47. A. Dalla Cort, L. Mandolini, and B. Masci, *J. Org. Chem., 45(19)*, 3923 (1980).
48. S. Pappalardo, F. Bottino, and R. Ronsisvalle, *J. Chem. Soc., Perkin Trans. 2, 1984 (6)*, 1001.
49. L. Wambach and F. Voegtle, *Tetrahedron Lett., 26(12)*, 1483 (1985).
50. R. S. Grewal and H. Hart, *Tetrahedron Lett., 31(30)*, 4271 (1990).
51. R. P. Sijbesma, W. P. Bosman, and R. J. M. Nolte, *J. Chem. Soc., Chem. Commun., 1991(13)*, 885; *J. Org. Chem., 56(10)*, 3199 (1991).
52. Zh. Li, H. Lin, M. Li, L. Shi, and R. Yu, *Geodeng Xuexio Huaxue Xuebao, 12(1)*, 30 (1991); *C.A., 115*, 235114 (1991).
53. C. Gennari, F. Molinari, M. Bartoletti, U. Piarulli, and D. Potenza, *J. Org. Chem., 56(10)*, 3201 (1991).
54. Y. Delaviz and H. W. Gibson, *Org. Prep. Proc., 23(3)*, 382 (1991).
55. M. Lofthagen, R. Vernon Clark, K. K. Baldridge, and J. S. Siegel, *J. Org. Chem., 57(1)*, 61 (1992).
56. T. Nagata, *Bull. Chem. Soc. Jpn., 65(2)*, 385 (1992); *C.A., 116*, 214206 (1992).
57. Y. Fujii, K. Kikuchi, K. Matsunati, K. Ota, M. Adachi, M. Syoji, I. Haneishi, and Y. Kuwana, *Chem. Lett., 1984 (9)*, 1487; *C.A., 102*, 132430 (1984).
58. M. C. Etter, Z. Urbanczyk-Lipowska, D. A. Jahn, and J. S. Frye, *J. Am. Chem. Soc., 108*, 5871 (1986).
59. Y. Aoyama, K. Mizokami, and H. Toi, *Chem. Lett., 1990 (4)*, 651; *C.A., 113*, 215540 (1990).
60. T. Motomura, K. Inoue, K. Kobayashi, and Y. Aoyama, *Tetrahedron Lett., 32(36)*, 4757 (1991).
61. C. J. Hawker and J. M. J. Frechet, *J. Am. Chem. Soc., 112(21)*, 7638 (1990).
62. T. Motomura and Y. Aoyama, *J. Org. Chem., 56(26)*, 7225 (1991).
63. K. W. Won and J. M. Prausnitz, *J. Chem. Thermodynamics, 7*, 661 (1975).
64. S. A. Kudchaker and B. J. Zwolinski, *J. Mol. Struct., 48*, 271 (1978).
65. S. A. Kudchaker, A. P. Kudchaker, R. C. Wilhoit, and B. J. Zwolinski, *Thermochim. Acta, 30(1–2)*, 319 (1979).
66. J. M. Spencer, R. S. Harner, L. I. Freed, and C. D. Penturelli, *J. Phys. Chem., 79(4)*, 332 (1975).
67. F. Shahidi, P. G. Farrell, and J. V. Westwood, *J. Chem. Res., Synop., 1980(10)*, 357.

68. B. L. Khandelwal, K. Kumar, and F. J. Berry, *Inorg. Chim. Acta, 47(2)*, 135 (1981).
69. A. Viikna, M. Gambino, G. Pouzard, and J. P. Bros, *Calorim. Anal. Therm., 13*, III. 16. 108 (1982); *C.A., 97*, 109493 (1982).
70. D. S. Veselinovic, D. A. Markovic, and M. V. Obradovic, *Glas. Hem. Drus. Beograd, 48(9)*, 619 (1983); *C.A., 101*, 72113 (1984).
71. S. N. Limaye, *J. Indian Chem. Soc., 59*, 916 (1982).
72. N. L. Egutkin, I. P. Malaya, and Yu. E. Nikitin, *Zh. Prikl. Khimii, 59(6)*, 1288 (1986); *C.A., 107*, 236162 (1987).
73. J. J. P. Furlong and N. S. Nudelman, *Acta Sud Am. Quim., 3(3–4)*, 129 (1983); *C.A., 101*, 229921 (1984).
74. S. D. Boyce, A. C. Barefoot, and J. F. Horning, *J. Labelled Compd. Radiopharm., 20(2)*, 243 (1983); *C.A., 99*, 52696 (1983).
75. J. Barbera, M. Marcos, E. Melendez, and J. L. Serrano, *Mol. Cryst. Liq. Cryst., 94(3–4)*, 367 (1983).
76. B. Trathnigg, K. Golob, H. Janek, J. Perne, and A. Popitsch, *Monatsh. Chem., 115(11)*, 1353 (1984).
77. H. Konshin, *THEOCHEM, 19(3–4)*, 267 (1984).
78. H. C. Box and E. E. Budzinski, *J. Chem. Phys., 83(11)*, 6059 (1985).
79. E. Haslam, T. H. Lilley, and J. P. McManus, *Thermochim. Acta, 86*, 183 (1985).
80. F. Dall-Acqua, S. Caffieri, and G. Rodighiero, *NATO ASI Ser., Ser. A, 85 (Primary Photo-Processes Biol. Med.)*, 259 (1985); *C.A., 103*, 195589 (1985).
81. V. J. Krukonis and R. T. Kurnik, *J. Chem. Eng. Data, 30*, 247 (1985).
82. R. E. Huie and P. Neta, *J. Phys. Chem., 89*, 3918 (1985).
83. G. Norwitz, N. Nataro, and P. N. Keliher, *Anal. Chem., 58(3)*, 639 (1986).
84. Y. Ebisuzaki, L. H. Askari, A. M. Bryan, and M. F. Nicol, *J. Chem. Phys., 87(11)*, 6659 (1987).
85. L. A. Fedorov, S. A. Sokolovskii, and A. N. Ermakov, *Zh. Strukt. Khim., 28(2)*, 50 (1987); *C.A., 107*, 235962 (1987).
86. I. B. Repinskaya, Vi. I. Mamatyuk, and V. A. Koptyug, *Zh. Org. Khim., 23(2)*, 350 (1987); *C.A., 107*, 197403 (1987).
87. F. J. Weigert, *J. Org. Chem., 52*, 921 (1987).
88. M. Moriyasu, A. Kato, and Y. Hashimoto, *Bull. Chem. Soc. Jpn., 61*, 2955 (1988); *C.A., 110*, 114335 (1989).
89. R. Yin and H. Shen, *Lanzhou Daxue Xuebao, Ziran Kexueban, 25(3)*, 45 (1989); *C.A., 113*, 97742 (1990).
90. D. A. Dixon, in *ACS Symposium Series No. 404, "Computer Applications In Applied Polymer Science II: Automation, Modeling and Simulation"* (T. Provder, ed.), p. 147ff, American Chemical Society, Washington, 1989.
91. C. Puebla and T.-K. Ha, *J. Mol. Struct. (Theochem), 204*, 337 (1990).
92. J. Suh and J. S. Heo, *J. Org. Chem., 55(20)*, 5531 (1990).
93. D. I. Lyamkin, K. G. Misyuk, A. G. Krieger, and E. V. Fronchek, *Izv. Vyssh. Uchebn. Zaved, Khim. Khim. Tekhnol., 34(7)*, 84 (1991); *C.A., 115*, 280986 (1991).
94. T. Shinegari, E. Kojima, Y. Ino, and K. Abe, *Phys. Rev. Lett., 66(16)*, 2112 (1991); *C.A., 114*, 257256 (1991).
95. A. Katrusiak, *Acta Crystallogr., Sect. B: Struct. Sci., B47(3)*, 398 (1991); *C.A., 115*, 18989 (1991).
96. M. V. Obradovic, D. S. Veselinovic, and D. A. Markovic, *J. Serb. Chem. Soc., 56(8–9), 513 (1991); C.A., 116*, 58643 (1992).
97. K. Mishima, S. Morii, and M. Nagatani, *Fukuoka Daigaku Kogaku Shuho, 47*, 197 (1991); *C.A., 116*, 68531 (1992).
98. M. C. Etter and G. Ranawake, *J. Am. Chem. Soc., 114(11)*, 4430 (1992).
99. M. K. Kokila, K. A. Nirmala, and S. N. Puttaraja, *Acta Crystallogr., Sect. C: Cryst. Struct. Commun., C48 (6)*, 1133 (1992); *C.A., 117*, 80558 (1992).
100. J. Basak, J. Penar, and K. Sykut, *Gazz. Chim. Ital., 122(1)*, 25 (1992); *C.A., 116*, 193437 (1992).
101. S. E. Branz and S. E. Fiore, Abstracts of the 203rd Natl. ACS Meeting, San Francisco, April 5–10, 1992, Organic Chemistry Section, Paper #326.

102. A. Katrusiak, *J. Mol. Struct., 269(3–4)*, 329 (1992); *C.A., 117*, 170544 (1992).
103. G. Pilcher, O. G. Parchment, I. H. Hillier, F. Heatley, D. Fletcher, M. A. V. Ribeiroda Silva, M. L. C. C. H. Ferrao, M. S. J. Monte, and J. Fang, *J. Phys. Chem., 97(1)*, 243 (1993).
104. M. N. Desai and M. H. Ghandi, *Record Chem. Progr., 30(3)*, 223 (1969).
105. V. Miketukova and R. W. Frei, *J. Chromatogr., 47*, 227 (1970).
106. V. Ulrich and P. Weber, *Hoppe-Seyler's Z. Physiol. Chem., 353*, 1171 (1972).
107. R. Anderson, A. H. Haines, and B. P. Stark, *Angew. Makromol. Chem., 62*, 101 (1977).
108. W. Dankelman and J. de Wit, *Angew. Makromol. Chem., 62*, 101 (1977).
109. E. J. Greenhow, R. Hargitt, and A. A. Shafi, *Angew. Makromol. Chem., 48*, 55 (1975); E. J. Greenhow and A. A. Shafi, *ibid., 53*, 187, (1976).
110. S. R. Bergman, E. Carlson, E. Dannen, and B. E. Sobel, *Clin. Chim. Acta, 104*, 53 (1980).
111. M. Nakamura, *Anal. Lett., 13 (A9)*, 771 (1980).
112. R. H. Dinius and J. M. Baker, *Microchem. J., 25(2)*, 209 (1980).
113. S. M. Dreijer-Van der Glas, T. Schalekamp, H. J. DeJong, and A. Bult, *Pharm. Weekbl., Sci. Ed., 5(2)*, 70 (1983); *C.A., 99*, 121913 (1983).
114. S. C. Lavlae and K. S. Pitre, *J. Indian Chem. Soc., 59(8)*, 965 (1982).
115. T. R. Oegama, Jr. and K. M. Cooper, *Anal. Biochem., 133(1)*, 233 (1983).
116. P. C. S. Feng and P. A. Hartman, *Appl. Environ. Microbiol., 43*, 1320 (1982).
117. R. Zhou and Z. Liu, *Gaodeng Xuexiao Huaxue Xuebao, 6(7)*, 601 (1985); *C.A., 103*, 210275 (1985).
118. E. Koller and O. S. Wolfbeis, *Monatsh. Chem., 116(1)*, 65 (1985).
119. N. Velghe and A. Claeys, *Analyst, 110*, 313 (1985).
120. A. Takadate, M. Irikura, T. Suehira, H. Fujino, and S. Goya, *Chem. Pharm. Bull., 33(3)*, 1164 (1985); *C.A., 111*, 126163 (1989).
121. M. R. P. Reddy, P. V. S. Kumar, J. P. Shyamasundar, and Y. Anjaneyulu, *J. Indian Chem. Soc., 66(11)*, 837 (1989); *C.A., 112*, 210174 (1990).
122. USSR Pat. SU 1,528, 794 (Oct. 30, 1989; M. A. Tishchenko, L. M. Shafran, A. P. Loburenko, V. B. Ryabalka, and A. I. Markino to Odessa Polytechnic Institute); *C.A., 112*, 191177 (1990).
123. J. L. Sotelo, F. J. Beltran, and M. Gonzalez, *Rev. R. Acad. Cienc. Exactas, Fis. Nat. Madrid, 83(2)*, 227 (1989); *C.A., 114*, 142456 (1991).
124. M. R. P. Reddy, P. V. S. Kumar, J. P. Shyamasundar, and Y. Anjaneyulu, *Proc. Indian Natl. Sci. Acad., Part A, 56(3)*, 255 (1990); *C.A., 114*, 74376 (1991).
125. Zh. Lu, *Diandu Yu Huanbao, 10(1)*, 20 (1990); *C.A., 113*, 164553 (1990).
126. J. C. Vidal and C. Vaquero, *Collect. Czech. Chem. Commun., 55(7)*, 1686 (1990).
127. Zh. Jinkui and R. Neeb, *Fresenius J. Anal. Chem., 238(1)*, 34 (1990); *C.A., 113*, 204127 (1990).
128. N. M. Malakhova, E. M. Nikipelova, and G. I. Savenko, *Khim. Tekhnol. Vody, 12(7)*, 627 (1990); *C.A., 113*, 243868 (1990).
129. M. Siroki and L. Maric, *Croat. Chem. Acta, 63(2)*, 255 (1990); *C.A., 114*, 54950 (1991).
130. R. M. Awadallah, A. E. Mohamed, M. A. El Meghraby, and M. M. A. E. Ramadan, *Transition Met. Chem. (London), 15(4)*, 274 (1990); *C.A., 113*, 108064 (1990).
131. Anon., *Chem. Eng. News, 68*, 40 (1990).
132. R. Purohit and S. Devi, *Talanta, 38(7)*, 753 (1991).
133. C. Lin, X. Zhang, G. Li, and S. Liang, *Huaxue Xuebao, 49(3)*, 249 (1991); *C.A., 114*, 258750 (1991).
134. C. Lin, X. Zhang, and X. Liu, *Analyst (London), 116(3)*, 277 (1991); *C.A., 114*, 156253 (1991).
135. P. A. M. Farias and A. K. Ohara, *Electroanalysis (N. Y.), 3(9)*, 985 (1991); *C.A., 116*, 186949 (1992).
136. Zh. Li, M. Li, R. Yu, F. Qian, and L. Shi, *Fenxi Huaxue, 19(10)*, 1150 (1991); *C.A., 116*, 98643 (1992).
137. G. B. Golubitskii, V. M. Ivanov, L. I. Gen, and V. N. Figurovskaya, *Vestn. Mosk. Univ., Ser. 2: Khim., 32(1)*, 42 (1991); *C.A., 115*, 84427 (1991).
138. Q. Liu, H. Zhang, and J. Cheng, *Talanta, 38(6)*, 669 (1991).
139. T. Okutani, A. Sakuragawa, and M. Murakami, *Anal. Sci., 7 (Suppl., Proc. Int. Congr. Anal. Sci., 1991, Pt. 1)*, 109 (1991); *C.A., 116*, 227156 (1992).

140. J. L. Sotelo, F. J. Betran, M. Gonzalez, and J. F. Garcia-Araya, *Ind. Eng. Chem. Res., 30(1)*, 222 (1991).
141. S. G. Kawatkar and S. J. Mane, *Acta Cienc. India, Chem., 17C(3)*, 275 (1991); *C.A., 117*, 82527 (1992).
142. S. Chakravarty and R. K. Mishra, *Anal. Sci., 8(5)*, 609 (1992); *C.A., 117*, 15565 (1993).
143. M. Petrovic, M. Kastelan-Macan, and A. J. M. Horvat, *J. Chromatogr., 607(1)*, 163 (1992).
144. S. Zareba, *Ann. Univ. Mariae Curie-Sklodowska, Sect. AA: Chem., 42-3*, 195 (1987–1988) [Publ. 1992]; *C.A., 118*, 93345 (1993).
145. H. Liu, L. Wang, L. Chen, and W. Liu, *Fenxi Ceshi Tongbao, 11(2)*, 78 (1992); *C.A., 118*, 160233 (1993).
146. N. N. Dutta, S. Borthakur, and G. S. Patil, *Sep. Sci. Technol., 27(11)*, 1435 (1992).
147. A. Ya. Strakov *et al., Khim. Geterosikl. Soedin., 1974(8)*, 1011 [196 references]; A. Ya. Strakhov, E. Gudriniece, and I. A. Strakova, *ibid., 1988(6)*, 723 [181 references].
148. G. Buechi and B. Egger, *J. Org. Chem., 36*, 2021 (1971).
149. C. J. Kowaski and K. W. Fields, *J. Org. Chem., 46(1)*, 197 (1981).
150. V. P. Makovetskii, I. B. Dzvinchuk, and A. A. Svishchuk, *Ukr. Khim. Zh., 46(8)*, 839 (1980); *C.A., 94*, 15463 (1981).
151. M. Bundule, E. Biesenieks, A. Kemme, J. Bleidelis, G. Duburs, and J. Uldrikis, *Khim. Geterotsikl. Soedin., 1980 (5)*, 666; *C.A., 93*, 168094 (1980).
152. L.-F. Tietze, G. V. Kiedrowski, K. Harms, W. Clegg, and G. Sheldrick, *Angew. Chem., 92(2)*, 130 (1980).
153. F. G. Gonzalez, M. G. Guillen, J. A. G. Perez, and E. R. Galan, *Carbohydr. Res., 78(1)*, 17 (1980).
154. W. Duerckheimer, W. Raether, H. Seliger, and H. Seidenath, *Arzneim.-Forsch., 30(7)*, 1041 (1980); *C.A., 93*, 220561 (1980).
155. Ger. Offen. DE 2,928,483 (Apr. 17, 1980; K. Lauer and E. Kiegel to Boehringer Mannheim G.m.b.H.); *C.A., 93*, 114325 (1980).
156. T. L. Macdonald and S. Mahalingam, *J. Am. Chem. Soc., 102*, 2113 (1980).
157. T. Umehara, Y. Inouye, and G. Kakisawa, *Bull. Chem. Soc. Jpn., 54(11)*, 3492 (1981); *C.A., 96*, 84779 (1982).
158. M. Azzaro, S. Geribaldi, and B. Videau, *Synthesis, 1981(11)*, 881.
159. E. Gudriniece, V. Barkane, and E. Liepins, *Khim. Geterotsikl. Soedin., 1981(4)*, 563; *C.A., 95*, 80863 (1981).
160. P. W. Scott, I. T. Harrison, and S. Bittner, *J. Org. Chem., 46(9)*, 1914 (1981).
161. I. Trummer, E. Ziegler, and O. S. Wolfbeis, *Synthesis, 1981(3)*, 225.
162. H. D. Kapadia and S. V. Sunthankar, *Indian J. Chem., Sect. B, 20B(11)*, 993 (1981).
163. Y. F. Zhou and N. Zh. Huang, *Synth. Commun., 12(10)*, 795 (1982).
164. M. M. Yousif, S. Saeki, and M. Hamana, *Chem. Pharm. Bull., 30(7)*, 2326 (1982); *C.A., 98*, 4468 (1983).
165. C. Dagher, R. Hanna, P. B. Terent'ev, Yu. G. Bundel, A. N. Kost, and B. I. Maksimov, *J. Heterocycl. Chem., 19(3)*, 645 (1982).
166. H. H. Wasserman and J. E. Pickett, *J. Am. Chem. Soc., 104*, 4695 (1982).
167. V. A. Nikolaev, O. V. Zhdanova, and I. K. Korobitsyna, *Zh. Org. Khim., 18(3)*, 559 (1982); *C.A., 96*, 217309 (1982).
168. J. H. Dodd, R. S. Garigipati, and S. M. Weinreb, *J. Org. Chem., 47*, 4045 (1982).
169. M. L. Bolte, W. D. Crow, and S. Yoshida, *Aust. J. Chem., 35*, 1421 (1982).
170. M. Gomez Guillen and J. P. Garcia Martin, *An. Quim., Ser. C, 79(2)*, 109 (1983); *C.A., 101*, 230985 (1984).
171. M. L. Jani and R. P. Soni, *Synthesis, 1983(11)*, 933.
172. N. S. Zefirov, T. S. Kuznetsova, and S. V. Kozhushkov, *Zh. Org. Khim., 19(8)*, 1599 (1983); *C.A., 100*, 22352 (1984).
173. S. Miyano, N. Abe, N. Mibu, K. Takeda, and K. Sumoto, *Synthesis, 1983(5)*, 401.

174. Jpn. Kokai Tokkyo Koho JP 59,144,758 [84, 144,758] (Aug. 18, 1984; to Ube Industries, Ltd.); *C.A.*, *102*, 62078 (1985).
175. R. O. Duthaler and U. H. U. Wegmann, *Helv. Chim. Acta, 67(7)*, 1755 (1984).
176. J. Yoshida, S. Yano, T. Ozawa, and N. Kawabata, *Tetrahedron Lett., 25(26)*, 2817 (1984).
177. J. V. Greenhill, M. A. Moten, and R. Hanke, *J. Chem. Soc., Perkin Trans. 1, 1984(6)*, 1213.
178. I. Zilniece and A. Zicmanis, *Latv. PSR Zinat. Akad. Vestis, Khim. Ser., 1984(5)*, 616; *C.A., 102*, 95604 (1985).
179. P. Dalla Croce and C. LaRosa, *Synthesis, 1984(11)*, 982.
180. J. V. Greenhill, M. A. Moten, and R. Hanke, *J. Chem. Soc., Perkin Trans. 1, 1984(2)*, 287.
181. D. W. Brooks, H. Mazdiyasni and S. Chakrabarti, *Tetrahedron Lett., 25(12)*, 1241 (1984).
182. A. Arnoldi, *Synthesis, 1984(10)*, 856.
183. G. Kneen and P. J. Maddocks, *Synth. Commun., 16(13)*, 1635 (1986).
184. M. P. Sammes, P. N. Maini, and A. R. Katritzky, *J. Chem. Soc., Chem. Commun., 1984(6)*, 354.
185. Jpn. Kokai Tokkyo Koho JP 60,109,579 [85, 109,579] (June 15, 1985; to Kureha Chemical Industry Co., Ltd.); *C.A., 103*, 196072 (1985).
186. Eur. Pat. Appl. EP 133,960 (March 13, 1985; R. Pelerin to Roure Betrand Dupont S. A.); *C.A., 105*, 226057 (1986).
187. K. Matano, M. Yoshida, M. Ohta, and K. Tanaka, *Chem. Pharm. Bull., 33(9)*, 4057 (1985); *C.A., 105*, 115034 (1986).
188. E. Reimann and J. M. Friesinger, *Arch. Pharm. (Weinheim, Ger.), 318 (10)*, 871 (1985); *C.A., 104*, 109395 (1985).
189. M. Suesse and S. Johne, *Z. Chem., 25(12)*, 432 (1985).
190. F. A. Lakhvich, T. S. Khlebnicova, and A. A. Akhrem, *Synthesis, 1985(8)*, 784.
191. A. R. Katritzky and C. M. Marson, *Tetrahedron Lett., 26(39)*, 4715 (1985); see also: A. R. Katritzky and C. M. Marson, *J. Org. Chem., 52*, 2726 (1987).
192. A. A. Maksimenko, B. B. Rivkin, I. D. Sadekov, and V. I. Minkin, *Khim. Getrotsikl. Soedin., 1985(6)*, 769; *C.A., 103*, 196056 (1985).
193. Y. Tamai, Y. Mizutani, H. Hagiwara, H. Uda, and N. Harada, *J. Chem. Res., Synop., 1985(5)*, 148; *C.A., 103*, 177984 (1985).
194. A. A. Akhrem, V. N. Pshenichnyi, O. V. Gulyakevich, and V. A. Khripach, *Khim. Getrotsikl. Soedin., 1985(8)*, 1130; *C.A., 104*, 19461 (1985).
195. R. M. Moriarty, B. R. Bailey, III, O. Prakash, and I. Prakash, *J. Am. Chem. Soc., 107(5)*, 1375 (1985).
196. J. Tsuji, H. Watanabe, I. Minami, and I. Shimizu, *J. Am. Chem. Soc., 107*, 2196 (1985).
197. Y. Houbrechts, P. Laszlo, and P. Pennetreau, *Tetrahedron Lett., 27(6)*, 705 (1986).
198. N. M. Berry, M. C. P. Darey, and L. M. Harwood, *Synthesis, 1986(6)*, 476.
199. M. Abou-Gharbia, *Heterocycles, 24(5)*, 1347 (1986).
200. N. M. Berry, M. C. P. Darey, and L. M. Harwood, *Tetrahedron Lett., 27(20)*, 2319 (1986).
201. S. T. Purrington, C. L. Bumgardner, N. V. Lazaridis, and P. Singh, *J. Org. Chem., 52(19)*, 4307 (1987).
202. J. C. Pelletier and M. P. Cava, *Synthesis, 1987(5)*, 474.
203. T. Asami, N. Takahashi, and S. Yoshida, *Agric. Biol. Chem., 51(1)*, 205; *C.A., 106*, 195923 (1987).
204. Y. Huang and X. Lu, *Tetrahedron Lett., 28(49)*, 6219 (1987).
205. O. I. Yurchenko, K. S. Pushkareva, G. A. Zheldubovskaya, N. V. Komarov, and G. A. Berkova, *Zh. Obshch. Khim., 57(4)*, 935 (1987); *C.A., 108*, 55513 (1988).
206. N. Zh. Hunag, M. V. Lakshmikantham, and M. P. Cava, *J. Org. Chem., 52(2)*, 169 (1987).
207. H. J. Monteiro, *Synth. Commun., 17(8)*, 983 (1987).
208. F. A. Lakhvich, I. I. Polozov, and A. A. Akhram, *Dokl. Akad. Nauk SSSR, 298(6)*, 1395 (1988); *C.A., 109*, 170076 (1988).
209. X.-P. Gu, T. Okuhara, I. Ikeda, and M. Okahara, *Synthesis, 1988(7)*, 535.
210. C. Glidewell, D. Lloyd, and S. Metcalfe, *Synthesis, 1988(4)*, 319.
211. J. Yoshida, K. Sakaguchi, and S. Isoe, *J. Org. Chem., 53*, 2525 (1988).

212. D. T. Manning, A. R. Cooke, and R. D. Gruenhagen, *J. Agric. Food Chem., 36(1)*, 169 (1988); *C.A., 108*, 51156 (1988).
213. M. Kocevar, S. Polanc, M. Tisler, and B. Vercek, *Synth. Commun., 19(9–10)*, 1713 (1989).
214. F. A. Lakhvich, L. G. Lis, D. B. Rubinov, I. L. Rubinova, and A. A. Akhrem, *Zh. Org. Khim., 25(7)*, 1417 (1989); *C.A., 112*, 157717 (1990).
215. S. Ohta, M. Sumino,T. Sasaki, and N. Yamagami, *Heterocycles, 29(8)*, 1455 (1989).
216. R. Martinez, E. Cortes, R. A. Toscano, and I. Linzaga, *J. Heterocycl. Chem., 27(2)*, 363 (1990).
217. J. E. Oliver, K. R. Wilzer, and R. M. Waters, *Synthesis, 1990(12)*, 1117.
218. U. Rose, *Arch. Pharm., 323(5)*, 281 (1990); *C.A., 113*, 97428 (1990).
219. M. A. Staley, R. L. Shone, and M. Miyano, *Synth. Commun., 20(12)*, 1869 (1990).
220. J. V. Greenhill, J. Hanaee, and P. J. Steel, *J. Chem. Soc., Perkin Trans. 1, 1990(7)*, 1869.
221. W. M. Best, R. F. C. Brown, P. D. Tiso, and K. G. Watson, *Aust. J. Chem., 43(2)*, 427 (1990); C.A., 113, 115214 (1990).
222. T. Ogawa, K. Hayami, H. Imaya, and H. Suzuki, *Chem. Lett., 1990(6)*, 937; *C.A., 113*, 171830 (1990).
223. J. Ozols, A. Pyrko, F. Lakvics, B. Vigante, R. Dubure, G. Dubur, and A. Ahrem, *Khim. Geterotsikl. Soedin., 1990(1)*, 66; *C.A., 113*, 58902 (1990).
224. N. Martin-Leon, M. Quinteiro, C. Seoane, and J. L. Soto, *Liebigs Ann. Chem., 1990(1)*, 101.
225. Reilly Industries, Inc., brochure, “POLYDMAP—A Guide to Its Successful Use,” 1990.
226. A. A. Zenyuk, L. I. Ukhova, and L. G. Lis, *Zh. Org. Khim., 26(10)*, 2232 (1990); *C.A., 115*, 158556 (1991).
227. A. N. Nyrko, *Zh. Org. Khim., 27(10)*, 2237 (1991); *C.A., 116*, 173607 (1992).
228. D. Sole and J. Banjoch, *Tetrahedron Lett., 32*, 5183 (1991).
229. J. E. Oliver, W. R. Lusby, and R. M. Waters, *J. Heterocycl. Chem., 28(6)*, 1565 (1991).
230. S. Arseniyadis, R. Rodriguez, E. Cabrera, A. Thompson, and G. Ourisson, *Tetrahedron, 47(34)*, 7045 (1991).
231. G. A. Kraus, B. E. Johnston, and J. M. Applegate, *J. Org. Chem., 56(19)*, 5688 (1991).
232. A. S. Kotnis, *Tetrahedron Lett., 32(29)*, 3441 (1991).
233. G. A. Kraus and Y. Wu, *Tetrahedron Lett., 32(31)*, 3803 (1991).
234. A. Krief, D. Surleraux, and H. Frauenrath, *Tetrahedron Lett., 29*, 6157 (1988); A. Krief, D. Surleraux, and M. J. Robson, *Synlett., 1991(4)*, 276.
235. V. A. Nikolaev, S. M. Korneev, I. V. Terent’eva, and I. K. Korobitsyna, *Zh. Org. Khim., 27(10)*, 2085 (1991); *C.A., 116*, 193750 (1992).
236. B. Wermeckes, S. Ye, and F. Beck, *Chem. Lett., 1992*, 609.
237. P. H. Nelson and J. T. Nelson, *Synthesis, 1992(12)*, 1287.
238. A. A. Zenyuk, A. V. Korchik, L. G. Lis, and L. I. Ukhova, *Zh. Org. Khim., 28(5)*, 947 (1992); *C.A., 118*, 101320 (1993).
239. O. V. Gulyakevich and A. L. Mikhal’chuk, *Zh. Obshch. Khim., 62(3)*, 715 (1992); *C.A., 117*, 234330 (1992).
240. D. Schinzer and M. Kalesse, *Tetrahedron Lett., 32(36)*, 4691 (1991).
241. P. M. Zavlin, D. A. Efremov, and A. L. Govorkov, *Zh. Obshch. Khim., 62(1)*, 221 (1992); *C.A., 117*, 171552 (1992).
242. Eur. Pat. Appl. EP 529,517 (March 3, 1993; T. Nakamatsu, Y. Nishida, and N. Kometani to Sumitomo Chemical Co., Ltd; Sumika Fine Chemicals Co., Ltd.); *C.A., 118*, 212666 (1993).

16

Postscript

In order to restrict the size of this book it became necessary to omit thousands of pertinent and useful references. Yet, it is hoped that the final selection has delivered the main message: Resorcinol, the industrial specialty chemical, is a versatile compound with a wide array of important applications in many fields, with wide occurrence in natural products, and useful in applied as well as academic projects. Resorcinol mirrors the world of chemistry, as it has a great many established uses, but also is involved in interdisciplinary ideas and at the frontiers of science and engineering.

Looking back at the preceding chapters, we believe that the facts refute the current, broad-brush assessment of chemistry by the (hopefully few) negativists: From the commercial side, an example of negativism would be the hard-nosed talk of some that research and development has become unfocused and expensive, has run out of ideas, and has seen its good days. From the academic side, an example of negativism is the really not so different view being heard, at least once, that we don't need more science, that chemical research has become sterile, and that university professors should take up applied research to survive. While it is true that a good portion of applied and academic work in chemistry is not likely to reap a profit and/or glory, it should be remembered that a prospector does not find gold or oil every day—once in a lifetime can be enough. Indeed, the best will do better. This book provides evidence that a certain portion, perhaps 10–20%, of both applied and academic work is so good that it does stand a chance of making money or a solid reputation. And 5% of this work is so useful/novel and/or exciting that it is likely to succeed on both counts—if the stars are in the right constellation (that is where management comes in).

The pace of this book is reason for optimism for the enterprise. The flow of ideas is so vigorous that those who take a longer-term view and remain flexible should flourish and get their payoff. This is particularly true for the academics, the truly gifted and dedicated people who work so very hard to generate and maintain the stream of exciting ideas. Surely, a great nation can understand that, can relearn to appreciate the promise of chemistry, and can support the university people generously, with a big return on their investment.

Looking back at the whole story, additional thoughts emerge on the business side of chemistry. Some of the customers, some of the public, even a few university people, feel that business puts profit above all, that chemical companies don't care if they poison the world and drown it in nondegradable waste, as long as they make obscene profits. It is hoped that this book can help to change that bias. We all applaud environmentalism, and the discovery of environmental issues by people of good will, most with idealistic fervor and motivation. This author's experience is that pollution was illegal and unethical long ago, but there were then and there are now some unethical characters who need to be punished according to the law. Let us remind the public that it was chemistry that developed the increasingly refined analytical methods that allow us to find compounds in the environment at the parts-per-trillion level, and that these methods were not available then. Even 40 years ago, range-finding toxicity tests of great value were conducted before a new chemical came to market, and in this area, too, much progress has been made, with chemistry a participant in developing improved tests.

Let us applaud as well the knowledgeable and committed managements that supported all this diversity, all these risky yet exciting research ideas. Nonetheless, it was demonstrated as a fact of life in this business that some of the best conceived and supported ideas, worked on intensively for years, have yet to pay off, if ever (see, e.g., the chapters on pharmaceuticals and on imaging technologies).

Looking back at the story of resorcinol, it comes to mind that we should remind the public that chemistry is one of the disciplines in the Division of Natural Sciences and, again, that nature's products are not all benign—as shown in the chapter on Natural Products. An analogy for the situation is mushroom picking, fondly remembered as a family activity in the author's childhood in Europe. We only picked two varieties, the Emperor (Kaiser) mushroom and the yellow egg (Eier Schwammerln) variety, both easily recognized and considered to be delicacies by many when prepared in a variety of ways. At the time, however, and most likely today as well, every year a good number of people died mistakenly by picking poisonous mushrooms and eating them. Yet, mushroom picking was never banned by the governments. The moral of the story is: know your product and how to handle it; educate people so they can pick and handle a product safely. Even a toxic product can be handled safely and can have beneficial uses.

Although countries of origin of a given piece of work were seldom mentioned in the text, the references show that good work is now coming not only from the old strongholds but from anywhere and everywhere, from Buenos Aires to Stockholm, from Asia, Africa, and Australia. The list of authors reads like a list of the nationals of the United Nations. Many of these scientists and specialists were trained in the U.S. This has played a part in the global competition in chemistry. The U.S. has lost its dominance, if not participation, in some areas. For instance, a careful reader of the references in Chapter 10 will note that the Japanese companies are heavily engaged in the area of imaging technology. Indeed, looking at the section of *Chemical Abstracts* (published by the American Chemical Society) titled "Radiation Chemistry, Photochemistry and Photographic and Other Reprographic Processes," for years there has been a long list of patents from Japan in every issue, even a dominance. This shows not only a strategic use of the patent system but also recognition of the importance of research and development in this area of rapid innovation. Here is a good example of global competition in chemistry and how to be in the forefront.

At present, the takeover/merger mania of the 1980s has moderated. In the 1990s so far the main business theme has been to cut costs in order to improve sagging profits.Thus, we now have the ongoing, agonizing for many, restructuring/rationalizing/reengineering—in part of need and, it would seem, in part of fashion. The chemical industry has been and is deeply involved. Swapping of businesses seems rational, splitting diversified companies into parts—usually separating businesses with good growth prospects from the older, slow-growth parts—is commonplace. Many more people are losing good jobs. There are signs that business is improving, that reason will prevail, and that competent managements will last.

The example of resorcinol has indeed reflected, on a smaller scale, the trends of the whole chemical enterprise. From that point of view, there is reason for considerable optimism as regards the future of chemistry.

Index

Acetone/formaldehyde resin, reaction with resorcinol, 92
N-Acetyl-m-aminophenol, 298
2-Acetyl-1,3-cyclohexanedione, 21
  alkylation, 476, 477
2-Acetylresorcinol (2,6-dihydroxyacetophenone), 21
4-Acetylresorcinol (2,4-dihydroxyacetophenone), 21
Acne preventative, 217
2-Acylresorcinols, 469, 473
Adhesion promoters, 336
Adhesives
  for wood, 85, 426, 428
  for metal, 245
Aerogel foams, 109–111
5-Alkadienylresorcinols, 408
2-Alkanoyl-1,3-cyclohexanediones, 471
7-Alkoxycoumarin aromas, 380
4′-Alkoxy-2′-hydroxy-4-butylazobenzene, 320
7-Alkoxyisoflavone fodder additives, 166
3-Alkoxy-4-nitrosophenol, 387
Alkylresorcinol/formaldehyde resins, 68, 92
Alkylresorcinol diglycidyl ethers, 241, 242
5-n-Alkyl-and n-alkenylresorcinols in nature, 407–409
Alloxydim, 137
7-Aminocoumarins, 282, 292, 295
  as light sensitizers in photoresists, 318
m-Aminophenol, 125–129, 132, 133
  N-alkyl
    by direct alkylation, 130, 131
    from resorcinol, 130–132
  allyl ether, 256
  N,N-dialkyl, 130–132, 284
m-Aminophenol (*cont.*)
  preparation
    from 3-amino-2-cyclohexen-1-one, 132
    from 1,3-cyclohexanedione, 132
    from resorcinol, 125–129
  producers from resorcinol, 398
  stabilization, 133
  use
    accelerator for resorcinol/formaldehyde resin cure, 91
    in adhesives, 91–94
    in agricultural chemicals, 139, 140, 168–171
    curing agent for epoxy resins, 247
    as a drug intermediate, 180, 183–185, 189, 192, 194, 196, 205, 217
    as a dye intermediate, 281–286, 288–290, 296, 299–309
    in formaldehyde resins in lithography, 116
    in hair dyes, 378, 379
    in maleimide/cyanate compound, 259
    in photographic emulsions, 314, 315
    for preparing diamine-type monomers, 234–236
    in skin-tanning compositions, 382
3-Amino-4′-(m-aminophenoxy)benzophenone, 235
7-Amino- and 7-hydroxy-2-phenoxazones, 417
4-(3-Aminophenoxy)-4′-(4-aminophenoxy)diphenylsulfone, 236
2-(3-Aminophenoxy)-6-chloropyridine, 236
4-Aminoresorcinol, 378

p-Aminosalicylic acid (PAS) and sodium salt
stimulator of oleoresin production in conifers, 164
tuberculostatic drug, 183
Aminoxanthene dyes, 284, 285
Angelicin, 423
Angelol B, 412
Aniline/resorcinol/formaldehyde resins, 67
Anticancer/antitumor agents, 194, 197, 199, 202, 203, 209, 210, 213, 215, 217, 406, 409, 411, 412, 430, 484
Anti-juvenile hormones, 166, 414–416
Antioxidants and heat stabilizers, 336–341
Aramids, 252, 253
Aromas, 53, 380–382
3-Aryl-4-hydroxycoumarins, 413
Aryl poly(m-phenylene-cyclohexylphosphonates), 345
Aspirochlorine, 431
Assay for optical purity of chiral compounds, 458
Atranorin, 419
Azo dyes, 280, 281, 293–295

Basic Pure Blue 3G (Rhodulin Pure Blue 3G), 286
Baskets (molecular), 437ff.
Benzotriazolylresorcinol, 352, 353, 361, 362
Benzoximate, 137
4-Benzoylresorcinol: *see* 2,4-dihydroxybenzophenone
Benzylresorcinol, 339
Bikaverin, 427
Bilobol, 407
1,3-bis-(3-aminophenoxy)benzene, 234, 253–255
1,3-bis-(4-aminophenoxy)benzene, 234, 255
1,3-bis-(3-aminophenoxy)benzophenone, 234, 254
1,3-bis-[4-(3-aminophenoxy)benzoyl]benzene, 235, 254
4,4′-bis-(3-aminophenoxy)biphenyl, 255, 257
1,3-bis-(3-aminophenoxy)-5-chlorobenzene, 234, 254
4,4′-bis-(3-aminophenoxy)diphenyl sulfone, 236, 253, 257
2,6-bis-(3-aminophenoxy)pyridine, 235, 257
m-bis-(2,4-diphenoxyphenoxy)benzene, 261
4,6-bis-(10-hydroxy-9-anthryl)resorcinol tetraacetate, 388
bis-(m-phenylene)-32-crown-10 macrocycle, 447
Blast furnace taphole mix, 108
Bonducellin, 429
Brake pad compositions, 108
Brasilein, 286
4-Bromomethyl-7-methoxycoumarin, 381
4-Bromoresorcinol in photography, 315
4-tert-Butylresorcinol, 298
5-tert-Butylresorcinol oxidation, 418

Calixarenes containing resorcinol, 437–445
antioxidants/antiozonants/aging inhibitors, 340, 341
binding of polar guests of biological origin, 444
binding glutaric acid, 444
flame retardants, 346, 347
in glycosidation of ribose, 444
in lactic acid recovery, 445
octaester/amide in composites, 258
phosphorylated, 443
in photoresist composition, 319
polyglycidyl ether, 244, 245
preparation from alkyl 2,4-dimethoxycinnamate, 443
redox responsive, 441
in rubber compounds, 340–341
as sensors of hydrogen binding organics (ribose, xylose, galactose), 388, 444–445
tetrasulfonated, 444
Calix[4]resorcinarenes: *see* Calixarenes containing resorcinol
Cannabinoid compounds as drugs, 185, 186, 191, 192, 205, 421, 422
Cannabinol and cannabidiol, 420
Cannabis constituents, 420
3-Carbethoxypsoralen, 424
Carbochromen, 181
Carbonless paper copying, 287–289
3,3′-Carbonyl-bis-(7-methoxycoumarin), 281
Carcerands, carciplexes, cavitands and caviplexes, 439–441
Cardanol and cardol, 410
Carenadiols, 205
Carteolol, 181
Cashew nutshell liquid, 410, 411
Catalyst for p-nitrophenylacetate hydrolysis, 449
Chemicarciplexes, 440, 445

Chlomethoxynil, 137
7-(Chlorocarbonylmethoxy)-4-methylcoumarin [CMMC], 281
3-Chloro-4-methyl-7-hydroxycoumarin, 166
Chlorophenol/resorcinol/formaldehyde resin, 60, 66, 70, 71
2- and 4-Chlororesorcinol, 21
Chromoglycolate sodium (Intal), 180
Chromophoric substrates for proteinases, 196, 208
cis-Chrysanthemic acid (and 2,2-di-desmethylchrysanthemic acid), 476
 from 2,2,5,5-tetramethyl-1,3-cyclohexanedione, 476
Chrysoin S (Tropaeolin O), 280
3-Cinnamoyl-7-diethylaminocoumarin, 291
Clavamine, 168
Cloproxydim, 139
Coatings for polymerization reactors, 114–115
Complexometric/colorimetric analysis of metal ions, 455–457
Concrete additives, 111, 112
Coriandrin, 423
Corrosion inhibitors, 343
Corrugated paper board, 117, 118
Cosmetics ingredients, 380–383
Coumaphos, 136
Cristacarpin, 426
Cresol or xylenol/resorcinol/formaldehyde resins, 65
 polyglyciyl ether, 243
Crystallization accelerators, 342
Cupric beta-resorcylate, 377
Cyclobisuberodiene, 413
1,3-Cyclohexanediol (hexahydroresorcinol), 18
1,3-Cyclohexanedione (dihydroresorcinol) and derivatives
 acylation, 471, 475
 4-alkylation, 470
 aromatization (dehydrogenation), 44, 45, 52, 381
 C- vs. O-alkylation, 460
 in catalyst composition, 342
 crystal phase/structure transition, 454, 455
 2-diazo, 464
 enthalpy of formation, 455
 hydroxylation, 473
 intermediate for agricultural chemicals, 137–139, 142–157
 intermediate for carteolol and pindolol, 181
 intermediate for indole, 208
1,3-Cyclohexanedione (dihydroresorcinol) and derivatives (*cont.*)
 intermediate for jasmone and methyl jasmonate, 460
 intermediate for karanjol, 467
 intermediate for meptazinol, 185
 intermediate for methicillin, 184
 intermediete for molindone, 183
 intermediate for 4-oxo-4,5,6,7-tetrahydroindole, 465
 intermediate for supermolecules, 448
 ketone/enol equilibrium, 453
 mesylates and their displacement reactions, 460
 in organic synthesis, 460–478
 oxidation, 476
 polymorphism, 455
 preparation, 17, 44, 45, 50
 2,2-thio-bis-compound, 471
Cyclohexanetrione, 476
Cyclohexylidene-bis-resorcinol, 337
Cycloxydim, 138
Cycolor color copying process, 290, 291
Cylindrocyclophane B, 431

Dandruff control, 185, 186, 383
Daidzein, 428
Decamerium bromide, 183
Demethoxyaflatoxin B2, 475
Demethylvignafuran, 426
Depsides of lichen, 419, 420
Desertorin, 413
Desmedipham, 139
Developers for carbonless and thermal printing paper, 310–313
4,6-Diacetylresorcinol crystal structure, 454
Diamines of oligomeric aromatic ether-sulfones and sulfide-sulfones, 236
3,4′-Diaminodiphenyl ether
 preparation, 232, 233
 use in aramids, 252, 253
 use in polyimides, 254
4,6-Diaminoresorcinol
 preparation, 233
 in poly(p-phenylenebenzobisoxazole), 267
2,6-Diaminotoluene, 384
1,3-Diamino-2,4,6-trinitrobenzene (DATB), 375
2-Diazo-4,6-di-tert-butyl-1,3-cyclohexanedione, 476
2-Diazo-1,3-cyclohexanedione, 464

2-Diazo-5,5-dimethyl-1,3-cyclohexanedione, 471
Diazotype reprography, 286
4,6-Di-tert-alkylresorcinols, 388
4,6-Di-tert-butylresorcinol
  antioxidant, 338, 339
  fungicide, 168
  oxidation, 418
  preparation, 19
4,6-Dichlororesorcinol, 21, 386
Dicyanates, 266
4,6-Di-(alpha,alpha-dimethylbenzyl)resorcinol, 339
3-[2-(Diethylamino)ethyl]-4-methyl-7-hydroxycoumarin, 385
Dihydroresorcinol: *see* 1,3-cyclohexanedione
Dihydroxyacetophenone oxime, 383
2,4-Dihydroxybenzaldehyde (beta-resorcaldehyde)
  intermediate for metochalcone, 182
  nitrones from, 72
  phenylhydrazone in diazotypy, 299
  preparation, 22
2,4-Dihydroxybenzoic acid (beta-resorcylic acid)
  adhesion promoter, 336
  and derivatives as couplers in diazotypy, 287, 312
  occurrance in plants, 406
  preparation, 14
2,6-Dihydroxybenzoic acid (gamma-resorcylic acid)
  adhesion promoter, 336
  component of drugs, 183–185, 209, 216
  n-decyl ester, 314
  derivatives as insecticides and herbicides, 136, 137, 163, 164
  4-dimethylaminopyridine salt as epoxy cure accelerator, 342
  occurrance in plants, 406
  preparation, 16
  triethylenediamine salt as epoxy cure accelerator, 342
  use in electrolytes, 343, 344
  use in photographic materials, 315
3,5-Dihydroxybenzoic acid (alpha-resorcylic acid)
  butyl ester as developer for leuco dyes, 311
  ethanolamide, 338
  preparation, 15
2,4-Dihydroxybenzophenones (4-benzoylresorcinol)
  antioxidant, 338
  fungicides, 167
  hydrazide, 358
  light screens for skin protection, 187, 214
  oxime, 343
  in photographic materials, 316
  methylene acceptor in rubber compounds, 69
  preparation, 355, 356
  use in lithography, 317
  UV light absorber, 331, 332
1,3-Dihydroxy-4,4′-bis-4′-(2′,2′,6′,6′-tetramethylpiperidyl)benzene, 354
2,4-Dihydroxy-4′-(2-carboxyvinyl)benzophenone, 354
2,2′-Dihydroxy-4,4′-diacryloxybenzophenone, 167
2′,6′-Dihydroxy-9-(2,5-dihydroxyphenyl)octylphenone, 337
2,2′-Dihydroxy-4,4′-dimethoxybenzophenone, 333
  disulfonate, 334
  UV light stabilizer, 331, 332
2,4-Dihydroxy-3,6-dimethylbenzonitrile, 53
3,8-Dihydroxy-5a,10b-diphenylcoumarano-2,3,2′,3′-cumaran
  preparation, 237
  use in polyarylates, 250
2,4-Dihydroxy-4′-fluorobenzophenone, 350
4,6-Dihydroxyisophthalic acid: *see* resorcinol-4,6-dicarboxylic acid
2,2′-Dihydroxy-4-methoxybenzophenone, 333
2,4-Dihydroxy-6-n-pentadecylbenzoic acid, 411
3-(2′,4′-Dihydroxyphenylazo-1′-)-1,2,4-triazole (TRIAR), 459
2,4-Dihydroxy-6-(10-phenyldecyl)acetophenone, 409
2,4-Dihydroxyphenyl-N-phenyl nitrone, 72
4,6-Diiodoresorcinol, 22
m-Diisopropylbenzene preparation, 32
Dimedone: *see* 5,5-dimethyl-1,3-cyclohexanedione
Dimeflin, 180
2,6-Dimethoxybenzoic acid, 16
4-(alpha,alpha-dimethylbenzyl)resorcinol, 386
5,5-Dimethyl-1,3-cyclohexanedione (dimedone)
  acylation, 475

5,5-Dimethyl-1,3-cyclohexanedione (dimedone) (*cont.*)
  alkylidene-bis-compounds (NMR analysis), 454
  2-diazo derivative, 471
  hardening accelerator for polyester resins, 342
  monofluorination, 470
  silicon rubber cure, 365
5,5-Dimethyl-1,2,3-cyclohexanetrione, 464
2,2-Dimethyl-7-hydroxychroman, 416
1,3-Dimethoxybenzene: *see* resorcinol dimethyl ether
2,6-Dimethoxybenzoic acid, 16
3,5-Dimethoxytoluene, 380
4-(alpha, alpha-Dimethylbenzyl)resorcinol, 386
2,5-Dimethyl-6-cyano-1,3-cyclohexanedione, 380
Di-O-methyl-12-oxocurvularin, 426
5,5-Dimethyl-3-phosphoryloxy-2-cyclohexene-1-one, 477
2,3-Dimethylpyridine, 47
Dimethylresorcinols, 476
4,6-Dinitroresorcinol
  in explosives, 376
  preparation, 233
4,6-Dinitrosoresorcinol in metal ion analysis, 458
Dioxybenzone, 187
7-Diphenylaminocoumarin, 324
m-Di-(propargyloxy)benzene, 168
Diresorcyl sulfide: *see* thio-bisresorcinol
Dithio-bis-resorcinol, 76
4-Dodecylresorcinol, 314
  diglycidyl ether, 241
Doubly-strapped porphyrin, 448

Edrophonium chloride, 184
Efloxate, 181
Electrolytes, 321, 343, 344
Electrophotographic photoreceptors, 324
Electroviscous fluid, 385
Emeguisin A, 429
Encecalin, 414
Environmental and health regulations, 401
Eosin G
  additive in cellulose pulp bleaching, 387
  dye, 283, 382
  preparation, 292
Epiphorellic acid, 420
Epoxy resins, 238
Eremomycin, 220
Erythrosine in dental diagnostics, 187
Erythrosine J, 284
3-Ethoxycarbonyl-7-hydroxycoumarin, 359
7-Ethoxycoumarin for assay of monooxygenase activity, 456
Ethyl 2-methyl-4,6-dihydroxybenzoate, 53
Ethylresorcinol, 298
Explosives/propellant components, 375–378

Fast Acid Violet B, 285
Fenocinol, 185
Fenoterol, 179
Ferroelectric liquid crystal composition, 320
Fire-resistant wooden doors, 94, 96
Fire-retardent phenolic resin composites, 96, 243
Fireworks compositions, 117
Flame/fire-retardent additives for polymers, 261, 262, 335, 344–348
Fluoran dyes, 285, 288, 297, 299–309
Fluoresceine (sodium salt)
  derivatives in lithography, 319
  in detergent compositions, 386
  diagnostic aid in eye injuries, 187
  in polyarylates, 250
  in poly(cyanoaryl)ethers, 263
  preparation, 283
  xanthene dyes, 283
Fluorescence (polarization) immunoassay, 190, 195, 296
Fluorescent dyes, 282
  in security paper, 295
Fluorescent probes,
  amino acid analysis, 211
  for carboxylic acids and hydroxyl compounds, 281, 457
  DNA sequence analysis, 188
Fluorimetric assays
  for amines and alcohols, 297
  for cytochrome oxidases, 207
  for E.coli, carboxylesterases, phosphatases and sulfatases, 457
  for hydrolases, 199
  for microsomal monooxygenase activity, 456
  substrates for hydrolases, 192, 199, 217
m-Fluorophenol, 389
4-Fluororesorcinol, 22
Foundry molding compounds, 108
Food additives, 337, 364

Formetanate, 140
Functional fluids and additives therefor, 260, 337–341, 348, 349, 365
Furanochromones, 424
Furocoumarins, 423, 424

Geipaverin, 412
Glycidyl ethers of resorcinol, derivatives and resins, 239–245
Goal (herbicide), 137
Golf ball compositions, 73
Guangdougen (Chinese herbal medicine), 181
Gypsum additives, 117

Haloxon, 136
Hair dyes, 378, 379
Hair growth stimulating composition, 214
Hashish, 420
Heat-generating (thermogenous) composition, 385
Heat/processing stabilizers for polymers, 336–341
Hemiacertal cleavage catalyst, 447
Hericenone, 431
2-Hexadecanoylresorcinol, 409
Hexahydroresorcinol: *see* 1,3-cyclohexanediol
7-Hexoxycoumarin, 380
2-n-Hexyl-5-n-pentyl- and 2-n-hexyl-5-n-heptylresorcinol, 407
4-n-Hexylresorcinol
  anthelmintic, 182
  anticancer agent, 210
  in antifouling coatings, 367
  antiseptic agent,182
  bacteriostat and anthelmintic, 182
  browning inhibitor in food, 364
  preparation, 19
HM-50 aramid, 252
Hologram material, 297
Hormothamnione, 430
Host-guest chemistry, 437–450
2-Hydroxy-4-acroyloxybenzophenone, 333
4-(2-Hydroxy-4-alkoxybenzylideneamino)cinnamate esters, 320
4-Hydroxybenzo(b)furan (karanjol), 467
2-Hydroxycarbazole, 292
4-Hydroxycarbazole, 461
7-Hydroxychromanone esters, 165
7-Hydroxychromenes, 165
7-Hydroxycoumarin, 295, 384, 411
  derivatives as drugs, 181, 182, 185, 197, 199, 201, 209, 213, 219
  ester, in assay of carboxyesterases etc., 457
  uses, 136, 166, 167, 281–282
5-Hydroxy-2,2-dimethyl-4-chromanone, 466, 467
m-Hydroxydiphenylamine, 384
m-(2-Hydroxyethoxy)phenoxyacetic acid
  preparation, 232
  use in polyesters, 248
2-Hydroxy-4-glycidyloxybenzophenone, 357
2-Hydroxy-4-(2′-hydroxyethoxy)benzophenone
  preparation, 359, 363
  uses, 333, 357
2-Hydroxy-4-(methacroyloxybutoxy)benzophenone, 359
2-Hydroxy-4-methallyloxyacetophenone, 385
2-Hydroxy-4-methoxybenzophenone, 333
3-Hydroxy-4′-methoxydiphenylamine, 383
7-Hydroxy-4-methylcoumarin (umbelliferone) preparation, 383,
2-Hydroxy-4-n-octoxy- and 4-n-dodecyloxybenzophenone, 333
3-Hydroxyphenoxytetrazole, 364
Hydroxyphenyl ether-type agricultural chemicals, 137, 157–162
N-(m-Hydroxyphenyl)maleimide, 257
Hydroxyphenyltellurium(IV) halides, 451
8-Hydroxyquinoline/resorcinol/formaldehyde resins, 458
Hydroxystilbenes from smartweed, 198
Hymechromon, 182

2-(2-Indolyl)-1,3-cyclohexanedione, 475
Intal, 180
Intraocular lenses, 359, 363
2- and 4-Iodoresorcinol, 22
Ion-exchange resins, 102–106, 263, 266
Ipriflavone, 185
4-Isoamylresorcinol, 210
4-Isopropylresorcinol acne preventative, 217
7-Isobutoxycoumarin, 380
Isodidymic acid, 419
Isoflavanoids, 413
Isoliquiridigenine, 203
Isopterofuran, 426
Isoxaben, 137

Jasmone and methyl jasmonate, 460

Kairomones, 471
Karanjol, 467
Karbutylate, 139
Kuwanons G and H, 426

Lacmoid (Resorcin blue), 416
Laser dyes, 282
Lasierin, 412
Lasiodiplodin, 424, 425
Lead dinitroresorcinate, 376
Lead beta-resorcylate, 376, 377
Lead styphnate, 375–376
Lead 2,4,6-trinitroresorcinate, 375–376
Leuco dyes for carbonless and thermal copying processes, 288–289, 299–309
Leukotriene inhibitors, 188, 189
Lichen constituents, 416–420
Light valves for high definition television (HDTV), 323
Liquid crystal
  compounds, 320
  polymers (LCPs), 251
Litmus (lacmus), 416
Lithography, 316

Macrolides, 424–426
Magnetic recording material, 321
Malabaricone A, 409
Marianol (4-methyl-7-ethoxycoumarin), 380
Marijuana constituents, 420
Melamine/urea/resorcinol/formaldehyde resins, 93
Mepronil, 140
Meptazinol, 185
m-Mercaptophenol, 383
N-Mesylaminoalkyl-m-aminophenols, 378
Metacyclophanes, 445–446
Metal ion analysis and extraction, 387, 455–459
N-[m-(Meth)allyloxyphenyl]maleimides, 256
Methicillin, 184
6-Methoxybenzoxazoline-2(3H)-one (MBOA), 429
7-Methoxycoumarin, 380
  carbonyl azides as fluorescent tags, 457
7-Methoxy-2,2-dimethylchromene, 415
N-Methyl-m-aminophenol, 379
2-Methylcardol, 410
4-Methyl-1,3-cyclohexanedione, 470
Methyl 2,4-dihydroxy-3-n-butyl-6-methoxybenzoate, 380
Methyl 2,4-dihydroxy-3,6-dimethylbenzoate, 380, 382, 467
Methyl 2,4-dihydroxy-3-isopropyl-6-methylbenzoate, 381
5,5′-Methylene-bis-(2,4-dihydroxybenzophenone), 351
5,5′-Methylene-bis-(2-hydroxy-4-methoxybenzophenone), 349, 356
Methylene-bis-resorcinol, 318
4-Methyl-7-ethoxycoumarin (Marianol), 380
2-Methyl-6-hydroxybenzoxazole, 383
4-Methyl-7-hydroxycoumarin (Coumarin 4), 282
4-Methylolresorcinol and its diglycidyl ether, 241
2-Methylpyridine, 47
2-Methylresorcinol
  dibenzoate, 355
  for epoxy resin cure, 245
  in hair dyes, 378
  preparation, 19, 47, 383, 386, 477
4-Methylresorcinol, 41, 47, 298, 386
5-Methylresorcinol (orcinol), 39, 47, 51, 54
4-Methylumbelliferone-beta-glucuronide for detection of *E. coli*, 457
Metochalcone, 182
Metoclopramide, 185
Mexenone, 187
Microcapsules, 112–114, 288, 290, 291
Microminutin, 413
Miriquidic acid, 419
Molecular recognition, 437
  nucleotides, 449
  D-phenylalanin, 448
Molindone, 183
Monocillin 1, 426
Moncut, 140
Monorden, 424, 426
Mulberrofuran A, 213

Nabazenil, 185
Nabilon, 185, 421
Naboctate, 185
2-Naphthol, 78
Nedocril (Tilade), 180
Neostigmine bromide, 184
Nitrogen fixation by microbes, 428
Nitroresorcinol phosphoramidates, 167
4-Nitroso-2,5-dimethylresorcinol, 387
4-Nitroso-5-methylresorcinol, 387
Nonylphenol/resorcinol/formaldehyde resin, 70
Novobiocin, 185

m-Octadecylaminophenol, 339
Oligo(benzyl)resorcinols, 323
Olivetol: *see* 5-pentylresorcinol
Optical bleaches, 282, 297
Optical recording material for laser discs, 322
Orcein, 416
Orcinol: *see* 5-methylresorcinol
Orciprenaline, 179
Orseille, 416
o-Orsenillic acid, 406
Oxazine dyes, 286
Oxine(8-hydroxyquinoline)/resorcinol/formaldehyde resins, 458
4-Oxo-4,5,6,7-tetrahydroindole, 195
Oxybenzone, 187
Oxyfluorfen (Goal), 137

PAR: *see* 4-(2-pyridylazo)resorcinol
PAS: *see* p-aminosalicylic acid
Pasiniazid,183
Patalenes
  heterocyclic, 165
PBO resins: *see* poly[p-phenylenebenzo-bis-(oxazole)]
Peniolactol, 427
5-Pentylresorcinol (olivetol), 20
Perfume components, 380–382
Phenisopham, 140
Phenmedipham, 139
Phenol assay in presence of resorcinol, 456
Phenolic resin foams, 109
Phenol/resorcinol/formaldehyde resins
  in adhesives, 85–102
  analysis by NMR spectroscopy, 456
  in composite materials, 95–102
  in molded brake pads, 108
  free phenol and resorcinol content by thermometric titrimetry, 456
  polyglycidyl ethers, 242
  in rubber, 67
Phenol/resorcinol/formaldehyde/furfural or furfuryl alcohol resins, 96
m-Phenoxyphenol, 261
3-(3-Phenoxy)phenoxyphenol, 261
Phentolamine, 183
m-Phenylaminophenol
  styrenated, 339
(m-Phenylenedioxy)diacetic acid
  preparation, 231
  use in polyesters, 248, 249
2,2′-(m-Phenylenedioxy)diethanol (HER)
  dipropenyl ether, 268
  oxidation, 232
  preparation, 230, 231
  use in polyesters, 249
  use in polyurethanes, 237
  use in UV-light curable resins, 268
Phenyl ether herbicides, fungicides, insecticides, 137, 157–162
4-Phenylsulfonylresorcinol, 314, 315
Phoenicine, 418
Photocopying, 322
Photographic materials, 313–316
Photoinitiators and photosensitizers, 281, 291, 296
Photoresists, 316
Photosensitive compositions, 115, 116
Photosynthetic electron transport (PET) inhibitors, 470
Phytoalexins, 405
Pindolol, 181
Pinosylvine, 487
Plant extracts
  as drugs, 181, 213, 217, 219
  as dyes, 286, 416
Plasticizers, 344
Polyamides, 252, 253
Polyamide/imides, 258–260
Poly(aryl ethers), 260
Poly(aryl ether/ketones), 265
Poly(aryl ether/sulfones), 265
Polyarylates, 249, 250
Polyazomethines, 268
Polybenz-1,3-oxazine-2,4-diones, 267
Poly(cyanoaryl ethers), 262–265
Polyesters, 248–252
  from dimethyl resorcinol-4,6-dicarboxylate, 248
  from m-(2-hydroxyethoxy)phenoxyacetic acid, 248
  from resorcinol: *see* polyarylates
Poly(ester crown ether), 249
Poly(ester/imides), 258, 259
Polyimides, 253
  ethynyl-endcapped, 255
Polymaleimides, 256, 257
Polyphenolic plant extracts
  use for adhesives, 93, 95
Poly[p-phenylenebenzo-bis-(oxazole)] (PBO), 267
Poly(phenylene ethers), 260–262

Polyurethanes, 237, 238, 365
Porphyrin, double strapped, 448
Precocene-1
dimer, 414
epoxidation, 414, 415
preparation, 414, 416
Processing stabilizers for polymers, 336–341
Prohexadione, 138
Propellants (double base), 377, 378
2-Propylresorcinol preparation, 19, 473
2-Propyl-4-acetylresorcinol, 21, 206
Psoralens, 423, 424
PVC reactor
antistick coatings, 114, 115, 389
Pyoluetin, 407
2-(2-Pyridinyl)-5,5-dimethyl-1,3-cyclohexanedione, 472
4-(2-Pyridylazo)resorcinol (PAR)
in analysis, 455
corrosion inhibitor, 343
Pyridyl-bis-resorcinol compounds, 450
Pyrocatechinoazo-triazole (TRIAP), 459
Pyronine B, 283

Quality standards, 402

Raclopride, 183
Ralgro, 424
Ramoxipride, 183
Reproterol, 179
Resorantel, 137
beta-Resorcaldehyde: *see* 2,4-dihydroxybenzaldehyde
Resorcin blue (lacmoid), 416
Resorcinol
adsorption from organic solvents, 386
acetate octanoate, 388
in acne preventatives, 217
adhesion promoter, 336
analysis, 23
in presence of phenol, 459
in analysis for nitrite/nitrate, 456–458
in analysis for sialic acids, 457
in analysis for zinc, 456
antiseptic and keratolytic agent, 185, 202
(aut)oxidation in ammoniacal solution, 418
binary system with catechol, 454
in binder for refractory brick, 389
binding by host, 446
broad categories of use, 399

Resorcinol (*cont.*)
C^{13}-labelled, 451
catalyst, 341, 342, 366
color stabilization, 384
complex with poly(vinylpyrrolidone), 454
composition with salicylic acid, camphor, menthol and thymol, 210
concrete additive, 365
conformational analysis, 450
conformation and rotational barriers, 451, 454
as corrosion inhibitor, 343
coupler in diazotypy, 286
alpha/beta crystal phase transition, 452, 454
and derivatives and resins
as curing agents/accelerators for epoxy resins, 245–247, 336, 341, 342
uses in analytical chemistry, 455–459
diacetate
preparation, 40, 54, 230
in polyarylate preparation, 252
stabilizer for color images, 310
dibenzoate, 367
di-(3,4-dicarboxyphenyl)ether, 254
dicyanate, 266
dicyclopentadiene condensation resin, 268
diglycidyl ether (RDGE), 239–241, 313
dihexanoate, 386
di-(2-hydroxyethyl)ether (HER): *see* 2,2′-(m-phenylenedioxy)diethanol
di-(hydroxypropyl)ether, 238
dimethyl ether
occurrance in nature, 380, 406
preparation, 22
discovery of, 406
distribution coefficient in water/butyl acetate or MIBK, 450
in explosives, 378
extraction with dialkyl and diaryl sulfoxides, 451
formaldehyde scavenger, 92, 93, 114
glycosides in topical preparations, 214
H-bonding with proton acceptors, 450
H/D exchange, 446, 451, 453
in hair growth-stimulating composition, 214, 383
heat of dilution in water,452
heat of solution in water, 451
in heat-generating composition, 385
hydroxylation, 387
ideal gas thermodynamic properties, 450

Resorcinol (*cont.*)
- inclusion complex in beta-cyclodextrin, 447
- isomerization catalyzed by zeolite, 453
- methylation, 386
- monoacetate, 186
- monobenzoate
  - developer for leuco dyes, 310
  - latent UV stabilizer, 332
  - preparatioon, 22
  - use in rubber compounds, 66
- monomethyl ether, 22
  - occurrance in Indian tobacco, 406
  - preparation, 22
- mono-(2-hydroxyethyl)ether, 287
- occurrance in plants, 406
- oxidation, 418
- ozonation (for effluent treatment), 457, 459
- as plant growth promoter, 168
- in photographic materials, 314, 315
- photometric assay with diazonium salt, 457
- physical properties, 6, 450
- in polyarylates, 249–251
- preparation/processes
  - from acetone/acrylic compound condensates, 42
  - from alkyl-substituted propiolates and dialkyl 3-oxoglutarates, 51
  - from anisole by acetoxylation, 54
  - from benzene disulfonation/fusion, 27
  - from coal pyrolyzates, 47
  - from m-cresol by sulfonation/fusion, 54
  - from 1,3-cyclohexanedione, 44, 45, 52, 473, 475, 476
  - from cyclohexanone, 49, 51
  - from 2-cyclohexenone, 49
  - from m-diisopropylbenzene by hydroperoxidation/cleavage, 31
  - from diketene and ketones, 52, 382
  - from disodium 3,3′-diphenylsulfonedisulfonate, 54
  - from 2,3-epoxycyclohexanone, 49, 51
  - from haloaromatic precursors, 49
  - from 3-hydroxycyclohexanone, 51
  - from m-isopropylphenol, 54
  - from oxygenated cyclohexanes, 49–51
  - from phenols by hydroxylation, 54
  - from m-phenylenediamine, 40
  - from phenylsulfinylacetate esters/unsaturated ketones, 52
  - from phthalic anhydride, 54
  - from plant sources, 48, 406

Resorcinol (*cont.*)
- preparation/processes (*cont.*)
  - from shale oil, 48
  - from wattle tree extracts, 48
- prices, 397
- producers and capacities, 394, 395
- propargyl ethers, 168
- properties and spectra, 5
- protonation in (super)acid, 451, 453, 454
- rate of reaction with formaldehyde, 86
- separation from catechol and/or hydroquinone, 36, 39, 459
- solubility in supercritical CO_2, 452
- in stabilizers for hydrogen peroxide, 366
- stabilizer for propellants, 378
- steam distillation, 452
- stillresidue, neat or formaldehyde condensate
  - coupler in diazotypy, 287
  - polyglycidyl ether, 242
  - sulfomethylated (superplasticizer for concrete), 112
  - taphole mix, 108
  - use in molding sands and clays, 107, 108
  - use in rubber compounds, 59–83
- sulfonation and sulfonic acids, 13
- thiocyanation kinetics, 454
- toxicity, 6
- UV spectra in sulfuric acid solution, 451, 454
- waste disposal, 11
- in wood pulp bleaching, 366
- as wood pulping additive, 366

Resorcinol/acetaldehyde resins, 61, 317

Resorcinol/acetone condensation products, 384, 386, 388

Resorcinol/aldehyde resins
- for hemodialysis, 117
- as tanning agents, 112

Resorcinol/amine/formaldehyde resins
- for enzyme fixation, 117

Resorcinol/aniline/formaldehyde resins, 67

Resorcinol/chlorophenol/formaldehyde resins, 60, 71

Resorcinol-containing pultrusion resins, 100–102

Resorcinol/cyanuric chloride condensation polymer, 266

Resorcinol-4,6-dicarboxylic acid (4,6-dihydroxyisophthalic acid)
- preparation, 17

Resorcinol-4,6-dicarboxylic acid (4,6-dihydroxyisophthalic acid) (*cont.*)
  use in making polybenz-1,3-oxazine-2,4-diones, 267
  use in polyesters, 248
Resorcinol/dicyclopentadiene resin, 268
Resorcinol/dimethylolcarbamide/N-methylolcaprolactam resin, 91
Resorcinol/ 4,4′-diphenylmethane diisocyanate adduct, 365
Resorcinol/diepoxide adducts in rubber, 73
Resorcinol-4,6-disulfonic acid sodium salt, 342
Resorcinol/formaldehyde resins
  additives for concrete, 111
  as adhesives for wood, 85
  analysis by GLC, HPLC, NMR, 456
  binders for foundry cores and friction materials, 107
  binders for refractory brick, 389
  in body armor making, 118
  for corrugated paper board, 117, 118
  curing agent for epoxy resins, 245, 247
  foams, 109
  in gypsum, 117
  in lithography, 115
  in microcapsules, 112–114
  modified
    as binders for explosives, 375
    concrete additives, 111–112
    coatings for polymerization reactors, 114–115
    in photosensitive compositions, 316
    in rubber compounds, 68, 70
  polyglycidyl ethers, 242, 243
  in polyurethane resins, 238
  in pyrotechnic compositions, 116, 117
  in rubber compounds, 60
  as soil stabilizers, 106, 107
  styrenated, 70, 71
  surgical adhesives, 116, 119
Resorcinol/formaldehyde/latex (RFL) dips, 60
Resorcinol/furfural resin, 96
Resorcinol/hexamethylenetetramine adduct: *see* resotropin
Resorcinol/ heximinium salt condensation product, 382
Resorcinol/nonylphenol/formaldehyde resin, 70
Resorcinol/phosphonitrilic resin 266
Resorcinol/polyphenolic plant extract/formaldehyde resin, 93, 95
Resorcinol/sulfur dichloride condensation product, 389
beta-Resorcylaldehyde: *see* 2,4-dihydroxybenzaldehyde
alpha-Resorcylic acid: *see* 3,5-dihydroxybenzoic acid
beta-Resorcylic acid: *see* 2,4-dihydroxybenzoic acid
gamma-Resorcylic acid: *see* 2,6-dihydroxybenzoic acid
Resorcylic acid derivatives in agricultural chemicals, 136, 137, 141, 163, 164
Resorcylic acid lactones (RALs) and derivatives, 141
Resorufin aminoglycosides, 208
Resothiomycin, 215
Resotropin, 75
Rhodamine B preparation, 284
Rhodamine B and 6G, 284
Rose Bengal, 296
Rotenone, 141
Ruscodibenzofuran, 426

Sanggenone O, 428
Schizopeltic acid, 419
Security paper, 295
Semecarpuflavanone,428
Sensitizers
  for carbonless and thermal printing paper, 310–313
  photobleachable, 291
Seselin, 412
Sethoxydim, 137
Siccanin, 185
Silver halide photographic materials, 313–316
Skin care compositions, 202
Sofalcon, 181
Soil stabilizers and drilling aids, 106
Solar cells, 322
Sparasol, 406
Spider toxins (NSTX-3, JSTX-3 and argiotoxin 636), 168, 428, 429
Squaraine compounds, 322, 524
Stabilizers for PVC, 242
Stains (biological), 280, 283, 286, 416
Starburst polyether, 449
Styphnic acid (2,4,6-trinitroresorcinol),
  complexes with 1- and 2-subtituted naphthalenes, 450
  preparation and use in explosives, 376, 377
Styrenated resorcinol/formaldehyde resin, 70, 71

5-Styrylresorcinol, 407
Sulforhodamine B, 285
Sulibenzone, 187
Sunscreen formulations, 355, 357, 360, 362
Superplasticizer for concrete, 112
Surgical adhesives, 117
Sweeteners for foods, 364

Tanning agents for hides, 112
Tannins (condensed) and tannin extracts in adhesives, 426, 428, 430
Teicoplanin, 220
Terbutalin, 179
Tetrafluoron, 140
Tetrahydrocannabinols (THCs), 420–422
2,2′,4,4′-Tetrahydroxybenzophenone
  copper chelate, 377
  dilithium salt, 385
  preparation, 356, 363
  UV stabilizer, 333, 355, 360
1,8,15,22-Tetraphenyl[1_4]metacyclophane-3,5,10,12,17,19,24,26-octol, 438, 439
Tetraphenyl resorcinol diphosphate, 335, 344–348
3-Thenoyl-7-diethylaminocoumarin, 296
Thermographic direct and transfer recording, 289, 290
4-(2-Thiazolyl) resorcinol (TAR)
  in metal ion analysis, 455
  in assay of free fatty acids in biological fluids, 456
Thickening agent for greases, 348
Thielavins A,B,C, 427
2,2′-Thio-bis-1,3-cyclohexanedione, 471
Thio-bis-resorcinol
  coupler in diazotypy, 287
  dandruff fighter, 383
  use in epoxy resins, 244
  in hair dyes, 379
  use in lithography, 317
  use in tall oil rosin esters, 366
  use in rubber, 76
  vulcanization accelerator, 77
Thioxolone, 187
Tilade, 180
Tire construction, 59, 60
Tobacco aromas, 380, 381
Tralkoxydim, 138
2,4,6-Trichlororesorcinol, 21
Tricornan, 447
Tri-(2,6-dialkoxyphenyl)phosphine
  preparation, 387
  use in a toner for electrophotography, 322
m-(Trifluoroethylamino)phenol, 379
4-Trifluoromethyl-7-hydroxycoumarin (Coumarin 152), 282
Triglycidyl-m-aminophenol, 244, 365
2,4,6-Triiodoresorcinol, 22, 193, 194
Trimethylresorcinol, 389
Trimethyl-trihydroxyflavan-containing resins, 65
2,4,6-Trinitroresorcinol (styphnic acid), 395, 396
Trioxatricornan, 447
Trioxsalen
  analog, 424
  dermatological agent, 187
  preparation, 204
Triplet sensitizers and photoinitiators, 281
2,4,6-Tris-(2,4-dihydroxybenzyl)resorcinol, 336
Tris-(3-hydroxy-4,6-di-tert-butylphenyl)phosphite, 339
Tropaeolin O (Chrysoin S), 280

Ultraviolet light absorbers/stabilizers, 331–335, 349–364
Umbelliferone, 411
Urea/resorcinol/formaldehyde resins 92, 93, 95

Velcrands and velcraplexes, 441
Vermistatin hydrogenation, 430
Visnadin, 181
Vulcanizing agents for silicone rubber, 365

Wairol, 412
Water desalting, 259
Wood pulping additive, 366

Xanthenes, 282
Xanthyletin, 412
p-Xylylene-bis-2,4-dihydroxybenzoates, 311

Zearalenone, 424–425
Zeranol and zearanol, 424